McGRAW-HILL SERIES IN WATER RESOURCES AND ENVIRONMENTAL ENGINEERING

VEN TE CHOW,
ROLF ELIASSEN,
and RAY K. LINSLEY,
Consulting Editors

BAILEY and OLLIS: Biochemical Engineering Fundamentals
BOCKRATH: Environmental Law for Engineers, Scientists, and Managers
CANTER: Environmental Impact Assessment
CHANLETT: Environmental Protection
GRAF: Hydraulics of Sediment Transport
HALL AND DRACUP: Water Resources Systems Engineering
JAMES AND LEE: Economics of Water Resources Planning
LINSLEY AND FRANZINI: Water Resources Engineering
LINSLEY, KOHLER, AND PAULHUS: Hydrology for Engineers
METCALF AND EDDY, INC.: Wastewater Engineering: Collection, Treatment, Disposal
NEMEROW: Scientific Stream Pollution Analysis
RICH: Environmental Systems Engineering
SCHROEDER: Water and Wastewater Treatment
TCHOBANOGLOUS, THEISEN, AND ELIASSEN: Solid Wastes:
    Engineering Principles and Management Issues
WALTON: Groundwater Resource Evaluation
WIENER: The Role of Water in Development: An Analysis of Principles of
    Comprehensive Planning

BIOCHEMICAL ENGINEERING FUNDAMENTALS

James E. Bailey
Professor of Chemical Engineering
University of Houston

David F. Ollis
Associate Professor of Chemical Engineering
Princeton University

McGRAW-HILL BOOK COMPANY

New York   St. Louis   San Francisco   Auckland   Bogotá   Düsseldorf
Johannesburg   London   Madrid   Mexico   Montreal   New Delhi   Panama
Paris   São Paulo   Singapore   Sydney   Tokyo   Toronto

BIOCHEMICAL ENGINEERING FUNDAMENTALS

234567890 KPKP 783210987

This book was set in Times New Roman. The editors were B. J. Clark,
Barbara Tokay, and James W. Bradley; the cover was designed by
Nicholas Krenitsky; the production supervisor was Charles Hess.
The drawings were done by J & R Services, Inc.
Kingsport Press, Inc., was printer and binder.

Library of Congress Cataloging in Publication Data

Bailey, James Edwin, date
    Biochemical engineering fundamentals.

    (McGraw-Hill series in water resources and environmental engineering)
(McGraw-Hill chemical engineering series)
    Includes index.
    1. Biochemical engineering.   I. Ollis, David F., joint author.   II. Title.
TP248.3.B34   660'.63   76-40006
ISBN 0-07-003210-6

*To Mary K and Marcy,
and our parents.*

CONTENTS

PREFACE

Biochemical engineering is the domain of microbial and enzyme processes, natural or artificial. The field encompasses many topics including industrial fermentation, enzyme utilization, and wastewater treatment. Our intention has been to treat the majority of these topics by covering the biochemical and engineering principles upon which these processes are based. We have provided an inclusive base of fundamentals and applications in the hope that the broad subject of biochemical engineering may soon be incorporated into many more engineering curricula than the few presently offering this course.

Biochemical engineering courses for juniors, seniors, and graduate students in chemical, environmental, civil, or food engineering may be taught from appropriate portions of this text. The authors, both chemical engineers, have presented undergraduate and graduate versions of a biochemical engineering course on quarter or semester bases over the last five years at Princeton University and the University of Houston. Portions of our notes from which the text has been developed have been used in biochemical engineering courses at the University of California at Berkeley, Iowa State University, the University of Massachusetts, and the University of Virginia.

Topics covered by the text include biochemistry, microbiology, reactor design and analysis, and transport phenomena. Our approach varies from traditional presentations in several ways. We have tried to interweave descriptive material on the life sciences together with engineering processes and analytical techniques. The implications of life science fundamentals for engineering processes are frequently discussed in sections dealing with biological principles. For example, the introduction of molecular genetics, viruses, mutation and genetic manipulation is followed by a discussion of recent applications of microbial genetics in developing especially productive microorganisms for several fermentations. The appropriate analytical techniques are presented after initial descriptive and background material has been given. Thus, enzyme kinetics and technology are introduced after the description of biochemicals including proteins; pure culture reactor dynamics are analyzed prior to the introduction of multiple species interactions.

Both text examples and end-of-chapter problems provide the student with opportunities to apply the concepts presented and to broaden understanding of the subject. More than 130 problems, spanning a range of difficulty, require discussions, derivations, and/or calculations from the student. A teacher's manual for this text, which provides many solutions and additional suggestions, is available.

Acknowledgments are due to Peter Reilly and Murray Moo-Young for their criticisms, to Elmer Gaden and Harold Bungay for copies of their own course notes, and to George Tsao, whose 1970 Chemical Engineering Education articles identified the need for a broad ranging text such as the present effort. Peter Reilly also kindly donated several homework problems. In addition, we are indebted to J. F. Andrews, E. L. Cussler, A. C. Payatakes, W. Phillips, D. A. Saville, and C. J. Shearer for valuable discussions. The authors take full responsibility for any errors and welcome comments and suggestions from readers.

Former graduate students Y. K. Cho, J. Fazel-Madjlessi, D. Chinloy, R. Marinangeli, H. Hager, M. Fish, H. Altman, E. Jacobs, and R. Datta have commented on sections of the text and have assisted in numerous details involved in its publication. To Betty Bixby, Carol Field, Nancy Brown, Linda Faught, Lorraine Harden, Charlotte Hennessey, and Juanita Lazard, who have typed the several drafts, many thanks.

We are pleased to thank several established contributors in biochemical engineering including Elmer Gaden, Arthur E. Humphrey, and Daniel I. C. Wang for warmly encouraging our entry in this field several years ago. Unrestricted financial support provided by the Camille and Henry Dreyfus Foundation to both authors freed the time needed to complete this text.

James E. Bailey
David F. Ollis

ONE

A LITTLE MICROBIOLOGY

Small living creatures called *microorganisms* interact in numerous ways with human activities. On the large scale of the biosphere, which consists of all regions of the earth containing life, microorganisms play a primary role in the capture of energy from the sun. Their biological activities also complete critical segments of the cycles of carbon, oxygen, nitrogen, and other elements essential for life. Microbes are also responsible for many human diseases. It has been argued that plagues caused by microorganisms have altered the course of history.

In this text, however, we concentrate on the purposeful utilization of microorganisms. A wide spectrum of profitable examples can be cited, including food processing, the manufacture of alcoholic beverages, and production of such complex organic molecules as vitamins and hormones. Moreover, microbial action provides an indispensable contribution in the treatment of sewage and many industrial wastes. Our principal objective is to understand and analyze such processes so that we can design and operate them in a rational way.

To reach this goal, however, a basic working knowledge of microbial growth and function is required. These factors and others peculiar to biological systems usually dominate biochemical process engineering. Consider for a moment that a living microorganism may be viewed in an approximate conceptual sense as an expanding chemical reactor which takes in chemical species called *nutrients* from its environment, grows, reproduces, and releases products into its surroundings. In instances such as sewage treatment, consumption of nutrients (here the organic sewage material) is the engineering objective. When microbes are grown for food sources or supplements, it is the mass of microbial matter produced which is desired. For a sewage-treatment process, on the other hand, this microbial matter produced by nutrient consumption constitutes an undesirable solid waste, and its amount should be minimized. Finally, the products formed and released during biological activity are of major concern in many industrial and natural contexts, including penicillin and ethanol manufacture. The relative rates of nutrient utilization, growth, and release of products depend strongly on the type of microorganism involved and on the temperature and composition of its environment. Understanding these interactions requires a foundation built upon biochemistry, biophysics, and microbiology. Since study of these subjects is not traditionally included in engineering education, a substantial portion of our efforts must be dedicated to them.

Whenever possible we shall extend our study of biological processes beyond qualitative understanding to determine quantitative mathematical representations. These mathematical models will often be extremely oversimplified and idealized, since even a single microorganism is a very complicated system. Nevertheless, basic concepts in microbiology will serve as a guide in formulating models and checking their validity, just as basic knowledge in fluid mechanics is useful when correlating the friction factor with the Reynolds number.

1.1. BIOPHYSICS AND THE CELL DOCTRINE

Microbiology is the study of living organisms too small to be seen clearly by the naked eye. As a rough rule of thumb, most microorganisms have a diameter of 0.1 mm or less. For many years, microbiology and other avenues of biological science were considered disciplines distinct from the physical sciences. It was thought that living things contained a "vital force" not governed by the laws of physics and chemistry.

In retrospect this is not surprising, for present knowledge indicates that even the simplest microorganism houses chemical reactors, information and control systems, and mass-transfer operations of amazing sophistication, efficiency, and organization. These conclusions have been reached in numerous experimental studies involving methods adapted from the physical sciences. Since this approach has proven so fruitful, the applicability of the principles of chemistry and physics to biological systems is now a widespread working hypothesis within the life sciences. The term *biophysics* is sometimes used to indicate explicitly the union of the biological and physical sciences.

A development critical to the understanding of living systems started in 1838, when Schleiden and Schwann first proposed the *cell theory*. This theory stated that all living systems are composed of cells and their products. Thus, the concept of a basic module, or building block, for life emerged. This notion of a common denominator permits an important decomposition in the analysis of living systems: first the component parts, the cells, can be studied, and then this knowledge is used to try to understand the complete organism.

The value of this decomposition rests on the fact that cells from a wide variety of organisms share many common features in their structure and function. In many instances this permits successful extrapolation of knowledge gained from experiments on cells from one organism to cells of other types. This existence of common cellular characteristics also simplifies our task of learning how microorganisms behave. By concentrating on the apparently universal features of cellular function, a basic framework for understanding all living systems can be established.

We should not leave this section with the impression that all cells are alike, however. Muscle cells are clearly different from those found in the eye or brain. Equally, there are many different types of single-celled organisms. These in turn can be classified in terms of the two major types of cellular organization described next.

1.2. THE STRUCTURE OF CELLS

Observations with the electron microscope have revealed two markedly different kinds of microbial cells. Although still linked by certain common features, these two classes are sufficiently distinct in their organization and function to warrant individual consideration here. So far as is known today, all cells belong to one of these groups.

1.2.1. Procaryotic Cells

Procaryotic cells, or *procaryotes*, are relatively small and simple cells. They usually exist alone, not associated with other cells. The typical dimension of these cells, which may be spherical, rodlike, or spiral, is from 0.5 to 3 μm.† In order to gain a qualitative feel for such dimensions, it is instructive to compare the relative sizes of cells with other components of the universe. As Fig. 1.1 reveals, the size of a procaryote relative to a man is approximately equal to the size of a man relative to the earth and less than the size of the hydrogen atom compared with that of a cell. These size relationships are very significant considerations when the details of cell function are investigated as we shall see later. The volume of procaryotes is on the order of 10^{-12} ml per cell, of which 50 to 80 percent is water. As a rough estimate, the mass of a single procaryote is 10^{-12} g.

Cells of this type grow rapidly and are widespread in the biosphere. Some, for example, can double in size, mass, and number in 20 min. Typically, procaryotes are biochemically versatile; i.e., they often can accept a wide variety of nutrients and further are capable of selecting the best nutrient from among several available in their environment. This feature and others to be recounted later make procaryotic cells adaptable to a wide range of environments. Since procaryotes usually exist as isolated single-celled organisms, they have little means

---

† 1 m (meter) = 10^3 mm (millimeter) = 10^6 μm (micrometer, formerly known as micron) = 10^{10} Å (angstrom units).

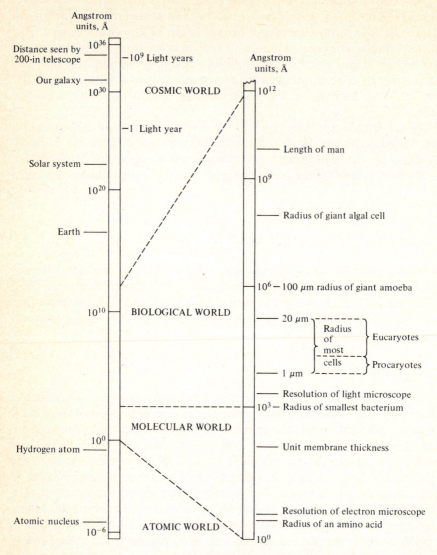

Figure 1.1 Characteristic dimensions of the universe. The biological world encompasses a broad spectrum of sizes. (*From "Cell Structure and Function," 2d ed., p. 35, by Ariel G. Loewy and Philip Siekevitz. Copyright* © *1963, 1969 by Holt, Rinehart and Winston Inc. Reprinted by permission of Holt, Rinehart and Winston.*)

of controlling their surroundings. Therefore the nutrient flexibility they exhibit is an essential characteristic for their survival. The rapid growth and biochemical versatility of procaryotes make them obvious choices for biological research and biochemical processing.

In Fig. 1.2 the basic features of a procaryotic cell are illustrated. The cell is surrounded by a rigid *wall*, approximately 200 Å thick. This wall lends structural

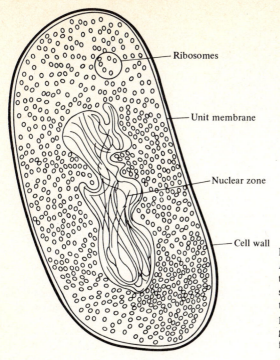

Ribosomes

Unit membrane

Nuclear zone

Cell wall

Figure 1.2 A sketch of a procaryote, *Escherichia coli*. This bacterium is native to man's intestinal tract and is sometimes simply called the intestinal bacterium. It is the most thoroughly investigated cell at present. Much of our knowledge of genetics at the molecular level is derived from studies of *E. coli*.

strength to the cell so that it can withstand a wide variety of external surroundings. Immediately inside this wall is the *cell membrane*, which typically has a thickness of about 70 Å. This membrane has a general structure common to membranes found in all cells. It is sometimes called a *plasma membrane*. These membranes play a critical role: they largely determine which chemical species can be transferred between the cell and its environment as well as the net rate of such transfer. Within the cell is a large, ill-defined region called the *nuclear zone*, which is the dominant control center for cell operation. The grainy dark spots apparent in the cell interior are the *ribosomes*, the sites of important biochemical reactions. The *cytoplasm* is the fluid occupying the remainder of the cell. Finally the *cytosol* is a colloidal suspension of large organic molecules. Not apparent here but visible in other photographs are clear, bubblelike regions called *storage granules*. We shall explore the composition and function of these structures within the procaryotic cell in greater detail after establishing the necessary background and defining some terms.

In order to bring out the similarities and differences between procaryotes, we consider another member of this family in Fig. 1.3. Again ribosomes, a nuclear zone, and cell wall are evident. This microorganism, however, is equipped with the biochemical machinery to utilize sunlight as an energy source, a capability illustrated by the presence of photosynthetic membranes.

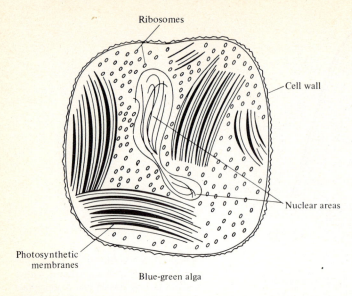

Figure 1.3 Another procaryote, a blue-green alga. This cell contains membranes capable of conducting *photosynthesis*, a complex process which captures light energy from the sun, provides the cell with organic molecules suitable for its reactions, and releases oxygen into the atmosphere.

1.2.2. Eucaryotic Cells

Eucaryotic cells, or *eucaryotes*, make up the other major class of cell types. As a rule these cells are 1000 to 10,000 times larger than procaryotes. All cells of higher organisms belong to this family. In order to meet the many specialized needs of animals, for example, eucaryotic cells exist in a wide variety of forms, as illustrated in Fig. 1.4. By coexisting and interacting in a cooperative manner in a higher organism, these cells can avoid the necessity for biochemical flexibility and adaptability so essential to procaryotes. Eucaryotic cells are not confined to plants and animals, however. In the next section we shall see several examples of eucaryotes which exist as single-celled organisms.

The internal structure of eucaryotes is considerably more complex than that in procaryotic cells, as can be seen in Figs. 1.5 and 1.6. Here there is a substantial degree of spatial organization and differentiation. The internal region is divided into a number of distinct compartments, which we shall explore in greater detail later; they have special structures and functions for conducting the business of the cell. At this point we shall only consider the general features of eucaryotic cells.

The cell is surrounded by a plasma, or unit, membrane similar to that found in procaryotes. On the exterior surface of this membrane may be a cell coat, or wall. The nature of the outer covering depends on the particular cell. For example, cells of higher animals usually have a thin cell coat. The specific adhesive properties of this coat are important in binding like cells to form specialized tissues and organs such as the liver. Plant cells, on the other hand, are often enclosed in a very strong,

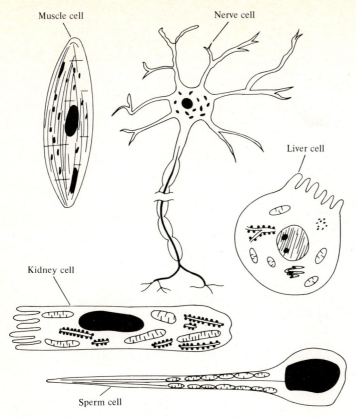

Figure 1.4 Several varieties of eucaryotes found in man. (*From "Cell Structure and Function,"* 2d ed., *p. 6, by Ariel G. Loewy and Philip Siekevitz. Copyright* © *1963, 1969 by Holt, Rinehart and Winston Inc. Reprinted by permission of Holt, Rinehart and Winston.*)

thick wall, which can be seen in Fig. 1.7. Wood consists for the most part of the walls of dead tree cells.

Important to the internal specialization of eucaryotic cells is the presence of unit membranes within the cell. A complex, convoluted membrane system, called the *endoplasmic reticulum*, leads from the cell membrane into the cell. The *nucleus* here is surrounded by a porous membrane. *Ribosomes*, reaction sites seen before in procaryotes, are embedded in the surface of much of the endoplasmic reticulum. (Ribosomes in procaryotes are smaller, however.) Ribosomes are in a sense analogous to the metal crystallites impregnated within porous supports to catalyze reactions in the classical process industry. A highly convoluted and twisting endoplasmic reticulum serves the same end as a very porous support for catalytic metals: it increases the available surface area per unit volume. The resemblance with such fabricated systems cannot be extended much further, however, because the living cell is more complex and sophisticated by many orders of magnitude.

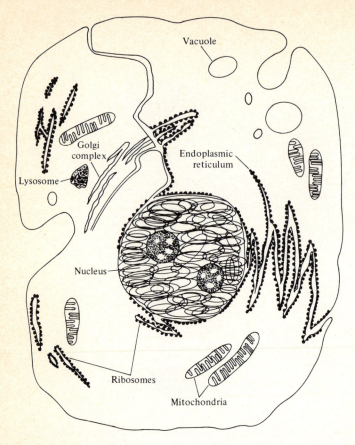

Figure 1.5 A typical eucaryotic cell. Such a typical cell is an imaginary construct, for there are wide variations between different eucaryotes. Many of these cells share common features and components, making the typical eucaryote a convenient and useful concept.

For example, a major function of the nucleus is to control the catalytic activity at the ribosomes. Not only are the reaction rates regulated, but the particular reactions which occur are determined by chemical messengers manufactured in the nucleus.

The nucleus is one of several interior regions surrounded by unit membranes. These specialized membrane-enclosed domains are known collectively as *organelles*. Catalyzing reactions whose products are the major energy supply of the cell, the *mitochondria* are organelles with an extremely specialized and organized internal structure. They are found in all eucaryotic cells which utilize oxygen in the process of energy generation. In *phototrophic cells*, which are those using light as a primary energy source, the *chloroplast* (see Fig. 1.7) is the organelle serving as the major cell powerhouse. Chloroplasts and mitochondria are the sites of many other important biochemical reactions in addition to their role in energy production.

The Golgi complex, lysosomes, and vacuoles are the remaining organelles illustrated in Figs. 1.5 to 1.7. In general, they serve to isolate chemical reactions or

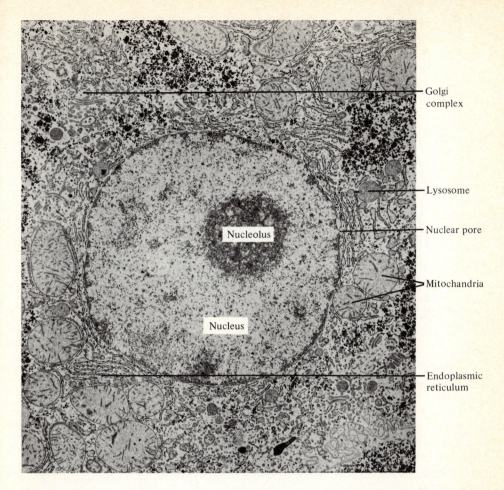

Golgi complex

Lysosome

Nuclear pore

Mitochandria

Endoplasmic reticulum

Nucleolus

Nucleus

Figure 1.6 Electron micrograph of a rat liver cell (× 11,000). (*Courtesy of George E. Palade, Yale University.*)

certain chemical compounds from the cytoplasm. This isolation is desirable either from the standpoint of reaction efficiency or protection of other cell components from the contents of the organelle.

Other interior features of eucaryotic cells include components involved in cell division and cell motion. Since such aspects of cell operation are not central to our purposes, we leave details on these matters to the references listed at the end of the chapter.

The discovery of similar organelles in a wide variety of cells allows a refinement of the major working advantages of the cell doctrine. The activities of the cell itself can now be decomposed conceptually into the activities of its component organelles, which in turn can be studied in isolation. In the absence of contrary evidence, similar organelles are assumed to perform similar operations and functions, regardless of the type of cell in which they are found.

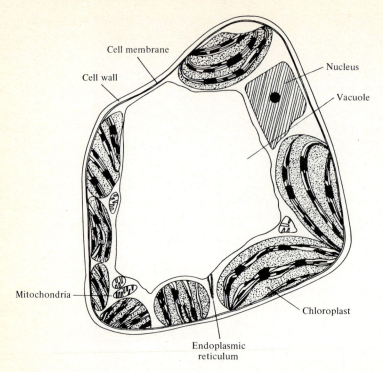

Figure 1.7 A leaf cell from a higher plant. The chloroplast is a specialized organelle for conducting photosynthesis. Other distinguishing features are rigid cell walls and large vacuoles.

Determination of the chemical composition, structure, and biochemical activities of organelles is a major goal of cell research. Much of the present knowledge of cell biochemistry came from investigation of these questions. Consequently we shall briefly enter the domain of the bioscientist and examine the centrifugation techniques widely employed to isolate components of the cell.

1.2.3. Cell Fractionation

A major problem in analyzing the characteristics of a particular organelle from a given type of cell is obtaining a sufficient quantity of the organelle for subsequent biochemical analysis. Typically this requires that a large number of organelles be isolated from a large number of cells, or a *cell population*. Let us follow a common procedure for this purpose, using liver cells as an example.

First a piece of liver tissue is minced in a blender. The resulting cell suspension is homogenized in a special solution using a rotating pestle within a tube or ultrasonic sound. Here an attempt is made to break the cells apart without significantly disturbing or disrupting the organelles within. Fractionation of the suspension, which now ideally contains a variety of isolated organelles, is the next step.

As process engineers, we know that any separation process is based upon exploitation of differences in the physical and/or chemical properties of the components to be isolated. The standard centrifugation techniques for fractionating cell organelles rely upon physical characteristics: size, shape, and density. A rudimentary analysis of centrifugation is presented in the following example.

Example 1.1. Analysis of particle motion in a centrifuge Suppose that a spherical particle of radius R and density ρ_p is placed in a centrifuge tube containing fluid medium of density ρ_f and viscosity μ_c. If this tube is then spun in a centrifuge at angular velocity ω (see Fig. 1E1.1), then after the particle has stopped accelerating, we have

$$\text{Drag force on particle} = \text{buoyancy force}$$

$$6\pi\mu_c R u_r = \frac{4\pi R^3}{3} G(\rho_p - \rho_f) \tag{1E1.1}$$

where u_r is the particle velocity in the r direction

$$u_r = \frac{dr}{dt} \tag{1E1.2}$$

and G is the acceleration due to centrifugal forces

$$G = \omega^2 r \tag{1E1.3}$$

Stokes' law has been used in Eq. (1E1.1) to express the drag force since particle velocities (and therefore particle Reynolds numbers) are usually very low in this situation. The usual gravitational-force term does not appear in Eq. (1E1.1) because the r direction in Fig. 1E1.1 is normal to the gravity

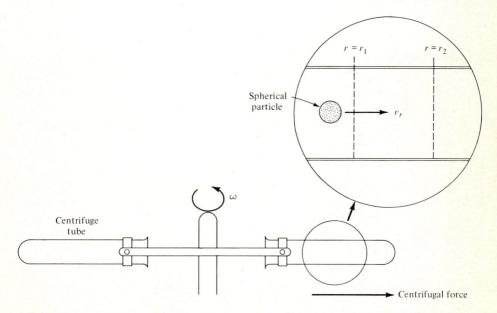

Figure 1E1.1 When a centrifuge is spun at high speed, particles suspended in the centrifuge tubes move away from the centrifuge axis. Since the rate of movement of these particles depends on their size, shape, and density, particles differing in these properties can be separated in a centrifuge.

force. Rotation of the centrifuge at high speed, however, produces an acceleration G usually many times larger than the acceleration of gravity.

Integration of this expression gives the time required for movement of the particle from position r_1 to r_2:

$$t = \frac{9}{2} \frac{\mu_c}{\omega^2 R^2 (\rho_p - \rho_f)} \ln \frac{r_2}{r_1} \tag{1E1.4}$$

Spheres with different sizes and/or densities will take different times to traverse the same distance in the centrifuge tube. This is the basis for the method of *differential centrifugation*. Since the larger, heavier particles such as nuclei and unbroken cells sediment more rapidly, they can be collected as a precipitate by spinning the suspension for a limited time at relatively low velocities. The supernatant suspension is then subjected to additional centrifugation at higher rotor speeds for a short time, and another precipitate containing mitochondria is isolated. By continuing this practice, known as *differential centrifugation*, a series of cell fractions can be obtained. The overall process is illustrated schematically in Fig. 1E1.2.

More sophisticated centrifugation methods employ liquid media with density gradients along the centrifuge tube. These techniques are also applicable for continued subdivision and fractionation of smaller cell constituents such as particular types of macromolecules in an organelle. Distinctions in chemical properties, also very valuable in such fine-scale separations, will be investigated in greater detail in Chap. 4.

There are several limitations in the application and interpretation of centrifugal cell-fractionation results. For an excellent summary the text of Mahler and Cordes† should be consulted. One difficulty, however, will plague us at almost

† H. R. Mahler and E. H. Cordes, "Biological Chemistry," 2d ed., Harper & Row, Publishers, Incorporated, New York, 1971.

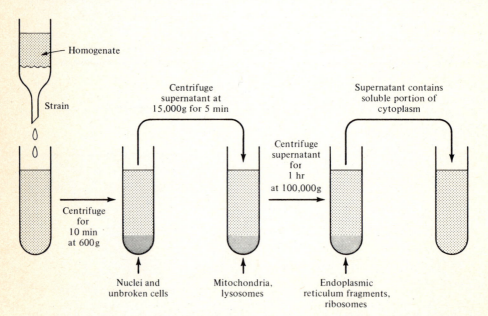

Figure 1E1.2 The steps in a typical differential centrifugation separation of cell constituents. Smaller components are isolated as the process proceeds.

every turn in investigating and utilizing microorganisms. In order to obtain a sufficient quantity of cells, organelles, biological molecules, or the like for analysis we are compelled to use a *population*, or a large number, of individual objects. It is common to assume that this population is *homogeneous*, i.e. that all its members are alike. In such a case the population serves only to amplify the characteristics of the individual so that it can be more conveniently observed.

Usually, however, the members of the population are different; the population is *heterogeneous*. For example, when the liver is minced, blood cells as well as liver cells are dispersed. Moreover, cells of the same general type within the liver are not identical. On a finer scale, similar organelles such as mitochondria within a single cell are generally different in some respects. Consequently, a cell fraction containing mitochondria, for example, is a heterogeneous population. When such a mixture of different components is analyzed, properties representing some kind of average over properties of its components are obtained. Therefore, the measured properties will depend upon the makeup of the population and will change if the population composition changes.

1.3. IMPORTANT CLASSES OF MICROBES

In this section we shall briefly review the kingdom of *protists*, which consists of all living things with a very simple biological organization relative to plants and animals. All unicellular (single-celled) organisms belong to this kingdom, and organisms containing multiple cells which are all of the same type are also classed as protists. Plants and animals, on the other hand, are distinguished by a diversity of cell types.

Table 1.1 shows a breakdown of the protist kingdom into groupings convenient for our purposes. These classifications show differences in several characteristics including the following: energy and nutritional requirements, growth and

Table 1.1. Classifications of microorganisms belonging to the kingdom of protists

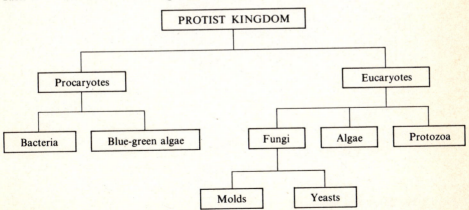

product-release rates, method of reproduction, and capability and means of motion. All these factors are of great practical importance in applications. Also significant in classification are differences in the *morphology*, or the physical form and structure, between these various types of organisms. The morphology of a microorganism has an influence on the rate of nutrient mass transfer to it and also can profoundly affect the fluid mechanics of a suspension containing the organism. Obviously then we must examine each class in Table 1.1 individually.

1.3.1. Bacteria

As seen earlier in our discussion of procaryotes, bacteria are relatively small organisms usually enclosed by rigid walls. In many species the outer surface of the cell wall is covered with a slimy, gummy coating called a *capsule* or *slime layer*. Bacteria are typically unicellular, but they may exist in three basic morphological forms (Fig. 1.8). Most cannot utilize light energy, are capable of motion (motile), and reproduce by division into two daughter cells (binary fission). Still, many exceptions to each of these rules are known.

There are many subdivisions of bacteria: some of the general groups of bacteria and some of their distinguishing characteristics are given in Table 1.2. The column labeled "Gram reaction" refers to the response of the bacteria to a relatively straightforward and rapid staining test. Cells are first stained with the dye crystal violet, then treated with an iodine solution and washed in alcohol.

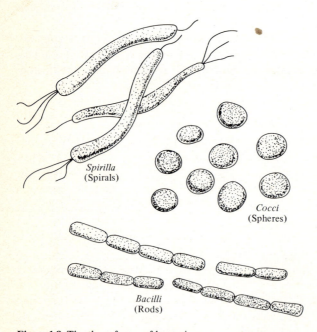

Figure 1.8 The three forms of bacteria.

Bacteria retaining the blue crystal-violet color after this process are called *gram-positive*; loss of color indicates a *gram-negative* species. Many characteristics of bacteria correlate very well with this test, which also indicates basic differences in cell-wall structure.

Whether or not oxygen is supplied to the cells is especially important in commercial exploitation of microorganisms (Chaps. 8, 10, and 12). In an *aerobic* process, oxygen is provided, usually as air, for use by the microorganisms. Manufacture of vinegar, some antibiotics, and animal-feed supplements are among the important microbial applications which employ aeration. The sparing solubility of oxygen in the aqueous media typical of these systems has major implications in process design (Chap. 8). The protists in an *anaerobic* process such as production of some alcohols or digestion of solid wastes function without oxygen.

Especially important in commercial utilization and control of bacteria is their ability to form *endospores* under adverse conditions. Endospores are dormant forms of the cell, capable of resisting heat, radiation, and poisonous chemicals. When the endospores are returned to surroundings suitable for cell function, they can germinate to give normal, functioning cells. This normal, biologically active cell state is often called the *vegetative form* in order to distinguish it from the spore form. As Table 1.2 indicates, there are two major groups of sporeforming bacteria. The aerobic *Bacillus* species are extremely widespread and adaptable. Several *Clostridium* species, which normally function under anaerobic conditions, die in the presence of oxygen in the vegetative state but form spores unaffected by oxygen. Some bacteria whose vegetative forms are rapidly killed at 45°C can form spores which survive boiling in water for several hours. Therefore, when we are attempting to kill microorganisms by heating (*heat sterilization*) and other means, the sporeforming capability of bacteria can create serious problems.

The blue-green algae will not be discussed here since they are not of great commercial significance. They are important, however, in the overall operation of natural aquatic systems since they participate in the nitrogen cycle (Chap. 12).

1.3.2. Yeasts

Yeasts form one of the important subclasses of fungi. Fungi, like bacteria, are widespread in nature although they usually live in the soil and in regions of lower relative humidity than bacteria. They are unable to extract energy from sunlight and usually are free-living. Although most fungi have a relatively complex morphology, yeasts are distinguished by their usual existence as single, small cells about 8 μm long and 5 μm in diameter.

The various paths of reproduction of yeasts are asexual (budding and fission), as shown in Fig. 1.9, and sexual. In budding, a small offspring cell begins to grow on the side of the original cell; physical separation of mature offspring from the parent may not be immediate, and formation of clumps of yeast cells involving several generations is possible. Fission occurs by division of the cell into two *equal* new cells. Sexual reproduction occurs by union of two *haploid* cells (each having a single chromosome) with dissolution of the adjoining wall to form a *diploid*

Table 1.2. Some major groups of bacterial species and some of their distinguishing characteristics

Class	Dominant morphological form	Some nutritional habits	Common habitat	Most members require O_2? (aerobic?)	Photosynthetic?	Forms endospores?	Gram reaction
Bacillus	Rods	Versatile; exists on many different nutrients	Soil	Yes	No	Yes	Positive
Clostridium	Rods	Many varieties have special nutrient requirements	Soil	Most species cannot tolerate O_2	No	Yes	Positive
Lactic acid (*Lactobacillus, Streptococcus, Leuconostoc*)	Rods or cocci	Lactic acid is a major end product of nutrient utilization; species acid-tolerant	Plants	No	No	No	Positive
Pseudomonas	Rods	Some extremely versatile and live on very wide range of nutrients	Soil, water	Yes	No	No	Negative

16

Rhizobium	Rods	Fixes nitrogen in association with legumes	Soil; in nodules of leguminous plants	Yes	No	No	Negative
Acetic acid bacteria (*Acetobacter, Gluconobacter*)	Rods; some *Acetobacter* species form extensive slime layers	Often consume alcohol; acid-tolerant	Decaying plants	Yes	No	No	Negative
Enteric or coliform bacteria (*E. coli* is one)	Rods	Simple organic compounds	Some naturally reside in intestine of higher animals	No, but can use O_2 if present	No	No	Negative
Rhodospirillum	Rods, spirals	Can fix N_2 or produce H_2	Specialized aqueous environments	No	Yes	No	Negative
Corynebacterium	Irregular form; reproduction not by binary fission; often nonmotile	Simple requirements	Soil, human body	No, but use O_2 if present	No	No	Positive

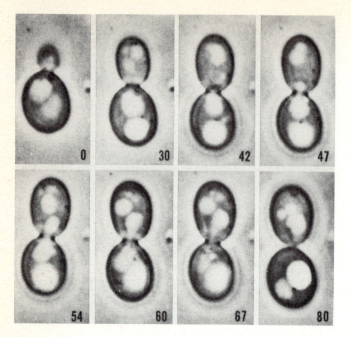

Figure 1.9 Reproduction of yeast by asexual budding is shown in the lower series of photographs. Numbers denote elapsed time in minutes. As illustrated in the upper sketch, sexual reproduction also occurs in the yeast life cycle. (*Photographs courtesy of C. F. Robinow*)

cell (two chromosomes per cell). The nucleus in the diploid cell undergoes one or several divisions and forms *ascospores;* each of these eventually becomes an individual new haploid cell, which may then undergo subsequent reproduction by budding, fission, or sexual fusion again. The ascospores, which here are a normal stage in the reproductive cycle of these organisms, should not be confused with the endospores, discussed above, which are a defense mechanism against unnatural surroundings.

In the production of alcoholic beverages, yeasts are the only important industrial microbes. In addition to supplying the consumer market for beer and wine, anaerobic yeast activities produce industrial alcohol and glycerol. The yeasts themselves are also grown for baking purposes and as protein supplements to animal feed (Chap. 10).

1.3.3. Molds

Molds are higher fungi with a vegetative structure called a *mycelium.* As illustrated in Fig. 1.10, the mycelium is a highly branched system of tubes. Within these enclosing tubes is a mobile mass of cytoplasm containing many nuclei. The mycelium may consist of more than one cell of related types. The long, thin filaments of cells within the mycelium are called *hyphae.* In some cases the

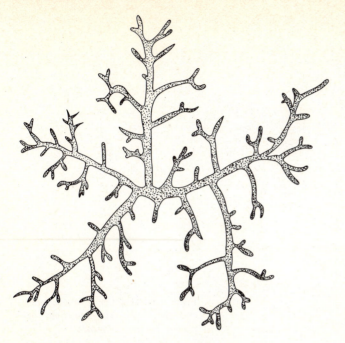

Figure 1.10 The mycelial structure of molds. A dense mycelium can cause conditions near its center to differ considerably from those at the outer extremities.

mycelium may be very dense. This property, coupled with molds' oxygen-supply requirements for normal function, can cause difficulties in their cultivation, since the mycelium can represent a substantial mass-transfer resistance. This problem and the unusual flow properties of suspensions of mycelia will be explored in further detail in Chaps. 7 and 8.

The most important classes of molds industrially are *Aspergillus* and *Penicillium* (Fig. 1.11). Major useful products include antibiotics (biochemical compounds which kill certain microorganisms or inhibit their growth), organic acids, and biological catalysts.

The strain *Aspergillus niger* normally produces oxalic acid (HO_2CCO_2H). Limitation of both phosphate nutrient and certain metals such as copper, iron, and manganese results in a predominant yield of citric acid [$HOOCCH_2COH(COOH)CH_2COOH$]. This limitation method is the basis for the commercial biochemical citric acid process. Thus *A. niger* is an interesting example differentiating approaches to biochemical-reactor design and optimization from those of nonbiological reactors: a much greater selectivity can sometimes be achieved in the biological system by minor alteration of feed composition to the reactor.

This example (as well as that of penicillin below) should motivate us to learn the essentials of cell structure, metabolism, and function which are woven into the

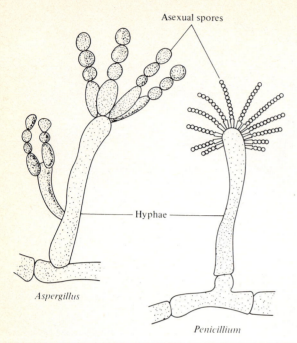

Asexual spores

Hyphae

Aspergillus

Penicillium

Figure 1.11 The hyphae of *Aspergillus* and *Pencillium*, two industrially important varieties of molds.

following chapters. Without this background in cellular processes and characteristics, our skills as process engineers, which are well suited to design and analysis of many aspects of biochemical processes, could be wasted because key biological features of the system would be ignored.

Penicillin production offers an example of a second fundamental difference between microbiological and nonbiological reactors. Major improvement in production of penicillin has arisen by use of ultraviolet irradiation of *Penicillium* spores to produce *mutants* of the original *Penicillium* strain (Fig. 1.12). Cell alteration by various techniques may result in orders-of-magnitude improvements in desired yield; it serves to indicate the central importance to the process design engineer of microbial genetics. This latter area occupies a large fraction of the present efforts in both university and industrial biological research. The pragmatic importance of designing mutations (or, in other situations, of avoiding such mutations) emphasizes the need for close cooperation between engineers, biologists, and biochemists in biochemical process design and evaluation. The history of penicillin production is in itself a story of joint development of new techniques, including deep-submerged production, solvent extraction of a delicate product on a large scale, air-sterilization procedure for high volumetric flow rates, and achievement of particular mutations with high penicillin yields.

Before leaving bacteria and fungi, we should mention the *actinomycetes*, a group of microorganisms with some properties of both fungi and bacteria. These organisms are extremely important for antibiotic manufacture. Although formally

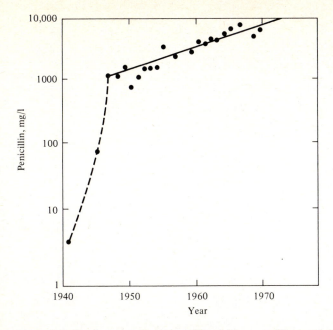

Figure 1.12 Maximum attainable pencillin yields over a 30-yr period. Development of special mold strains by mutation has produced an exponential increase in yields for the past 25 yr. A similar trend also holds for yields of the antibiotic streptomycin. (*Reprinted by permission of A. L. Demain, Overproduction of Microbial Metabolites due to Alteration of Regulation, in T. K. Ghose and A. Fiechter (eds.), "Advances in Biochemical Engineering 1," p. 129, Springer-Verlag, New York, 1971.*)

classified as bacteria, actinomycetes resemble fungi in their formation of long, highly branched hyphae. Also, design of antibiotic production processes utilizing actinomycetes is very similar to those involving molds. One difference, however, is the susceptibility of actinomycetes to infection and disease by biological agents which also can attack bacteria. These agents will be examined briefly later in Chap. 6.

1.3.4. Algae and Protozoa

These relatively large eucaryotes have sophisticated and highly organized structures. Two types of algae are shown in Fig. 1.13. *Euglena* has flagella for locomotion, lacks a rigid wall, and has an eyespot sensitive to light. The cell, guided by the eyespot, moves in response to stimulus by illumination—clearly a valuable asset since most algae require energy in the form of light. Many diatoms (another kind of algae) have exterior skeletons of complex architecture which are impregnated with silica. These skeletons are widely employed as filter aids in industry.

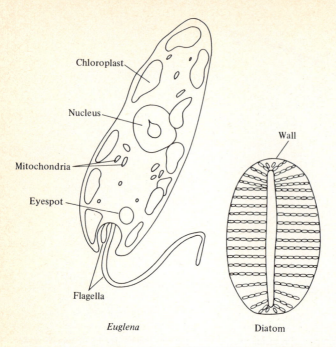

Chloroplast

Nucleus

Wall

Mitochondria

Eyespot

Flagella

Euglena   Diatom

Figure 1.13 Two varieties of algae. *Euglena* possesses flagella which, guided by the eyespot, propel the alga toward the light, where energy can be harnessed by the chloroplasts. The silica impregnated wall of diatoms exhibits detailed patterns and symmetries.

Current commercial interest in algae is concentrated on their possible exploitation as foodstuffs and food supplements. In Japan, several processes for algae food cultivation are in operation today. Also important in Asia is use of seaweed in the human diet. While not microorganisms, many seaweeds are actually multicellular algae. Like the simpler blue-green algae, eucaryotic algae serve a vital function in the cycles of matter on earth (Chap. 12).

Just as algae may be viewed as primitive plants, *protozoa*, which cannot exploit sunlight's energy, are in a sense primitive animals. The great variety of life forms characterized as protozoa is evident from the examples in Fig. 1.14. Although protozoa are not now employed for industrial manufacture of either cells or products, their activities are significant among the microorganisms which participate in biological waste treatment (Chap. 12). These processes, widely employed in urban communities and large industrial plants throughout the world, are surprisingly complicated from a microbiological viewpoint. Since a wide variety of nutrients are present in sewage or industrial wastes, a correspondingly large collection of different protists are present and indeed necessary in treatment operations. These diverse organisms compete for nutrients, devour each other, and interact in numerous ways characteristic of a small-scale ecological system. A survey and analysis of interactions between different species will be considered in Chap. 11.

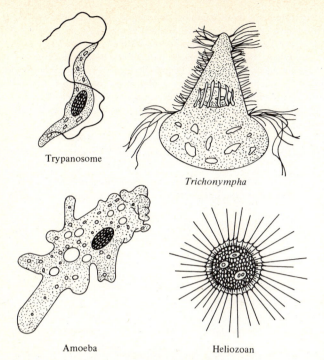

Trypanosome

Trichonympha

Amoeba

Heliozoan

Figure 1.14 A sampling of protozoa. Some trypanosomes carry serious disease, including African sleeping sickness. The *Trichonympha* inhabit the intestines of termites and assist them in digesting wood. While the amoeba has a changing, amorphous shape, the heliozoa have an internal skeleton and definite form.

1.4. A PERSPECTIVE FOR FURTHER STUDY

In our brief sojourn through microbiology in this chapter, we have seen the basic importance of the cell doctrine. The importance of basic biology for understanding biochemical process systems has been emphasized. In the next few chapters, our attention will continue to be concentrated on fundamental cell biology.

Chapter 2 reviews the chemicals which the cell must synthesize for survival and reproduction; the catalysts used to facilitate reactions within the cell are examined in Chaps. 3 and 4. Reaction sequences necessary for cellular function are then studied in Chap. 5, with control of these reactions and genetics the primary topics of Chap. 6. After investigating the kinetics of microbial systems in Chap. 7, we direct our remaining efforts to analysis, design, control, and optimization of biological systems.

PROBLEMS

1.1. Microbiologists Read a biography of one of the early influential men in microbiology, e.g., Robert Hooke, Anton van Leeuwenhoek, Lazzaro Spallanzani, Louis Pasteur, Walter Reed, D. Iwanowski, P. Rous, Theodor Schwann, or M. J. Schleiden. Prepare a short summary indicating what technical and social obstacles were or were not overcome, what technical achievements did or did not result from careful quantitation, and the place of induction and deduction in the studies of these men.

1.2. Experimental microbiology Many techniques in microbiology are simple and relatively well established but unfamiliar to those who have completed general, organic, and physical chemistry. As observation and measurement underlie any useful analysis, some appreciation of techniques and their accuracy is indispensible. Take a microbiology laboratory course in parallel or following this course if you have not previously done so. Lacking the opportunity, read carefully through a short laboratory paperback on this topic as you follow the chapters of this text, e.g., Ref. 5. As you do this exercise, remember Claude Bernard: "to experiment without a preconceived idea is to wander aimlessly." For the laboratory course or paperback, summarize the purpose(s) of each experiment. For each such experimental setup, what other information could you obtain?

1.3. Observation Read a brief account of microscope techniques including dark-field, phase-contrast, fluorescence, and electron microscopy. Develop a list of relative advantages and limitations for each technique.

1.4. Definition Define the following terms and when there is more than one, compare and contrast their general characteristics with those of others in the same group:

(*a*) Procaryotes, eucaryotes
(*b*) Cell wall, plasma membrane, endoplasmic reticulum
(*c*) Cytoplasm
(*d*) Nucleus, nuclear zone
(*e*) Ribosome, mitochondria, chloroplast
(*f*) Morphology
(*g*) Spirilla, cocci, bacilli
(*h*) Budding, sexual fusion, fission, sporulation
(*i*) Protozoa, algae, mycelia, amoeba

1.5. Identification and classification (*a*) Sketch the diagram showing the kingdom of the protists (from memory).

(*b*) The taxonomy of microbial species is based largely on visual observation with the optical microscope. Using either Bergey's "Manual of Determinative Bacteriology" or "A Guide to the Identification of the Genera of Bacteria," by Skerman, locate the organisms *Escherichia coli*, *Staphylococcus aureus*, *Bacillus cerus*, and *Spirillum serpens*. For any two, list the distinguishing characteristics which would lead to ultimate idenification. Begin most generally, passing from family through subgroups (tribe, genus) to species.

1.6. "The view from the ground floor" Pick a microbial topic of interest (beer fermentation, antibiotic production, yeast growth, waste-water treatment, soil microbiology, vaccines, pickling, cheese manufacture, lake ecology, salt-water microbiology, etc.). Read a descriptive account of the topic in a text such as the Kirk-Othmer "Encyclopedia of Chemical Technology," etc. Sketch a process or natural flow scheme indicating important microbial species, food sources, and waste or exit streams. Each week during the course, add one or two pages to the description indicating the (lack of) importance of the chapter topic to the process. At the end of the course, prepare a short paper on your topic and present it to the class.

REFERENCES

1. W. R. Sistrom: "Microbial Life," 2d ed., Holt, Rinehart, and Winston, Inc., N.Y., 1969. The first three chapters of this introductory microbiology text cover most of the topics in this chapter. Also included is material on metabolism, growth, and genetics, which are treated later in this text.
2. M. Frobisher: "Fundamentals of Microbiology," 8th ed., W. B. Saunders Company, Phila., 1968. A descriptive presentation of microbial life forms including cell classification, viruses, sterilization, immunology, and microbial applications. The ever-present connections between microorganism and man are emphasized.

3. R. Y. Stanier, M. Doudoroff, and E. A. Aldelberg: "The Microbial World," 4th ed., Prentice-Hall, Inc., Englewood Cliffs, N.J., 1975. More advanced than the previous texts and very well written. Covers microbiology history, classes of microbes, symbiosis, and disease, as well as microbial metabolism. Genetics at the molecular level, including mutation and regulation, is presented in detail.
4. "The Living Cell," readings from *Scientific American*, W. H. Freeman and Company, San Francisco, 1965. Reprints of *Scientific American* articles including levels of cell complexity, energetics, synthesis, division and differentiation, and special activities such as communication, stimulation, and muscle action. While some of the material is now somewhat outdated, this wide-ranging and profusely illustrated collection is still very worthwhile reading.

Problems

5. K. T. Crabtree and R. D. Hinsdill: "Fundamental Experiments in Microbiology," W. B. Saunders Co., Philadelphia, 1974.

CHAPTER
TWO

CHEMICALS OF LIFE

The viable organism must synthesize all the chemicals needed to operate, maintain, and reproduce the cell. In the following chapters we investigate the kinetics, energetics, and control of the major biochemical pathways for such syntheses. A necessary prerequisite for such studies is familiarity with the reactants, products, catalysts, and chemical controllers which participate in reaction networks of the cell.

The present chapter is concerned with the predominant cell polymeric chemicals and the smaller monomer units from which the larger polymers are derived. The four main classes of polymeric cell compounds are the fats and lipids, the polysaccharides (cellulose, starch, etc.), the information-encoded polydeoxyribonucleic and polyribonucleic acids (DNA, RNA), and proteins. The physicochemical properties of these compounds are important both in understanding cellular function and in rationally designing processes incorporating living microorganisms.

The various biological polymers may be usefully regarded as being either repetitive or nonrepetitive in structure. *Repetitive biological polymers* contain one kind of monomeric subunit; distinctions between different types of the same polymers are primarily due to differences in molecular weight and the degree of branching of the polymer chains. The major function of repetitive polymers in the cell is to provide *structures* with the desired mechanical strength, chemical inertness, and permeability. Repetitive polymers also provide a means of *nutrient storage*. In the latter function, for example, a 1 M glucose solution can be stored as the polymer glycogen, a cell polysaccharide reserve, with a concurrent reduction in molarity by a factor of 10,000 or greater. As cells may need to store excess food supplies without seriously disrupting the intracellular ionic strength, commodity storage as polymer is a useful asset.

Nonrepetitive polymers may contain from several up to 20 *different* monomer species. Further, each of these biological polymers has a fixed molecular weight and monomer composition, and the monomers are linked together in a fixed, genetically determined sequence.

The elemental pool from which the polymers are constructed is exemplified by the *E. coli* composition given in Table 2.1. The predominant elements (hydrogen, oxygen, nitrogen, and carbon) form chemical bonds by completing their outer shells with one, two, three, and four electrons, respectively. They are the lightest elements in the periodic table with such properties, and (except for hydrogen) they can form multiple bonds as well. The variety of chemicals which can be assembled from these four elements include, if we add a little sulfur, the four major biopolymer classes.

In addition to variety, the biochemical compounds assembled from these elements are quite stable, reacting only slowly with each other, water, and other cellular compounds. Chemical reactions involving such compounds are catalyzed by biological catalysts: proteins which are called *enzymes* (recall that a catalyst is a substance which allows an increase in a reaction rate without itself undergoing a permanent change). Consequently, by controlling both the number and type of enzymes which the cell contains, the cell regulates both the type and rate of chemical reactions which occur within it. Details of these control mechanisms are considered in Chap. 6.

While phosphorus and sulfur occur in the organic matter of all living things, they are present in relatively small amounts. The ionized forms of sodium, potassium, magnesium, calcium, and chlorine are always present, and trace amounts of

Table 2.1. The composition of *E. coli*

Element	Percentage of dry weight
Carbon	50
Oxygen	20
Nitrogen	14
Hydrogen	8
Phosphorus	3
Sulfur	1
Potassium	1
Sodium	1
Calcium	0.5
Magnesium	0.5
Chlorine	0.5
Iron	0.2
All others	~0.3

Data for *E. coli* assembled by S. E. Luria, in I. C. Gunsalus and R. Y. Stanier (eds.), "The Bacteria," vol. 1, chap. 1, Academic Press Inc., New York, 1960.

manganese, iron, cobalt, copper, and zinc are necessary for proper activation of certain enzymes. Some organisms also require miniscule amounts of boron, aluminum, vanadium, molybdenum, iodine, silicon, fluorine, and tin. Thus, subject to further revision, the present count of elements necessary for life is 24.

The solvent within which cells live is, of course, water. In addition to its relatively unusual properties (a high heat of vaporization, a high dielectric constant, the ability to ionize into acid and base, and propensity for hydrogen bonding), water is an extremely important reactant which participates in many enzyme-catalyzed reactions. Also, the properties which biopolymers exhibit depend strongly on the properties of the solvent within which they are placed; this fact provides the means of many separation process designs. The biological fitness of water and other common cell chemicals is discussed by Blum [8].†

2.1. LIPIDS

By definition, lipids are biological compounds which are soluble in nonpolar solvents such as benzene, chloroform, and ether and practically insoluble in water. Consequently, these molecules are diverse in their chemical structure and their biological function. Their relative insolubility leads to their presence predominantly in the nonaqueous biological phases, especially the plasma and organelle membranes. Fats, which simply serve as polymeric biological fuel storage, are lipids, as are several important mediators of biological activity. Lipids also constitute portions of more complex molecules such as lipoproteins and liposaccharides, which again appear predominantly in biological membranes of cells and the external walls of some viruses.

2.1.1. Fatty Acids and Related Lipids

Saturated fatty acids are relatively simple lipids with the general formula

$$CH_3-(CH_2)_n-C\overset{\displaystyle O}{\underset{\displaystyle OH}{\Big\langle}}$$

The hydrocarbon chain is constructed from identical two-carbon monomers, so that fatty acids may be regarded as noninformational biopolymers with a terminal carboxylic group. The value of n is typically between 12 and 20 (even numbers) in biological systems.

Unsaturated fatty acids are formed upon replacement of a saturated $(-C-C-)$ bond by a double bond $(-C=C-)$. For example, oleic acid is the unsaturated counterpart of stearic acid $(n = 16)$:

$$CH_3-(CH_2)_{16}-COOH \qquad CH_3-(CH_2)_7-HC=CH-(CH_2)_7-COOH$$

Stearic acid   Oleic acid

† Numbers in brackets indicate the references listed at the end of the chapter.

The hydrocarbon chain is nearly insoluble in water, but the carboxyl group is very hydrophilic. When a fatty acid is placed at an air-water interface, a small amount of the acid forms an oriented monolayer, with the polar group hydrated in the water and the hydrocarbon tails out on the air side (Fig. 2.1). The same phenomenon occurs in the action of soaps, which are fatty acid salts. The soap-monolayer formation greatly lowers the air-water interfacial tension, and the ability of the solution to wet and cleanse confined regions is greatly increased.

$$Na^+ \; ^-O-\overset{\overset{\displaystyle O}{\|}}{C}-(CH_2)_7-HC=CH-(CH_2)_7-CH_3$$

A soap: sodium oleate

These hydrophilic-hydrophobic lipid molecules possess very small solubilities; elevation of the solution concentration above the monomolecular solubility limit results in the condensation of excess solutes into larger ordered structures called *micelles* (Fig. 2.1). This spontaneous process occurs because the overall free energy of the resultant (micelle plus solution) mixture is lower than that of the original solution. The structure of the micelle is dictated by the favorable increase in the number of hydrophobic-hydrophobic and hydrophilic-hydrophilic contacts and concurrent diminution of hydrophilic-hydrophobic associations. Such interactions between hydrophilic and hydrophobic portions of the *same* biopolymer are also known to favor the folding of such polymer chains into a single preferred configuration. This behavior will be discussed shortly with respect to DNA and proteins.

Important as reservoirs of fuel, *fats* are esters formed by condensation of fatty acids with glycerol:

$$
\begin{array}{lll}
CH_2OH & HO-OC(CH_2)_{n_1}-CH_3 & CH_2O-\overset{\overset{\displaystyle O}{\|}}{C}-(CH_2)_{n_1}-CH_3 \\[6pt]
| & + & | \quad\;\; O \\[6pt]
CHOH \;+ & HO-OC(CH_2)_{n_2}-CH_3 \;\xrightarrow{-3H_2O}\; & CHO\; -\overset{\|}{C}-(CH_2)_{n_2}-CH_3 \\[6pt]
| & + & | \quad\;\; O \\[6pt]
CH_2OH & HO-OC(CH_2)_{n_3}-CH_3 & CH_2O-\overset{\|}{C}-(CH_2)_{n_3}-CH_3 \\[4pt]
\text{Glycerol} & \text{Fatty acids} & \text{A fat}
\end{array}
$$

Fats and other lipids discussed in this section are hydrolyzed to glycerol and soap by heating in alkaline solution, the historical method for making soap from animal fats. The reverse of the fat synthesis reaction shown above is catalyzed by fat-splitting enzymes at body temperatures in the digestive tract of animals; microbes also secrete such enzymes to hydrolyze particulate fats into smaller fragments, which can then be taken in through the cell membrane.

Closely related to the fats in structure but not function are the phosphoglycerides. In these molecules, phosphoric acid replaces a fatty acid esterified to one end

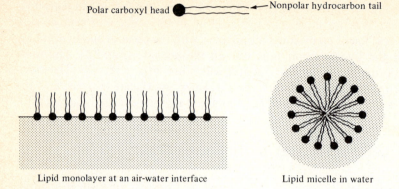

Figure 2.1 Some stable configurations of fatty acids in water.

of glycerol. The result is again a molecule with strongly hydrophilic and hydrophobic portions; thus micelle formation is again observable at sufficiently large phosphoglyceride concentrations. A flat double-molecular layer structure may be formed across a small aperture in a sheet submerged in a phosphoplipid solution (Fig. 2.2a). The resulting planar lipid bilayer has a thickness of about 70 Å (7×10^{-7} cm). Biological plasma membranes typically contain appreciable concentrations of phospholipids and other polar lipids (Table 2.2). Also, plasma membranes show an apparent molecular bilayer (Fig. 2.2b) of thickness similar to

Table 2.2. Some major polymer components of cell walls and membranes

Component	Occurrence
Cell walls:	
Homopolysaccharides:	
Chitin (poly-*N*-acetylglucosamine) (cellulose structure, linear)	Fungi, some yeasts
Polymannans (highly branched)	Yeasts
Polymannuronic acid	Algae
Heteropolysaccharides:	
Murein (poly-*N*-acetylglucosamine-*N*-acetyl-muramic acid) cross-linked by oligopeptides	Bacteria
Techoic acids (ribitol or glycerol polymers with phosphate groups in polymer chain)	Gram-positive bacteria
Lipopolysaccharides	Gram-negative slime-forming bacteria
Plasma membranes:	
Lipids (including phospholipids), protein, and carbohydrate	

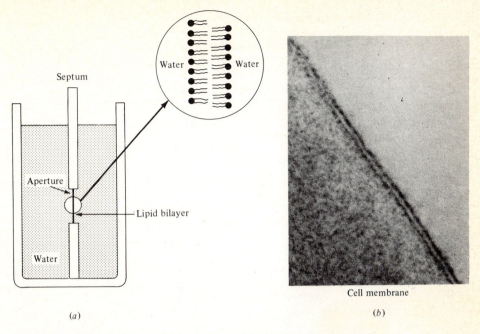

Figure 2.2 (*a*) The spontaneous formation of a stable phosphoglyceride bilayer in the aperture between two compartments filled with water and lipid. (*b*) This structure strongly resembles the bilayer appearance of cell membranes in electron micrographs. (*Electron micrograph reprinted by permission from J. B. Robertson.*) *Membrane Models: Theoretical and Real, in "The Nervous System, vol. 1: The Basic Neurosciences," D. B. Tower (ed.), p. 43, Raven Press, New York.*

the spontaneously formed phosphoglyceride double layer in Fig. 2.2*a*. Consequently, it appears that the bilayer lipid membranes might serve as convenient synthetic systems for fundamental characterization of thin membrane processes.

Several physical properties of lipid bilayer membranes are similar to those of plasma membranes. Both lipid and plasma membranes have high passive electrical resistance and capacitance. The resultant impermeability of natural membranes to such highly charged species as phosphorylated compounds is largely a result of this property. The membrane thereby allows the cell to contain a reservoir of charged nutrients and metabolic intermediates, as well as maintaining a considerable difference between the internal and external concentrations of small cations such as H^+, K^+, and Na^+.

There are at least six means by which cells can exchange material with the surrounding aqueous environment; they are summarized in Table 2.3. The last three paths are all clearly unique to systems with access to an external energy source since in each case work must be done in order to acquire the nutrient in the intracellular volume. Methods 3 and 4 implicate specific compounds in the cell membranes, often called *permeases*, which are probably protein (Sec. 2.4).

Table 2.3. Intra- and extracellular transport mechanisms of microbes

1. Passive diffusion	Solute moves through membrane by Fickian diffusion in direction of decreasing concentration
2. Solvent drag	Water movement across membrane entrains some solutes (ultrafiltration, Chap. 4)
3. Facilitated transport	Sparingly soluble substance passes through membrane by specific complexation with membrane-confined species of greater solubility (Chap. 5)
4. Active transport	Energy-coupled process in which cell does work by concentrating solute *against* its concentration gradient (Chap. 5)
5. Pinocytosis	Solid nutrient particles diffuse or are drawn into tiny digestive channels leading through cell wall
6. Phagocytosis	Especially for some protozoa; cell extends two or more "fingers," which surround food particle and eventually internalize it

The importance of solvent drag does not appear to be established. However, we note that a growing bacterium, for example, must double its volume and thus its water content over the course of the cell cycle.

An intriguing similarity between bilayer lipid membranes and plasma membranes is their ability to be modified in their selective ion permeabilities by the addition of small amounts of various substances. In particular, several antibiotics and other cation-complexing molecules have been found to markedly increase passive ion transport in both types of membranes. In more complex processes, the cell walls of viable organisms can be rendered leaky by mild chemical or heat treatment. This has been used advantageously in the microbial production of metabolic intermediates (Example 2.1) and in the treatment of cells to decrease their nucleic acid content before use as animal foodstuffs.

Example 2.1. Modification of Biomembrane Permeability Microbial populations have historically been used to produce natural end products of their metabolism such as ethanol and acetic acid (Chap. 10). Many metabolic intermediates, such as amino acids, are also valuable; these are normally confined within the cell by the plasma membrane and cell wall. Fukui and Ishida [9, p. 33] indicate in general that the normal cell membrane properties can be altered: "amino acid production by fermentation requires the use of techniques to stimulate excretion of product. These include (1) limitation of the factors required for complete cell wall synthesis, (2) the use of an inhibitor of cell wall synthesis and (3) addition of an agent that partially impairs the function of the cell wall." Knowledge of the cell-wall composition (Table 2.2) and biosynthetic pathways (Chap. 5) clearly aids the rational evolution of a culture-medium composition which utilizes method 1 or 2. An example of the large influence of added chemical agents (method 3) is seen in the effect (Table 2E1.1) of the surface-active compound cetylpyridinium chloride (CPC) on aspartic acid production by the intestinal bacterium *E. coli.*

Table 2E1.1

Time, h	Free L-aspartic acid concentration, g/l†	
	Without CPC	With CPC
1	1	12
2	2	22
3	3	30
4	4	35

† Data from Fukui and Ishida, "Microbial Production of Amino Acids," Fig. 5.5, Kodandsha Ltd., Tokyo, and John Wiley & Sons, Inc., New York, 1972.

We return to further discussion of membrane transport for the first four varieties in Table 2.3 in subsequent chapters; the topics of pinocytosis and phagocytosis are beyond the scope of this book since they involve massive collective motion of the cell membrane. In this context, we should emphasize that biological membranes are considerably more complex than the model lipid bilayer structure of Fig. 2.2a.

The cell membrane and cell walls are the only division between the internal environment needed for cell function and the external medium, often far different from the former. As Table 2.2 indicates, three of the four biopolymer classes (lipid, polysaccharide, and protein) are found in biological membranes and walls, as are complexes and cross-linked combinations of these molecules such as lipoproteins and lipopolysaccharides. The specific composition of a given wall will determine its resistance to change by mechanical disruption, chemical or enzymatic hydrolysis, or other techniques used in cell harvesting and product recovery.

2.1.2. Fat-soluble Vitamins, Steroids, and Other Lipids

A *vitamin* is an organic substance which is required in trace amounts for normal cell function. The vitamins which cannot be synthesized internally by an organism are termed *essential vitamins;* in their absence in the external medium, the cell cannot survive. (This fact has been used advantageously by growing microbes in test media as a probe for the presence or absence of a particular vitamin.) The water-soluble vitamins such as vitamin C (ascorbic acid) are not lipids by definition. However, vitamins A, E, K, and D are water-insoluble and dissolve in organic solvents. Consequently these vitamins are classified as lipids.

The ultimate role of the lipid-soluble vitamins appears obscure, vitamin A being an isolated counterexample. The hydrophobic nature of this vitamin (necessary to prevent night blindness in human beings) is evident from its polyisoprene

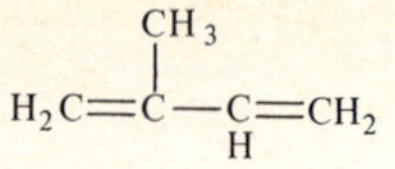

Isoprene

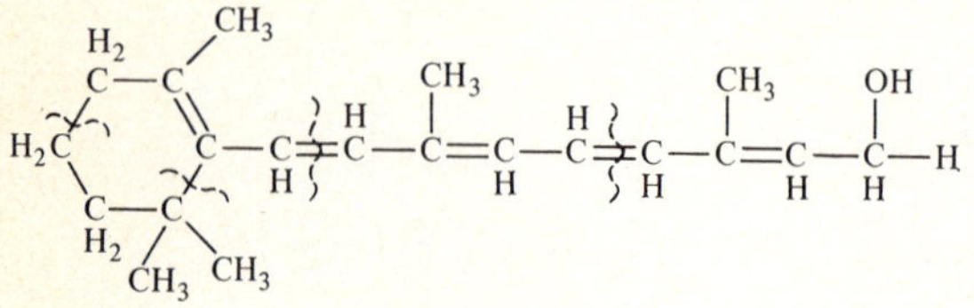

One form of vitamin A
(vitamin A$_1$)

Figure 2.3 Isoprene, a building block of several lipids and lipid components. One of these is the fat-soluble vitamin A.

structure, shown in Fig. 2.3. Interest in vitamin supply from microbial and other foods derives largely from the fact that the water-soluble vitamins thiamine, riboflavin, niacin (nicotinic acid), pantothenic acid, biotin, folic acid, and choline and the lipid vitamins A, D, E, and K are all essential (or probably essential) for children and/or adults. Many microorganisms can synthesize a number of these compounds, yeast, for example, providing the precursor ergosterol, which is converted by sunlight to vitamin D$_2$ (calciferol). The fat-soluble vitamin K is synthesized by microbes in animal and human digestive tracts, an excellent example of mutually assisting populations (commensalism, considered further in Chap. 11). Several water-soluble vitamins are known to be necessary for activity of specific enzymes.

Steroids are a class of lipid biochemicals with the general structure shown in Fig. 2.4a. Of these, a subgroup (hormones) constitutes some of the extremely potent controllers of biological reaction rates; hormones may be effective at levels of 10^{-8} M in human tissue. Microbes are currently used to carry out relatively minor transformations of such active steroids to yield more valuable products. For example, progesterone can be converted into *cortisone* in a two-step process (microbial, then chemical) (Fig. 2.4b). Evidently, the complexity of the reactant is such that only the action of an enzyme, produced by perhaps only one or several kinds of microbes, carries out the reaction with a useful *selectivity* (minimal side-product generation). The familiar steroid *cholesterol* (Fig. 2.4c) occurs almost exclusively in membranes of animal tissues. Related sterol compounds have been shown to alter cell plasma-membrane permeabilities.

Recent research indicates that another kind of lipid known as *prostaglandins* may be of utmost importance in mammalian biology. These substances are formed by oxidation of fatty acids to give a ringed structure in the middle of their hydrocarbon tail. In the sense that they control the level of hormone activity, prostaglandins appear to regulate biological function at a higher level than hormones. Better understanding of these powerful biochemical controllers promises new techniques in medicine.

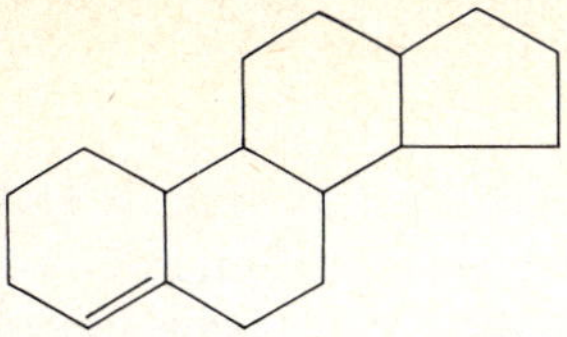

(*a*) General steroid base: perhydroxycyclopentano phenanthrene

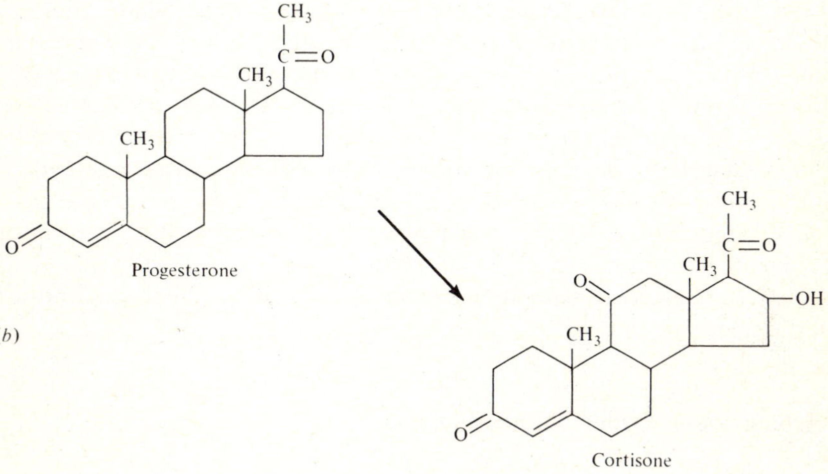
CH3
C=O
CH3
CH3
O
Progesterone
CH3
C=O
CH3
OH
O
CH3
O
Cortisone

(*b*)

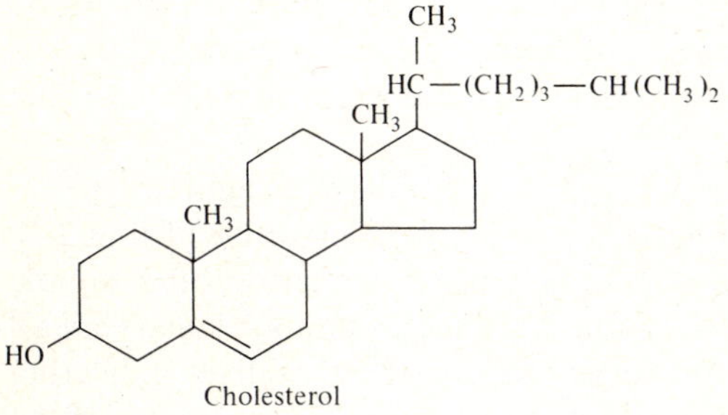
CH3
HC—(CH2)3—CH(CH3)2
CH3
CH3
HO
Cholesterol

(*c*)

Figure 2.4 Some examples of steroid structure.

An important food-storage polymer for some bacteria is poly-β-hydroxybutyric acid (PHB). The repeating unit is

$$\begin{array}{c} \quad CH_3 \qquad\quad O \\ \quad | \qquad\qquad \| \\ -CH-CH_2-C-O- \end{array}$$

The polymer occurs as granules within the cells. In the absence of sufficient food supply, the cell depolymerizes this reserve to yield the soluble, easily metabolized β-hydroxybutyric acid.

2.2. SUGARS AND POLYSACCHARIDES

The *carbohydrates* are organic compounds with the general formula $(CH_2O)_n$, where $n \geq 3$. These compounds are found in all animal, plant, and microbial cells; the higher-molecular-weight polymers serve both structural and storage functions. The formula $(CH_2O)_n$ is sufficiently accurate to be useful in calculating overall elemental balances and energy release in biochemical oxidations.

In the biosphere, carbohydrate matter (including starches and cellulose) exceeds the combined amount of all other organic compounds. When photosynthesis occurs, carbon dioxide is converted to simple sugars (C_3 to C_9) in reactions driven by the incident sunlight (considered further in Chap. 5, bioenergetics). These sugars are then polymerized into forms suitable for structure (cellulose) or sugar storage (starches). By these processes, radiant solar energy is stored in chemical form for subsequent utilization. The magnitude of this energy transformation is estimated to be 10^{18} kcal/yr, corresponding to storage of 0.1 percent of the annual incident radiant energy. Much of the annual 10^{18} kcal stored is of course ultimately released in subsequent oxidation (largely cellular respiration) to carbon dioxide.

2.2.1. D-Glucose and Other Monosaccharides

Monosaccharides, or *simple sugars*, are the smallest carbohydrates. Containing from three to nine carbon atoms, monosaccharides serve as the monomeric blocks for noninformational biopolymers with molecular weights ranging into the millions.

D(+)-Glucose, the optical isomer which rotates polarized light in the + direction, like other simple sugars, is a polyhydroxyalcohol derivative (Table 2.4).

Although D-glucose is by far the most common monosaccharide, other simple sugars are also found in living organisms (Table 2.4). These common sugars are all either aldehyde or ketone derivatives. In sugar nomenclature, prefixes indicating these functional groups are often combined with a name fixing the length of the carbon chain. Thus, glucose is an aldohexose; the notation D referring to a particular optical isomer occurring almost exclusively in living systems (see optical activity, Sec. 2.4).

In solution D-glucose is present largely as a ring structure, *pyranose* (note the standard numbering scheme for six carbons and the α, β labels for the position of the —OH group on the number 1 carbon).

Table 2.4. Monosaccharides commonly found in biological systems

	Aldoses (aldehyde derivatives; prefix *aldo-*)			Ketoses (ketone derivatives; prefix *keto-*)
Triose (three-carbon)	CHO | HCOH | CH_2OH D-Glyceraldehyde			CH_2OH | C=O | CH_2OH Dihydroxyacetone
Pentose (five-carbon)	CHO | HCOH | HCOH | HCOH | CH_2OH D-Ribose			CH_2OH | C=O | HCOH | HCOH | CH_2OH D-Ribulose
Hexose (six-carbon)	CHO | HCOH | HOCH | HCOH | HCOH | CH_2OH D-Glucose	CHO | HOCH | HOCH | HCOH | HCOH | CH_2OH D-Mannose	CHO | HCOH | HOCH | HOCH | HCOH | CH_2OH D-Galactose	CH_2OH | C=O | HOCH | HCOH | HCOH | CH_2OH D-Fructose

The five membered sugars D-ribose and deoxyribose are major components of the nucleic acid monomers of DNA and RNA and other biochemicals to be discussed shortly.

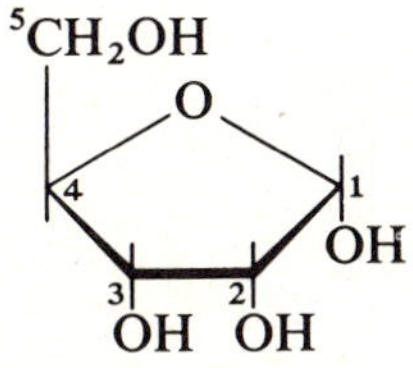

D-Ribose

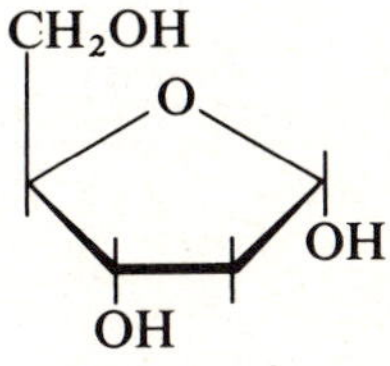

Deoxyribose

2.2.2. Disaccharides to Polysaccharides

Because the ringed form of many simple sugars predominates in solution, they do not exhibit the characteristic reactions of aldehydes or ketones. In the D-glucose ring above, the —OH attached to position 1 is relatively reactive. As shown below, this hydroxyl group can condense with an —OH on the 4 carbon of another sugar to eliminate a water molecule and form an α-1,4-*glycosidic bond*:

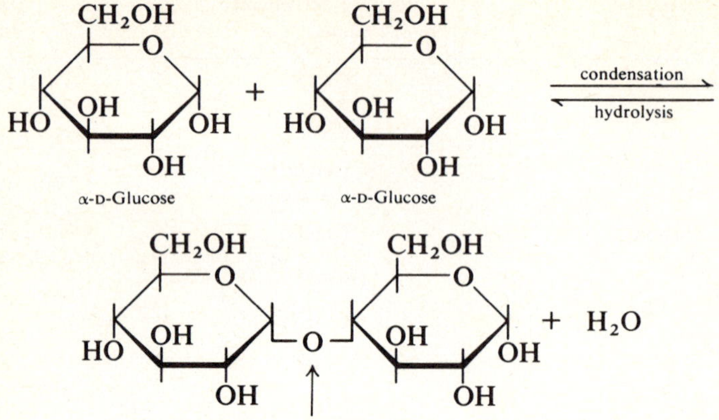

α-D-Glucose   α-D-Glucose

condensation →
← hydrolysis

α-1,4-glycosidic bond

α-Maltose

The condensation product of two monosaccharides is a *disaccharide*. In addition to maltose, which is formed from two D-glucose molecules, the following disaccharides are relatively common:

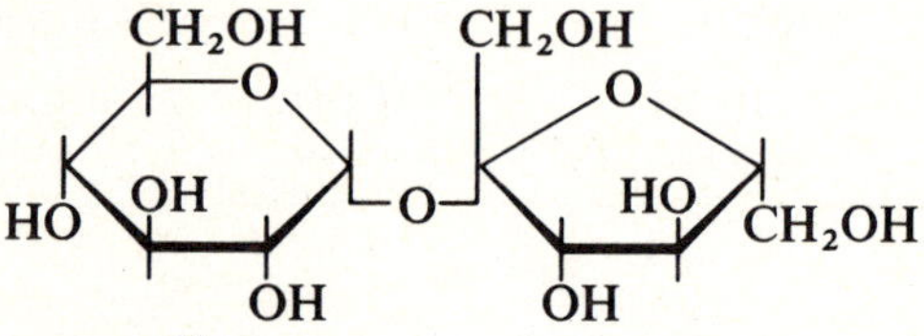

α-D-Glucose   β-D-fructose

Sucrose

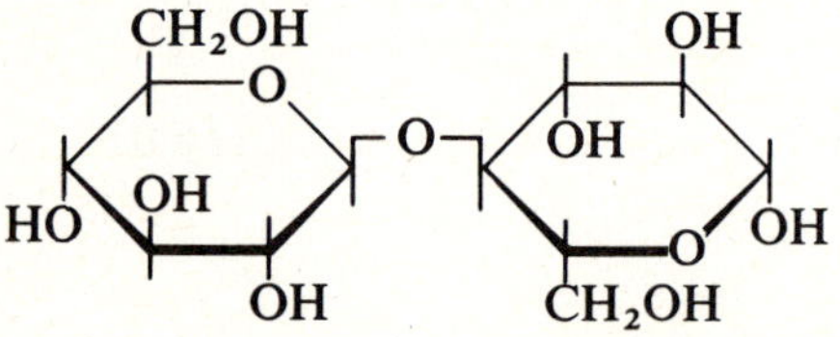

β-D-Glucose   β-D-galactose

Lactose

Table sugar is *sucrose*, a major foodstuff which is found in all photosynthetic plants. Among all disaccharides, sucrose is the easiest to hydrolyze: the resulting mixture of glucose and fructose monosaccharides is called *invert sugar*. Found only in milk, *lactose* is a relatively rare but important disaccharide. Since many people are lactose-intolerant and therefore cannot digest milk, enzyme processes to hydrolyze milk lactose are currently under development.

Continued polymerization of glucose can occur by formation of new 1,4-glycosidic bonds. *Amylose* is a straight-chain polymer of glucose subunits whose molecular weight may vary from several thousand to half a million:

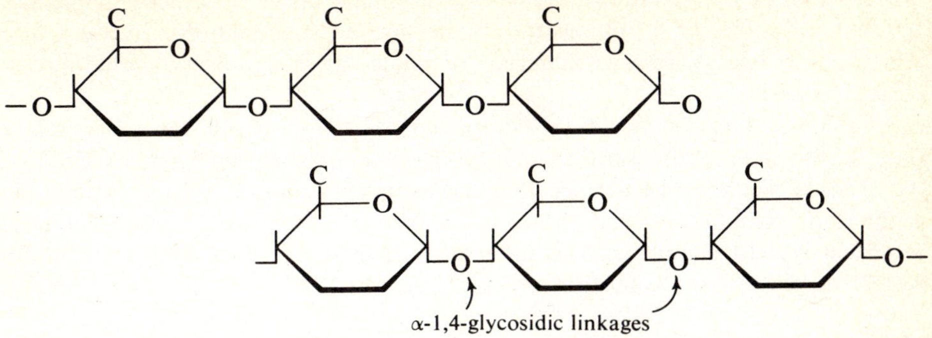

α-1,4-glycosidic linkages

Portion of an amylose chain (−OH groups omitted for clarity)

Amylose typically constitutes about 20 percent of *starch*, the reserve food in plants, although this percentage varies widely. Granules of starch are large enough to be seen in many plant cells examined through a microscope.

While the amylose fraction of starch consists of straight-chain, water-insoluble polymers[†], the bulk of starch is *amylopectin*. Amylopectin, also a D-glucose polymer, is distinguished by a substantial amount of branching. Branches occur from the ends of amylose segments averaging 25 glucose units in length. Such structures arise when condensation occurs between the glycosidic −OH on one chain and the 6 carbon on another glucose:

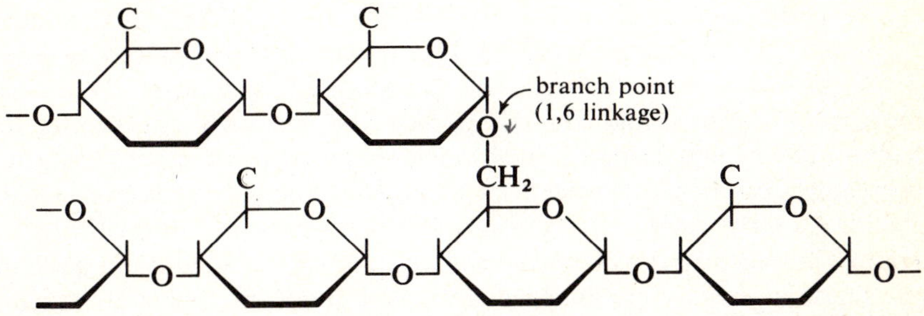

branch point
(1,6 linkage)

Amylopectin molecules are typically larger than amylose, with molecular weights ranging up to 1 to 2 million. Amylopectin is soluble in water, and it can form gels by absorbing water. After partial hydrolysis of starch, by acid or certain enzymes, many branched remnants of amylopectin called *dextrins* remain. Dextrins are used as thickeners and in pastes. Naturally, glucose, maltose, and other relatively small sugars are also obtained by partial hydrolysis. Corn syrup is derived from corn starch in this manner.

The glucose reservoirs in animals, especially numerous in liver and muscle cells, are granules of *glycogen*, a polymer which resembles amylopectin in that it is

† M. M. Green, G. Blankenhorn, and H. Hart, Textbook Errors. 123: Which Starch Fraction is Water-Soluble, Amylose or Amylopectin, J. Chem. Educ., **52**: 729, 1975.

highly branched. Here, the degree of branching is greater as there are typically only 12 glucose units in the straight-chain segments between branches. Glycogen molecular weights of 5 million and greater are not uncommon. Glycogen also serves as an energy reserve material for some microorganisms, including the enteric bacteria.

Cellulose, a major structural component of all plant cell walls from algae (Fig. 2.5) to trees, is the most abundant organic compound on earth. Cotton and wood are two common examples of materials rich in cellulose. Estimates place the amount of cellulose formed in the biosphere at 10^{11} tons/yr. Each cellulose molecule is a long, unbranched chain of D-glucose subunits with a molecular weight ranging from 50,000 to over 1 million.

The glucose chain of cellulose

Although the glycosidic linkage in cellulose occurs between the 1 and 4 carbons of successive glucose units, the subunits are bonded differently than in amylose (compare the following structural formula with the preceding one for amylose). This difference in structure is extremely significant. While many microorganisms, plants, and animals possess the necessary enzymes to break (hydrolyze) the α-1,4-glycosidic bonds which are found in starch or glycogen, few living creatures can hydrolyze the β-1,4 bonds of cellulose. (If cellulose were liable to attack by a large variety of organisms, it would not be a very useful structural material.) The exceptions which can utilize cellulose as foodstuffs are quite interesting, since *symbiotic* relationships are often involved. Bacteria which can decompose cellulose live in the rumen of the cow and in the intestines of the termite. Consequently, in return for a steady diet of grass or wood (already masticated), these bacteria assist their hosts' digestion. No such cellulytic bacteria inhabit the human gut. Development of economical processes for cellulose degradation, which would ease waste-disposal problems and provide valuable simple sugars, is a major current activity in biochemical engineering.

The complex cell-wall structures which enclose bacteria are composed of murein (Table 2.2), a network of chains of glucose derivatives which are cross-linked by chains of amino acids (Sec. 2.4.1). The cross-linking is believed to be so extensive that the cell wall of bacteria is essentially one giant macromolecular bag.

The outer membranes and walls of microbes must be resistant to degradation by many extracellular (present outside the cell) enzymes if the organisms are to survive. Existence of such a protective cell wall permits, for example, the existence

Figure 2.5 Electron micrograph of lamellae from the wall of a green marine alga, *Chaetomorpha melagonium*, viewed from outside the cell. Shadowed, Pd/Au., magnification 30,000 ×. (*By courtesy of Dr. Eva Frei and Professor R. D. Preston, F. R. S., Astbury Department of Biophysics, University of Leeds U.K.*)

of cellulytic bacteria in ruminants and vitamin K producers in human digestive tracts. As microbes are currently being extensively considered as an additional protein food source, it appears in general that for human consumption the maximum digestible protein value is obtained only after pretreatment to rupture or hydrolyze cell membranes and walls. Techniques for this purpose as well as enzyme recovery are considered in Chap. 4.

2.3. FROM NUCLEOTIDES TO RNA AND DNA

The *informational* biopolymer DNA (deoxyribonucleic acid) contains all the cell's hereditary information. When cells divide, each daughter cell receives a complete copy of its parent's DNA, which allows offspring to resemble parent in form and operation. The information which DNA possesses is found in the *sequence* of the subunit nucleotides along the polymer chain. The mechanism of information transfer and the timing of DNA replication in the cell cycle are discussed in Chap. 6, along with other aspects of cellular controls. The subunit chemistry, the structure of the DNA polymer, and the information coded in the subunit sequence are now considered.

2.3.1. Building Blocks, the Energy Carrier, and Coenzymes

In addition to their presence in nucleic acids, nucleotides and their derivatives are of considerable biological interest on their own. Three components make up all nucleotides: (1) phosphoric acid, (2) ribose or deoxyribose (five-carbon sugars), and (3) a nitrogenous base, usually derived from either purine or pyrimidine (see below).

Purine   Pyrimidine

These three components are joined in a similar fashion in two nucleotide types (Fig. 2.6) which are distinguished by the five-carbon sugar involved.
Ribonucleic acid (RNA) is a polymer of ribose-containing nucleotides, while deoxyribose is the sugar component of nucleotides making up the biopolymer deoxyribonucleic acid (DNA). Also shown in Fig. 2.6 are the four nitrogenous bases found in DNA and RNA nucleotide components. Three of the bases, adenine (A), guanine (G), and cytosine (C), are common to both types of nucleic acids. Thymine (T) is found only in DNA, while uracil (U), a closely related pyrimidine base, is unique to RNA. Both series of nucleotides are strong acids because of their phosphoric acid groups.

Of particular biological significance is the nucleotide *adenosine*, made with ribose and adenine. Sketched in Fig. 2.7, along with its important derivatives, is *adenosine monophosphate* (AMP). One or two additional phosphoric acid groups can condense with AMP to yield ADP (adenosine diphosphate) and ATP (adenosine triphosphate), respectively. The phosphodiester bonds connecting the phosphate groups have especially useful free energies of hydrolysis. For example, the reaction of ATP to yield ADP and phosphate is accompanied by a Gibbs free-energy change of -7.3 kcal/mol at 37°C and pH 7 [recall that pH $= -\log_{10}$ (molar H^+ concentration)].

Although we are accustomed to thinking primarily in terms of thermal energy, the cell is essentially an isothermal system where *chemical* energy transformations are the rule. As examined in considerable additional detail in Chap. 5, ATP is the major carrier of chemical energy in all cells. That is, ATP is the means by which the cell temporarily stores the energy derived from nutrients or sunlight for subsequent use such as biosynthesis of high polymers, transport of materials through membranes, and cell motion. While adenosine phosphates are the predominant forms of energy carriers, the diphosphate and triphosphate derivatives of the other nucleotides also serve related functions in the cell's chemistry.

The significance of the cyclic form of AMP, so called because of an intramolecular ring involving the phosphate group (Fig. 2.7), is only beginning to emerge. It serves as a regulator in many cellular reactions, including those that

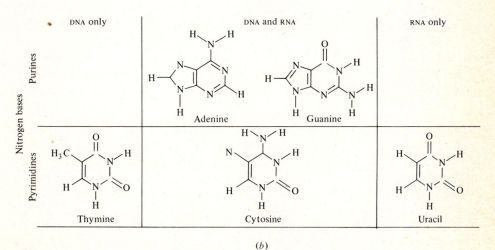

Figure 2.6 (a) The general structure of ribonucleotides and deoxyribonucleotides. (b) The five nitrogenous bases found in DNA and RNA.

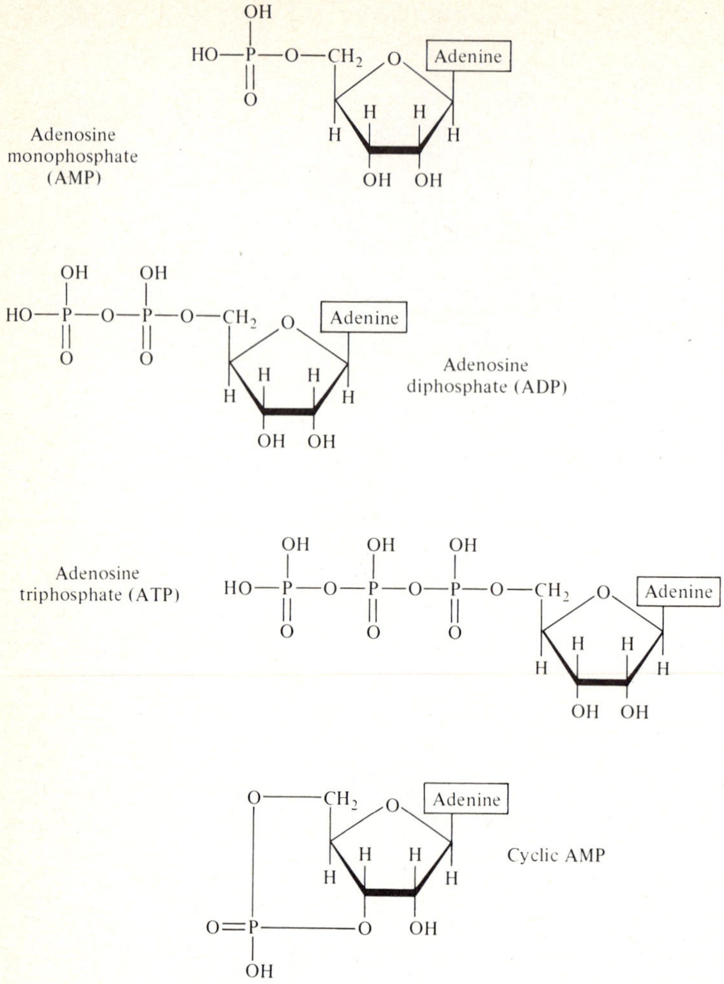

Figure 2.7 The phosphates of adenosine. AMP, ADP, and ATP are important in cellular energy transfer processes, and cyclic AMP serves a regulatory function.

form sugar and fat-storage polymers. An inadequacy of cyclic AMP in tissue is related to one kind of cancer, a condition of relatively uncontrolled cell growth.

In addition to providing components for nucleic acids, the adenine-ribose monophosphate is a major portion of the *coenzymes* shown in Fig. 2.8. Enzyme kinetics are considered in the following chapter; it suffices here to note that the *coenzyme* is the additional organic moiety which is necessary for the activation of certain enzymes to the catalytically useful form. As essentially all reactions within the cell are catalyzed by enzymes, variations of the coenzyme level provide a convenient short-term method for cellular regulation of *active* enzyme and thus of the rate of intracellular reactions.

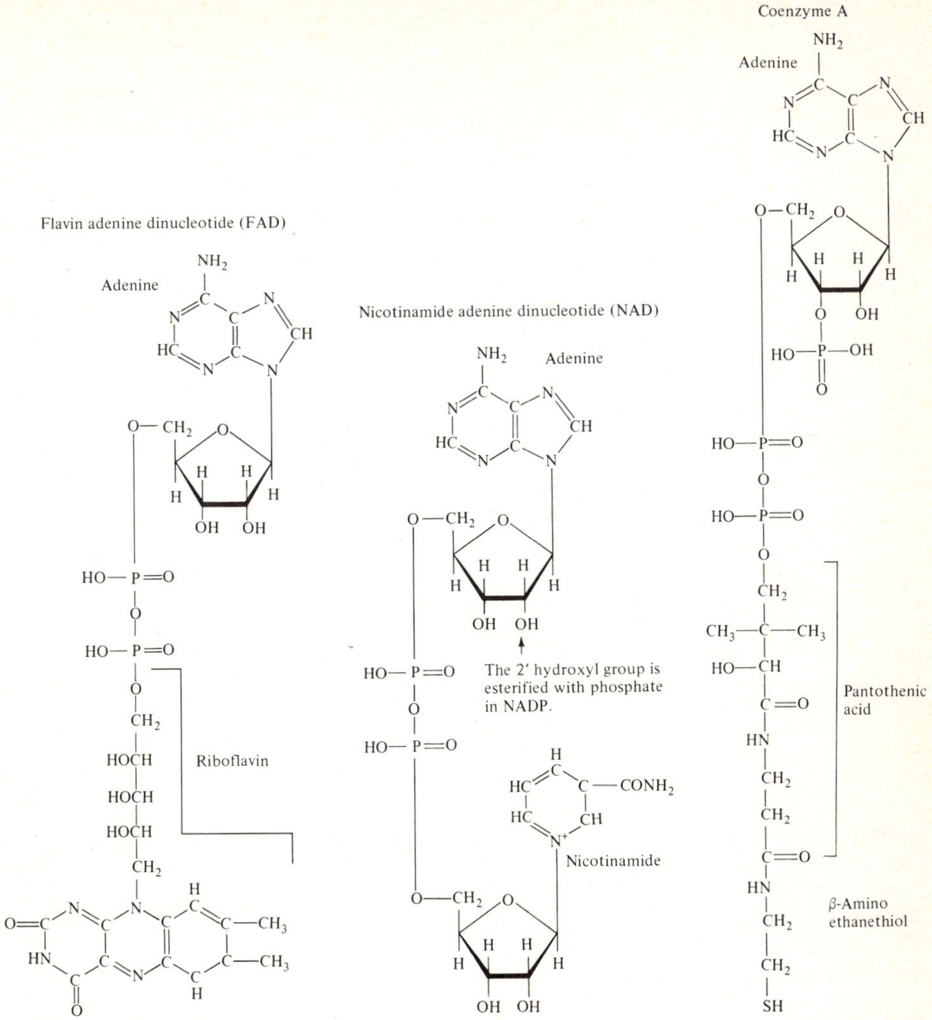

Coenzyme A

NH_2

Adenine

Flavin adenine dinucleotide (FAD)

NH_2

Adenine

Nicotinamide adenine dinucleotide (NAD)

NH_2 Adenine

$O-CH_2$

$HO-P=O$

$HO-P=O$

CH_2

HOCH   Riboflavin

HOCH

HOCH

CH_2

$O=C$ N N C—CH_3

HN C N C—CH_3

C H

O

The 2' hydroxyl group is
esterified with phosphate
in NADP.

HC C—$CONH_2$

HC CH

Nicotinamide

$O-CH_2$

OH OH

$O-CH_2$

OH OH

$HO-P-OH$

$HO-P=O$

$HO-P=O$

CH_2

CH_3-C-CH_3

$HO-CH$ Pantothenic
                     acid
$C=O$

HN

CH_2

CH_2

$C=O$

HN   β-Amino
                    ethanethiol
CH_2

CH_2

SH

Figure 2.8 Three important coenzymes derived from nucleotides. (*From A. L. Lehninger, "Bio-chemistry," 1st ed., pp. 250, 248, Worth Publishers, N.Y. 1970.*)

2.3.2. Biological Information Storage: DNA and RNA

As with the polysaccharides, the polynucleotides are formed by condensation of its monomers. For both RNA and DNA, the nucleotides are connected between the 3' and 5' carbons of successive sugar rings. Fig. 2.9 illustrates a possible sequence in a trinucleotide. The DNA molecules in cells are enormously larger: all the hereditary information in procaryotes is contained in one DNA molecule with a molecular weight of the order of 2×10^9. In eucaryotes, the nucleus may contain several large DNA molecules. The negative charges on DNA are balanced by

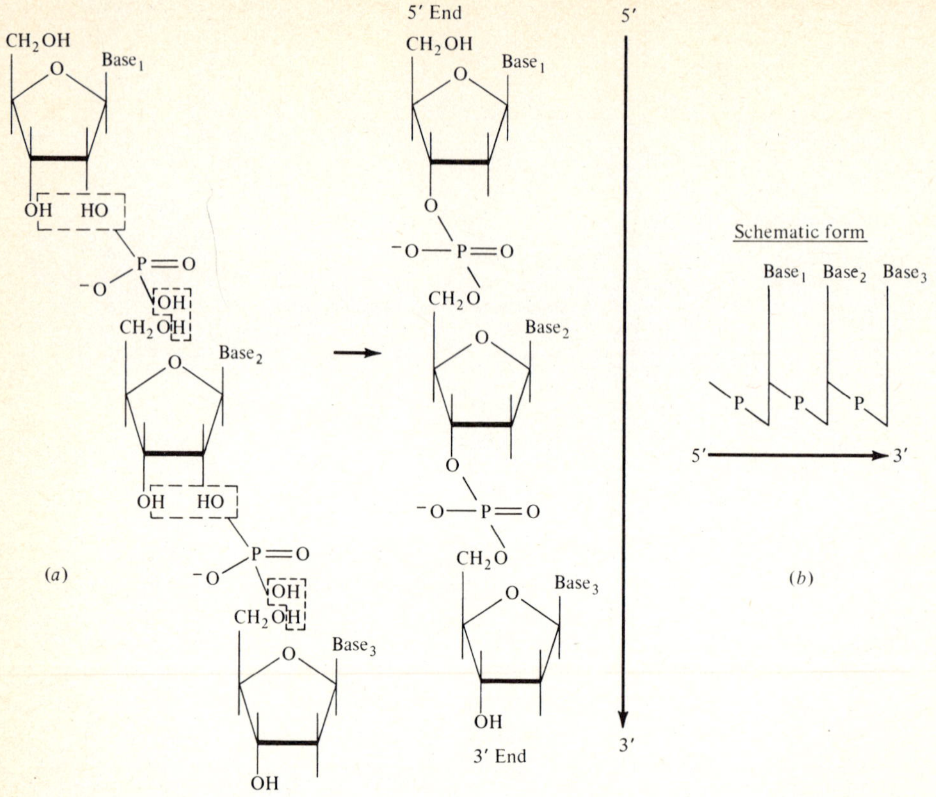

Figure 2.9 (a) Condensation of several nucleotides to form a chain linked by phosphodiester bonds. Notice the chain has a direction, there is a 5′ sugar carbon on one end and a hydroxyl group on the 3′ carbon at the other. In a listing sequence of nucleotide bases, it is conventional to begin from the 5′ end. (b) The standard schematic for a nucleotide chain.

divalent ions (procaryotes) or basic amino acids (eucaryotes). Recently it has been found that organelles such as mitochondria and chloroplasts also contain DNA and that these organelles reproduce independently when a eucaryote divides. This has led to speculation that these relatively sophisticated organelles have evolved from small procaryotes which long ago established a symbiotic relationship with a larger cell. Other cell components containing DNA will be introduced in Chap. 6.

In the two intertwined helical molecules which make up DNA (Fig. 2.10), the phosphate groups are found on the outer surface and the bases point toward the chain center. The neighboring bases of the two chains are always paired in the same manner: adenine with thymine and cytosine with guanine. Fig. 2.11 illustrates this pairing, which provides molecular stabilization by hydrogen bonding between complementary bases. While the hydrogen bonds are relatively weak

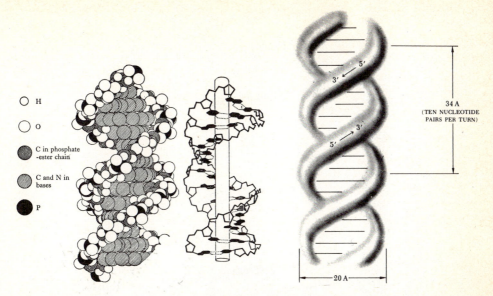

H

O

C in phosphate -ester chain

C and N in bases

P

5′
3′
3′
5′

34 A
(TEN NUCLEOTIDE
PAIRS PER TURN)

← 20 A →

Figure 2.10 Several views of the double-helical structure of DNA. The sketch on the right reveals some geometrical parameters of the molecule. (*Left-hand and center diagrams reprinted with permission from M. Yudkin and R. Offord, "A Guidebook to Biochemistry," p. 52, Cambridge University Press, London*, 1971.

(from several to 12 kcal/mol), the large number formed in DNA and other informational biopolymers represents a major structural determinant.

Base pairing within DNA (Fig. 2.11) also provides a means for duplication of DNA (Fig. 2.12). If the two complementary DNA strands are separated and double helices are constructed from each strand following the base-pairing rules, the end products are two new molecules, each identical to the original double-stranded DNA and each containing one new strand and one old strand. Therefore, base pairing provides a chemical reader for the biological message coded in the DNA nucleotide sequence.

There are three distinct types of ribonucleic acid (RNAs) in all cells and a fourth type found in some viruses. Like DNA, each RNA molecule has a definite sequence of nucleotides. The nucleotides in RNA, however, come from the series with ribose rather than deoxyribose as the sugar component, and uracil substitutes for thymine in pairing with adenine. The various RNAs which participate in normal cell function serve the purpose of reading and implementing the genetic message of DNA. *Messenger RNA* (mRNA) is formed in the nucleus and is complementary to a base sequence from DNA. Each mRNA molecule carries a message from DNA to another part of the cell. Since the length of these messages

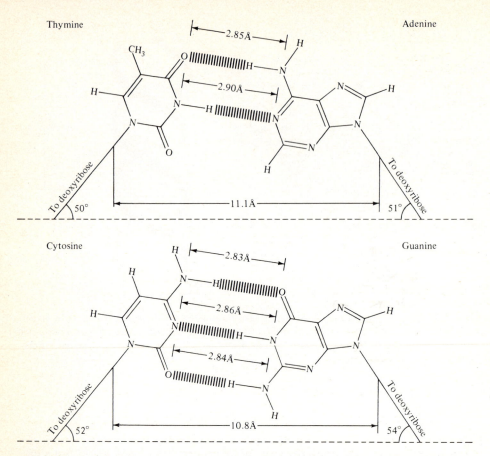

Figure 2.11 The complementary structures of the adenine-thymine and guanine-cytosine pairs provide for hydrogen bonding while maintaining very similar pair geometries. This feature permits base pairing within the DNA molecule double helix. (*From "Cell Structure and Function," 2d ed., p. 141, by Ariel G. Loewy and Philip Siekevitz. Copyright © 1963, 1969 by Holt, Rinehart and Winston. Reprinted by permission of Holt, Rinehart, and Winston.*)

varies considerably, so does the size of mRNA; a chain of 10^4 nucleotides is a typical value.

The message from mRNA is read in the *ribosome* (Figs. 1.2, 1.3, and 1.6). Up to 65 percent of the ribosome is *ribosomal RNA* (rRNA), which in turn can be separated in a centrifuge into three different varieties of nucleic acids denoted as 23S, 16S, and 5S and with characteristic chain lengths of 3×10^3, 1.5×10^3, and 10^2 nucleotides, respectively.

Transfer RNA (tRNA), the smallest type, with only 70 to 80 nucleotide components, is found in the cell's cytoplasm and assists in the translation of the genetic code at the ribosome. Table 2.5 characterizes the various RNAs in the *E. coli* bacterium.

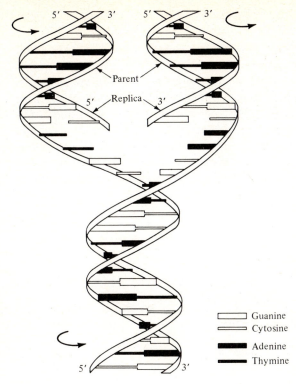

5' 3'   5'   3'

Parent

Replica   3'

5'

5'   3'

⎕⎕⎕⎕ Guanine
═══ Cytosine
▬▬▬ Adenine
▬▬▬ Thymine

Figure 2.12 A simplified diagram of DNA replication. As the parent strands separate, complementary strands are added to each parent, resulting in two daughter molecules identical to the parent. Notice that each daughter molecule contains one strand from the parent. (*From "Cell Structure and Function,"* 2d ed., p. 145, by Ariel G. Loewy and Philip Siekevitz. Copyright © 1963, 1969 by Holt, Rinehart and Winston, Inc. Reprinted by permission of Holt, Rinehart and Winston.)

The end result of this intricate information transmitting and translating system is a protein molecule. Proteins are the tangible biochemical expression of the information and instructions carried by DNA. In the terminology of classical process control, they are the final control elements which implement the DNA controller messages to the cellular process. The dynamics of these controllers is reconsidered in Chaps. 3 and 6.

Table 2.5. Properties of *E. coli* RNAs†

Type	Sedimentation coefficient	mol wt	No. of nucleotide residues	Percent of total cell RNA
mRNA	6S–25S	25,000–1,000,000	75–3000	~2
tRNA	~4S	23,000–30,000	75–90	16
rRNA	5S	~35,000	~100 ⎫	
	16S	~550,000	~1500 ⎬	82
	23S	~1,100,000	~3100 ⎭	

† From A. L. Lehninger, "Biochemistry," 2d ed., table 12-3, Worth Publishers, Inc, New York, 1975.

2.4. AMINO ACIDS INTO PROTEINS

If inorganic molecules including water are excluded, *proteins* are the most abundant organic molecules within the cell: typically 30 to 70 percent of the cell's dry weight is protein. All proteins contain the four most prevalent biological elements: carbon, hydrogen, nitrogen, and oxygen. Average weight percentages of these elements in proteins are C (50 percent), H (7 percent), O (23 percent), and N (16 percent). In addition, sulfur (up to 3 percent) contributes to the three-dimensional stabilization of almost all proteins by the formation of disulfide (S—S) bonds between sulfur atoms at different locations along the polymer chain. The size of these nonrepetitive biopolymers varies considerably, molecular weights ranging from 6000 to over 1 million.

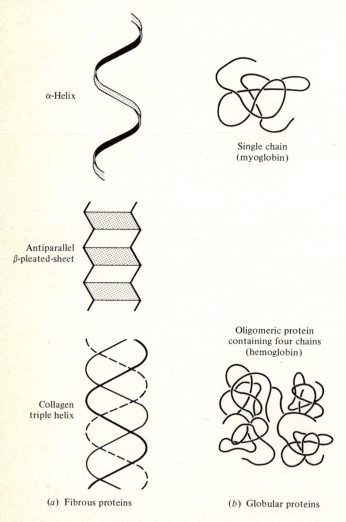

α-Helix

Single chain
(myoglobin)

Antiparallel
β-pleated-sheet

Oligomeric protein
containing four chains
(hemoglobin)

Collagen
triple helix

(*a*) Fibrous proteins

(*b*) Globular proteins

Figure 2.13 The two major types of protein structure with variations. (*a*) fibrous and (*b*) globular.

In view of the great diversity of biological functions which proteins can serve (Table 2.6), the prevalence of proteins within the cell is not surprising. The two major types of protein conformation, fibrous and globular, are shown in Fig. 2.13. Some proteins contribute to the structure of cell membranes (Table 2.2). Others can serve a motive function. Some single-celled organisms possess small, hairlike

Table 2.6. The diverse biological functions of proteins†

Protein	Occurrence or function
Enzymes (biological catalysts):	
Ribonuclease	Hydrolyzes RNA
Cytochrome *c*	Transfers electrons
Trypsin	Hydrolyzes some peptides
Storage proteins:	
Ovalbumin	Egg-white protein
Casein	Milk protein
Zein	Seed protein of corn
Transport proteins:	
Hemoglobin	Transports O_2 in blood of vertebrates
Myoglobin	Transports O_2 in muscle
Serum albumin	Transports fatty acids in blood
β_1-Lipoprotein	Transports lipids in blood
Contractile proteins:	
Dynein	Cilia and flagella
Protective proteins in vertebrate blood:	
Antibodies	Form complexes with foreign proteins
Thrombin	Component of blood clotting mechanism
Toxins:	
Clostridium botulinum toxin	Causes bacterial food poisoning
Diphtheria toxin	Bacterial toxin
Snake venoms	Enzymes which hydrolyze phosphoglycerides
Hormones:	
Insulin	Regulates glucose metabolism
Growth hormone	Stimulates growth of bones
Structural proteins:	
Glycoproteins	Cell coats and walls
Membrane-structure proteins	Component of membranes
α-Keratin	Skin, feathers, nails, hoofs
Sclerotin	Exoskeletons of insects
Collagen	Fibrous connective tissue (tendons, bone, cartilage)
Elastin	Elastic connective tissue (ligaments)

† From A. L. Lehninger, "Biochemistry," table 3-3, Worth Publishers, Inc., New York, 1970.

appendages, called *flagella*, whose motion serves to propel the cell; the flagella are driven by contractile proteins.

A critical role of certain proteins is catalysis. Protein catalysts called *enzymes* determine the rate of chemical reactions occurring within the cell and are also implicated in selective transport of solutes through cell and organelle membranes. Enzymes are found in a variety of cell locales: some are suspended in the cytoplasm and thus dispersed throughout the cell interior. Others are localized by attachment to membranes or association with larger molecular assemblies.

Enzymes are isolated, purified, and characterized by physical and chemical techniques often applicable to other proteins as well. Protein separation methods (Chap. 4) are based on differences in molecular properties, which in turn are partly due to the characteristics of the constituent amino acids, our next topic.

2.4.1. Amino Acid Building Blocks and Polypeptides

The monomeric building blocks of polypeptides and proteins are the α-amino acids, whose general structure is

$$\text{H}_2\text{N}\cdots\overset{\overset{\displaystyle \text{H}}{|}}{\underset{\underset{\displaystyle \text{R}}{|}}{\text{C}}}\cdots\text{COOH}$$

Thus, amino acids are differentiated according to the R group attached to the α carbon (the one closest to the —COOH group). Since there are generally four different groups attached to this carbon (the exception being glycine, for which R = H), it is asymmetrical.

Many organic compounds of biological importance, sugars and amino acids included, are optically active: they possess at least one asymmetric carbon and therefore occur in two isomeric forms, as shown below for the amino acids:

$$\text{H}_2\text{N}\cdots\overset{\overset{\displaystyle \text{H}}{|}}{\underset{\underset{\displaystyle \text{R}}{|}}{\text{C}}}\cdots\text{COOH} \qquad\qquad \text{HOOC}\cdots\overset{\overset{\displaystyle \text{H}}{|}}{\underset{\underset{\displaystyle \text{R}}{|}}{\text{C}}}\cdots\text{NH}_2$$

$$\text{L form} \qquad\qquad\qquad\qquad \text{D form}$$

Solutions of a pure isomer will rotate plane polarized light either to the right (dextro- or *d*-) or left (levo- or *l*-). This isomerism is extraordinarily important since viable organisms can usually utilize only one form, lacking the isomerization catalysts needed to interconvert the two. Such a feature has been used to remove the undesired form from a racemic (*dl*) mixture of the same amino acid: a microbe utilizing only the undesired form is grown in the solution, leaving the desired form untouched. (What new separation cost must now be paid?)

The enzyme catalysts themselves are typically capable of catalyzing reactions involving one of the two forms only. This property has been used commercially in Japan to resolve a racemic mixture of amino acids, converting one isomer only, so that the final solution consists of two considerably different and hence more easily separable components. Additional details on this process follow in Chap. 4. As direct physical resolution of optical isomers is expensive and slow, such microbial and enzymatic (as well as other chemical) aids for resolution may become more important.

It is intriguing to note that only the L-amino acid isomers are found in proteins. Appearance of D-amino acids in nature is rare: they are found in the cell walls of some microorganisms and also in some antibiotics.

The acid ($-COOH$) and base ($-NH_2$) groups of amino acids can ionize in aqueous solution. The acid is positively charged (cation) at low pH and negatively charged (anion) at high pH. At an intermediate pH value, the amino acid acts as a dipolar ion, or zwitterion, which has no net charge. This pH is the *isoelectric point*, which varies according to the R group involved (Table 2.7). An amino acid at its isoelectric point will not migrate under the influence of an applied electric field; it also exhibits a minimum in its solubility. These properties of amino acids allow utilization of such separation techniques as ion exchange, electrodialysis, and electrophoresis for mixture resolution (Chap. 4).

Illustrated in Fig. 2.14 are the 20 amino acids commonly found in proteins. In addition to the carboxyl and amine groups possessed by all amino acids (except proline), the R groups of some acids can ionize. Several of the acids possess nonpolar R groups which are hydrophobic, others being considerably hydrophilic in character. The nature of these side chains is important in determining both the ultimate role played by the protein and the structure of the protein itself, as discussed in the next section.

Table 2.7. pK ($= -\log K$) **values of terminal amino, terminal carboxyl, and R groups for several amino acids**†

Amino acid	pK_1 α-COOH	pK_2 α-NH$_3$	pK_R R group
Glycine	2.34	9.6	
Alanine	2.34	9.69	
Leucine	2.36	9.60	
Serine	2.21	9.15	
Threonine	2.63	10.43	
Glutamine	2.17	9.13	
Aspartic acid	2.09	9.82	3.86
Glutamic acid	2.19	9.67	4.25
Histidine	1.82	9.17	6.0
Cysteine	1.71	10.78	8.33
Tyrosine	2.20	9.11	10.07
Lysine	2.18	8.95	10.53
Arginine	2.17	9.04	12.48

† From A. L. Lehninger, "Biochemistry," 2d ed., table 4-2, Worth Publishers, Inc., New York, 1975

Number of carbon atoms in R-group

0	1	2	3	4	Cyclic

H—

CH$_3$—

$$\begin{array}{c} CH_3 \\ \diagup \\ CH- \\ \diagdown \\ CH_3 \end{array}$$

$$\begin{array}{c} CH_3 \\ \diagup \\ CH-CH_2- \\ \diagdown \\ CH_3 \end{array}$$

Glycine
(Gly)

Alanine
(Ala)

Valine
(Val)

Leucine
(Leu)

Tryptophan
(Trp)

$$CH_3-S-CH_2-CH_2-$$

Methionine
(Met)

—CH$_2$—

Phenylalanine
(Phe)

$$\begin{array}{c} CH_3-CH_2-CH- \\ | \\ CH_3 \end{array}$$

Isoleucine
(Ile)

Hydrophobic

Figure 2.14 The twenty amino acids found in proteins.

Simple proteins are condensation products of amino acid chains. In protein formation, the condensation reaction occurs between the amino group of one amino acid and the carboxyl group of another, forming a *peptide bond*:

Since the peptide bond has some double-bond character, the six atoms in the dashed-box region lie in a plane. Note that every amino acid links with the next via a peptide bond, thus suggesting that a single catalyst (enzyme) can join all the subunits provided that some other mechanism orders the subunits in advance of peptide-bond formation (Chap. 6).

The amino acid fragment remaining after peptide-bond formation is the *residue*, which is designated with a *-yl* ending, e.g., glycine, glycyl; alanine, alanyl. The list of a sequence of residues in an oligopeptide is begun with the end residue containing the free amino group.

Polypeptides are formed by further condensation reactions (Fig. 2.15). A little reflection suggests that as the length of the chain increases, the physicochemical characteristics of the polymer will be increasingly dominated by the R groups of

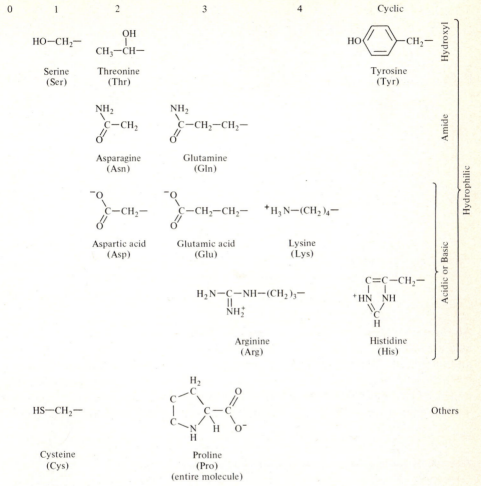

Figure 2.14 (*cont.*)

the residues while the amino and carboxyl groups on the ends will shrink in importance. By convention, the term polypeptide is reserved for these relatively short chains. These molecules have considerable biological significance. Many hormones are polypeptides.

Larger chains are called *proteins*, with the diffuse dividing line between these two classifications ranging from 50 to 100 amino acid residues. Since the average molecular weight of a residue is about 120, proteins have molecular weights larger than about 10,000, and some protein molecular weights exceed 1 million.

Determination of the amino acid content of a particular protein or a protein mixture can be accomplished in an automated analyzer. Total hydrolysis of the protein can be achieved by heating it in 6 N HCl for 10 to 24 h at 100 to 120°C.

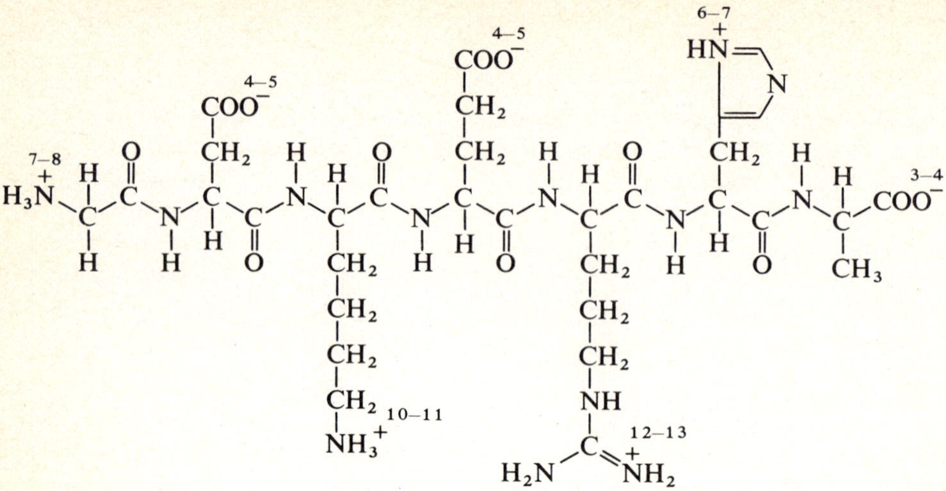

Gly-Asp-Lys-Glu-Arg-His-Ala

Figure 2.15 A hypothetical polypeptide, showing the ionizing groups found among the amino acids. Notice that the listing sequence for the amino acid residues starts at the N-terminal end.

All the amino acids with the exceptions of tryptophan, glutamic acid, and aspartic acid are recovered intact in the hydrochloride form. These can then be separated and measured. Alkali hydrolysis can be used to estimate the tryptophan content. The result of such experiments on the proteins from *E. coli*, an intestinal bacterium, are listed in Table 2.8. From these and other data, it has been found that *not all* 20 amino acids in Fig. 2.14 are found in all proteins. Amino acids do not occur

Table 2.8. Proportions of amino acids contained in the proteins of *E. coli*†

Amino acid	Relative frequency (Ala = 100)	Amino acid	Relative frequency (Ala = 100)
Ala	100	Thr	35
Glx(Gln + Glu)	83	Pro	35
Asx(Asn + Asp)	76	Ile	34
Leu	60	Met	29
Gly	60	Phe	25
Lys	54	Tyr	17
Ser	46	Cys	14
Val	46	Trp	8
Arg	41	His	5

† From A. L. Lehninger, "Biochemistry," 2d ed., table 5.3, Worth Publishers, Inc., New York, 1975.

in equimolar amounts in any known protein. For a given protein, however, the relative amounts of the various amino acids are fixed.

Amino acids are not the only constituents of all proteins. Many proteins contain other organic or even inorganic components. The other part of these *conjugated proteins* is a *prosthetic group*. If only amino acids are present, the term *simple protein*, already introduced above, is used. *Hemoglobin*, the oxygen-carrying molecule in red blood cells, in a familiar example of a conjugated protein: it has four heme groups, which are organometallic complexes containing iron. A related, smaller molecule, *myoglobin*, has one heme group. As discussed above, the prosthetic portion of ribosomes is RNA.

2.4.2 Protein Structure

The wide variety of specific tasks served by proteins (Table 2.6) is in large part due to the variety of forms which proteins may take. The structure of many proteins is conveniently described in three levels. A fourth structural level must be considered if the protein consists of more than a single chain. As seen from Table 2.9, each level of structure is determined by different factors, hence the great range of protein structures and functions.

2.4.3. Primary and Secondary Structure

The *primary structure* of a protein is its sequence of amino acid residues. The first protein structure completely determined was that of insulin (Sanger and coworkers in 1955). It is now generally recognized that every protein has not only a

Table 2.9. Protein structure

1. Primary	Amino acid sequence, joined through peptide bonds
2. Secondary	Manner of extension of polymer chain, due largely to hydrogen bonding between residues not widely separated along chain
3. Tertiary	Folding, bending of polymer chain, induced by covalent disulfide, hydrogen, and salt bonds as well as hydrophobic and hydrophilic interactions
4. Quaternary	How different polypeptide chains fit together; structure stabilized by same forces as tertiary structure

definite amino acid content but also a *unique sequence* in which the residues are connected. As we shall explore further in Sec. 6.1, this sequence is prescribed by a sequence of nucleotides in DNA.

While a detailed discussion of amino acid sequencing would take us too far from our major objective, we should indicate the general concept employed. By a variety of techniques, the protein is broken into different polypeptide fragments. For example, enzymes are available which break bonds between specific residues within the protein. The resulting relatively short polypeptide chains can be sequenced using reactions developed by Sanger and others. The fragment sequences, some of which partially overlap, are then shuffled and sorted until a consistent overall pattern is obtained and the complete amino acid sequence deciphered. The sequence of amino acid residues in several globular proteins is now known. So far no repeating pattern of residues has been found, although the available evidence suggests some patterns may exist in fibrous proteins.

In many instances, proteins with related functions in one organism or proteins with the same function in different organisms are very closely related at the molecular level. For example, the enzyme cytochrome *c* is found in all plants, animals, and aerobic microorganisms. Of its 104 residues, 35 do not vary in any of the species studied to date. As Table 2.10 reveals, the number of differences between widely different species is surprisingly small. (Why might this be so?)

The amino acid sequence, which is one-dimensional, can have a profound influence on three-dimensional structure. In fact, it has now been conclusively determined (see Fig. 2.16) for several proteins that their three-dimensional conformation is completely determined by the amino acid sequence. A unique conformation is assumed because of various interactions between the residues. When this fact is coupled with knowledge of the cell's coding system for the amino acid sequence (Sec. 6.1), the significance of primary protein structure emerges. It is the

Table 2.10. Variations in the structure of cytochrome *c*†

Cytochrome compared with human cytochrome	No. of variant amino acid residues	Cytochrome compared with human cytochrome	No. of variant amino acid residues
Chimpanzee	0	Tuna fish	21
Rhesus monkey	1	Dogfish	23
Kangaroo	10	Moth (*Samia cynthia*)	31
Dog	11	Wheat	35
Horse	12	Neurospora	43
Chicken	13	Yeast	44
Rattlesnake	14		

† Adapted from more complete table in A. White, P. Handler, and E. L. Smith, "Principles of Biochemistry," 5th ed., p. 787, McGraw-Hill Book Company, New York, 1973.

Native form

Random coil

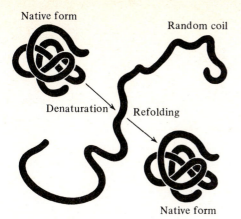

Denaturation   Refolding

Native form

Figure 2.16 After a protein has been denatured by exposure to an adverse environment, it will often return to the native, biologically active conformation following restoration of suitable conditions. This result indicates that primary protein structure determines secondary and tertiary structure. (*From A. L. Lehninger, "Biochemistry," 2d ed., p. 62, Worth Publishers, New York, 1975.*)

link between the central DNA control center of the cell and the elaborate, highly specific protein molecules which mediate and regulate the diverse biochemical activities essential for cell survival.

2.4.4. Three-dimensional Conformation: Secondary and Tertiary Structure

Secondary structure of proteins refers to the way in which the polypeptide chain is extended, especially in fibrous proteins. As noted earlier in Fig. 2.13, there are two general structural types: helixes and sheets. We recall that the partial double-bond character of the peptide bond holds its neighbors in a plane. Consequently, rotation is possible only about two of every three bonds along the protein backbone (Fig. 2.17). This restricts the possible shapes which the chain can assume.

The configuration believed to occur in hair, wool, and some other fibrous proteins arises because of hydrogen bonding between atoms in closely neighboring residues. If the protein chain is coiled as shown in Fig. 2.18, hydrogen bonding

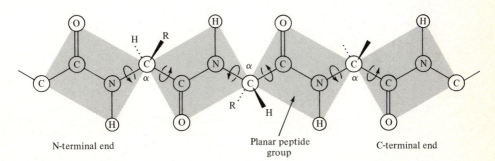

N-terminal end   Planar peptide group   C-terminal end

Figure 2.17 The planar nature of the peptide bond restricts rotation in the peptide chain. (*From A. L. Lehninger, "Biochemistry," 2d ed., p. 128, Worth Publishers, New York, 1975.*)

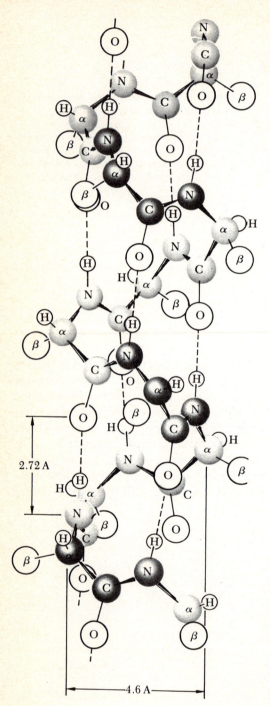

Figure 2.18 The α-helical structure of fibrous proteins. In this type of secondary structure, hydrogen bonds form between all α-amino and α-carboxyl groups. (*Reprinted from F. J. Reithel, "Concepts in Biochemistry," p. 219 McGraw-Hill Book Company, New York, 1967.*)

can occur between the $-C=0$ group of one residue and the $-NH$ group of its neighbor four units down the chain. This configuration, an α *helix*, is the only one which allows hydrogen bonding and also involves no deformation of bonds along the molecular backbone. Collagen, the most abundant protein in higher animals, is thought to consist of three such α helixes, which are intertwined in a *superhelix* (Fig. 2.19). Rigid and relatively stretch-resistant, collagen is a biological cable. It is found in skin, tendons, the cornea of the eye, and numerous other parts of the body.

Because of the relative weakness of individual hydrogen bonds, the α-helical structure is easily disturbed. Proteins can lose this structure in aqueous solution because of competition of water molecules for hydrogen bonds. If wool is steamed and stretched, it assumes a different *pleated-sheet structure*. This arrangement is

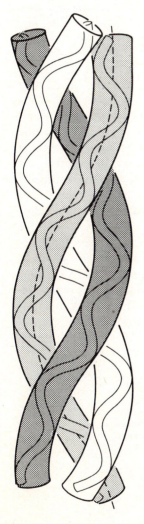

Figure 2.19 A drawing of collagen illustrating a triple helix composed of strands which themselves have helical structure (*Redrawn from R. E. Dickerson and I. Geis, "The Structure and Action of Proteins," p. 41, W. A. Benjamin, Inc., Menlo Park. Copyright © 1969 by Dickerson and Geis.*)

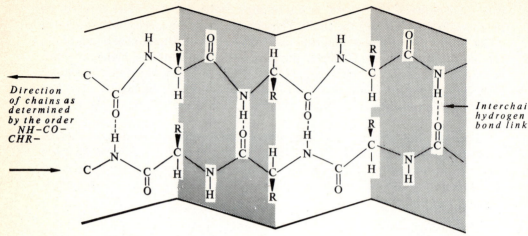

Figure 2.20 The pleated sheet secondary structure of silk. The orientation of each sheet is the reverse of its two nearest neighboring sheets so that only Ala-Ala and Gly-Gly contacts occur. (*Reprinted from N. A. Edwards and K. A. Hassall, "Cellular Biochemistry & Physiology," p. 55, McGraw-Hill Publishing Company Ltd., London, 1971.*)

the naturally stable one in silk fibers (Fig. 2.20). The pleated-sheet configuration is also stabilized by hydrogen bonds which exist between neighboring parallel chains. While the parallel sheets are flexible, they are very resistant to stretching.

Achievement of the secondary structure of fibrous proteins is of importance in the food industry. While plant products such as soybeans are valuable sources of essential amino acids, they lack the consistency and texture of meat. Consequently, in the preparation of "textured protein," globular proteins in solution are spun into a more extended, fibrous structure.

Protein conformation is also influenced by interactions between R groups widely separated along the chains. These interactions determine *tertiary structure*, i.e., how the chain is folded or bent to form the compact, relatively rigid configuration typical of globular proteins. Especially important in this regard are covalent bonds which often exist between two cysteinyl residues. By eliminating two hydrogen atoms, a disulfide bond is formed between two —SH groups:

Such bonds are found as cross-links within a polypeptide chain, and they sometimes hold together two otherwise separate chains: insulin is actually two subchains of 21 and 30 residues which are linked by disulfide bonds (Fig. 2.21). There is also an internal cross-link in the shorter subchain.

Several relatively weak interactions, including ionic effects, hydrogen bonds, and hydrophobic interactions between nonpolar R groups, also influence tertiary structure (Fig. 2.22). Like micelle formation by lipids, examined earlier, a protein structure with hydrophobic groups concentrated in the molecule's interior and relatively hydrophilic functions outside is probably the most stable. It is sometimes called the *oil-drop model* of protein structure.

The fact that tertiary structures are reasonably well established for only eight proteins to date is indicative of the enormous experimental difficulties involved. After crystals of pure protein have been prepared, x-ray crystallography is employed. Each atom is a scattering center for x-rays; the crystal diffraction pattern contains information on structural details down to 2 Å resolution. In order to obtain this detail, however, it was necessary in the case of myoglobin (Fig. 2.23) to calculate 10,000 Fourier transforms, a task which had to wait for the advent of the high-speed digital computer.

Protein conformation seems consistent with its biological function. There is considerable evidence that in many cases a definite three-dimensional shape is

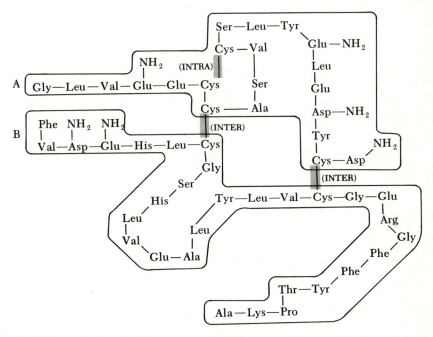

Figure 2.21 Disulfide bonds link two peptide subchains in the insulin molecule. Note the disulfide linkage between two residues of the shorter subchain. (*Reprinted from F. J. Reithel, "Concepts in Biochemistry," p. 237, McGraw-Hill Book Company, New York, 1967.*)

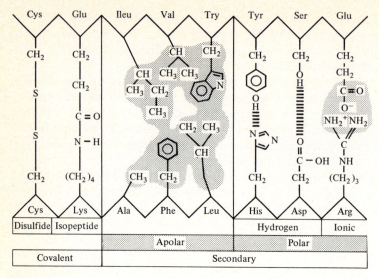

Figure 2.22 A summary of the bonds and interactions which stabilize tertiary and quaternary protein structure. (*From "Cell Structure and Functions," p. 210 by Ariel G. Loewy and Philip Siekeritz. Copyright © 1963, 1969 by Holt, Rinehart and Winston, Inc., Reprinted by permission of Holt, Rinehart and Winston.*)

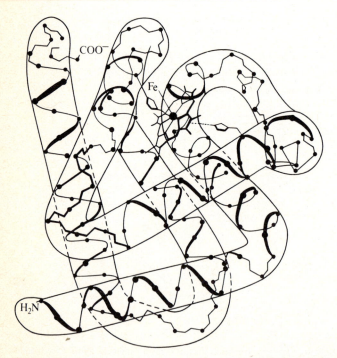

Figure 2.23 The structure of myoglobin. (*Reprinted from N. A. Edwards and K. A. Hassall, "Cellular Biochemistry & Physiology," p. 57, McGraw-Hill Publishing Company Ltd., London, 1971.*)

essential for the protein to work properly. This principle is embodied in the *lock-and-key model* for enzyme specificity. This simple model suggests why the proteins which catalyze biochemical reactions are often highly specific. That is, only definite reactants, or *substrates*, are converted by a particular enzyme. For example, many enzymes will break the 1,4-glycosidic linkage in starch or glycogen, while very few cause hydrolysis of the type of 1,4 linkage found in cellulose. The lock-and-key model (Fig. 2.24) holds that the enzyme has a specific site (the "lock") which is a geometrical complement of the substrate (the "key") and that only substrates with the proper complementary shape can bind to the enzyme so that catalysis occurs.

The available information on tertiary enzyme structure has not only confirmed this hypothesis but also has provided insights into the catalytic action of enzymes. Lysozyme, an enzyme found in tears, mucous, and egg white, is the most thoroughly explored example. By hydrolyzing the murein cell wall and thereby killing gram-positive bacteria, lysozyme serves a protective function. It will form a stable unreactive complex with a trimer derived from murein polymer, and this complex provides an intimate view of substrate-enzyme interaction. Examination of Fig. 2.25 shows that both helical and sheet secondary structures may be found within globular proteins. These observations indicate that no single rule governs the architecture of the protein. The variety of functions and constituent-group interactions also suggests such a conclusion.

When protein is exposed to conditions sufficiently different from its normal biological environment, a structural change, called *denaturation*, typically leaves the protein unable to serve its normal function (Fig. 2.16). Relatively small changes in solution temperature or pH, for example, may cause denaturation, which usually occurs without severing covalent bonds. If a heated dilute solution of denatured protein is slowly cooled back to its normal biological temperature, the reverse process or, *renaturation*, with restoration of protein function may occur.

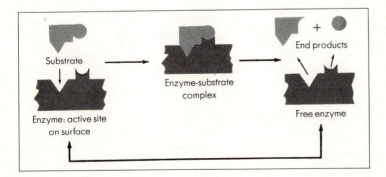

Figure 2.24 Simplified diagram of the lock-and-key model for enzyme action. Here the shape of the catalytically active site (the lock) is complementary to the shape of the substrate key. (*Reprinted from M. J. Pelczar, Jr. and R. D. Reid, "Microbiology," 3d ed., p. 158, McGraw-Hill Book Company, New York, 1972.*)

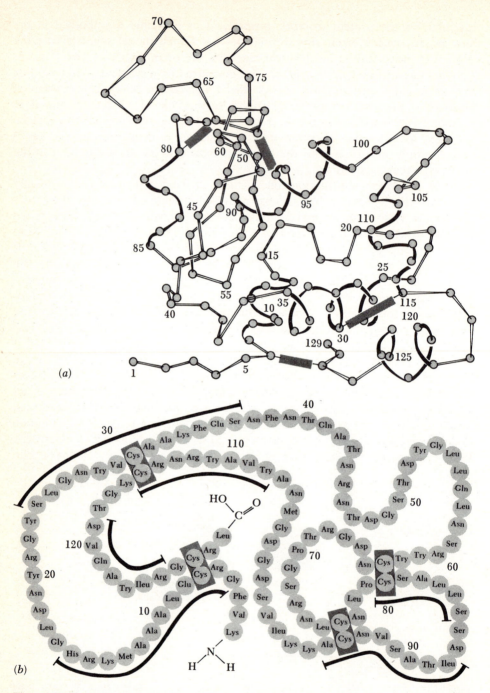

Figure 2.25 The tertiary structure of cystalline egg white lysozyme reveals subdomains of helical and sheet secondary structure (*By permission of C. C. F. Blake and the editor of Nature. From "Structure of Hen Egg-White Lysozyme," Nature,* **206***; 757, 1965.*)

As with other transitions, higher temperatures favor the state of greater entropy (disorder) hence extendedness; cooling favors the greatest number of favorable interactions leading to compactness and renaturation.

The relative ease of thermally or chemically inducing denaturation of many proteins reminds us that the biochemical process engineer may operate microbial processes only over a fairly narrow range of, for example, pH, temperature, and ionic strength. This knowledge is also useful in other subjects such as protein recovery (Chap. 4) and sterilizer design (Chaps. 7 and 9).

2.4.5. Quaternary Structure and Biological Regulation

Proteins may consist of more than one polypeptide chain, hemoglobin being perhaps the best known example. How these chains fit together in the molecule is the *quaternary structure*. As Table 2.11 indicates, a great many proteins, especially enzymes (usually indicated by the *-ase* suffix), are oligomeric and consequently possess quaternary structure. The forces stabilizing quaternary structure are believed to be the same as those which produce tertiary structure. While disulfide bonds are sometimes present, as in insulin, most of the proteins in Table 2.11 are held together by relatively weak interactions (Table 2.9). Many oligomeric proteins are known to be *self-condensing*; e.g., separate hemoglobin α chains and β chains in solution rapidly reunite to form intact hemoglobin molecules. This feature is of great significance, for it indicates that at least in some cases the one-dimensional biochemical coding from DNA which specifies primary protein structure ultimately determines all higher structural levels and hence the specific biological role of the protein. An example of quaternary-structure influence appears in Prob. 2.11.

Evidence to date suggests that the subunit makeup of some protein molecules is especially suited for at least two important biological functions: control of the catalytic activity of enzymes and flexibility in construction of related but different molecules from the same collection of subunits. Different proteins known as *isozymes* or *isoenzymes* demonstrate the latter point. Isozymes are different molecular forms of an enzyme which catalyze the same reaction within the same species. While the existence of isozymes might seem a wasteful duplication of effort, the availability of parallel but different catalytic processes provides an essential ingredient in some biochemical control systems (Chaps. 3 and 5). Several isozymes are known to be such *oligomeric* proteins; in one instance isoenzymes are composed of five subunits, only two of which are distinct. This might be viewed as an economical device since five different proteins may be constructed from the two polypeptide components.

2.5. The Hierarchy of Cellular Organization

In the previous sections we have reviewed the major smaller biochemicals and the biopolymers constructed from them. Although the relationships between chemical structure and cellular functions have already been emphasized, it will provide a

Table 2.11. Characteristics of several oligomeric proteins†

Protein	Molecular weight	No. of chains	No. of —S—S— bonds
Insulin	5,798	1 + 1	3
Ribonuclease	13,683	1	4
Lysozyme	14,400	1	5
Myoglobin	17,000	1	0
Papain	20,900	1	3
Trypsin	23,800	1	6
Chymotrypsin	24,500	3	5
Carboxypeptidase	34,300	1	0
Hexokinase	45,000	2	0
Taka-amylase	52,000	1	4
Thioredoxin reductase	65,800	—	
Bovine serum albumin	66,500	1	17
Yeast enolase	67,000	2	0
Hemoglobin	68,000	2 + 2	0
Liver alcohol dehydrogenase	78,000	2	0
Alkaline phosphatase	80,000	2	4
Hemerythrin	107,000	8	0
Glyceraldehyde-phosphate dehydrogenase	140,000	4	0
Lactic dehydrogenase	140,000	4	0
γ-Globulin	140,000	2 + 2	25
Yeast alcohol dehydrogenase	150,000	4	0
Tryptophan synthetase	159,000	2 + 2	
Aldolase	160,000	4(?)	0
Phosphorylase b	185,000	2	
Threonine deaminase (*Salmonella*)	194,000	4	
Fumarase	200,000	4	0
Tryptophanase	220,000	8	4
Formyltetrahydrofolate synthetase	230,000	4	
Aspartate transcarbamylase	310,000	4 + 4	0
Glutamic dehydrogenase	316,000	6	0
Fibrinogen	330,000	2 + 2 + 2	
Phosphorylase a	370,000	4	
Myosin	500,000	2 + 3	0
β-Galactosidase	540,000	4	
Ribulose diphosphate carboxylase	557,000	24	

† From "Cell Structure and Function" 2d ed., p. 221, by Ariel G. Loewy and Philip Siekevitz. Copyright © 1963, 1969 by Holt, Rinehart, and Winston. Reprinted by permission of Holt, Rinehart and Winston.

useful perspective to reconsider the dynamic nature and spatial inhomogeneity of the structures within which such functions occur. As viable cells function and grow, the biopolymers of this chapter must be repeatedly synthesized by the cells. Usually, the nutrient medium of the cell consists of entities such as sugar, carbon dioxide, a few amino acids, water, and some inorganic ions, but typically significant amounts of the biopolymer starting materials are absent. Thus, from these simplest of precursors, the cell must synthesize the remainder of the needed amino acids, nucleic acids, lipids, proteins, etc. The energy-consuming processes inherent in precursor synthesis and biopolymer formation are considered in Chap. 5.

There are many *supramolecular* assemblies within the cell. The cell membrane is a grand collection of from one to a range of molecular varieties (Table 2.2). The ribosomes are a separable combination of protein and several different nucleic acids. In *multienzyme complexes*, several different enzymes catalyzing a sequence

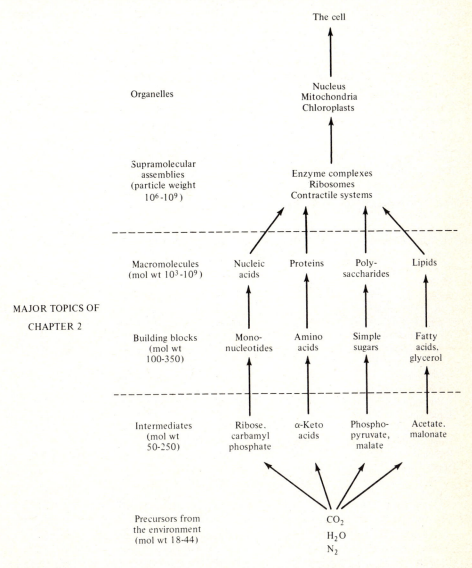

Figure 2.26 Ascending order of increasing complexity and organization of the location of the biomolecules described in this chapter. (*From A. L. Lehninger, "Biochemistry," 1st ed., p. 19, Worth Publishers, N.Y. 1970.*)

of chemical reactions are maintained in close proximity, presumably to maximize utilization of the reaction intermediates. The organelles such as mitochondria, ribosomes, and chloroplasts represent the final level of organizational sophistication before the cell itself. The different levels of complexity, from elemental components to the cell, are presented in Fig. 2.26.

The locations of the various biochemicals of this chapter within a procaryotic microorganism are summarized in Fig. 2.27. Small molecules like amino acids and simple sugars are dispersed throughout the cytoplasm, as are some larger molecules including certain enzymes and tRNA. Other biopolymers are localized within the cell or near a surface such as the cell membrane.

The overall chemical composition of *E. coli* is outlined in Table 2.12. The ions, nutrients, intermediates, biopolymers and other constituents are linked together through reaction networks (Chaps. 5 and 6). Individual reactions within such networks are catalyzed by enzymes under the mild conditions conducive to maintenance of the native protein forms of the previous section. The organelles and membranes are assembled by the combination of energy-consuming reactions (Chap. 5) and the natural (spontaneous) tendency of their constituent molecules to form spatially inhomogeneous structures rather than a uniform solute sea.

Table 2.12. Composition of a rapidly growing *E. coli* cell†

Component	Percent of total cell weight	Average mol. wt.	Approximate no. per cell	No. of different kinds
H_2O	70	18	4×10^{10}	1
Inorganic ions (Na^+, K^+, Mg^{2+}, Ca^{2+}, Fe^{2+}, Cl^-, PO_4^{4-}, SO_4^{2-}, etc.)	1	40	2.5×10^8	20
Carbohydrates and precursors	3	150	2×10^8	200
Amino acids and precursors	0.4	120	3×10^7	100
Nucleotides and precursors	0.4	300	1.2×10^7	200
Lipids and precursors	2	750	2.5×10^7	50
Other small molecules‡	0.2	150	1.5×10^7	200
Proteins	15	40,000	10^6	2000–3000
Nucleic acids:				
DNA	1	2.5×10^9	4	1
RNA	6			
16S rRNA	...	500,000	3×10^4	1
23S rRNA	...	1,000,000	3×10^4	1
tRNA	...	25,000	4×10^5	40
mRNA	...	1,000,000	10^3	1000

† J. D. Watson, "Molecular Biology of the Gene," table 3.3, W. A. Benjamin, Inc., New York, 1970.

‡ Heme, quinones, breakdown products of food molecules, etc.

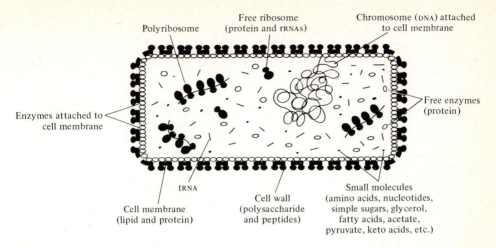

Figure 2.27 A schematic diagram of the intestinal bacterium *E. coli*, showing the location of the biomolecules described in this chapter.

Chapters 3 and 4 examine the kinetics of enzyme-catalyzed reactions, enzyme applications, and protein recovery methods. Chapters 5 to 7 consider energy flows and mass balances in cellular processes, genetics and the control of cellular networks, and finally mathematical representations of cell-population kinetics. This collection is the prelude to biochemical-reactor analysis, which occupies the balance of the text.

PROBLEMS

2.1 Chemical structure (*a*) Write down the specific or general chemical structure, as appropriate, for the following substances. Comment on the (possible) functional importance of various groups on the molecule.

1. Fat, phospholipid, micelle, vitamin A
2. D(+)-glucose, pyranose, lactose, starch, amylopectin, cellulose
3. Ribonucleotide, adenosine monophosphate (AMP), adenosine triphosphate (ATP), nicotinamide adenine dinucleotide (NAD), deoxyribonucleic acid (DNA)
4. Amino acid, lysine, tripeptide, protein
5. Globular and fibrous proteins

(*b*) Sketch the hierarchy of protein configuration and indicate the important features of each level of protein configuration (primary through quaternary).

2.2 Spectrophotometry Measurements of the amount of light of a specific wavelength absorbed by a liquid solution are conducted with a spectrophotometer, an instrument of basic importance in biochemical analyses. Spectrophotometric measurements are used to determine solute concentrations and to identify solutes through their absorption spectra. Interpretation of the former class of spectrophotometric data is based on the Beer-Lambert law

$$A = \log \frac{I_0}{I} = a_m bc$$

where A = absorbancy, or optical density (OD)

I_0 = intensity of incident light

I = intensity of transmitted light

a_m = molar extinction coefficient

b = path length

c = molar concentration of solute

Usually the incident-light wavelength is adjusted to the value λ_{max} at which the solute of interest absorbs most strongly. Table 2P2.1 gives λ_{max} values for a few substances of biological interest and the corresponding molar extinction coefficients (for 1-cm light path and pH 7). Additional data are available in handbooks; notice that many of the λ_{max} values lie in the ultraviolet range.

Table 2P2.1†

Compound	λ_{max}, mμ	$a_m \times 10^{-3}$, 1/(cm · mol)
Tryptophan	218	33.5
ATP	259	15.4
NAD$^+$	259	18
Riboflavin	260	37
NADH	339	6.22
FAD	450	11.3

† See Chap. 5 for conventions regarding NAD$^+$, NADH nomenclature.

(a) An ATP solution in a 1-cm cuvette transmits 70 percent of the incident light at 259 mμ. Calculate the concentration of ATP. What is the optical density of a 5×10^{-5} M ATP solution?

(b) Suppose two compounds A and B have the molar extinction coefficients given in Table 2P2.2. If a solution containing both A and B has an OD (1-cm cuvette) of 0.35 at 340 mμ and an OD of 0.220 at 410 mμ, calculate the A and B concentrations.

Table 2P2.2

Wavelength, mμ	A, 1/(cm · mol)	B, 1/(cm · mol)
340	14,000	7100
410	2,900	6600

2.3 Molecular weights Sedimentation and diffusion measurements are often used to provide molecular-weight estimates based on the Svedberg equation

$$M = \frac{RTs}{\mathscr{D}(1 - \bar{v}\rho)}$$

where R = gas constant

T = absolute temperature

s = sedimentation coefficient, s

$\mathscr{D}$ = diffusion coefficient, cm^2/s

$\bar{v}$ = partial specific volume of substance, cm^3/g

ρ = solution density, g/cm^3

Verify the molecular weights of the substances in Table 2P3.1 ($T = 20°$C). Assume $\rho = 1.0$.

Table 2P3.1

Substance	$s \times 10^{13}$	$\mathscr{D} \times 10^7$	$\bar{v}$	M
Lysozyme	1.91	11.2	0.703	14,400
Fibrinogen	7.9	2.02	0.706	330,000
Bushy stunt virus	132	1.15	0.74	10,700,000

2.4 pH and buffering capacity of polymer (*a*) Calculate the pH of a 0.2 M solution of serine; serine·HCl; potassium serinate.

(*b*) Sketch the titration profile for addition of the strong base KOH to the serine·HCl solution; use scales of pH vs. moles KOH added per mole serine.

(*c*) Repeat part (*a*) for the polypeptides serine$_5$, serine$_{20}$, and serine$_{100}$ for solutions of the same weight percent solute as 0.2 M serine·HCl.

(*d*) Repeat part (*c*) for lysine·HCl, assuming that protonation of the amino NH_2 has an association equilibrium constant $K_{eq} = 8.91 \times 10^{-6}$ l/mol and the side-chain amino acid an association constant of 2.95×10^{-11} l/mol.

(*e*) Estimate the pH of a 1 wt% solution of cellulose. Repeat for starch (take molecular weight to be very large, $> 10^6$). What influence do storage polymers in general have on cell pH?

2.5 Amino acid separations (*a*) Answer the following statements *true* or *false*; justify your responses:

1. The pK of a simple amino acid is given by

$$pK = \tfrac{1}{2}(pK_{COOH} + pK_{NH_2})$$

2. Amino acids are neutral between pK_{COOH} and pK_{NH_2}.

(*b*) In amino acid chromatography, useful chromatographic media are sulfonic acid derivatives of polystyrene. The basic components of amino acids interact with the negative sulfonic acid groups, and the hydrophobic portions interact favorably with the aromatic structure of the polymer. At pH values of 3, 7, and 10, give the (approximate) sequence of amino acids which would elute (appear in order) from a column containing:

1. Only sulfonic acid groups on inert support
2. Only polystyrene
3. Equal quantities of accessible sulfonic acid and polystyrene

(*c*) On the same chromatography matrices as in part (*b*) above, would you expect hydrophobic or hydrophilic, e.g., acid or base, interactions to be relatively more important in separating mixtures of polypeptides? Of proteins?

2.6 Macromolecular physical chemistry From your own knowledge, a chemistry text, etc., describe the general reactions and/or physical conformation changes involved in:

(*a*) Frying an egg, burning toast, curdling milk, setting Jell-O, blowing a soap bubble, crystallizing a protein, tanning of leather, whipping meringue.

(*b*) In the development of substitute foods from microbial or other sources, e.g., single-cell protein, soyburgers, texture and rheology (flow response to stress) are clearly important parameters. Outline how you would organize a synthetic-food testing laboratory. (What backgrounds would you look for in personnel; what kinds of tests would you organize?)

2.7 Mass balances: stoichiometry The major elemental demands of a growing cell are carbon, oxygen, hydrogen, and nitrogen. Assuming the carbon source to be characterized by (CH_2O) and nitrogen by NH_4^+

$$\alpha CH_2O + \beta NH_4^+ + \zeta O_2 \longrightarrow \underset{\substack{\text{New cellular} \\ \text{material}}}{C_a H_b O_c N_d} + \eta CO_2 + \gamma H_2O$$

(a) Write down the restrictions between the coefficients α, β, ... from consideration of elemental conservation. Given the "composition" of a cell, what additional information is needed to solve for the coefficient values?

(b) Repeat part (a) for photosynthetic systems:

$$\alpha CO_2 + \beta NH_4^+ + \gamma H_2O \longrightarrow C_a H_b O_c N_d + \varepsilon O_2$$

(c) Describe an explicit series of experiments by which you would evaluate the stoichiometric formula of the cell (a, b, c, d).

2.8 Stoichiometry: aerobic vs. anaerobic species In order to conveniently calculate nutrient consumption, aeration, and heat-transfer rates (Chap. 8), a general stoichiometric equation can be formulated by assigning to the cell an apparent molecular formula.

(a) Calculate the cell formula for the growth of yeast on sugar when it is observed to give the following balance (assume 1 mol of cell is produced):

$$100 \text{ g } C_6H_{12}O_2 + 5.1 \text{ g } NH_3 + 46.63 \text{ g } O_2 \longrightarrow 43.23 \text{ g cell mass} + 41.08 \text{ g } H_2O + 67.42 \text{ g } O_2$$

(b) If the above reaction produces 0.30017 "mol" of cell, what cell formula results?

(c) Anaerobic fermentations typically produce a variety of partially oxygenated compounds in addition to cell mass. Keeping the cell "molecular formula" of part (b) above, calculate the unknown coefficients for the following typical equation (coefficients are mole quantities, not mass):

$$0.55556 \text{ glucose} + \alpha \text{ ammonia}$$
$$\rightarrow \beta \text{ cell mass}$$
$$+ 0.05697 \text{ glycerol}$$
$$+ \gamma \text{ butanol, (a typical alcoholic product)}$$
$$+ \varepsilon \text{ succinic acid, (a typical acid product)}$$
$$+ 0.01164 \text{ water} + 1.03076 \text{ (carbon dioxide)}$$
$$+ 1.00380 \text{ ethanol}$$

Kinetically, the rates of processes (a) and (c) can conveniently be controlled by changing the aeration rate, giving yeast and ethanol at any intermediate desired level (*Vienna process*) [10].

2.9 Plasma and biomolecular lipid membranes Tien summarized certain properties of natural and bilayer lipid membranes as shown in Table 2P9.1.

Table 2P9.1†

	Natural membranes	Lipid bilayers
Thickness, Å:		
Electron microscopy	40–130	60–90
X-ray diffraction	40–85	
Optical methods	—	40–80
Capacitance technique	30–150	40–130
Resistance, $\Omega \cdot cm^2$	10^2–10^5	10^3–10^9
Breakdown voltage	100	100–550
Capacitance, $\mu F/cm^2$	0.5–1.3	0.4–1.3

† H. T. Tien, Bilayer Lipid Membranes: An Experimental Model for Biological Membranes, in M. L. Hair (ed.), "Chemistry of Biosurfaces," vol. 1, p. 239, Marcel Dekker, New York, 1971.

(a) Why do these membranes have a common range of thicknesses? (Be as specific as possible.)

(b) What cell functions are (possibly) served by membrane resistance and capacitance?

(c) Compare the breakdown field of the order of 1 V/Å with the ionization potential per length of any C_{10} or higher hydrocarbon. Comment.

2.10 Molecular mobility in plasma membranes In the fluid-mosaic model of plasma membranes, the specific functional molecules such as permeases, antigens, etc., are thought to be loosely set *in* a phospholipid bilayer. In experiments with plasma membranes from chicken red blood cells and mouse white blood cells, two different plasma-membrane-encased cells were induced to form a *single* large hybrid cell. By means of fluorescent tags on specific membrane molecules [*antibodies* which adsorbed on plasma-bound *antigens* (Sec. 10.4)], it was observed that 5 min after hybrid formation, the new plasma membrane was *initially* approximately two distinct hemispheres, each containing only the marked antibody-antigen for mouse or chicken membrane, but 40 min later, the distribution of both markers was uniform over the entire hybrid plasma cell surface.

(a) From Einstein's equation for the mean square displacement $\langle x^2 \rangle$ in time t, $\mathscr{D} = \langle x^2 \rangle / t$, estimate the antigen-antibody surface diffusion coefficient $\mathscr{D}$ (state your assumptions).

(b) Antigens are mostly protein and may be relatively small. It is known that three major groups of antibodies exist, each a different gamma globulin group, denoted γA, γG, γM (other types, γY and γE are known). The molecular weight of these two-unit structures is typically about 72,000. The "Chemical Engineer's Handbook"† gives diffusion coefficients for large molecules in nonelectrolytes as follows: $\mathscr{D} = 1.1 \times 10^{-6}$, 5×10^{-7}, and 2.5×10^{-7} cm^2/s for molecular weights of 10^4, 10^5, 10^6 respectively. What is the upper limit of $\mathscr{D}$ for an antibody-antigen complex in a nonelectrolyte bulk fluid?

(c) In measurements of surface diffusion of adsorbed species *on* solid surfaces, it has always been difficult to eliminate the parallel effect of desorption into the adjacent bulk fluid, bulk diffusion therein, and readsorption onto the solid surface. How would you design an experiment to prove the absence of this parallel mechanism in the experiment under discussion?

2.11 Concerted-transition model for hemoglobin O_2 Binding The concerted-transition model envisions an inactive form of protein T_0 in equilibrium with an active form R_0:

$$R_0 \underset{}{\overset{L}{\rightleftharpoons}} T_0 \qquad \text{equilibrium constant } L$$

In turn, the active form, which in the case of hemoglobin has four subunits, may bind one molecule of substrate S per subunit:

$$R_0 \underset{-S}{\overset{+S}{\rightleftharpoons}} R_1 \underset{-S}{\overset{+S}{\rightleftharpoons}} R_2 \underset{-S}{\overset{+S}{\rightleftharpoons}} R_3 \underset{-S}{\overset{+S}{\rightleftharpoons}} R_4$$

Let K_{DS} denote the dissociation constant for S binding in each step above

$$K_{DS} = \frac{[S][R_{i-1}]}{[R_i]} \qquad i = 1, 2, 3, 4$$

and evaluate the ratio

$$y = \frac{\text{total subunit concentration}}{\text{concentration of S-occupied subunits}}$$

in terms of K_{DS}/S and L. Plot y vs. K_{DS} for $L = 9000$. Look up a graph of hemoglobin-bound O_2 vs. O_2 partial pressure in a biochemistry text, and compare it with your plot. Read further in the reference you have found about cooperative phenomena in biochemistry.

2.12 Multiunit enzymes: mistake frequency Multiunit proteins, e.g., some allosteric enzymes, may seem to be more complicated structures than single-chain proteins, yet the percentage of mistakes in the final structures may be fewer.

---

† R. Perry and C. Chilton, "Chemical Engineer's Handbook," 5th ed., McGraw-Hill Book Company, New York, 1973.

(*a*) Consider the synthesis of two proteins, each with 850 amino acid residues. One is a single chain; the second is a three-unit structure of 200, 300, and 350 residues. If the probability of error is the same for each residue and is equal to 5×10^{-9} per residue, evaluate the fraction of complete structures with one or more errors when the presence of two mistakes in any chain prevents it from achieving (or being incorporated in) an active structure.

Convincing as part (*a*) may be, it has also been observed that (1) small portions of some enzymes may be removed without loss of activity and that (2) the same function may be accomplished by a considerable variety of protein sequences as illustrated in Table 2.10 for cytochrome *c*. What other function may be served in the cell by these replaceable or removable residues? (Look back at the internal cell structure.)

2.13 Mass flows Analysis of correct mass and energy balances in cellular processes requires a careful appreciation of the *sequence* of cellular chemical events in time and in space. (More detailed examples of various metabolic paths appear in subsequent chapters.) For the major biopolymer and higher structures of this chapter, form composite sketches of Figs. 2.26 and 2.27 which follow, by arrows, the time-space movement involved in cell-material synthesis from precursors, for example, O_2, H_2O, NH_4^+ and glucose. Comment where appropriate on the (lack of) restriction in space for particular species.

2.14 Fitness of biochemicals This chapter has surveyed the major existing biochemicals and biopolymers in microbial systems. For an interesting account of *why* these particular biochemicals may have arisen, given the history of the earth's development, read Ref. 8. Summarize Blum's major arguments relating to the fitness of water, glucose, and ATP for their biological roles.

REFERENCES

1. M. Yudkin and R. Offord: "A Guidebook to Biochemistry," Cambridge University Press, London, 1971. A readable introductory text which describes the essentials of biochemistry with a minimum amount of detail.
2. A. G. Loewy and P. Siekevitz: "Cell Structure and Function," Holt, Rinehart and Winston, Inc., New York, 1969. An excellent introductory short course in biochemistry is formed in Part Three. Emphasis on the experimental verification of biological theories and models is an especially strong feature.
3. A. L. Lehninger: "Biochemistry" 2d ed., Worth Publishers, New York, 1975. An excellent, more advanced biochemistry text. Continued reference to examples from cell and human physiology helps the reader in appreciating, remembering, and organizing the material.
4. H. R. Mahler and E. H. Cordes: "Biological Chemistry," Harper & Row, Publishers, Incorporated, New York, 1971. Greater chemical and mathematical detail than Lehninger; the sections on experimental techniques in biochemistry and microbiology go farther in modeling and analysis in a transport-phenomena vein than most other texts.
5. R. M. Dowben: "General Physiology: A Molecular Approach," Harper & Row, Publishers, Incorporated, New York, 1969. A book on cell physiology drawing upon the most fundamental principles of physical chemistry; challenging but enjoyable reading.
6. R. W. Dickerson and I. Geis: "The Structure and Action of Proteins," Harper & Row, Publishers, Incorporated, New York, 1969. A joint project of a biologist and a science artist, this monograph provides a refreshing survey of the various protein functions with special emphasis on relations with protein structure. Most useful with one of the above references.
7. J. D. Watson: "The Double Helix," Atheneum, New York, 1968. The story of the discovery of DNA structure, a turning point in biology. Also a candid glimpse at the process of discovery in modern research.
8. H. F. Blum: "Time's Arrow and Evolution," 3d ed., Princeton University Press, Princeton, N.J., 1968. An inquiry into the biological fitness of elements and chemicals from an evolutionary viewpoint.

9. Fukui and Ishida: "Microbial Production of Amino Acids," Kodansha Ltd., Tokyo, and John Wiley and Sons, Inc., New York, 1972. An interesting survey of devices to release metabolic intermediates, in this case amino acids.

Problems

10. J. B. Harrison: *Adv. Ind. Microb.*, **10**: 129, 1971.

THREE

THE KINETICS OF
ENZYME-CATALYZED REACTIONS

In the previous chapter we learned that there are a great variety of chemical compounds within the cell. How are they manufactured and combined at sufficient reaction rates under relatively mild temperatures and pressures? How does the cell select exactly which reactants will be combined and which molecules will be decomposed? The answer is catalysis by enzymes, which, as we already know, are globular proteins.

A breakthrough in enzymology occurred in 1897 when Büchner first extracted active enzymes from living cells. Büchner's work provided two major contributions: first, it showed that catalysts at work in a living organism could also function completely independently of any life process. As we shall explore in the next chapter, isolated enzymes enjoy a wide spectrum of applications today. Also, efforts to isolate and purify individual enzymes were initiated by Büchner's discoveries. The first successful isolation of a pure enzyme was achieved in 1926 by Sumner. Studies of the isolated enzyme revealed it to be protein, a property now well established for enzymes in general.

Since Sumner's time the number of known enzymes has increased rapidly (Fig. 3.1), and the current total is in excess of 1500. It is likely that the trend evidenced in Fig. 3.1 will continue, for in 1970 the known reactions in the relatively simple bacterium *E. coli* were estimated to represent only one-third to one-fifth of all reactions actually occurring within the cell. As each biochemical reaction is in general catalyzed by a different enzyme, these estimates can be directly translated into numbers of known and total enzymes.

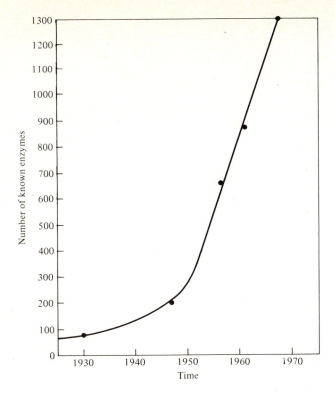

Figure 3.1 The rapidly increasing number of known enzymes.

Since we shall have an occasion to mention a variety of enzymes in this chapter and later, a few words on enzyme nomenclature are in order. Unfortunately, there is no scheme which is consistently applied to all enzymes. In most cases, however, enzyme nomenclature derives from what the enzyme *does* rather than what the enzyme *is:* the suffix *-ase* is added either to the substrate name (urease is the enzyme which catalyzes urea decomposition) or to the reaction which is catalyzed (alcohol dehydrogenase catalyzes the oxidative dehydrogenation of an alcohol). Exceptions to this nomenclature system are long familiar enzymes such as pepsin and trypsin in the human digestive tract, rennin (used in cheese making), and the "old yellow" enzyme, which causes browning of sliced apples.

There are six major classes of reactions which enzymes catalyze. These units form the basis for the Enzyme Commission (EC) system for classifying and assigning index numbers to all enzymes (Table 3.1). Although the common enzyme nomenclature established by past use and tradition is often employed instead of the "official" names, the EC system provides a convenient tabulation and organization of the variety of functions served by enzymes.

Table 3.1. Enzyme Commission classification system for enzymes (class names, Enzyme Commission type numbers, and type of reactions catalyzed)†

1. Oxidoreductases (oxidation-reduction reactions)	4. Lyases (addition to double bonds)
1.1 Acting on $-\overset{\mid}{C}H-OH$	4.1 $-\overset{\mid}{C}=\overset{\mid}{C}-$
1.2 Acting on $-\overset{\mid}{C}=O$	4.2 $-\overset{\mid}{C}=O$
1.3 Acting on $-CH=CH-$	4.3 $-\overset{\mid}{C}=N-$
1.4 Acting on $-\overset{\mid}{C}H-NH_2$	....
1.5 Acting on $-\overset{\mid}{C}H-NH-$	5. Isomerases (isomerization reactions)
1.6 Acting on NADH; NADPH	5.1 Racemases
....	....
2. Transferases (transfer of functional groups)	6. Ligases (formation of bonds with ATP cleavage)
2.1 One-carbon groups	6.1 C—O
2.2 Aldehydic or ketonic groups	6.2 C—S
2.3 Acyl groups	6.3 C—N
2.4 Glycosyl groups	6.4 C—C
....	....
2.7 Phosphate groups	
2.8 S-containing groups	
3. Hydrolases (hydrolysis reactions)	
3.1 Esters	
3.2 Glycosidic bonds	
....	
3.4 Peptide bonds	
3.5 Other C—N bonds	
3.6 Acid anhydrides	
....	

† A. L. Lehninger, "Biochemistry," 2d ed., table 8-1, Worth Publishers, Inc., New York, 1975.

To be sure there is no confusion, let us remember what a catalyst is. A catalyst increases the rate of a chemical reaction without undergoing a permanent chemical change. Also, a catalyst influences the *rate* of a chemical reaction but does not affect reaction equilibrium (Fig. 3.2). Equilibrium concentrations can be calculated using only the thermodynamic properties of the substrates (remember that substrate is the biochemist's term for what we usually call a reactant) and products. Reaction kinetics, however, involve molecular dynamics and are presently impossible to predict accurately without experimental data.

For design and analysis of a reacting system, we must have a mathematical formula which gives the *reaction rate* (moles reacted per unit time per unit volume) in terms of composition, temperature, and pressure of the reaction mixture. If you have studied catalyzed reactions before, you have seen a strategy for obtaining a reasonable reaction rate expression. First an educated guess is

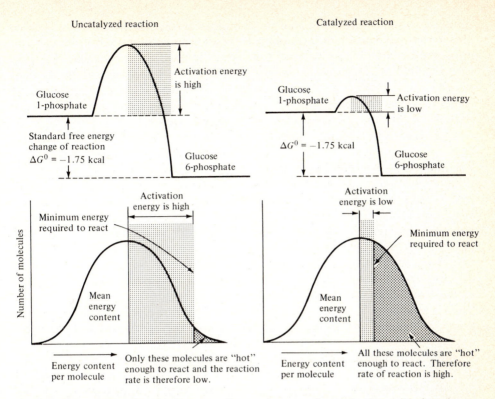

Figure 3.2 By lowering the activation energy for reaction, a catalyst makes it possible for substrate molecules with smaller internal energies to react. (*Reprinted by permission from A. L. Lehninger,* "*Bioenergetics,*" *2d ed. p. 35, W. A. Benjamin, Inc., Menlo Park, Ca., 1971.*)

made about what elementary reactions go on at the molecular scale. Then, assuming that one of these molecular events is slow relative to all the rest, an expression for obtaining a reasonable reaction rate expression. First an educated guess is tions. We shall follow precisely this approach in this chapter's quest for rate expressions for enzyme-catalyzed reactions.

The similarities between synthetic catalysts and enzymes go beyond the use of a common technique for modeling reaction kinetics. The rate expressions eventually obtained for both types of catalysts are very similar and sometimes of identical forms. This arises because, in both cases, it is known that the reacting molecules form some sort of complex with the catalyst. This general phenomenon in catalysis accounts for such similarity in rate expresssions. We shall have more to say on this shortly.

Although some have already been mentioned, it is important that we remember the differences between enzymes and synthetic catalysts. Most synthetic catalysts are not *specific;* i.e., they will catalyze a variety of reactions involving many different kinds of reactants. While some enzymes are not very specific, many will catalyze only one reaction involving only certain substrates. Usually the degree of *specificity* of an enzyme is related to its biological role. Obviously it is not desirable for an enzyme with the task of hydrolyzing proteins into smaller peptides and amino acids to be highly specific. On the other hand, an enzyme catalyzing an isomerization of one particular compound must be very discriminating. As mentioned earlier in Sec. 2.4.3, enzyme specificity is thought to be a consequence of its elaborate three-dimensional conformation which allows formation of the *active site* responsible for the catalytic ability of the enzyme.

Another distinguishing characteristic of enzymes is their frequent need for *cofactors*. A cofactor is a nonprotein compound which combines with an otherwise inactive protein (the *apoenzyme*) to give a catalytically active complex. While this complex is often called a *holoenzyme* by biochemists, we shall often simply call it an *enzyme*. Two distinct varieties of cofactors exist. The simplest cofactors are *metal ions* (Table 3.2). On the other hand, a complex organic molecule called a *coenzyme* may serve as a cofactor. We have already encountered the coenzymes NAD, FAD, and coenzyme A (CoA) in the last chapter, and ATP serves as a cofactor in some instances. Often these cofactors are bound relatively loosely to the enzyme, and there is equilibrium between enzyme, apoenzyme, and cofactor. In the strict sense of the word, tightly bound nonprotein structures such as the heme group found in cytochrome *c* are coenzymes. However, the name prosthetic group, introduced previously, is usually reserved for such irreversibly attached groups.

Both synthetic and biological catalysts can gradually lose activity as they participate in chemical reactions. As Fig. 3.3 illustrates, some of these deactivation processes are conceptually and physically similar. Still, enzymes are in general far

Table 3.2. Some enzymes containing or requiring metal ions as cofactors†

Zn^{2+}	Alcohol dehydrogenase
	Carbonic anhydrase
	Carboxypeptidase
Mg^{2+}	Phosphohydrolases
	Phosphotransferases
Mn^{2+}	Arginase
	Phosphotransferases
Fe^{2+} or Fe^{3+}	Cytochromes
	Peroxidase
	Catalase
	Ferredoxin
Cu^{2+} (Cu^{+})	Tyrosinase
	Cytochrome oxidase
K^{+}	Pyruvate phosphokinase
	(also requires Mg^{2+})
Na^{+}	Plasma membrane ATPase
	(also requires K^{+} and Mg^{2+})

† A. L. Lehninger, "Biochemistry," 2d ed., table 8-2, Worth Publishers, Inc., New York, 1975.

Phenomenon	Cause
Solid Catalysts	
(*a*) Product inhibition (reversible)	Accumulating product occupies active sites, decreasing overall reaction rate.
(*b*) Catalyst poisoning (irreversible)	The active site reacts essentially irreversibly with some species to yield an inactive catalyst site.
(*c*) Catalyst sintering (irreversible)	A high surface area catalyst loses active surface area resulting in a decrease in the number of active sites per catalyst volume.
Enzyme Catalysts	
(*a*) Product inhibition	Same as above
(*b*) Catalyst poisoning	Same as above
(*c*) Catalyst denaturation (irreversible and/or reversible)	The enzyme molecule conformation changes sufficiently to result in a loss of the catalyst activity; a first order process in enzyme concentration thus again active site concentration decreases in time.

Note: The characterizations reversible and irreversible are qualitative indicators

Figure 3.3 Some related deactivation phenomena in catalysis by enzymes or solid surfaces.

more fragile. While their complicated, contorted shapes in space often endow enzymes with unusual specificity and activity, it is relatively easy to disturb the native conformation and destroy the enzyme's catalytic properties. We shall return to denaturation in Sec. 3.6; inhibition, another means of enzyme deactivation, will be treated in Sec. 3.5.

It is often asserted that enzymes are more active, i.e., allow reactions to go faster, than nonbiological catalysts. A common measure of activity is the *turnover number*, which is the net number of substrate molecules reacted per catalyst site per unit time. To permit some comparison of relative activities, turnover numbers for several reactions are listed in Table 3.3, which shows that at the ambient temperatures where enzymes are most active they are able to catalyze reactions faster than the majority of artificial catalysts. When the reaction temperature is increased, however, solid catalysts may become as active as enzymes. Unfortunately, enzyme activity does not increase continuously as the temperature is raised. Instead, the enzyme typically denatures at quite a low temperature, often only slightly above that at which it is typically found.

Table 3.3. Some turnover numbers for enzyme- and solid-catalyzed reactions
N = molecules per site per second†

Enzyme	Reaction	Range of reported N values at 0–37°C
Ribonuclease	Transfer phosphate of a polynucleotide	2–2×10^3
Trypsin	Hydrolysis of peptides	3×10^{-3}–1×10^2
Papain	Hydrolysis of peptides	8×10^{-2}–1×10^1
Bromelain	Hydrolysis of peptides	4×10^{-3}–5×10^{-1}
Carbonic anhydrase	Reversible hydration of carbonyl compounds	8×10^{-1}–6×10^5
Fumarate hydratase	L-Malate $\rightleftharpoons$ fumarate + H_2O	1×10^3 (forward) 3×10^3 (reverse)

Solid catalyst	Reaction	N	Temp., °C
SiO$_2$–Al$_2$O$_3$ (impregnated)	Cumene cracking	3×10^{-8}	25
		2×10^4	420
Decationized zeolite	Cumene cracking	$\sim 10^3$	25
		$\sim 10^8$	325
V$_2$O$_3$	Cyclohexene dehydrogenation	7×10^{-11}	25
		10^2	350
Treated Cu$_3$Au	HCO$_2$H dehydrogenation	2×10^7	25
		3×10^{10}	327
AlCl$_3$–Al$_2$O$_3$	n-Hexane isomerization (liquid)	1×10^{-2}	25
		1.5×10^{-2}	60

† R. W. Maatman: Enzyme and Solid Catalyst Efficiencies and Solid Catalyst Site Densities, *Catal. Rev.*, **8**: 1, 1973).

A truly unique aspect of enzyme catalysis is its susceptibility to control by small molecules. Several enzymes are "turned off" by the presence of another chemical compound, often the end product of a sequence of reactions in which the regulated enzyme participates. This feature of some enzymes is essential for normal cell function. Some aspects of the kinetics of such enzymes will be explored in Sec. 3.6, and in Chap. 6 we shall learn how altering the normal channels of cellular control can improve industrial biological processes tremendously.

Before beginning our efforts to model the kinetics of enzyme-catalyzed reactions, we must review the available experimental evidence on molecular events which actually occur during reaction. With this foundation, we shall be able to make reasonable hypotheses about the reaction sequence and from them to derive useful reaction rate expressions.

3.1. THE ENZYME-SUBSTRATE COMPLEX AND ENZYME ACTION

There is no single theory currently available which accounts for the unusual specificity and activity of enzyme catalysis. However, there are a number of plausible ideas supported by experimental evidence for a few specific enzymes.

Probably, then, all or some collection of these phenomena acting together combine to give enzymes their special properties. In this section we shall outline some of these concepts. Our review will necessarily be brief, and the interested reader should consult the references for further details. Since all the theories mentioned here are at best partial successes, we must be wary of attempting to synthesize a single theory of enzyme activity.

Verified by numerous experimental investigations involving such diverse techniques as x-ray crystallography, spectroscopy, and electron-spin resonance is the existence of a *substrate-enzyme complex*. The substrate binds to a specific region of the enzyme called the *active site*, where reaction occurs and products are released. Binding to create the complex is sometimes due to the type of weak attractive forces outlined in Sec. 2.4.3 although covalent attachments are known for some cases. As shown schematically in Figs. 2.29, 2.30, and 3.4, the complex is formed when the substrate key joins with the enzyme lock. Especially evident in Fig. 3.4 are the hydrogen bonds which form between the substrate and groups widely separated in the amino acid chain of the enzyme.

This example also nicely illustrates the notion of an active site. The protein molecule is folded in such a way that a group of reactive amino acid side chains in the enzyme presents a very specific site to the substrate. The reactive groups encountered in enzymes include the R group of Asp, Cys, Glu, His, Lys, Met, Ser, Thr, and the end amino and carboxyl functions. Since the number of such groups near the substrate is typically 20 (far less than the total number of amino acid residues present), only a small fraction of the enzyme is believed to participate directly in the enzyme's active site. Large enzymes may have more than one active site. Many of the remaining amino acids determine the folding along a chain of amino acids (secondary structure) and the placement of one part of a folded chain next to another (tertiary structure), which help create the active site itself (Fig. 3.7).

While some of the ideas described below are still somewhat controversial, we should emphasize that the notions of active sites and the enzyme-substrate complex are universally accepted and form the starting point for most theories of enzyme action. These concepts will also be the cornerstone of our analysis of enzyme kinetics.

Two different aspects of current thinking on enzyme activity are shown schematically in Fig. 3.5. Enzymes can hold substrates so that their reactive regions are close to each other and to the enzyme's catalytic groups. This feature, which quite logically can accelerate a chemical reaction, is known as the *proximity effect*. Consider now that the two substrates are not spherically symmetrical molecules. Consequently, reaction will occur only when the molecules come together at the proper orientation so that the reactive atoms or groups are in close juxtaposition. Enzymes are believed to bind substrates in especially favorable positions, thereby contributing an *orientation effect*, which accelerates the rate of reaction. Also called *orbital steering*, this phenomenon has qualitative merit as a contributing factor to enzyme catalysis. The quantitative magnitude of its effect, however, is still difficult to assess in general.

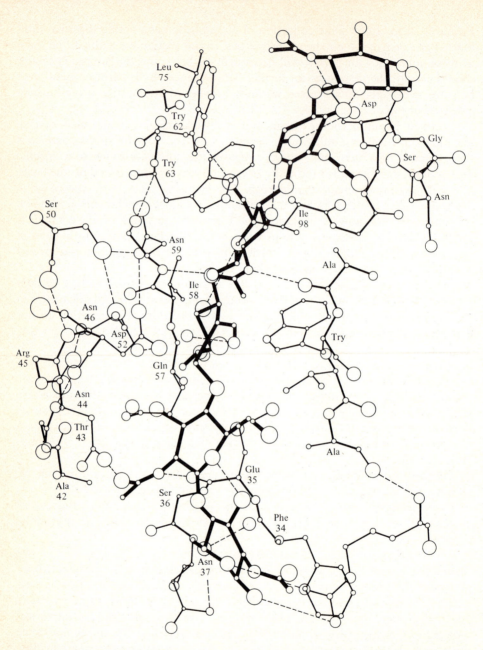

Figure 3.4 Another view of the active site of lysozyme, showing a hexasaccharide substrate in heavy lines. Larger and smaller circles denote oxygen and nitrogen atoms, respectively, and hydrogen bonds are indicated by dashed lines. (*Reprinted by permission from M. Yudkin and R. Offord, "A Guidebook to Biochemistry," p. 48, Cambridge University Press, London, 1971.*)

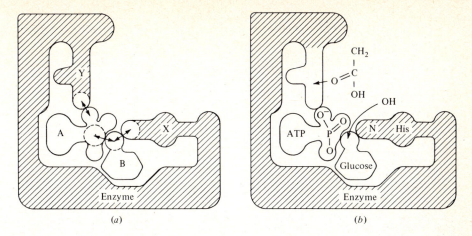

Figure 3.5 An enzyme may accelerate reaction by holding two substrates close to each other (proximity effect) and at an advantageous angle (orientation effect). (*Reprinted by permission of D. E. Koshland.*)

Before a brief foray into the chemistry of enzyme action, we should mention one other hypothesis related to enzyme geometry. It is thought that for some enzymes the binding of substrate causes the shape of the enzyme to change slightly (Fig. 3.6). As illustrated with a simple example in Fig. 3.6, the *induced fit* of substrate to enzyme may add to the catalytic process. There are also more sophisticated extensions of the induced-fit model, in which a number of different intermediate enzyme-substrate complexes are formed as the reaction progresses. A slightly elastic and flexible enzyme molecule would have the ability to make delicate adjustments in the position of its catalytic groups to hasten the transformation of each intermediate. Studies of the three-dimensional structure of the

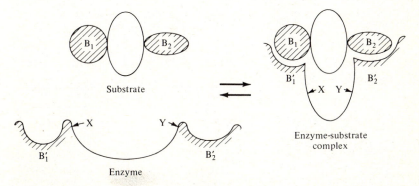

Figure 3.6 In the induced-fit model, binding of substrate causes a change in enzyme shape which facilitates reaction.

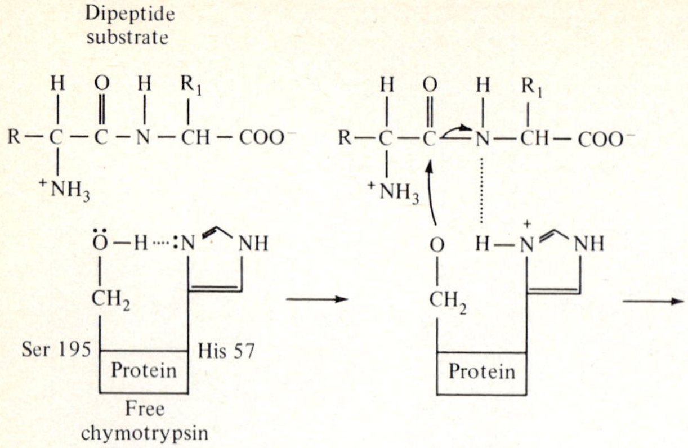

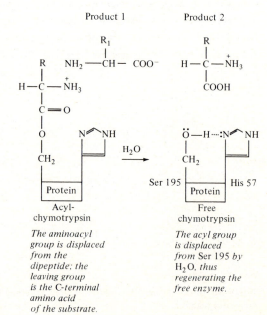

Dipeptide substrate

The hydroxyl group of Ser 195 is hydrogen-bonded to the imidazole nitrogen of His 57.

The proton is very rapidly transferred from Ser 195 to the imidazole nitrogen of His. 57.
A hydrogen bond between imidazole and substrate forms transiently and positions the dipeptide for the nucleophilic attack of the serine oxygen on the carboxyl carbon atom of the substrate.

Product 1   Product 2

The aminoacyl group is displaced from the dipeptide; the leaving group is the C-terminal amino acid of the substrate.

The acyl group is displaced from Ser 195 by H_2O, thus regenerating the free enzyme.

(a)

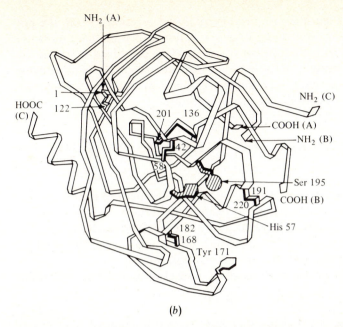

NH₂ (A)

1
HOOC
(C)
122
201   136
42
58
NH₂ (C)
COOH (A)
NH₂ (B)
Ser 195
191
220
COOH (B)
182
168
His 57
Tyr 171

(b)

Figure 3.7 (*a*) A proposed reaction sequence for chymotrypsin action on a peptide bond, and (*b*) structure of chymotrypsin. (*From A. L. Lehninger, "Biochemistry," 2d ed., pp. 229, 231. Worth Publishers, New York, 1975. Structure redrawn from B. W. Matthews, et al., "Three-Dimensional Structure of Tosyl-α-Chymotrypsin, Nature,* **214:** *652, 1967.*)

enzymes lysozyme and carboxypeptidase A with and without substrates have shown a change in enzyme conformation upon addition of the substrate. Since these are the only two enzymes for which data of this type are available, it is possible that a substrate-induced change in active-site configuration may be a general characteristic of enzyme catalysis.

Catalytic processes well known to the organic chemist also appear to be at work in some enzymes. One of these is *general acid-base catalysis*, where the catalyst accepts or donates protons somewhere in the overall catalytic process. In one of the few enzymes for which a reasonably complete catalytic sequence is proposed (Fig. 3.7) this mode of catalysis is present. *Chymotrypsin*, derived from the pancreas, is a *proteolytic* (protein-hydrolyzing) enzyme with specificity for peptide bonds where the carbonyl side is a tyrosine, tryptophan, or phenylalanine residue. Water is believed to serve as a proton-transfer agent in both the general acid- and general base-catalyzed portions of the chymotrypsin mechanism. Several reactions important to cellular chemistry, including carbonyl addition and ester hydrolysis, are in principle amenable to general acid-base catalysis.

Participating in enzyme catalysis may be a number of other phenomena such as covalent catalysis, strain, electrostatic catalysis, multifunctional catalysis, and solvent effects (recall the oil-drop structure from Chap. 2). Details on these mechanisms available in the references reveal that, like the notions reviewed here, they are probably involved in some enzyme-catalyzed reactions. From the possible involvement of these factors and others, it is not surprising that no one has yet devised a simple general scheme for assessing their combined influence and relative importance. Fortunately, armed only with the basic idea of an enzyme-substrate complex as an essential reaction intermediate, we shall be able to formulate useful rate expressions for biochemical reactions.

3.2. SIMPLE ENZYME KINETICS WITH ONE AND TWO SUBSTRATES

Our objective in this section is to develop suitable mathematical expressions for the rates of enzyme-catalyzed reactions. Naturally, the crucial test of a reaction rate equation is comparison with experimentally determined rates. Since there are some pitfalls awaiting the unwary, we should first briefly outline some of the experimental methods employed to measure reaction rates.

Let us be quite clear on what we mean by a reaction rate. If the reaction is

$$S \longrightarrow P$$

the reaction rate v, in the quasi-steady state approximation (Sec. 3.2.1), is defined by

$$v = \frac{-ds}{dt} = \frac{dp}{dt} \tag{3.1}$$

where lowercase letters denote molar concentration. The dimensions of v are consequently moles per unit volume per unit time. The reaction rate is an *intensive* quantity, defined at each point in the reaction mixture. Consequently, the rate will vary with position in a mixture if concentrations or other intensive state variables change from point to point. In experimental kinetic studies, well-mixed reactors are often used to that the reaction rate is spatially uniform.

As usual in our engineering modeling work, the term *point* in the rate definition is not used in a strict geometrical sense. Instead, a point is a volume large enough to contain many molecules but very small relative to the entire reacting system. It is very important to remember this seemingly minor detail when applying engineering analysis to biological systems. Since cells may contain only a few molecules of a given compound A_1, the concepts of A_1 concentration and reaction rate at a point really cannot be used rigorously in modeling molecular processes of an isolated single cell.

A typical experiment to measure enzyme kinetics (shorthand for kinetics of reactions catalyzed by enzymes) might proceed as follows. At time zero, solutions of substrate and of an appropriate purified enzyme are mixed in a well-stirred, closed isothermal vessel containing a buffer solution to control pH. The substrate and/or product concentrations are monitored at various later times. Techniques for following the time course of the reaction may include spectrophotometric, manometric, electrode, polarimetric, and sampling-analysis methods. Typically, only *initial-rate* data are used. Since the reaction conditions including enzyme and substrate concentrations are known best at time zero, the initial slope of the substrate or product concentration vs. time curve is estimated from the data:

$$\left.\frac{dp}{dt}\right|_{t=0} = -\left.\frac{ds}{dt}\right|_{t=0} = v\left.\right|_{t=0} \qquad \text{where } e = e_0\,, s = s_0\,, \text{ and } p = 0 \qquad (3.2)$$

Figure 3.8 shows actual data for such an experiment. Notice that some product is present initially, indicating that the reported zero reaction time is not the actual time when the reaction started. This is one of the inherent difficulties of the initial-rate method, and others are described in texts on chemical kinetics. Still, initial concentrations as well as enzymatic activity are reasonably reproducible, and these considerations favor the initial-rate approach.

The problem of reproducible enzyme activity is an important one which also has serious implications in design of reactors employing isolated enzymes. Since we have already mentioned that proteins may denature when removed from their native biological surroundings, it should not surprise us that an isolated enzyme in a "strange" aqueous environment can gradually lose its catalytic activity (Fig. 3.9). Although this gradual deactivation is known to occur for many enzymes, the phenomenon is scarcely mentioned in many treatments of enzyme kinetics. While perhaps not important in vivo (in the intact living organism), where enzyme synthesis compensates for any loss of previously active enzymes, enzyme deactivation in vitro (removed from a living cell) cannot be overlooked in either kinetic studies or enzyme reactor engineering. This topic will surface again in Sec. 3.7.

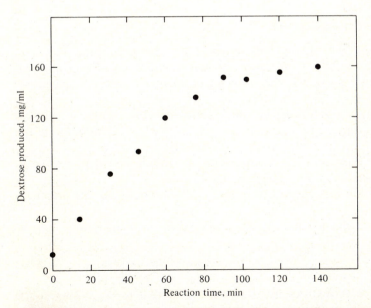

Figure 3.8 Batch reactor data for hydrolysis of 30 percent starch by glucoamylase (60°C, $e_0 = 11,600$ units, reactor volume = 1 l).

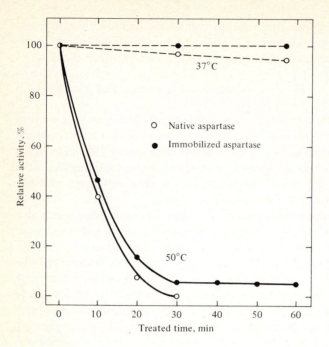

Figure 3.9 Loss of activity in solution. At 37°C, aspartase in solution loses more than 5 percent of its initial activity within 1 hr. (*Reprinted from T. Tosa et al., Continuous Production of L-Aspartic Acid by Immobilized Aspartase, Bioctech. Bioeng.,* **15:** 69, 1973.)

3.2.1. Michaelis-Menten Kinetics

Now let us suppose that armed with a good set of experimental rate data (Fig. 3.10), we face the task of representing it mathematically. From information of the type shown in Fig. 3.10 the following qualitative features often emerge:

1. The rate of reaction is first order in concentration of substrate at relatively low values of concentration. [Recall that if $v = (\text{const})(s^n)$, n is the order of the reaction rate.]
2. As the substrate concentration is continually increased, the reaction order in substrate diminishes continuously from one to zero.
3. The rate of reaction is proportional to the total amount of enzyme present.

Henri observed this behavior and in 1902 proposed the rate equation

$$v = \frac{v_{\max}s}{K_m + s} \qquad \text{where } v_{\max} = \alpha e_0 \qquad (3.3)$$

which exhibits all three features listed above. Notice that $v = v_{\max}/2$ when s is equal to K_m. To avoid confusion of the type found in some of the literature, we should strongly emphasize that s is the concentration of free substrate in the

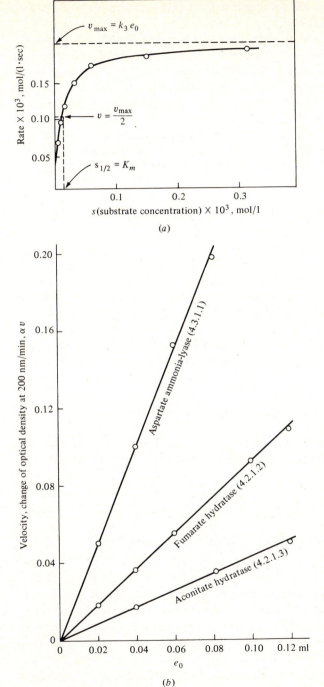

Figure 3.10 Kinetic data for enzyme-catalyzed reactions. (*a*) Enzyme concentration is held constant when studying substrate concentration dependence; (*b*) the converse holds for investigation of the influence of enzyme concentration. [(*a*) *reprinted from K. J. Laidler, "The Chemical Kinetics of Enzyme Action," p.* 64, *The Clarendon Press, Oxford,* 1958, *Data of L. Ouellet, K. J. Laidler, M. F. Morales, Molecular Kinetics of Muscle Adenosine Triphosphate, Arch. Biochem. Biophys.,* **39:** 37, 1952), (*b*) *reprinted by permission from M. Dixon and E. C. Webb, "Enzymes," 2d ed., p.* 55, *Academic Press, Inc., New York,* 1964).]

reaction mixture, while e_0 is the total amount of enzyme present in both the free and combined forms (see below).

Although Henri provided a theoretical explanation for Eq. (3.3) using a hypothesized reaction mechanism, his derivation and the similar one offered in 1913 by Michaelis and Menten are now recognized as not rigorous in general. However, the general methodology of the Michaelis-Menten treatment will find repeated useful (although still not rigorously justified) application in derivation of more complicated kinetic models later in this chapter. Consequently, we shall provide a brief summary of their development before proceeding to others.

As a starting point, it is assumed that the enzyme E and substrate S combine to form a complex ES, which then dissociates into product P and free (or uncombined) enzyme E:

$$S + E \underset{k_{-1}}{\overset{k_1}{\rightleftharpoons}} ES \qquad (3.4a)$$

$$ES \xrightarrow{k_2} P + E \qquad (3.4b)$$

This mechanism includes the intermediate complex discussed above, as well as regeneration of catalyst in its original form upon completion of the reaction sequence. While perhaps considerably oversimplified, Eqs, (3.4) are certainly reasonable.

Henri and Michaelis and Menten assumed that reaction (3.4a) is in equilibrium, which, in conjunction with the mass-action law for the kinetics of molecular events, gives

$$\frac{se}{(es)} = \frac{k_{-1}}{k_1} = K_m = \text{dissociation constant} \qquad (3.5)$$

Here s, e and (es) denote the concentrations of S, E, and ES, respectively. Decomposition of the complex to product and free enzyme is assumed irreversible:

$$v = \frac{dp}{dt} = k_2(es) \qquad (3.6)$$

Since all enzyme present is either free or complexed, we also have

$$e + (es) = e_0 \qquad (3.7)$$

where e_0 is the total concentration of enzyme in the system. This is known from the amount of enzyme initially charged into the reactor. Equation (3.3) with v_{max} equal to $k_3 e_0$ can now be obtained by eliminating (es) and e from the three previous equations. We should note here that a reaction described by Eq. (3.3) is commonly referred to as having *Michaelis-Menten kinetics*, although certainly other investigators made equal contributions to its development and justification. The parameter v_{max} is called the *maximum* or *limiting velocity*, and K_m is known as the *Michaelis constant*. While the Michaelis-Menten equation successfully describes the kinetics of many enzyme-catalyzed reactions, it is not universally valid. We shall explore the extensions and modifications necessary for certain enzymes and reaction conditions in later sections and in the problems.

Briggs and Haldane have provided the derivation of Eq. (3.3) which later kinetic studies and mathematical analyses have shown to be the most general. For reaction in a well-mixed closed vessel we can write the following mass balances for substrate and the ES complex:

$$v = -\frac{ds}{dt} = k_1 se - k_{-1}(es) \tag{3.8}$$

and
$$\frac{d(es)}{dt} = k_1 se - (k_{-1} + k_2)(es) \tag{3.9}$$

Using Eq. (3.7) in the previous two equations gives a closed set: two simultaneous ordinary differential equations in two unknowns, s and (es). The appropriate initial conditions are, of course,

$$s(0) = s_0 \qquad (es)(0) = 0 \tag{3.10}$$

These equations cannot be solved analytically but they can readily be integrated on a computer to find the concentrations of S, E, ES, and P as functions of time. In calculated results illustrated in Fig. 3.11, it is evident that to a good approximation

$$\frac{d(es)}{dt} = 0 \tag{3.11}$$

after a brief start-up period. In the key step in their analysis Briggs and Haldane assumed condition (3.11) to be true. This assumption, commonly called the *quasi-steady-state approximation*, can be proved valid for the present case provided that e_0/s_0 is sufficiently small. If the initial substrate concentration is not large

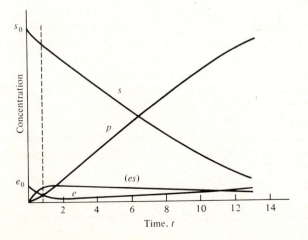

Figure 3.11 The time course of the reaction $E + S \underset{k_{-1}}{\overset{k_{+1}}{\rightleftharpoons}} ES \xrightarrow{k_{+2}} P + E$ with $k_{+1} \cong k_{-1} \cong k_{+2}$.

compared with the total enzyme concentration, the assumption breaks down. In many instances, however, the amount of catalyst present is considerably less than the amount of reactant, so that (3.11) is an excellent approximation after start-up.

Proceeding from Eq. (3.11), as Briggs and Haldane did, it is possible to use Eqs. (3.7) and (3.9) to eliminate e and (es) from the problem, leaving

$$v = -\frac{ds}{dt} = \frac{k_2 e_0 s}{[(k_{-1} + k_2)/k_1] + s} \tag{3.12}$$

Consequently the Michaelis-Menten form results, where Eq. (3.12) shows now

$$v_{max} = k_2 e_0 \tag{3.13}$$

and

$$K_m = \frac{k_{-1} + k_2}{k_1} \tag{3.14}$$

Notice that K_m no longer can be interpreted physically as a dissociation constant.

With the Michaelis-Menten rate expression at hand, the time course of the reaction can now be determined analytically by integrating

$$\frac{ds}{dt} = \frac{-v_{max}s}{K_m + s} \qquad \text{with } s(0) = s_0 \tag{3.15}$$

to obtain

$$v_{max} t = s_0 - s + K_m \ln \frac{s_0}{s} \tag{3.16}$$

Naturally this equation is most easily exploited indirectly by computing the reaction time t required to reach various substrate concentrations, rather than vice versa.

It is instructive to compare the behavior predicted by Eq. (3.16) with that obtained without the quasi-steady-state approximation. As Fig. 3.12 reveals, the deviation is significant when the total enzyme concentration approaches s_0. Consequently the Michaelis-Menten equation should really not be used in such cases. As another example, we show the dependence of esterase activity on enzyme concentration in Fig. 3.13. The linear dependence predicted by the Michaelis-Menten model is followed initially but does not hold for large enzyme concentrations. Remember that the slope at the linear portion is equal to $k_2 s/(K_m + s)$.

We should consider whether or not the quasi-steady-state approximation can be justified in some cases of large e_0/s_0, perhaps with different rate constants. This could be done if the ES complex tended to disappear much faster than it was formed, i.e., if K_m were very large relative to unity. Such a situation, however, does not arise for any known enzyme. On the contrary, the Michaelis constant is quite small, with typical values in the range 10^{-2} to 10^{-5} M.

Consequently, if the enzyme concentration is comparable to s_0, we have no proper justification for simplifying the kinetic model with the quasi-steady-state approximation. Situations with relatively large enzyme concentration can occur in

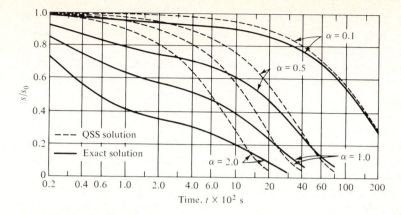

Figure 3.12 Computed time course of batch hydrolysis of acetyl 1-phenylalanine ether by chymotrypsin. Considerable discrepancies between the exact solution and the quasi-steady-state solution arise when e_0/s_0 $(=\alpha)$ is not sufficiently small. (*Reprinted by permission from H. C. Lim, On Kinetic Behavior at High Enzyme Concentrations, Am. Inst. Chem. Eng., J.,* **19:** 659, 1973.)

several instances. If an enzyme reactor is operated under conditions where most of the substrate is converted, s falls to a value comparable to e_0 and the Michaelis-Menten equation may be inappropriate for the final stages of reaction. Fortunately, the reaction has become arbitrarily slow here, and this situation is consequently seldom of interest. Also in the case of enzyme reactions at interfaces, which we shall investigate in Sec. 3.7, the substrate concentration in the neighborhood of the enzyme may be quite small. These possible pitfalls in the Michaelis-Menten model have come to light only very recently, and consequently it is still commonly employed for analyzing and designing enzyme reactors.

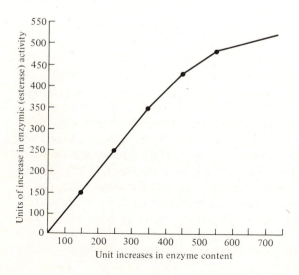

Figure 3.13 These data on esterase activity show deviations from Michaelis-Menten kinetics at large values of initial enzyme content. (*Reprinted by permission from M. Frobisher, "Fundamentals of Microbiology,"* 8th ed., p. 61, W. B. Saunders Co., Philadelphia, 1968.)

Although they lack rigorous justification in some situations, the equilibrium assumption and the Michaelis-Menten rate equation have proven to be valuable working tools. We shall often employ them in this spirit.

While on the subject of simple Michaelis-Menten kinetics, we should emphasize that exactly the same mathematical form is widely used for expressing the rates of many solid-catalyzed reactions. In chemical engineering practice, kinetics described by Eq. (3.3) is called *Langmuir-Hinshelwood* or *Hougen and Watson kinetics*. Since reactions involving synthetic solid catalysts are widespread throughout the chemical and petroleum industries, a large effort has been devoted to the design and analysis of catalytic reactors. Much of this work has employed Langmuir-Hinshelwood kinetics and is therefore directly applicable to enzyme-catalyzed reactions described by Michaelis-Menten kinetics. In the remainder of our studies, we shall continue to observe analogies, similarities, and identities between classical chemical engineering and biochemical technology. Also, of course, we shall encounter many novel features of biological processes.

3.2.2. Evaluation of Parameters in the Michaelis-Menten Equation

The Michaelis-Menten equation in its original form [Eq. (3.3)] is not well suited for estimation of the kinetic parameters v_{max} and K_m. As Fig. 3.10 shows, it is quite difficult to estimate v_{max} accurately from a plot of v vs. s. By rearranging Eq. (3.3) we can derive the following options for data plotting and graphical parameter evaluation:

$$\frac{1}{v} = \frac{1}{v_{max}} + \frac{K_m}{v_{max}}\frac{1}{s} \tag{3.17}$$

$$\frac{s}{v} = \frac{K_m}{v_{max}} + \frac{1}{v_{max}}s \tag{3.18}$$

$$v = v_{max} - K_m\frac{v}{s} \tag{3.19}$$

Each equation suggests an appropriate linear plot. In evaluation of the kinetic parameters of the model using such plots, however, several points should be noted. Plotting Eq. (3.17) as $1/v$ vs. $1/s$ (known as *Lineweaver-Burk plot*) cleanly separates dependent and independent variables (Fig. 3.14a). The most accurately known rate values, near v_{max}, will tend to be clustered near the origin, while those rate values which are least accurately measured will be far from the origin and will tend most strongly to determine the slope K_m/v_{max}. Thus, a linear least-squares fitting should *not* be used with such a plot. The second equation, (3.18), tends to spread out the data points for higher values of v so that the slope $1/v_{max}$ can be accurately determined. The intercept often occurs quite close to the origin, so that accurate measure of K_m by this method is subject to large errors. The third method uses the *Eadie-Hofstee plot*: v is graphed against v/s (Fig. 3.14b). Both coordinates contain the measured variable v, which is subject to the largest errors.

Table 3.4. Kinetic parameters for several enzymes†

Enzyme	Substrate	Temp °C	pH	k_2, s^{-1}	$1/K_m$, M^{-1}
Pepsin	Carbobenzoxy-L-glutamyl-L-tyrosine ethyl ester	31.6	4.0	0.00108	530
	Carbobenzoxy-L-glutamyl-L-tyrosine	31.6	4.0	0.00141	560
Trypsin	Benzoyl-L-argininamide	25.5	7.8	27.0	480
	Chymotrypsinogen	19.6	7.5	2,900	< 770
	Sturin	24.5	7.5	13,100	400
	Benzoyl-L-arginine ester	25.0	8.0	26.7	12,500
Chymotrypsin	Methyl hydrocinnamate	25.0	7.8	0.026	256
	Methyl dl-α-chloro-β-phenylpropionate	25.0	7.8	0.135	83.3
	Methyl d-β-phenyllactate	25.0	7.8	0.139	28.6
	Methyl-l-β-phenyllactate	25.0	7.8	1.38	100
	Benzoyl-L-phenylalanine methyl ester	25.0	7.8	51.0	217
	Acetyl-L-tryptophan ethyl ester	25.0	7.8	30.7	588
	Acetyl-L-tyrosine ethyl ester	25.0	7.8	193.0	31.2
	Benzoyl-L-o-nitrotyrosine ethyl ester	25.0	7.8	3.27	90.9
	Benzoyl-L-tyrosine ethyl ester	25.0	7.8	78.0	250
	Benzoyl-L-phenylalanine ethyl ester	25.0	7.8	37.4	167
	Benzoyl-L-methionine ethyl ester	25.0	7.8	0.77	1,250
	Benzoyl-L-tyrosinamide	25.0	7.8	0.625	23.8
	Acetyl-L-tyrosinamide	25.0	7.8	0.279	12.3
Carboxy-peptidase	Carbobenzoxyglycyl-L-tryptophan	25.0 / 25.0	7.5 / 8.2	89 / 94	196 / 164
	Carbobenzoxyglycyl-L-phenylalanine	25.0	7.5	181	154
	Carbobenzoxyglycyl-L-leucine	25.0	7.5	10.6	37
Adenosine triphosphatase	ATP	25.0	7.0	104	79,000
Urease	Urea	20.8 / 20.8	7.1 / 8.0	20,000 / 30,800	250 / 256

† K. J. Laidler, "The Chemical Kinetics of Enzyme Action," p. 67, Oxford University Press, London, 1958.

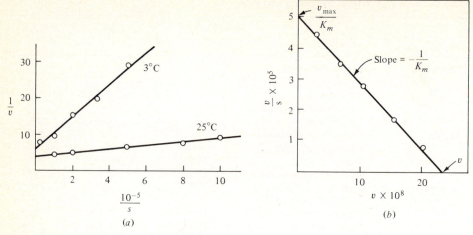

Figure 3.14 (*a*) Lineweaver-Burk plot of experimental data for pepsin. (*b*) Eadie-Hofstee plot of data for hydrolysis of methyl hydrocinnamate catalyzed by chymotrypsin. (*Reprinted by permission from K. J. Laidler, "The Chemical Kinetics of Enzyme Action," pp. 65–66, The Clarendon Press, Oxford, 1958.*)

These considerations suggest the following strategy for evaluating v_{max} and K_m: determine v_{max} from a plot of Eq. (3.17) (find intercept accurately) or Eq. (3.18) (find slope accurately). Then return to a graph of v vs. s and find $s_{1/2}$, the substrate concentration where v is equal to $v_{max}/2$. Recalling the comment following Eq. (3.3), we see that K_m is equal in magnitude and dimension to $s_{1/2}$.

It is important to have a feel for the magnitudes of these kinetic parameters. Table 3.4 shows the range of values encountered for several different enzymes. Notice that almost all the experiments reported were performed at moderate temperatures and pH values. The exception is pepsin, which has the task of hydrolyzing proteins in the acid environment of the stomach. Consequently, the enzyme has the greatest activity under the acidic conditions employed in the experimental determination of its kinetic parameters. Models for representing the effects of pH and temperature on enzyme kinetics will be explored in Sec. 3.5.

3.2.3. Kinetics for Two-Substrate Reactions and Cofactor Activation

The vast majority of enzyme-catalyzed reactions involve at least two substrates. In many, however, one of the substrates is water, whose concentration is essentially constant and typically 1000 or more times larger than the concentration of the other substrates. As our analysis in this section will reveal, in such a case the reaction may be treated as in the preceding section, where S is the only substrate considered. Moreover, the kinetic models developed here can be applied in some instances to explain the influence of cofactors on enzyme reaction rates.

In many two-substrate reactions it appears that a ternary complex may be formed with both substrates attached to the enzyme. One possible sequence in this situation is

Dissociation
equilibrium constant

$$
\begin{aligned}
E + S_1 &\rightleftharpoons ES_1 & K_1 \\
E + S_2 &\rightleftharpoons ES_2 & K_2 \\
ES_1 + S_2 &\rightleftharpoons ES_1S_2 & K_{12} \\
ES_2 + S_1 &\rightleftharpoons ES_1S_2 & K_{21} \\
ES_1S_2 &\xrightarrow{\ k\ } P + E &
\end{aligned}
\tag{3.20}
$$

so that

$$
v = k(es_1 s_2) \tag{3.21}
$$

As before, lowercase symbols are used to denote concentration, and a lowercase symbol in parentheses is the concentration of one species, often a complex. Assuming equilibria in the first four reactions in (3.20) leads to

$$
v = \frac{ke_0}{1 + \dfrac{K_{21}}{s_1} + \dfrac{K_{12}}{s_2} + \dfrac{1/2(K_2 K_{21} + K_1 K_{12})}{s_1 s_2}} \tag{3.22}
$$

Derivation of this expression is straightforward so long as we remember that the total concentration of enzyme e_0 must be equal to the sum of the concentration of free enzyme e plus the concentrations of the *three* complexes ES_1, ES_2, and ES_1S_2. As for the one-substrate situation, an analysis based on the quasi-steady-state approximation can be undertaken. This leads in the general case to a rather unwieldy equation with too many parameters for practical application. In some special situations the equation reduces to (3.22).

Equation (3.22) can be shortened slightly by noting that we always have

$$
K_1 K_{12} = K_2 K_{21} \tag{3.23}
$$

Rearrangement of (3.22) results in the now familiar form

$$
v = \frac{v_{max}^* s_1}{K_1^* + s_1} \tag{3.24}
$$

where

$$
v_{max}^* = \frac{ke_0 s_2}{s_2 + K_{12}} \tag{3.25}
$$

and

$$
K_1^* = \frac{K_{21} s_2 + K_1 K_{12}}{s_2 + K_{12}} \tag{3.26}
$$

From the previous three equations it is apparent that if s_2 is held constant and s_1 is varied, the reaction will follow Michaelis-Menten kinetics. However, Eqs. (3.25) and (3.26) show that the apparent maximum rate and Michaelis constant are functions of S_2 concentration.

Assuming that the above sequence and the rate expression for a two-substrate reaction are correct, we can verify the original Michaelis-Menten form (3.3) when one substrate is in great excess. In this case, v_{max}^* becomes $= ke_0$, while K_1^* approaches the constant value K_{21}. Therefore, the two-substrate reaction can be treated as though s_1 were the only substrate when $s_2 \gg K_{12}$.

The participation of a cofactor, whether metal ion or coenzyme, in a one-substrate enzymatic reaction (or a two-substrate reaction with $s_2 \gg K_{12}$) can be modeled as in Fig. 3.15. Since this situation resembles so closely the two-substrate mechanism just analyzed, the details of the derivation based on the equilibrium assumption are not necessary. The final result is

$$v = \frac{kecs}{cs + K_s(c + K_c)} \tag{3.27}$$

where c is the cofactor concentration. If substrate concentration s is considered fixed, this rate expression shows a Michaelis-Menten dependence on cofactor concentration c. Thus, if there is little cofactor present ($c \ll K_c$), the reaction velocity is first order in c. On the other hand, for $c \gg K_c$, the single-substrate equation is recovered, and the rate is independent of cofactor concentration.

If s_1 and s_2 or c and s_1 must bind to e in an obligatory order, the appropriate kinetic equation is obtained by letting K_{ij} of the forbidden binding reaction approach infinity.

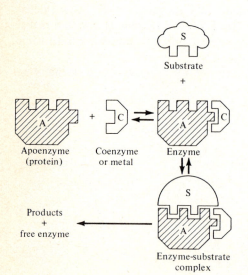

Figure 3.15 Schematic illustration of a plausible mechanism for enzyme catalysis requiring a cofactor.

3.3. DETERMINATION OF ELEMENTARY-STEP RATE CONSTANTS

In Sec. 3.2.2, we explored convenient plotting techniques for determining the parameters v_{max} and K_m which appear in the Michaelis-Menten rate equation. As the Briggs-Haldane derivation reveals, however, these parameters depend on the elementary-step rate constants k_1, k_{-1}, and k_2. Examination of the specific relationships in Eqs. (3.13) and (3.14) shows that knowledge of K_m and v_{max} is not sufficient to determine the elementary rate constants.

These more fundamental kinetic parameters are of interest for at least two reasons: (1) They provide a better picture of what occurs during the process of enzyme catalysis. We can learn, for example, how rapidly substrate combines with free enzyme and how this rate compares with the reverse process of ES complex decomposition. (2) We saw earlier that the simplification leading to the Michaelis-Menten equation is rigorously justified only when the total enzyme concentration is relatively small. In cases where this is not true, a careful treatment of enzyme reaction kinetics would involve integration of the differential equations (3.8) and (3.9) for s, p, and (es). Such an analysis cannot be undertaken, however, unless the elementary-step rate constants k_1, k_{-1}, and k_2 are known. With these motivations, we next explore two different experimental techniques for determining these parameters.

3.3.1. Pre-Steady-State Kinetics

We have already seen in Fig. 3.11 that there is a short period just after reaction is initiated when the quasi-steady-state approximation does not apply. In Fig. 3.16 the initial behavior in a batch reactor has been magnified. Notice that initially the concentration (es) of the enzyme-substrate complex is zero, and some time is required for this concentration to coincide with the prediction of the quasi-steady-state approximation. Application of this approximation gives the dashed line labeled $(es)_s$ in Fig. 3.16. In particular, the initial complex concentration from the quasi-steady-state solution is

$$(es)_{s_0} = \frac{e_0 s_0}{K_m + s_0} \tag{3.28}$$

Next we shall seek an approximate solution for the transient behavior of (es) during the initial period before the steady-state-approximation concentration is achieved. If we assume that, relative to (es), substrate concentration s changes little during the pre-steady-state period, using

$$s(t) = s_0 \qquad \text{for small } t \tag{3.29}$$

in the mass balance (3.9) for (es) gives

$$\frac{d(es)}{dt} = k_1 s_0 [e_0 - (es)] - (es)(k_{-1} + k_2) \tag{3.30}$$

Solution of this differential equation with the initial condition

$$(es)(0) = 0 \tag{3.31}$$

yields

$$(es)(t) = (es)_{s_0}\{1 - \exp[-k_1 t(K_m + s_0)]\} \tag{3.32}$$

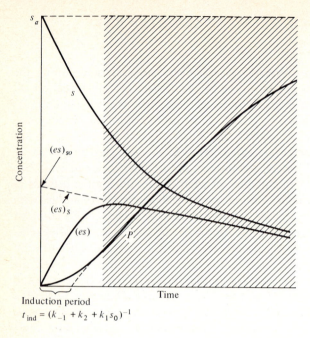

Figure 3.16 The solid curves show batch substrate (s), product (p), and enzyme-substrate complex (es) concentrations vs. time as computed from Eqs. (3.8), (3.9), and (3.33). The subscript s denotes approximate values obtained with the quasi-steady state approximation.

In order to evaluate the time course of product concentration p, Eq. (3.32) is substituted into

$$\frac{dp}{dt} = k_2(es) \tag{3.33}$$

Assuming no product is present initially, the resulting expression for $p(t)$ is

$$p(t) = k_2(es)_{s_0} t + \frac{1}{k_{-1} + k_2 + k_1 s_0}\{1 - \exp\left[-k_1 t(K_m + s_0)\right]\} \tag{3.34}$$

where the definition (3.14) of the Michaelis constant has been employed. For sufficiently large times, the exponential term in Eq. (3.34) is negligible. Then the product concentration follows the straight line given by

$$p_s(t) = k_2(es)_{s_0}\left(t - \frac{1}{k_{-1} + k_2 + k_1 s_0}\right) \tag{3.35}$$

The *induction time* t_{induc} for the reaction is defined by extrapolating $p_s(t)$ back to zero product concentration (see Fig. 3.16). From Eq. (3.35), the induction time is related to the kinetic parameters of the reaction system according to

$$t_{induc} = \frac{1}{k_{-1} + k_2 + k_1 s_0} = \frac{1}{k_1(K_m + s_0)} \tag{3.36}$$

Experimental evaluation of t_{induc} for several s_0 values gives the sum $k_{-1} + k_2$ and k_1. As noted earlier, k_2 can be determined separately by measuring v_{max} at different enzyme concentrations e_0. With this information, k_{-1} follows, and all the elementary-step rate constants are known.

There are other ways in which pre-steady-state data can be exploited. For very short times, the

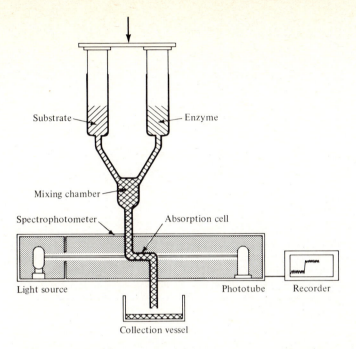

Figure 3.17 Stopped-flow apparatus for measurement of pre-steady-state enzyme kinetics. The experiment is initiated by rapidly mixing substrate and enzyme solutions and pumping the mixture quickly into an absorption cell in a spectrophotometer. The flow is then stopped and the progress of reaction monitored via changes in the absorbance of the reaction mixture. (*From "Cell Structure and Function," p. 240, by Ariel G. Loewy and Philip Siekevitz. Copyright © 1963, 1969 by Holt, Rinehart and Winston, Inc. Reprinted by permission of Holt, Rinehart and Winston.*)

exponential term in Eq. (3.34) can be approximated by the first few terms in a series expansion. This approach yields the approximate solution

$$p(t) = k_1 k_2 e_0 s_0 t^2 \qquad (3.37)$$

Hence a plot of p vs. t^2 gives the product $k_1 k_2$. Combined with values for v_{max} and K_m and their definitions, this information allows determination of all three desired parameters.

We should not be misled by these mathematical derivations into thinking that it is easy to obtain k_1, k_{-1}, and k_2. The pre-steady-state period is quite short (~ 1 s or less) for enzyme-catalyzed reactions, so that a considerable number of data must be accumulated in a very short time. Also, the zero time when reaction initiates must be very precisely controlled. One popular experimental technique for collecting such kinetic data is shown schematically in Fig. 3.17. Perfected by Chance and coworkers, this device can be operated on a continuous or stopped-flow basis. In a stopped-flow experiment, rapid continuous flow of substrate and enzyme is briefly maintained to establish uniform conditions in the absorption cell. When flow is stopped, the absorption cell is a tiny batch reactor; its behavior can be continuously monitored with the spectrophotometer.

Actual data obtained with this procedure are presented in Fig. 3.18. It is interesting to note that, using elementary rate constants obtained in the study, the calculated value of K_m for peroxidose was 4.4×10^{-7} M. Independent evaluation of K_m by the methods described in Sec. 3.2.2 gave the value 5.0×10^{-7} M. In view of the mathematical manipulations and difficult experiments involved in obtaining the first value, such agreement is quite satisfactory and increases confidence in the pre-steady-state method.

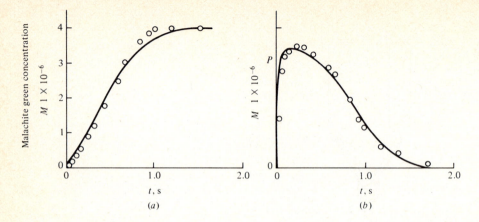

Figure 3.18 Results of stopped-flow experiments (circles) on peroxidase kinetics: (*a*) product concentration and (*b*) enzyme-substrate complex concentration. The solid lines are computed results. (Experimental initial conditions: horseradish peroxidase 10^{-6} *M*, H_2O_2 4×10^{-6} *M*, leucomalachite green 15×10^{-6} *M*; in acetate buffer at pH 4. (*Reprinted by permission from B. Chance, The Kinetics of the Enzyme-Substrate Compound of Peroxidase, J. Biol. Chem.,* **151**: 553 1943.)

3.3.2. Relaxation Kinetics

The pre-steady-state method can be used only on a limited class of fast chemical reactions. With a flow apparatus like that in Fig. 3.17 it is difficult to follow composition changes over time intervals smaller than the millisecond range. The *relaxation methods*, developed and widely applied by Eigen and coworkers, can explore reactions which occur on a time scale of nanoseconds.

Although there are several variations of the method, all are based on the same principles: an external perturbation of some reaction condition such as temperature, pressure, or electric field density is applied to a reaction mixture at equilibrium (or steady state). The resulting response of the reaction system is then monitored continuously. As illustrated schematically in Fig. 3.19, the response following a sudden step change in reaction conditions is a transition to a new, nearby equilibrium or steady state.

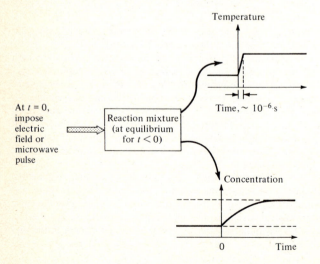

Figure 3.19 In a relaxation experiment, a small step change in reaction conditions, e.g., in temperature, causes a transient approach to a new steady state.

Although the theory and practice of relaxation methods is developed for oscillatory perturbations, we shall explore briefly only the step-change technique. Let us consider for the moment only the equilibrium between substrate, enzyme, and enzyme-substrate complex:

$$E + S \underset{k_{-1}}{\overset{k_1}{\rightleftharpoons}} ES \tag{3.38}$$

For this reaction, we know that

$$s + (es) = s_0 \tag{3.39}$$

and

$$e + (es) = e_0 \tag{3.40}$$

so that in a batch reactor, the mass balance for s

$$\frac{ds}{dt} = k_{-1}(es) - k_1 se \tag{3.41}$$

becomes

$$\frac{ds}{dt} = k_{-1}(s_0 - s) - k_1 s(e_0 + s - s_0) \equiv f(s) \tag{3.42}$$

If we let s^* denote the s concentration at equilibrium *after* the step change in reaction conditions, s^* can be determined by solving

$$f(s^*) = 0 \tag{3.43}$$

where $f(s)$ is the right-side of Eq. (3.42) and is equal to ds/dt. In any relaxation experiment s will be close to s^*, and so we can approximate $f(s)$ by the first terms in its Taylor expansion

$$f(s) \approx f(s^*) + \frac{df(s^*)}{ds}(s - s^*) + \text{terms of order } (s - s^*)^2 \text{ and higher} \tag{3.44}$$

zero because of Eq. (3.43)

Letting χ be the deviation from the equilibrium concentration

$$\chi = s - s^* \tag{3.45}$$

and remembering that s^* is time-invariant, we can combine Eqs. (3.42) to (3.45) to obtain the *linearized* mass balance [squared and higher-order terms neglected in (3.44)]

$$\frac{d\chi}{dt} = -[k_{-1} + k_1(e_0 - s_0 + 2s^*)]\chi \tag{3.46}$$

If the system is initially at the old equilibrium corresponding to conditions before the step perturbation (see Fig. 3.20),

$$\chi(0) = \Delta\chi_0 \tag{3.47}$$

and consequently

$$\chi(t) = \Delta\chi_0 e^{-t/\tau} \tag{3.48}$$

where

$$\frac{1}{\tau} = k_{-1} + k_1(e_0 - s_0 + 2s^*) = k_{-1} + k_1(e^* + s^*) \tag{3.49}$$

In Eq. (3.49) e^* represents the enzyme concentration at the final equilibrium state. Figure 3.20 shows how τ can be determined from the experimentally measured time course of $\chi(t)$. After τ, χ^*, and e^* (or

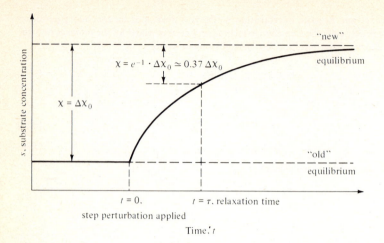

Figure 3.20 Relaxation response to a step-function perturbation. Actually it is sufficient to measure any quantity proportional to s. The relaxation time τ is the time required for the deviation χ from the new equilibrium to decay to 37 percent of its initial value.

e_0 and s_0) have been measured, Eq. (3.49) provides a relationship between k_1 and k_{-1} which is independent of the equilibrium equation. Consequently, both k_1 and k_{-1} can be calculated.

This technique has enjoyed wide application in chemical-kinetics research, including investigations of enzyme catalysis which we now review.

3.3.3. Some Results of Transient-Kinetics Investigation

Listed in Table 3.5 are k_1 and k_{-1} values for a variety of enzymes and substrates. These numbers were obtained by the methods just outlined as well as other specialized approaches described in the references. Examining the rate constant k_1 for combination of enzyme and substrate, we see that measured

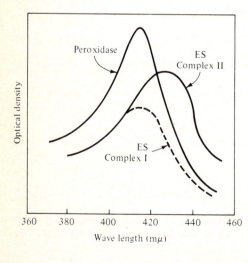

Figure 3.21 Absorption spectra measured in a continuous-flow apparatus. (See *Fig. 3.18*; in a continuous-flow experiment substrate and enzyme solutions are fed continuously through the absorption cell.) The data show the presence of two enzyme-substrate complexes ES_I(green) and ES_{II}(red). For this system a suitable sequence is

$$E + S \rightleftarrows ES_I \rightleftarrows ES_{II} \rightarrow E + P$$

(*After B. Chance, 1949.*)

Table 3.5. Rate constants for several elementary substrate-enzyme reactions†‡

Protein or enzyme	Substrate	k_1, $(M \cdot s)^{-1}$	k_{-1}, s^{-1}
Fumarase	Fumarate	$> 10^9$	$> 4.5 \times 10^4$
	L-Malate	$> 10^9$	$> 4 \times 10^4$
Acetylcholinesterase	Acetylcholine	$> 10^9$	
Urease	Urea	$> 5 \times 10^6$	
Myosin ATPase	ATP	$> 8 \times 10^6$	
Hexokinase	Glucose	3.7×10^6	1.5×10^3
	MgATP	$> 4 \times 10^6$	$> 6 \times 10^2$
β-Amylase	Amylose	$> 5.8 \times 10^7$	
Liver alcohol dehydrogenase	NAD	5.3×10^5	74
	NADH	1.1×10^7	3.1
Liver alcohol dehydrogenase– NAD	C_2H_5OH	$> 1.2 \times 10^4$	> 74
	CH_3CHO	$> 3.7 \times 10^5$	> 3.1
Malate dehydrogenase	NADH	6.8×10^8	2.4×10^2
Old yellow enzyme	FMN	1.5×10^6	$\sim 10^{-4}$
Catalase	H_2O_2	5×10^6	
Catalase-H_2O_2	H_2O_2	1.5×10^7	
Peroxidase	H_2O_2	9×10^6	< 1.4
Peroxidase-H_2O_2(II)	Hydroquinone	2.5×10^6	
Peroxidase	Cytochrome c	1.2×10^8	
Glutamic-aspartic trans- aminase, aldehydic	Glutamate	3.3×10^7	2.8×10^3
	Aspartate	$> 10^7$	$> 5 \times 10^3$
	NH_2OH	3.7×10^6	38
	Ketoglutarate, oxalacetate	$> 5 \times 10^8$	$> 5 \times 10^4$
Aminic	Oxalacetate	7×10^7	1.4×10^2
	Ketoglutarate	2.1×10^7	70
BSA	NR′	2×10^6	35
	NSR′	3.5×10^5	2.5
Anti-R antibodies§	NR′	2.2×10^7	50
Anti-DNP antibodies§	DNP-lysine	8.1×10^7	1.1
$Hb(O_2)_3$	O_2	2×10^7	36
Globin	Carboxyheme	7×10^7	

† H. R. Mahler and E. H. Cordes, "Biological Chemistry," 2d ed., p. 322, Harper & Row, Publishers, Incorporated, New York, 1971.

‡ FMN, flavin mononucleotide; BSA, bovine serum albumin; NR′, 1-naphthol-4-(4′-azo-benzene azophenyl)arsonic acid; NSR′, [naphthol-3-sulfonic acid 4-(4′-azobenzene azo)]phenyl-arsonic acid; DNP, 2,4-dinitrophenyl; Hb, hemoglobin.

§ Rabbit γ-globulin.

values are all very near the theoretical maximum, since the absolute bimolecular-collision rate constant in solution is 10^9 to 10^{11} $[(mol \cdot s)/l]^{-1}$. On the other hand, the reverse process is relatively slow and exhibits considerably wider variation from reaction to reaction. The same comments apply to decomposition of the intermediate complex into product and free enzyme (see k_2 values in Table 3.4).

Another interesting feature of enzyme catalysts which has been revealed by pre-steady-state and relaxation studies is the presence of more than one simple reaction intermediate ES (Fig. 3.21). Thus the intermediate complex passes through a series of different configurations which may be regarded as isomerizations. Steady-state experiments are incapable of distinguishing whether more than one reaction intermediate exists when the dissociation rate (to products) of the last such intermediate complex is slow compared with that of earlier complexes. Transient relaxation studies provide a valuable probe for such situations.

3.4 OTHER PATTERNS OF SUBSTRATE CONCENTRATION DEPENDENCE

Not all rates of enzyme-catalyzed reactions follow Michaelis-Menten kinetics, as described in Sec. 3.2. In this section we shall examine the more common forms of deviation. The first kind of unusual behavior, often associated with regulatory enzymes, can be explained by reaction mechanisms where more than one substrate molecule can bind to the enzyme. In the second case, we shall consider the possibility that the substrate is actually a mixture of different reacting species with different kinetic properties. This analysis will reveal some of the problems encountered when working with impure or poorly defined mixtures.

3.4.1. Substrate Activation and Inhibition

The sigmoidal dependence of reaction rate on substrate concentration exhibited by some enzymes is evidence of a type of activation effect (Fig. 3.22). At low substrate concentrations, the binding of one substrate molecule enhances the binding of the next one (mathematically stated, the increment in v resulting from an increment in s, dv/ds, is increasing). Such behavior can be modeled by assuming a concerted transition of protein subunits: we assume that the enzyme, which is probably oligomeric, has multiple binding sites for substrate and that the first substrate molecule bound to the enzyme alters the enzyme's structure so that the remaining sites have a stronger affinity for the substrate. Since the mathematical

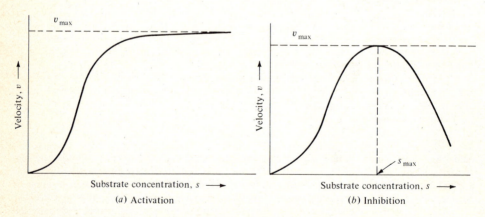

Figure 3.22 Substrate activation (*a*) and inhibition (*b*).

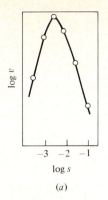

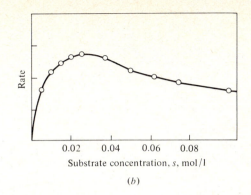

(a)   (b)

Figure 3.23 Experimental evidence of substrate inhibition. (*K. J. Laidler, "The Chemical Kinetics of Enzyme Action," p. 71, The Clarendon Press, Oxford, 1958. (a) Data of K. -D. Augustinsson, Substrate Concentration and Specificity of Choline Ester Splitting Enzymes, Arch. Biochem., **23**: 111, 1949; (b) data of R. Lumry, E. L. Smith, and R. R. Glantz, Kinetics of Carboxypeptidase Action. I. Effect of Various Extrinsic Factors on Kinetic Parameters, J. Am. Chem. Soc., **73**: 4330, 1951.*)

analysis for this model parallels the hemoglobin model outlined in Prob. 2.11, we need not pursue it here.

Sometimes when a large amount of substrate is present, the enzyme-catalyzed reaction is diminished by the excess substrate (Fig. 3.23). This phenomenon is called *substrate inhibition*. We should note from Fig. 3.23 that the reaction rate v passes through a maximum as substrate concentration is increased. When s is larger than s ($v = v_{max}$), a decrease in substrate concentration causes an increase in reaction rate. The behavior of substrate-inhibited reactions can be modeled quite nicely using the Michaelis-Menten approach. In the assumed mechanism here, however, a second substrate molecule can bind to the enzyme. When S joins the ES complex, an unreactive intermediate results:

Reaction steps at equilibrium:

Dissociation constant

$$E + S \underset{k_{-1}}{\overset{k_1}{\rightleftharpoons}} ES \qquad K_1 \left(\text{for example, } \frac{k_{-1}}{k_1}\right) \qquad (3.50)$$

$$ES + S \rightleftharpoons ES_2 \qquad K_2$$

Slow step: $$\qquad\qquad\qquad ES \overset{k}{\longrightarrow} E + P \qquad\qquad\qquad (3.51)$$

When the two dissociation-equilibrium relationships and an enzyme balance analogous to Eq. (3.7) are used, straightforward algebraic manipulations yield

$$v = \frac{ke_0}{1 + K_1/s + s/K_2} \qquad (3.52)$$

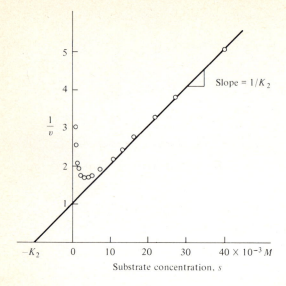

Figure 3.24 K_2 can be determined from the slope or intercept of a v vs. s plot. Data for hydrolysis of ethyl butyrate by sheep liver carboxylesterase. (*Reprinted by permission from M. Dixon and E. C. Webb, "Enzymes," 2d ed., p. 78, Academic Press, Inc., New York, 1964.*)

It is not difficult to deduce the necessary parameters in this equation from experimental data. First, by plotting $1/v$ against s, a straight line with slope $1/K_2$ is obtained at high substrate concentrations (Fig. 3.24). Next we can evaluate K_1 from Eq. (3.53) where $s_{\max}$ satisfies $dv/ds = 0$ using Eq. (3.52).

$$s_{\max} = \sqrt{K_1 K_2} \tag{3.53}$$

Finally k can be determined by varying e_0.

The unconventional dependences on substrate concentration just described are believed to be important in controlling the rates and relative importance of different reaction pathways within the cell. If enzymes with these properties are isolated and employed commercially, their unusual behavior can also be of special significance. These matters will be explored briefly in the problems.

3.4.2. Multiple Substrates Reacting on a Single Enzyme

Many enzymes can utilize more than a single substrate. In such cases, the various reactants compete with each other for the active site, which is here presumed to be the same for each reaction. The most obvious examples of such enzymes are hydrolytic depolymerases, which act on many identical bonds, often regardless of the length of the polymer segment. When the substrate takes the form of a dimer, trimer, or oligomer, the kinetic parameters characterizing the reaction may change from the values applicable to large polymeric units. An example of such a system is lysozyme, the enzyme found in egg whites, mucus, and tears. Lysozyme hydrolyses complex mucopolysaccharides, including the murein cell wall, thereby killing gram-positive bacteria. The previous comments are reflected in Table 3.6, which summarizes the measured rate and dissociation constants for each

Table 3.6. Kinetic parameters for lysozyme action on bacterial cell wall (GlcNAc–MurNAc copolymer)†
GlcNAc = N-acetylglucosamine, MurNAc = N-acetylmurein

No.	Reaction	Type
1.	$S_i = (\text{GlcNAc-MurNAc})_i = i$ unit oligomer	Association
	$E + S_i \xrightarrow{\ K_{ij}\ } C_{ij} \quad j = 0, \ldots, i+1$	
2.	$C_{ij} \xrightarrow{\ k_c\ } A_j + S_{i-j} \quad j = 1, \ldots, i-1$	Hydrolysis of complex
3.	$C_{ii} \xrightarrow{\ k_v\ } A_i + H_2O$	Dissociation of intermediate $(j = i)$
4.	$A_i\,(+H_2O) \xrightarrow{\ k_H\ } E + S_i$	Hydrolysis of intermediate
5.	$A_i + S_j \xrightarrow{\ k_T\ } E + S_{i+j}$	Transglycosylation to yield $i + j$ unit

$$K_{11} = 0.5\ M^{-1} \qquad K_{21} = 6.310^{-4}\ M^{-1} \qquad K_{22} = 2000\ M^{-1} \qquad K_{32} = 3.5 \times 10^4\ M^{-1}$$

$$k_c = 1.75\ s^{-1} \qquad k_v = 100\ s^{-1} \qquad k_T = 2.77 \times 10^5 (M \cdot s)^{-1}$$

† After D. M. Chipman, A Kinetic Analysis of the Reaction of Lysozyme with Oligosaccharides from Bacterial Cell Walls, *Biochemistry*, **10:** 1714, 1971).

lysozyme substrate when it is the only substrate present. The parameter variations are seen to be greatest with the monomer and oligomers.

In the above examples, as in the following one, each condensation linkage in the oligomer or polymer can be viewed conceptually as a separate substrate. Also, a polymeric substrate like starch is not really a single substrate in the normal sense, for the polymer is a mixture of long chains of different lengths, which may also contain various monomer-monomer links (Sec. 2.2.2). Shown in Fig. 3.25 are the differences in the course of enzymatic hydrolysis of several starches. The kinetic behavior of a better-defined system is illustrated in Fig. 3.26. In commercial processes for thinning of starch solutions by amylases, these reactions are

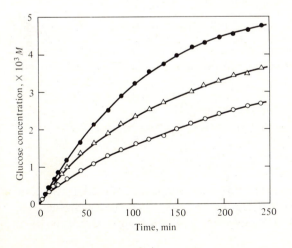

Figure 3.25 Reaction-time curves for hydrolysis of different amylose fractions (25°C, pH 4.6, $s_0 = 0.1$ g/100 ml, $e_0 = 6.8 \times 10^{-8} M$) (on amyloglucosidase)

○ mol wt ≈ 1,650,000
△ mol wt ≈ 1,100,000
● mol wt ≈ 360,000

[*Reprinted by permission from K. Hiromi et al., A Kinetic Method for the Determination of Number-Average Molecular Weight of Linear High Polymer by Using an Exo-Enzyme, J. Biochem. (Tokyo),* **60:** 439, 1966.]

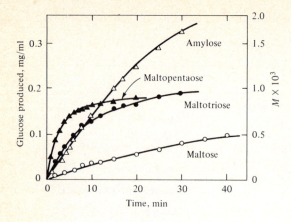

Figure 3.26 Reaction-time curves for glucose oligomers and polymers (15°C, pH 5.15, $s_0 = 0.04$ M, $e_0 = 2.82 \times 10^{-7}$ M) The enzyme is amyloglucosidase. [*S. Ono et al., Kinetic Studies of Gluc-amylase, J. Biochem. (Tokyo)*, **55**: 315, 1964.]

important. Similar effects are to be expected in analogous reactions, e.g., hydrolysis of proteins with proteolytic enzymes. Before proceeding to a quantitative treatment of competition between different substrates, we should point out that other phenomena may be involved in the above examples, including variations in physical size and differences in the number of reactive bonds in the diverse substrates in the reaction mixture.

Suppose that the sequence below describes the reactions of two different substrates catalyzed by one enzyme:

Reaction steps at equilibrium:

Disassociation constant

$$E + S_1 \rightleftharpoons ES_1 \qquad K_1 \qquad (3.54)$$
$$E + S_2 \rightleftharpoons ES_2 \qquad K_2$$

Slow steps:

$$ES_1 \xrightarrow{k_1} E + P_1 \qquad ES_2 \xrightarrow{k_2} E + P_2 \qquad (3.55)$$

Here, each substrate binds a certain fraction of the enzyme present. When we employ the overall enzyme balance and equilibria, as in our earlier analysis, the rate expression is

$$-\frac{ds_1}{dt} = v_1 = \frac{k_1 e_0 s_1 / K_1}{1 + s_1/K_1 + s_2/K_2} \qquad (3.56)$$

and

$$-\frac{ds_2}{dt} = v_2 = \frac{k_2 e_0 s_2 / K_2}{1 + s_1/K_1 + s_2/K_2} \qquad (3.57)$$

Equations (3.56) and (3.57) indicate that if two reactions are catalyzed by the same enzyme, the individual velocities are slower in the presence of both substrates than in the absence of one of the substrates. This fact has been used advantageously to determine whether the same enzyme in an undefined sample acts upon both substrates or each substrate's reaction is catalyzed by a separate enzyme.

Often in experiments and plant operation, it is neither possible nor practical to measure and monitor the concentrations of all species present in the reaction mixture. For example, in Fig. 3.25, the plotted glucose production is actually the net overall effect of many simultaneous reactions, including hydrolysis of maltose, maltotriose, and so forth. Started in other terms, the overall reaction considered in the kinetic analysis is actually

$$\text{Glucose polymers} \xrightarrow{\text{amyloglucosidase}} \text{glucose}$$

Thus, all the species containing glucose subunits have been *lumped* together into one imaginary species whose concentration is relatively easy to measure. Lumping of this type is extremely widespread in all branches of chemical kinetics. Consider, for example, catalytic cracking of gas oil, which is often usefully represented as a three-species reaction

$$\text{Gas oil} \longrightarrow \text{gasoline} \longrightarrow \text{gases}$$

when designing or analyzing the reactor although each "species" is clearly a mixture of compounds. Although not often explicitly recognized in the literature of the life sciences and biotechnology, lumping is a pervasive practice when investigating, modeling, or designing biological processes. In this book, we shall watch carefully for this and other simplifying assumptions employed, usually implicitly, in treating biological systems. Any assumption can fail and cause a breakdown in the analysis.

One possible pitfall of lumping can be illustrated with the present example. If s_T is the total concentration of all substrates present

$$s_T = s_1 + s_2 \tag{3.58}$$

then the overall rate of disappearance of s_T is

$$v_T = -\frac{ds_T}{dt} = -\left(\frac{ds_1}{dt} + \frac{ds_2}{dt}\right) = \frac{e_0(k_1 s_1/K_1 + k_2 s_2/K_2)}{1 + s_1/K_1 + s_2/K_2} \tag{3.59}$$

By varying the relative values of s_1 and s_2, this rate can be changed. For example, if the total substrate concentration has a specified value s_{T0}, the overall rate v_T does *not* in general have a unique value. Instead, it may lie anywhere in the range indicated below, depending on the s_1/s_2 ratio:

$$v_T\bigg|_{\substack{s_1 = s_{T0} \\ s_2 = 0}} = \frac{e_0 k_1 s_{T0}}{K_1 + s_{T0}} \xrightarrow[\substack{\text{increase} \\ s_2}]{\substack{\text{decrease} \\ s_1}} v_T\bigg|_{\substack{s_1 = 0 \\ s_2 = s_{T0}}} = \frac{e_0 k_2 s_{T0}}{K_2 + s_{T0}} \tag{3.60}$$

Clearly, then, the values of $v_{\max}$ and K_m we would measure in a kinetic experiment would depend on the detailed composition (s_1/s_2 ratio) as well as the total substrate concentration. Equally, the performance of an enzyme reactor fed a mixture of S_1 and S_2 will change if their relative amounts do, even though the total feed reactant concentration is maintained constant. As we shall discuss later, difficulties of this nature are magnified many-fold when working with living cells rather than isolated enzymes.

3.5. MODULATION AND REGULATION OF ENZYMATIC ACTIVITY

Chemical species other than the substrate can combine with enzymes to alter or modulate their catalytic activity. Such substances, called *modulators* or *effectors*, may be normal constituents of the cell. In other cases they enter from the cell's environment or act on isolated enzymes. Although most of our attention in this section will be concentrated on *inhibition*, where the modulator decreases activity, cases of enzyme *activation* by effectors are also known.

Modulators such as lead salts can combine with some enzymes, and loss of catalytic activity results. In this case and many others, the substances we commonly call *poisons* are strong enzyme inhibitors. By shutting off one reaction in a long sequence necessary for life, small quantities of inhibitors from the environment can have a profound effect on living organisms.

The combination of an enzyme with an effector is itself a chemical reaction and may therefore be fully reversible, partially reversible, or essentially irreversible. Known examples of irreversible inhibitors include cyanide ions, which deactivate xanthine oxidase, and a group of chemicals collectively termed *nerve gases*, which deactivate cholinesterases (enzymes which are an integral part of nerve transmission and thus of motor ability). If the inhibitor acts irreversibly, the Michaelis-Menten approach to inhibitor-influenced kinetics cannot be used since this method assumes equilibrium between the free and complexed forms. Often, irreversible inhibition increases with time as more and more enzyme molecules are gradually deactivated. Other cases, more difficult to detect, involve the partial deactivation of enzyme. In such an instance, the inhibited enzyme retains catalytic activity although at a level reduced from the pure form.

Section 2.5 mentioned the remarkable chemical capabilities of most microorganisms, which, supplied with only a few relatively simple precursors, can manufacture a vast array of complex molecules. In order to perform this feat, it is necessary for the supply of precursors to be proportioned very efficiently among the many synthesis routes leading to many end products. Optimum utilization of the available chemical raw materials in most instances requires that only the necessary amount of any end product be manufactured. If enough of one monomer is present, for example, its synthesis should be curtailed and attention devoted to making other compounds which are in relatively short supply.

Reversible modulation of enzyme activity is one control mechanism employed by the cell to achieve efficient use of nutrients. (The other major control device is discussed in Sec. 6.1.) The most intriguing examples of enzyme regulation involve interconnected networks of reactions with several control loops, but their analysis must wait until we have studied the basic chemistry of cellular operation in Chap. 5. For present purposes, we shall employ control of a single sequence of reactions as an example. Figure 3.27 shows a five-step sequence for the biosynthesis of the amino acid L-isoleucine. Regulation of this sequence is achieved by *feedback inhibition:* the final product, L-isoleucine, inhibits the activity of the first enzyme in the path. Thus, if the final product begins to build up, the biosynthesis process will be stopped.

Since the reactions catalyzed by enzymes E_2 through E_5 are essentially at equilibrium and the first reaction is irreversible, the response of this control device

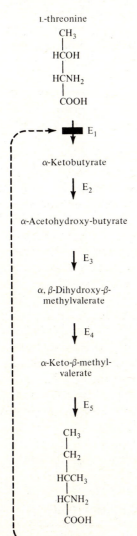

Figure 3.27 In this feedback-inhibition system, the activity of enzyme E_1 (L-theonine deaminase) is reduced by the presence of the end product, L-isoleucine. (*From A. L. Lehninger, Biochemistry, 1st ed., p. 180, Worth Publishers, Inc., New York, 1970.*)

is especially fast. Indeed, it is a general property of most regulatory enzymes that they catalyze irreversible reactions. Also, it should be obvious from the biological context that such "natural" enzyme modulation must be reversible. For example, if L-isoleucine is depleted by its use in protein synthesis, enzyme E_1 must be disinhibited to a higher activity so that the required supply of the amino acid is restored.

In the case of reversible inhibition, the approaches described in Sec. 3.2 prove quite fruitful. Since many isolated enzyme-substrate systems exhibit Michaelis-Menten kinetics, it is traditional to classify inhibitors by their influence on the Michaelis-Menten equation parameters v_{max} and K_m. The Michaelis constant K_m in its simplest quasi-steady-state form is the ratio of rate constants $(k_{-1} + k_2)/k_1$.

Reversible inhibitors are termed *competitive* if their presence increases the value of K_m but does not alter v_{max}. Their effect can be countered or reversed by increasing the substrate concentration. On the other hand, by rendering the enzyme or the enzyme-substrate complex inactive, a *noncompetitive* inhibitor decreases the v_{max} of the enzyme but does not alter the K_m value. Consequently, increasing the substrate concentration to any level cannot produce as great a reaction rate as possible with the uninhibited enzyme. Common noncompetitive inhibitors are heavy-metal ions, which combine reversibly with the sulfhydryl (—SH) group of cysteine residues.

Several different combinations and variations on the two basic types of reversible inhibitors are known. Some of these are listed in Table 3.7. Experimental discrimination between these possibilities will be discussed shortly. First, however, we shall briefly review some of the current theories and available data on the mechanisms of modulator action.

Table 3.7. Partial classification of reversible inhibitors†

	Type	Description	Result
I*a*	Fully competitive	Inhibitor adsorbs at substrate binding site	Increase in apparent value of K_m
b	Partially competitive	Inhibitor and substrate combine with different groups; inhibitor binding affects substrate binding	Increase in apparent value of K_m
II*a*	Noncompetitive	Inhibitor binding does not affect ES affinity, but ternary EIS complex does not decompose into products	No change in K_m, decrease of v_{max}
b	Noncompetitive	Same as II*a* except that EIS decomposes into product at a finite rate different from that of ES	No change in K_m, decrease of v_{max}
III	Mixed inhibitor		Affects both K_m and v_{max}

† M. Dixon and E. C. Webb, "Enzymes," 2d ed., table VIII.1, Academic Press, Inc., New York, 1964.

3.5.1. The Mechanisms of Reversible Enzyme Modulation

Many known competitive inhibitors, called *substrate analogs*, bear close relationships to the normal substrates. Thus, it is thought that these inhibitors have the key to fit into the enzyme active site, or lock, but that the key is not quite right so the lock does not work; i.e., no chemical reaction results. For example, consider the inhibition of succinic acid dehydrogenation by malonic acid:

$$
\begin{array}{ccc}
\text{COOH} & \text{COOH} & \text{COOH} \\
| & | & | \\
\text{CH}_2 \xrightarrow[\substack{\text{inhibited competitively} \\ \text{by malonic acid}}]{\substack{\text{catalyzed by} \\ \text{succinic dehydrogenase}}} & \text{CH} & \text{CH}_2 \\
| & \| & | \\
\text{CH}_2 & \text{CH} & \text{COOH} \\
| & | & \\
\text{COOH} & \text{COOH} & \text{Malonic acid} \\
\text{Succinic acid} & \text{Fumaric acid} &
\end{array}
$$

The malonic acid can complex with succinic dehydrogenase, but it does not react.

The action of one of the sulfa drugs, sulfanilamide, is due to its effect as a competitive inhibitor. Sulfanilamide is very similar in structure to *p*-aminobenzoic acid, an important vitamin for many bacteria. By inhibiting the enzyme which causes *p*-aminobenzoic acid to react to give folic acid, the sulfa drug can block the biochemical machinery of the bacterium and kill it.

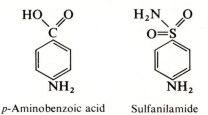

p-Aminobenzoic acid   Sulfanilamide

Another mechanism, called *allosteric control*, yields behavior typical of non-competitive inhibition and is thought to be the dominant mechanism for noncompetitive inhibition and activation. The name allosteric (other shape) was originally coined for this mechanism because many effectors of enzymic activity have structures much different from the substrate. From this fact it has been concluded that effectors work by binding at specific regulatory sites distinct from the sites which catalyze substrate reactions. An enzyme which possesses sites for modulation as well as catalysis has consequently been named an *allosteric enzyme*.

Allosteric control may either inhibit (reduce) or activate (increase) the catalytic ability of the enzyme. One schematic view of this process is depicted in Fig. 3.28. Notice that the enzyme here is visualized as having two pieces. Many allosteric enzymes are known to be oligomeric; we have already mentioned the allosteric (nonenzyme) protein hemoglobin. Experimental studies on one oligomeric allosteric enzyme, aspartyl transcarbamylase, have provided the most dramatic evidence to date in support of the allostery theory (Fig. 3.29).

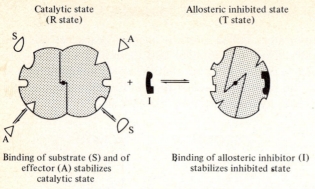

Catalytic state
(R state)

Allosteric inhibited state
(T state)

Binding of substrate (S) and of
effector (A) stabilizes
catalytic state

Binding of allosteric inhibitor (I)
stabilizes inhibited state

"Symmetry principle" excludes a mixed T-R state

Figure 3.28 In the symmetry model for allosteric control of enzyme activity, binding of an activator A and substrate S gives a catalytically active R enzyme configuration, while binding of an inhibitor changes all subunits in the oligomeric protein molecule to an inactive T state. (*From "Cell Structure and Function," 2d ed., p. 265, by Ariel G. Loewy and Philip Siekevitz. Copyright © 1963, 1969 by Holt, Rinehart and Winston, Inc. Reprinted by permission of Holt, Rinehart and Winston.*)

Reaction sequence in cytidine triphosphate (CTP) biosynthesis*

Feedback inhibition

Aspartase

Carbamyl
phosphate

ATCase E_2 E_3 E_4 E_5

CTP

*Note dissimilarity in structures of substrates and inhibitor

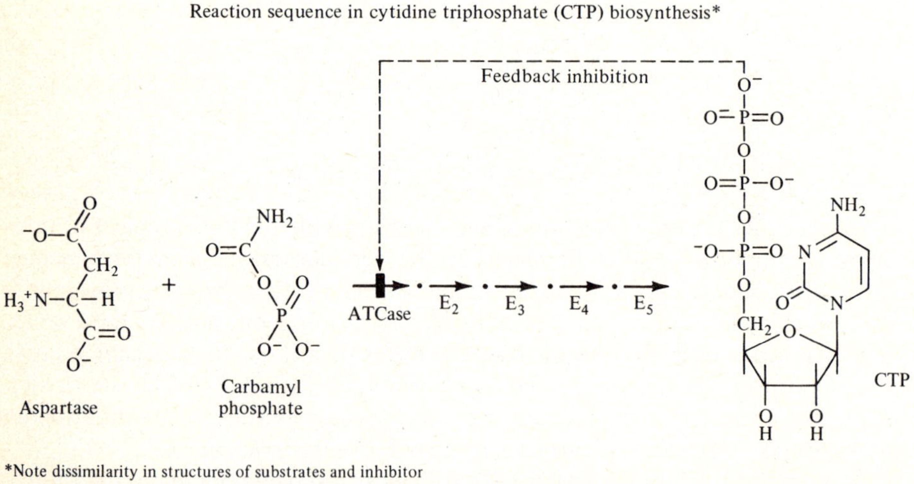

Figure 3.29 Some of the experimental findings for the allosteric enzyme aspartate transcarbamylase (ATCase). (*a*) While the native ATCase exhibits the sigmoidal rate curve often observed for allosteric enzymes, the desensitized enzyme follows conventional Michaelis-Menten form and is also insensitive to inhibition by CTP. The catalytic activity of the desensitized enzyme strongly suggests that the modular and catalytic sites are separate as shown in Fig. 3.28. (*b*) ATCase is separated into subunits and then centrifuged, with fractions then taken from the top of the centrifuge tube (abscissa). The O.D. data

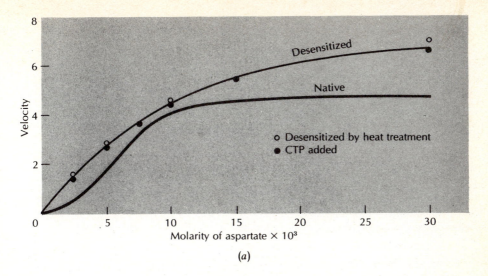

(a)

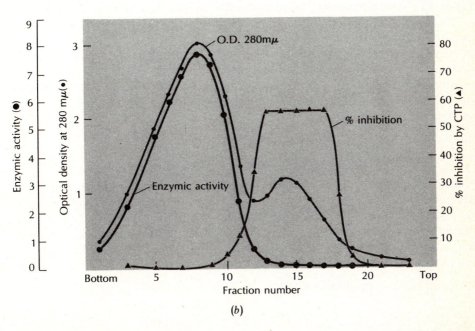

(b)

show two peaks of protein; separate measurements of ATCase activity show that the heavier fragments contain all the catalytic subunits. However, only the fractions associated with the smaller right hand peak are inhibited by CTP. [(a) *Reprinted from J. C. Gerhart and A. B. Pardee, The Enzymology of Control by Feedback Inhibition, J. Biol. Chem.,* **237:** 891, 1962; (b) *reprinted with permission from J. C. Gerhart and H. K. Schachman, Distinct Subunits for the Regulation and Catalytic Activity of Aspartate Transcarbamylase, Biochemistry,* **4:** 1054, 1965. Copyright by the American Chemical Society.)*

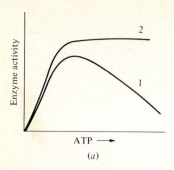 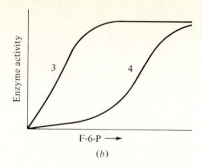

(a) (b)

Figure 3.30 Dependence of phosphofructokinase activity on (a) ATP and (b) fructose 6-phosphate (F 6-P) concentrations. Curves 1 and 2 are for low and high F 6-P levels, respectively, and curves 3 and 4 are for low and high ATP levels, respectively. (*Reprinted by permission from E. R. Stadtman, Allosteric Regulation of Enzyme Activity, Adv. Enzymol. Rel. Sub. Biochem.,* **28:** 41, 1966.)

From this interesting experiment, however, we should not conclude that the kinetics of regulatory enzymes is always straightforward. As an illustration of some of the unusual possibilities, examine the behavior of phosphofructokinase, where ATP can apparently bind at both catalytic and regulatory sites (Fig. 3.30).

In closing, we should note that our assumed sequence [(3.50) and (3.51)] for substrate inhibition could describe an allosteric enzyme. In the absence of detailed information of the type presented in Fig. 3.29 we cannot be certain that an effector does *not* bind at the active site. All that matters from the kinetic viewpoint is that substrate affinity and the rate of complex breakdown are changed by presence of the modulator. In the following section, we shall seek mathematical formulas suitable for representing these effects.

3.5.2. Analysis of Modulator Effects on Kinetics

A major contribution of the Michaelis-Menten approach to enzyme kinetics is accounting quantitatively for the influence of modulators. To begin our analysis, we shall assume that the following sequence is a reasonable representation for partially competitive inhibition. Here E and S have the usual significance, and I is the inhibitor. Also, EI is an enzyme-inhibitor complex; the ternary enzyme-inhibitor-substrate complex is denoted EIS.

Reaction steps at equilibrium:

Dissociation
constant

$$E + S \rightleftharpoons ES \qquad K_s$$

$$E + I \rightleftharpoons EI \qquad K_i \qquad\qquad (3.61)$$

$$EI + S \rightleftharpoons EIS \qquad K_{is}$$

$$ES + I \rightleftharpoons EIS \qquad K_{si}$$

Slow steps:

$$\text{ES} \xrightarrow{k} \text{E} + \text{P} \qquad \text{EIS} \xrightarrow{k} \text{EI} + \text{P} \tag{3.62}$$

If all four equilibria in Eq. (3.61) are written out algebraically and all variables except *total* enzyme concentration e_0 and *free* substrate and inhibitor concentrations (s and i respectively) are eliminated, there results

$$v = \frac{ke_0}{\left(1 + \dfrac{K_s}{s}\right)\dfrac{1 + i/K_i}{1 + (i/K_i)(K_s/K_{is})}} \tag{3.63}$$

In the limiting case of low inhibitor concentration, we obtain the original Michaelis-Menten form

$$\lim_{i \to 0} v = \frac{ke_0}{1 + K_s/s} \tag{3.64}$$

while if the inhibitor concentration i is very large relative to both K_i and $K_s/K_i K_{is}$, a different form holds:

$$\lim_{i \to \infty} v = \frac{ke_0}{1 + K_{is}/s} \tag{3.65}$$

Writing Eq. (3.63) as

$$v = \frac{ke_0 s}{s + K_m{}^{\text{app}}} \tag{3.66}$$

where the apparent Michaelis constant $K_m{}^{\text{app}}$ is given by

$$K_m{}^{\text{app}} = K_s \frac{1 + i/K_i}{1 + (i/K_i)(K_s/K_{is})} \tag{3.67}$$

we can see that $v_{\max}$ is unaffected by the magnitude of inhibitor concentration and that for the partial-inhibition case under consideration $K_m{}^{\text{app}}$ lies between the values K_s and K_{is}. Thus, the inhibitor reduces binding between free substrate and enzyme but does not totally prevent it.

The reaction rate for total competitive inhibition can be deduced as a special case of our preceding analysis. A totally competitive inhibitor completely blocks the active site, and, correspondingly, it cannot form a complex with the enzyme if the substrate is already complexed. Consequently, no ternary EIS complex exists, and K_{is} is infinity (the tendency of an EIS complex to dissociate is infinite). Thus, for total competition, Eq. (3.63) simplifies to

$$v = \frac{ke_0 s}{s + K_s(1 + i/K_i)} \tag{3.68}$$

Note that in this situation the apparent Michaelis constant $[K_s(1 + i/K_i)]$ increases without bound as i is increased, where there are bounds on K_m^{app} for partially competitive inhibition.

The noncompetitive-inhibition cases mentioned in Table 3.7 arise when inhibitor binding does not affect the binding of the substrate (K_s unchanged). However, binding of inhibitor does partially slow or completely halt the breakdown of the EIS complex into products. The appropriate equations for partially and totally noncompetitive inhibition are:

Partially noncompetitive:

$$v = \frac{ke_0 + k^i ie_0/K_i}{\left(1 + \dfrac{K_s}{s}\right)\left(1 + \dfrac{i}{K_i}\right)} = \frac{ke_0}{\left(1 + \dfrac{K_s}{s}\right)} \frac{1 + i/K_i}{1 + (i/K_i)(k^i/k)} \tag{3.69}$$

Totally noncompetitive:

$$v = \frac{ke_0}{(1 + K_s/s)(1 + i/K_i)} \tag{3.70}$$

where k^i is the reaction-rate constant for Eq. (3.62) ($k^i = k$ for competitive inhibition and $k^i = 0$ for totally noncompetitive inhibition). Notice that here $K_m = K_s$, as in the original Michaelis-Menten treatment, but now the apparent maximum velocity v_{max}^{app} is, for example,

$$v_{max}^{app} = \frac{ke_0}{1 + i/K_i} \tag{3.71}$$

for totally noncompetitive inhibition.

Enzyme activation is readily analyzed beginning from a slightly modified version of the sequence shown in Eqs. (3.61) and (3.62). Only (3.62) need be modified to

$$ES \xrightarrow{k_1} E + P \tag{3.72}$$

$$EIS \xrightarrow{k_2} E + P$$

where $k_2 > k_1$. Consequently I activates the enzyme since the EI complex is a more active catalyst than E alone. Quantitative treatment of activation will be left as an exercise for the reader.

As a summary and guide to the use of these equations for evaluating their parameters and discriminating between the various types of inhibition, Fig. 3.31 shows how inhibition affects substrate-concentration–reaction-rate data plotted in various ways. Table 3.8 lists the ordinate and abscissa intercepts for Lineweaver-Burk plots for these inhibited systems.

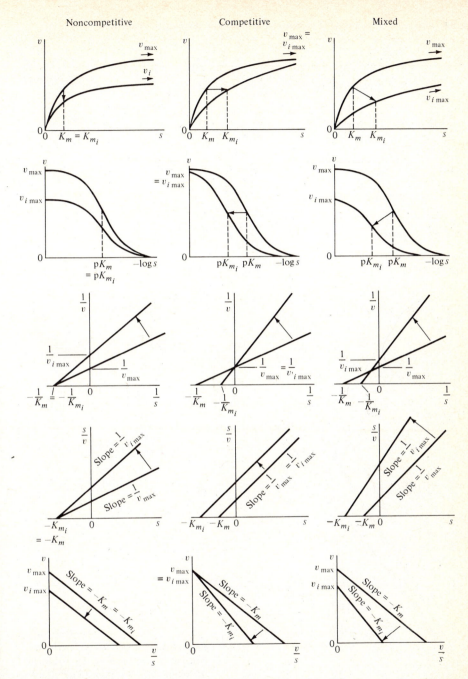

Figure 3.31 Effects of various types of inhibition as reflected in different rate plots. (*Reprinted by permission from M. Dixon and E. C. Webb, "Enzymes," 2d ed., p. 326, Academic Press, Inc., New York, 1964.*)

Table 3.8. Intercepts of reciprocal (Lineweaver-Burk) plots in presence of inhibitor

No.	Type		
Ia	Purely competitive	$\dfrac{1}{v_{max}}$	$\dfrac{1}{K_s(1 + i/K_i)}$
Ib	Partially competitive	$\dfrac{1}{v_{max}}$	$\dfrac{1 + iK_s/K_iK_{is}}{K_s(1 + i/K_i)}$
IIa	Purely noncompetitive	$\dfrac{1 + i/K_i}{v_{max}}$	$\dfrac{1}{K_s}$
IIb	Partially noncompetitive	$\dfrac{1 + i/K_i}{v_{max} + k^i e_0/K_i}$	$\dfrac{1}{K_s}$
Ib and IIa	Mixed	$\dfrac{1 + iK_s/K_iK_s}{v_{max}}$	$\dfrac{1 + iK_s/K_iK_{is}}{K_s(1 + i/K_i)}$

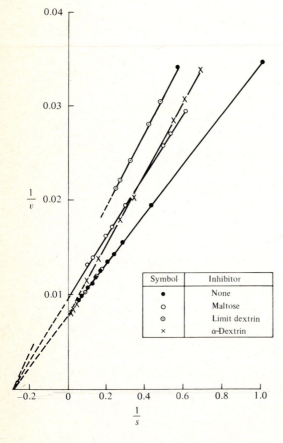

Symbol	Inhibitor
●	None
○	Maltose
⊙	Limit dextrin
×	α-Dextrin

Figure 3.32 Competitive (α-dextrin) and noncompetitive (maltose, limit dextrin) inhibition of α-amylase. (*Reprinted by permission from S. Aiba, A. E. Humphrey, and N. Millis, "Biochemical Engineering," 2d ed., p. 101, University of Tokyo Press, Tokyo, 1974.*)

The hydrolysis of starch with α-amylase offers examples for both types of reversible inhibition. By comparing the experimental data illustrated in Fig. 3.32 with Fig. 3.31, we see that α-dextrin is a competitive inhibitor, while noncompetitive inhibition is caused by both maltose and limit dextrin.

3.6. OTHER INFLUENCES ON ENZYME ACTIVITY

Before proceeding, it may be helpful to recall the major objective of this chapter: to be able to represent the rate of an enzyme-catalyzed reaction mathematically. Without suitable rate expressions, we cannot design reactors or experiments employing isolated enzymes. Moreover, when we reach the subject of cell growth kinetics, we shall discover that many aspects of enzyme kinetics can be applied. Therefore a thorough exploration of the variables which affect enzyme catalysis and a quantitative analysis of their influence are essential.

We have already explored how different chemical compounds which bind with enzymes can influence the rate of enzyme-catalyzed reactions. Many other factors can influence the catalytic activity of enzymes, presumably by affecting the enzyme's shape or ionization state. Included among these factors are:

1. pH
2. Temperature
3. Fluid forces (shear stress and hydrostatic pressure)
4. Chemical agents (such as alcohol and urea)
5. Irradiation (light, sound, ionizing radiation)

In this section we shall concentrate on the first three effects.

Sometimes the change in catalytic activity caused by a shift in pH, for example, is undone by returning to original reaction conditions. This situation is in a sense analogous to the reversible-inhibition cases considered above. The equilibrium (or quasi-steady-state) conditions prevailing in the original environment are merely shifted slightly by a small change in one of the factors above. In general, the amount of change in the environment from the enzyme's native biological habitat must be relatively small or deactivation of the enzyme will be likely to occur.

Denaturation normally involves a cooperative phase transition with destruction of part or all of the hydrogen bonds and at times even the sulfur linkages maintaining the enzyme in its active configuration (Fig. 2.22). Thus there is a large structural change in the molecule itself, in contrast to deactivation by ionization (neutralization) or inhibition. Ionization or inhibitor binding is known to cause changes in enzyme configuration resulting in reversible transitions from active to inactive: this change is normally relatively small in a physical sense. As a general rule, denaturation is reversible only when conditions are relatively mild. Environments far removed from the native one tend to cause irreversible damage to the protein. Also, even when denaturation can be reversed, the return of the enzyme to its original activity is often slow compared with reversal of ionization or reversible inhibitor binding.

3.6.1. The Effect of pH on Enzyme Kinetics in Solution

Figure 2.14 lists the various amino acids from which all proteins are constructed. These biochemical units possess basic, neutral, or acidic side groups. Consequently, the intact enzyme may contain both positively or negatively charged groups at any given pH. Such ionizable groups are often apparently part of the active site since acid- and base-type catalytic action has been closely linked to several enzyme mechanisms. For the appropriate acid or base catalysis to be possible, the ionizable groups in the active site must often each possess a particular charge; i.e., the catalytically active enzyme exists in only *one* particular ionization state. Thus, the catalytically active enzyme may be a large or small fraction of the total enzyme concentration, depending upon the pH. Figure 3.33 illustrates the influence of pH on several enzymes. In several cases shown there, catalytic activity of the enzyme passes through a maximum (at the optimum pH) as pH is increased.

At the optimum pH the maximum possible amount of enzyme is in the active form. The most interesting parameters are the two pH values at which the first groups necessary for the active site ionize (or neutralize) as the pH shifts to the basic or acidic side from the optimum pH. In the following acid-base reactions, E^- denotes the active enzyme form while E and E^{2-} are inactive forms obtained by protonation and deprotonation of the active site of E, respectively:

$$ E \underset{\substack{+H^+ \\ K_1}}{\overset{-H^+}{\rightleftharpoons}} E^- \underset{\substack{+H^+ \\ K_2}}{\overset{-H^+}{\rightleftharpoons}} E^{2-} \tag{3.73} $$

Further ionizations away from the E^- state of the enzyme presumably do not need to be considered since the first ionization is assumed here to deactivate the enzyme completely.

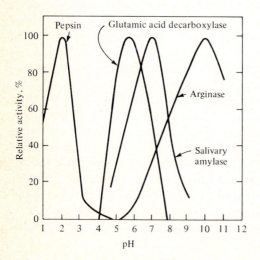

Figure 3.33 Enzyme activity is strongly pH dependent, and the range of maximum activity differs greatly depending on the enzyme involved. (*Redrawn with permission from J. S. Fruton and S. Simmonds, "General Biochemistry," p. 260, John Wiley & Sons, Inc., New York, 1953.*)

Acid-base equilibrium between the three species E, H$^+$, and E$^-$, for example, requires

$$\frac{h^+ e^-}{e} = K_1 \qquad (3.74)$$

Notice that when E and E$^-$ are present in equal amounts ($e^- = e$), the pH is equal to pK_1, since

$$pK_1 \equiv -\log K_1 = -\log \left(\frac{e^-}{e}\right)^0 - \log h^+ \equiv pH \qquad (3.75)$$

After writing the equilibrium relation for the second ionization

$$\frac{h^+ e^{2-}}{e^-} = K_2 \qquad (3.76)$$

we can determine the fraction of total enzyme present which is active. When we let e_0 denote the total enzyme concentration,

$$e_0 = e + e^- + e^{2-} \qquad (3.77)$$

the active fraction y^- is e^-/e_0 and is given by

$$y^- \equiv \frac{e^-}{e + e^- + e^{2-}} = \frac{e^-}{h^+ e^-/K_1 + e^- + K_2 e^-/h^+}$$

or
$$y^- = \frac{1}{1 + h^+/K_1 + K_2/h^+} \qquad (3.78)$$

Similarly, the enzyme fractions in the acid and base forms are given by

$$y = \frac{1}{1 + K_1/h^+ + K_1 K_2/(h^+)^2} \qquad (3.79)$$

and
$$y^{2-} = \frac{1}{1 + h^+/K_2 + (h^+)^2/K_1 K_2} \qquad (3.80)$$

The functions y, y^-, and y^{2-} are commonly known as the *Michaelis pH functions*.

Figure 3.34a and b illustrates the three functions y, y^-, and y^{2-} vs. pH when the pK's are 5 and 10 (Fig. 3.34a) or 7 and 8 (Fig. 3.34b). Taking the logarithm of Eq. (3.78) gives

$$\log y^- = -\log \left(1 + \frac{h^+}{K_1} + \frac{K_2}{h^+}\right) \qquad (3.81)$$

Since p$K_1 <$ pK_2, it follows that $K_1 > K_2$. If K_1 and K_2 are well separated, so that $\Delta pK = |pK_1 - pK_2| > 2.0$, there will be three linear regions in a plot of $\log y^-$ vs. pH:

pH $<$ pK_1:
$$\frac{h^+}{K_1} > 1 \qquad \frac{K_2}{h^+} \ll 1$$

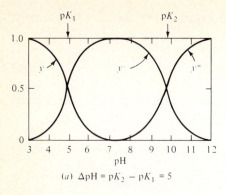
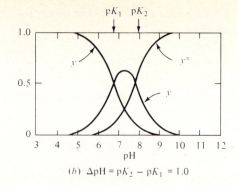

(a) $\Delta\mathrm{pH} = \mathrm{p}K_2 - \mathrm{p}K_1 = 5$ (b) $\Delta\mathrm{pH} = \mathrm{p}K_2 - \mathrm{p}K_1 = 1.0$

Figure 3.34 The functions of y, y^-, y^{2-} for (a) $\Delta\mathrm{pH} = 5$ and (b) $\Delta\mathrm{pH} = 1.00$. (*Reprinted by permission from M. Dixon and E. C. Webb, "Enzymes," 2d ed., p. 120, Academic Press, Inc., New York, 1964.*)

and
$$\log y^- \approx -\log h^+ + \log K_1 = \mathrm{pH} - \mathrm{p}K_1 \tag{3.82}$$

When $\mathrm{p}K_1 < \mathrm{pH} < \mathrm{p}K_2$, $h^+/K_1 < 1$ and $K_2/h^+ < 1$, so that

$$\log y^- \approx \log 1 = 0 \tag{3.83}$$

If $\mathrm{pH} > \mathrm{p}K_2$, $h^+/K_1 \ll 1$ and $K_2/h^+ > 1$. Consequently

$$\log y^- \approx -\log \frac{K_2}{h^+} = \mathrm{p}K_2 - \mathrm{pH} \tag{3.84}$$

Plotting these linear approximations vs. pH in Fig. 3.35 yields two intersections of adjoining segments at $\mathrm{pH} = \mathrm{p}K_1$ and $\mathrm{pH} = \mathrm{p}K_2$, respectively. This extrapolation method allows determination of the two $\mathrm{p}K$'s of interest which will not generally be the $\mathrm{p}K$'s of the individual isolated ionizable groups. (Why?) As an exercise, show that if the two $\mathrm{p}K$'s are not well separated, the K's can conveniently be determined by finding the pH of the maximum value of y^- and either pH corresponding to $y^- = y^-_{\max}/2$.

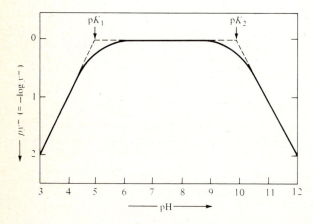

Figure 3.35 Plot of $py^- = -\log y^-$ vs. pH. (*Reprinted by permission from M. Dixon and E. C. Webb, "Enzymes," 2d ed., p. 123, Academic Press, Inc., New York, 1964.*)

Protonation and deprotonation are very rapid processes compared with most reaction rates in solution. Therefore, it can be assumed that the fraction of enzyme in the active ionization state is y^- even when the enzyme is serving as a catalyst. Consequently, the influence of ionization on the maximum reaction velocity v_{max} is obtained by replacing the total enzyme concentration e_0 with the total active form concentration $e_0 y^-$:

$$v_{max} = ke_0 y^- = \frac{ke_0}{1 + h^+/K_1 + K_2/h^+} \tag{3.85}$$

Hence plots of log v_{max} vs. pH should exhibit the same behavior as log y^- vs. pH; some data for fumarate hydratase which appear to confirm this point are presented in Fig. 3.36. Notice the similarity of these curves to those shown in Fig. 3.33.

Although it has been assumed that only one form of the enzyme is active, the value of the Michaelis constant is also affected by pH since K_m is defined with respect to the free-enzyme, substrate, and enzyme-substrate-complex concentration and not the concentrations of the active forms of these species. Suppose the true dissociation constant for the active form of the enzyme is K_s, given by

$$K_s = \frac{e_{act} s_{act}}{(es)_{act}} \tag{3.86}$$

and K_m is defined as usual:

$$K_m = \frac{e \cdot s}{(es)} \tag{3.87}$$

As above, we let $y_{s,\,act}$ and $y_{(es),\,act}$ be the active fractions of substrate and ES complex, respectively. Notice that now we are also considering S to be an ionizable compound having only one active form. From these definitions we obtain

$$K_m = \frac{(e_{act}/y^-)(s_{act}/y_{s,\,act})}{(es)_{act}/y_{(es),\,act}} = K_s \frac{y_{(es),\,act}}{y^- y_{s,\,act}}$$

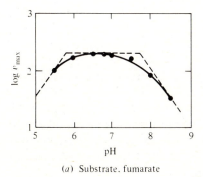

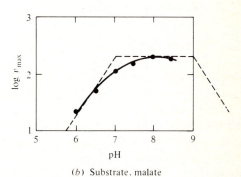

(a) Substrate, fumarate   (b) Substrate, malate

Figure 3.36 Plots of v_{max} vs. pH with (a) fumarate and (b) malate as substrates. (*Reprinted by permission from M. Dixon and E. C. Webb, "Enzymes," 2d ed., p. 132, Academic Press, Inc., New York, 1964.*)

so that

$$\log K_m = \log K_s + \log y_{(es),\ act} - \log y^- - \log y_{s,\ act} \tag{3.88}$$

If the substrate does not have more than one form ($y_{s,\ act} = 1.0$) but binding of the substrate changes the ease of ionization of different active-site groups, then $y_{(es),\ act} \neq y^-$ and

$$\log K_m = \log K_s + \log y_{(es),\ act} - \log y^- \tag{3.89}$$

Notice that the signs of $\log y_{(es),\ act}$ and $\log y^-$ are opposite: if ionization of a group on the free enzyme produces an increase in K_m, ionization of the same group on the enzyme-substrate complex will produce the opposite effect. On the other hand, if the pK's of active-site ionizable groups are unaffected by formation of the ES complex, then y^- is equal to $y_{(es),\ act}$, and K_m is the true dissociation constant. In this situation the influence of pH on enzyme kinetics is only through its effect on v_{max}, as shown in Eq. (3.85). In current practice, Eq. (3.85) alone is commonly used to represent the dependence of reaction rates on pH.

We should always remember that the theory just described may not apply for pH values far removed from the optimum pH. Under these circumstances, the force stabilizing the native protein conformation may be so disturbed that denaturation occurs and the equilibria in reaction (3.73) are altered. If this occurs, we cannot expect the normal enzymatic activity to return quickly if the pH is restored to its optimal value.

3.6.2. Enzyme Reaction Rates and Temperature

In any study of chemical kinetics, a recurring theme is the Arrhenius form relating temperature to a reaction-rate constant

$$k = Ae^{-E_a/RT} \tag{3.90}$$

where E_a = activation energy
R = gas-law constant
A = frequency factor
T = absolute temperature

In an Arrhenius plot, $\log k$ is graphed against $1/T$ to give a straight line with a slope of $-E_a/R$ [assuming, of course, that Eq. (3.90) holds]. The Arrhenius dependence on temperature is indeed satisfied for the rate constants of many enzyme-catalyzed reactions, as exemplified by the data in Fig. 3.37.

It should be noted, however, that the temperature range of Fig. 3.37 is quite limited. No temperatures significantly larger than the usual biological range were considered. What would happen if we attempted to push the enzyme farther and try for higher rates via higher temperatures? The result would in most cases be disastrous, as the example of Fig. 3.38 illustrates.

For many proteins, denaturation begins to occur at 45 to 50°C and is severe at 55°C. One physical mechanism for this phenomenon is obvious: as the temperature increases, the atoms in the enzyme molecule have greater energies and a

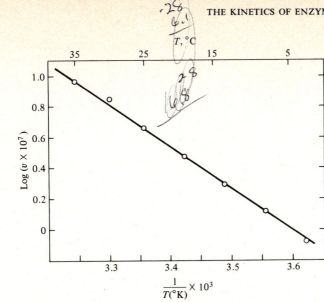

Figure 3.37 Arrhenius plot for an enzyme catalyzed reaction (myosin catalyzed hydrolysis of ATP). (*Reprinted from K. J. Laidler, "The Chemical Kinetics of Enzyme Action," p. 197, The Clarendon Press, Oxford, 1958. Data of L. Quellet, K. J. Laidler, M. F. Morales, Molecular Kinetics of Muscle Adenosine Triphosphate, Arch. Biochem. Biophys., **39**: 37, 1952.*)

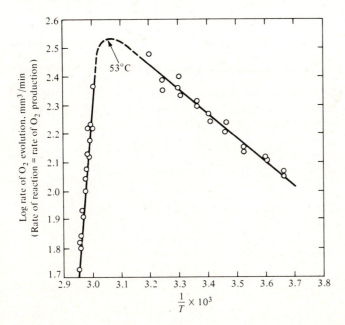

Figure 3.38 Arrhenius rate dependence breaks down at high temperatures, above which enzyme deactivation predominates (H_2O_2 decomposition catalyzed by catalase). (*Reprinted by permission from I. W. Sizer, Temperature Activation and Inactivation of the Crystalline Catalase–Hydrogen Peroxide System, J. Biol. Chem., **154**: 461, 1944.*)

greater tendency to move. Eventually, they acquire sufficient energy to overcome the weak interactions holding the globular protein structure together, and denaturation follows.

Thermal denaturation of enzymes may involve both reversible and irreversible processes. If the enzyme denatures *reversibly* with temperature, then at equilibrium the proportion of enzymes in the denatured (*d*) and active (*a*) form is simply given by

$$\frac{e_d}{e_a} = \exp\left(\frac{-\Delta G_d}{RT}\right) \tag{3.91}$$

where the free energy of denaturation is

$$\Delta G_d = \Delta H_d - T\,\Delta S_d \tag{3.92}$$

Although individual hydrogen bonds are quite weak, typically with bond energies of 5 to 7 kcal/mol, the enthalpy of denaturation of enzymes ΔH_d is quite high: 68 and 73.5 kcal/mol for free trypsin and hen egg white lysozyme, respectively. The entropy changes upon denaturation for these enzymes are $+213$ cal/(mol · K).

Due to the large heats of denaturation, the proportion of active enzyme is quite sensitive to small changes in temperature, as seen in Fig. 3.39. Here the enzyme denatures almost totally over a range of 30 Celsius degrees.

When the temperature is held constant, a number of enzyme systems are found to denature *irreversibly* in time, often following approximately a first-order decay law

$$\frac{de}{dt} = -k_d e \tag{3.93}$$

so that

$$e(t) = e(0)\exp\left(-k_d t\right) \tag{3.94}$$

The decay constant k_d depends on temperature, often in an Arrhenius fashion. Listed in Table 3.9 are the activation energies E and entropies ΔS^* for k_d for several common enzymes. (Recall from transition state theory that

$$k_d = (k_B T/h)\exp\left(\Delta S^*/R\right)\exp\left(-E/RT\right)$$

where k and h are Boltzmann's and Planck's constants, respectively.)

Table 3.9. Energies and entropies of activation for enzyme denaturation†

Enzyme	pH	Energy of activation, kcal/mol	Entropy of activation ΔS^* (e.u./mol)
Pancreatic lipase	6.0	46.0	68.2
Trypsin	6.5	40.8	44.7
Pepsin	4.83	56–147	unknown
ATPase	7.0	70	150.0

† K. J. Laidler and P. S. Bunting, "The Chemical Kinetics of Enzyme Action," 2d ed., p. 430, Oxford University Press, London, 1973.

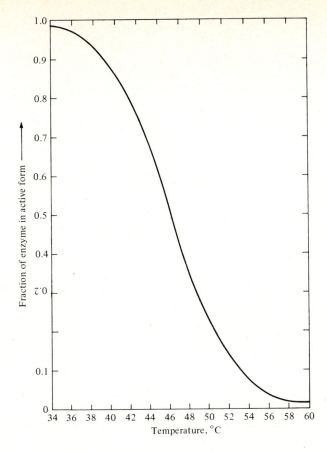

Figure 3.39 Thermodynamic arguments indicate that the fraction of total enzyme in an active form decreases significantly over a narrow temperature range.

Without enzyme activity, cells cannot function. In some cases, destruction of a very small fraction of the cell's enzymes results in its death. Viewed in this light, the preceding discussion reveals why heat can be used to *sterilize* gases, liquids, or solids, i.e., to rid them of microbial life. Very few enzymes (and consequently few microbes) can survive prolonged heating, as we shall explore in greater detail in Chap. 7.

3.6.3 Mechanical Forces Acting on Enzymes

Mechanical forces can disturb the elaborate shape of an enzyme molecule to such a degree that denaturation occurs. Included among such forces are shear forces created by flowing fluids. The apparatus shown schematically in Fig. 3.40a can be used to measure denaturation caused by shear forces. Let γ denote the shear rate averaged over the test cross section and θ the time the enzyme is exposed to shear. Experiments on shear denaturation of catalase showed that activity loss can be correlated as a function of the product $\gamma\theta$ (Fig. 3.40b).

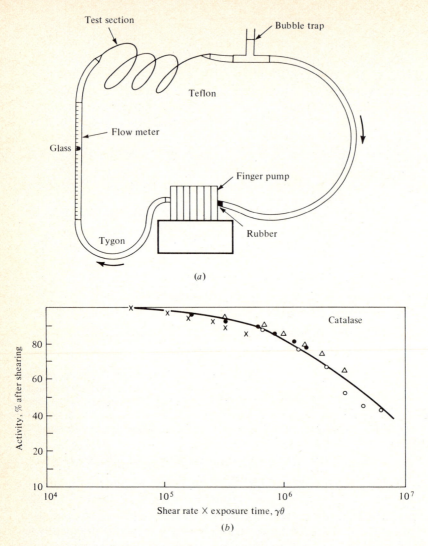

Figure 3.40 (*a*) System for experimental study of shear denaturation of enzymes. (*b*) Deactivation of catalase caused by exposure to shear. (*Reprinted with permission from S. E. Charm and B. L. Wong, Enzyme Inactivation with Shearing, Biotech. Bioeng.,* **12**: 1103, 1970.)

This characteristic mechanical fragility of enzymes may impose limits on the shear forces which can be tolerated in enzyme reactors being stirred to increase substrate mass-transfer rates or in an enzyme ultrafiltration system in which increasing membrane throughput causes increased fluid shear just in front of the membrane and passing through it.

Another mechanical force, surface tension, often causes denaturation of proteins and consequent inactivation of enzymes. Since the surface tension of the interface between air and pure water is 80 dyn/cm, foaming or frothing in protein solutions invariably causes denaturation since protein adsorbs at the air-water interface. On the other hand, liquid-liquid interfaces normally have considerably lower surface tensions. Similarly, the plasma membrane of the cell is believed to have a surface tension of the order of 1 dyn/cm or less. As active proteins are known to exist in such plasma membranes, evidently these very low surface tensions do not deactivate the enzymes. Foam fractionation is a separation technique whereby molecules are concentrated at surfactant-air interfaces without deactivation; here the surfactant lowers the air-liquid surface tension to the order of 1 dyn/cm.

Still other phenomena can disturb enzymes so that their catalytic activity is reduced. We must leave these to the references, however. As a general rule of thumb, the following may suffice. If the enzyme is surrounded *in vitro* with essentially the same environment it enjoys *in vivo*, it will be active. If *any* parameter of its environment is altered significantly, loss of activity is likely to occur.

There is a corollary of this idea which finds numerous applications in biochemical technology. If an enzyme is required which is active at extreme temperature or pH values, we should look for a microorganism which normally lives under these conditions. It will often contain enzymes especially adapted for use in an unusual environment. The development of alkaline-stable enzymes for use in laundry detergents is a good example of this practice, as wash-water pH is typically 9.0 to 9.5.

3.7. ENZYME REACTIONS IN HETEROGENEOUS SYSTEMS

Our attention to this point has been concentrated on enzymes in solution acting on substrates in solution. This is not a universal situation, as we have already hinted in Fig. 2.27. In that diagram of a procaryote, it was indicated that some of the cell's enzymes are attached to the cell membrane. Similar features are found in eucaryotes. In mitochondria, for example, the enzymes for a very complicated chain of reactions are bound to a convoluted internal-membrane system (see Chap. 5).

Many other combinations of enzyme and substrate physical states arise in nature and technology, as Fig. 3.41 illustrates. While we cannot investigate all these particular cases now, we shall consider two different classes of heterogeneous systems. First, the kinetics of enzymes in solution acting on insoluble substrates will be examined, and then, in the concluding section of this chapter, we shall analyze reactions of soluble substrates catalyzed by enzymes attached to surfaces. This treatment is extended in Chap. 7.

3.7.1. Reactions Involving Insoluble Substrates

Sometimes only the soluble form of a substrate which may also exist as a separate phase is suitable for enzyme catalysis. An example is evident in the data shown in Fig. 3.42. Since all molecular species have a finite aqueous solubility, a small

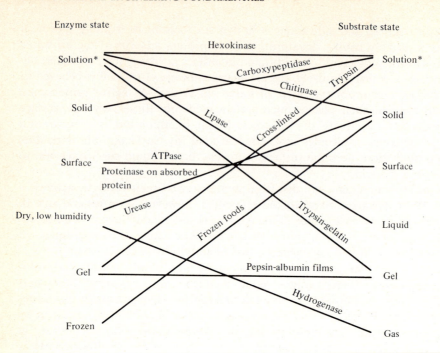

Figure 3.41 Enzymes in several different states catalyze reactions of substrates in various forms. The classical solution case is only one of a broad spectrum of possible enzyme-substrate interactions. (*Reprinted from A. D. McLaren and L. Packer, Some Aspects of Enzyme Reactions in Heterogeneous Systems, Adv. Enzymol. Rel. Sub. Biochem.* **33**: 245, 1970.)

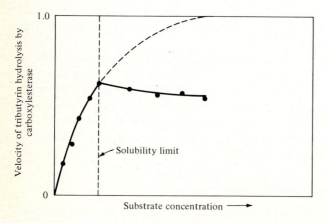

Figure 3.42 Reaction only of soluble-substrate form. (*Reprinted by permission from M. Dixon and E. C. Webb, "Enzymes," 2d ed., p. 90, Academic Press, Inc., New York, 1964.*)

amount of substrate will always be in solution to supply the enzyme. This may be so slow a process, however, that the rate is nil for practical purposes. Let us turn next to some examples of the opposite sort.

One of these is the hydrolysis of methyl butyrate by pancreatic lipase, a fat-splitting enzyme which is secreted in the human digestive tract. In contrast to the previous example, here the reaction does not occur until an insoluble form (liquid droplets) of the substrate is available (Fig. 3.43). Apparently the enzyme is active only at the liquid-liquid interface. Recalling the possibility of enzyme denaturation by interfacial tension, it is interesting to note that bile salts, which are also secreted into the digestive tract, may adsorb on fat droplets and reduce the interfacial forces, just as with surfactant foam fractionation mentioned above.

Other enzymes are active toward both the soluble and insoluble form of the substrate. Trypsin, a protease, is found to digest both free lysozyme and lysozyme adsorbed on the surface of kaolinite. A second enzyme which hydrolyses both soluble and "insoluble" substrates is lysozyme itself. As noted earlier, lysozyme is active in splitting the murein bacterial cell wall. However, it also catalyzes the breakdown of soluble oligomers derived from the cell-wall polymer (Sec. 3.4.2).

An interesting variation on the kinetic equations derived earlier arises when enzyme in solution acts on an insoluble substrate by adsorbing onto the surface of the substrate. In contrast to previous cases, where the reaction rate increases in direct proportion to total enzyme concentration, a limiting rate is approached as enzyme concentration is increased. This behavior is evident in kinetic data for hydrolysis of a solid cube of protein (specifically thiogel, a cross-linked gelatin) under the action of trypsin (Fig. 3.44).

In order to develop a reasonable model for such kinetics, we begin by turning the tables: the enzyme now adsorbs on the substrate. Letting A denote a vacant

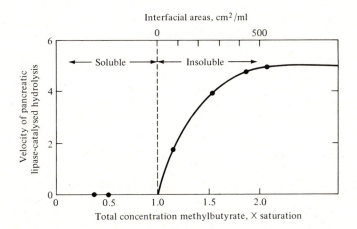

Figure 3.43 Reaction at liquid-liquid interface. (*Reprinted by permission from L. Sarda et P. Desnuelle, Action de la Lipase Pancréatique sur les Esters en Émulsion, Biochim. Biophys. Acta,* **30:** 513, 1958.)

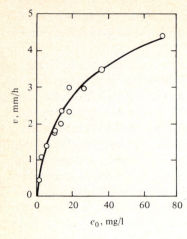

Figure 3.44 Dependence of the rate of disappearance of solid substrate (thiogel) on the concentration e_0 of enzyme in solution. [*Data of A. G. Tsuk and G. Oster, Determination of Enzyme Activity by a Linear Measurement, Nature (London),* **190**: 721, 1961.]

site on the substrate surface, we assume the following equilibrium:

$$E + A \underset{k_{des}}{\overset{k_{ads}}{\rightleftharpoons}} EA \qquad (3.95)$$

If a_0 is the total number of adsorption sites on the substrate surface per unit volume of the reaction mixture, we have

$$a_0 = a + (ea) \qquad (3.96)$$

Combining this with the equilibrium relation for enzyme adsorption (3.95) gives

$$(ea) = \frac{a_0 e}{K_A + e} \qquad \text{with } K_A = \frac{k_{des}}{k_{ads}} \qquad (3.97)$$

The development is completed by assuming that the hydrolysis is accomplished by irreversible decomposition of the EA complex. Consequently

$$v = k_3(ea) = \frac{k_3 a_0 e}{K_A + e} \qquad (3.98)$$

Recalling our notation conventions, e is the concentration of free enzyme and is related to the total concentration e_0 at the start of the experiment by

$$e_0 = e + (ea) \qquad (3.99)$$

If the initial concentration of enzyme is much larger than that of substrate $(e_0 \gg a_0)$, we may assume to an excellent degree of approximation that

$$e_0 \approx e \qquad (3.100)$$

so that

$$v = \frac{k_3 a_0 e_0}{K_A + e_0} \qquad (3.101)$$

The situation $e_0 \gg a_0$ is not at all uncommon for reactions involving solid substrates. For example, the data in Fig. 3.44 were obtained in an experiment for which e_0/a_0 was approximately 4000. This case contrasts sharply with reactions in solution, where s_0 is typically much larger than e_0.

Rate equation (3.101) reveals that a Lineweaver-Burk double-reciprocal plot ($1/v$ vs. $1/e_0$) should be linear. This indeed occurs, as Fig. 3.45 demonstrates. In Fig. 3.45a, the data shown in Fig. 3.44 have been replotted. The other part of the figure illustrates a similar result for another soluble-enzyme–solid-substrate system.

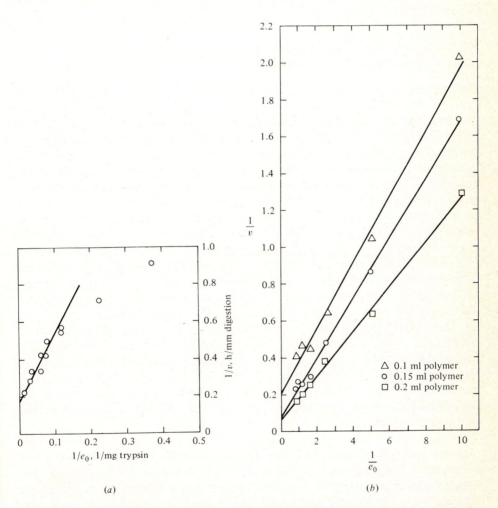

Figure 3.45 (a) Double-reciprocal plot of the data in Fig. 3.44. (b) Another double-reciprocal plot for digestion of an insoluble substrate (poly-β-hydroxybutyrate particles) by an enzyme (depolymerase of *P. lemoignei*) in solution. (*Reprinted from A. D. McLaren and L. Packer, Some Aspects of Enzyme Reactions in Heterogeneous Systems, Adv. Enzymol. Rel. Sub. Biochem.*, **33**: 245, 1970.)

Many potential microbial nutrients begin as solid particles (in waste streams, lakes, compost piles, etc.). Hydrolysis of these particles by extracellular enzymes is clearly necessary before the cell can absorb or pump the then solubilized nutrient through its own membrane. Hence we can expect the enzyme kinetics in this section to be of importance in such particulate-substrate reactors.

Before leaving this topic, we should emphasize two further points: (1) The development above is not necessarily restricted to solid substrates. It has also been applied to a dispersed phase of immiscible liquid substrate, e.g., the system shown in Fig. 3.43; (2) we have ignored possible differences in concentrations between bulk fluid phases and interfaces. These will be examined explicitly in the next section.

3.7.2. Chemical-Reaction–Mass-Transfer Interaction for Reactions on Immobilized Enzymes

We have already commented on the presence of enzymes bound to plasma membranes within the cell. In recent years, there has been considerable interest in *in vitro* enzyme immobilization. With a variety of techniques, enzymes isolated from living organisms are immobilized on or within solid supports, e.g., polystyrene, porous glass, or silica-alumina. One motivation for such studies is to elucidate the behavior of bound enzymes in nature. By experimenting with a relatively simple synthetic model of a membrane-bound enzyme system, important distinctions between this configuration and soluble enzymes can be explored.

Moreover, in several actual or proposed commercial applications of enzymes, immobilized-enzyme systems offer several distinct advantages relative to free enzymes in solution. One of these is increased longevity of enzyme activity (Fig. 3.46). Further, an enzyme bound to a solid phase is more easily recovered from the reaction mixture than its soluble counterpart. Some of the promising avenues for use of immobilized enzymes will be explored in the next chapter, along with the most important methods of enzyme immobilization.

Our major theme in this section is the influence of mass-transfer resistance on reactions catalyzed by immobilized enzymes. Because substrate is consumed at a solid surface covered with enzymes, we can expect the substrate concentration at the surface to be smaller than the concentration in the bulk fluid far from the interface (Fig. 3.47). Diffusive and convective mass-transport processes supply substrate for the reaction by moving substrate from the high bulk concentration to the lower surface value.

For such a situation, the observed rate of reaction depends on the rate of these mass-transport phenomena as well as the true, *intrinsic* kinetics of the reactions occurring at the solid-liquid interface. This is well appreciated in most reactor design courses: a tremendous amount of research on the problems of interaction between mass transfer and reaction in porous synthetic catalysts has been published in recent years. Until recently much of this existing knowledge was overlooked or ignored in investigations of immobilized-enzyme catalysis. After trivial changes in nomenclature and notation, much of the groundwork already estab-

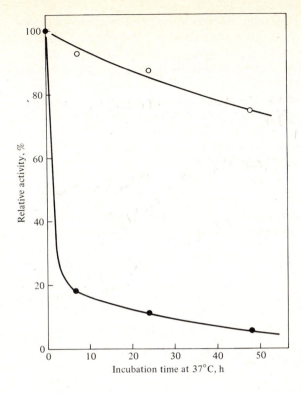

Figure 3.46 β-D-Glucosidase in solution ($\bullet$) loses activity at 37°C much more rapidly than the immobilized form [(β-D-Glucosidase–cellulose carbonate, ($\circ$)]. (*Reprinted from S. A. Barker et al., β-D-Glucosidase Chemically Bound to Microcrystalline Cellulose, Carbohydr. Res.,* **20:** 1, 1971.)

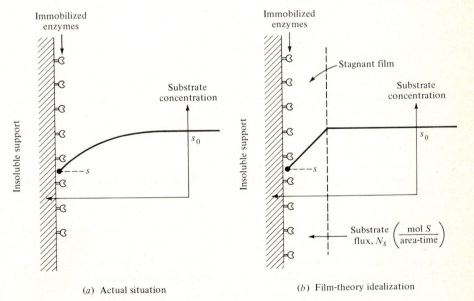

(*a*) Actual situation

(*b*) Film-theory idealization

Figure 3.47 Mass transfer of substrate from bulk solution to an interface where an enzyme-catalyzed reaction occurs in (*a*) an actual situation and (*b*) a film-theory idealization.

lished can be directly applied to enzymes. Several features of some immobilized-enzyme systems, however, have no obvious analogs in better-studied areas of heterogeneous catalysis. In the following paragraphs, we shall try to bring out these aspects without neglecting the important basic concepts of chemical-reaction–mass-transfer coupling.

One traditional model, called the *Nernst diffusion layer* in the biochemistry literature and a *stagnant film* in engineering, leads to the following equation for the flux N_s (moles per unit time per unit area) of substrate from the bulk fluid (sometimes called the *pool* by biochemists) to the interface (see Fig. 3.47):

$$N_s = k_s(s_0 - s) \tag{3.102}$$

Here s is the substrate concentration at the interface, and k_s is the mass-transfer coefficient. A function of physical properties as well as hydrodynamic conditions near the interface, k_s can usually be evaluated from available correlations (discussed in Chap. 8).

In steady state, substrate cannot accumulate at the catalyst interface. Consequently the rate of substrate supply by mass transfer must be exactly counterbalanced by the rate of substrate consumption in the reaction at the interface. Assuming that Michaelis-Menten kinetics applies at the surface, and letting $\bar{v}$ denote the *surface* reaction rate (moles per unit time per unit area), we have

$$k_s(s_0 - s) = \bar{v} = \frac{\bar{v}_{max} s}{K_m + s} \tag{3.103}$$

The number of parameters necessary to specify the system can be reduced from four (k_s, s_0, $\bar{v}_{max}$, and K_m) to two (Da and κ) by introducing the dimensionless variables

$$x = \frac{s}{s_0} \qquad Da = \frac{\bar{v}_{max}}{k_s s_0} \qquad \kappa = \frac{K_m}{s_0} \tag{3.104}$$

In terms of these quantities, the substrate mass balance (3.103) becomes

$$\frac{1 - x}{Da} = \frac{x}{\kappa + x} \tag{3.105}$$

where $0 \leq x \leq 1.0$.

The physical significance of Da, the *Damköhler number*, should be emphasized:

$$Da = \frac{\bar{v}_{max}}{k_s s_0} = \frac{\text{maximum reaction rate}}{\text{maximum mass-transfer rate}} \tag{3.106}$$

Thus, if Da is much less than unity, the maximum mass-transfer rate is much larger than the maximum rate of reaction (low mass-transfer resistance). When the mass-transfer resistance is large, mass transfer is the limiting process and Da is much greater than 1. These cases are known as the *reaction-limited regime* and the *diffusion-limited regime*, respectively.

Algebraic manipulation of Eq. (3.105) gives a quadratic equation for s, so that an analytical solution is available:

$$x = \frac{\beta}{2}\left(\pm\sqrt{1 + \frac{4\kappa}{\beta^2}} - 1\right)$$

(3.107)

where

$$\beta \equiv Da + \kappa - 1$$

(3.108)

The $+$ and $-$ signs are taken for $\beta > 0$, $\beta < 0$ respectively. When $\beta = 0$, $x = \sqrt{\kappa}$. Using this value for s/s_0, either the right- or left-hand side of Eq. (3.105) can be used to evaluate the dimensionless observed reaction rate $\bar{v}_{obs}$ ($\equiv \bar{v}/\bar{v}_{max}$). It should be clear that $\bar{v}$ will not in general exhibit Michaelis-Menten dependence on s_0. Also, it is no longer correct to equate K_m to the substrate concentration $s_{1/2}$ where $\bar{v}$ is equal to one-half the observed maximal reaction rate $\bar{v}_{max}{}^{app}$. In general, $s_{1/2}$ varies with Da.

In spite of this fact, the parameter $s_{1/2}$ is frequently cited as the apparent $K_m(K_m{}^{app})$, and this value is employed as a measure of mass-transfer influence. While this practice may be convenient for the experimentalist who has only to measure $s_{1/2}$, it runs serious risk of misinterpretation. One may be tempted to use the *incorrect* equation

$$\bar{v}_{obs} = \frac{\bar{v}_{max}{}^{app}s_0}{K_m{}^{app} + s_0}$$

(3.109)

for the observed rate. Moreover, even if this form provides an adequate empirical representation of the observed kinetics in a given situation, it is not generally correct since it ignores the dependence of $K_m{}^{app}$ on fluid properties and hydrodynamics near the interface. The experimental results shown in Fig. 3.48 make this dependence evident.

By tradition in chemical engineering, the influence of mass transfer on the overall reaction process is represented using the *effectiveness factor* η, which is defined physically by

$$\eta \equiv \frac{\text{observed reaction rate}}{\substack{\text{rate which would be obtained} \\ \text{with no mass-transfer resistance,} \\ \text{i.e., surface concentration} \\ s = \text{bulk concentration } s_0}}$$

(3.110)

Consequently, for the problem in question,

$$\eta = \frac{x/(\kappa + x)}{1/(\kappa + 1)}$$

(3.111)

so that $\eta \leq 1$ and, in general, the effect of increasing mass-transfer resistance is to reduce the observed activity of the catalyst. An experimental demonstration of this phenomenon is related in Fig. 3.49.

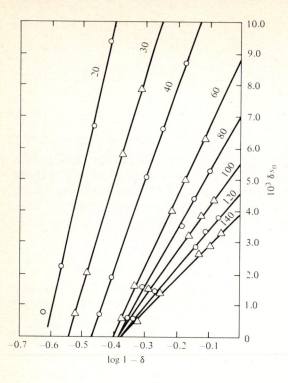

$\log 1 - \delta$

Figure 3.48 Experimental relationships between substrate conversion $\delta[= (s_0 - s_e)/s_0$, where s_0 and s_e are feed and exit substrate concentrations, respectively] and flow rate (numbers on curves, in milliliters per hour) in a packed bed immobilized enzyme reactor (hydrolysis of benzoyl L-arginine ethyl ester by CM-cellulose-ficin). If plug flow and Michaelis-Menten kinetics are assumed, the slope of all the lines should be constant and equal to K_m. The change in slope with flow rate shows a significant mass-transfer effect on overall kinetics. (*Reprinted by permission from M. D. Lilly, W. E. Hornby, and E. M. Crook, The Kinetics of Carboxymethylcellulose-Ficin in Packed Beds, Biochem. J.,* **100**: 718, 1966.)

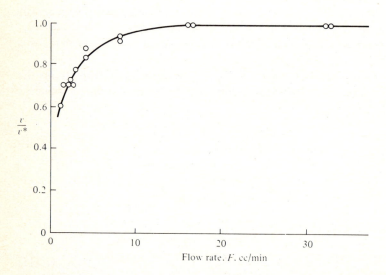

Flow rate. F. cc/min

Figure 3.49 Experimental test for external diffusion effects; v^* *is* the rate obtained at high flow rates through a packed column (hydrolysis of 4×10^{-4} M D,L-N-benzoyl DL-arginine-p-nitroanilide at pH 8, $T = 25°C$ by immobilized trypsin). [*Reprinted by permission from J. R. Ford et al., Recirculation Reactor System for Kinetic Studies of Immobilized Enzymes, in L. B. Wingard, Jr. (ed.), "Enzyme Engineering," Wiley-Interscience, New York, 1972.*]

For Da approaching zero (very slow reaction relative to maximum mass-transfer rate), Eq. (3.105) shows that x must approach unity, and so for the reaction-limited regime (Da $\rightarrow 0$)

$$\eta = 1 \qquad \bar{v} = \frac{\bar{v}_{max} s_0}{K_m + s_0} \tag{3.112}$$

Here the *observed* reaction kinetics is the same as the true, *intrinsic* kinetics at the fluid-solid interface. If a new immobilized-enzyme surface has been prepared, it is important to determine whether the immobilization procedure has changed the catalytic behavior of the enzyme. The evaluation of $\bar{v}_{max}$ and K_m needed for such a determination must occur in experiments for which Da $\ll 1$ in order to avoid disguise by significant mass-transfer resistance.

Intrinsic kinetics is also required for engineering design of immobilized-enzyme reactors, since only with this information can fluid properties, shape of the enzyme support, and mixing characteristics be adequately taken into account. Many experimental reactors designed to operate in the reaction-limited regime have been developed for ordinary heterogeneous catalysis. Some have been adapted for immobilized-enzyme kinetics studies (Fig. 3.50). They all share the common concept of high fluid flow rates near the catalyst to minimize mass-transfer resistance (larger k_s, smaller Da). There are several possible drawbacks to this approach which are peculiar to immobilized enzymes. Shear forces might cause partial or complete denaturation of the attached enzymes (see earlier discussion in Sec. 3.6.3). Also, relative motion of catalyst particles may give enzyme loss by particle-particle abrasion.

The diffusion-limited regime of systems with coupling between chemical reaction and mass transfer arises when $\bar{v}_{max}$ is much larger than $k_s s_0$, so that Da $\ll 1$. By expanding the square root in Eq. (3.107) and subsequent manipulation, we can derive for the diffusion-limited regime (Da $\rightarrow \infty$, κ finite)

$$\eta = \frac{1 + \kappa}{\text{Da}} \qquad \bar{v} = k_s s_0 \tag{3.113}$$

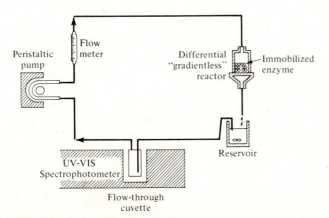

Figure 3.50 Recirculation reactor for study of external mass transfer effects on immobilized kinetics.

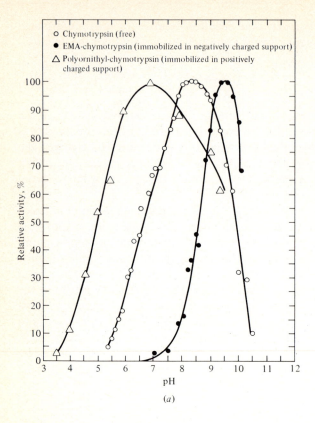

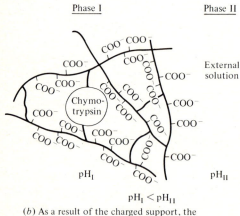

Phase I   Phase II

(b) As a result of the charged support, the local enzyme environment is changed

Figure 3.51 Use of a charged support gives an immobilized enzyme different pH dependence than the free enzyme. Substrate is acetyl-L-tyrosine ethyl ester. [(a) *Reprinted with permission from L. Goldstein, et al., A Water-insoluble Polyanionic Derivative of Trypsin. II. Effect of the Polyelectrolyte Carrier on the Kinetic Behavior of Bound Trypsin,* Biochemistry, **3**: 1913, 1964. Copyright by the American Chemical Society. (b) Reprinted by permission from L. Goldstein and E. Katchalski, Z. Anal. Chem., **243**: 375, 1968.]

Thus, so long as Da is very large, the observed rate of reaction $\bar{v}$ is *first order* in bulk substrate concentration and totally independent of the intrinsic rate parameters $v_{\max}$ and K_m. This behavior disguises the true kinetic parameters of the immobilized enzyme. In the diffusion-limited regime, for example, the observed activity for given s_0 is constant even though the enzymes at the interface may be losing activity because of adverse temperature, pH, or other conditions. Studies aimed at determining activity retention or denaturation rates of immobilized enzymes should therefore also be conducted as close as possible to the reaction-limited regime.

More facets of immobilized-enzyme kinetics will be examined in the problems, and diffusion-reaction interactions in porous immobilized pellets will be considered in Chap. 7. As a final example here, consider the experimental data shown in Fig. 3.51. The pH-activity curves for enzymes in the free and immobilized states can shift over several pH units, depending on the ionic properties of the support. Several phenomena may be involved. First, a charged surface will have either an attraction or repulsion for H^+ ions, so that the pH near the surface which the enzyme actually sees may be shifted from the bulk value used in plotting Fig. 3.51. Also, the charged support may alter the ionization state of groups within the enzyme. Finally, the reaction itself may produce or consume acid, again creating a pH difference between the bulk fluid and the solid surface. Of course, some of these same considerations apply when there are modulators present in the reaction mixture.

The kinetics of enzyme-catalysed reactions have now been examined in some detail. Although the examples cited above and the problems reveal some of the applications of enzymes, we have only scratched the surface. The following chapter will add to the theme of applications. Besides reviewing common uses of enzymes, we shall endeavor to present a balanced view of enzyme technology by also discussing enzyme production and isolation techniques.

PROBLEMS

3.1. Determination of K_m and v_m Initial rates of an enzyme-catalyzed reaction for various substrate concentrations are listed in Table 3P1.1.

Table 3P1.1

s, mol/l	v, mol/(l·min) $\times 10^6$
4.1×10^{-3}	177 1.77×10^{-4}
9.5×10^{-4}	173
5.2×10^{-4}	125
1.03×10^{-4}	106
4.9×10^{-5}	80
1.06×10^{-5}	67
5.1×10^{-6}	43

(a) Evaluate $v_{\max}$ and K_m by a Lineweaver-Burk plot.
(b) Using an Eadie-Hofstee plot, evaluate v_m and K_m.
(c) Calculate the standard deviation of the slope and intercept for each method.

3.2. Substrate inhibition Derive an expression for the reaction velocity v in terms of S, E, and reaction constants:

$$E + S \underset{}{\overset{K_s}{\rightleftharpoons}} ES \qquad ES + S \underset{}{\overset{K_s'}{\rightleftharpoons}} ESS \qquad ES \overset{k}{\longrightarrow} E + P$$

State your assumptions.

3.3. Activation kinetics Using the reaction sequences given in Eqs. (3.61) and (3.72), develop a rate expression which includes the influence of an activator.

3.4. Multiple enzyme-substrate complexes Figure 3.21 revealed that multiple complexes are involved in some enzyme-catalyzed reactions. Assuming the reaction sequence

$$S + E \underset{k_2}{\overset{k_1}{\rightleftharpoons}} (ES)_1 \underset{k_4}{\overset{k_3}{\rightleftharpoons}} (ES)_2 \overset{k_5}{\longrightarrow} P + E$$

develop suitable rate expressions using (a) the Michaelis equilibrium approach and (b) the quasi-steady-state approximations for the complexes.

3.5. Relaxation kinetics with sinusoidal perturbations As Fig. 3P5.1 indicates, when a reaction system at equilibrium (or steady state) is perturbed slightly in a sinusoidal fashion, the concentrations of the reacting species also become sinusoidal. Develop the equations necessary to relate the observed concentration fluctuations to the kinetic parameters for the reaction

$$A \underset{k_2}{\overset{k_1}{\rightleftharpoons}} B \qquad v = k_1 a - k_2 b$$

when the medium temperature is oscillated according to

$$T = T_0(1 + \alpha \sin wt) \qquad \text{where } \alpha \ll 1$$

Indicate how you would determine k_1 and k_2 from the response of a and b to the periodic perturbations.

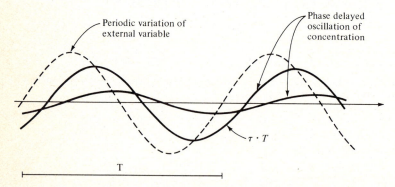

Figure 3P5.1 Response to small-amplitude periodic environmental perturbations.

3.6. Reversible reactions For the reversible reaction

$$E + S \underset{k_{-1}}{\overset{k_1}{\rightleftharpoons}} (ES) \underset{k_{-2}}{\overset{k_2}{\rightleftharpoons}} E + P$$

show that:
   (a) The reaction will proceed far to the right only if $k_1 k_2 \gg k_{-1} k_{-2}$.
   (b) The time variation of s and p, in the quasi-steady-state approximation (Sec. 3.2.1), is given by

$$-\frac{ds}{dt} = \frac{dp}{dt} = \frac{(v_s/K_s)s - (v_P/K_P)p}{1 + s/K_s + p/K_P}$$

where K_s and K_P have the usual forms.
   (c) The parameters v_s, v_P, K_s, and K_P are not independent.
   (d) Under what conditions will a Lineweaver-Burk plot of the equation in part (b) yield useful results?
   (e) Integrate dp/dt above to obtain $p(t)$ in terms of t and the value of p at equilibrium

$$At = Bp + C \ln \left(1 - \frac{p}{p_{eq}}\right)$$

   (f) If $K_s = K_P$, show that $B = 0$.
3.7. Enzyme deactivation An enzyme irreversibly denatures according to Eq. (3.94).
   (a) Show that (in the absence of mass-transfer influences) only v_{max} and not K_m is affected by such a change.
   (b) For enzymes acting on insoluble substrates [Eq. (3.101)], show that the converse of part (a) will appear to be correct if the investigator is unaware that the active enzyme concentration is changing.
   (c) If an enzyme is immobilized on a solid surface and at $t = 0$ the Dahmköhler number (3.104) is large, plot the variation of the effectiveness factor η and of the reaction rate vs. time. If due allowance was not made for mass-transfer effects, how might initial rate data of this system be misleading? (Assume soluble substrate.)
3.8. pH dependence An enzyme-catalyzed reaction irreversibly generates protons according to the equation

$$H_2O + E + S^+ \longrightarrow E + SOH + H^+$$

If the active form of the enzyme is e^-, and $e^-/e = 1.0$ at $pH = 6.0 = pK_1$, $e^{2-}/e^- = 1.0$ at $pH = 10 = pK_2$:
   (a) Show that the reaction velocity at pH 7.0 is given approximately by

$$v = v_{max} \frac{s}{K_s + s} \frac{K_1}{K_1 + h} \qquad \text{where } h = [H^+]$$

   (b) Integrate the previous equation to show that the time variation of s obeys

$$\alpha \ln \frac{s}{s_0} + \beta(s_0 - s) + \delta(s_0^2 - s^2) = v_m t$$

where α and β depend on the initial pH.
   (c) The concentration dependence of a charged species (substrate, inhibitor) in a charged matrix, e.g., that of an enzyme membrane, may be expressed as

$$[C^+]_{membrane} = [C^+]_{bulk\ solution} e^{ze\psi/RT}$$

where z = ion charge of species
      e = charge of one electron
      ψ = electrostatic potential of membrane ($\psi_{soln} = 0$)

Show that K_s and K_I have apparent values given by

$$K^{app} = K_{(true)} e^{-ze\psi/RT}$$

if bulk solution concentrations are used. Making allowance for the existence of e, e^-, and e^{2-} in the membrane, show that the maximum initial rate (ignoring mass-transfer effects) is obtained by choosing an enzyme support matrix with a ψ such that

$$1 + \frac{h_0}{K_1 \alpha} + \frac{K_2 \alpha}{h_0} = \left(\alpha + \frac{s_0}{K_s} \right) \left(\frac{K_2}{h_0} - \frac{h^0}{K_2 \alpha^2} \right)$$

where $\alpha = e^{ze\psi/RT}$ and $h_0 = [H^+]$ in bulk solution.

(d) An enzyme which dissolves a solid substrate has a net charge $+n$. As the potential ψ at the surface of the solid is varied, how will the observed reaction rate change if the bulk-enzyme concentration e_0 is much larger than the substrate surface area S_0?

3.9. Inhibitor kinetics A pesticide inhibits the activity of a particular enzyme A, which can therefore be used to assay for the presence of the pesticide in an unknown sample.

(a) In the laboratory, the initial rate data shown in Table 3P9.1 were obtained. Is the pesticide a competitive or noncompetitive inhibitor? Evaluate K_I, v_{max}, and K_m.

Table 3P9.1

	v, mol/(l·min) $\times 10^6$	
s, mol/l	No inhibitor	10^{-5} M inhibitor
3.3×10^{-4}	56	37
5.0×10^{-4}	71	47
6.7×10^{-4}	88	61
1.65×10^{-3}	129	103
2.21×10^{-3}	149	125

(b) After 50 ml of the same enzyme solution in part (a) is mixed with 50 ml of 8×10^{-4} M substrate and 25 ml of sample, the initial rate observed is 43 μmol/min. What is the pesticide concentration in the unknown *assuming* no other inhibitors or substrates are present in the sample?

(c) Evaluate K_I, v_{max}, and K_m from Fig. 3.23b (carboxypeptidase example).

3.10. Simultaneous kinetic- and mass-transfer-parameter evaluations An enzyme is immobilized on the surface of a nonporous solid. Assuming that the external mass-transfer resistance for substrate is not negligible and that the Michaelis-Menten equation describes the intrinsic kinetics.

(a) Derive an expression which indicates the explicit form of the coefficients in a Lineweaver-Burk plot. From this result, what is the *apparent* maximal velocity v_m^{app} and Michaelis constant K_m^{app} in terms of the real variables v_m, K_m, and k (mass-transfer coefficient)?

(b) If a sufficient range of substrate concentrations is examined, show graphically (sketch) how the parameters v_m, K_m, and k can be evaluated.

3.11. Microencapsulation: β-galactosidase The kinetics of lactose conversion by β-galactosidase includes a product-inhibition term:

$$v = \frac{v_m s}{K_m(1 + p/K_i) + s}$$

To allow easy recovery of β-galactosidase from milk, the enzyme solution was encapsulated in thin microcapsular cellulose nitrate membranes of diameter $d \approx 30$ μm (microcapsules).

(a) If $v_m = k_p e_0$, $k_p = 0.57$ μmol/(mg enzyme · min), $K_m = 0.54$ mM, $K_i = 1.5$ mM, construct Lineweaver-Burk plots for $p = 0$, 0.5, 1.5, 5, and 10 mM ($e_0 = 75$ mg per 100 ml).

(b) Without evaluating the formula above, use the graph in part (a) to sketch the trajectory of a batch reaction ($1/v$ vs. $1/s$ for this plot) when the initial values are

s_0, mM	0.5	5.0	10.0	5.0
p_0, mM	0	0	0	5.0

(c) Show that even under the maximum possible rate of reaction in the microcapsule, the system in part (a) is reaction-rate-limited, i.e., diffusion of substrate and product is rapid compared to reaction [6].

3.12. Heat generation in enzyme conversions The maximum temperature rise in a cylindrical plug-flow reactor can be estimated by a closed heat balance around the conversion of substrate on a single flow through the reactor.

(a) Writing heat generated ≥ heat gained by flowing medium, calculate the maximum temperature rise obtainable for a single enzyme-catalyzed reaction in terms of heat of reaction per mole reactant ΔH_r, fractional conversion of reactant δ, reactant inlet concentration s_0, liquid heat capacity C_p, and $\Delta T \equiv T_{\text{outlet}} - T_{\text{inlet}}$.

(b) For 80 percent hydrolysis of 20% lactose solution ($\Delta H_r = -7100$ cal/g mol), show that the maximum temperature rise is only a few degrees Celsius.

(c) When a cooling jacket is applied to a thin enzyme reactor, the exit centerline temperature is approximately given by

$$T^* = \frac{T_{\text{outlet}} - T_{\text{wall}}}{T_{\text{inlet}} - T_{\text{wall}}} = \text{erf} \frac{z}{2\sqrt{\alpha t}}$$

where z = column length, t = residence time of fluid, and $\alpha = k_t/\rho C_p$, where k_t = overall thermal conductivity, ρ = packed bed density, and erf is the error function (tabulated for the argument, $z/2\sqrt{\alpha t}$, in any handbook). Show that for the conditions of part (b), T^* approaches unity rapidly (calculate T^* when $z = 1, 2, 3, 4$ in) [7].

3.13. Multiple immobilized enzymes Employing the film-theory idealization, investigate what happens when consecutive reactions

$$\text{S} \xrightarrow{\text{E}_1} \text{R} \xrightarrow{\text{E}_2} \text{P}$$

occur in an immobilized-enzyme system. The enzymes E_1 and E_2 are both attached to a nonporous support surface, the enzymes are intimately mixed and uniformly distributed on the support, and the rates of both reactions can be approximated using first-order kinetics.

(a) What is the rate of disappearance of S? How is the fraction of S which is converted into P affected by the mass-transfer coefficient of R?

(b) How would you expect the answer to the last question to change as the average distance between E_1 and E_2 molecules increases?

3.14. Hysteresis: substrate inhibition An enzyme which exhibits substrate-inhibited kinetics [Eq. (3.52)] is immobilized on an impermeable solid surface. Assuming that the stagnant-film model adequately describes substrate transport from the bulk fluid to the surface, determine observed dimensionless reaction rate(s) for each of the following sets of parameter values:

$$\frac{v_{\text{max}}}{s_0 K_s} = 2 \quad \text{and} \quad \frac{K_1}{s_0} = 0.01 \quad \text{for} \quad \frac{K_2}{s_0} = \begin{cases} 0.4 & (a) \\ 0.1 & (b) \\ 0.01 & (c) \end{cases}$$

For parameter set (b), plot in one figure both the dimensionless rate of mass transfer to the surface and the dimensionless surface reaction rate vs. the surface substrate concentration. What physical interpretation can be attached to the intersection(s) of these curves? Discuss the possible significance of your result with regard to design and operation of immobilized-enzyme reactors. For an experimental demonstration of the phenomenon you have discovered, see Ref. 8.

REFERENCES

All the biochemistry text books listed at the end of Chap. 1 contain introductory material on enzyme kinetics. Mahler and Cordes (Ref. 4 of Chap. 2) is especially strong on some detailed and modern aspects of this subject. For more detailed information, the following books and articles may be consulted.

1. A. Bernhard: "Structure and Function of Enzymes," W. A. Benjamin, New York, 1968. A short, very readable introduction to enzyme fundamentals. Available in paperback.
2. M. Dixon and E. C. Webb: "Enzymes," 2d ed, Academic Press, Inc., New York, 1964. The most complete single-volume reference on enzymes. In addition to a 112-page chapter on enzyme kinetics, this handbook of the enzymologist considers many other aspects and includes an extensive table of enzymes, the reactions they catalyze, and appropriate references.
3. K. J. Laidler and P. S. Bunting: "The Chemical Kinetics of Enzyme Action," 2d ed., Oxford University Press, London, 1973. A more detailed treatment of enzyme kinetics. Extensive presentations of experimental evidence for various rate equations.
4. O. R. Zaborsky: "Immobilized Enzymes," CRC Press, Cleveland, Ohio, 1973. A valuable summary and review of artificial immobilized enzymes and their kinetics; an excellent guide to the literature through 1973.
5. A. D. McLaren and L. Packer: Some Aspects of Enzyme Reactions in Heterogeneous Systems, *Advan. Enzymol.*, **33**: 245 (1970). A fascinating review paper covering the entire gamut of heterogeneous enzyme reactions. While less detail is provided on artificial systems than in the previous reference, this work surveys an important but relatively untraveled domain including bound enzymes in the cell and on soil and insoluble substrates.

Problems

6. D. T. Wadiak and R. G. Carbonell: Kinetic Behavior of Microencapsulated β-Galactosidase, *Biotech. Bioeng.*, **17**: 1157 (1975).
7. W. H. Pitcher, Jr.: p. 151 in R. A. Messing (ed.), "Immobilized Enzymes for Industrial Reactors," Academic Press, Inc., New York, 1975.
8. H. Degn: Bistability Caused by Substrate Inhibition of Peroxidase in an Open Reaction System, *Nature*, **217**: 1047 (1968).

FOUR

ISOLATION AND UTILIZATION OF ENZYMES

The purpose of this chapter is twofold; to survey some of the applications of enzymes and to review means of obtaining enzymes suitable for these applications. With the singular exception of solid-phase enzyme synthesis (Sec. 4.2.4), all enzymes are derived from living sources (Table 4.1). Although all living cells produce enzymes, one of the three sources—plant, animal, or microbial—may be favored for a given enzyme or utilization.

For example, some enzymes may be available only from animal sources. Enzymes obtained from animals, however, may be relatively expensive, e.g., rennin from calf's stomach, and may depend on other markets, e.g., demand for lamb or beef, for their availability. While some plant enzymes are relatively easy to obtain (papain from papaya), their supply is also governed by food demands. Microbial enzymes are produced by methods which can easily be scaled up. Moreover, due to the rapid doubling time of microbes compared with plants or animals, microbial processes may be more easily attuned to the current market demands for enzymes. On the other hand, for use in food or drug processes, only those microorganisms certified as safe may be exploited for enzyme production.

While all enzymes used today are derived from living organisms, in this chapter we consider only enzymes which are utilized in the absence of life. Such biological catalysts include *extracellular enzymes*, secreted by cells in order to degrade polymeric nutrients into molecules small enough to permeate cell walls. Grinding, mashing, lysing or otherwise killing and splitting whole cells open frees *intracellular enzymes*, which are produced and normally confined within individual cells.

Table 4.1. Some enzymes of industrial importance†

Name	Source	Application	Notes	Commercial importance
		Starch-liquifying amylases		
Diastase	Malt	Digestive aid; supplement to bread; syrup	α-Amylase activity, β-amylase activity	+ + +
Takadiastase	*Aspergillus oryzae*	Digestive aid; supplement to bread; syrup	Contains many other enzymes, protease, RNase	+ + +
Amylase	*Bacillus subtilis*	Desizing textiles; syrup; alcohol fermentation industry; glucose production	Crude preparation contains protease	+ + +
Acid-resistant amylase	*Aspergillus niger*	Digestive aid	Optimum pH 4-5	+
		Starch-saccharifying amylases		
Amyloglucosidase	*Rhizopus niveus, A. niger, Endomycopsis fibuliger*	Glucose production		+ + +
Invertase	*Saccharomyces Cerevisiae*	Confectionaries, to prevent crystallization of sugar; chocolate; high-test molasses		
Pectinase	*Sclerotina libertina Coniothyrium diplodiella, Aspergillus oryzae, Aspergillus niger, Aspergillus flavus*	Increase yield and for clarifying juice Removal of pectin; coffee concentration	Scrase (Sankyo); Pectinol (Rohm and Hass); Takamine Pectinase Clarase (Takamine); Filtragol (I. G. Farben)	+ + +

† Adapted from K. Arima, Microbial Enzyme Production, in M. P. Starr (ed.), "Global Impacts of Applied Microbiology," pp. 278–279, John Wiley & Sons, Inc., New York, 1964.

156

Animal and vegetable proteases

Name	Source	Use	Activity	Remarks
Trypsin	Animal pancreas	Medical uses; meat tenderizers; beer haze removal	+++	
Pepsin	Animal stomach	Digestive aid; meat tenderizer	+++	
α-Chymotrypsin	Animal stomach	Medical uses	+++	
Rennet	Calf stomach	Cheese manufacture	++	
Pancreas protease	Animal pancreas	Digestive aid; cleaning; leather-bating; dehairing; feed improvement		
Papain	Papaya	Digestive aid; medical uses; beer haze removal; meat tenderizer	+++	
Bromelain, ficin	Pineapple, fig	Digestive aid; medical uses; beer haze removal; meat tenderizer	++	

Microbial proteases

Name	Source	Use	Activity	Remarks
Protease	*A. oryzae*	Flavoring of sake; haze removal in sake	+	
Protease	*A. niger*	Feed, digestive aid	++	Acid resistant protease; optimum pH 2–3
Protease	*B. subtilis*	Detergents; removal of gelatin from film (recovery of silver); fish solubles; meat tenderizer	++	Optimum pH 7.0
Protease	*Streptomyces griseus*	Detergents; removal of gelatin from film (recovery of silver); fish solubles; meat tenderizer	++	Optimum pH 8.0
Varidase	*Streptococcus* sp.	Medical use	++	Lederle
Streptokinase	*Streptococcus* sp.		++	Profibrinolysin

Table 4.1. Some enzymes of industrial importance (*Continued*)

Enzyme	Source	Other commercial enzymes		
Penicillinase	B. subtilis, Bacillus cereus	Removal of penicillin	Takamine Schenley	+
Glucose oxidase	Aspergillus niger, Dee O, Dee G	For removal of oxygen or glucose from various foods; dried-egg manufacture	Takamine	+
Hyaluronidase	Penicillium chrysogenum Animal, bacteria	For glucose determination Medical use	Nagase Co.	+ +
Lipase	Pancreas, mold (Rhizopus)	Digestive aid; flavoring of milk products		+
Cytochrome c	Yeast (Candida)	Medical use	Sankyo Co.	+
Catalase		Sterilization of milk		
Keratinase	Streptomyces fradiae	Removal of hair from hides	Merck Co.	+
		Nucleolytic and other new enzymes		
5'-Phosphodiesterase	Penicillium citrinum, S. griseus, B. subtilis	Inosinic acid and guanylic acid manufacture. (5'-nucleotides)	Yamasa Co., Takeda Co.	+ + +
Adenylic acid deaminase	A. oryzae	AMP → IMP	In Takadiastase	+
Microbial rennet	Mucor sp.	Cheese manufacture	Meito Sangyo Co	+
Naringinase. hesperidinase	Aspergillus niger	Removal of bitter taste from citrus juice	Rohm and Haas	+
Glucose isomerase	Lactobacillus brevis	Glucose → fructose	Optimum pH 6-7	
Laccase	Coriolus versicolor	Drying of lacquer		
Cellulase	Tricoderma koningi	Digestive aid	Optimum pH 4.6	

Certain applications of enzymes demand the use of a relatively pure extract. Two such examples are mentioned by Underkofler [2]: (1) glucose oxidase for desugaring of eggs must be free of any protein-splitting enzymes, and (2) proteolytic enzymes injected into animals for meat tenderization just before slaughter must not contain any compounds which would cause a serious physiological reaction. Other examples requiring relative purity include enzymes in clinical diagnosis and some enzymes in food processing.

Studies of enzyme kinetics have been generally carried out with the purest possible enzyme preparations. As indicated in Chap. 3, such research also involves the fewest possible number of substrates (one if achievable) and a controlled solution with known levels of activators (Ca^{2+}, Mg^{2+}, etc.), cofactors, and inhibitors. The results of these studies provide the clearest picture of enzyme kinetics.

Many useful industrial enzyme preparations are not highly purified. They contain a number of enzymes with different catalytic functions and, under most conditions, are not used with anything approaching either a pure substrate or a completely defined synthetic medium in the sense defined in Chap. 3. Also, the simultaneous use of several different enzymes may be more efficient than sequential catalysis by a separated series of the enzymes. In spite of this added complexity, such enzyme preparations are kinetically more simple than the integrated living organisms from which they are produced and are thus logically considered before the chapters dealing with the bioenergetics of microbial metabolism and industrial microbiological routes of product synthesis.

4.1. Production of Crude Enzyme Extracts

Crude extracts are usually produced by a brief processing, typically mechanical, of a source which is inherently relatively rich in the desired enzyme(s). Concentrating now on microbial sources, we note that two different types of processes are used to produce enzymes, the *solid-culture*, or *Koji method*, and the *submerged-liquid culture*. In Fig. 4.1 the first technique is illustrated schematically: the Koji culture unit (*H*) shown in this process contains trays of substrate, here wheat bran mixed with water and other ingredients such as buffers and supplementary nutrients. Extracellular-enzyme preparations can then be obtained by water or buffer washes of the surface culture, which undergo further processing as indicated in the figure. Figure 4.2 illustrates the usual process sequence for solid-culture enzyme production.

A more recent technique for producing microbial enzymes, emanating from aerated-culture reactor designs of World War II for penicillin production, is submerged-liquid culture growth (Fig. 4.3). Here compact biological reactor vessels, called *fermentors*, typically from 1000 to 30,000 gal, are used instead of the more spacious surface-culture processes. Sterilized air (freed from microorganisms) is pumped through the reactor liquid or *broth;* such variables as temperature, pH, and concentration level of some key nutrients in the broth are continually monitored, and some are controlled (Fig. 4.4).

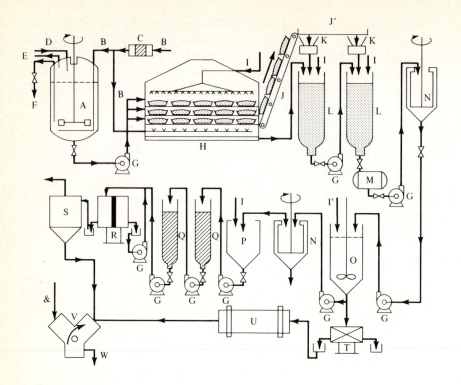

Figure 4.1 Cellulase production using the Koji technique. *A*—submerged seed culture of *T. viride*; *B*—oil-free compressed air; *C*—air filter; *D*—inoculum; *E*—exhaust air; *F*—sample collection; *G*—centrifugal pump; *H*—automatic wheat bran culture of *T. viride*; *I*—water spray; *I'*—ammonium sulphate, acetone, alcohol and water; *J*—belt conveyor; *J'*—screw conveyor; *K*—hopper; *L*—extraction column; *M*—storage tank; *N*—centrifuge; *O*—precipitation tank; *P*—mixing tank; *Q*—ion-exchange column; *R*—membrane concentrator; *S*—spray dryer; *T*—filter press; *U*—rotary dryer; *V*—mixer; *W*—cellulase preparation; &—salt stabiliser. (*Reprinted by permission from T. K. Ghose and A. N. Pathak, Cellulases I: Sources, Technology, Process Biochem.,* **April, 1973:** 35.)

The length of culture time needed and the control of such factors as temperature and feed and air sterility depend on both the particular microbe and the particular product desired. Firm generalizations regarding the point in the reaction when a particular product may be expected to appear are few. For example, in the production of bacterial amylase (Fig. 4.5) the broth is simply sampled until two successive amylase analyses of the broth show the same concentration. The reaction process is then halted by cooling the fermenter. Notice that the production of amylase bears no obvious relation to either pH or sugar level. From the final submerged-culture broth, a simple extracellular enzyme extract is derived by filtration or centrifugation; an intracellular-enzyme extract is obtained by breaking up the cells in the filter cake or centrifuge precipitate. Breakup or lysis of cells can be achieved by a variety of techniques, including homogenation in a blender, mechanical grinding, or ultrasonication.

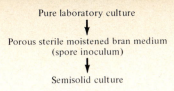

Pure laboratory culture

↓

Porous sterile moistened bran medium
(spore inoculum)

↓

Semisolid culture

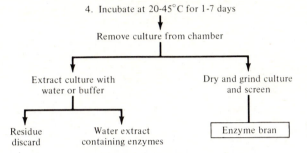

1. Sterilize moistened bran medium with steam

2. Cool and inoculate with spore inoculum

3. Spread on enclosed trays or tumble in
   enclosed drums

4. Incubate at 20-45°C for 1-7 days

↓

Remove culture from chamber

Extract culture with
water or buffer

Dry and grind culture
and screen

Residue
discard

Water extract
containing enzymes

Enzyme bran

Figure 4.2 A typical processing sequence for production of enzymes using the solid-culture (Koji) method. (*Reprinted from L. A. Underkofler, Manufacture and Uses of Industrial Microbial Enzymes in Bioengineering and Food Processing, CEP Symposium No. 69,* **62:** 12, 1966.)

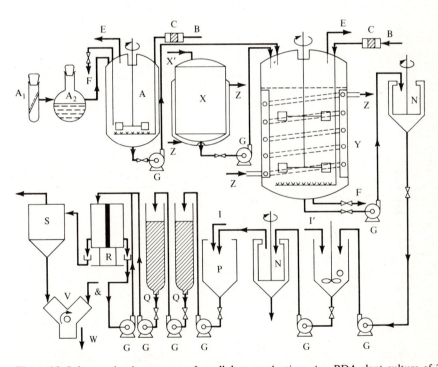

Figure 4.3 Submerged-culture process for cellulase production. A_1—PDA slant culture of *T. viride;* A_2—spore suspension of *T. viride; A*—seed; *B*—oil-free compressed air; *C*—air filter; *E*—air exhaust; *F*—sample collection; *G*—centrifugal pump ; *X*—substrate storage tank; *X'*—feed; *Z*—water line; *Y*—main fermentor; *N*—centrifuge; *O*—precipitation tank; *I'*—ammonium sulphate, acetone, alcohol and water; *I*—water; *P*—mixing tank; *Q*—ion-exchange column; *R*—membrane concentrator; *S*—spray dryer; *V*—mixer; *W*—cellulase preparation; *&*—salt stabiliser (*Reprinted by permission from T. K. Ghose and A. N. Pathak, Cellulases I: Sources, Technology, Process Biochem.,* **April, 1973:** 35.)

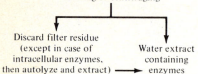

Pure laboratory culture
↓
Sterile liquid shaken flask culture
↓
Seed tank culture
↓
Submerged liquid culture

1. Sterilize liquid nutrient in closed vessel

2. Cool and inoculate from seed tank

3. Grow at 25-40°C with aeration and agitation for 1-5 days
↓
Separate microorgranisms and insolubles by filtering or centrifuging

Discard filter residue (except in case of intracellular enzymes, then autolyze and extract) ⟶

Water extract containing enzymes

Figure 4.4 A typical processing sequence for production of enzymes using submerged culture. (*Reprinted from L. A. Underkofler, Manufacture and Uses of Industrial Microbial Enzymes in Bioengineering and Food Processing, CEP Symposium Series No. 69,* **62:** 13 1966.)

With development of improved sterilization procedures, instrumentation, and controls, submerged-culture reactors are progressively displacing surface-culture techniques. Still, surface culture may be superior in some applications, as the list of relative features in Fig. 4.6 suggests. The kinetics of microbial growth for producing enzymes or other chemicals in surface or submerged culture are considered in Chap. 7; we shall explore further details of microbial technology in Chaps. 9, 10, and 12.

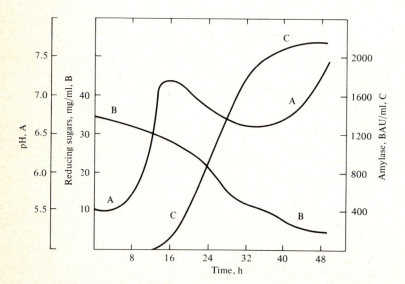

Figure 4.5 Time course of pH, reducing-sugar, and amylase concentrations during bacterial amylase production. (*Reprinted from L. A. Underkofler, Manufacture and Uses of Industrial Microbial Enzymes in Bioengineering and Food Processing, CEP Symposium Series No. 69,* **62:** 16, 1966.)

Surface	Submerged
Requires much space for trays	Uses compact closed fermentors
Requires much hand labor	Requires minimum of labor
Uses low-pressure air blower	Requires high-pressure air
Little power requirement	Needs considerable power for air compressors and agitators
Minimum control necessary	Requires careful control
Little contamination problem	Contamination frequently a serious problem
Recovery involves extraction with aqueous solution, filtration or centrifugation, and perhaps evaporation and/or precipitation	Recovery involves filtration or centrifugation, and perhaps evaporation and/or precipitation

Figure 4.6 Distinguishing features of solid and submerged culture processes. (*Reprinted from L. A. Underkofler, Manufacture and Uses of Industrial Microbial Enzymes in Bioengineering and Food Processing, CEP Symposium Series No. 69, 62: 13 1966.*)

Crude enzyme extract solutions are often given further treatment, which may include flocculation of particulate matter, adjustment of pH, and addition of preservatives such as toluene, organic acids and salts, phenolic compounds, or sodium fluoride. Then, concentration of the enzyme extract can be obtained by spray drying or enzyme precipitation (Fig. 4.7).

Operations such as mixing, grinding, centrifugation, gravity and vacuum filtration, and evaporation are described in chemical engineering texts on unit operations and separations. While we shall treat aeration and particle filtration in Chap. 8 and outline some other general techniques for product recovery in Chap. 10, we shall focus in the remainder of this chapter on some of the relatively specialized techniques used in the production of high-purity proteins such as enzymes. These include paper, gel, adsorption, ion-exchange, and affinity chromatography as well as electrophoresis, ultracentrifugation, fractional precipitation, freeze-drying (or lyophilization), dialysis, and crystallization. Some of these techniques (electrophoresis, chromatography, and ultracentrifugation) are difficult to scale up beyond laboratory dimensions. Many (fractional precipitation, dialysis, ultracentrifugation, and affinity chromatography) are usually operated batchwise. The design of large-scale continuous or semicontinuous versions of these various techniques is a major engineering challenge.

4.2. Enzyme Purification

On several occasions in the preceding chapters we have emphasized that, because of their complex structure, many proteins possess very specific biological activity. It should not be surprising, therefore, that proteins also have distinct physical properties which determine their binding at an interface or their solubilities. By

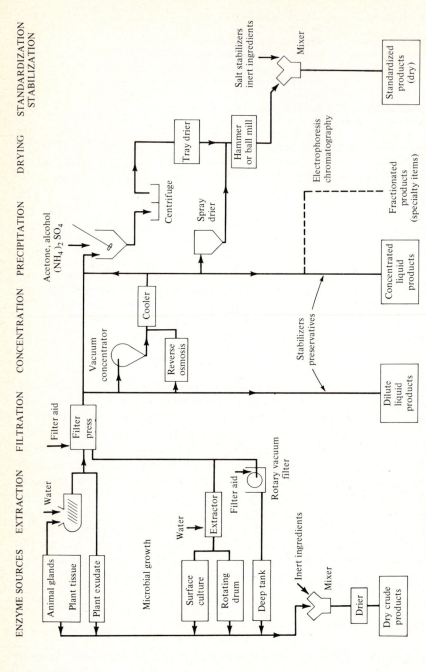

ENZYME SOURCES   EXTRACTION   FILTRATION   CONCENTRATION   PRECIPITATION   DRYING   STANDARDIZATION
STABILIZATION

Figure 4.7 A summary of processing operations used in manufacture of commercial enzymes. (*Reprinted from W. T. Faith et al., Production and Application of Enzymes, in T. K. Ghose and A. Fiechter (eds.), "Advances in Biochemical Engineering," vol. 1, p. 92, Springer-Verlag, New York, 1971.*)

exploiting the unique properties of a particular protein, we can separate it from a multicomponent mixture containing many proteins and other biochemicals. Sumner was the first to isolate an enzyme sucessfully when, in 1926, he crystallized urease from a jack bean. At the present time, more than 100 enzymes have been crystallized, and many more have been highly purified by other methods.

To provide an overview of this section and to place the techniques described here in perspective, two different enzyme-purification recipes for biochemical experiments are listed in Examples 4.1 and 4.2. The first is a lengthy procedure involving ammonium sulfate fractionation in order to concentrate the enzyme alcohol dehydrogenase from isolated yeast cells. In the second, simpler scheme in Example 4.2, aldolase enzyme can be obtained from a tissue, rabbit muscle. Most of the individual steps in these examples will be discussed below. These are organized according to the enzyme properties which are exploited to achieve a given separation.

Example 4.1. Procedures for isolation of enzymes from isolated cells (yeast alcohol dehydrogenase from dried bakers' yeast)†

1. Grind yeast in ball mill.
2. Mix 100 g powdered dried yeast with 100 ml 0.066 M disodium phosphate buffer which is also 0.001 M in ethylene diamine tetraacetate (EDTA).
3. Add 200 ml buffer, stir 2 h at 37°C, and let stand at ambient 3 h.
4. Remove yeast-cell debris by centrifugation (28,000g for 20 min) (assay).
5. Remove thermally labile protein by heating at ~ 55°C for 15 min (alcohol dehydrogenase is relatively heat-stable) (assay).
6. Add 50 ml acetone per 100 ml heat-treated extract (mixing at -2 to -4°C) and discard precipitate. Add additional 55 ml acetone, centrifuge (28,000g), and discard supernatant (assay).
7. Resuspend precipitate in 50 ml of 10^{-3} M phosphate buffer (pH 7.5) + 10^{-3} M EDTA cold buffer, dialyze twice with 3 l fresh phosphate buffer for 1 h.
8. At 0°C gradually add 36 g ammonium sulfate per 100 ml dialyzed solution and centrifuge at 0°C (assay).
9. Suspend the first $(NH_4)_2SO_4$ precipitate in 30 ml of 50% saturated $(NH_4)_2SO_4$ solution. Separate supernatant and remaining precipitate and repeat precipitate suspension with 45, 40, 35, and 25% saturated ammonium sulfate solutions (assay solutions).
10. Combine high-activity supernatant and add saturated ammonium sulfate solution until incipient turbidity is observed. Let crystallization proceed for several days at 0°C. Dissolve precipitate and recrystallize in 10 ml of 10^{-3} M EDTA (assay).

The results are

	Total units, $\times 10^{-6}$	Specific activity	Units
First extract	25	4,000	
After heat	30	8,000	100
Acetone precipitate	20	45,000	80
First $(NH_4)_2SO_4$ precipitate	14	70,000	56

† From S. Chaykin, "Biochemistry Laboratory Techniques," pp. 22–32, John Wiley & Sons, Inc., New York, 1966.

Saturation $(NH_4)_2SO_4$ %	Protein in solution	
	Specific activity	Total protein, %
50	180,000	14.9
45	200,000	12.5
40	650,000	12.0
35	440,000	6.6
25	430,000	19.5

Example 4.2. Procedure for isolation of enzyme from animal tissue (rabbit-muscle aldolase)‡

1. Grind 1 lb thawed rabbit muscle at 0°C.
2. Extract with 1 lb of solution (10^{-2} M KOH and 10^{-3} M EDTA) for 10 min; squeeze extract through cheesecloth; repeat extraction of muscle with a second solution volume.
3. Combine extracts, filter through glass-wool plug, and adjust pH to 7.5 with cold 0.1 N KOH (assay).
4. Add equal volume of saturated ammonium sulfate solution (pH adjusted to 7.5 with ammonia).
5. Cool to 0 to 5°C. Separate precipitate by gravity filtration or by refrigerated centrifugation ($13,000\,g$) for 30 min (assay supernatant).
6. Bring supernatant to 0.53 saturation in ammonium sulfate and store at 0°C for 1 week to effect crystallization.
7. Suspend crystals in 10^{-3} M EDTA, remove insoluble precipitate, recrystallize by addition of ammonium sulfate to 0.45 saturation, crystallize, and assay.

4.2.1. Differences in Ionic Properties: Precipitation, Electrophoresis, and Ion-Exchange Chromatography

The ionic properties of proteins have already been mentioned (see Table 2.7). To recall an important item introduced earlier, the *isoelectric point* is the pH at which there is no net charge on the protein. This important pH value, often denoted by pI, is listed in Table 4.2 for several enzymes and other proteins.

Table 4.2. Isoelectric points for several enzymes and other proteins†

pI = isoelectric pH			
Pepsin	~ 1.0	Hemoglobin	6.8
Egg albumin	4.6	Myoglobin	7.0
Serum albumin	4.9	Chymotrypsinogen	9.5
Urease	5.0	Cytochrome c	10.65
β-Lactoglobulin	5.2	Lysozyme	11.0
γ_1-Globulin	6.6		

† Adapted from A. L. Lehninger, "Biochemistry," 2d ed., table 7.1, Worth Publishers, Inc., New York, 1975.

‡ From S. Chaykin, "Biochemistry Laboratory Techniques," pp. 36–38, John Wiley & Sons, Inc., New York, 1966.

It is not surprising that the solubility of a protein depends on the pH of the solution and is usually smallest at the isoelectric pH. Since the protein molecules have no net charge at pI and some charge at different pH, electrostatic repulsive forces between solute molecules are minimized at pI. This behavior suggests a *fractional-precipitation* procedure for separating proteins with different isoelectric points: at a given pH the proteins with the nearest pI will tend to precipitate, other things (such as molecular weight) being equal. By varying pH, fractions containing different proteins are separated.

However, imposing wide pH variations on a protein solution runs the risk of denaturing many of its components. Consequently, another precipitation technique commonly called *salting out* is more prevalent. Protein solubility is markedly affected by the salt concentration of the solution. In 1889, Hofmeister observed that more negatively charged anions of salts were most effective in causing precipitation of proteins; a similar trend is evident with cations. The Hofmeister (or lyotropic) series of ions, in order of approximate *diminishing* effectiveness in salting out proteins is as follows:

Anions: citrate^{3-}, tartrate^{2-}, SO_4^{2-}, F^-, IO_3^-, $H_2PO_4^-$, acetate$^-$, $B_2O_3^-$, Cl^-, ClO_3^-, Br_2^-, NO_3^-, ClO_4^-, I^-, CNS^-
Cations: Th^{4+}, Al^{3+}, H^+, Ba^{2+}, Sr^{2+}, Ca^{2+}, Mg^{2+}, Cs^+, Rb^+, NH_4^+, K^+, Na^+, Li^+

As ammonium sulfate is very soluble in aqueous solutions, it is a common salt for such biochemical isolations, although the Hofmeister series reveals the potential usefulness of many other salts.

The major variable governing salting out of proteins is the ionic strength μ_{is}, defined by

$$\mu_{is} = \frac{1}{2} \sum_{i=1}^{n} c_i Z_i^2 \qquad (4.1)$$

where c_i and Z_i are the concentration and charge of the ith ion, respectively, and n is the total number of *ions* in solution. It is believed that the addition of various salts tends to reduce the water concentration available for hydration of ionized or polarizable protein groups, thereby diminishing the protein solubility. As the data in Fig. 4.8 show, protein solubility at high ionic strength can often be correlated by the relation

$$\log(\text{solubility, g/l}) = \beta' - \kappa_s \mu_{is} \qquad (4.2)$$

where κ_s is called the *salting-out constant*.

Precipitation of enzymes and other proteins can be accomplished by several other means. Some organic solvents such as acetone are good protein precipitants. Presumably such solvents, like salts, cause a reduction in water concentration. Proteins which denature at higher temperatures more easily than the component of interest can be removed from solution by "cooking" them. After denaturation, the protein is less soluble (remember the oil-drop model: unfolding a globular

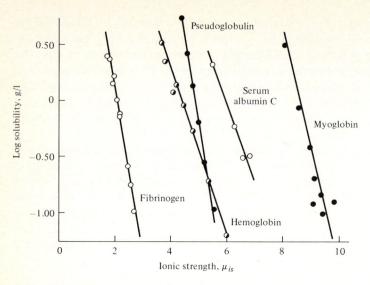

Figure 4.8 Dependence of protein solubility on ionic strength of ammonium sulfate solutions. (*Reprinted from E. Cohn and J. T. Edsall, "Proteins, Amino Acids, and Peptides," p. 602, Academic Press, New York, 1943.*)

protein will contact hydrophobic groups with the aqueous phase) and it precipitates. Another precipitation mechanism is to use casein, diatomaceous earth (containing diatoms, the porous skeletons of many algae, Fig. 1.14), or gelatin: proteins settle out of solution by absorption to these large particles.

A combination of these precipitation techniques is evident in the purification recipes given above. As shown in Example 4.1 the process begins with enzyme extraction in a buffered solution. This is followed by slight heating to remove, by precipitation, those proteins or components which denature more easily than the dehydrogenase of interest. Most of the enzyme is precipitated (along with other molecules of similar solubility) with acetone addition, which separates the enzyme from many of the low-molecular-weight extraction components; a subsequent resuspension and precipitation in ammonium sulfate effects a further purification in terms of enzyme activity per gram of precipitate. This purification is typically achieved, however, at the cost of some of the total amount of enzyme in each step (bottom of Example 4.1).

A further purification to separate the dehydrogenase from other proteins of similar solubility is effected by suspending the precipitate from step 8 (Example 4.1) in progressively weaker solutions of ammonium sulfate. Since the solubility of a protein increases as salt concentration decreases below saturation values, resolution of the enzyme is accomplished with each progressively weaker contact (Example 4.1). The supernatants from these resuspensions have specific activities (enzyme activity per total soluble protein) of the order of 100 times that of the first crude buffer extract. In addition, a great deal of extraneous material has been removed.

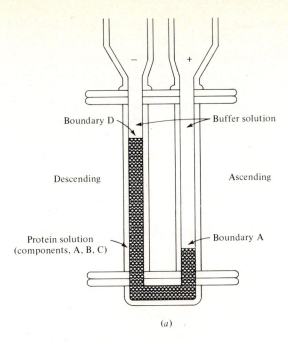

(a)

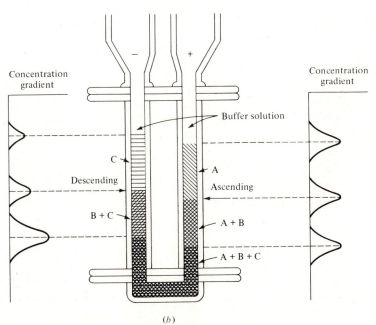

(b)

Figure 4.9 Location of protein components in an electrophoresis cell (a) before and (b) after application of an electric field. (*Courtesy of Beckman Instruments, Inc.*)

Electrophoresis is the movement of a charged species in an electric field. The steady velocity v_E achieved by a particle of charge q in a fluid under the influence of an electric field of strength E is found from the momentum balance

$$-\text{Fluid drag on particle} = \text{electrostatic force} = qE \tag{4.3}$$

Since most enzymes are globular proteins, use can be made of Stokes' law for the drag on a sphere of radius r_p moving in a newtonian liquid of viscosity μ_c.

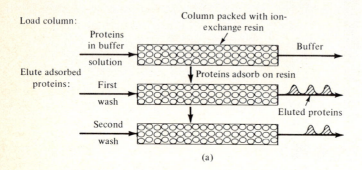

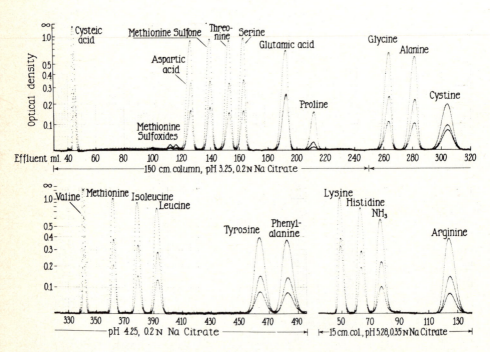

Figure 4.10 Ion exchange chromatography: (*a*) the procedure; (*b*) the results of an application to amino acid analysis. Labels above the chromatogram identify the feed conditions of the wash stream. [(*b*) *Reprinted with permission from D. H. Spackman, W. H. Stein, and S. Moore, Automatic Recording Apparatus for Use in the Chromatography of Amino Acids, Anal. Chem.,* **30**: 1190, 1958. *Copyright by the American Chemical Society.*]

$$-\text{Fluid drag on particle} = 6\pi\mu_c r_p v_E \qquad (4.4)$$

so that
$$v_E = \frac{qE}{6\pi\mu_c r_p} \qquad (4.5)$$

Since in general each enzyme species will have a different net charge q, application of an electric field will result in different protein velocities. Thus a sample containing several enzymes can be separated into its component parts (Fig. 4.9). By varying the pH of the electrophoretic medium, the velocity of a protein can be altered. If for a given protein the pI is smaller than the pH, its charge and velocity will be negative. Protein components with pI > pH will move in the opposite direction. This technique can be utilized to measure pI: the pI of proteins or other molecules can easily be determined by placing the molecules in a pH gradient and noting where in the gradient the electrophoretic velocity v_E becomes zero.

A general problem in considering scale-up of electrophoresis techniques is ohmic heating of the solution. As solutions which stabilize proteins are often quite conductive, high current densities are typical and the achievement of large-scale heat-exchange capabilities for such systems is not yet at hand. Small-scale electrophoretic equipment has been developed for continuous operation.

In *ion-exchange chromatography* the protein solution is passed over a fixed-bed column containing ion-exchange resin (Fig. 4.10). One common resin for protein purification is CM-cellulose, a cation-exchange resin obtained by linking negatively charged carboxymethyl groups to a cellulose backbone. Cationic (positively charged) proteins will bind to this resin by electrostatic forces, the strength of the attachment depending on the net positive charge at the pH of the column feed. After protein has been deposited in this manner, the column is developed by washing it with buffers of increasing pH and/or ionic strength. Such changes in the carrier solution cause weakly bound proteins to detach from the resin first, followed by more tightly attached molecules as conditions become far removed from those in the depositing step. Consequently the wash, which is protein-free when fed to the ion-exchange column, picks up a certain characteristic portion of the bound protein. The effluents (eluate) obtained by washing the column are then collected in small fractions under different conditions. Similar procedures are employed with anionic exchangers, which typically employ diethylaminoethyl (DEAE) cellulose ion-exchange resins.

4.2.2. Differences as Adsorbates: Adsorption and Affinity Chromatography

Proteins have varying tendencies to adsorb on such materials as starch, diatomaceous earth, and polyacrylamide gel. As a result, when a solution containing proteins is passed over such a phase, each protein moves at an effective velocity which decreases as the protein's propensity to adsorb increases. The following analogy is often used to explain the basic principle of chromatography. Imagine a number of coachmen who start together down a road lined with pubs. Obviously those with the greatest thirst will complete the journey last, while those with no taste for ale will progress rapidly.

The simplest model of chromatography assumes local equilibrium between protein in solution and adsorbed protein

$$K_i = \frac{w_i}{x_i} \tag{4.6}$$

where K_i = equilibrium partition coefficient
w_i = mole fraction of protein i in adsorbent
x_i = mole fraction of ith protein in solution

In a typical application, a small amount of sample protein solution is introduced into a buffered carrier stream, which then passes through a packed column of adsorbent (Fig. 4.11). If the carrier fluid travels through the bed at velocity v, the velocity v_i of the ith protein component is given by

$$\frac{v_i}{v} = \left(1 + \frac{1 - \varepsilon}{\varepsilon} K_i\right)^{-1} \tag{4.7}$$

The symbol ε denotes the void fraction of the bed and is equal to the volume occupied by fluid divided by the total column volume. It is important to recognize that v in Eq. (4.7) is the true fluid velocity [= (fluid volumetric flow rate)/(ε)(column cross-sectional area)].

In accordance with the coachman analogy, Eq. (4.7) indicates that protein with greater affinities for the second phase (large K_i's) will move through the column relatively slowly whereas compounds with very small tendencies to partition into or onto the second phase (small K_i) will move at velocities approaching that of the solvent.

An intriguing chromatographic technique based on the natural chemical specificity of some biopolymers is *affinity chromatography* (Fig. 4.12). A number of proteins and other biological macromolecules (A) complex with some other molecular entity (B) with a high degree of specificity: if species B is attached to a solid which is then used to pack a column, the K for component A is very large while the K's for other solution components are essentially zero. Examples of very

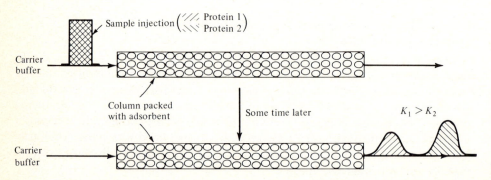

Figure 4.11 Schematic diagram of a chromatographic separation of proteins.

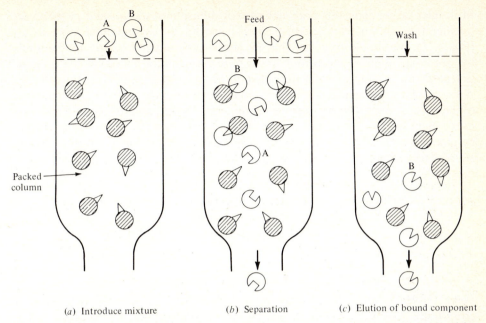

(a) Introduce mixture   (b) Separation   (c) Elution of bound component

Figure 4.12 Schematic diagram of affinity chromatography, where fine separations are achieved by specific interactions between an immobilized agent and a component in the liquid phase. Three steps are shown: (a) introduction of mixture, (b) separation, and (c) elution of bound component.

specific pair interactions include enzyme-inhibitor, antigen-antibody, and lectin-cell:

$$\text{Enzyme + inhibitor} \quad \rightleftharpoons \quad \text{enzyme-inhibitor complex}$$

$$\text{Antibody + antigen} \quad \longrightarrow \quad \text{antibody-antigen precipitate}$$

$$\text{Lectin + cell wall} \quad \longrightarrow \quad \text{lectin–cell-wall complex}$$

For example, an affinity-chromatography column for isolation of DNA depolymerase can be prepared by chemically binding DNA depolymerase (DNase) inhibitor to the surface of some convenient bead material, e.g., agarose or polyacrylamide. Flow of a lysed-cell centrifugate (such as that obtained after step 4 of Example 4.1) through such a column could result in an essentially quantitative binding of the enzyme DNA depolymerase and passage of nearly all other materials. From Eq. (4.7) $v_i \approx v$ for all other solution components. Subsequently, a solution containing DNase inhibitor or some other appropriate eluant can be passed through the column to release and recover the bound enzyme for further use.

4.2.3 Differences in Size: Molecular-Sieve Chromatography, Dialysis, Ultrafiltration, and Ultracentrifugation

Another isolation technique known as *molecular-sieve* (or gel-filtration or gel-permeation) *chromatography* separates proteins of different sizes. In this instance the column is packed with gel particles with pores of a characteristic size.

Molecules larger than this pore size cannot diffuse into the gel and pass rapidly through the column, while smaller species penetrate the gel and move more slowly (Fig. 4.13). We should be aware that the separation may not be governed entirely by molecular exclusion according to the pore diameter (or fiber diameter if the gel is regarded as a tangled collection of solid fibers). Other possible influential parameters include the effective protein diffusivity within the material, the tendency of the protein to adsorb on the internal material surface, and the osmotic pressure of the material. Currently the exact relative importance of these factors appears to be unknown. However, molecular exclusion often predominates. In this case, an equivalent equilibrium constant K_{av} for use in Eq. (4.7) can be calculated from

$$K_{av,\ i} = \exp\left[-\pi L(r_g + r_i)^2\right] \tag{4.8}$$

where $K_{av,\ i}$ = fraction of total internal material volume available
for spherical molecule of radius r_i
L = concentration of gel fiber, expressed in length per
volume, cm/cm^3
r_g = radius of gel fiber
r_i = radius of spherical molecule of ith species

Some typical r_g and r_i values are listed in Table 4.3. By varying the degree of cross-linking between chains in the gel the value of the effective gel-fiber radius can be altered. A new gel support can be characterized by plotting $-(\ln K_{av})^{0.5}$ vs. r_s. From Eq. (4.8), the slope is $(\pi L)^{0.5}$, and the intercept is $(\pi L)^{0.5}r_r$. The K_{av} values for a Sephadex and several Sephadex-agarose (Sepharose) gels are shown in Fig. 4.14. Such plots using K_{av} vs. molecular weight rather than radius r_i may be used when the radii of interest are not available.

Figure 4.15 illustrates the results of a protein separation achieved by molecular-sieve chromatography. Notice that the various proteins do not appear all at once but instead emerge as diffuse peaks. Several factors can contribute to peak spreading, including molecular diffusion, mass-transfer resistances, and variation in axial velocity over the column cross section. The response in Fig. 4.15 is typical of all the chromatographic methods we have reviewed in this chapter. The fractionation of very large proteins, of viruses, cell fragments, and even different cell types in a mixture of cells can be carried out in an analogous chromatographic fashion provided (as in the methods surveyed above) that support material has a tendency to discriminate between various mixture components.

Dialysis and *ultrafiltration* are separation processes based on the ability of a membrane to pass some solution components and retard or prevent passage of

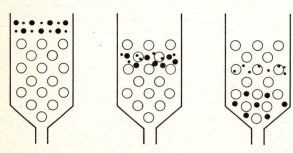

Figure 4.13 In molecular-sieve chromatography, larger molecules pass more rapidly through the column, while smaller molecules are retarded by occasional excursions into the gel particles.

Table 4.3. Some radii estimated from diffusion studies for several molecules†

Protein	Mol wt	Diffusity, $D \times 10^7$, cm^2/s	r_i, Å
Ribonuclease	13,683	11.9	18.0
Lysozyme	14,100	10.4	20.6
Chymotrypsinogen	23,200	9.5	22.5
Serum albumin	65,000	5.94	36.1
Catalase	250,000	4.1	52.2
Urease	480,000	3.46	61.9
Typical fiber radii in gel		r_g, Å	
Sephadex		7	
Agarose		25	

† Selected from summary in C. Tanford, "Physical Chemistry of Macromolecules," table 21.1, John Wiley & Sons, Inc., New York, 1961.

others. Several synthetic membrane materials exhibit the essential traits of interest: the tendency to pass small molecules easily (by diffusion and/or coupled flow from a high-concentration side to a low-concentration side) and to retain large molecules.

The object in dialysis is to remove various undesirable salts and small solutes from a solution, as in step 7 of the alcohol dehydrogenase isolation procedure (Example 4.1). To prevent the loss of desirable small ions or molecules (as may be

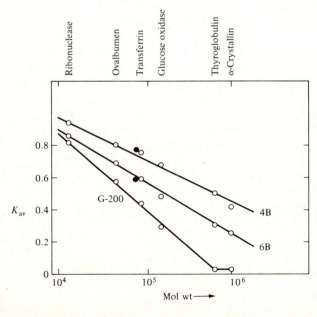

Figure 4.14 Protein vs. molecular-weight selectivity curves of Sephadex G-200, and Sepharose 6B (∼6 percent agarose gel, experimental) and Sepharose 4B (∼4 percent gel). (*Reprinted from M. K. Joustra, Gel Filtration on Agarose Gels, in "Progress in separation and Purification," vol. 2, p. 183, Wiley-Interscience, New York, 1969.*)

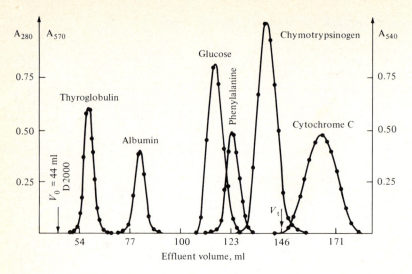

Figure 4.15 Chromatogram resulting from molecular-sieve chromatography of a solute mixture (packing is 6 percent cross-linked desulfated agar. Buffer is pH 7.5, 0.05 M tris-HCl). (*Reprinted by permission from J. Porath, Chromatographic Methods in Fractionation of Enzymes, in L. B. Wingard, Jr. (ed.), "Enzyme Engineering," p. 154, John Wiley & Sons, New York, 1972.*)

needed to ensure protein stability, for example) the composition of the *dialysate fluid* in which the dialysis bag containing the sample is placed must be such that no finite concentration difference of these particular dialyzable components exists across the dialysis membrane, e.g., the use of 10^{-3} M phosphate buffer as dialysate in step 7 of Example 4.1. When there are many chemicals which we wish to retain in the sample (such as blood), the cost of the necessary dialysate fluid may be very large. If a low concentration of certain dialyzable compounds is ultimately needed in the sample, relatively larger volumes of dialysate fluid must be used, for example, 6 l of phosphate buffer to dialyze 50 ml of sample (Example 4.1). This is necessary so that the concentration in the dialysate fluid will remain very small and there will always be a diffusive driving force. Dialysis is essentially a diffusion-controlled process.

Ultrafiltration is based upon the ability of a membrane, under a hydrostatic pressure head, to reject relatively high molecular weight components while passing both solvent and low molecular weight solutes. Thus, while dialysis generally removes low molecular weight molecules and ions from the original solution volume, ultrafiltration *concentrates* the original protein solution by removing much solvent as well as small solutes.

With ultrafiltration, the fluxes of solvent and of solute (protein) are often reasonably represented by

$$\mathcal{N}_1 \text{ (solvent)} = L_p(\Delta p - \sigma \, \Delta \pi) \tag{4.9}$$

$$\mathcal{N}_2 \text{ (solute)} \approx \bar{C}_2(1 - \sigma)\mathcal{N}_1 + P \, \Delta C_2 \tag{4.10}$$

where Δp = pressure difference applied across membrane

$\Delta \pi$ = osmotic-pressure difference

ΔC_2 = protein-concentration difference across membrane

$\bar{C}_2$ = average solute concentration in solution

The quantities L_p and P are membrane permeabilities indicative of the sustainable mass flux per membrane surface area per unit-pressure driving force. For ultrafiltration of protein solutions, the osmotic pressure difference is generally small compared with Δp. Thus

$$\mathcal{N}_1 \approx L_p \, \Delta p \qquad (4.11)$$

and

$$\mathcal{N}_2 \approx \bar{C}_2(1 - \sigma)\mathcal{N}_1 + P \, \Delta C_2 \qquad (4.12)$$

The quantity σ in the above equations is the reflection coefficient, and it indicates the tendency of solute to be transported with solvent. When the reflection coefficient is small ($\sigma \approx 0$), much of the solute may be transported *with* the solvent, as evidenced by the term $\bar{C}_2(1 - \sigma)\mathcal{N}_1$ in Eq. (4.12). When σ is large ($\sigma \approx 1.0$), the solute will tend to be transported at a rate *independent* of pressure, as seen from the second term in Eq. (4.12).

Typical permeabilities and reflection-coefficient values for several artificial membrane materials are listed in Table 4.4 The dialysis tubing indicated in the table could be used in an artificial kidney since it can pass urea (molecular weight 60) easily while retaining essentially all protein, as characterized by bovine serum albumin. Also, the numerical values given in Table 4.4 show that in actual operation, solvent fluxes are rather small; e.g., in 1 day and at a Δp of 10 atm the wet gel

Table 4.4. Flow characteristics of three types of cellophane membranes†

Substance	Mol wt	Molecular radius, Å	Reflection coefficient Dialysis tubing‡	Reflection coefficient Cellophane§	Reflection coefficient Wet gel¶
D_2O	20	1.9	0.002	...	0.001
Urea	60	2.7	0.024	0.006	0.004
Glucose	180	4.4	0.20	0.044	0.016
Sucrose	342	5.3	0.37	0.074	0.028
Raffinose	595	6.1	0.44	0.089	0.035
Inulin	991	12	0.76	0.43	0.23
Bovine serum albumin	66,000	37	1.02	1.03	0.73
Calculated pore radius, Å	...	...	23	41	82
Membrane constant, L_p 10^{-5} g $\cdot$ (cm$^2 \cdot$ s $\cdot$ atm)$^{-1}$	...	...	1.7	6.5	25

† Data from R. P. Durbin, *J. Gen. Physiol.*, **44**: 315 (1960).

‡ Visking cellulose.

§ DuPont 450-PT-62 cellophane.

¶ Sylvania 300 viscose wet gel.

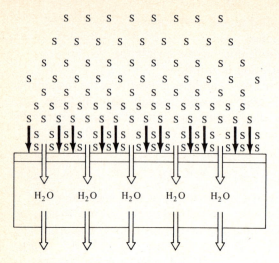

Figure 4.16 Concentration polarization caused by buildup of solute(s) near the upstream membrane surface.

membrane will pass a flux of

$$\mathcal{N}_1 = L_p \, \Delta p = (25)[10^{-5}\,\text{g}/(\text{cm}^2 \cdot \text{s} \cdot \text{atm})](86{,}400\,\text{s/day})(10\,\text{atm})$$

$$= 216\,\text{g}/(\text{day} \cdot \text{cm}^2\,\text{of area}) \tag{4.13}$$

or the order of 80 gal/(day·ft²).

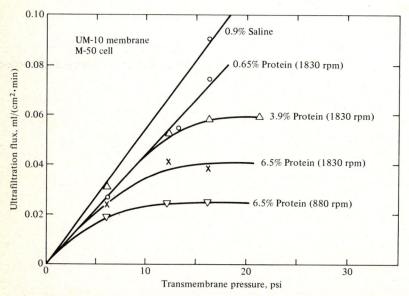

Figure 4.17 The influence of concentration polarization on deviations from linearity and approach to a constant flux as transmembrane pressure is increased. [*Reprinted by permission from M. C. Porter, Applications of Membranes to Enzyme Isolation and Purification, in L. B. Wingard, Jr. (ed.), "Enzyme Engineering," p. 119, John Wiley & Sons, Inc., New York, 1972.*]

In practice, the direct proportionality between $\mathcal{N}_1$ and Δp as given in Eq. (4.11) breaks down at high pressures, largely due to concentration polarization. As shown in Fig. 4.16, solute molecules S are carried to the membrane surface by the solvent, and, at steady state, this convective flux of S is exactly counterbalanced by an equal and opposite purely diffusive flux. Eventually, this "gel layer" of impermeable solute limits the attainable solvent flux, as revealed by experimental findings in Fig. 4.17. These data, which were taken using a stirred cell, also indicate that concentration polarization can be reduced by increasing mixing near the membrane surface.

Also, shear denaturation, mentioned in the last chapter, can be a significant problem in membrane processing of protein solutions. Moreover, in any enzyme-isolation procedure, some precautions in pump selection and piping layout must be taken to minimize shear damage to the protein component of interest.

Further details of dialysis and ultrafiltration and related techniques (reverse osmosis, electromembrane processing) are available in Lacey and Loeb [14]. In closing our discussion of these methods, we should emphasize that they are useful only for concentrating or removing small molecules from a protein solution. Unless the proteins have very different molecular weights, membranes are not especially helpful for separating one protein from another, a problem usually called *fractionation*. All the other methods surveyed earlier, however, can be applied to achieve fractionation.

4.2.4. Other Techniques, Including Solid-Phase Enzyme Synthesis

We begin this section with a brief account of isolation practices not yet mentioned and conclude with an unusual method for total synthesis of proteins and polypeptides.

Enzymes can be *crystallized*, usually from ammonium sulfate solution, by supersaturating a concentrated enzyme solution via increasing salt concentration or varying pH or temperature. The process of crystal formation and growth may take days or weeks and does not necessarily yield pure enzyme. On the contrary, the first crystal masses are often only about 50 percent pure and contain several enzymes. Repeated recrystallization is usually effective in isolating relatively pure enzyme.

Foaming can also be employed to fractionate and concentrate enzymes. The most surface-active components will concentrate at the interface between the solution and gas bubbled through it, and these species can then be collected in the foam at the top of the vessel. When protein is collected in the foam, denaturation may occur if interfaces have large surface-tension values. The same techniques can also be used, however, to remove impurities from solution.

In several processes for *concentrating* enzyme solutions, the fundamental concept is removal of solvent. This can be achieved in several ways besides the ultrafiltration method already described. One is *lyophilization*, or freeze-drying. To remove solvent from a solution frozen in an acetone—dry-ice bath, a vacuum is maintained so that solvent continually evaporates. Other variants of this method include flash evaporation, vacuum evaporation, and freeze-thaw procedures.

While peptides have been synthesized using the classical approaches of organic chemistry, the most promising method for achieving complete synthesis of enzymes is the *solid-phase* technique developed by R. Bruce Merrifield and coworkers. The basic concept, depicted schematically in Fig. 4.18, is sequential addition of amino acids to a growing chain attached to a solid. The advantage of the solid phase is that the growing protein product can easily be separated from the reaction mixture following each amino acid addition. Using this procedure, the 124-residue chain of ribonuclease has been synthesized. One limitation of making enzymes this way is that the primary structure must be known: unless the amino acid sequence is known, it cannot be reconstructed. Automated devices for con-

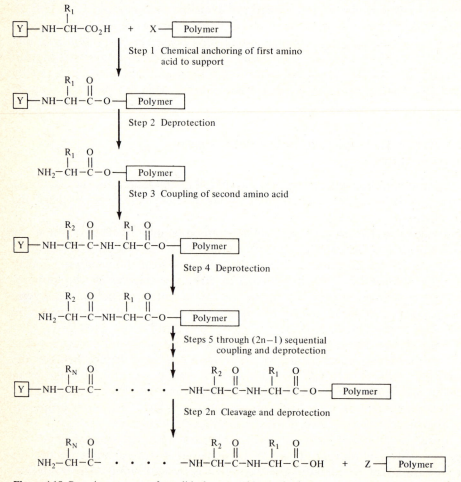

Figure 4.18 Reaction sequence for solid phase peptide synthesis.

ducting the synthesis have been developed. Solid-phase enzyme synthesis can be used, for example, to investigate the function of the large number of residues which are not in the active site. Also, attempts can be made to construct relatively small synthetic enzymes which have active sites similar to natural enzymes.

However, solid-phase synthesis is not without its problems. If one coupling reaction does not occur at a growing chain, a failure sequence is produced. As the number of residues increases, any inefficiencies in coupling at each step are compounded. If the coupling efficiency of each individual step were 99 percent for example, only 29 percent of the 124 residue chains in the ribonuclease synthesis would have the correct primary structure. With a 90 percent coupling efficiency, which has been observed on some occasions, 99.9998 percent of the chains would be failure sequences in a 124-residue synthesis. Mitigating these difficulties is the possibility that some failure sequences will be close enough to the specified structure to possess some degree of catalytic activity. (Recall Table 2.10.)

4.2.5. Concluding Remarks

Several general points should be brought out in relation to previous discussions in Sec. 4.2. First, many of the methods described here have also been applied to mixtures of other biological molecules besides proteins. For example, one automated amino acid analyzer employs ion-exchange chromatography, with which the amino acid content of a protein can readily be determined.

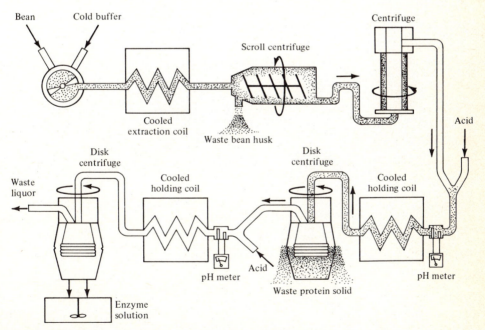

Figure 4.19 Flowsheet for continuous production of prolyl-tRNA synthetase from mung bean. (*Reprinted from M. D. Lilly and P. Dunnill, Sci. J.,* **5:** 59, 1969.)

We have not considered the criteria of purity employed in enzyme isolation. How do we decide when a homogeneous enzyme preparation of sufficient purity has been obtained? This question is addressed in the references.

A problem of general concern with all separation techniques is the choice between batch and continuous operation. Normally, continuous operation is advantageous as any process increases in magnitude; however, with biological separations or reactors, the possibility of microbial contamination of either the feed or the device itself is an inconvenience which can lessen this advantage. One recently developed continuous enzyme-isolation process is illustrated in Fig. 4.19.

Both small-scale chromatographic and electrophoretic separations have been operated on a continuous basis. In the continuous liquid chromatograph illustrated in Fig. 4.20, the feed is introduced at one point in the top of a slowly rotating annular chromatographic column. Carrier solution is introduced at other points in the top. Components with larger K values are carried further around the annulus periphery by column rotation before being eluted into the fixed collection tubes. It is evident that relatively large carrier-solution volumes are required for

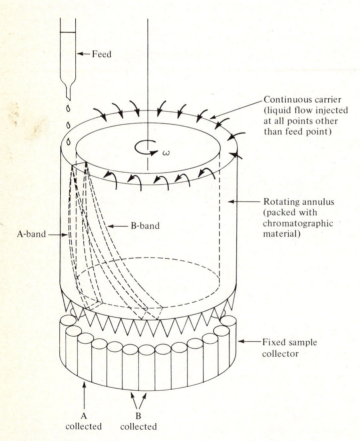

Figure 4.20 Schematic of continuous-liquid-chromatography apparatus.

such a technique; the same disadvantage arises with continuous electrophoresis. There similar separating bands are formed by applying an electric field perpendicular to the downward flow but in the plane of the electrophoresis support medium.

However, at least in the current scale of isolated-enzyme technology, the lack of large-scale separation devices for protein purification is not a serious problem. The present annual consumption of purified enzyme does not approach that of any common or even fine chemical.

4.3. Enzyme Immobilization

Immobilization of an enzyme means that it has been confined or localized so that it can be reused continuously. There are several reasons why immobilization may be desirable: for processing with isolated enzymes, an immobilized form can be retained in the reactor. With enzymes in solution, on the other hand, some enzymes will leave the reactor with the final product. Not only do new enzymes have to be introduced to replace the lost ones, but the enzymes in the product may be undesirable impurities which must be removed. Also, as we saw in Chap. 3, immobilized enzymes may retain their activity longer than those in solution (Figs. 3.9 and 3.46). Other special features of immobilized-enzyme systems will be discussed in a later section on their uses. Much of the current interest in immobilized enzymes may be classified within the science of *bionics*, which is the study of the unique features of biological systems in order to use or copy them in synthetic devices. As we have already noted, many enzymes in the cell are immobilized by attachment to, or confinement in, water-insoluble materials.

The various methods devised for enzyme immobilization can be subdivided into two general classes: chemical methods, where covalent bonds are formed with the enzyme, and physical methods, where weaker interactions or containment of the enzyme are involved (Fig. 4.21). In one of the foremost chemical methods, enzymes are covalently bonded directly to a water-insoluble support. Some common support materials are listed in Table 4.5. The reactions involved in establishing the covalent linkages and the enzymes which have been immobilized by this method are described in detail in Zaborsky (Ref. 4 of Chap. 3). Immobili-

Table 4.5. Insoluble supports useful for covalent enzyme attachments†

Synthetic supports	Natural supports
Acrylamide-based polymers	Agarose (Sepharose)
Maleic anhydride-based polymers	Cellulose
Methacrylic acid-based polymers	Dextran (Sephadex)
Polypeptides	Glass
Styrene-based polymers	Starch

† From O. Zaborsky, "Immobilized Enzymes," p. 9, CRC Press, Cleveland, 1973.

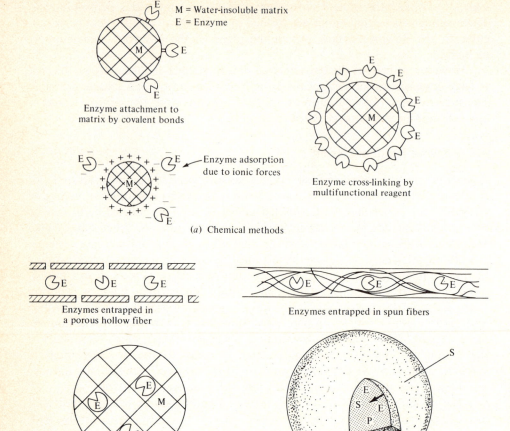

(a) Chemical methods

(b) Physical methods

Figure 4.21 A summary of techniques for enzyme immobilization. (a) chemical methods and (b) physical methods.

zation by this method is relatively straightforward experimentally. A disadvantage is that residues participating in the enzyme active site should not be involved in the attachment reaction. This requires luck or some knowledge of the amino acids participating in the catalytic site.

Multifunctional reagents can both bind enzyme and react with themselves to form polymers. Several strategies of immobilized-enzyme preparations are based on this property. In one, the enzyme is mixed only with the reagent, so that after polymerization bound enzyme is distributed throughout the polymer. As we saw in Chap. 3, enzymes in the interior of such a particle may be only partially utilized due to depletion of substrate by reaction near the exterior. Consequently, it may be preferable to keep the enzyme on the outer surface of an insoluble support. This can be achieved using multifunctional reagents as follows: the enzyme is first adsorbed on a surface active insoluble support. Then, the multifunctional reagent is introduced to cross-link the adsorbed enzymes and form a net around the support. Again, the preparation of immobilized enzymes using this method is simple, although precise controls of pH and concentrations are necessary. However, the gelatinous nature of these preparations hinders their use in some packed-column applications.

The adsorbents typically employed in this procedure, which are given in Table 4.6, can be used by themselves without the subsequent cross-linking step. Care must be exercised, however, because adsorption may be reversible and in some cases can be reversed by a small change in pH, temperature, or ionic strength.

A promising physical immobilization procedure is the fiber-entrapment scheme shown schematically in Fig. 4.21. In this method, an aqueous enzyme solution is emulsified with a solution of fiber-forming polymer. Subsequent spinning of this emulsion into a hardening or coagulation bath yields the fiber-entrapped enzyme. More than 30 enzymes have been immobilized in this fashion with large enzyme loadings and good activity retention.

In the remaining immobilization procedures, the enzyme stays in solution, but this solution is physically confined so that the enzyme cannot escape. There is a biological analog of these techniques: the lysosome within the cell contains hydrolytic enzymes, which would kill the cell immediately if they were released into it. As dramatically revealed in Fig. 4.22, the lysosomes are digestive organelles. Unnecessary, undesirable, or malfunctioning biopolymeric materials, like the mitochondrion in Fig. 4.22, are decomposed in the lysosome into smaller molecules, which can be reutilized by the cell's biochemical machinery.

In the artificial immobilization method known as *microencapsulation*, enzymes are entrapped in small capsules with diameters ranging from 1 to 300 μm. The capsules are surrounded by spherical membranes which have pores permitting small substrates and product molecules to enter and leave the capsule. The pores are too small, however, for enzymes and other large molecules to penetrate. Thus, the membrane acts as a dialyzer wall.

Two different types of semipermeable microcapsules can be made. The first

Table 4.6. Insoluble supports for enzyme immobilization by adsorption†

Alumina	Clays
Anion-exchange resins	Collagen
Calcium carbonate	Collodion
Carbon	Conditioned metal or glass plates
Cation-exchange resins	Diatomaceous earth
Celluloses	Hydroxylapatite

† From O. Zaborsky, "Immobilized Enzymes," p. 75, CRC Press, Cleveland, 1973.

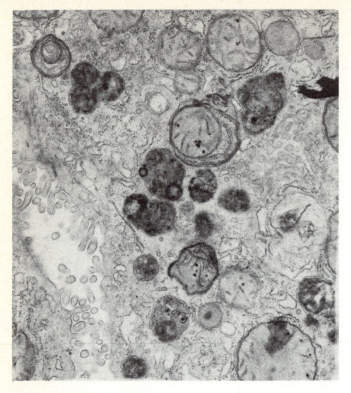

Figure 4.22 Lysosomes digest (presumably) defective cell organelles. (*Courtesy of Dr. Keith R. Porter*)

has a permanent polymeric membrane. To form this membrane, a copolymerization reaction can be conducted at the interface between an organic phase and a dispersed aqueous phase containing the enzyme. By choosing a water-insoluble monomer as one of the reactants and a monomer slightly soluble in both phases for the other, the copolymerization will occur only in the vicinity of the interface. Another method for making permanent microcapsules involves coacervation, a phase separation in a polymer solution which concentrates polymer at the interface of a microdroplet.

Nonpermanent microcapsules can be made by emulsifying an aqueous enzyme solution with a surfactant. This forms liquid-surfactant membranes, as illustrated in Fig. 4.23. The microcapsules so produced can then be added to an aqueous substrate solution.

Figure 4.23 (*a*) A drop consisting of aggregated liquid surfactant membrane microcapsules. (*b*) Drop structure and function. (*From O. Zaborsky, "Immobilized Enzymes," © CRC Press, Inc., 1973. Used by permission of CRC Press, Inc.*)

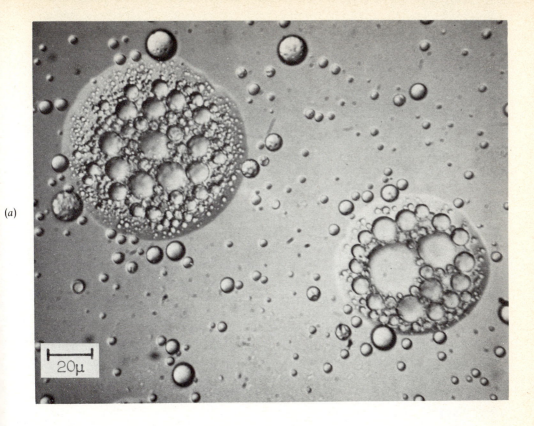

(a)

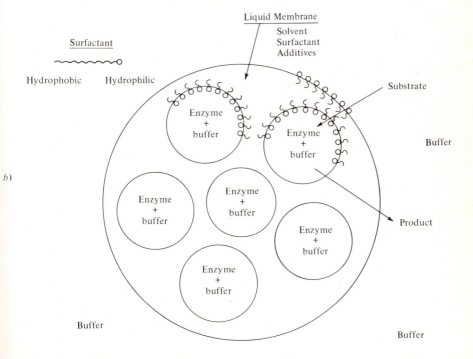

b)

187

Both microencapsulation methods have the potential of offering a very large surface area (for example, 2500 cm² per millimeter of enzyme solution) and the possibility of added specificity: the membrane can be made in some cases to admit some substrates selectively and exclude others. In principle these methods should be applicable to a large variety of enzymes. However, the membrane is a significant mass-transfer barrier, so that the "effectiveness factor" for the enzymes may be quite small. Also, these techniques are not applicable when the size of the substrate molecule approaches that of the enzyme.

We have already discussed (Sec. 4.2) how semipermeable-membrane filtration devices can be used to contain enzyme while allowing interchange of smaller molecules with a neighboring solution. A continuous flow ultrafiltration concept illustrated schematically in Fig. 4.24, also contains the enzyme as in microencapsulation. The surface areas separating the two solutions are substantially lower when ultrafiltration is used, and shear denaturation may occur. The mass-transfer rates available through membranes are small and may limit the overall rate. However, almost any enzyme or combination of enzymes can be immobilized in this way. It is advantageous when the substrate has very high molecular weight or is insoluble, situations where polymer-bound enzymes typically may have a small efficiency. Also, the semipermeable membrane will not permit relatively large product molecules from polymer-hydrolysis reactions to escape from the enzyme solution. This provides an interesting method of controlling the molecular-weight distribution of the products which ultimately leave the reactor.

In Sec. 4.6, we shall return to immobilized enzymes and examine some of their present and possible future applications. In current technology, however, enzymes in solution enjoy far greater use, as outlined in the next two sections.

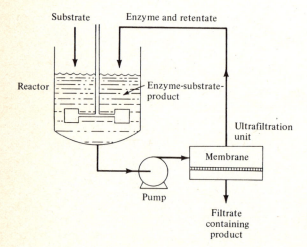

Figure 4.24 Enzyme entrapment on a macroscale. An ultrafiltration membrane is used to retain enzyme and other large molecules in the reactor.

4.4. Applications of Hydrolytic Enzymes

The action of hydrolytic enzymes is important not only in obvious macroscopic degradations such as food spoilage, starch thinning, and waste treatment but also in the chemistry of ripening picked green fruit, self-lysis of dead whole cells (autolysis), desirable aging of meat, curing cheeses, preventing beer haze, texturizing candies, treating wounds, and desizing textiles. These and other uses, the enzymes involved, and their sources will occupy our attention in this section.

A general classification of the major hydrolytic enzymes is given in Table 4.7. The three grouping of enzymes are those involved in the hydrolysis of ester, glucose, and various nitrogen bonds, respectively. Finer classifications of all such enzyme groupings are available and in common use. For the present, we shall simply point out that some enzymes catalyze the hydrolysis of a large variety of glucose linkages, for example, whereas other enzymes may catalyze the hydrolysis of only one glucose oligomer. Thus, the name of the enzyme by itself is not necessarily indicative of the precise substrate specificity.

Table 4.7. Hydrolytic enzymes†

Enzyme	Substrate	Catabolic product
Esterases:		
Lipases	Glycerides (fats)	Glycerol + fatty acids
Phosphatases:		
Lecithinase	Lecithin	Choline + H_3PO_4 + fat
Pectin esterase	Pectin methyl ester	Methanol + poly-galacturonic acid
Carbohydrases:		
Fructosidases	Sucrose	Fructose + glucose
α-Glucosidases		
(maltase)	Maltose	Glucose
β-Glucosidases		
(cellobiase)	Cellobiose	Glucose
β-Galactosidases		
(lactase)	Lactose	Galactose + glucose
Amylases	Starch	Maltose or glucose + maltooligosaccharides
Cellulase	Cellulose	Cellobiose
Cytase	....	Simple sugars
Polygalacturonase	Polygalacturonic acid	Galacturonic acid
Nitrogen-carrying compounds		
Proteinases	Proteins	Polypeptides
Polypeptidases	Proteins	Amino acids
Desaminases:		
Urease	Urea	$CO_2 + NH_3$
Asparaginase	Asparagine	Aspartic acid + NH_3
Deaminases	Amino acids	NH_3 + organic acids

† H. H. Weiser, "Practical Food Microbiology and Technology," p. 37, Avi Publishing Co., Westport, Conn., 1962.

We should note here that most enzymes have been named according to the chemical reactions they are observed to catalyze rather than their structure.†
Since a one-enzyme–one-reaction uniqueness does not generally exist, enzymes from different plant or animal sources which catalyze a given reaction will not always have the same molecular structure or necessarily the same kinetics. Consequently, maximum reaction rate, Michaelis constant, pH of optimum stability or activity, and other properties will depend on the particular enzyme source used. Most hydrolytic enzymes used commercially are extracellular microbial products.

Since water is an omnipresent substrate at roughly 55 M concentration, and since hydrolytic enzymes are invariably associated with degradative reactions, e.g., conversion of starch to sugar, proteins to polypeptides and amino acids, and lipids to their constituent glycerols, fatty acids, and phosphate bases, the following discussion is organized around the various possible polymer substrates.

4.4.1. Esterase Applications

This group of enzymes carries out the cleavage or synthesis of various ester bonds to yield an acid and an alcohol:

$$R_1COOR_2 + H_2O \longleftrightarrow R_1COOH + R_2OH \qquad (4.14)$$

The most important subgroups of these enzymes are the lipases, which hydrolyze fats into glycerol and fatty acids:

$$
\begin{array}{ccccc}
H_2C{-}OR_1 & & H_2COH & & R_1OOH \\
| & & | & & + \\
HC{-}OR_2 & \xrightarrow{\text{lipase}} & HCOH & + & R_2OOH \qquad (4.15) \\
| & & | & & + \\
H_2C{-}OR_3 & & H_2COH & & R_3OOH \\
\text{Lipid} & & \text{Glycerol} & & \text{Fatty acids}
\end{array}
$$

Pancreatic lipase, secreted into the digestive tract following neutralization of stomach-imparted acidity, splits ingested fats into fatty acids as well as the intermediate products of mono- and diglycerides. As mentioned in Chap. 3, the insolubility of higher-molecular-weight fats apparently requires the enzyme to act at the water-fat interface in order to yield an appreciable rate of hydrolysis (see Fig. 3.43).

The specificity of lipases and a second group of esterases known as aliesterases is not extreme. Lipases are active toward hydrolysis of high-molecular-weight fats and inactive toward the hydrolysis of fats formed from short-chain fatty acids: the reverse applies to aliesterases. The activity of the former is increased both by surfactants, e.g., bile salts, which tend to stabilize high-surface-area emulsions, and by calcium ions, which apparently precipitate the fatty acid hydrolysis products or

---

† The international classification scheme of the Enzyme Commission was outlined in Table 3.1. For further information on classification, see M. Dixon and E. C. Webb, "Enzymes," chap. 5. Academic Press, Inc., New York, 1964.

complex with them. This removes the free fatty acids, which otherwise tend to act on the lipase in an inhibitory manner.

In aerobic waste-digestion processes, like those occurring in natural water and in activated-sludge treatments (Chap. 12), mass transfer of sufficient oxygen is necessary to maintain the desired life forms. Since oxygen transport is relatively slow through fats, thin layers of such fats must often be continuously removed from the surface of aerated oxidation tanks in treatment plants; the skimmed fat-rich liquid is then digested by cells which are able to manufacture extracellular lipases. Formation of a continuous oil or fat layer over natural water will often quickly cut off the supply of dissolved oxygen with the resulting death of macroscopic and microscopic oxygen-requiring life which cannot pierce the layer.

In the meat-processing industry, the production of relatively fat-free meats can be carried out by partial fat hydrolysis of the meat cuts using a lipase preparation. Drains in food processing or in domestic or industrial waste treatment may be clogged by a mixture of biological materials containing proteins, carbohydrates, and fats. Synthetic mixtures of proteases, various carbohydrases, and lipases have been successfully used to control or remedy such problems.

The second class of esterase enzymes is known collectively as phosphatases; they are intricately involved in the biosynthetic metabolic pathways internal to cells which utilize the chemical free energy of fuel oxidation; practical applications outside living systems are not known.

4.4.2. Carbohydrases

All these enzymes cleave glucoside linkages. Since such glucose polymers as glycogen, starch, cellulose, and maltose are common carbon sources for macroscopic and microscopic life, the universal presence of such enzymes is apparent.

We have already remarked that the most plentiful polysaccharide is cellulose. The extracellulose enzyme system denoted as cellulase, which splits the polymer into dimer units of cellobiose, is thought to depend on three molecules: an S factor and two enzymes, C_1 and C_x. The first, or S, factor apparently solubilizes or removes the noncellulosic casing of lignin which surrounds native cellulose wood fibers; two enzymes which serve this function have been identified. The enzyme C_1 is able to attack insoluble or crystallized forms of cellulose, and the final enzyme C_x effects the conversion of soluble cellulose into cellobiose. This process is shown schematically in Fig. 4.25. Cellulase preparations find numerous uses, which are expected to expand in the future. Some of them are listed in Table 4.8.

The conversion of cellulose to carbon dioxide by microorganism digestion is normally a slow process. On one of Thor Heyerdahl's raft voyages across the Atlantic Ocean, he encountered undigested cellulose fibers (claimed to result from former paper) almost continuously distributed in the surface water. However, in some controlled microbial situations, e.g., the aerated tank in activated-sludge treatment of sewage, and in compost piles, where certain aerobic thermophilic (adapted to high temperatures) organisms thrive, the digestion of wood, paper, and other cellulose forms is relatively rapid. To date, almost all biochemical

1) Crystalline Cellulose

$\xrightarrow{\quad C_1 \quad}$ Soluble cellulose (may result from cleavage of hydrogen bonds or also some chain hyrdolysis)

2) Soluble cellulose $\xrightarrow{\quad E \quad}$ cellobiose and oligomers

$E = \beta\text{-}1 \rightarrow 4$ Glucanases:

Exoglucanase: splits single glucose residues from chain

Endoglucanase: splits chain randomly

3) β-Glucosidases: split oligomers and dimers (cellobiose) into glucose

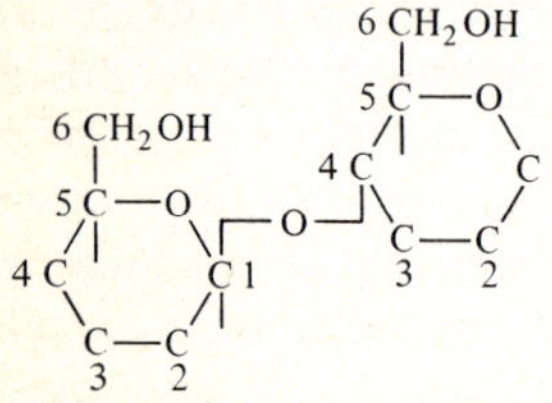

$\beta\text{-}1 \rightarrow 4$ Carbon linkage of cellobiose; same linkage occurs in all celluloses

Figure 4.25 Degradative sequence of enzyme attack of cellulose.

cellulose-disposal methods utilize these microbial methods rather than free enzymes. Digestion of cellulose to nourish microbial cells for possible food value or to provide substrate for industrial chemical manufacture and a high-pressure metal-catalyzed reaction involving carbon monoxide and cellulosic wastes to produce synthetic oil are some of the potential processes which may eventually provide better answers to cellulose-disposal problems.

Table 4.8. Applications of cellulases†

Increasing tensile strength of high-z sulfite pulp	Isolating soybean and coconut proteins, sweet potato starch and corn starch
Recovering agar-agar; production of seaweed jelly	Modifying rice tissue, carrot, or other foodstuffs
Processing fibrous materials as ruminant feed from delignified bark	Producing fruit vinegar or pectin from citrus pulp
Saccharifying delignified woody waste associated with xylanase	Improving solubility of raw materials in brewing
Treating rice grains for sake brewing in Japan	Digesting excreta in septic tanks
Removing soybean seedcoat	Supplementing enzymes in feed for poultry and pigs
Extracting green-tea component	Producing protoplasts from higher plants

† From T. K. Ghose and A. N. Pathak, Cellulases—2: Applications, *Process Biochem.*, **May, 1973:** 21.

Amylases are more common enzymes which can hydrolyze the glycosidic bonds in starch and related glucose-containing compounds. To appreciate the distinction between the two major types of amylases, we should recall that starch contains straight-chain glucose polymers called amylose and a branched component known as amylopectin. The branched structure is relatively more soluble than the linear amylose and also is effective in rapidly raising the viscosity of starch solutions. The action of α-amylase reduces the solution viscosity by acting *randomly* along the glucose chain: α-amylase is often called the *starch-liquefying enzyme* for this reason. β-Amylase can attack starch only on the nonreducing ends of the polymer and always results in maltose when a linear chain is hydrolyzed. Because of the characteristic production of the sugar maltose, β-amylase is also called a *saccharifying enzyme*. A soluble mixture of starch and β-amylase yields maltose and a remainder of dextrins, starch remnants with 1, 6 linkages on the end. β-Amylase cannot hydrolyze these bonds.

Another saccharifying enzyme, *amyloglucosidase* (also called glucoamylase, among other names) attacks only the nonreducing ends of starch, glycogen, dextrins, and maltose. Sequential treatment with α-amylase and glucoamylase or enzyme mixtures are utilized where pure glucose rather than maltose is desired: in distilleries (as opposed to breweries) and in the manufacture of glucose syrups (corn syrup) and crystalline glucose. It is estimated that 1.35 billion pounds of glucose were produced by this method in the United States in 1971.

Amylase preparations are used in brewing (to thin a potential polysaccharide source for use as a carbon nutrient) and in paper and textile manufacture (to thin starch solutions for size, the starchy coating used to strengthen fibers before weaving). These and other applications given in Table 4.9 makes these enzymes one of the most important groups in commerical use. The relative proportions of α- and β-amylase selected in various applications depends on the result desired.

The sources of amylases (or *diastases*, as these enzymes are often called) are very numerous. This is not surprising since starch is a common form of carbon fuel for many life forms. Amylases are produced by a number of bacteria and molds; an important example is the amylase produced by *Clostridium acetylbutyli-cum* which is clearly involved in the microbial conversion of polysaccharides to butanol and acetone. Commercial amylase preparations used in human foods are normally obtained from grains, notably barley, wheat, rye, oats, maize, sorghum, and rice. The ratio of saccharifying to liquefying enzyme activity depends not only on the particular grain but also upon whether the grain is or is not germinated. Germinated grain normally has considerably greater α-amylase activity, as shown in Table 4.10.

The other carbohydrases listed in Table 4.7 also involve cleavages of sugar polymers. The enzyme or enzymes shown as cytase are formed during germination of cereal grains and act to dissolve the cell wall (made up of protein and complex carbohydrates) surrounding the starch granules, so that the starch can be utilized as an energy source for the growing seed. In the production of malt (softened, germinated barley) for brewing, the ungerminated seeds are exposed to a favorable temperature and humidity so that rapid germination occurs, with resulting

Table 4.9. Common applications of amylase preparations†

Industry	Use
Glucose and syrup	Total or partial hydrolysis of corn starch by amyloglucosidase or α-amylase to give a large quantity of sweeteners
Brewing	Conversion of crushed grain starch to maltose (a suitable dissacharide substrate for yeast fermentation)
Breadmaking	*Leavening* Conversion of sufficient starch to fermentable saccharides needed for carbon dioxide generation
Fruit juice	Hydrolysis of starch causing turbidity due to insolubility
Papermaking	α-Amylase action to liquefy starch coatings to a desired viscosity for application to fibers (variable-weight papers)
Textiles	*Sizing* α-Amylase activity to liquefy starch; resulting solution used to strengthen warp threads before weaving
	Desizing α,β-Amylase action to remove size from woven material so that all threads will dye uniformly and fabric will have desired texture
Candy	Production of candy of desired softness

† Adapted from H. H. Weiser, "Practical Food Microbiology and Technology," p. 37, Avi Publishing Co., Westport, Conn., 1962.

Table 4.10. Comparison of amylase activity of various grain sources†
—Below sensitivity of test

	Total relative activity			
	β-Amylase		α-Amylase	
Cereal	Ungerminated	Germinated	Ungerminated	Germinated
Barley	29.8	34.4	0.058	94.0
Wheat	25.1	23.7	0.063	214.7
Rye	17.8	17.6	0.111	119.8
Oats	2.4	—	0.297	60.3
Maize	—	—	0.249	35.6
Sorghum	—	—	0.127	75.6
Rice	—	0.2	—	2.3

† Adapted from H. Tauber, "Chemistry and Technology of Enzymes," pp. 77–78, John Wiley & Sons, Inc., New York; data of E. Kneen, *Cereal Chem.*, **21**: 304 (1944).

large increase in amylase activity (Table 4.10). The germinated barley is then kiln-dried slowly; this halts all enzyme activity without irreversible inactivation. The dried malt preparation is then ground, and its enormous liquefying and saccharifying power (to convert starches to fermentable sugars) is utilized in the subsequent yeast fermentation (Chap. 10).

As noted in Chap. 2, milk sugar, lactose, is a combination of glucose and galactose. Splitting lactose into these sweeter components by *lactase* is common in the manufacture of ice cream products. The enzyme *maltase* hydrolyzes the maltose linkage; maltase is typically produced by the same organisms which produce β-amylase.

A related enzyme, invertase, hydrolyzes sucrose and polysaccharides containing a β-D-fructofuranosyl linkage. The enzyme name derives from the early observation that the hydrolyzed sucrose solution containing fructose and glucose rotates a polarized light beam in the direction opposite that of the original solution. The partially or completely hydrolyzed solution allows two properties desirable in syrup and candy manufacturing: a slightly sweeter taste than sucrose and a much higher sugar concentration before hardening.

4.4.3. Proteolytic Enzymes

The variety and uses of enzymes which selectively attack nitrogen-carrying compounds, especially proteins, is quite large. As with the amylases, the mode of attack on polyamino acids is either on terminal groups (exopeptidases) or internal linkages (endopeptidases).

Since enzymes, the essential catalysts of living organisms, are themselves protein, it is not surprising that protein-splitting enzymes are often initially synthesized in an inactive form. Activation of these proteolytic enzymes is then accomplished in one of two ways (Fig. 4.26). The enzyme is synthesized in an inactive form suitable either for storage or for transport from the site of synthesis to the desired site of activity, as is the case for pepsin, trypsin, chymotrypsin, and carboxypeptidase. It is interesting to note that the activation of pepsin and trypsin is *autocatalytic*: the inactive enzyme precursor is a substrate for the active form of the enzyme, the reaction product being more of the activated enzyme. Equally interesting is the fact that further proteolytic attack is not observed after the initial conversion of inactive trypsin or pepsin to the active form. A second group of enzymes, typically exopeptidases, require one of perhaps several specific metal ions for activation. Dialysis of the enzyme-containing solution with resultant loss of metal ions is a standard form of determining specific metal ion requirements of enzymes and microbial life in general.

The commercial sources of proteases include animals (pancreas) and large plants (saps, juices) as well as yeast, mold, and bacteria. Some of them are listed in Table 4.1, along with the corresponding applications. According to 1972 data and projections, proteolytic enzymes are the most important group from an economic viewpoint (see Sec. 4.7).

(*a*) From a precursor

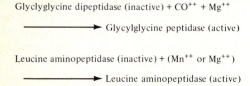

(*b*) By presence of a metal ion

Glyclyglycine dipeptidase (inactive) + CO^{++} + Mg^{++}

$\longrightarrow$ Glycylglycine peptidase (active)

Leucine aminopeptidase (inactive) + (Mn^{++} or Mg^{++})

$\longrightarrow$ Leucine aminopeptidase (active)

Figure 4.26 Activation of proteolytic enzymes. (*a*) from a precursor and (*b*) by presence of a metal ion.

The major uses of free proteases occur in dry cleaning, detergents, meat processing (tenderization), cheese making (rennin only), tanning, silver recovery from photographic film (pepsin), production of digestive aids, and certain medical treatments of inflammations and virulent wounds. Enzymes were used in laundry aids as early as 1913. During the late 1960s an explosive increase in protease utilization in detergents occurred. The enzymes used facilitate spot removal; they are a mixture of bacterial neutral and alkaline proteases which are active over the pH range of 6.5 to 10 and temperatures from 30 to 60°C. A peak in this enzyme application occurred in 1969, when 30 to 75 percent of all European detergents and about 40 percent of detergents in the United States contained enzymes. However, subsequent warnings from the U.S. Federal Trade Commission caused concern about health hazards from these preparations, and this enzyme market plummeted in 1970 and 1971. Following retraction of the Trade Commission warning, a partial recovery followed, with 1980 United States demand for detergent bacterial protease estimated at 6 million dollars.

The tenderization of individual meat pieces by commercial tenderizer products depends on proteolytic action of the relatively inexpensive and non-heat-labile plant proteases papain and bromelain. Aging of whole meat carcasses prior to cutting and packaging is normally accomplished by controlled partial self-digestion (autolysis) of the bled meat at about 15°C in the presence of ultraviolet light, which acts as a germicidal agent preventing the concurrent surface growth of undesirable microorganisms.

Ground pancreas preparations from different animal sources contain all the digestive proteases, including typsin, as well as lipases and amylases. These obviously digestive mixtures are useful in dehairing animal hides and the simultaneous removal of other noncollagen protein from the hides. Since pepsin itself attacks collagen, the fibrous skin protein which is converted into leather, this proteolytic enzyme is useless in tanning.

In the dairy industries, rennin (or rennet) is the single most important enzyme. It acts by removing a glycopeptide from soluble calcium casein to yield a relatively insoluble calcium paracaseinate, which precipitates to form the desired curd. Other proteases are also effective in converting calcium caseinate to calcium paracaseinate. However, these enzymes normally continue the proteolysis much further, and thus curd formation is prevented since the later degradation products are more soluble. Shortages of animal rennin have stimulated development of suitable microbial rennin enzymes, and a few of them have been used commercially.

Clinical and medicinal applications of proteolytic enzymes include both digestive aids and cleansers of serious wounds. Since enzymes are proteins, digestive-enzyme aids are suitably coated to protect the enzyme during passage through the stomach, where the acid environment could cause protein denaturation. The sources of enzymes shown in Table 4.1 are all nonhuman. Injection of some foreign proteases (pig trysin differs from human trysin) into human beings has been used to reduce tissue inflammation: the highly purified crystallizable form of the enzyme minimizes immune system response. The natural defenses of live cells against protease attack are usually inactivated in dead cells. This convenient difference allows application of solutions of proteases to virulent or oozing wounds: selective liquefaction of the dead tissue and cells in achieved, which facilitates wound drainage and thereby decreases the time needed for healing.

Proteolytic enzymes, especially trypsin, apparently reduce inflammation and swelling associated with internal injuries and infections by dissolving blood clots and extracellular-protein precipitates, by locally activating other body defenses which do the same thing, or both. Some severe lung infections resulting in accumulation of viscous lung deposits have been successfully reduced or eliminated by proteolytic-enzyme administration.

4.4.4. Enzyme Mixtures, Pectic Enzymes, and Additional Applications

Mixtures of enzymes, either of the same general type, for example, α- and β-amylase, or trypsin and chymotrypsin, or of different types such as found in pancreas extracts, e.g., trypsin, lipase, and amylase, are often more conveniently or more successfully used than single enzyme preparations. Thus, blends of different amylases achieve large yields of saccharified starch suitable for yeast fermentations yielding alcohol; a combination with less β-amylase achieves the desired thinning of starches (as in textile sizing) without too great a saccharification. Similarly, pancreas' extracts from different animals contain a mixture of digestive enzymes which together carry out many of the same functions as enzymes in human pancreatic fluids. At least one of the enzyme preparations marketed in detergents in this country contained a combination of both bacterial amylase and proteases, claimed to be more effective in removing certain stains than the other additions consisting only of protease.

Oxidation of galactose to galacturonic acid followed by dehydration and resulting polymerization yields polygalacturonic acid molecules. Naturally occurring plant molecules containing such polygalacturonic acid species as a major portion or fraction are collectively termed *pectins*. Often the acid forms of the molecules are found esterified with methanol. The two major pectin-hydrolyzing enzymes are pectin methylesterase (pectin esterase, hydrolyzing pectase) and polygalacturonase (pectinase, pectolase), which hydrolyze the methyl ester and the glycosidic linkages, respectively. Major sources are fruits for the former and fungi for the latter.

There are two main applications for *pectic enzymes*. Crushing fruits and vegetables yields juices which have high viscosity, desirable in the production of tomato and citrus juices but often not so for apple cider and other fruit juices. A controlled partial pectin hydrolysis of these juices yields a free-flowing product which retains enough viscosity to prevent undesirable settling of particulate matter. A greater hydrolysis is effected with apple juice: the hydrolyzed product is much more easily filtered, to yield a clear juice. If the juices are to be used in jelly manufacture, only the pectin esterase is added. The resulting polyacid hydrolysis product is then gelled by precipitation with calcium ions.

The second important application of pectic enzymes is wine production. Addition of the pectic enzyme mixture to the crushed grapes tends to increase the weight yield of juice, allow extraction of greater color from the grape skin, and permit faster filtering and pressing. Later addition to the fermented product again gives a faster subsequent separation of the wine from the yeast and grape sediment and yields a clearer wine with an increased stability (largely resulting from reduction of the suspended protein concentration in the final wine).

In both these cases, a major use of pectic enzymes is thus the development of a process stream with a desirable viscosity and filterability. The application includes benefits to process economics and the appearance of the product.

Future utilization of pectic enzymes to treat soft woods is a possibility. When trees of this type, e.g., Norway spruce, are felled, they resist penetration by chemical preservatives. Pretreatment with pectic enzymes has been shown to improve the efficiency of preservative treatment by rendering the wood more permeable.

Several additional hydrolytic enzymes enjoy small-scale applications, and promising new processes using hydrolases are being developed (Table 4.11).

In the food industry, the enzyme melibiase is now being used to hydrolyze raffinose, a trisaccharide of fructose, glucose, and galactose which is found in sugar beets. Since raffinose inhibits beet-sugar crystallization, its removal by enzyme hydrolysis increases the yield and quality of the sucrose product.

Selective removal of undesirable compounds in foods is the common purpose of many hydrolase uses. Relatively recent developments are enzyme processes for eliminating bitter-tasting naringin from fruits and juices

$$\underset{\substack{\text{(Bitter}\\\text{flavor)}}}{\text{Naringin}} \xrightarrow{\text{naringinase}} \text{prunin} \xrightarrow[\substack{\text{glucosidase}}]{\text{a flavonoid}} \underset{\text{(Nonbitter)}}{\text{naringenin}}$$

Table 4.11. Hydrolytic enzyme applications which are expanding or in development

Enzyme	Process
Dextranase	Removal of tooth plaque
Lactase	Removal of lactose from whey, milk
Isoamylase	Production of maltose from starch
Keratinase	Modification of wool, hair, leather
Tannase	Removal of tannic acid from foods
Penicillin amidase	Production of semi-synthetic penicillin core from natural penicillin G
Ribonuclease	Production of 5'-nucleotides from RNA

and transforming hesperidine, a source of turbidity in canned mandarin orange syrup

$$\text{Hesperidine} \xrightarrow{\text{hesperidinase}} \text{hesperetin}$$
$$\text{(White turbidity)} \qquad\qquad \text{(Nonturbid)}$$

4.5. OTHER ENZYME APPLICATIONS

While the hydrolytic-enzyme applications considered above dominate past and present enzyme technology, other enzyme processes currently serve important functions in the food, pharmaceutical, and biochemical industries. Moreover, many new applications are emerging, as we shall see in the remainder of this chapter.

4.5.1. Medical Applications of Enzymes

A rapidly growing area of medicine now involves the use of free or extracellular enzymes. A brief listing of some of the more common or promising enzymes in diagnosis, therapy, and treatment is given in Table 4.12.

Table 4.12. Some enzymes of importance in medicinal applications

Enzyme	Typical application
Trypsin	Anti-inflammation agent, wound cleanser
Glucose oxidase	Glucose test in blood or urine
Lysozyme (though protein, not affected by trypsin, chymotrypsin, or papain)	Recommended in treatment of certain ulcers, measles, multiple sclerosis, some skin diseases, and postoperative infections (antibacterial agent)
Hyaluronidase (from beef testicles)	Hydrolyses polyhyaluronic acid, a relatively impermeable polymer found between human cells; administered to increase diffusion of coinjected compounds, e.g., antibiotics, adrenaline, heparin, and local anaesthetic in surgery and dentistry
Digestive enzyme (mixtures of amylase, lipase, protease, and cellulases)	Digestive aids
Streptokinase	Anti-inflammatory agent
Streptodornase	Anti-inflammatory agent; also digests DNA, reducing viscosity of wound exudates
Penicillinase	Removal of allergenic form of penicillin from allergic individuals
Asparaginase	Anticancer agent

An enzyme which is present in protective body fluids such as nasal mucus and tears is *lysozyme*, which hydrolyzes the mucopolysaccharides of a number of (gram-positive) bacterial cell walls. The enzyme is used as an antibacterial agent (it apparently has other catalytic functions as well).

Initial stages of certain diseases and the presence of internal injuries give rise to elevated or depressed levels of enzyme concentrations in the easily sampled body fluids of lymph, blood, and urine. The presence of a given enzyme can usually be detected quite simply using an appropriate substrate test. Consequently assay of enzyme activity in body fluids can be a useful diagnostic tool. As Table 4.13 indicates, serum enzyme levels provide useful input in diagnosing a number of cardiac, pancreatic, muscular, bone, and malignant disorders.

The enzyme L-*asparaginase* catalyzes the hydrolysis of L-asparagine

$$\text{L-Asparagine} + H_2O \longrightarrow \text{L-aspartate} + NH_3$$

Table 4.13. Enzymes whose levels in human blood serum provide useful diagnostic indicators†

Serum enzymes of diagnostic value	Principal clinical conditions in which enzyme determination is of diagnostic value
Transferases:	
Glutamate-oxalacetate transaminase	Myocardial infarction liver disease
Glutamate-pyruvate transaminase	Liver disease
Creatine phosphokinase	Myocardial infarction, muscle disease
Oxidoreductase (dehydrogenase):	
Lactate dehydrogenase	Myocardial infarction
α-Hydroxybutyrate dehydrogenase	Myocardial infarction
Malate dehydrogenase	Myocardial infarction
Isocitrate dehydrogenase	Acute hepatitis
Hydrolases:	
Amylase	Acute pancreatitis
Lipase	Acute pancreatitis
Alkaline phosphatases	Bone disease, liver disease
5-Nucleotidase	Obstructive jaundice, hepatic metastases
Acid phosphatase	Carcinoma of prostate
Leucine aminopeptidase	Obstructive jaundice, hepatic metastases,
Cholinesterase	Liver disease, Organophosphorus poisoning, suxamethonium sensitivity
Lyases:	
Aldolase	Muscle disease, myocardial infarction

† Sidney B. Rosalki, "Diagnostic Enzymology," p. 9, 2d., Dade Division, American Hospital Supply Corporation, Miami, Fla., 1969.

Since some cancer cells require L-asparagine, their growth can be inhibited by using L-asparaginase to remove this essential nutrient. *E. coli* is known to produce two different asparaginases, only one of which exhibits antilymphoma activity.

As noted earlier, the cell membranes and the intermembrane materials of macroscopic animals are relatively impermeable to many substances; in the latter example this property is due to the presence of a very viscous polyhyaluronic acid. Effective administration of certain drugs and local anaesthetics, e.g., in dental work, is enhanced by simultaneous injection of *hyaluronidase*, which partially hydrolyses this cellular diffusion barrier.

A number of single-cell species can manufacture *penicillinase* and thus protect themselves from otherwise lethal levels of the antibiotic penicillin. Human beings do not produce penicillinase; the appearance of allergic symptoms in a patient to whom penicillin has been administered can be treated with injection of penicillinase solutions which, ideally, convert the drug into a nonallergenic form.

Glucose oxidase catalyzes the oxidation of glucose to gluconic acid

$$\text{Glucose} + \tfrac{1}{2}O_2 + H_2O \longrightarrow \text{gluconic acid} + H_2O_2$$

Since the resulting hydrogen peroxide is easily noted with an appropriate reducible indicator, glucose oxidase provides a sensitive specific test for the presence of glucose in blood and urine (as in diabetes). Additional medical applications of enzymes will be mentioned later during our discussion of immobilized enzymes.

4.5.2. Nonhydrolytic Enzymes in Current and Developing Industrial Technology

Glucose oxidase, the very useful enzyme just mentioned, finds applications whenever glucose or oxygen removal is desirable. Dried egg powders undergo an undesirable darkening due to a reaction between glucose and protein. This reaction, commonly called a *Browning reaction* or a *Maillard reaction*, can be prevented by addition of glucose oxidase. In the production and storage of orange soft drinks, canned beverages, dried food powders, mayonnaise, salad dressing, and cheese slices, the presence of oxygen (which would otherwise eventually form other oxygenated products associated with undesirable flavors) is usually avoided by addition of glucose oxidase and, in the case of cheese wrapping, glucose itself. Since enzyme activity is normally maintained for a long time at storage temperatures, such enzyme additions also increase the shelf life of food products by continually removing the oxygen which diffuses through the food packaging.

The hydrogen peroxide produced in the glucose oxidase-catalyzed reaction has an antibacterial action; if the presence of hydrogen peroxide is undesirable in the product, *catalase* is added, which catalyzes the reaction

$$2H_2O_2 \longrightarrow 2H_2O + O_2 \tag{4.16}$$

With catalase present this reaction proceeds very quickly $[v_{max}/e_0 = 1.2 \times 10^7 (\text{s} \cdot M)^{-1}, \; K_m = 10^{-7} \, M]$. Because of its rapid action to decompose peroxide, *catalase* is used in the rinse for some hair dyes.

Besides these established applications, those listed in Table 4.14 are now being studied, and some are being implemented commercially. One of the commercially important cases is use of glucose isomerase to produce fructose from glucose. Since this process and some others in this list are often accomplished with enzymes in an immobilized form, we consider them in greater detail in the next section.

Table 4.14. Some recent applications or promising future uses for nonhydrolytic enzymes†

Production of L-malate from fumarate by fumarase

Production of L-aspartate from fumarate by aspartase

Production of ATP from adenine by microbial enzymes

Production of NAD from adenine and nicotinamide by microbial enzymes

Production of fructose from glucose by glucose isomerase

Production of fructose from sorbitol by sorbitol dehydrogenase

Resolution of DL-amino acids by immobilized aminoacylase

Production of L-tyrosine and L-dopa by tyrosine phenol-lyase

Production of L-dopa from the corresponding α-keto acid by transaminase

Enzymatic production of L-tryptophan from indole, pyruvate, and ammonia

Enzymatic production of L-lysine from α-amino-ε-caprolactam

Transformation of saturated fatty acids to polyunsaturated fatty acids by fatty acid desaturase

Production of galactonolactose from whey using glucose oxidase and galactose oxidase

† From E. K. Pye and L. B. Wingard (eds.), "Enzyme Engineering," p. 365, Plenum Press, New York, 1973.

4.6. IMMOBILIZED-ENZYME TECHNOLOGY

Compared with their counterparts in solution, immobilized enzymes possess the advantage of easy recovery for continuous reuse. In addition, immobilized enzymes frequently permit continuous processing of a substrate stream with relatively small losses in catalytic activity. Finally, when we immobilize an enzyme, we have an opportunity to fix its position near other enzymes participating in a catalytic sequence say, on one side of a membrane or in the same porous pellet.

These characteristics make immobilized enzymes attractive if a very large throughput of substrate is required and/or the enzymes involved are expensive. Moreover, the ability to confine an enzyme in a well-defined, predetermined space provides opportunities for applications unique to immobilized enzymes. In the next section we examine some present and potential applications of immobilized enzymes.

4.6.1. Industrial Processes

Several large-scale industrial processes already in operation employ immobilized-enzyme catalysts at some point. Two notable examples are production of high-fructose syrups from corn starch and manufacture of L-amino acids by resolution of racemic amino acid mixtures (containing both D and L optical isomers). Also, immobilized penicillin amidase has been used commercially in manufacture of semisynthetic penicillins. The first application mentioned is probably the most important economically at present, but unfortunately many aspects of the process details are proprietary. Enzyme utilization in sweetener manufacture is considerably ahead of published information, so that our discussion of this topic is necessarily abbreviated and incomplete.

Immobilized enzymes can participate in sweetener production in at least two different ways. First, after the cornstarch has been thinned by acid or enzyme hydrolysis or a combination of the two, conversion of the soluble cornstarch to D-glucose (dextrose) can be achieved on immobilized amyloglucosidase. Pilot-plant experiments at Iowa State University with amyloglucosidase immobilized on porous silica have demonstrated that adequate conversions can be obtained in less than 20 min. Although the conventional soluble-enzyme process requires a residence time of 75 h, the immobilized-amyloglucosidase process is not yet commercially employed.

A second immobilized-enzyme application arises because D-glucose is not an ideal end product. It usually cannot be substituted directly for sugar (sucrose) because glucose is less sweet. Also, glucose crystallization in concentrated solutions can make subsequent handling and processing difficult. These problems are alleviated considerably by isomerizing some of the glucose to fructose, using the enzyme glucose isomerase

D-Glucose   D-Fructose

The equilibrium constant of this reaction at 50°C is approximately unity, and this value does not change greatly with temperature since estimates of the heat of the isomerization reaction are of the order of 1 kcal/mol. Consequently, the equilibrium product contains roughly a 1 : 1 ratio of glucose to fructose. Such a mixture has greater sweetness than glucose alone and is well suited to substitute for sugar in many applications including soft drinks. Frustose-glucose syrups may eventually replace invert sugar, the market volume of which in 1972 was 2 billion pounds.

An intracellular enzyme, glucose isomerase, is produced by a number of microorganisms, *Streptomyces* sp. being among the preferred sources. The need to disrupt cells without destroying the enzyme makes glucose isomerase substantially more costly that, say, the extracellular hydrolases. Also, glucose isomerase is very sensitive to several inhibitors. Both these factors suggest that retention of the enzyme in immobilized form under well-controlled reaction conditions is a desirable strategy. Several forms of immobilized glucose isomerase have been prepared, including enzyme fixed in whole cells, which in turn are held in collagen.

At least one manufacturer of cornstarch-derived sweeteners, Clinton Corn Processing Company, has been using immobilized glucose isomerase for a number of years, and several other companies have licensed the Clinton technology. Their process flowsheet is shown in Fig. 4.27. Many of the separation and treatment needs between the saccharification and isomerization units are dictated by basic enzymology: in order to maximize heat stability of the α-amylase used in the liquefaction step (which typically operates at about 105°C) calcium ions are required. These ions, on the other hand, inhibit glucose isomerase, and they are removed by ion exchange before the dextrose liquor is fed to the isomerization reactor.

Demand for L-amino acids for food and medical applications has been growing rapidly. While microbial processes for L-amino acid have been developed, considerable efforts have also been devoted to a production scheme based on chemical synthesis. Chemically synthesized amino acids suffer the disadvantage of being racemic mixtures. The D isomer in this mixture is generally of no nutritive

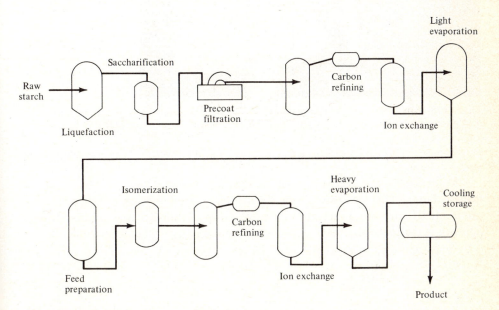

Figure 4.27 Flowsheet summarizing the glucose isomerase process for production of high-fructose corn syrups. (*Redrawn from J. D. Harden, On-Line Control Optimizes Processing, Food Eng.,* **44**(12); 59, 1972.)

Figure 4.28 Immobilized enzyme columns used in the Tanabe Seiyaku Co., Ltd. process for L-amino acid production. (*Photograph courtesy of Dr. I. Chibata, Tanabe Seiyaku Co., Ltd.*)

value: it is desirable to obtain a product consisting strictly of the physiologically active L isomer.

A process achieving this goal is that of Tanabe Seiyaku Co., Ltd., of Osaka, Japan. This was the first publically announced commercial process using immobilized enzymes. Essential to the optical resolution scheme used by Tanabe Seiyaku is the following reaction, which is catalyzed by the enzyme aminoacylase:

$$H_2O + \underset{\substack{\text{DL-Acylamino}\\\text{acid}}}{\text{DL-R}-\text{CH}-\text{COOH}\ |\ \text{NHCOR}'} \xrightarrow{\text{aminoacylase}}$$

$$\underset{\substack{\text{L-Amino}\\\text{acid}}}{\text{L-R}-\text{CH}-\text{COOH}\ |\ \text{NH}_2} + \underset{\substack{\text{D-Acylamino}\\\text{acid}}}{\text{D-R}-\text{CH}-\text{COOH}\ |\ \text{NHCOR}'}$$

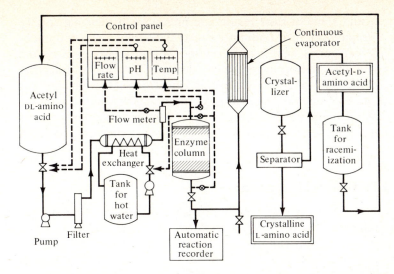

Figure 4.29 Flowsheet for the Tanabe Seiyaku process, which uses immobilized aminoacylase. [*Reprinted by permission from I. Chibata et al., Preparation and Industrial Application of Immobilized Aminoacylases, in G. Terui (ed.), "Fermentation Technology Today," p. 387, Society of Fermentation Technology, Japan, Osaka, Japan, 1972.*]

Following this step, which is carried out in a column reactor containing immobilized aminoacylase (Fig. 4.28), the desired L-amino acid is separated from the unhydrolyzed acyl D-amino acid based on solubility differences. The D-acylamino acid is then racemized to the DL-acylamino acid, which is recycled to the aminocylase column. A flowsheet for the process is shown in Fig. 4.29.

    I. Chibata, T. Tosa, and coworkers at Tanabe Seiyaku spent considerable effort in an attempt to develop an optimal immobilized-enzyme formulation for

Table 4.15. Properties of different forms of aminoacylase (acetyl-DL-methionine as substrate)†

		Immobilized aminoacylases		
Property	Native aminoacylase	Ionic binding (DEAE-Sephadex)	Covalent binding (Iodoacetylcellulose)	Entrapped by polyacrylamide
Optimum pH	7.5–8.0	7.0	7.5–8.5	7.0
Optimum temperature, °C	60	72	55	65
Activation energy, kcal/mol	6.9	7.0	3.9	5.3
Optimum Co^{2+}, mM	0.5	0.5	0.5	0.5
K_m, mM	5.7	8.7	6.7	5.0
v_{max}, μmol/h	1.52	3.33	4.65	2.33
Preparation		Easy	Difficult	Difficult
Binding force		Weak	Strong	Strong
Regeneration		Possible	Impossible	Impossible

    † From T. Mori, T. Sato, T. Tosa, and I. Chibata, Studies on Immobilized Enzymes X: Preparation and Properties of Aminoacylase Entrapped into Acrylamide Gel-Lattice, *Enzymologia*, **43:** 213 (1972).

use in this process, and some of their results are presented in their paper [21]. Table 4.15 gives a summary of their findings for several different forms of immobilized enzymes. The categories in this table remind us that a variety of factors besides initial activity must be considered when selecting a catalyst for industrial use. In the present case, aminoacylase ionically bound to DEAE-Sephadex was chosen because of its high activity, ease of preparation, regeneration capability, and stability. In 1973 Tanabe Seiyaku reported that this formulation had been used for more than 2 yr with no physical decomposition or reduction in binding activity.

4.6.2. Medical and Analytical Applications

The number of known inborn human metabolic disorders now exceeds 120, and many of them are related to the absence of one particular enzyme normally found in the body. For example, phenylketonuria, a disease leading to mental retardation, is thought to be caused by a deficiency in the enzyme which converts phenylalanine into tyrosine. A possible alternative to the current therapy (a phenylalanine-free diet) is to replace the missing enzyme. However, an enzyme with the same function from a nonhuman source cannot be introduced directly into the body because it would trigger an adverse immunological response. A possible approach to this problem involves isolation of the enzyme within a microcapsule, fiber, or gel. In this manner, the enzyme may not cause an adverse response but the small substrate can reach it through the gel, hollow fiber, or microcapsule membrane. While membrane-contained enzymes are not susceptible to antibody attack, protein buildup on the in vivo membrane surfaces adds mass-transfer resistance and causes decreased efficiency of substrate utilization.

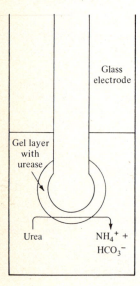

Figure 4.30 A urea electrode constructed by immobilizing urease in a gel attached to a glass electrode.

A variant of this idea has been proposed for construction of a compact artificial kidney. In this device, urease and an adsorbent resin or charcoal are encapsulated together, so that ammonia produced by urea decomposition is adsorbed within the microcapsule:

$$\text{Urea} \xrightarrow[\substack{\text{into} \\ \text{microcapsule}}]{\text{diffusion}} \text{urea} \xrightarrow{\text{urease}} HCO_3^- + NH_4^+ \xrightarrow[\substack{\text{on resin} \\ \text{or charcoal}}]{\text{adsorption}}$$

Among the small-scale trials of immobilized enzymes in reaction processes are steroid (see Sec. 2.1.2) conversions. Cortisol, a useful drug in arthritis treatment, can be made from the cheap precursor 11-deoxycortisol in a column of immobilized 11-β-hydroxylase, following which cortisol can be converted into a superior therapuetic component, prednisolone, using immobilized Δ^1 dehydrogenase in a subsequent packed-bed reactor:

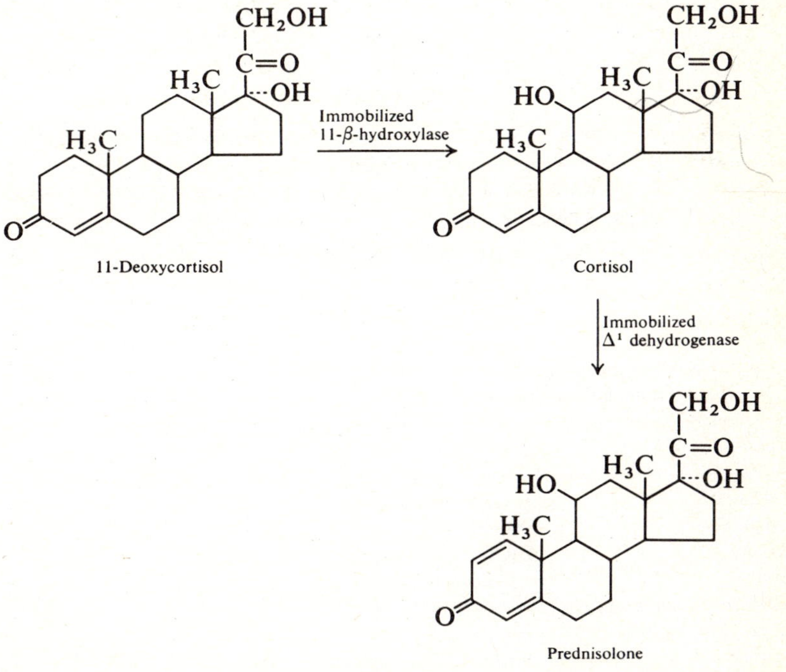

11-Deoxycortisol

Cortisol

Immobilized 11-β-hydroxylase

Immobilized Δ^1 dehydrogenase

Prednisolone

Note the extreme specificity of these enzymes. Most commercial steroid transformations are currently achieved by microbial processes. We shall examine some of them in Chap. 10.

Immobilized enzymes have already made important contributions to analytical biochemistry, and others are certain to follow. One example is the immobilized-enzyme electrode, which permits continuous monitoring of small concentrations of a specific biochemical. In the urea electrode (Fig. 4.30) immobilized urease decomposes urea into ions which can be detected using standard electrochemical techniques. Also, using immobilized-enzyme electrodes, standard biochemical tests can be automated, as in the schematic diagram in Fig. 4.31. Such

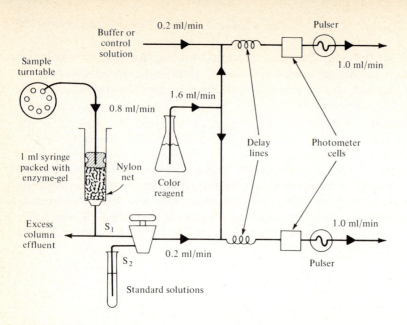

Figure 4.31 Schematic diagram of a system for automated analysis of glucose (using immobilized glucose oxidase) and lactate (using immobilized lactate dehydrogenase). (*Reprinted with permission from G. T. Hicks and S. J. Updike, The Preparation and Characterization of Lyophilized Polyacrylamide Enzyme Gels for Chemical Analysis, Anal. Chem.*, **38**; 726, 1966. *Copyright by the American Chemical Society.*)

Table 4.16. Selected compounds whose concentrations can be monitored using immobilized-enzyme electrodes

Compound	Compound
Acetaldehyde	D-Galactose
Acetylcholine	D-Glucose
D-Alanine	D-Glutamate
L-Alanine	L-Gulono-λ-lactone
Aliphatic nitro compounds	Hypoxanthine
Alkaline phosphatase	D-Lactose
L-Arginine	Lactate dehydrogenase
D-Aspartate	L-Lactose
Benzaldehyde	NADH
Cholinesterase	Penicillin
Creatine	Some pesticides
Creatine phosphokinase	L-Phenylalanine
L-Cysteine	Phosphate
Dehydrogenases	Sulfate
Diamines	L-Tryptophan
2-Deoxy-D-glucose	L-Tyrosine
Ethanol	Urea
Formaldehyde	Uric acid
L-Galactonolactone	

a system can be used, for example, to determine glucose or lactate levels by employing immobilized glucose oxidase and lactate dehydrogenase, respectively.

Based on similar strategies, immobilized-enzyme electrodes have now been constructed for a wide variety of biologically important compounds (Table 4.16). Also, a solid-surface fluorometric method has recently been developed for observation of enzyme reactions in membranes. With this approach, direct assay of the concentration of many enzymes, substrates, and cofactors is possible. A wealth of additional material on this subject will be found in the references.

Immobilization of biochemicals for affinity chromatography has already been discussed as a protein-isolation tool. This principle—immobilization of one species which has an extremely high affinity for the material to be removed from solution—can readily be extended to permit purification or analysis of enzyme inhibitors, cofactors, antigens, antibodies, and other substances.

4.6.3. Utilization and Regeneration of Cofactors

As just mentioned, immobilized cofactors such as AMP, cyclic AMP, and NAD can be used in affinity-chromatography columns to isolate enzymes requiring these cofactors. This complementary biochemical interaction can also be used in reverse: the immobilized cofactor-requiring enzymes can be used to assist in isolating their respective cofactors.

In addition to these separations applications, cofactors are also of interest in the capacity of functioning coenzymes. Only two of the six general classes of enzymes are catalytically active without cofactors. If the other four groups of enzymes are to enjoy industrial use, efficient large-scale methods must be developed for production, separation, and isolation of cofactors. Moreover, special reactor and catalyst designs are required for effective coenzyme utilization and regeneration. Many of these problems are not satisfactorily resolved at present. However, the potential commercial and scientific rewards of a well-developed cofactor technology are sufficiently large for considerable research in this area to be currently in progress. Here we shall touch upon some of the concepts for enzyme-coenzyme reactors: additional information on this and other aspects of cofactor applications is available in the references.

By retaining both enzyme and coenzyme in hollow fibers which substrate and product can permeate, the desired reaction can be carried out without loss of enzyme or coenzyme. Fig. 4.32 shows schematically a laboratory reactor design based on this concept. One reaction system which has been studied in this apparatus is oxidation of ethanol to acetate in a two-step sequence of reactions which require NAD:

$$\text{Ethanol} + \text{NAD} \xrightarrow[\text{dehydrogenase}]{\text{alcohol}} \text{acetaldehyde} + \text{NADH}_2$$

$$\text{Acetaldehyde} + \text{NAD} \xrightarrow[\text{dehydrogenase}]{\text{aldehyde}} \text{acetate} + \text{NADH}_2$$

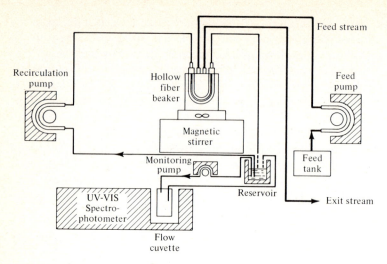

Figure 4.32 A hollow fiber system permitting continuous operation of an enzyme-cofactor reactor.

Clearly these reactions will take place continuously only if $NADH_2$ is continuously oxidized, a process which has been accomplished by adding the enzyme pig-heart diaphorase to the catalytic mixture inside the hollow fibers. With this enzyme, O_2 serves as the $NADH_2$ electron acceptor, and H_2O_2 is formed along with NAD as the coenzyme regeneration reaction occurs. Since hydrogen peroxide causes inactivation of many enzymes, catalase is also included in the enzyme-coenzyme solution. Extension of this scheme to other reactions seems straightforward, although it does not appear adaptable to cases where the substrate molecular weight approaches those of the coenzyme(s) and enzymes involved (why?).

Let us consider next possible enzyme-coenzyme reaction schemes where enzymes and/or their respective coenzymes are bound to insoluble supports. Any such formulation must allow reversible physical contact of enzyme and coenzyme and must provide opportunities for cofactor regeneration. At least three different approaches to this problem can be identified:

1. Immobilize the coenzyme and put enzyme and all necessary substances for regeneration in solution. All participants in the catalytic system except coenzyme will then be contained in the reactor effluent.
2. A complementary idea to 1: the enzyme is immobilized with cofactor in solution and thus in the effluent. Enzymes and additional substrates required for coenzyme regeneration may be passed through this reactor in solution or contained in a separate regeneration reactor, which may itself employ immobilized enzymes.
3. Link the enzyme and coenzyme to each other using a long, flexible molecule as a connection. Such a coenzyme-on-a-string structure is then immobilized.

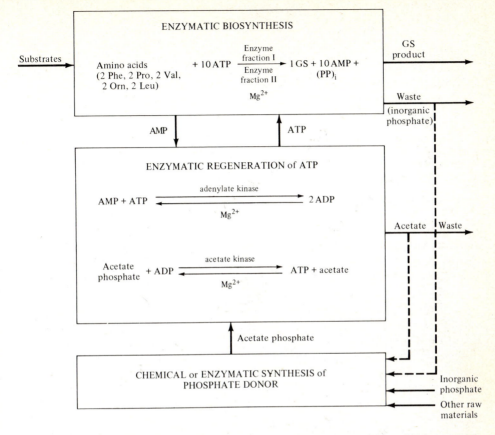

Figure 4.33 Schematic diagram of an enzymatic process for *in vitro* synthesis of the antibiotic gramicidin S. An essential feature of the process is regeneration of ATP.

Concepts 1 and 2 have been demonstrated on a laboratory scale, while the feasibility of approach 3 is still under study.

Choice between schemes 1 and 2 for a possible large-scale process depends on a variety of technological and economic factors. Critical among these is the ease, efficiency, and expense with which enzyme, coenzyme, and/or regeneration systems can be removed from the product stream and recycled into the reactor. If recovery of none of these species is feasible, for example, concept 1 will have an advantage since most coenzymes are much more expensive than any other part of the catalytic system.

The second design concept underlies one approach to large-scale, cell-free enzymatic synthesis. In this process, hydrolysis of ATP to AMP provides the energy necessary for the synthesis reaction, which involves peptide-bond formation. The desired product of this particular study is gramicidin S, a cyclic decapeptide antibiotic containing two identical pentapeptides

$$
\begin{array}{ccc}
& \text{Leu} & \\
\text{Phe} & & \text{Phe} \\
\text{Val} & & \text{Pro} \\
\text{Pro} & & \text{Val} \\
\text{Phe} & & \text{Orn} \\
& \text{Leu} &
\end{array}
$$

Gramicidin S

Figure 4.33 gives an overall view of the synthesis process. This system can be studied in further detail in Ref. 7.

4.7. THE SCALE OF ENZYME TECHNOLOGY

In our discussion of enzyme-purification procedures, we mentioned that the amounts of enzymes produced and used in today's technology are very small relative even to fine chemicals. Here we shall briefly examine the economic magnitude of the enzyme industry.

Table 4.17 gives some of the results of an enzyme market study completed in 1972. Projected as well as then current data are shown. The authors of that report concluded that the markets for glucose oxidase, glucose isomerase, mibrobial rennin, and immobilized enzymes in general had the greatest growth possibilities if then current development programs were successful. In addition, several enzymes including lactase, isoamylase, lipases, nonspecific proteases, and flavor enzymes in the early stages of development in 1972 were found to have relatively significant market potential. For example, the potential United States market for lactase alone was estimated at 11 million dollars per year.

While the figures given above may become rapidly outdated, they show that the size of the market for isolated enzymes was relatively small in 1972. For example, the total United States enzyme market in 1971 was estimated at about 35 million dollars. In comparison, the total chemical sales of Du Pont in 1972 were 3.55 billion dollars, and total United States 1971 drug sales approached 5 billion dollars. Nevertheless, enzyme catalysts serve many unique needs, and their economic impact is expected to increase.

Table 4.17. Present and projected markets for enzymes†

	Millions of dollars		
	1971	1975	1980
Amylases:			
Amylglucosidase	1.70	2.10	2.60
Fungal amylase	2.01	2.40	3.10
Bacterial (α-amylase)	4.60	8.00	8.50
Total	8.31	12.50	14.20

(*continued*)

	Millions of dollars		
	1971	1975	1980
Proteases:			
Fungal	0.76	0.82	0.91
Bacterial	0.83	1.00	1.30
Detergent	1.00	2.00	4.70
Pancreatin	0.80	0.75	0.75
Rennins:			
Animal plus			
microbial	7.50	8.10	8.90
Pepsin	2.75	2.80	2.85
Papain	3.58	3.87	4.27
Total	17.22	19.34	23.68
Other enzymes:			
Glucose oxidase	0.35	0.60	0.90
Cellulase	0.10	0.15	0.20
Invertase	0.10	0.10	0.10
Glucose isomerase	1.00	3.00	6.00
Pectinases	1.56	1.75	2.10
Medical diagnostics,			
research, etc.	5.50	7.30	9.80
Total	8.61	12.90	19.10
Total U.S. markets:			
Amylases	8.31	12.50	14.20
Proteases	18.34	20.77	24.51
Others	8.61	12.15	19.00
Total	35.26	45.42	57.71
World-wide markets			
for selected enzymes:			
Pentinases	2.81	3.12	3.66
Cellulase	1.80	‡	‡
Pancreatin	12.80	‡	‡
Bromelain	0.8	2.4	‡
Papain	6.0	6.49	7.17
Fungal protease	1.37	1.60	1.95
Pepsin	4.30	‡	‡
Bacterial protease	2.83	5.0	10.0
Glucose oxidase	0.62	0.908	1.464
Glucose isomerase	1.00	4.00	8.00
Rennin	23.0	27.53	35.12

† Bernard Wolnak and Associates, Present and Future Technological and Commercial Status of Enzymes, report prepared for the National Science Foundation, *U.S. Dept. Commerce Natl. Tech. Inf. Serv. Doc.* PB-219 636, **December, 1972.**

‡ Not available.

PROBLEMS

4.1. Enzymatic desugaring of egg whites We have 1000 lb of egg white, containing 2.7 percent glucose; it is to be desugared by adding 34,000 units of glucose oxidase, which also contains some catalase. Next 6 lb of 30 wt% hydrogen peroxide is added slowly to the mixture, which is maintained at pH 7 and 24°C. How long will it take to reduce the glucose level to 0.1 percent?

4.2. Particulate substrates A suspension of uniform-sized gelatin (polyglycine) particles is mixed with a proteolytic enzyme powder to give a final mixture of α volume fraction gelatin, and e moles of enzyme per liter of suspension.

(a) If the particle size is initially d_0, derive expressions for the time-varying production rate of the (assumed) sole product glycine and for the time rate of change of particle size. Assume that the rate of reaction is given by the Michaelis-Menten form for $s_0 \ll e_0$, that is,

$$\frac{\text{Rate}}{\text{Volume}} = \frac{ke_0 S_a}{e_0 + K_E}$$

where S_a is the surface area of substrate per unit volume of reactor.

(b) If silver particles in developed photographic film are to be recovered by enzyme digestion of the gelatinous binder, what is the shortest length of contact time between film and enzyme solution needed for total binder hydrolysis?

4.3. Chromatography In chromatography, each solute has an equilibrium partition coefficient $K_i = s_i/c_i$, where s_i is the adsorbed species concentration and c_i the bulk concentration. If a sample of width d at the injection point is to be separated into peaks of at least this peak-to-peak spacing, show that:

(a) The mean residence time of any single component i is given by

$$t_i = \frac{L}{v\varepsilon}(1 - \varepsilon)(1 + K_i)$$

where n is fluid velocity and L the column length.

(b) The required maximum column length L is the maximum value of

$$L = \frac{d(K_j - K_i)}{(1 + K_j)^2(1 + K_i)} \qquad \text{where} \quad K_j > K_i$$

(c) In liquid chromatography, the diffusion coefficient is typically 10^{-6} cm²/s for larger molecules (Table 4.3). If the chromatography column is packed with particles of 10^{-2} cm diameter, use the Einstein equation for the root-square (rms) distance traversed by a molecule in time t, $\langle z^2 \rangle^{1/2} = (Dt)^{1/2}$, to show that flow velocities larger than the order of 10^{-4} cm/s will require a longer column than indicated in part (b).

(d) Charm and coworkers [24] have used an affinity-chromatography column to separate serum-hepatitis antigen from pooled blood plasma. For such specific separation $K_i \to \infty$ (i = antigen), $K_j \approx 0$ ($j \neq i$). If the column is packed with nonporous beads of radius R and the antigen concentration is n particles per liter, show that the maximum number of blood volumes which can be cleared continuously of antigen by a single unit of column volume is given by

$$\frac{3(1 - \varepsilon)}{n\pi r^2 R}$$

where ε is the column void volume and r the antigen radius.

4.4. Chromatography From Fig. 4.13 and a correlation from the data in Table 4.3, deduce the values of L and r_g for the three chromatographic materials of Fig. 4.13.

4.5. Enzyme-isolation efficiency (a) Assuming a 50 percent yield in step 4 of Example 4.1, calculate the fraction of initial enzyme recovered in the procedure.

(b) Assuming that each reaction step in Fig. 4.19 achieves 99 percent completion, calculate the fraction of the first peptide (step 1) which arrives in the final active enzyme product for an enzyme with 100 peptides.

(c) What fractional completion per step is needed in part (b) to reach a final enzyme yield which is 1 percent of the concentration of starting peptide?

(d) How would you rate prospects of enzyme solid-phase synthesis vs. the traditional procedures such as Example 4.1 for providing enzymes in the future?

4.6. Autocatalysis The batch-autocatalytic activation of pepsin [Eq. (4.27)] may be presumed to follow Michaelis-Menten kinetics.

(a) Show that the maximum rate is reached when

$$s = \frac{K_s(s_0 + e_0 - s)}{K_s + s}$$

(b) Show that the course of the reaction in time is

$$t = \frac{1}{k_3}\left[\frac{K}{s_0 + e_0}\ln\left(\frac{e_0}{s_0}\frac{s}{s_0 + e_0 - s}\right) - \ln\left(1 + \frac{s_0 - s}{e_0}\right)\right]$$

(c) Sketch the Lineweaver-Burk or $1/v$ vs. $1/s$ behavior of such an autocatalytic system. What other form of plotting data would be more useful?

4.7. Multiple substrates In industrial enzyme reactors, more than one substrate is often present in the reaction mixture.

(a) Show that for a two-substrate solution, the rate of conversion of S_1 and S_2 is given by

$$v_j = \frac{v_{max, j}}{(1 + K_j/s_j)(1 + s_i/K_i)} \qquad (j, i) = (1, 2), (2, 1)$$

and thus each substrate acts as a competitive inhibitor for the other.

(b) Derive a general form for the total reaction velocity when m substrates are present simultaneously.

4.8. Cellulose-hydrolysis kinetics The enzymes which degrade cellulose include a solubilizing enzyme, an enzyme producing the dimer cellobiose, and the cellobiose hydrolyzer β-glucosidase (Fig. 4.25). As the earlier two enzymes are inhibited by cellobiose (taken to represent all inhibitory oligomers), a simplified reaction network can be written

$$[G_2] + E_1 \xrightarrow{k_1} E_1[G_2]$$

$$E_1 + G_2 \xrightarrow{k_3} E_1 G_2$$

$$E_1[G_2] \longrightarrow E_1 + G_2$$

where $[G_2]$, G_2 are insoluble cellulose and soluble cellobiose and E_1 is indicative of the enzyme involved in the slowest step leading to cellobiose.

(a) In a β-glucosidase-deficient system, glucose production can be ignored. Derive appropriate forms giving the explicit dependence of G_2 upon time for competitive inhibition. Repeat for non-competitive inhibition.

(b) Show graphically that a plot of G_2/t vs. $(1/t)\ln(s_0/G_2)$ should provide a means of distinguishing between uninhibited, competitive, and noncompetitive inhibited systems. Why is a plot of G_2 rather than remaining substrate useful here? (Experimentally, the kinetics is noncompetitive [22]).

4.9. Capillary-driven separations In paper and thin-layer chromatography, the carrier fluid velocity is not constant in time. The pressure differential due to the advancing solvent-solid interface in the paper capillaries is given by $\Delta P = 4\sigma \cos(\theta)/d_c$, where σ is surface tension, d_c is capillary diameter, and θ is fluid-solid contact angle. The viscous resistance to a flow rate $\dot{m}$ is proportional to μz (viscosity times length of liquid-filled portion of capillary) and inversely proportional to the capillary cross section ($\propto d_c^2$).

(a) Assuming $\dot{m}$ proportional to dz/dt, show that z_s^2 and t are proportional, where z_s is the position of solvent front.

(b) Prove that the distance between any two solute peaks also has the same $z - t$ dependence.

(c) To represent actual chromatographic supports more realistically suppose that the normalized fraction of capillaries of diameter d varies about the size d_m of maximum frequency as $N(d) = A \exp -(d - d_m)^2$. Develop an expression giving the relative water content of the capillary material vs. time and distance.

(d) Ignoring other dispersion effects (see Chap. 9), derive an expression for the apparent peak-to-peak distance of any two solutes.

4.10. Enzyme electrode

Well-mixed bulk fluid:

Fluid	enzyme		electrode
film	membrane	$\longleftarrow$	surface
$z = L$		$z = 0$	

A urease enzyme electrode is to be used to detect the presence of urea, e.g., in blood. We consider a one-dimensional model. If the bulk-solution concentration u_b is much less than the Michaelis constant for urea hydrolysis, the local reaction velocity at a point in the enzyme membrane is given by

$$v = \frac{keu}{K_u} = k'eu$$

(a) If e is uniform in the membrane, derive expressions for the concentration profile of both urea and product ammonia when the reaction is $H^+ + 2H_2O + NH_2CONH_2 \rightarrow 2NH_4^+ + HCO_3^-$ in a buffered solution. As the electrode does not consume NH_4^+, the boundary condition at $z = 0$ is $d(NH_4^+)/dz = 0$.

(b) If the membrane effectiveness factor η is defined as

$$\eta = \frac{\text{rate of observed conversion of urea}}{\text{rate if } u = u_b \text{ everywhere in membrane}}$$

show that $1/\eta = \phi + \phi^2/\text{Bi}$, where the Biot number Bi is the ratio of external (film) to internal (membrane) mass transfer

$$\text{Bi} = k_l \frac{L}{\mathscr{D}_u}$$

(c) The electrode response follows the Nernst equation if we use the NH_4^+ level at $z = 0$:

$$\Delta V = V - V_{ref} = \frac{RT}{Fe} \ln \frac{(NH_4^+)_{z=0}}{(NH_4^+)_{ref}}$$

where R = gas constant
T = absolute temperature
F = Faraday's constant = number electron charges per mole
e^- = electron charge

From the solutions to parts (a) and (b), sketch the variation of electrode response ΔV vs. bulk urea concentration for small and large values of ϕ and Bi. In which design regime(s) can the electrode be used to measure u_b?

4.11. High-pressure homogenizer Dunnill and Lilly [23] observed that protein recovery from passage of bakers'-yeast cell suspensions through a large pressure drop confined to a small volume (homogenizer) followed the equation

$$\ln \frac{R_m}{R_m - R} = KN$$

where K = first-order rate constant

N = number of passes through homogenizer

R_m = maximal achievable protein release

R = protein release after N passes

$K = kp^a$, where k, a depend on microbe and p is operating pressure

(a) Show that a first-order rate law gives the expression observed, treating N as a continuous variable.

(b) If the power input per pass W varies linearly with operating pressure p, show that a maximum exists in the curve of Q (percent protein released per kilowatt power input) vs. pressure.

(c) If $a = 2.9$ for bakers' yeast and the maximum in Q is 7.0 at 570 kg_f/cm^2 for $N = 1$, evaluate k and W/P and calculate the profiles of Q vs. p for $N = 1, 2$.

REFERENCES

Most of the references given for Chap. 3 have information pertinent to this chapter. Dixon and Webb (Ref. 2 of Chap. 3) is probably the best single reference for isolation, and Zaborsky (Ref. 4 of Chap. 3) should be consulted for extensive discussions of immobilized enzyme production and uses. Several general sources covering many aspects of enzyme technology are available, including:

1. H. Tauber: "The Chemistry and Technology of Enzymes," John Wiley & Sons, Inc., New York, 1949.
2. L. A. Underkofler, Manufacture and Use of Industrial Microbial Enzymes, *Bioeng. Food Process. CEP Symp. Ser.* No. 69, **62**: 11 (1966).
3. W. T. Faith, C. E. Neubeck, and E. T. Reese: Production and Applications of Enzymes, *Advan. Biochem. Eng.* **1**: 77 (1971).
4. K. Arima: Microbial Enzyme Production, in M. P. Starr (ed.), "Global Impacts of Applied Microbiology," John Wiley & Sons, Inc., New York, 1964.
5. Bernard Wolnak and Associates: The Present and Future Technological Status of Enzymes, *U.S. Dept. Commerce Doc.* PB-219636, *Natl. Tech. Inf. Serv.*, December 1972.
6. L. B. Wingard, Jr. (ed.), "Enzyme Engineering," Interscience Publishers, New York, 1972.
7. E. K. Pye and L. B. Wingard, Jr. (eds.): "Enzyme Engineering," vol. 2, Plenum Press, New York, 1974.

Diverse techniques for concentrations, isolation, and separation of proteins are considered in the following references:

8. William B. Jakoby (ed.): "Methods in Enzymology," vol. 22, "Enzyme Purification and Related Techniques," Academic Press, Inc., New York, 1971. Chapter 37, on scale-up of protein isolation, is especially good.
9. P. Alexander and R. J. Block (eds.): "A Laboratory Manual of Analytical Methods of Protein Chemistry," Pergamon Press, New York, 1961. A multiple-volume edition dealing with laboratory methods in considerable detail.
10. H. A. Sober, R. W. Hartley, Jr., William R. Carroll, and E. A. Peterson: Fractionation of Proteins, p. 1, in H. Neurath (ed.), "The Proteins," 2d ed., vol. 3, Academic Press, Inc., 1965. A concise review of experimental literature to 1964.
11. C. J. van Oss: Separation and Purification of Plasma Proteins: Analytical and Preparative Separation, Purification, and Concentration Methods, p. 187, in E. S. Perry (ed.), "Progress in Separation and Purification," vol. 1, Interscience Publishers, New York, 1968. General review of electrophoresis, membrane, centrifugation, chromatography, solubility, and chemical-specificity separations.
12. T. Gerritsen (ed.): Modern Separation Methods of Macromolecules Ref. 11, vol. 2, p. 1 (1969). A detailed survey of recently available techniques including disc and free-flow electrophoresis, gradient centrifugation and resolubilization methods, partition and countercurrent distribution, and agarose-gel filtration.
13. C. Tanford: "Physical Chemistry of Macromolecules," John Wiley & Son, Inc., New York, 1961. An excellent discussion of the fundamental macromolecular properties upon which all protein separations are based.

14. R. E. Lacey and S. Loeb: "Industrial Processing with Membranes," Wiley-Interscience, New York, 1972. A thorough discussion of theory and applications of reverse osmosis, ultrafiltration, and electromembrane processes.

15. M. Florkin and E. H. Storz: "Comprehensive Biochemistry," vol. 4, "Separation Methods," American Elsevier Publishing Company, Inc., New York, 1962. Detailed surveys for various biochemical separations, predominently by chromatography.

16. F. X. Pollio and R. Kunin, Use of Macroreticular Ion Exchange Resins for the Fractionation and Purification of Enzymes and Related Proteins, *CEP Symp. Ser.*, No. 108, **67**: 66 (1971). Discussion of ionic adsorbents with bimodal (macroreticular) pore-site distributions for protein purification.

In addition to Zaborsky's book and Refs. 6 and 7, the following books concentrate on immobilized enzymes and their applications:

17. G. G. Guilbault: "Enzymatic Methods of Analysis," Pergamon Press, New York, 1970.

18. M. Salmona, C. Saronio, and S. Garatlini (eds.): "Insolubilized Enzymes," Raven Press, New York, 1974.

19. A. C. Olson and C. L. Cooney, "Immobilized Enzymes in Food and Microbial Processes," Plenum Press, New York, 1974.

20. R. A. Messing, ed.: "Immobilized Enzymes for Industrial Reactors," Academic Press, Inc., New York, 1975.

Details on the Tanabe Seiyaku L-amino acid process are given in the following paper and its references:

21. I. Chibata, T. Tosa, T. Sato, T. Mori, and Y. Matsuo: Preparation and Industrial Application of Immobilized Aminoacylases, p. 383 in G. Terui (ed.), "Fermentation Technology Today," Society of Fermentation Technology, Japan, 1972.

Problems

22. J. A. Howell and J. D. Stuck: Kinetics of Solka-floc Cellulose Hydrolysis by *Trichoderma viride* Cellulase, *Biotech. Bioeng.*, **17**: 873, 1975.

23. P. Dunnill and M. Lilly: Protein Extraction and Recovery from Microbial Cells, pp. 179–207 in "Single Cell Protein II" S. Tannenbaum and D. I. C. Wang (eds.), MIT Press, Cambridge, Mass 1975.

24. S. F. Charm and B. L. Wong: An Immunoadsorbent Process for Removing Hepatitis Antigen from Blood and Plasma, *Biotech. Bioeng.*, **16**: 593, 1974.

METABOLIC PATHWAYS AND
ENERGETICS OF THE CELL

In other phases of engineering, we consider transport of momentum, energy, and mass. While momentum transfer is involved in cell motion, the exchanges of material and energy between the cell and its environment are of greater importance for our purposes. We shall discuss these exchanges and their interconnections in some detail. Before proceeding, however, some refocusing of our engineering outlook may be necessary. In most processes we deal with, thermal-energy changes are by far the most important. Consequently, the notion of a *heat balance* is a useful approximation although in principle it is incorrect. When examining biological systems, the proper procedure, which considers total energy, is essential. Heat transfer is relatively unimportant in living cells compared with the transport, storage, and utilization of *chemical energy*. A little reflection on our previous studies reveals why: enzymes, the specific but fragile catalysts essential for operation of living systems, denature outside a narrow temperature range. Therefore, the cell would die or sporulate if its temperature changed significantly.

Consideration of energy flows within the cell serves several purposes. It provides an excellent conceptual framework for the entire subject of *cell metabolism*, which is defined as the sum total of all the chemical reactions which occur in the cell. By analyzing the intricate procedures which living organisms use for transferring chemical energy, we can learn how cells construct and maintain a definite structure. The science of biophysics had to wait for understanding of the principles of cell energetics, since before their discovery, living systems appeared to violate the second law of thermodynamics. A cell produces order (itself and its offspring) from its disorderly surroundings. We now know in some detail how energy from the environment is used to drive this process. Finally, and perhaps of greatest importance in microbial process engineering, the study of energy exchanges helps explain the major distinction between cell function in the presence and absence of oxygen. As we have already commented, these two conditions are called aerobic and anaerobic, respectively. While some cells (*obligate* or *strict anaerobes*) do not use free oxygen and other microorganisms require it (*obligate aerobes*), a third class of cells can grow in either environment. Yeast is a familiar example of this metabolically versatile third group, known collectively as *facultative anaerobes*.

Two different kinds of energy, light and chemical, are tapped by inhabitants of the microbial world. Organisms which rely on light energy are called *phototrophs*, while *chemotrophs* extract energy by breaking down certain nutrients. Further subdivision of the chemotrophs is possible according to the nature of the energy-yielding nutrient. In particular, *lithotrophs* oxidize inorganic material and *organotrophs* employ organic nutrients for energy production. Some examples will be mentioned shortly after we have introduced cellular nutrition.

The energy obtained from the environment is typically stored and shuttled in convenient high-energy intermediates such as ATP (Sec. 2.3.1). The cell uses this energy to perform three types of work: the chemical synthesis of large or complex molecules (growth), the transport of ionic and neutral substances into or out of the cell or its internal organelles, and the mechanical work required for cell division and motion. All these processes are, by themselves, nonspontaneous and result in an *increase* of free energy of the cell. Consequently, they occur only when simultaneously coupled to another process which has a negative free-energy change of greater magnitude. We shall return to this theme in Sec. 5.1.

The chemical work is performed with relatively high efficiency of free-energy utilization, typically greater than about 20 percent. The transport work also involves ATP consumption in a process unique to living systems. Small molecules and ions can be moved through membranes against a concentration gradient to achieve a ratio of concentrations on the two sides of the membrane as great as 10^5. Mechanical work is evident during cell division and bacterial and protozoal movement. Animal muscle activity and sperm swimming also imply ATP participation; the resulting direct conversion of chemical free energy into mechanical work without such intermediates as electricity or heated gases is also unique to life.

Since natural systems contain energy-utilization and chemical-conversion schemes largely without parallel in the man-made sphere of processes, examination of biological systems suggests ideas which may lead to new developments in the synthetic areas of man's endeavor. The probing of natural schemes for energy conversion and information organization in cells in order to improve synthetic counterparts is another application of bionics. As Lehninger suggests, the living cell may be viewed as an ultimate form of the current trends toward miniaturization of electrical and information-storage devices; the cell achieves on a molecular scale what still requires macroscopic elements in man-made systems.

In order to grow and reproduce, microorganisms must ingest the raw materials necessary to manufacture membranes, proteins, walls, chromosomes, and other components. From our review of the atoms and compounds of life, four major

Table 5.1. Typical sources of elements

Element	Source
Carbon	CO_2, sugars, proteins, fats
Nitrogen	Proteins, NH_3, NO_3^-
Sulfur	Proteins, SO_4^{2-}
Phosphorus	PO_4^{3-}

requirements are evident: carbon, nitrogen, sulfur, and phosphorus (remember that water is always available within the cell to supply hydrogen and oxygen). Typical sources of these elements are listed in Table 5.1, and Fig. 5.1 shows a schematic view of their utilization in an imaginary bacterium. This organism would be called a *heterotroph* because its carbon comes from an organic compound. Other microorganisms can use simpler nutrients: *autotrophs* require only CO_2 to supply their carbon needs. By combining these classifications by carbon source with those based on energy source, we can construct appropriate adjectives which are descriptive of the cell's metabolism. Table 5.2 gives this nomenclature and example organisms of each type. The fact that different microorganisms employ different carbon and energy sources shows clearly that all cells do not possess the same internal chemical machinery. While the differences are important and will be discussed further, we shall concentrate in this chapter on aspects of metabolism which are common to many varieties of living cells.

In Fig. 5.1 the reactions within the cell have been subdivided into three classes: degradation of nutrients (class I), biosynthesis of small molecules (class II), and biosynthesis of large macromolecules (class III) from the class II reaction products. The number of different chemical reactions necessary for

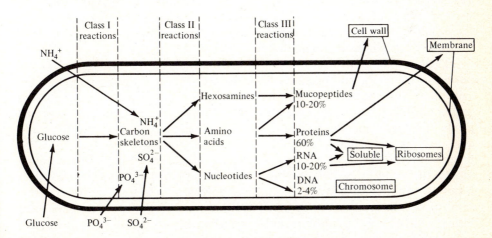

Figure 5.1 Schematic diagram showing synthesis of biological macromolecules from simple nutrients in a bacterium. [*Reprinted by permission from J. Mandelstam and K. McQuillen (eds.), "Biochemistry of Bacterial Growth," 2d ed., p. 4, Blackwell Scientific Publications, Oxford, 1973.*]

Table 5.2. Classification of organisms by carbon and energy source

Carbon source	Energy source	
	Chemical	Light
Organic compounds	Chemoheterotrophs (higher animals, protozoa, fungi, and most bacteria)	Photoheterotrophs (some bacteria, some eucaryotic algae)
CO_2	Chemoautotrophs (some bacteria)	Photoautotrophs (higher plants, eucaryotic algae, blue-green algae, and some bacteria)

sustenance of cell life is of the order of 1000 or more. As we have already commented, each reaction is catalyzed by an enzyme. The enzymes serve the essential function of determining which reactions occur and their relative rates.

In order to appreciate the importance of regulating relative reaction rates, a different view of the cell's chemistry is useful. Many reacting substances within the cell (*metabolites*) can be attacked simultaneously by several different enzymes—perhaps to oxidize them, reduce them, or couple them to other molecules. Thus, the sequences of reactions occurring in the cell intersect and overlap in complex ways, as Fig. 5.2 illustrates. Notice, for example, how pyruvate, the final product of the Embden-Meyerhof (also called Embden-Meyerhof-Parnas or EMP) pathway, can be used in five different pathways. Although the reaction patterns shown in Fig. 5.2 may appear very complex, we should recognize that it is an extremely abbreviated version of the entire story: nowhere near 1000 steps are shown. Also, Fig. 5.2 does not indicate that portions of these metabolic pathways are reversible, so that they help accomplish either synthesis or degradation of a biomolecule. Later in this chapter we shall examine some of these pathways and the enzymatic machinery for controlling them in greater detail.

There is still another serious omission from Fig. 5.1. We have not shown the release of metabolic end products from the cell. Many of these compounds are substances unnecessary or useless for the cell's function; others, e.g., antibiotics and extracellular enzymes, serve a purpose. From man's perspective, these end products (organic and amino acids, antibiotics, and many others) are often valuable, making it profitable to produce them by growing microorganisms. For microbial processes, we usually seek a cellular species which is inefficient from a biological point of view in that it produces much more of some biochemical than it can use. Another objective in this chapter, therefore, is to trace the interconversions of various chemical compounds within the cell so that we shall have some idea of what we may be able to manufacture using microorganisms. While we shall only scratch the surface of these topics here and in Chaps. 10 and 12, we should be able to gain an appreciation of the tremendous complexity and variety of reaction processes which abound in the microbial world.

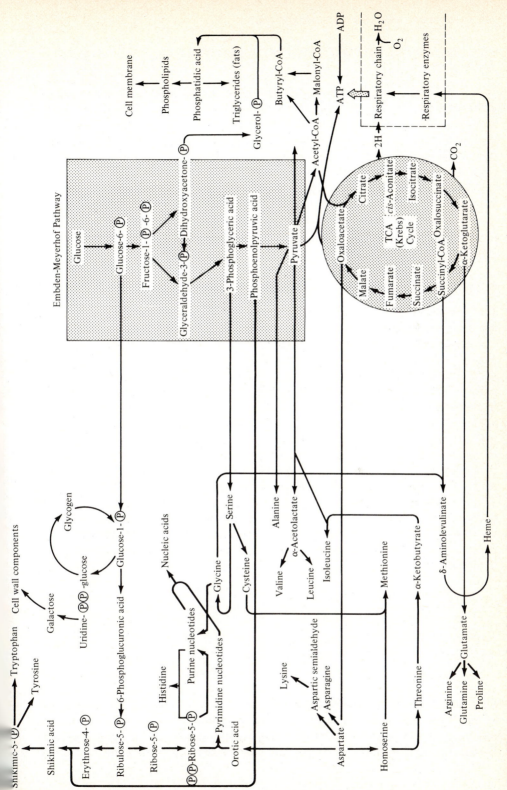

Figure 5.2 A summary of a few of the major metabolic pathways in the bacterium *E. Coli.* (*Reprinted by permission from J. D. Watson, "Molecular Biology of the Gene," 2d ed., pp. 96–97, W. A. Benjamin, Inc., New York, 1970.*)

5.1. THE CONCEPT OF ENERGY COUPLING: ATP AND NAD

In Chaps. 3 and 4 the kinetics of isolated enzyme systems and the industrial utilization of free enzymes and enzyme preparations have been considered. The majority of the kinetic studies mentioned in Chap. 3 and of the current important usages cited in Chap. 4 involve hydrolytic enzymes, which, by definition, split or degrade larger molecules into smaller components by consuming water as a second substrate. Such degradations proceed with a decrease in free energy of the system and an increase in entropy and are therefore spontaneous in a closed system. The present section introduces the mechanisms by which an open system, the living cell, is able to couple energy-yielding (*exergonic*) reactions with chemical reactions and other functions which do not occur unless energy is supplied (*endergonic* processes).

5.1.1. Open Systems and Some Thermodynamics

Living systems continuously exchange mass with their surroundings and therefore are *open systems* in the usual thermodynamic sense. An open system typically does not achieve an equilibrium with its surroundings but instead strives for a *steady-state* existence. Use of irreversible thermodynamics and the concept of the steady state should, theoretically, replace use of equilibrium thermodynamics and the concept of equilibrium for a true description of such systems. For example, using irreversible thermodynamics, we can show that the steady state is characterized by achievement of the *minimal possible rate of entropy production* for a given nutrient-utilization rate. However, the equilibrium properties of chemical systems are more easily measured and quantified than the fundamental variables of irreversible thermodynamics. The following discussion will therefore use equilibrium free-energy data in considering the efficiency of biochemical steady-state processes. This approach, and much of this chapter's presentation, is inspired by Lehninger's book "Bioenergetics" [1], which should be consulted for additional details.

The free-energy change of a chemical reaction represents the maximal possible work performable by that reaction under the condition of constant temperature

$$A \rightleftharpoons B \tag{5.1}$$

$$\Delta G = G_B - G_A = W_{max} \tag{5.2}$$

Also, by the well-known relationship

$$\Delta G = -RT \ln K_{eq} \qquad \text{where } K_{eq} = \frac{b_{eq}}{a_{eq}} \tag{5.3}$$

the free-energy change is also related to the concentration of A and B at equilibrium. If the free-energy change is negative, the reaction can *do* work. In this case the reaction will proceed (from left to right) spontaneously in a closed system. However, only the input of energy can cause a reaction with a positive ΔG to occur for a closed system. As just noted, we shall routinely assume that the same considerations are good approximations for the open system of the living cell. Another approximation will greatly simplify our discussion without interfering with our understanding of the principles of cell energetics. The standard free-energy change $\Delta G°$ used here applies to a standard situation at 25°C, pH 7, and all species at 1.0 M concentration. While biological systems usually operate near the standard temperature of 25°C, the pH may not be 7 and concentrations are typically much smaller than 1.0 M. We shall consider corrections for these deviations from the standard state in problems, but we shall liberally use standard-free-energy data in the following discussions. Such data are readily available from thermodynamic tables.

In coupled-reaction nets of the kind indicated in Fig. 5.2, the direction of a major reaction path in cell metabolism is often not properly indicated by examining an isolated reaction. For example, consider the reaction between two isomers in the Embden-Meyerhof pathway for carbohydrate breakdown

$$
\begin{array}{lll}
\begin{array}{l}
\text{CHO} \\
| \\
\text{CHOH} \\
| \\
\text{CH}_2\text{O}-\text{\textcircled{P}}
\end{array}
&\longrightarrow
\begin{array}{l}
\text{CHO} \\
| \\
\text{C}=\text{O} \\
| \\
\text{CH}_2\text{O}-\text{\textcircled{P}}
\end{array}
& \Delta G° = -1830 \text{ cal/mol} \qquad (5.4)
\end{array}
$$

3-Phosphoglyceraldehyde   Dihydroxyacetone—$\text{\textcircled{P}}$

where $\text{\textcircled{P}}$ denotes phosphate. Because of the negative free-energy change, equilibrium favors the dihydroxyacetone by a 22 : 1 ratio. However, as Fig. 5.2 indicates, when this reaction occurs *within* the glycolysis sequence, glyceraldehyde phosphate is continually removed by reactions leading ultimately to pyruvate. As the glyceraldehyde phosphate is tapped off, reaction (5.4) is forced to proceed from right to left in an attempt to maintain equilibrium.

In the next section, we shall consider possible coupling between several reactions. While we still cannot analyze the thermodynamics of the entire system of cellular chemical reactions, examination of small sets of linked reactions will reveal the basic principles of energy transport in the cell's chemical engine.

5.1.2. ATP and Other Phosphate Compounds

The structure and some properties of adenosine triphosphate (ATP) have already been mentioned (see Sec. 2.3.1 and Fig. 2.7). As noted there, the enzymatic hydrolysis of ATP to yield ADP and inorganic phosphate has a large negative free-energy change

$$\text{ATP} + \text{H}_2\text{O} \longrightarrow \text{ADP} + \text{P}_i \qquad \Delta G° = -7.3 \text{ kcal/mol} \qquad (5.5)$$

where P_i denotes inorganic phosphate. Thus, a substantial amount of free energy may be released by the hydrolysis, and, by reversing the reaction and adding phosphate to ADP, energy can be stored for later use. Let us next examine how the latter situation can occur using coupled chemical reactions.

Another example from the Embden-Meyerhof glycolytic pathway serves to illustrate the concept of a *common chemical intermediate*. Conversion of an aldehyde in an aqueous medium to a carboxylic acid results in a free-energy decrease of about 7000 cal/mol, as summarized in Fig. 5.3. In an isolated solution, this chemical free energy would be completely dissipated. This does not occur in the living cell. In biochemical glucose oxidation, when 3-phosphoglyceraldehyde is converted into a carboxylic acid (3-phosphoglycerate), ATP is simultaneously regenerated from ADP (see reaction 2 of Fig. 5.3). The free-energy decrease resulting from aldehyde oxidation is coupled by the cell enzymes to the simultaneous regeneration of ATP. Since reaction 2 results in approximately no free-energy change, the free energy released in oxidation of 3-phosphoglyceraldehyde has been transformed into a so-called *high-energy phosphate bond* in adenosine triphosphate.

The elementary reaction sequence by which the conversions actually occur is

1 Isolated oxidation of aldehyde to carboxylic acid (aqueous solution),

$$RCHO + H_2O \longrightarrow 2H + RCOO^- + H^+, \qquad \Delta G_1{}^0 \simeq -7 \text{ kcal/mol}$$

2 Same reactions, coupled to ATP generation (glucose oxidation),

$$RCHO + HPO_4^- + ADP^{3-} \longrightarrow 2H + RCOO^- + ATP^{4-},$$

$$\Delta G_2{}^0 \simeq 0 \text{ kcal/mol}$$

3 Evidently, *2 − 1* yields,

$$ADP^{3-} + HPO_4{}^- + H^+ \longrightarrow ATP^{4-} + H_2O, \qquad \Delta G_3 \simeq +7 \text{ kcal/mol}$$

The elementary reactions occurring in *2* are,

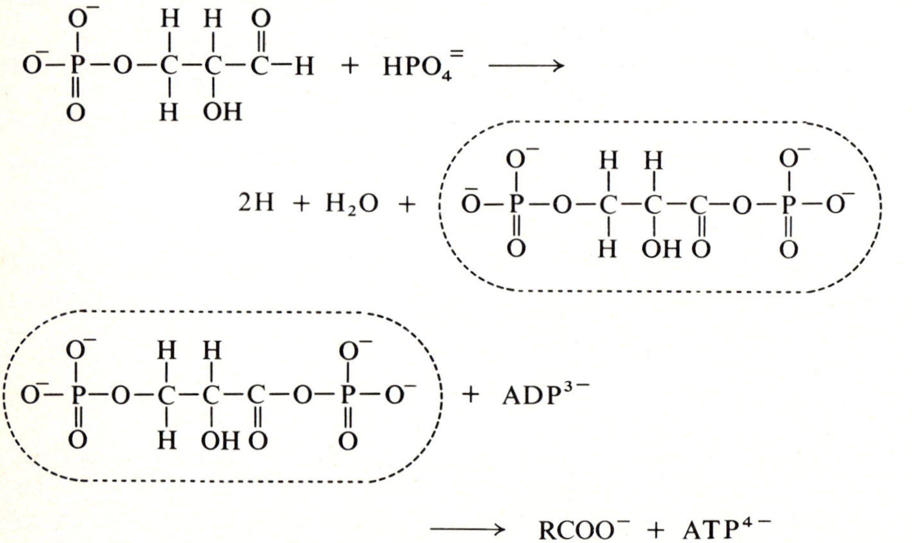

Figure 5.3 Example of reactions coupled via a common intermediate.

shown at the bottom of Fig. 5.3 in reactions 4 and 5. The central important features of these last two reactions are: (1) the appearance of a common intermediate; the same compound is a product of the first reaction and a reactant in the second and (2) the fact that the phosphorylated intermediate formed and consumed has a larger free energy of hydrolysis (with phosphate removal) than ATP. The equilibrium of reaction 5 lies to the right, or product, side: thus as the cell consumes ATP and produces ADP in biosynthetic reactions and other endergonic processes, this part of glucose metabolism is one of several points at which the cell regenerates the needed ATP. This regeneration is accomplished by the conversion of a partially metabolized nutrient into a high-energy phosphorylated intermediate, which then donates a phosphate to ADP via an enzyme-catalyzed reaction.

The phosphorylation of various compounds, including ATP, serves several functions. It provides a useful means of storing considerable fractions of the free energy of fuel oxidation. Since the free energies of hydrolysis of these phosphorylated compounds are all different (as seen in Fig. 5.4), the phosphorylated

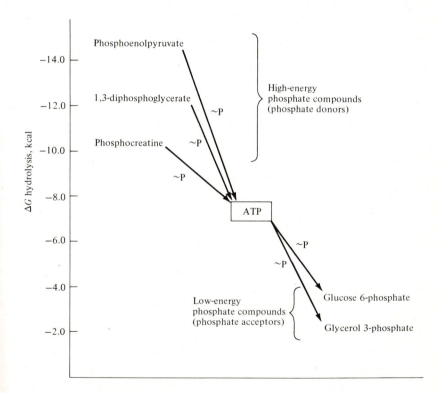

Figure 5.4 The free energy of ATP hydrolysis lies between those for the groups of high- and low-energy phosphate compounds. (*Reprinted by permission from A. Lehninger, "Bioenergetics," 2d ed., p. 45, W. A. Benjamin, Inc., Palo Alto, Ca., 1974.*)

derivatives provide a means of either regenerating or consuming the chemical free energy of the phosphate-ADP bond in ATP. Thus, hydrolysis of some of those phosphate compounds with a large free-energy change is used in ATP formation; at other reaction locations, ATP is consumed, e.g., in order to activate glucose by converting it to glucose 6-phosphate.

Yet another function is served by phosphorylation. Highly ionized organic substances are virtually unable to permeate the cell's plasma membranes. The charged phosphorylated compounds which serve as metabolic intermediates may therefore be contained within the cell. In this manner, the maximum amounts of energy and chemical raw materials can be extracted from a nutrient. Typically, the last reaction in production of a metabolic end product is a dephosphorylation. The uncharged waste material can then escape from the cell's interior.

5.1.3. Oxidation and Reduction: Coupling via NAD

We have just reviewed the role of ATP as a shuttle for phosphate groups bound with rather high energies. In this section we shall consider briefly how oxidation and reduction reactions are conducted biologically and introduce the connection between these mechanisms and ATP metabolism.

To begin with, we should recall that oxidation of a compound means that it loses electrons and that addition of electrons is reduction of a compound. When an organic compound is oxidized biochemically, it usually loses electrons in the form of hydrogen atoms: consequently, oxidation is synonymous with *dehydrogenation*. Similarly, *hydrogenation* is the usual way of adding electrons, or reducing a compound, e.g., the reduction of pyruvic acid and oxidation of lactic acid

$$
\begin{array}{ccc}
CH_3 & & CH_3 \\
| & +2H \ (\text{reduction of pyruvic acid}) & | \\
C{=}O & \longrightleftharpoons & CHOH \\
| & -2H \ (\text{oxidation of lactic acid}) & | \\
COOH & & COOH \\
\text{Pyruvic acid} & & \text{Lactic acid}
\end{array}
$$

(Pyruvic acid, as just mentioned, is really the same compound as pyruvate in Fig. 5.2. Pyruvate refers to the ionized form CH_3COCOO^-, which predominates at biological pH.)

Pairs of hydrogen atoms freed during oxidations or required in reductions are carried by nucleotide derivatives, especially nicotinamide adenine dinucleotide (NAD) (see Fig. 2.8) and its phosphorylated form NADP. These compounds were classified as coenzymes earlier since they usually must be present when an oxidation or reduction is conducted. When hydrogen atoms are needed, for example, the nicotinamide group of reduced NAD can contribute them by undergoing the oxidation

Reduced form   Oxidized form

This oxidation is readily reversible, so that NAD can also accept electrons (H atoms) when they are made available by oxidation of other compounds. For convenience in writing oxidation-reduction reaction stoichiometry, we shall denote the oxidized and reduced forms of NAD as NAD and $NADH_2$, respectively. The actual molecular species present are NAD^+ and NADH, a notation found in many biochemistry texts.

In its role as electron shuttle, NAD serves two major functions. The first is analogous to one of ATP's jobs: *reducing power* (= electrons ≈ H atoms) made available during breakdown of nutrients is carried to biosynthetic reactions. Such a transfer of reducing power is often necessary because the oxidation state of the nutrients to be used for constructing cell components is different from the oxidation state of biosynthesis products. As already observed, the oxidation state of carbon within the cell is approximately the same as carbohydrate (CH_2O). Thus, autotrophic organisms, which employ CO_2 as their carbon source, use considerable reducing power when assimilating carbon

$$CO_2 + 4H \longrightarrow (CH_2O) + H_2O$$

While carbon dominates reducing-power requirements, nitrogen and sulfur assimilation often also demand some of the cell's carrier-bound hydrogen. To estimate these needs, we may assume that cell material contains nitrogen at the oxidation level of ammonia (NH_3), while sulfur's oxidation state is approximately that of sulfide (S^{2-}). For example, use of sulfate as a sulfur source requires considerable reducing power, as suggested by

$$SO_4^{2-} + 8H \longrightarrow S^{2-} + 4H_2O$$

NAD and related pyridine nucleotide compounds carrying hydrogen also participate in ATP formation in aerobic metabolism. As we shall investigate in greater detail in Sec. 5.3, the hydrogen atoms in $NADH_2$ are combined with oxygen in a cascade of reactions known as the *respiratory chain*. The energy released in this oxidation is sufficient to form three molecules of ATP from ADP.

It is an intriguing fact that all biological systems known, whether anaerobic, aerobic, or photosynthetic, utilize ATP as a central means of accumulating oxidative or radiant energy in a form convenient for driving the endergonic processes of the cell. The remainder of the chapter will follow ATP utilization and electron transfer through progressively more complex pathways of anaerobic, aerobic, and photosynthetic systems. Consideration of oxidation-reduction balances will allow us to anticipate connections between nutrient composition, metabolic pathways, and end products. Since the free energy of fuel oxidation is eventually stored in ATP, the efficiency of ATP utilization in different cell processes gives a useful reflection of relative energy demands by the various cell functions. Further, study of ATP logically allows accounting for both the progress of carbon skeletons and the utilization of the free energy released during their formation. Finally, comparison of the energetics of these different biological systems gives insights into the origin of the relatively high thermodynamic efficiency of these chemical engines.

5.2. ANAEROBIC METABOLISM: FERMENTATION

Breakdown of nutrients to obtain energy is called *catabolism*. Among the many catabolic reaction pathways which operate in nature, the simplest occur under *anaerobic* (oxygen-free) conditions. Such conditions may predominate in soil, mud, and deep seas and at the inner portions of a dense mycelium or a deep wound. The microorganisms known as strict anaerobes require such conditions: for them oxygen is poisonous. Included among the strict anaerobes are several microbes which cause disease (*pathogens*). For example, the anaerobic bacterium *Clostridium tetani*, responsible for tetanus, or lockjaw, is widespread in nature in the form of spores. Consequently, when a deep wound introduces the spores into the body and also cuts off blood circulation, the anaerobic conditions necessary for germination of the spores and growth of vegetative cells are created.

The biological reaction mechanisms for extracting energy under anaerobic conditions are called *fermentations*, following Pasteur, who called fermentation *life without air*. Since most early microbiological processes such as wine making were anaerobic, the term fermentation was also used to describe these processes. Unfortunately for the novice in biotechnology, this practice has persisted, and microbial conversion processes are still usually called fermentations in industry even though many of these processes are aerobic. Since biochemists reserve the term for anaerobic operation, the exact meaning of fermentation must often be inferred from the context. In this chapter we shall adopt the biochemist's language, but when we delve into microbial technology in later chapters, we shall follow convention and use the word fermentation in the usual industrial sense.

Carbohydrates are by far the most important class of nutrients for fermentations, although some microbial species can also ferment amino acids. As an illustration of the diversity of the microbial world, we note that almost any carbohydrate or related compound is fermented by some microorganism. Most microorganisms which can employ carbohydrate in a fermentation are capable of fermenting the simple sugar glucose. In the following section we shall examine the major processes for glucose breakdown. There are at least seven different glucose fermentation pathways, and the particular one used and the end products produced depend on the microorganism involved.

5.2.1. Embden-Meyerhof Glycolysis

The Embden-Meyerhof pathway already seen in Fig. 5.2 dominates the glucose fermentation called *glycolysis*, in which the nutrient glucose is broken down into two molecules of end product, lactic acid (lactate). In this section we examine this best-known fermentation pathway with emphasis on energetics and reducing power transfer.

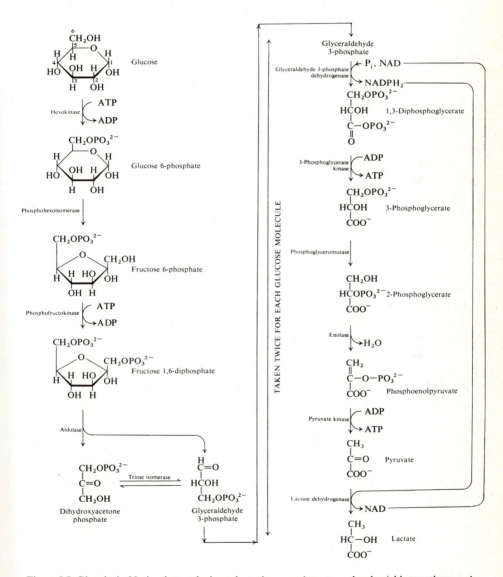

Figure 5.5 Glycolysis. Notice that each six-carbon glucose substrate molecule yields two three-carbon intermediates, each of which passes through the reaction sequence on the right-hand side.

Figure 5.5 gives all the steps in Embden-Meyerhof glycolysis, including the enzymes involved. All these enzymes are found free in the cell's cytoplasm. As in most reactions in class I of Fig. 5.1, each step is quite simple and involves isomerization, ring splitting, or transfer of a small group such as hydrogen or phosphate. The overall process can be divided into two parts which are characteristic of all fermentations of glucose. First the glucose chain is split and oxidized by removal of hydrogen atoms. In the reductive portion of the fermentation which follows, the oxidized products from the first stage are reduced using the hydrogen atoms which were extracted during the first stage. Figure 5.5 shows the role of NAD as a shuttle connecting these two processes. As in any fermentation, all the hydrogen atoms produced are used again within the fermentation pathway. Consequently, no net reducing power is made available to the cell. This has an important bearing on anaerobic end products, as we shall see shortly.

Considering now the participation of ATP in glycolysis, we see that two molecules of ATP are converted to ADP during the first part of the fermentation. However, at two points during the second portions (which is followed by both three-carbon species resulting from one glucose molecule), ATP is regenerated from ADP. The net result is production of two molecules of ATP from each glucose molecule so that the overall reaction for Embden-Meyerhof glycolysis is

$$\text{Glucose} + 2P_i + 2\text{ADP} \longrightarrow 2 \text{ lactate} + 2\text{ATP}$$

$$\Delta G° = -32,400 \text{ cal/mol} \tag{5.6}$$

The ATP produced in this anaerobic pathway requires 2 moles of free phosphate nutrient per mole of glucose metabolized. The source of this phosphate may be external to the cell or may derive from later reactions which consume ATP and generate monophosphate.

Comparing the free-energy change for reaction (5.6) with the corresponding quantity for glucose breakdown alone

$$\text{Glucose} \longrightarrow 2 \text{ lactic acid} \qquad \Delta G° = -47,000 \text{ cal/mol} \tag{5.7}$$

shows that a free-energy total of 14.6 kcal, or 7.3 kcal for each mole of ATP generated, has been conserved by the pathway as high-energy phosphate compounds. The apparent efficiency of free-energy transfer is thus about $\frac{14}{47} \times 100 \approx$ 31 percent. Correction of standard-free-energy data for the concentrations and pH prevailing *in vivo* (see Table 5.3) suggests that this estimate is quite low and that the true efficiency is about 53 percent. The reason for this high efficiency is apparent from the free-energy data given in Fig. 5.6: almost all the reactions in the glycolysis sequence are readily reversible.

Glycolysis is sometimes called the *homolactic* fermentation because lactic acid is the only end product. Its occurrence in muscle during exercise (which creates anaerobic conditions locally) causes pain and muscle fatigue as lactic acid accumulates. Some species of the *lactic acid bacteria*, a group of significant economic importance, also conduct the homolactic fermentation. By virtue of their ability to produce lactic acid and survive in its presence, these bacteria can eliminate other microorganisms which cannot tolerate low pH. The lactic acid bacteria are found

Table 5.3. In the human erythrocyte, the concentrations of glycolysis intermediates differ significantly from the 1 M value used in standard-free energy calculations†

Intermediate	Concentration, μM
Glucose	5000
Glucose 6-phosphate(G6P)	83
Fructose 6-phosphate(F6P)	14
Fructose 1,6-diphosphate(FDP)	31
Dihydroxyacetone phosphate(DHP)	138
Glyceraldehyde 3-phosphate(GAP)	18.5
3-Phosphoglycerate (3PG)	118
2-Phosphoglycerate (2PG)	29.5
Phosphoenolpyruvate (PEP)	23
Pyruvate (Pyr)	51
Lactate (Lact)	2900
ATP	1850
ADP	138
Phosphate	1000

† A. L. Lehninger, "Biochemistry," 2d ed., table 16.1, Worth Publishers, Inc., New York, 1975.

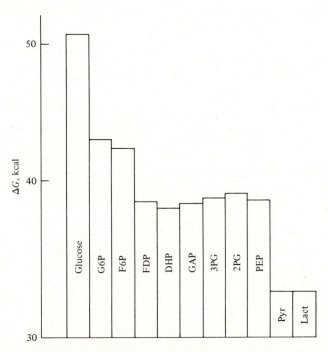

Figure 5.6 Free energies of glycolysis. Three steps involve relatively large free energy changes while the remainder are near equilibrium. Abbreviations for glycolysis intermediates are defined in Table 5.3. (*Reprinted by permission from A. L. Lehninger, "Biochemistry," p. 327, Worth Publishers, New York, 1970.*)

in and on plants, animals, and milk. In each habitat, activities beneficial or detrimental to people can occur. For example, milk souring is caused by certain streptococci, while other members of the lactic acid bacteria family are important in the production of cheese, yogurt, and other fermented-milk products. Lactic acid bacteria are found in plant products such as pickles, sauerkraut, and beer and wine. In both these fermented beverages the growth of these bacteria sometimes produces an undesired acidity.

Products other than lactic acid result from other fermentation pathways present in numerous microorganisms. Some of the possibilities will be explored in the following section and in Chap. 10.

5.2.2. Other Fermentation Pathways and Their End Products

A mass balance between nutrients and the resulting metabolic end products will be helpful in our survey of other fermentations: the molar ratios of total carbon, hydrogen, and oxygen summed over all products must be equal to the corresponding quantities for the substrate. For example, the alcohol fermentation produces ethanol (C_2H_5OH) and CO_2 in a 1:1 ratio from glucose ($C_6H_{12}O_6$). A strict oxidation-reduction balance applies for fermentations because they generate no net change in the cell's reducing power. We should *not* conclude from this, however, that a strict stoichiometric relationship applies between the *amount* of consumed nutrient (say glucose) and the *amounts* of end products formed by fermentation since some of the C, H, O provided by the glucose nutrient will be used in synthesis of new cellular material (recall Probs. 2.7 and 2.8).

Many types of glucose fermentation rely on the Embden-Meyerhof pathway; it is defined as the reaction sequence in Figs. 5.2 and 5.5 which leads from glucose to pyruvic acid (pyruvate). Consequently, these fermentations differ from homolactic fermentation only in the ultimate fate of pyruvate. For example, Fig. 5.7 compares the *alcoholic fermentation* with the homolactic one. By far the most economically important fermentation, the alcoholic fermentation yields ethanol and carbon dioxide from glucose. It is carried out in some plant tissues and by yeasts and other fungi.

Some other end products obtained through the Embden-Meyerhof pathways are given in Table 5.4, along with the different bacteria which produce them. Since meeting the cell's energy needs usually requires more glucose consumption than is needed to provide three-carbon skeletons for biosynthesis of cell components, a substantial portion of the end products is released into the cell's environment. Consequently, any of the end products listed above could be manufactured by growing their respective microorganisms. A different summary of the spectrum of possible end products is provided in Fig. 5.8.

We should not leave this cursory look at diverse catabolic pathways for glucose with the impression that the Embden-Meyerhof scheme is the only route to the key intermediate, pyruvate. Two of the known alternatives which also produce pyruvate contain the typical reaction of the *hexose monophosphate* shunt;

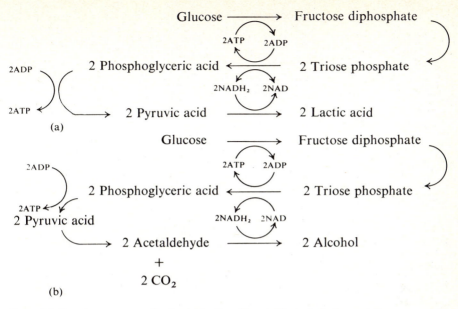

Figure 5.7 Reaction sequences for the (*a*) lactic acid and (*b*) alcoholic fermentations.

Table 5.4. Bacterial sugar fermentations which proceed through the Embden-Meyerhof pathway†

Fermentation class	Principal products from pyruvic acid	Bacterial group(s) where found
1. Homolactic	Lactic acid	Lactic acid bacteria of genera *Streptococcus*, *Pediococcus*, *Lactobacillus* (some species)
2. Mixed acid	Lactic acid, acetic acid, succinic acid, formic acid (or CO_2 and H_2), ethanol	Many enteric bacteria, e.g., *Escherichia, Salmonella, Shigella, Proteus, Yersinia*
a. Butanediol	As in 2 but also 2,3-butanediol	*Aerobacter, Serratia, Aeromonas, Bacillus polymyxa*
3. Butyric acid	Butyric acid, acetic acid, CO_2 and H_2	Many anaerobic sporeformers (*Clostridium*); some non-spore-forming anaerobes (*Butyribacterium*)
a. Butanolacetone	As in 3 but also butanol, ethanol, acetone, and isopropanol	Certain anaerobic sporeformers (*Clostridium* spp.)
4. Propionic acid	Propionic acid, acetic acid, succinic acid, CO_2	*Propionibacterium, Veillonella*

† R. Y. Stanier, M. Doudoroff, and E. A. Adelberg, "The Microbial World," 3d ed., p. 183, Prentice-Hall, Inc., Englewood Cliffs, N.J., 1970.

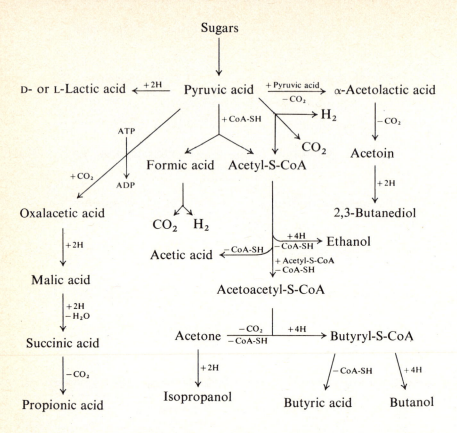

Figure 5.8 Another view of fermentation pathways from pyruvate and the resulting end products. *(From R. Y. Stanier, M. Doudoroff, and E. A. Adelberg, "The Microbial World," 3d ed., p. 184. Reprinted by permission of Prentice-Hall, Inc., Englewood Cliffs, N. J.)*

the glucose 6-phosphate formed by glucose phosphorylation is immediately oxidized via the reaction

$$\text{Glucose 6-phosphate} + \text{NAD} \longrightarrow \text{6-phosphogluconate} + \text{NADH}_2 \quad (5.8)$$

In the heterolactic fermentation carried out by some lactic acid bacteria, the pathway to the end products of lactic acid, ethanol, and CO_2 proceeds via *pentose phosphate* (see Fig. 5.9). Another variation of the hexose monophosphate shunt is the *Entner-Doudoroff pathway*, also shown in Fig. 5.9.

Notice that in both reaction sequences only 1 mole of ATP is generated per mole of glucose consumed, in contrast to the Embden-Meyerhof scheme, which yields twice this amount. The hexose monophosphate shunt pathways have the advantage of providing phosphorylated intermediates containing from three to

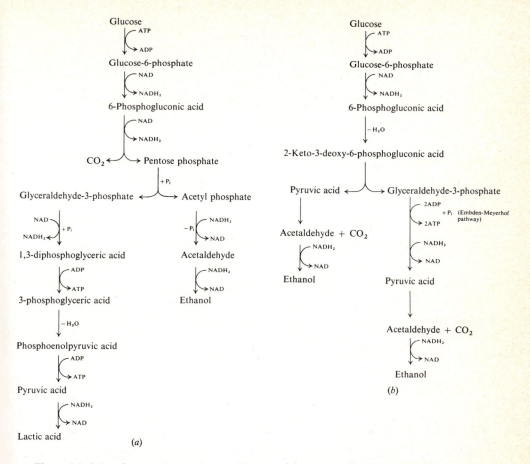

Figure 5.9 Other fermentation pathways. Through (*a*) pentose phosphate and (*b*) the Entner-Doudoroff pathway. (*From R. Y. Stanier, M. Doudoroff, and E. A. Adelberg, "The Microbial World," 3d ed., pp. 189, 190. Reprinted by permission of Prentice-Hall, Inc., Englewood Cliffs, N. J.*)

seven carbon atoms. Thus, they offer the cell's biosynthetic machinery a wider spectrum of raw materials. Figure 5.10 illustrates the most energetically favorable fermentation pathway currently known. With overall stoichiometry given by

$$2 \text{ Glucose} \longrightarrow 3 \text{ acetic acid} + 2 \text{ lactic acid} \qquad (5.9)$$

this scheme yields five ATP molecules for every two glucose molecules processed. Generation of ATP from ADP by these fermentation pathways is often called *substrate-level phosphorylation*. Our study of respiration below will reveal a much more efficient mechanism for extracting energy from nutrients.

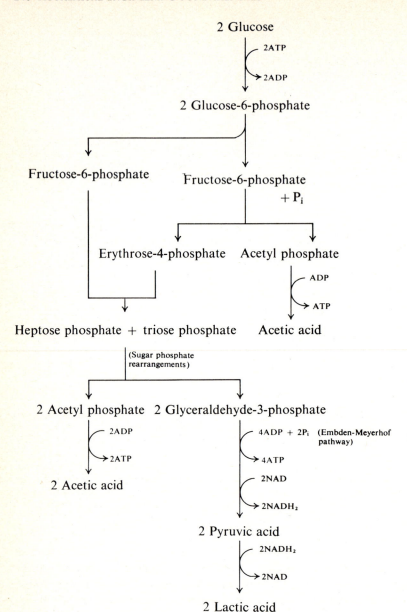

Figure 5.10 Lactic and acetic acids are produced in fermentation of glucose by *Bifidobacterium*. This pathway, which yields 2.5 ATP molecules for each molecule of glucose consumed, results in greater net ATP synthesis than any other known fermentation route.

5.2.3. Enzyme Regulation of Fermentation Rates

At the end of Sec. 3.5.1, allosteric control of enzyme activity was introduced and explained. One allosteric enzyme, phosphofructokinase, participates in glycolysis in the reaction

$$\text{ATP} + \text{fructose 6-phosphate} \xrightarrow{\text{phosphofructokinase}} \text{ADP} + \text{fructose 1,6-diphosphate} \quad (5.10)$$

We have already seen in Fig. 3.30 that ATP serves not only as a substrate but also as a modulator for phosphofructokinase. In terms of the allostery model with active and inactive enzyme configurations, ATP inhibits the enzyme by stimulating reversion to an inactive state. It is also known that ADP is an activator for phosphofructokinase, presumably promoting transition to the active form

$$\text{Phosphofructokinase (active)} \underset{\text{ADP}}{\overset{\text{ATP}}{\rightleftharpoons}} \text{phosphofructokinase (inactive)} \quad (5.11)$$

If we regard ATP as one of the end products of glycolysis, we can see that a feedback-inhibition control scheme (examine Fig. 3.27 and related discussion) is at work. As ATP accumulates, an enzyme appearing in the early stages of the glycolysis is "shut off." On the other hand, if ATP concentration within the cell is low, ADP concentration must be high since their sum is constant. In this instance, phosphofructokinase is activated, and ATP regeneration via substrate-level phosphorylation proceeds relatively rapidly.

Returning to Fig. 5.6, we can see that phosphofructokinase mediates one of three reactions in the glycolysis sequence which is not near equilibrium. We have already mentioned that this is a common characteristic of reactions catalyzed by allosteric enzymes. But what about the other two reactions with large free-energy changes? These are not regulated in glycolysis, and so they seem to violate our rule of thumb that uncontrolled reactions are usually poised near equilibrium.

A brief foray into the domain of biosynthesis provides a partial answer to this difficulty and reveals other interesting features. In the synthesis of glucose from pyruvate, all the intermediates found in Embden-Meyerhof glycolysis participate. More important, all enzymes catalyzing reactions near equilibrium in glycolysis also catalyze reactions near equilibrium in glucose biosynthesis. In biosynthesis, however, different enzymes catalyze the phosphorylation and dephosphorylation reactions, which are somewhat different from their reverse counterparts in glycolysis. These biosynthetic reactions (going upward in Fig. 5.11) proceed spontaneously with a decrease of free energy; Fig. 5.11 shows that two of these are also allosteric, so that the rate of glucose synthesis is suitably controlled. Notice the significant economy achieved by using the same enzymes for most of the degradative and biosynthetic steps, with independent control of the two processes achieved using a few allosteric enzymes.

5.3. RESPIRATION AND AEROBIC METABOLISM

Respiration is an energy-producing process in which organic or reduced inorganic compounds are oxidized by inorganic compounds. As Table 5.5 indicates, various bacteria conduct respiration using a wide variety of reductants and oxidants. When an oxidant other than oxygen is involved, the process is called *anaerobic respiration*, the term *aerobic respiration* being reserved for the more usual situation where O_2 is the oxidant. Recall that lithotrophs are organisms employing inorganic reductants. Several lithotrophs are evident in Table 5.5. So far as is known today, all lithotrophs are also autotrophs; they obtain carbon from CO_2.

In the most common form of respiration, an organic compound is oxidized using oxygen. We consider only this case in the remainder of this section, and the term respiration will be used to describe this process. It is convenient to decompose the overall process of respiration into two phases. In the first,

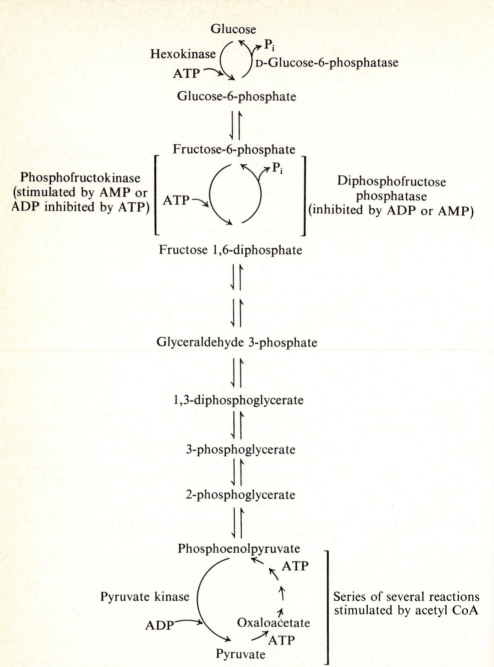

Figure 5.11 Glucose degradation via glucolysis (descending reactions) and glucose biosynthesis (ascending reactions) share many common reactions, namely those near equilibrium in both pathways. Reactions far from equilibrium are catalyzed by different enzymes (mostly allosteric) in the two pathways, thereby allowing independent control of glucose synthesis and degradation. (*Adapted from A. Lehninger, "Bioenergetics," 2d ed., p. 129, W. A. Benjamin, Inc., Palo Alto, Ca., 1974.*)

Table 5.5. Reductants and oxidants in bacterial respirations†

Reductant	Oxidant	Products	Organism
H_2	O_2	H_2O	Hydrogen bacteria
H_2	SO_4^{2-}	$H_2O + S^{2-}$	*Desulfovibrio*
Organic compounds	O_2	$CO_2 + H_2O$	Many bacteria, all plants and animals
NH_3	O_2	$NO_2^- + H_2O$	Nitrifying bacteria
NO_2^-	O_2	$NO_3^- + H_2O$	Nitrifying bacteria
Organic compounds	NO_3^-	$N_2 + CO_2$	Denitrifying bacteria
Fe^{2+}	O_2	Fe^{3+}	*Ferrobacillus* (iron bacteria)
S^{2-}	O_2	$SO_4^{2-} + H_2O$	*Thiobacillus* (sulfur bacteria)

† From W. R. Sistrom, "Microbial Life," 2d ed., table 4-2, p. 53, Holt, Rinehart, and Winston, Inc., New York, 1969.

organic compounds are oxidized to CO_2 and pairs of hydrogen atoms (electrons) are transferred to NAD. Next, the hydrogen atoms are passed through a sequence of reactions, during which ATP is regenerated from ADP. At the end of their journey, the hydrogen atoms are combined with oxygen to give water. These two phases of biological oxidation will now be examined in greater detail.

5.3.1. The TCA Cycle and Respiratory Chain

For the moment let us continue the story of carbohydrate metabolism started in Sec. 5.2. All the pathways to pyruvate described there can also operate during respiration. The reactions peculiar to respiration start with pyruvate. To begin with, pyruvate is not reduced to some end product using the hydrogen atoms obtained during glucose breakdown. Instead, this reducing power is saved for other uses, to be discussed shortly. Moreover, additional reducing power is generated from pyruvate by converting it to an acetic acid derivative (acetyl CoA)

$$CH_3COCOOH + NAD + CoA^\dagger{-}SH \xrightarrow{\text{pyruvic dehydrogenase complex}} CH_3CO{-}S{-}CoA + CO_2 + NADH_2 \quad (5.12)$$
Acetyl CoA

Acetyl CoA is also a key intermediate in the catabolism of amino acids and fatty acids. Consequently, through acetyl CoA three classes of biomolecules can be oxidized. The first phase of this oxidation is carried out in a cyclic reaction sequence with many different names: *tricarboxylic acid* (TCA) *cycle*, Krebs cycle, and citric acid cycle. Following Krebs, who played a major role in elucidating the cycle, we shall use the first name when referring to the reactions given in Fig. 5.12.

† For coenzyme A see Fig. 2.8.

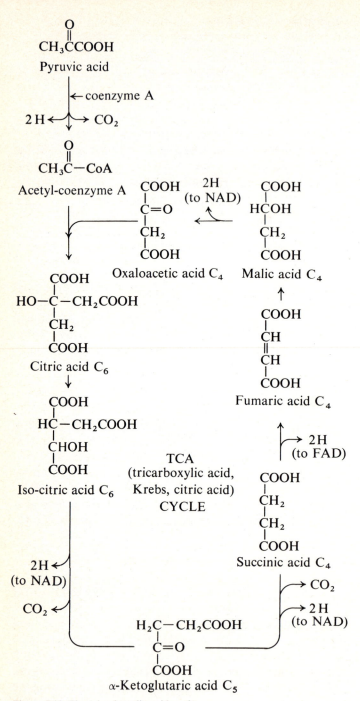

Figure 5.12 The tricarboxylic acid cycle.

Notice that the two remaining carbon atoms from pyruvate (one was lost as CO_2 in the reaction to acetyl CoA) enter the cycle to create a six-carbon acid from one with four carbons. In one pass around the TCA cycle, however, two carbon atoms are expelled as CO_2. Thus, this first phase of respiration consumes all the carbon atoms from the original glucose nutrient.

The TCA cycle as shown in Fig. 5.12 appears to serve a strictly catalytic function: no input of carbon other than the substrate is indicated. As Fig. 5.2 shows, however, the TCA cycle serves a very important function as a pool of precursors for biosynthetic reactions. Consequently, some intermediates in the cycle are constantly being tapped off and must be replaced. This is accomplished by synthesis of oxaloacetic acid from pyruvate or another three-carbon acid.

Leaving carbon metabolism, let us next follow the reactions in which hydrogen atoms are oxidized to water, the process from which aerobic cells derive most of their energy. In each pass around the TCA cycle, four pairs of hydrogen are liberated. Three pairs are transferred to NAD and, as outlined in Fig. 5.12, the hydrogen atoms resulting from succinic acid dehydrogenation are transferred to flavin adenine dinucleotide (FAD) (see Fig. 2.8). Some of the reducing power derived from the TCA cycle may be needed in biosynthesis reactions; the rest is used to generate ATP.

This last feature of aerobic metabolism can quite useful in conjunction with the information on biological oxidation states given in Sec. 5.1.3. For example, a chemoheterotroph living aerobically on glucose requires no reducing power for carbon assimilation and consequently will generate the maximum amount of ATP during respiration. Since one of ATP's major functions is driving biosynthesis, we can tentatively conclude that under such conditions the cell would either grow most rapidly or consume the minimum amount of nutrient for a given growth rate. On the other hand, the same organism with a more oxidized carbon substrate would have to divert reducing power for substrate assimilation and consequently would produce less ATP.

In the following discussion of ATP generation during respiration, we shall concentrate on such a chemoheterotroph example: all the hydrogen atoms obtained during glucose breakdown are available for reaction in the *respiratory chain*. Figure 5.13 shows an abbreviated diagram of this reaction sequence and how it ties in with the ultimate breakdown of pyruvate via acetyl CoA and the TCA (Krebs) cycle. In this diagram, FP_1 and FP_2 denote two different flavoproteins, which are enzymes containing FAD for transport of electrons (hydrogen atoms). Electrons from NAD are funneled by FP_1 to coenzyme Q (designated as Q in the figure), and in this process one molecule of ADP is regenerated to ATP for each pair of electrons passed. The electrons obtained from succinate dehydrogenation in the TCA cycle are carried by FAD in FP_2 directly to coenzyme Q.

From there, all electrons enjoy the common fate of passing through a sequence of *cytochromes*, proteins containing heme groups which are designated b, c, a, and a_3 in Fig. 5.13. Along the way, ATP is generated twice for each electron pair. The process of ATP regeneration in the respiratory chain is called *oxidative phosphorylation*. Ultimately, the hydrogen atoms are combined with dissolved

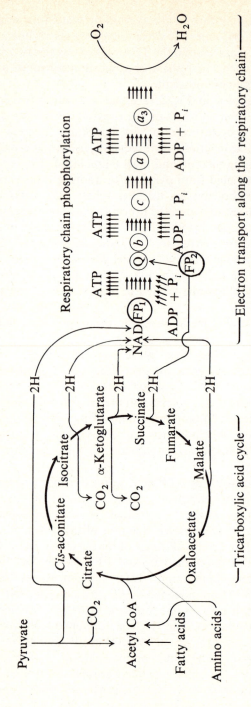

Figure 5.13 High-energy electrons released in oxidation of carbohydrates, fatty acids, and amino acids as they move through the respiratory chain: $FP_1(NADH_2)$, FP_2(succinate dehydrogenase), Q (coenzyme Q), b, c, a, and a_3 (the cytochromes). *(Reprinted by permission from A. Lehninger, "Bioenergetics," 2d ed., p. 74, W. A. Benjamin, Inc., Palo Alto, Ca., 1974.)*

oxygen to yield water as the second final product of the oxidation. If we examine the free-energy changes along the respiratory chain, we find that ATP is regenerated at each point where there is a sufficiently large decrease in electron free energy to more than offset the 7.3 kcal/mol needed to phosphorylate ADP.

5.3.2. Respiration Energetics and Growth

The end products of respiration, carbon dioxide and water, have far less utilizable free energy than fermentation products; e.g., alcohol has some nutritional value but water does not. Consequently respiration potentially makes available much more energy for use by the cell than glycolysis since $\Delta G°$ is -686 kcal/mol for the reaction

$$\text{Glucose} + 6\text{O}_2 \longrightarrow 6\text{CO}_2 + 6\text{H}_2\text{O} \qquad (5.13)$$

Let us examine how efficiently respiring living systems tap this large source of energy.

We can conveniently estimate the actual energy yield of respiration by counting the number of ATP molecules regenerated per glucose molecule. Since an exact count requires several biochemical details [1, sec. 5.6], we shall simply state the final result:

$$\text{Glucose} + 36\text{P}_i + 36\text{ADP} + 6\text{O}_2 \xrightarrow{\text{respiration}} 6\text{CO}_2 + 42\text{H}_2\text{O} + 36\text{ATP} \quad (5.14)$$

Since ATP hydrolysis has a standard free-energy change of -7.3 kcal/mol, the free-energy change of reaction (5.14) is approximately

$$\Delta G° \approx (36 \text{ mol ATP/mol glucose})(7.3 \text{ kcal/mol ATP}) = -263 \text{ kcal/mol glucose} \qquad (5.15)$$

This is *18 times* the energy which the cell captured during glycolysis. As in glycolysis, energy retention is very efficient:

$$\text{Energy capture efficiency} = \tfrac{263}{686} \approx 38\% \qquad (5.16)$$

If this figure is corrected for the nonstandard concentrations within the cell, a rather astounding efficiency estimate of greater than 60 percent results. Most of the remaining energy is dissipated as heat, which must be removed in some fashion to keep the temperature in the physiologically suitable range.

Since *chemical* engines are not commonly studied in engineering thermodynamics, it may be helpful to draw an analogy with the classical example of compressing a gas in a cylinder with a piston. If the gas is compressed rapidly, much energy is lost as heat and therefore cannot be recovered in a subsequent expansion. An analogous process would be burning glucose in air, which is quite inefficient. On the other hand, the reversible compression ideal can be approached by pushing the piston very slowly, so that heat generation is minimized. Similarly, by carrying out

Table 5.6. Weight and ATP yields for respiration and fermentation of glucose by different microorganisms†

Organism	Kind of energy metabolism	Grams dry wt		Moles ATP per mole glucose
		Per mole glucose	Per mole ATP	
Yeast	Glycolysis	20	10	2
Yeast	Respiration	~ 100	?	High
Lactic acid bacterium	Glycolysis	20	10	2
	Hexose monophosphate fermentation	10	10	1
Zymomonas	Entner-Doudoroff fermentation	10	10	1

† From W. R. Sistrom, "Microbial Life" 2d ed., table 4-3, p. 61, Holt, Rinehart, and Winston, Inc., New York, 1969.

glucose oxidation in many steps, where each has a relatively small free-energy change, the living cell is able to approach reversibility and to maximize efficiency for extraction of energy.

The relative yields of ATP from fermentation and respiration of glucose have an important practical implication: facultative anaerobes demand far less glucose nutrient in the presence of oxygen than without it. As already mentioned, this is due to the greater availability of ATP in respiring cells. The data presented in Table 5.6 dramatically underscore this fact. About five times more yeast is produced under aerobic conditions than is obtained from glycolysis. Also, glycolysis, which we know yields twice as much ATP as the pentose phosphate (PP) or Entner-Doudoroff (ED) fermentation, produces exactly twice the yield of dry cellular material. The information in Table 5.6 indicates that the amount of cellular substance formed per mole of ATP regenerated is constant at 10 g (dry weight) for anaerobic growth of many microorganisms. Using this information, we can convert observations of growth yield based on substrate into moles ATP regenerated per mole of substrate consumed. This in turn suggests which metabolic pathway might be operating. Such calculations are not foolproof, however, since some bacteria do not yield 10 g of dry weight per mole of regenerated ATP in anaerobic growth. As Table 5.6 suggests, growth yields based on ATP are more difficult to estimate and more variable for aerobic growth.

5.3.3. Partial Oxidation and Its End Products

Normally water and carbon dioxide are the metabolic end products of respiration for most aerobic microorganisms. Under abnormal conditions or with a few aerobic microbes, however, the oxidation of organic nutrient is not carried to completion, and end products then accumulate. Since some partial oxidations are of economic importance, we shall briefly review some of them here. Other applications will be mentioned in Chap. 10.

Some partial oxidations are determined by the microorganism's environment. One of these, production of citric acid by the mold *A. niger*, was cited in Sec. 1.3.3. By keeping sugar concentration very high and the concentration of iron (a cofactor for an enzyme which uses citric acid as a substrate) low, the yield of citric acid can be greatly increased (Table 5.7). The total volume of the citric acid business was estimated at 30 million dollars per year in 1967. This suggests the large relative importance of processes using living organisms, for this figure for *one* such microbial process is of the same order as the current total market for *all* isolated enzymes. By growing other fungi and aerobic bacteria under abnormal conditions, other intermediates of the TCA cycle such as α-ketoglutaric acid and fumaric acid are produced.

The acetic acid bacteria are well known for their tendency for incomplete oxidation. Both the *Acetobacter* and *Gluconobacter* genera oxidize ethanol to

Table 5.7. Iron concentration vs. citric acid yield†‡ *Aspergillus niger* **utilizing sucrose; submerged culture**

Iron§ ($FeCl_3$), mg/l	Weight yield; %¶
0.0	67.0
0.05	73.0
0.50	88.0
0.75	79.0
1.00	76.0
2.00	71.0
5.00	57.0
10.00	39.0

† H. J. Peppler (ed.), "Microbial Technology," table 8-1, Reinhold Publishing Corporation, New York, 1967.

‡ Medium composition: sucrose solution purified by ion exchange 3.6 MΩ resistance at 40% concentration, diluted to 14.2% sugar content, KH_2PO_4, 0.014%; $MgSO_4 \cdot 7H_2O$, 0.1%; $(NH_4)_2CO_3$, 0.2%; HCl to pH 2.6. (From U.S. patent 2,970,084.)

§ Supplied as $FeCl_3$.

¶ (g critic acid produced)/(g hexose moiety supplied) × 100.

acetic acid, but in the absence of ethanol *Acetobacter* bacteria can oxidize acetic acid to CO_2. The *Gluconobacter*, on the other hand, cannot metabolize acetic acid; some species are known to lack the enzymes necessary for conducting the TCA cycle. The acetic acid end product secreted by *Acetobacter* growing in ethanol is defined to be *vinegar*. The vinegar products shipped in 1962 had a total value of 60 million dollars. In Table 5.8, other end products available from the acetic acid bacteria are summarized. More on these commercially important processes will follow in Chap. 10.

Table 5.8. The variety of useful end products obtainable by partial oxidations conducted by the acetic acid bacteria[†]

$$CH_3CH_2OH + O_2 \longrightarrow CH_3COOH + H_2O$$
$$\text{Ethanol} \qquad\qquad \text{Acetic acid}$$

$$CH_3CH_2CH_2OH + O_2 \longrightarrow CH_3CH_2COOH + H_2O$$
$$\text{Propanol} \qquad\qquad \text{Propionic acid}$$

$$(H_3C)_2CHOH + \tfrac{1}{2}O_2 \longrightarrow (H_3C)_2C{=}O + H_2O$$
$$\text{Isopropanol} \qquad\qquad \text{Acetone}$$

$$\begin{array}{c} CH_2OH \\ | \\ CHOH + \tfrac{1}{2}O_2 \\ | \\ CH_2OH \\ \text{Glycerol} \end{array} \longrightarrow \begin{array}{c} CH_2OH \\ | \\ C{=}O + H_2O \\ | \\ CH_2OH \\ \text{Dihydroxyacetone} \end{array}$$

$$\begin{array}{c} CH_3 \\ | \\ (CHOH)_2 + \tfrac{1}{2}O_2 \\ | \\ CH_3 \\ \text{2,3-Butanediol} \end{array} \longrightarrow \begin{array}{c} CH_3 \\ | \\ CHOH + H_2O \\ | \\ C{=}O \\ | \\ CH_3 \\ \text{Acetoin} \end{array}$$

$$\begin{array}{c} \;\;\;\;H\;H\;OHOH \\ \;\;\;\;|\;\;|\;\;|\;\;| \\ HOCH_2-C-C-C-C-CH_2OH + \tfrac{1}{2}O_2 \\ \;\;\;\;|\;\;|\;\;|\;\;| \\ \;\;\;\;OHOHH\;H \\ \text{Mannitol} \end{array} \longrightarrow \begin{array}{c} \;\;\;\;H\;\;OHOH \\ \;\;\;\;|\;\;\;\;|\;\;| \\ HOCH_2-C-C-C-C-CH_2OH + H_2O \\ \;\;\;\;\|\;\;|\;\;|\;\;| \\ \;\;\;\;O\;\;OHH\;H \\ \text{Fructose} \end{array}$$

$$\begin{array}{c} \;\;\;\;H\;H\;OHH \\ \;\;\;\;|\;\;|\;\;|\;\;| \\ HOCH_2-C-C-C-C-CHO + \tfrac{1}{2}O_2 \\ \;\;\;\;|\;\;|\;\;|\;\;| \\ \;\;\;\;OHOHH\;OH \\ \text{Glucose} \end{array} \longrightarrow \begin{array}{c} \;\;\;\;H\;H\;OHH \\ \;\;\;\;|\;\;|\;\;|\;\;| \\ HOCH_2-C-C-C-C-COOH + H_2O \\ \;\;\;\;|\;\;|\;\;|\;\;| \\ \;\;\;\;OHOHH\;OH \\ \text{Gluconic acid} \end{array}$$

$$\begin{array}{c} \;\;\;\;H\;H\;OHH \\ \;\;\;\;|\;\;|\;\;|\;\;| \\ HOCH_2-C-C-C-C-COOH + \tfrac{1}{2}O_2 \\ \;\;\;\;|\;\;|\;\;|\;\;| \\ \;\;\;\;OHOHH\;OH \\ \text{Gluconic acid} \end{array} \longrightarrow \begin{array}{c} \;\;\;\;H\;\;OHH \\ \;\;\;\;|\;\;\;\;|\;\;| \\ HOCH_2-C-C-C-C-COOH + H_2O \\ \;\;\;\;\|\;\;|\;\;|\;\;| \\ \;\;\;\;O\;\;OHH\;OH \\ \text{5-Ketogluconic acid} \end{array}$$

[†] R. Y. Stanier, M. Doudoroff, and E. A. Adelberg, "The Microbial World," 3d ed., p. 211, Prentice-Hall, Inc., Englewood Cliffs, N.J., 1970.

5.3.4. Regulation and Localization of Respiration

Earlier we saw how phosphofructokinase modulates the rate of glycolysis by a type of feedback inhibition. One major control step in respiration is the dehydrogenation of isocitric acid to α-ketoglutaric acid in the TCA cycle:

$$
\text{Isocitric acid} + \text{NAD} \xrightarrow[\text{dehydrogenase}]{\text{isocitric}} \alpha\text{-ketoglutaric acid} + \text{NADH}_2 \qquad (5.17)
$$

with **ADP** accelerating and **ATP** inhibiting the reaction.

As indicated, ATP inhibits the allosteric enzyme isocitric dehydrogenase, while ADP accelerates the reaction rate. Thus, the pace of the TCA cycle and the overall respiration process is adjusted according to the availability of ATP.

As Fig. 5.14 shows, this control mechanism is integrated and coupled with the glycolysis controls mentioned earlier, so that the rates of glycolysis and the TCA cycle are coordinated by the cell's demand for ATP. Notice also that citric acid is also an inhibitor of the glycolysis sequence through phosphofructokinase. This helps to keep the rate of pyruvate supply in line with the needs of the TCA cycle reactions.

Another example of integrated control of respiration and glycolysis is the Pasteur effect. When facultative anaerobes switch from fermentation to respiratory metabolism, two changes rapidly occur: (1) the rate of glucose consumption falls rapidly, and (2) the production of lactate falls to zero. The second phenomenon is believed to be due to the high affinity of the respiratory chain for $NADH_2$. This makes very little reducing power available for lactate formation from pyruvate. Another type of competitive effect is apparently responsible for the slowdown in glucose consumption. The reactions for oxidative phosphorylation have a tendency to push the ATP/ADP ratio far higher than during glycolysis. As the amount of ATP is pushed up by these reactions with their strong affinity for ADP, phosphofructokinase is more strongly inhibited and the rate of the glycolysis sequence is reduced. Some of these results derive from experiments with cancer cells, which accumulate lactate while they respire. Consequently, they consume an unusually large amount of glucose. For a more complete account of these and other control mechanisms for respiration, see Lehninger [1, Chaps. 17 and 19].

It is quite possible that the greater affinity of the respiratory enzymes for ADP and $NADH_2$ involves mass transfer as well as chemical-reaction effects, for these enzymes are known to be immobilized within the cell on plasma membranes. In procaryotes, the enzymes are found attached to the inside of the cell membrane (recall Fig. 2.27) or to infoldings of this membrane called *mesosomes*. Since the various enzymes are bound in nearby vicinities on the membrane, an intermediate molecule is likely to find the next enzyme in the respiratory sequence much sooner than if the enzymes were in solution (see Prob. 3.13).

Even more dramatic evidence of a beautifully orchestrated spatial organization of enzymes is found in mitochondria, the organelles which carry out respiration in eucaryotic cells (Fig. 5.15). Both illustrations in Fig. 5.15 reveal a smooth outer membrane and an inner membrane with numerous folds, called *cristae*. The enzymes catalyzing the electron-transport reactions of the respiratory chain are located in this inner membrane. Protruding from the inner membrane are many knoblike particles, which are known to be the sites of oxidative phosphorylation of ADP to ATP. From substantial evidence that all the enzymes in the respiratory chain are held together in a single cluster, or *respiratory assembly*, it has been hypothesized that they are arranged in the reaction sequence Fig. 5.16 shows. Inside the mitochondrial membranes is a gellike material called the *matrix*, where the enzymes of the TCA cycle are found.

Even on a larger scale, mitochondria provide evidence of spatial organization. They are usually found near energy users, e.g., flagella, or energy sources e.g., fat droplets. Such an arrangement clearly reduces the mass-transfer resistance for transport of chemical-energy carriers from or to the mitochondria.

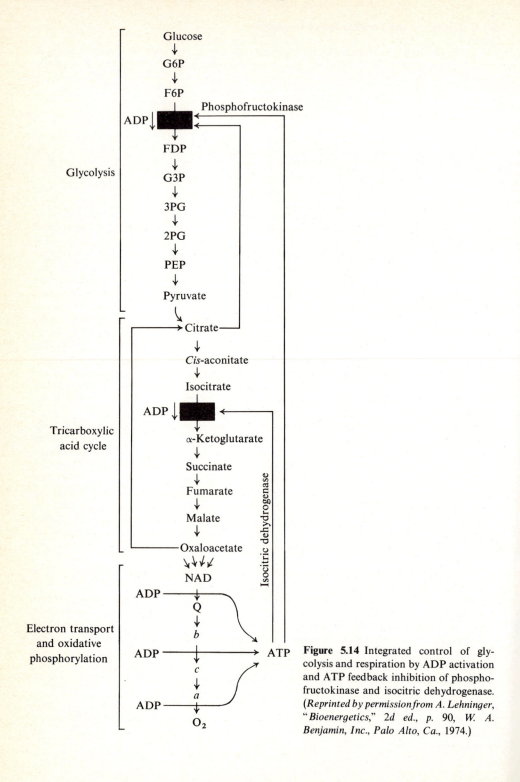

Figure 5.14 Integrated control of glycolysis and respiration by ADP activation and ATP feedback inhibition of phosphofructokinase and isocitric dehydrogenase. (*Reprinted by permission from A. Lehninger, "Bioenergetics," 2d ed., p. 90, W. A. Benjamin, Inc., Palo Alto, Ca., 1974.*)

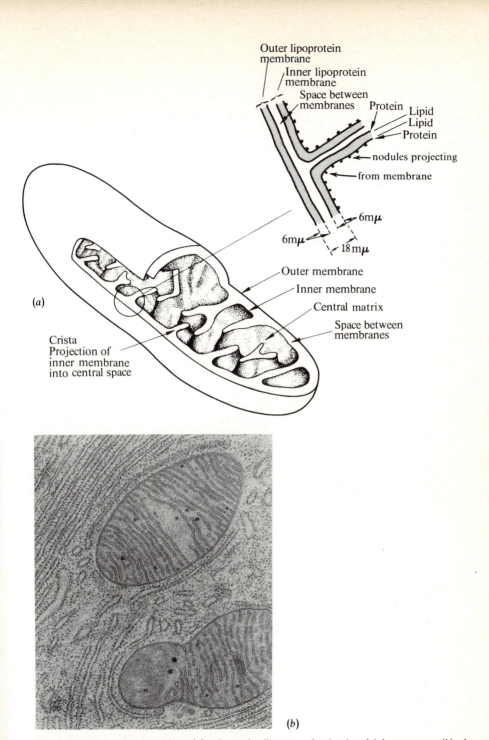

Outer lipoprotein membrane

Inner lipoprotein membrane

Space between membranes

Protein

Lipid

Lipid

Protein

nodules projecting

from membrane

6mμ

6mμ

18mμ

Outer membrane

Inner membrane

Central matrix

Space between membranes

(a)

Crista
Projection of
inner membrane
into central space

(b)

Figure 5.15 The mitochondrion. (a) schematic diagram of mitochondrial structure; (b) electron micrograph. [(a) *Reprinted by permission from N. A. Edwards and K. A. Hassall, "Cellular Biochemistry and Physiology," p. 27, McGraw-Hill Publishing Company, Ltd., London, 1971; (b) Electron micrograph courtesy of Dr. Keith R. Porter.*]

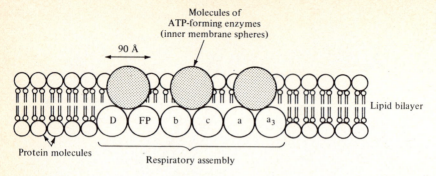

Figure 5.16 Hypothesized organization of electron-transfer enzymes in the inner membrane of the mitochondrion. (*Reprinted by permission of A. Lehninger, "Bioenergetics," 2d ed., p. 95, W. A. Benjamin, Inc., Palo Alto, Ca., 1974.*)

5.4. PHOTOSYNTHESIS: TAPPING THE ULTIMATE SOURCE

In most respirations, hydrogen atoms are continuously transferred from the fuel to oxygen, with simultaneous release of energy. Photosynthesis is largely the reverse of respiration: energy in the form of light is captured and used for conversion of carbon dioxide to glucose and its polymers. As we will reexamine in Chap. 12, photosynthesis is the prime supplier of energy for the biosphere. It extracts energy from the only significant source external to our planet, the sun. Also, photosynthesis plays a vital role in closing the carbon and oxygen cycles by reducing the carbon oxidized by respiration:

$$\text{Photosynthesis:} \quad 6CO_2 + 6H_2O + \text{light} \longrightarrow C_6H_{12}O_6 + 6O_2 \quad (5.18)$$

In procaryotes, photosynthesis occurs in stacked membranes, while the organelle called the *chloroplast* conducts photosynthesis for eucaryotes (Fig. 5.17). Both systems contain *chlorophyll*, a complex molecule which strongly absorbs visible light (Fig. 5.18). Light absorption results in an electronic excitation of the chlorophyll molecule, as shown schematically in Fig. 5.19. If isolated, an excited chlorophyll molecule would eventually return to the ground state with emission of light quanta in a process known as *fluorescence*. In photosynthetic systems, the excited chlorophyll can *donate* an electron to a sequence of enzymes which function analogously to the cytochromes of the oxidative phosphorylation pathway of the previous section. Fig. 5.19 illustrates an extremely simplified picture of ATP regeneration by this process, which is called *photophosphorylation*.

For the enzyme series of Fig. 5.19 to be able to produce ATP, the increase of free energy of the electron in the process of chlorophyll photoexcitation must be sufficiently great to drive the phosphorylation of ADP: 7300 cal/mol of free

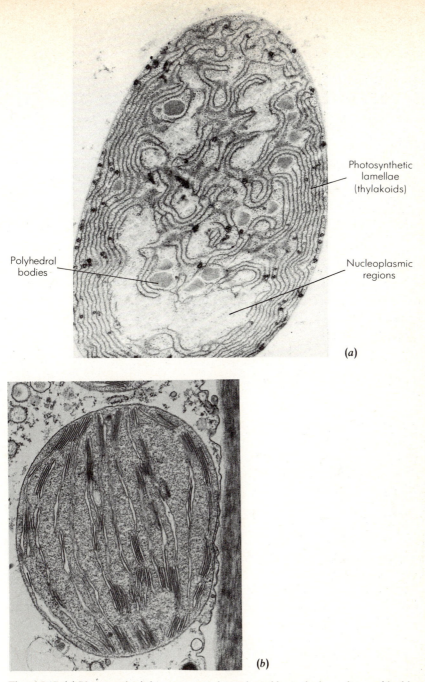

Photosynthetic
lamellae
(thylakoids)

Polyhedral
bodies

Nucleoplasmic
regions

(a)

(b)

Figure 5.17 (a) Photosynthesis in procaryotes is conducted in stacked membranes (the blue-green alga *Oscillatoria rubescens*). (b) Eucaryotic photosynthesis occurs in a specialized organelles, the chloroplast (a *Nitella* sp.) [(a) *Electron micrograph from Norma J. Lang, J. Phycol.*, **1** : 127, 1965; *used by permission of the Journal of Phycology;* (b) *Electron micrograph by M. C. Ledbetter, Brookhaven National Laboratory.*]

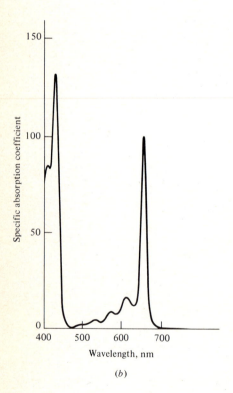

(*a*) Chlorophyll *a*

(*b*)

Figure 5.18 (*a*) The structure of a chlorophyll and (*b*) its light-absorption characteristics.

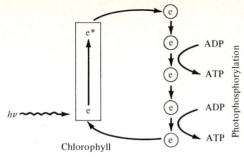

Figure 5.19 Recovery of photon energy in chloroplasts. The first electron carrier is ferredoxin; at least two cytochromes have also been identified.

energy is required. The energy content E_p of 1 mole of photons depends on the frequency (or wavelength), according to

$$E_p = hv = h\frac{c}{\lambda} \qquad (5.19)$$

where h = Planck's constant
v = frequency of photon
λ = wavelength of photon
c = speed of light in the medium

The two major absorption peaks of Fig. 5.19 correspond to 43,480 cal/mol (6500 Å) and about 67,000 cal/mol (4300 Å) of photons, respectively. Both these absorption bands evidently correspond to free-energy promotions in excess of that needed to drive a single phosphorylation of ADP.

5.4.1. Light Reactions and the Calvin Cycle

While the complete details of the common chlorophyll system are not yet elucidated, a more complete schematic of the system is shown in Fig. 5.20. This hypothetical but reasonable scheme includes functions for both kinds of chlorophyll known to coexist in plants. In addition to a cyclic photophosphorylation, noncyclic photophosphorylation is shown, in which electrons promoted from chlorophyll b travel through ferredoxin and the cytochrome system and finally pass to chlorophyll a. The charge restoration for chlorophyll b is thought to derive from the splitting of water into protons, electrons, and eventually molecular oxygen (shown at the lower right of the figure). Finally, the ubiquitous electron carrier NAD appears here in a phosphorylated form, NADP. NADP is specific to the hydrogen-removing or hydrogen-adding enzymes (dehydrogenases) which are needed in *biosynthetic* reactions, whereas the NAD form is specific to enzymes in the degradative reactions of the TCA cycle and the aerobic respiration sequence of Sec. 5.3. This part of photosynthesis, which is concerned with electron transport in the chlorophyll-cytochrome paths, contains the *light reactions*.

The ultimate synthesis of the major building block glucose arises in a complex cycle of catalyzed reactions which bears both a similarity and a dissimilarity to the TCA cycle (Fig. 5.21). The similarity is that the photosynthetic carbon-reduction cycle involves repetitive addition of carbon dioxide onto a regenerable carrier, *ribulose 1,5-diphosphate*, just as oxaloacetate is a continuously regenerated carrier in the TCA cycle. However, in the photosynthetic carbon-reduction, or *Calvin* cycle, the net result is production of glucose and *consumption* of the reduced form of NADP ($NADPH_2$) and of ATP. The TCA cycle produces reduced intermediates $NADH_2$ and $FADH_2$, which can be used to generate ATP. These *dark reactions* proceed in the absence of light as long as sufficient ATP and $NADPH_2$ are present. Though only glucose production is discussed here, some of the other carbon skeletons participating in the dark reactions are also being continuously removed into other biosynthetic pathways.

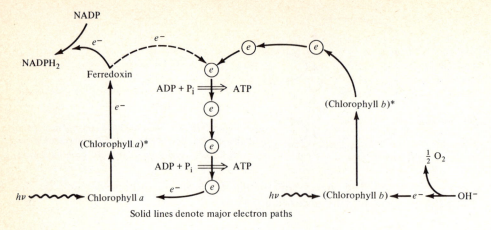

Figure 5.20 Schematic of electron flow in generation of reducing power ($NADH_2$) and oxygen in chloroplast photosynthesis.

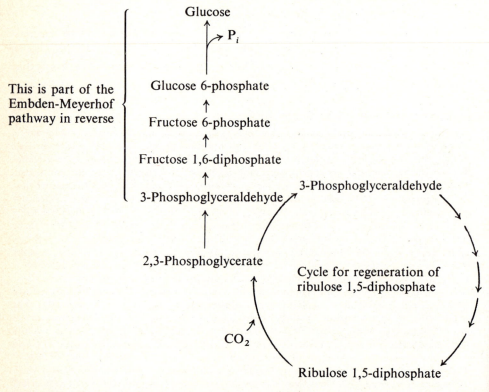

Figure 5.21 The Calvin cycle accomplishes the reduction of CO_2 necessary for glucose synthesis, an endergonic process. (*Reprinted by permission of A. L. Lehninger, "Bioenergetics," 2d ed., p. 117, W. A. Benjamin, Inc., Palo Alto, Ca., 1974.*)

5.4.2. Chloroplasts, Bacterial Photosynthesis, and Bioluminescence

All the light-phase reactions and most of the dark-phase reactions in photosynthesizing eucaryotic cells occur in organelles known as *chloroplasts*. These enclosed, solid-state, light-powered biochemical factories are similar in size and function to the mitochondria of aerobic cells. The number of chloroplasts per cell may vary from a few, as in the algae, to several hundred in more complex macroscopic plants. While molecular details of chloroplast structure are not considered here, it is interesting to note again that these organelles possess outer membranes impermeable to large phosphates and that the inner organization contains enzymes fixed on the surfaces of highly folded ultrathin membranes (Fig. 5.17b). Clearly a major purpose of such design is to maintain the highest possible local concentration of intermediates for the successive reactions. We should also note that in order to power the cell in the dark most photosynthetic eucaryotes contain mitochondria as well as chloroplasts.

Bacterial photosynthesis occurs in the two major groups of *green* and *purple* species. Both these bacterial families utilize light with the aid of chlorophyll and at least one carotenoid (a second class of light-excitable compounds useful as electron donors). A number of bacteria in these groups utilize sulfur in a manner chemically analogous to oxygen utilization in plants:

Bacteria: $$CO_2 + 2H_2S \xrightarrow{hv} (CH_2O) + H_2O + 2S \qquad (5.20)$$

Plants: $$CO_2 + 2H_2O \xrightarrow{hv} (CH_2O) + H_2O + O_2 \qquad (5.21)$$

Whereas plants have no use for the oxygen formed, some anaerobic purple bacteria, in the absence of further sulfur such as H_2S, can utilize the previously formed elemental sulfur to continue photosynthesis:

$$3CO_2 + 2S + 5H_2O \xrightarrow{hv} 3(CH_2O) + 2H_2SO_4 \qquad (5.22)$$

A few other bacteria can utilize certain organic molecules as electron donors in place of water or hydrogen sulfide.

Some bacteria, as well as some fungi, fishes, and other organisms, possess the ability to emit light (Fig. 5.22). This phenomenon, called *bioluminescence*, is in a sense the reverse of photoexcitation:

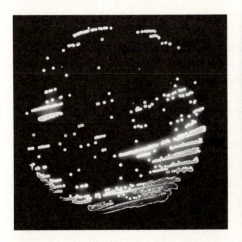 

Figure 5.22 (*a*) Streaks of luminous bacteria in a petri dish. (*b*) In submerged culture, luminous bacteria luminesce only near the surface, where sufficient dissolved oxygen is present, in an unaerated vessel. They light the entire medium in an aerated flask. (*From R. Y. Stanier, M. Doudoroff, and E. A. Adelberg, The Microbial World, 3d ed., p. 204, 1970. Reprinted by permission of Prentice-Hall, Inc., Englewood Cliffs, New Jersey.*)

chemical energy excites a molecule, which emits light as it returns to its ground state. Bacteria liminesce only in an aerobic environment, and they consume oxygen in the process. Thus, bioluminescence is also related to respiration. In fact, it is a shunt for some of the electrons produced during respiration. Instead of passing through the cytochrome system, some electrons are used to excite a light-emitting enzyme known as *luciferase*, as the following schematic of electron flow indicates:

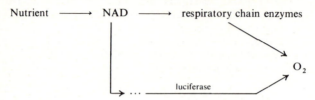

Electron flow in respiring luminescent bacteria

Besides providing another example of the energy-conversion processes conducted by living creatures, bioluminescence has been put to use by man. A very sensitive test for dissolved oxygen involves a suspension of luminous bacteria. Figure 5.22b reveals the basic idea: light is produced wherever free oxygen is present. More recently, the reactions of bioluminescence have been adapted to permit rapid quantitative analysis for ATP, as discussed in Chap. 7.

This completes our review of energy-yielding metabolism. In the remaining sections of this chapter and the first part of Chap. 6, we shall pursue cellular processes which consume energy. Since life requires all these processes, they are all clearly important. From our focus on the human use of microorganisms, biosynthesis is certainly one of the most significant energy-requiring activities.

5.5. BIOSYNTHESIS

In the opening pages of this text, we discussed three phenomena which characterize most microbial processes: substrate or nutrient utilization, cell growth, and product release. Biosynthesis influences all three. Nutrient requirements are in part dictated by the cell's need for precursor molecules. Some biosynthesis products are released into the environment of the cell. Finally, the rate of cell growth is essentially equivalent to the net rate of biosynthesis, the rate at which new cellular materials are formed. Cell growth rates vary widely. While the *E. coli* bacterium can double in 20 min, rat-liver cells reproduce over a 2- to 3-month cycle, and nerve cells do not multiply in adult human beings. Even in the last instance, however, some biosynthesis is necessary for maintenance and repair.

5.5.1. Overview of Biosynthesis Rates and ATP Utilization

Since protists are the organisms of major interest, we shall employ the bacterium *E. coli* as a model biosynthesizing cell. As Table 5.9 reveals, this tiny creature is a fantastic chemical plant. Consider protein synthesis, for example. On the average about 1400 protein molecules *per second* are manufactured within the cell. Since proteins are large biopolymers with an average of over 300 covalent bonds, peptide bonds are formed at the rate of 420,000 bonds per second. Further, since proteins are informational biopolymers, their monomeric units must be connected

Table 5.9. Biosynthetic activity during a 20-min cell-division cycle of *E. coli*†

Chemical component	Percent of dry weight	Approximate mol wt	Number of molecules per cell	Number of molecules synthesized per second	Number of molecules of ATP required to synthesize per second	Percent of total biosynthetic energy required
DNA	5	2,000,000,000	1	0.00083	60,000	2.5
RNA	10	1,000,000	15,000	12.5	75,000	3.1
Protein	70	60,000	1,700,000	1,400	2,120,000	88.0
Lipids	10	1,000	15,000,000	12,500	87,500	3.7
Polysaccharides	5	200,000	39,000	32.5	65,000	2.7

† A. L. Lehninger, "Bioenergetics," 2d ed., p. 123, W. A. Benjamin, Inc., Palo Alto, Ca, 1965.

in a definite, predetermined sequence. To put this feat in perspective, the first total synthesis of *one* protein was accomplished only recently by chemists after several months of successive chemical reactions.

The living cell is able to achieve such prodigious rates of protein synthesis at the expense of a very substantial portion of its metabolically derived energy. Bacteria devote almost all their energy to biosynthesis and, of this, roughly 88 percent is channeled into protein synthesis. Table 5.9 shows that approximately $2\frac{1}{2}$ million ATP molecules per second are invested in biosynthesis. Since the total ATP content of *E. coli* is approximately 5 million molecules, the total cell inventory is sufficient for only 2 seconds of work. This gives some indication of the large rates of ATP regeneration which must be maintained in order to sustain the cell.

5.5.2. Synthesis of Small Molecules

In these reactions, which are called class II reactions in Fig. 5.1, the monomeric building blocks are constructed. Approximately 70 different compounds are required for this purpose (Table 5.10). Also, ATP, NAD, other carriers, and coenzymes must be manufactured in class II reactions. The products of these reactions are known collectively as the *central intermediary metabolites*.

Glucose synthesis has already been described. Beginning with the modified sugar ribose phosphate, nucleotides are formed by stepwise addition of CO_2 and amino acid components. Deoxyribose nucleotides are then synthesized by reduction of ribose. Some features of these syntheses are given in Fig. 5.2. Since detailed knowledge of these class II reactions is not necessary for our purposes, we shall not pursue them. It is worthwhile to look deeper into amino acid synthesis, however. Not only do these reactions illustrate interesting metabolic control systems, but microbial production of various amino acids are important commercial processes.

Table 5.10. About 70 building blocks are used in synthesis of important biological polymers†

Class	Chemical nature of subunits	No. of different kinds used
Nucleic acids	Nucleotides	8
Deoxyribonucleic acids	Deoxyribonucleotides	4
Ribonucleic acids	Ribonucleotides	4
Proteins	Amino acids	20
Polysaccharides	Monosaccharides	~ 15‡
Complex lipids	Variable	~ 20‡

† R. Y. Stanier, M. Doudoroff, and E. A. Adelberg, "The Microbial World," 3d ed., p. 235, Prentice-Hall, Inc., Englewood Cliffs, N.J., 1970.

‡ The number of building blocks used for the synthesis of any given member of these classes is usually much smaller.

Living cells assimilate nitrogen by incorporating it into the amino acids glutamic acid and glutamine. First glutamic acid is formed by reaction between ammonia and α-ketoglutaric acid, one of the TCA cycle intermediates:

$$HOOC(CH_2)_2COCOOH + NH_3 + NADH_2 \longrightarrow$$
α-Ketoglutaric acid

$$HOOC(CH_2)_2CHNH_2COOH + NAD + H_2O \qquad (5.23)$$
Glutamic acid

Additional ammonia can be accepted by adding it to glutamic acid to give glutamine:

$$HOOC(CH_2)_2CHNH_2COOH + NH_3 + ATP \longrightarrow$$
Glutamic acid

$$HOOC(CH_2)CHNH_2CONH_2 + ADP + P_i + H_2O \qquad \Delta G^\circ = -3.9 \text{ kcal/mol}$$
Glutamine

$$(5.24)$$

The appearance of ATP should signal to us that reaction (5.24) occurs in multiple steps with common intermediates. Indeed, the standard-free-energy change for the reaction

$$\text{Glutamic acid} + NH_3 \longrightarrow \text{glutamine} + H_2O \qquad (5.25)$$

is 3.4 kcal/mol, and the reaction occurs *in vivo* through two steps

$$ATP + \text{glutamic acid} \longrightarrow \text{glutamyl phosphate} + ADP$$
$$\text{Glutamyl phosphate} + NH_3 \longrightarrow \text{glutamine} + P_i$$
$$(5.26)$$

the sum of which is given in (5.24). Some microorganisms take in nitrogen in the form of nitrate NO_3^- and free nitrogen N_2. but it is now known that these

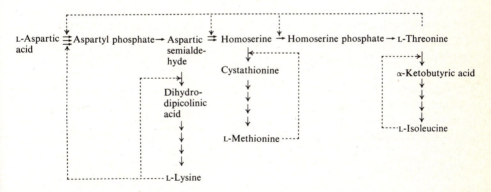

Figure 5.23 Control systems for the aspartate pathway in *E. coli*. Dotted arrows denote feedback inhibition, while solid arrows are reaction steps catalyzed by different enzymes. Note the use of isozymes, different enzymes which catalyze the same reaction, to achieve independent regulation of synthesis of each end product.

Table 5.11. Classification of amino acid synthesis†

Precursors	Products	Family
α-Ketoglutarate ⟶ glutamate	Glutamine Arginine Proline	Glutamate
Oxalacetate ⟶ aspartate	Asparagine Methionine Threonine Isoleucine (in part) Lysine (in part)‡	Aspartate
Phosphoenolpyruvate + erythrose-4-phosphate	Phenylalanine (in part) Tyrosine (in part) Tryptophan (in part)	Aromatic
3-Phosphoglycerate ⟶ serine	Glycine Cysteine	Serine
Pyruvate	Alanine Valine Leucine (in part)	Pyruvate
Phosphoribosyl pyrophosphate + ATP	Histidine (in part)	

† R. Y. Stanier, M. Doudoroff, and E. A. Adelberg, "The Microbial World," 3d ed., p. 241, Prentice-Hall, Inc., Englewood Cliffs, N.J., 1970.

‡ In all procaryotic and some eucaryotic protists, lysine has an alternative derivation.

nutrients are first transformed into ammonia before being assimilated in either reaction (5.23) or (5.24).

Other amino acids are synthesized by branched pathways, some of which appear in Fig. 5.2. These can be organized into five families, shown in Table 5.11, among which the aspartate family and its control mechanisms has been studied extensively. Figure 5.23 summarizes the various pathways and the intricate pattern of control employed by *E. coli*. This is the first example we have seen of isozymes, which are enzymes catalyzing the same reaction. As the three arrows from L-aspartic acid to aspartyl phosphate indicate, three different aspartokinase enzymes participate in this step. Each one, however, is regulated by different end products of the diverging pathways. In this way, the end product of one branch cannot shut off synthesis of other end products. The first and third enzymes are allosterically inhibited by threonine and lysine, respectively, while the second enzyme is not regulated allosterically but controlled by another method considered in Sec. 6.1.

5.5.3. Macromolecule Synthesis

In the next step in the buildup of cell components (class III in Fig. 5.1) monomeric subunits are joined to form large biopolymers. Energy must be invested in order to assemble these macromolecules; Fig. 5.24 shows schematically

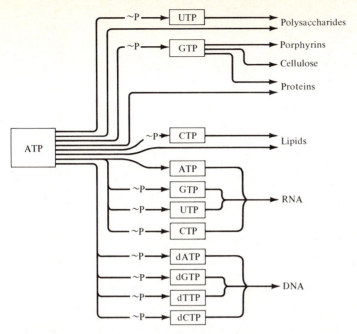

Figure 5.24 The high energy triphosphate forms of the nucleosides participate in various biosynthetic routes. (*Reprinted by permission from A. Lehninger, "Bioenergetics," 2d ed., p. 136, W. A. Benjamin, Inc., Palo Alto, Ca., 1974.*)

how energy stored in the phosphate bonds of ATP is mobilized in the form of other nucleoside triphosphates for constructing the four classes of biomolecules.

At this stage of biosynthesis, a different breakdown of ATP is quite common. Instead of splitting into $ADP + P_i$ (orthophosphate), a *pyrophosphate* group is severed leaving AMP

$$\text{(A)}-\text{(R)}-P{\sim}P{\sim}P + H_2O \longrightarrow \text{(A)}-\text{(R)}-P + P{\sim}P \qquad (5.27)$$

$$\underset{\text{ATP}}{} \qquad\qquad\qquad \underset{\text{AMP}}{} \quad \underset{\text{Pyrophosphate}}{}$$

Following this step, a *second* high-energy phosphate bond is broken by hydrolyzing pyrophosphate to orthophosphate

$$\begin{matrix} O^- & & O^- \\ | & & | \\ ^-O-P-O-P-O^- + H_2O \longrightarrow & 2HO-P-O^- \\ \| & \| & \| \\ O & O & O \end{matrix} \qquad (5.28)$$

$$\underset{\text{Pyrophosphate}}{} \qquad\qquad\qquad \underset{\text{Orthophosphate}}{}$$

Thus, approximately double the free-energy driving force (~ 14 kcal/mol) of a split to ADP is available. Some version of this mechanism operates in the synthesis of lipids, RNA, DNA, and glycogen. The common-intermediate principle is again involved, although in a more complicated form. For example, the addition of one glucose unit to glycogen proceeds in six steps linked by overlapping intermediates (Table 5.12).

Table 5.12. Reaction sequence for glycogen biosynthesis†

1. Glucose + ATP $\xrightarrow{\text{hexokinase}}$ glucose 6-phosphate + ADP

2. Glucose 6-phosphate $\xrightarrow{\text{phosphoglucomutase}}$ glucose 1-phosphate

3. Glucose 1-phosphate + UTP $\xrightarrow[\text{pyrophosphorylase}]{\text{UDP·glucose}}$ UDP·glucose + pyrophosphate

4. UDP·glucose + glycogen$_n$ $\xrightarrow[\text{synthetase}]{\text{glycogen}}$ glycogen$_{n+1}$ + UDP

5. ATP + UDP $\xrightarrow[\text{diphosphokinase}]{\text{nucleoside}}$ ADP + UTP

6. Pyrophosphate + H_2O $\xrightarrow{\text{pyrophosphatase}}$ 2 phosphate

Sum: glucose + 2ATP + glycogen$_n$ $\longrightarrow$ 2ADP + 2P + glycogen$_{n+1}$

† A. L. Lehninger, "Bioenergetics," 2d ed., p. 140, W. A. Benjamin, Inc., Palo Alto, Ca., 1971.

Naturally, synthesis of informational polymers (RNA, DNA, and proteins) is a considerably more complex process. In both cases, however, the monomer is activated to permit its addition to the polymer chain. For RNA and DNA synthesis, the nucleotides enter the scheme as nucleoside triphosphates (a nucleotide + pyrophosphate). Activation of amino acids for protein construction can be represented by

$$\text{Amino acid} + \text{ATP} \longrightarrow \underset{\text{Active amino acid}}{\text{amino acyl·AMP}} + \text{P}\sim\text{P} \qquad (5.29)$$

Thus, in both cases, about 14 kcal/mol is invested to push the monomer addition reaction to completion.

While we shall defer the details on construction of the proper sequence of monomers for proteins until Sec. 6.1, the basic strategy will be outlined now. The following brute-force approach may come to mind first: since any given amino acid–amino acid bond may appear in a given protein, the order of (number of amino acids)^2/2, or about 200 enzymes, are needed to form the various bonds possible. The required number apparently must be larger than this first estimate, however, since the number of different proteins is also large (of the order of 1000 in a typical bacterium). Thus, if an amino acid sequence a—b—c—d— is to be produced, the enzyme forming the —c—d— bond must recognize the presence of a—b—c— and other enzymes forming c—x bonds (where x is any of the 19 other amino acids) must also not mistakenly add —x. The total number of enzymes needed would be very large, and their biosynthesis would require substantial energy. A different approach is found in living systems. Since the particular peptide bond joining adjacent amino acids is identical for every amino acid pair, the step·of *catalyzing the addition* is separated from the task of *preparing the sequence* of subunits to be added in the proper order. Information from DNA is vital to the latter process, as we shall explore further in Chap. 6.

5.6. TRANSPORT ACROSS CELL MEMBRANES

In order to maintain intracellular concentrations in a definite narrow range while exterior concentrations may vary widely, the cell controls the rate of exchange of matter with the environment. Basic to this control are the cell membranes, which act to hinder interchange of some substances and accelerate the transport rate of others. At least three different means of transport across cell membranes are known: passive diffusion, facilitated diffusion, and active transport. Regardless of the mechanism, the transport characteristics of a given membrane with a given substrate are often expressed in terms of the membrane permeability K, which is computed from

$$K = \frac{V}{At} \ln \frac{c_e - c_{i0}}{c_e - c_i(t)} \qquad (5.30)$$

where V = cell volume

A = cell external surface area

c_e = external concentration

c_{i0} = interior substrate concentration initially

$c_i(t)$ = interior substrate concentration after elapsed time t

As Eq. (5.30) indicates, the units of permeability are those of a velocity (centimeters per second). The assumptions underlying this equation are rather extreme, however, as we shall explore in the problems.

5.6.1. Passive and Facilitated Diffusion

In passive diffusion, material moves across the membrane from regions of high concentration to low concentration (Fig. 5.25). The diffusion rate is proportional to the overall *driving force*, the concentration difference across the membrane (Fig. 5.26). Thermodynamic considerations show that passive diffusion is spontaneous: the free-energy change ΔG accompanying the transport of material from a region of concentration c_2 to another locale where the concentration is c_1 is given by

$$\Delta G = RT \ln \frac{c_1}{c_2} \qquad (5.31)$$

Since c_1 is smaller than c_2 in passive diffusion, ΔG is negative.

Because of the structure of the plasma membrane which surrounds all cells and the organelles of eucaryotes, not all chemical species penetrate the membrane with equal ease. The diffusion rate of most large molecules is strongly correlated with the molecules' solubility in lipids (Fig. 5.27). This should not surprise us since the central core of the plasma membrane is believed to be mostly lipid, perhaps arranged in a bilayer (recall Fig. 2.2). A variation in the unit-membrane model has been developed to account for the abnormally rapid diffusion of water and other small molecules through the membrane: very small pores through the membrane are envisioned.

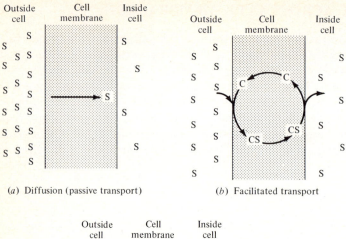

(a) Diffusion (passive transport)

(b) Facilitated transport

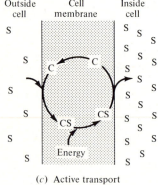

(c) Active transport

Figure 5.25 Various modes of membrane transport: (a) diffusion, (b) facilitated transport, and (c) active transport.

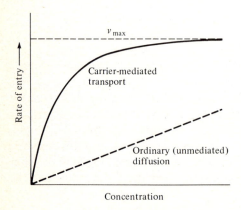

Figure 5.26 Mass transfer rates exhibit saturation when a carrier mediates membrane transport.

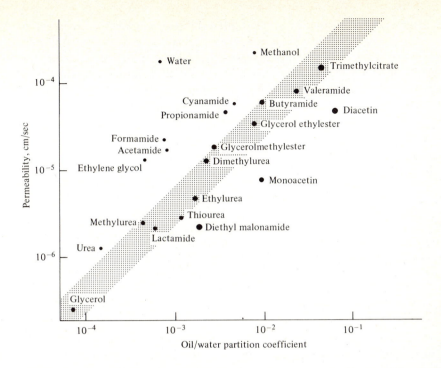

Figure 5.27 Permeability of many substances through the membrane of the alga *Chara* correlates well with lipid solubility (partition coefficient). The sizes of the dots are roughly proportional to the respective molecular sizes. (*After Collander, 1947.*)

Charged molecules and other polar substances have a very small lipid solubility. Thus, these species have little tendency to pass through the membrane by ordinary diffusion. The importance of this property has already been emphasized: most metabolic intermediates are charged and consequently stay within the cell. Still, there are some polar components (glucose, H^+, Na^+, K^+ and others) which move across the membrane barrier quite rapidly.

Facilitated diffusion provides one mechanism for this anomaly. As illustrated schematically in Fig. 5.25, substrate on the outside combines with a *carrier molecule*, which diffuses to the other side, where the complex splits, leaving the ferried molecule inside the cell. This mode of biological transport has several distinguishing characteristics. One of them is indicated in Fig. 5.26: instead of increasing linearly with the exterior solute concentration, the transport rate reaches a saturation level. Further increases in the overall mass-transfer driving force have no influence on the rate of transport. From our exposure to Michaelis-Menten and other models of enzyme kinetics, it should be obvious why the scheme shown in Fig. 5.25 has been postulated as a facilitated-diffusion mechanism.

The other essential properties of facilitated diffusion are also reminiscent of enzyme kinetics: only specific compounds are transported, and specific inhibitors slow the process. Because of this specificity and kinetic behavior, carrier molecules are believed to be proteins. These carriers are called *permeases*. Perhaps the best known example of facilitated diffusion is glucose transport in the human erythrocyte (red blood cell). Details on this system and additional information on facilitated transport are available in the references.

5.6.2. Active Transport

As already shown in Fig. 5.25, active transport has two distinguishing character-. istics: (1) it moves a component *against* its chemical (or electrochemical) gradient, from regions of low to high concentration; (2) this process requires metabolic energy, as Eq. (5.31) indicates. For example, if glucose is transported from the cell's environment, where its concentration is 0.001 M, to the interior of the cell, where glucose concentration is 0.1 M, then

$$\Delta G^\circ = [1.98 \text{ cal/(mol·K)}](298 \text{ K})(\ln 100) = 2.72 \text{ kcal/mol}$$

Consequently at least this amount of free energy must be expended to drive the processes.

Before examining two examples of active-transport systems and their hypothesized mechanisms, we should extend our energetic calculations to include transport of charged species. If the component being transferred is charged, Eq. (5.31) needs to be modified to read

$$\Delta G^\circ = RT \ln \frac{c_1}{c_2} + Z_1 F \, \Delta\psi \qquad (5.32)$$

where Z_1 = number of charges on transported molecules
$\quad\quad F$ = faraday = 96,493 C/g equiv
$\quad\quad \Delta\psi$ = potential difference across membrane, V

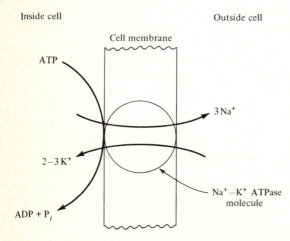

Inside cell

Outside cell

Cell membrane

ATP

3 Na⁺

2–3 K⁺

Na⁺–K⁺ ATPase molecule

ADP + P$_i$

Figure 5.28 In this model of the Na⁺–K⁺ active transport process, the two transport processes are coupled and driven by the ATP. (*Reprinted by permission from A. Lehninger, "Bioenergetics," 2d ed., p. 203, W. A. Benjamin, Inc., Palo Alto, Ca., 1974.*)

Membrane

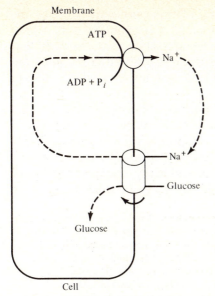

Figure 5.29 Na^+ external to the cell moves through the cell membrane in a passive carrier that also transports glucose. A high external concentration of Na^+ is maintained by active transport. (*Reprinted by permission from A. Lehninger, "Bioenergetics," 2d ed., p. 205, W A. Benjamin, Inc., Palo Alto, Ca., 1974.*)

Active transport of ions plays a central role in nerve action, but this process enjoys wider application: almost all cells have active-transport systems to maintain a proper balance between K^+, Na^+, and water within the cell. In particular, Na^+ is pumped out of the cell while K^+ is pumped in. The pumping action allows the cell to offset the simultaneous passive diffusion of these ions, which occurs continuously. In one model supported by several experiments, these ion-transport processes are coupled and driven by ATP (Fig. 5.28).

The second common class of active-transport systems pumps molecular nutrients such as glucose and amino acids into the cell at far greater rates than can be achieved by passive diffusion. In the cells of higher animals. glucose active transport is dependent on Na^+. Figure 5.29 shows the mechanism envisioned for this process. A different process is believed responsible for glucose active transport in bacteria. In this scheme, glucose is released into the cell's interior in the energized and relatively impermeable form of glucose 6-phosphate (Fig. 5.30). The rate of this process is believed to be the limiting step which determines the growth rate of some bacteria.

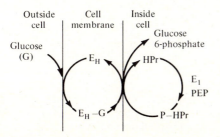

Figure 5.30 Phosphorylation of glucose inside the cell membrane helps maintain a large driving force for glucose membrane transport and also serves to keep the transported glucose moiety in the cell's interior. (*Reprinted by permission from A. L. Lehninger, "Biochemistry," 2d ed., p. 799, Worth Publishers, New York, 1975.*)

Eucaryotic cells must possess a hierarchy of transport systems, since the concentration of some components within organelles is maintained at a level other than that in the cell's cytoplasm. This feature serves as a reminder that the inside of a cell, especially a eucaryote, is not a uniform well-mixed pool. On the contrary, living cells are highly organized systems even down to the molecular level.

Many mathematical theories of active transport have been developed, although we shall not pursue them now. Further details are available in the references. As the fundamental molecular basis for active transport is not well known, it does not yet appear possible to choose between plausible theories. We shall close this section with a synthetic example of a facilitated- and active-transport movement of nitric acid through a liquid membrane in the presence of ferrous ion.

Example 5.1. Transport of nitric acid through a liquid membrane The facilitated-transport case is illustrated in Fig. 5E1.1. The carrier, ferrous ion, may be regarded as acting to increase the solubility of the neutral molecule in the membrane. Since the carrier solubility will often be much greater than that of the uncomplexed solute, NO, a 1 percent concentration gradient in the complex $FeNO^{2+}$ will carry a great deal more NO than a 1 percent gradient in the NO concentration profile. Additionally, the carrier ion may be very specific for a given solute, so that an aqueous ferrous chloride membrane will exhibit a much greater permeability toward NO than toward nitrogen or argon, neither of which forms ferrous complexes. Thus, carrier-facilitated transport exhibits both an *increased* membrane permeability and a *specificity* toward a particular solute. In various membranes of biological interest, examples of passive facilitated-transport systems are known for the solutes glucose, H^+, Na^+, K^+, and Ca^{2+} [4, 5].

The molecular coupling by which biological *active* transport occurs is not well understood: our thinking may be aided conceptually by considering the ferrous-ion transport of nitric oxide again. Recent research has indicated that the direction of net positive nitric oxide transport in Fig. 5E1.1 may be reversed by doing electrical work on the membrane, as illustrated in Fig. 5E1.2. Application of a

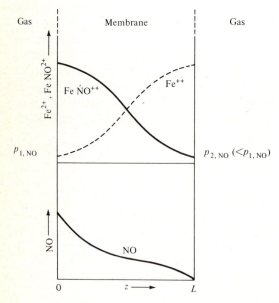

Figure 5E1.1 Passive or facilitated transport of NO by Fe^{2+} through an aqueous membrane of thickness L.

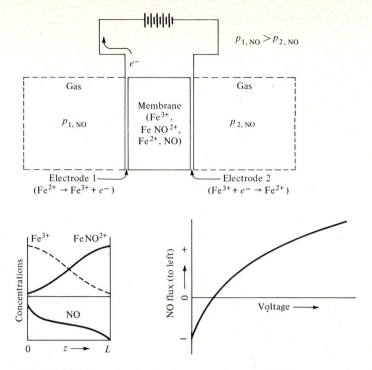

Figure 5E1.2 Schematic of *active-transport* scheme for NO transport.

voltage difference across the membrane allows oxidation of the carrier ion at one porous electrode and reduction of it at the other, according to

Electrode 1: $$Fe^{2+} \longrightarrow Fe^{3+} + e^-$$ (5E1.1)

Electrode 2: $$e^- + Fe^{3+} \longrightarrow Fe^{2+}$$ (5E1.2)

The ferric ion created and consumed does not complex with NO.

By continuously increasing the voltage drop across the membrane solution, the ferrous-ion concentration at the high-pressure side of the membrane can be driven below the average ferrous-ion concentration in the membrane, setting up a ferrous-ion gradient of a sign *opposite* that of the NO gradient. For thick enough membranes, the central portion of the film is nearly at equilibrium, given by

$$NO + Fe^{2+} \rightleftharpoons FeNO^{2+}$$ (5E1.3)

Thus $$[NO][Fe^{2+}] = K[FeNO^{2+}]$$

Differentiating gives

$$\frac{d(FeNO^{2+})}{dx} = \frac{1}{K}\left[\frac{d(NO)}{dx}Fe^{2+} + NO\frac{d(Fe^{2+})}{dx}\right]$$ (5E1.4)

The nitric oxide self-diffusion flux is given by

$$N_{NO} = \mathscr{D}_{NS}\frac{d(NO)}{dx}$$ (5E1.5)

The contribution due to nitric oxide transport by $FeNO^{2+}$ is

$$N_{FeNO^{2+}} = \mathscr{D}_{FeNO^{2+}} \frac{d(FeNO^{2+})}{dx} \tag{5E1.6}$$

and, using Eq. (5E1.4), we have for the net flux of NO

$$N = N_{NO} + N_{FeNO^{2+}} = \mathscr{D}_{NO} \frac{d(NO)}{dx} \left\{ 1 + \frac{\mathscr{D}_{FeNO^{2+}}}{\mathscr{D}_{NO}} \left[Fe^{2+} + NO \frac{d(Fe^{2+})/dx}{d(NO)/dx} \right] \right\} \tag{5E1.7}$$

Since $\mathscr{D}_{FeNO^{2+}}/\mathscr{D}_{NO} > 0$ and $d(Fe^{2+})/dx$ can be made continuously more negative by increasing the membrane voltage drop, evidently the net transport of NO from right to left (Fig. 5E1.2) can be made to proceed against an NO gradient. The data at the base of Fig. 5E1.2 illustrate such pumping action when $p_{1,NO} = p_{2,NO}$. From the data shown, a characteristic of irreversible processes appears: the greater the driving force, the lower the specific rate (rate per unit of driving force) and thus, in the present case, the lower the efficiency conversion of electrical work into a transport work.

5.7. CONCLUDING REMARKS

We now know that extremely complicated reaction processes abound in the living cell. In spite of the complexities, a few central features occur quite often and provide a helpful unifying link between all aspects of cell metabolism. Perhaps the most important common aspects of cell biochemistry are ATP's role as an energy carrier (Fig. 5.31) and the reducing power shuttle, NAD.

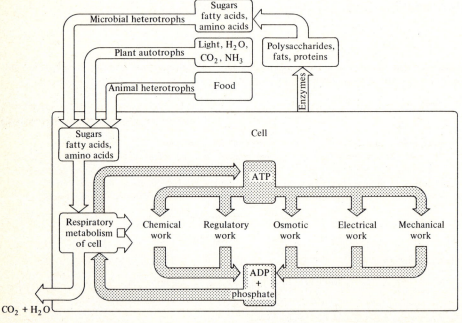

Figure 5.31 Overall schematic diagram of energy flows in the cell and its environment. (*From "Cell Structure and Function," 2d ed., p. 26, by Ariel G. Loewy and Philip Siekevitz, Copyright © 1963, 1969 by Holt, Rinehart and Winston, Inc. Reprinted by permission of Holt, Rinehart and Winston.*)

In the next chapter, we finish our introductory study of the cell's operation. After that, we shall tackle the difficult problem of analyzing, interpreting, and designing processes for exploitation of microorganisms.

PROBLEMS

5.1. Equilibrium The standard-free-energy change of the reaction phosphoenolpyruvate + ADP → pyruvate + ATP is -7.50 kcal. If the reaction is initiated with 6.0 mM ADP, 6.0 mM phosphoenolpyruvate, 6.0 mM ATP, and no pyruvate, what will the final concentrations be?

5.2. Mineral function For each element or species in Table 5P2.1 identify the important function(s) within the cell and the predominant chemical form for each function. Is it necessary in such an experiment for every element in the resulting ash to have a biochemical function? (See Ref. 6.)

Table 5P2.1. Approximate ash content from complete oxidation of yeasts

Species	Approximate content, %	Species	Approximate content, %
Phosphorus pentoxide	50	Silicon oxide	1
Potassium oxide	35	Sodium oxide	1
Magnesium oxide	5	Sulfur trioxide	0.5
Calcium oxide	1	Chlorine, iron	Trace

5.3. Standard-free-energy changes The free-energy change of a reaction ΔG is related to the half-cell potentials $E_1 - E_2$ of the individual half reactions

$$\Delta G = -nF\,\Delta E = -nF(E_1 - E_2)$$

where n is the number of electrons transferred per mole reactant converted, F is Faraday's constant $\approx$ 23 kcal/(V·mol), and E_i ($i = 1, 2$) is given by the Nernst equation

$$E_i = E_{i0} - \frac{RT}{nF}\ln\frac{s_{i,\,red}}{s_{i,\,ox}}$$

where s_i = concentration of a reactant in oxidized or reduced form, and E_{i0} = half-cell potential when $s_{i,\,red}/s_{i,\,ox} = 1.0$ compared to the half cell for H:

$$S_{i,\,red} \longrightarrow S_{i,\,ox} + ne^- \qquad E_0 \neq 0$$

$$H_2 \longrightarrow 2H^+ + 2e^- \qquad E_0 \equiv 0 \text{ by definition}$$

Some half-cell potentials of interest are given in Table 5P3.1.

From the standard half-cell potentials given, at standard conditions show that the following overall reactions are (not) possible:

(*a*) Electron transfer from excited chlorophyll
(*b*) Oxidation of β-hydroxybutyrate with oxygen
(*c*) Oxidation of acetaldehyde by acetaldehyde (dismutation).

Table 5P3.1

Half cell†	n	E_0, V
Acetate + CO_2 + 2H$^+$ $\longrightarrow$ pyruvate + 2H$_2$O	2	-0.70
Acetoacetate + 2H$^+$ $\longrightarrow$ β-hydroxybutyrate	2	-0.27
Pyruvate + 2H$^+$ $\longrightarrow$ lactate	2	-0.19
S + 2H$^+$ $\longrightarrow$ H$_2$S	2	-0.23
$\frac{1}{2}$O$_2$ + 2H$^+$ $\longrightarrow$ H$_2$O	2	0.82
NAD(P)$^+$ + 2H$^+$ $\longrightarrow$ NAD(P)H + H$^+$	2	-0.32
FAD + 2H$^+$ $\longrightarrow$ FADH$_2$ (free coenzyme)	2	-0.18
Chlorophyll$^+$ $\longrightarrow$ chlorophyll* + e^-	1	-0.2‡
Chlorophyll$^+$ $\longrightarrow$ chlorophyll	1	0.9‡
NO$_3^-$ + 2H$^+$ $\longrightarrow$ NO$_2^-$ + H$_2$O	2	0.42
SO$_4^{2-}$ + 2H$^+$ $\longrightarrow$ SO$_3^{2-}$ + H$_2$O	2	0.48
Acetate + 2H$^+$ $\longrightarrow$ acetaldehyde + H$_2$O	2	-0.60
Acetaldehyde + 2H$^+$ $\longrightarrow$ ethanol	2	-0.20

† Selected values from W. B. Wood, J. H. Wilson, R. M. Benbow, and L. E. Hood, "Biochemistry: A Problems Approach," pp. 190–191, W. A. Benjamin, Inc., Palo Alto, Calif., 1974.

‡ Eucaryotic photosynthesis, noncyclic portion of Fig. 5.25.

5.4. ΔG of cellular conditions (a) From Fig. 5.6 and Table 5.3, calculate the *standard*-free-energy change for each step in glycolysis in the human erythrocyte.

(b) From these data and the free energy of ATP hydrolysis to ADP under the same conditions, establish the need for ATP to drive the particular reaction steps in which it participates.

(c) From your calculations, verify the overall result that for glucose $\rightarrow$ 2 lactic acid, $\Delta G° = -52$ kcal/mol.

5.5. Network kinetics A cell carries out many reactions simultaneously, yet often the apparent behavior of a sequence of reactions is that of only a few.

(a) Show that, at steady state, the mathematical expression for the rate of P formation for the postulated Michaelis-Menten sequence

$$E + S \rightleftharpoons ES \longrightarrow E + P$$

is indistinguishable from that for the following equilibrated sequence:

$$S + E_1 \rightleftharpoons S_2 + E_1$$
$$S_2 + E_2 \rightleftharpoons S_3 + E_2$$
$$S_3 + E_3 \rightleftharpoons E_3S_3 \longrightarrow E_3 + P$$

(b) The 10 reactions in series in Fig. 5.6 can be represented by only three kinetic equations describing

$$A \xrightarrow{k_1} B \xrightarrow{k_2} C \xrightarrow{k_3} D \quad \text{(lactate)}.$$

Identify A, B, and C.

(c) The standard free energies of hydrolysis of phosphate compounds are given in Table 5P5.1. If the ATP/ADP ratio in the cell diminishes by a factor of 10 (with sum ATP + ADP = const), what happens to the free-energy differences calculated in Prob. 5.4 for the three steps (b) above? What (un)useful effect results for the cell?

Table 5P5.1

	$\Delta G°$, kcal/mol
Phosphoenolpyruvate	− 12.8
1,3-Diphosphoglycerate	− 11.8
Phosphocreatinine	− 10.5
Acetyl phosphate	− 10.1
ATP (terminal bond)	− 7.3
Glucose 1-phosphate	− 5.0
Fructose 6-phosphate	− 3.8
Glucose 6-phosphate	− 3.3
3-Phosphoglycerate	− 3.1
Glycerol 1-phosphate	− 2.3

5.6. ATP regeneration The cell drives many chemical reactions with unfavorable equilibrium constants by the coupled hydrolysis of ATP to ADP or AMP. The use of isolated enzymes for similar potential syntheses of interest would require (eventually) the transfer of a phosphate group from a system with a higher free energy of hydrolysis (in order to regenerate ATP). For the phosphate compounds listed in Prob. 5.5:

(a) What is the cost per pound of each compound with a higher free energy of hydrolysis? (Consult any biochemicals catalogue.)

(b) To produce ATP from ADP at the levels cited in Table 5.3, what concentration of the following phosphorylated species is needed: glucose 1-phosphate, fructose 6-phosphate, or glycerol 1-phosphate?

5.7. NAD regeneration in enzymatic steroid transformation A technique for steroid conversion with 20 β-hydroxysteroid dehydrogenase (20 β-HSDH) [7] consisted of three steps:

Steroid hydrogenation:

$$\text{Cortisone} + \text{NADH}_2 \underset{}{\overset{20\ \beta\text{-HSDH}}{\rightleftharpoons}} \text{NAD} + \text{H}_2 \cdot \text{cortisone} \qquad (1)$$

Ethanol oxidation:

$$\text{C}_2\text{H}_5\text{OH} + \text{NAD} \underset{\text{alcohol dehydrogenase}}{\rightleftharpoons} \text{CH}_3\text{CHO} + \text{NADH}_2 \qquad (2)$$

Acetaldehyde removal:

$$\text{CH}_3\text{CHO} + \text{NH}_2\text{NHCONH}_2 \longrightarrow \text{CH}_3\text{CHNNHCONH}_3 \qquad (3)$$
$$\text{Semicarbazide}$$

Reactions (1) and (2) together have an equilibrium constant near unity.

(a) Taking the equilibrium for (1) and (2) to be unity, evaluate the thermodynamically possible conversion of cortisone assuming that reaction (3) (which must also be reversible at equilibrium) has an equilibrium constant of 10, 10^3, 10^5, 10^{10} in units of reciprocal concentration. Plot the result as cortisone conversion vs. log K.

(b) Since steroids are relatively insoluble (saturation at 10^{-4} to 10^{-5} mol/l in water), the overall amount of cortisone in the system was increased by the addition of organic solvents, which formed a *two-phase* system, thus allowing the organic phase to hold relatively high concentrations of steroid (cortisone solubility $= 0.160$ g/100 ml in butyl acetate) which acted as a reservoir for the aqueous phase. Taking $\varepsilon =$ volume fraction of butyl acetate in water–butyl acetate emulsion, and 10^{-4} mol/l $=$ cortisone solubility in water, what is the maximum conversion possible if $H_2 \cdot$cortisone is insoluble in the solvent (assume same solubility in water as cortisone) and K for reaction (3) is 10, 10^3, 10^5? Repeat for $H_2 \cdot$cortisone solubility $= 0.160$ g/100 ml butyl acetate.

(c) Enumerate the thermodynamic devices used by these authors to drive the desired reaction (1).

5.8. Permeability. Derive Eq. (5.30). State all assumptions.

5.9. Network mass balance You have an absent minded friend who unfortunately is not a very good experimenter. He was asked to run a fermentation in a laboratory course but neglected to weigh the glucose he added and also forgot to analyze for ethanol. He has found that the fermentation produced the compounds shown in Table 5P9.1. He knows that all glucose ($C_6H_{12}O_6$) is oxidized to pyruvic acid ($CH_3COCOOH$) by glycolysis and that no other products except ethanol are formed. He comes to you for help in salvaging something out of this mess. How many moles of ethanol should you tell him have been formed? (State your assumptions.)

Table 5P9.1

	Moles
Lactic acid ($CH_3CHOHCOOH$)	10
Acetic acid (CH_3COOH)	5
Carbon dioxide (CO_2)	15
Hydrogen (H_2)	10

5.10. Definition of life Comment on the (lack of) necessity for each word in the following definition [8]: "Life is a potentially self-perpetuating open system of linked organic reactions, catalysed stepwise and almost isothermally by complex and specific organic catalysts which are themselves produced by the system." For a more recent discussion, see Ref. 9.

5.11. Free energy: electron-transfer basis The fact that intracellular precursors are typically ions and that electron transfer is intimately associated with free-energy transfer in glycolysis, TCA cycle, mitochondrial, and other typical metabolic processes suggests that energetics might most fundamentally be based on units of electron transfer. McCarty [10] suggests the use of a yield coefficient (see also Chap. 7) Y_e (grams of cell synthesized per electron equivalent) defined by

$$Y_e \equiv \frac{C}{hA}$$

where $C =$ grams of cells used to generate an electron equivalent
$h = 1.0$ (usually) $=$ number of electron moles actually transferred from a donor molecule divided by electron equivalents per mole (in proper half-cell reaction)
$A =$ electron equivalents of electron donor converted for energy per electron equivalent of cells synthesized

From the presumed cell formula $C_5H_7O_2N$, the value of C is calculated to be 5.65, assuming the half reaction for oxidation to be

$$\tfrac{1}{20}C_5H_7O_2N + \tfrac{9}{20}H_2O = \tfrac{1}{5}CO_2 + \tfrac{1}{20}HCO_3^- + \tfrac{1}{20}NH_4^+ + H^+ + e^-$$

(a) Verify the value of C above.
(b) Verify the values of A and Y_e in Table 5P11.1 from additional data in the table.

Table 5P11.1† Cell-yield coefficients estimated from energetics of substrate oxidation
$K = 0.6$, ammonia = nitrogen source

Electron donor	Electron acceptor	ΔG_r,‡ kcal	ΔG_p, kcal	A, calc	Y_e, calc, g/electron equiv
Acetate	O_2	-25.28	1.94	0.71	7.96
	NO_3^- ¶	-23.74	1.94	0.76	7.43
	SO_4^{2-}	-1.52	1.94	11.8	0.48
	CO_2	-0.85	1.94	21.1	0.27
Glucose	O_2	-28.70	-1.48	0.38	14.90
	CO_2	-4.26	-1.48	2.58	2.19
Ethanol	O_2	-26.27	0.95	0.58	9.76
	CO_2	-1.83	0.95	8.3	0.67

† P. L. McCarty, in R. Mitchell (ed.), "Water Pollution Microbiology," p. 107, Wiley-Interscience, New York, 1972.

‡ Products and reactants at unit thermodynamic activity except pH 7.

¶ Reduction to N_2.

Keeping the same value of $K = 0.60$, calculations of A were typically within 50 percent of measured values for 25 systems, about half anaerobic and half aerobic.

The value of A needed in the definition of Y_e above is estimated from

$$A = \frac{-\Delta G_p / K + \Delta G_c + \Delta G_n / K}{K \, \Delta G_r}$$

where ΔG_p = free energy required to convert carbon source used for cell synthesis to intermediate level

ΔG_n = free energy required to convert inorganic nitrogen source into ammonia, the oxidation state of nitrogen in cellular material

ΔG_c = free energy required to convert both intermediate-level carbon and ammonia into cellular material ≈ 7.5 kcal/electron equiv of cells (including inefficiencies)

K = average efficiency of free-energy transfer (range = 0.4 to 0.8 in heterotrophic or autotrophic bacteria)

ΔG_r = free-energy change per electron equivalent of substrate oxidized

Depending on the nitrogen source, the value of ΔG_n is 0 (ammonia), 3.25 (nitrite), 4.17 (nitrate) or 3.78 (N_2) kcal/mol (using assumed cell stoichiometry above).

5.12. Carried-mediated Transport Suppose component A is insoluble in a membrane of thickness L, and that its concentrations on either side of the membrane are a_1 and a_2. Carrier B, which exists only in the membrane, forms a complex AB with A at the membrane surfaces. Assuming that complex formation is always at equilibrium *at each surface* (not within membrane) and that the equilibrium constants are K_1 and K_2, what is the flux of A from side 1 to side 2 if AB has a diffusivity $\mathcal{D}$. Under what conditions (if any) can active transport occur?

REFERENCES

Reference 1 in Chap. 1 and Refs. 2 and 3 in Chap. 2 are recommended, as are the following sources:

1. Albert L. Lehninger, "Bioenergetics," 2d ed., W. A. Benjamin, Inc., New York, 1971. A superbly written paperback covering the central topic of the present chapter.

2. L. Peusner, "Concepts in Bioenergetics," Prentice-Hall, Inc., Englewood Cliffs, N.J., 1974. Another view of bioenergetics, this time combined with a more complete and self-contained treatment of thermodynamics.

3. J. Mandelstam and K. McQuillan (eds.): "Biochemistry of Bacterial Growth," 2d ed., Halsted Press, New York, 1973. A thorough chemical discussion of the reactions of interest in biosynthesis and ATP degradation. This collection, which is available in paperback, will also be a primary reference for Chap. 7.

The following references are rich in information concerning membrane transport:

4. A Kotyk and K. Janáček: "Cell Membrane Transport: Principles and Techniques," 2d ed., Plenum Press, New York, 1975.
5. E. E. Bittar (ed.): "Membranes and Ion Transport," vols. 1–3, Wiley-Interscience, New York, 1975.

Problems

6. C. N. Frey: History and Development of the Modern Yeast Industry, *Ind. Eng. Chem.*, **22:** 1154, 1930.
7. P. Cremonesi, G. Carrea, L. Ferrara, and E. Antonini: Enzymatic Preparation of 20β-Hydroxysteroids in a Two-Phase System, *Biotech. Bioeng.*, **17:** 1101, 1975.
8. J. Perrett: *New Biol.*, **12:** 68, 1952.
9. J. D. Bernal: "Theoretical and Mathematical Biology," p. 96ff., Blaisdell Publishing Company, New York, 1965.
10. P. L. McCarty: Energetics of Organic Matter Degradation, in R. Mitchell (ed.), "Water Pollution Microbiology," p. 91, Wiley-Interscience, New York, 1972.

CELLULAR GENETICS AND CONTROL SYSTEMS

The previous chapters have discussed enzyme kinetics and the gross features of some common microbial reaction networks. In a simpler chemical process plant, only a few products are synthesized, and the *steady-state* operation usually sought requires relatively minimal controls. The complex chemical plant which is the living cell is continually in a *transient* state; the cell alters the proportions of substrates, enzyme catalysts, RNA, cofactors, and other constituents as it procedes through the stages of its *life cycle*. Synthesis and regulation of the systems which control cellular dynamics are the subjects of this chapter. Important applications of these topics occur in viral infections of cultures, natural and induced mutations, genetic manipulations, and the versatility with which some cultures utilize multiple substrate feeds.

6.1. MOLECULAR GENETICS

To discuss cell genetics and expression of genetic information at the molecular level, we must return to the composition and structure of DNA, the three RNA varieties, and proteins presented in Chap. 2. In order to appreciate the significance of the mechanisms discussed below, we should remember that proteins, particularly in their role as enzymes, directly determine how the cell operates by determining which chemical reactions occur. Consequently, in a sense the enzyme makeup of a cell defines the cell's metabolic processes: specifying the concentration and kind of enzymes present determines the cell's reaction-network characteristics. Also, other proteins in the cell serve important functions (recall Table 2.6). Thus the inheritance which a dividing cell leaves to daughter cells must include the information for directing the production of the same protein constituents as found in the parent cell. The study of transmission of such information is called *genetics*. We shall next investigate how the nucleotide base sequence in deoxyribonucleic acid (DNA), a nonrepetitive and therefore informational biopolymer, determines the course of protein synthesis in the cell.

6.1.1. DNA Translation for Protein Synthesis

We already know that the DNA molecule is a very long double helix consisting of two chains. Each chain is a polymer constructed from four nucleotides: adenine (A), guanine (G), cytosine (C), and thymine (T). Moreover, the two chains in the DNA double helix have complementary sequences: A in one strand is always paired with T in the other, and G and C are likewise paired (Fig. 2.11).

As outlined in Sec. 2.3.2, the nucleotide sequence of DNA is translated into protein structure in the following manner. *Messenger ribonucleic acid* (mRNA) is synthesized using a DNA strand as a template. In order to achieve this, the DNA double helix unwinds as a result of attachment of RNA polymerase, a complex oligomeric enzyme containing five different peptide chains (Fig. 6.1). Then, a molecule of mRNA is assembled by the pairing of complementary nucleotides to each of the bases in a DNA segment (Fig. 6.2). Base pairing during this process resembles the pairing found in the DNA double helix except that uracil (U) rather than thymine appears in RNA (see Fig. 2.6). Consequently the pairs form as follows:

DNA (template)	RNA (being assembled)
T	A
A	U
C	G
G	C

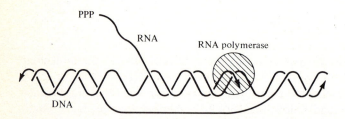

Figure 6.1 RNA polymerase opens a segment of the DNA double helix and moves to transcribe the DNA template onto an RNA stand. (*Reprinted by permission from J. D. Watson, "Molecular Biology of the Gene," 2d ed., p. 341, W. A. Benjamin, Inc., New York, 1970.*)

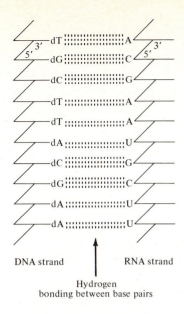

DNA strand   RNA strand

Hydrogen
bonding between base pairs

Figure 6.2 Complementary pairing between DNA and RNA bases provides the necessary information transfer during transcription.

The length of the mRNA molecule varies from about 300 nucleotide units to more than 3000. Stop and start signals in the form of a base sequence on the DNA strand determine the size of mRNA. Often, the mRNA corresponds to one *gene* on the DNA chain, where a gene is a nucleotide sequence containing the information for the synthesis of one protein. In some cases the mRNA molecule carries the genetic message from a group of related genes called an *operon* (see Sec. 6.1.3). Construction of mRNA with base ordering complementary to a DNA segment is known as *transcription*.

As we noted when discussing the energetics of biosynthesis in Chap. 5, an RNA chain is built up by splitting pyrophosphate from the triphosphate form of the nucleotides. Consequently, the equivalent of two high-energy phosphate bonds is invested in every RNA phosphodiester bond. This ensures that the synthesis reaction proceeds to "completion" and provides us with some indication of the importance of biopolymer synthesis to normal cell function.

The genetic information carried by mRNA is *translated* into the amino acid sequence of a protein by the ribosomes and *transfer ribonucleic acid* (tRNA). In a rough analogy with a magnetic tape player, the ribosome aligns and transports the tape (mRNA) while tRNA serves as the playback head which converts the information on mRNA into the desired output, an amino acid sequence.

Before reviewing the translation process, we must look more carefully at tRNA. It is known that each specific tRNA molecule can carry only one corresponding amino acid. Figure 6.3 shows the secondary structure for alanine tRNA from yeast. Attachment of an alanyl residue to its tRNA first requires activation to alanyl acyl-AMP, as depicted in Eq. (5.29). The activated residue is then bound to this particular tRNA by an enzyme specific for both the residue and tRNA. This ensures that only alanine is linked to alanine tRNA.

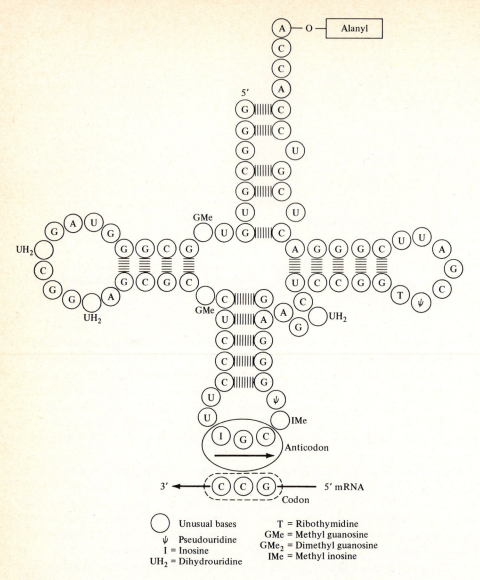

Figure 6.3 The anticodon of alanine tRNA recognizes the complementary three base codon in mRNA. (*Reprinted by permission from J. D. Watson, "Molecular Biology of the Gene," 2d ed., p. 361, W. A. Benjamin, Inc., New York, 1970.*)

Now we can turn to the specificity of RNA. At the base of the lower loop of the tRNA molecule is a sequence of three nucleotides called an *anticodon*. Such a structure is found in all tRNAs so far studied. This anticodon base sequence is complementary to a three-nucleotide segment of mRNA known as a *codon*. Biochemical research has revealed that each codon is a "word" in the genetic message; each codon specifies one amino acid. Since there are only four letters in the chemical alphabet of RNA (the four bases A, C, G, U) and there must be at least 20 words (one for each amino acid), a language or code is required for transmission of genetic information. Amazingly enough, the *genetic code* has been completely deciphered; further, it appears to be essentially universal in all living organisms—microbe, plant, and animal (Fig. 6.4).

Now we are ready to put the pieces together in an overview of the translation process, which is illustrated schematically in Fig. 6.5. After the mRNA has formed a complex with the ribosome, tRNAs carrying the amino acids corresponding to the first two mRNA codons are attached to the complex. With the amino acid sequence arranged, a peptide bond is formed between amino acid residues 1 and 2. The mRNA "tape" then shifts to the right by one codon, and tRNA for the first amino acid is released. The third amino acid, carried by its tRNA, then binds to the complex with subsequent attachment to the elongating peptide chain. This process continues until a stop codon is encountered (see Fig. 6.4), at which point the ribosome and mRNA separate.

Often more than one ribosome is attached to a single mRNA chain at a time.

	second				
first	U	C	A	G	third
U	Phe	Ser	Tyr	Cys	U
	Phe	Ser	Tyr	Cys	C
		Ser			A
	Leu	Ser		Try	G
C		Pro	His	Arg	U
	Leu	Pro	His	Arg	C
		Pro	Gln	Arg	A
	Leu	Pro	Gln	Arg	G
A	Ileu	Thr	Asn	Ser	U
	Ileu	Thr	Asn	Ser	C
		Thr	Lys	Arg	A
	Met	Thr	Lys		G
G	Val	Ala	Asp	Gly	U
	Val	Ala	Asp	Gly	C
	Val	Ala	Glu	Gly	A
	Val	Ala	Glu	Gly	G

Figure 6.4 The genetic code. Three codons (OCH, AMB, END) mean stop synthesis of the peptide chain. Examination of this chart reveals that a shift in one base in a codon often gives a similar amino acid, a "fail-safe" feature. [*Reprinted by permission of Dr. H. G. Khorana and the National Academy of Science (U.S.). Reprinted from D. Soll et al.; Studies on Polynucleotides: XLIX. Stimulation of the Binding of Aminoacyl s-RNA's to Ribosomes by Ribotrinucleotides and a Survey of Codon Assignments for 20 Amino Acids, Proc. Nat. Acad. of Sci. U.S.A.,* **54**: 1379, 1965.]

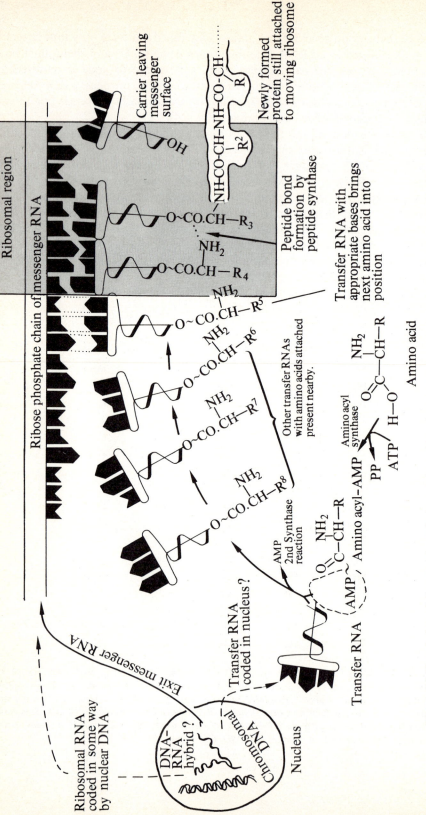

Figure 6.5 Translation of the genetic code into polypeptide primary structure occurs via specific interactions between mRNA and tRNA's at the ribosome. (*Reprinted by permission from N. A. Edwards and K. A. Hassall, "Cellular Biochemistry & Physiology," p. 342, McGraw-Hill Publishing Company Ltd., London, 1971.*)

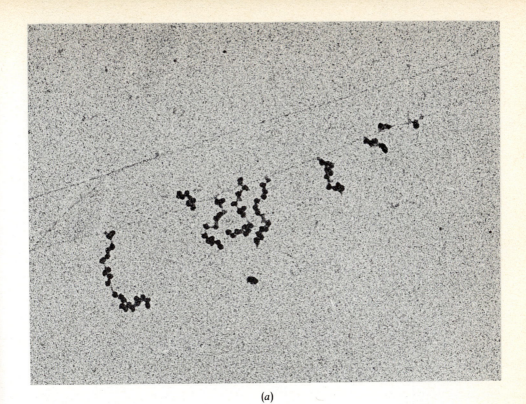

(a)

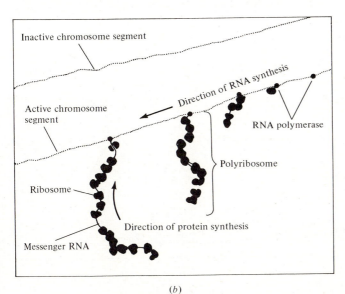

(b)

Figure 6.6 Chromosome segments, RNA polymerases, and polyribosomes shown in this 150,000 × electron micrograph (a) dramatically illustrate several basic principles of molecular genetics. The schematic drawing (b) aids in interpreting the photograph. [(a) *Electron micrograph by O. L. Miller, Jr., B. A. Hamkalo, and C. A. Thomas, Jr., Visualization of Bacterial Genes in Action, Science,* **169**, 392, 1970; (b) *From "The Visualization of Genes in Action," by O. L, Miller, Jr. Copyright © 1973 by Scientific American, Inc. All rights reserved.*]

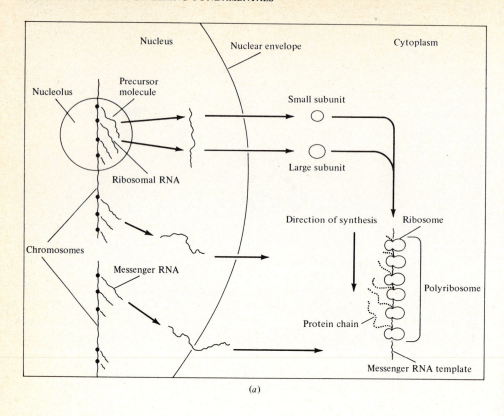

(a)

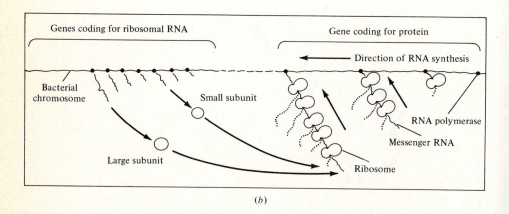

(b)

Figure 6.7 Spatial and temporal organization of molecular genetics in eucaryotes. (a) Transcription occurs in the nucleus, while translation is accomplished in the cytoplasm. As illustrated in part (b), a much more compact procedure prevails in procaryotes, where transcription and translation are localized and simultaneous. (*From " The Visualization of Genes in Action," by O. L. Miller, Jr. Copyright © (1973) by Scientific American, Inc. All rights reserved.*)

Polyribosomes are clusters containing several individual ribosomes. This arrangement, which can be highly organized, permits simultaneous translation of information carried by different portions of mRNA and thus lends speed and efficiency to protein synthesis. The electron micrograph in Fig. 6.6 provides direct, graphic evidence that the mechanism just described operates in *E. coli*. Elongating mRNA chains are seen extended from a DNA strand, and clusters of ribosomes are attached to each mRNA branch.

The molecules of *ribosomal ribonucleic acid* (rRNA) and tRNA, like mRNA, are synthesized using a portion of the DNA chain as a template. Thus, the information carried by DNA includes the structure of the translation devices for protein synthesis as well as the coded information for protein primary structure.

We should recall at this point that the interactions between the amino acids in the resulting sequence determines (Sec. 2.4) secondary, tertiary, and even the quaternary structure of proteins. The importance of weak interactions between atoms at all stages of genetic information storage, transmittal, and action should now be apparent.

Before discussing how genetic information is passed from generation to generation and how this process can be disturbed, we should recognize that the protein-synthesis mechanisms of eucaryotes and procaryotes are different. Figure 6.6 has already shown that transcription-synthesis of mRNA and translation of mRNA to proteins on the ribosomes occur at almost the same point within the nuclear region of the bacterial cell. Figure 6.7 reveals that in eucaryotes, the transcription and translation steps are conducted at different places in the cell. Additionally, the timing of the various events involved in protein assembly is different in eucaryotes and procaryotes. More on this will follow in Sec. 6.2.

6.1.2. DNA Replication and Mutation

Since DNA contains the information required for proper development and function of the living cell, it is essential that there be an almost foolproof mechanism for copying DNA. When a cell reproduces, each of the offspring must receive a complete set of genetic data in the form of DNA. We have already noted that the double-helical model of DNA structure provides a ready hypothesis for faithful reproduction of the molecule (Fig. 2.12). After the original strands unwind, new complementary chains are assembled, so that ultimately two intact DNA molecules result. As a result of this scheme, each offspring molecule contains one nucleotide chain from the parent molecule. This presumably helps minimize errors in the replication process.

Actually, DNA replication occurs in a slightly more complex fashion than envisioned in Fig. 2.12. It is not built up in one continuous sequential pass; instead an enzyme called *DNA polymerase* switches from one parent strand to another, adding about 1000 nucleotides to a daughter chain during each pass (Fig. 6.8). Such alternation is necessary because this enzyme can catalyze nucleotide addition only in the 5' to 3' direction (recall Fig. 2.9). The subchains constructed by

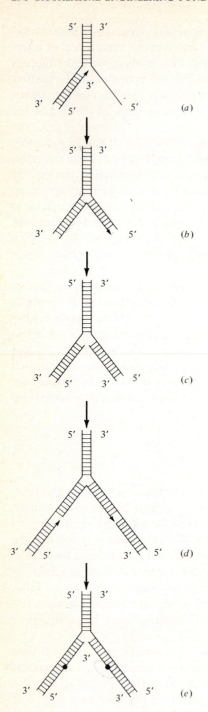

Figure 6.8 Replication of DNA requires the enzyme DNA polymerase and DNA ligase: (*a*) a new segment grows in 5′ → 3′ direction along one strand of parental double helix; (*b*) by switching to the other parent strand, new segment continues 5′ → 3′ growth; (*c*) at switch point, new strand is cleaved; (*d*) parental strands separate further, steps (*a*) and (*b*) are repeated; (*e*) cleavage again; new sections joined by ligase at (·). (*Reprinted by permission from A. L. Lehninger, "Bioenergetics," 2d ed., p. 159, W. A. Benjamin, Inc., Palo Alto, Ca., 1974.*)

DNA polymerase are linked by a second enzyme, *DNA ligase*, to obtain completed complementary base strands.

We shall next briefly outline some differences in DNA storage and duplication between procaryotes and eucaryotes. In the ill-defined nuclear region of procaryotes, there is a single *chromosome*, or carrier of genetic information, which consists of a circular double strand of DNA. This huge molecule is 1.2 mm long, about 20 Å thick, and has a molecular weight on the order of 2.8 billion. This provides enough storage capacity to code for about 2000 different proteins. Figure 6.9 shows an image of an actual *E. coli* chromosome as well as a schematic diagram showing how the circular DNA molecule is copied. The circular nature of *E. coli* DNA was first demonstrated by genetic studies aimed at finding the relative position of the genes. A chromosome in eucaryotes consists of a DNA molecule associated with proteins and possibly some RNA. As Table 6.1 reveals, eucaryotic cells typically contain several chromosomes.

A *mutation* is a change in a chromosome which is passed to succeeding generations. In molecular terms, mutation involves an alteration in the nucleotide sequence of DNA. Several possibilities are given in Fig. 6.10. To some extent, mutation is a spontaneous process which is constantly occurring. The rate of spontaneous mutation is rather low, however. A typical value is 1 error for every 10^6 gene duplications.

The importance of a gene-copying error depends on its nature. In *missense mutation*, a codon for one amino acid is altered so that a different amino acid is inserted into the protein. This type of mutation leads, for example, to the abnormal hemoglobin characteristic of sickle-cell anemia in human beings. In this instance the abnormal protein contains Val instead of Glu. Referring to the genetic code in Fig. 6.4, we see the codons GAA and GAG code for Glu while valine is called for by the codons GUA and GUG. Apparently then, the switch of only one base pair in a human chromosome can lead to serious genetic disease.

Table 6.1 Normal chromosome number in various species

Species	Number
Procaryotes (haploid: each chromosome present once):	
Bacteria	1
Eucaryotes (diploid: each chromosome except sex present twice per cell):	
Red clover	14
Honeybee	16
Frog	26
Hydra	30
Cat	38
Rat	42
Man	46
Chicken	78

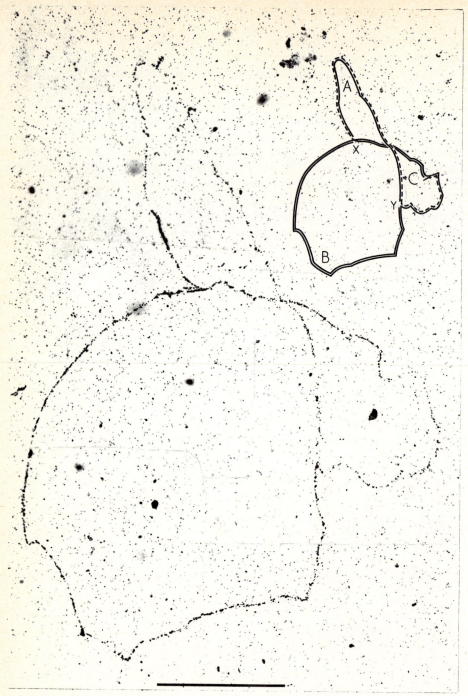

Figure 6.9 The autoradiograph in (*a*) provides an image of replication of an *E. coli* circular chromosome; (*b*) interpretation of the image. [(*a*) *Courtesy of J. Cairns.*]

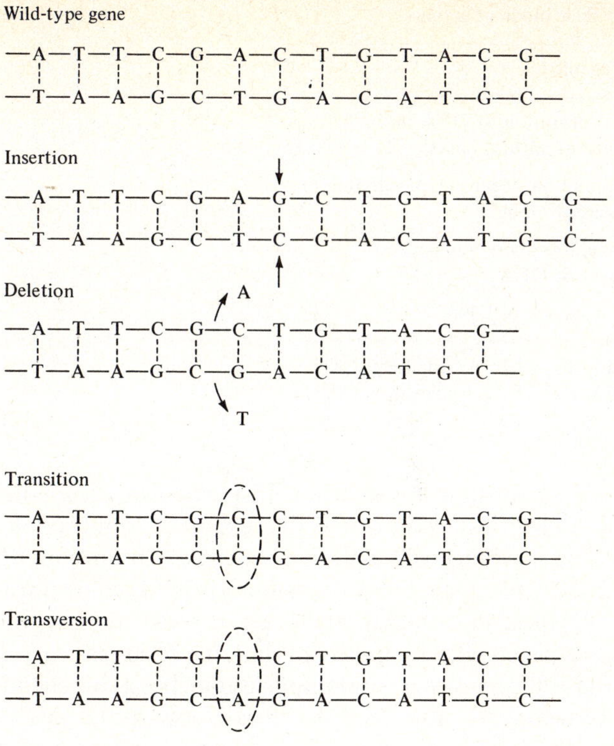

Wild-type gene

```
—A—T—T—C—G—A—C—T—G—T—A—C—G—
  |  |  |  |  |  |  |  |  |  |  |  |  |
—T—A—A—G—C—T—G—A—C—A—T—G—C—
```

Insertion

```
—A—T—T—C—G—A—G—C—T—G—T—A—C—G—
  |  |  |  |  |  |  |  |  |  |  |  |  |  |
—T—A—A—G—C—T—C—G—A—C—A—T—G—C—
```

Deletion

```
—A—T—T—C—G—C—T—G—T—A—C—G—
  |  |  |  |  |  |  |  |  |  |  |
—T—A—A—G—C—G—A—C—A—T—G—C—
```

Transition

```
—A—T—T—C—G—G—C—T—G—T—A—C—G—
  |  |  |  |  |  |  |  |  |  |  |  |  |
—T—A—A—G—C—C—G—A—C—A—T—G—C—
```

Transversion

```
—A—T—T—C—G—T—C—T—G—T—A—C—G—
  |  |  |  |  |  |  |  |  |  |  |  |  |
—T—A—A—G—C—A—G—A—C—A—T—G—C—
```

Figure 6.10 Different types of mutation in the base pair sequence of a DNA molecule.

Other alterations in codons can give rise to the OCH, AMB, or END stopping codons and cause premature termination of peptide synthesis; they are known as *nonsense mutations*.

There are several postulated mechanisms for spontaneous mutation. First, the nucleotide bases of DNA have several different structural forms, known as *tautomers*. Although the configurations given in Fig. 2.11 are believed to predominate, shifts to other tautomeric forms could cause errors in base pairing. Another possible cause of spontaneous mutation is interference with the enzymes necessary for DNA synthesis and repair. Finally, some intermediates of normal cellular metabolism, e.g., peroxides, nitrous acid, and formaldehyde, are *mutagens*, chemicals which interfere with normal DNA operation.

The action of chemical mutagens on DNA has been widely studied by growing cells in environments rich in such agents. Among the several classes of known mutagens are base analogs, which are compounds with structures similar to the bases normally found in DNA. Consequently, the analogs rather than the proper bases may be incorporated during synthesis of a DNA chain. Other types of mutagens and their mode of interference are listed in Table 6.2.

Table 6.2. Chemical mutagens and their mode of action†

Chemical agent	Mutagenic effect
Base analogs	Incorporation into DNA in place of natural bases
Nitrous acid	Deamination of purine, pyrimidine bases of DNA
Proflavin	Intercalation between stacked bases of DNA
Alkylating agents	Depurination of DNA

† From R. Y. Stanier, M. Doudoroff, and E. A. Adelberg: "The Microbial World," p. 418, Prentice-Hall, Inc., Englewood Cliffs, N.J., 1970.

Another common cause of mutation is radiation. In particular, ultraviolet light is strongly absorbed by DNA, to such an extent that exposure to ultraviolet light rapidly kills most cells. The surviving cells exhibit a high rate of mutation. All cells possess enzymatic machinery to repair DNA occasionally damaged by ultraviolet light. These repair enzymes, in a rather elaborate process, replace the damaged segment, which contains covalently linked pyrimidine residues.

The phenomenon of mutation occupies several important niches in biochemical engineering practice. Returning to Fig. 1.15, we recall how mutations in microorganisms can improve them for our use. Thus, mutagens and exposure to ultraviolet light have been employed in attempts to obtain mutated, more productive protists. Essential in any strain-development activities are effective means for identifying and isolating mutants with specific characteristics. Table 6.3 summarizes some of the basic techniques for this purpose. Recent *in vitro* procedures for recombinant DNA are discussed in Sec. 6.3.3 and in Probls. 6.10 and 8.20.

On the other hand, mutations can create processing difficulties. Successful operation of a microbial reactor often requires the availability of a pure strain of microorganisms with known characteristics. Consequently, *stock cultures* of the needed microbe must be maintained. Degradation of these cultures by mutation is always a possibility, so that regular testing of the stock is necessary to preserve its integrity. Other practical problems concerned with mutations will be explored in the following chapters.

6.1.3. Induction and Repression: Enzyme Control at the Gene Level

That living cells possess intricate control systems to ensure efficient use of material and energy resources is already known to us. We have previously explored how activation and inhibition of enzyme activity by metabolites helps channel these intermediates through the complex network of cellular reactions. In this section, we explore another level of control, which must be carefully distinguished from those discussed earlier. *Activation and inhibition influence the catalytic activity of*

Table 6.3. Methods for detection and selection of different bacterial mutants†

Type of mutant	Selection methods	Detection
Able to use as carbon source a compound not utilizable by the wild type	Plate on agar containing the compound in question as the only available carbon source	Plating method is absolutely selective; only the desired type will form colonies
Resistant to inhibitory chemical agents, such as penicillin, streptomycin, sulfonamides, dyes	Plate on agar containing the inhibitor	Plating method is absolutely selective; only the desired type will form colonies
Resistant to bacteriophage	Plate on agar previously spread with a suspension of phage	Plating method is absolutely selective; only the desired type will form colonies
Auxotrophic (requirement for one or more growth factors not required by wild-type cells)	Incubate in growth medium lacking the growth factor in question but containing penicillin; wild-type cells multiply and most are killed; auxotrophic mutants are unable to multiply without the growth factor and survive (penicillin kills only actively growing cells)	Plate on agar lacking the growth factor; mark the few wild-type colonies that appear, then add layer of agar containing growth factor; auxotrophs form colonies only after addition of growth factor (delayed-enrichment method)
Unable to ferment a given sugar	Apply penicillin technique as above, using sugar in question as only fermentable carbon source	Plate on agar containing sugar in question, plus chemical indicator that changes color in the presence of fermenting cells, e.g., acid-base indicator; mutant colonies appear a contrasting color
Temperature-sensitive DNA synthesis	Incubate cell suspension at 42°C for 15 min, add 5-bromouracil (5-BU), and continue incubation an additional 60 min; irradiate with light of 310 nm wavelength (cells that have incorporated 5-BU into their DNA are killed; cells that have failed to replicate their DNA at 42°C are spared)	Test colonies for ability to grow at 42°C, e.g., by replica plating
Sensitive to ultraviolet irradiation (unable to repair ultraviolet-light damage)	Infect culture with ultraviolet-light-inactivated bacteriophage; normal cells repair the ultraviolet-light lesions of the phage and are subsequently killed by the phage; repair-deficient cells are spared	Make suspensions from isolated colonies, measure survival at several ultraviolet light doses

† R. Y. Stanier, M. Doudoroff, and E. A. Adelberg, "The Microbial World," 3d. ed., p. 450, Prentice-Hall, Inc., Englewood Cliffs, N.J., 1970.

enzymes already present in the cell, but they do not alter the amount of enzyme present. Control occurs at the enzyme level. Enzyme *induction and repression,* the control devices of interest now, *cause change in the rates of enzyme synthesis* and act at the level of the gene. Perhaps the greatest similarity between these two kinds of control is their sensitivity to the concentration of relatively small molecules.

Genetic-level controls are most widespread in microorganisms, especially bacteria. Before delving into the details of the mechanisms, it may be helpful to place the matter in microbiological perspective. We should recall from Chap. 1 that bacteria are free-living, isolated cells which consequently have almost no

Table 6.4. The bacterium *Pseudomonas multivorans* can utilize any organic compound in this list as its sole carbon source†

Carbohydrates and carbohydrate derivatives (sugar acids and polyalcohols):	Caprylate	Aspartate
	Pelargonate	Glutamate
	Caprate	Lysine
Ribose	Dicarboxylic acids:	Arginine
Xylose	Malonate	Histidine
Arabinose	Succinate	Proline
Fucose	Fumarate	Tyrosine
Rhamnose	Glutarate	Phenylalanine
Glucose	Adipate	Tryptophan
Mannose	Pimelate	Kynurenine
Galactose	Suberate	Kynurenate
Fructose	Azelate	Other nitrogenous compounds:
Sucrose	Sebacate	Anthranilate
Trehalose	Other organic acids:	Benzylamine
Cellobiose	Citrate	Putrescine
Salicin	α-Ketoglutarate	Spermine
Gluconate	Pyruvate	Tryptamine
2-Ketogluconate	Aconitate	Butylamine
Saccharate	Citraconate	Amylamine
Mucate	Levulinate	Betaine
Mannitol	Glycolate	Sarcosine
Sorbitol	Malate	Hippurate
Inositol	Tartrate	Acetamide
Adonitol	Hydroxybutyrate	Nicotinate
Glycerol	Lactate	Trigonelline
Butylene glycol	Glycerate	Nitrogen-free ring compounds:
Fatty acids:	Hydroxymethylglutarate	Benzoyl formate
Acetate	Primary alcohols:	Benzoate
Propionate	Ethanol	*o*-Hydroxybenzoate
Butyrate	Propanol	*m*-Hydroxybenzoate
Isobutyrate	Butanol	*p*-Hydroxybenzoate
Valerate	Amino acids:	Phenyl acetate
Isovalerate	Alanine	Phenol
Caproate	Serine	Quinate
Heptanoate	Threonine	Testosterone

† R. Y. Stanier, M. Doudoroff, and E. A. Adelberg, "The Microbial World," 3d ed., p. 70, Prentice-Hall, Inc., Englewood Cliffs, N.J., 1970.

influence on their external environment. Therefore, they must be very adaptable: it may be essential for their survival that they function efficiently in variable surroundings.

Many bacteria indeed possess such versatility. They can synthesize enzyme systems for effective utilization of many different types of nutrients. For example, Table 6.4 lists over 90 different compounds which can serve *Pseudomonas multivorans* as its sole source of carbon. Different enzymes are generally required for each nutrient. Consequently, the bacterium must carry genetic information of them all. The sum total of all information carried in the chromosomes is called the cell's *genotype*.

However, an organism such as the bacterium of Table 6.4 will not require the same mix of enzymes for all these nutrients, and synthesis of any extra protein represents a waste of valuable energy and intermediary metabolites. Thus, for maximum efficiency in a particular environment, only a portion of the total genetic information should be expressed (actually synthesized) as protein. This indeed occurs in a number of instances. The term *phenotype* denotes the observable features of an organism. In view of the previous comments, the phenotype arises from a combination of two factors, the genotype and the organism's environment. Not all aspects of the phenotype are influenced by the environment: *constitutive enzymes* are synthesized at rates independent of the cell's surroundings.

The biosynthesis rate of *inducible enzymes*, however, are sensitive to environmental influence. A common example is β-galactosidase synthesis. This enzyme catalyzes the hydrolysis of lactose to its component parts (Fig. 6.11); the reaction

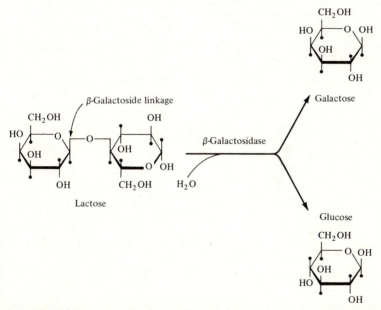

Figure 6.11 The enzyme β-galactosidase hydrolyzes lactose into galactose and glucose, substrates suitable for subsequent metabolic utilization by the cell.

is essential if the cell is to employ lactose as a nutrient since only the hydrolysis products can be used in subsequent reactions. As Fig. 6.12 illustrates, the amount of β-galactosidase in the cell is related to the need for this enzyme. When its substrate level rises, larger amounts of the enzyme are produced up to a maximum value. Thus, the substrate *induces* the formation of the enzyme.

An analogous situation arises with the *repressible enzymes*. For example, although *E. coli* can make all the enzymes necessary to synthesize all 20 amino acids from simpler precursors, the enzymes are not found in significant quantity when the amino acids are supplied to the cell as nutrients. In this instance the end product of a biosynthetic pathway *represses* synthesis of the enzymes for that pathway. This phenomenon is demonstrated in Fig. 6.12 using histidine synthesis as an example.

Useful models for induction and repression, developed by Monod and his coworkers in France, are given in Fig. 6.13, which reveals the basic similarities of the two mechanisms. In both, a regulator gene produces a protein involved in control of the operator gene. If the operator gene is blocked, the following structural gene (which codes for the enzyme of interest) is not transcribed.

The regulator gene in the induction model (Fig. 6.13*a*) produces a repressor molecule which can prevent enzyme production. When the inducer is present, it binds with the repressor to form an inactive complex, which does not interfere with subsequent DNA transcription. In the repression model of Fig. 6.13*b*, the regulator gene must complex with another molecule to yield a repressor. Without the corepressor (histidine in the example above), enzyme synthesis proceeds.

The regulator and operator gene may simultaneously control synthesis of several enzymes. For example, when production of β-galactosidase is induced, so are two other enzymes. One of these is a galactoside permease, which participates in active transport of β-galactosides. Obviously, there is a logical connection between the functions of these two enzymes, and coordinating their biosynthesis

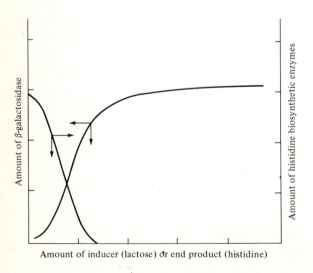

Figure 6.12 Cell content of β-galactosidase increases with increasing concentration of inducer (lactose) in the growth medium. Increased medium concentration of a repressor (histidine) decreases cellular content of enzymes catalyzing repressor biosynthesis. (*Reprinted by permission from J. D. Watson, "Molecular Biology of the Gene," 2d. ed., p. 439, W. A. Benjamin, Inc., New York, 1970.*)

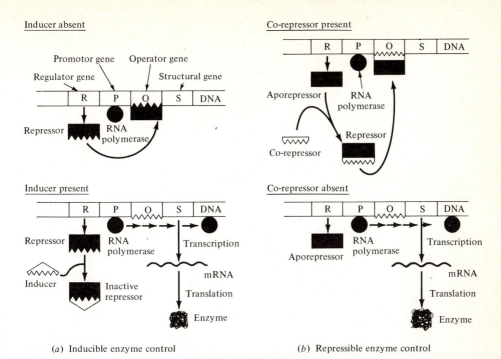

Figure 6.13 (*a*) An inducer acts by inactivating repressor so that the repressor does not bind to the operator gene and block transcription of the structural gene. (*b*) A (co-) repressor combines with an aporepressor to form an active repressor, which can block expression of the structural gene by binding at the repressor gene. In this manner synthesis of repressible enzymes is controlled. [*Reprinted from A. L. Demain, Theoretical and Applied Aspects of Enzyme Regulation and Biosynthesis in Microbial Cells, in L. B. Wingard, Jr. (ed.), "Enzyme Engineering," Interscience, New York, 1972.*]

therefore makes good sense. Such a collection of several genes which are regulated together is called an *operon*. The case just mentioned is the *lac operon*, whose features are summarized schematically in Fig. 6.14.

There are several variations on these themes of genetic level control. While we cannot pursue them all, the *glucose effect* should be mentioned. If *E. coli* bacteria are cultivated in a medium containing glucose and a less easily metabolized carbon source such as lactose, the glucose is preferentially consumed. Under these circumstances lactose does *not* induce β-galactosidase. In the cell growing rapidly on glucose, formation of cyclic AMP is inhibited. Cyclic AMP (Fig. 2.7) is apparently required for RNA polymerase to recognize the lac operon, so that a reduction in cyclic AMP concentration stops the induction process. Since the amount of cyclic AMP is dictated by catabolic products of glucose, this control device is often called *catabolite repression*.

It should be emphasized that catabolite repression can occur in the absence of glucose. In general, a bacterium growing in a mixture of carbon sources will selectively utilize the best (the one that gives the fastest growth rate) and catabolically repress utilization of the less useful nutrients. What would happen when the first carbon source is exhausted? (See Chap. 7, fermentation patterns.)

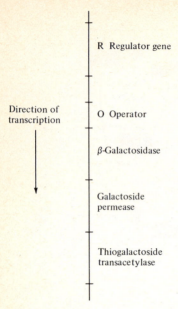

Direction of transcription

R Regulator gene

O Operator

β-Galactosidase

Galactoside permease

Thiogalactoside transacetylase

Figure 6.14 In the lac operon, a single regulator-operator pair of genes controls synthesis of three proteins with related biological functions.

6.1.4. Overview of Information Flow in the Cell

A familiar cliché asserts that one picture is worth a thousand words. Figure 6.15 proves the point: here in one compact schematic all the cell's control and information carrying channels are summarized. Their interrelationships are clearly revealed. Notice that a type of cascade control is embodied here: activation and inhibition (at the enzyme level) allow rapid adjustment to short-term changes in the cell's chemical balance, whereas induction and repression (at the gene level) readjust the entire metabolic pattern when a long-term disturbance in the cell's environment appears. To better understand this process we must consider the time sequence of events in a growing cell.

6.2. GROWTH AND REPRODUCTION OF A SINGLE CELL

We may define *growth* of an organism as an orderly increase in all its chemical constituents. When most single-celled organisms grow, they eventually divide. Consequently, growth of a population usually implies an increase in the *number* of cells as well as the *mass* of all cellular material: either parameter may be used to investigate cell growth quantitatively. Actually, several different measures of mass are in use. Figure 6.16 summarizes them and shows their relationships.

In the following chapter we shall attempt to understand growth when many cells coexist in a nutrient solution. A simpler objective faces us in this section. Here we shall review the events of the *cell cycle*, which is the time interval between the formation of *one* cell by division of its mother cell and the subsequent division

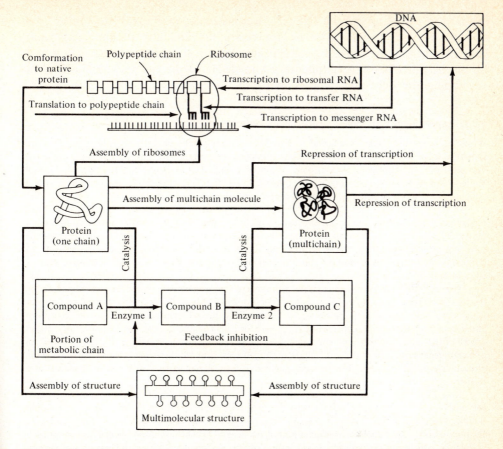

Figure 6.15 A hierarchy of control and information transmittal and processing systems is evident in this schematic illustration of information flow in the cell. (*From "Cell Structure and Function,"* 2d ed., p. 31, by Ariel G. Loewy and Philip Siekevitz. Copyright © 1963, 1969 by Holt, Rinehart and Winston, Inc. Reprinted by permission of Holt, Rinehart and Winston.)

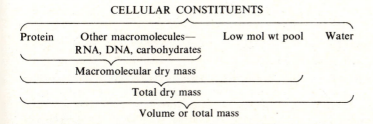

Figure 6.16 Different measures of biomass include different cellular components.

of that one cell. Study of the cell cycle is a necessary preliminary to Chap. 7 for two reasons: (1) it will make us appreciate the difficulty of analyzing what happens when many coexisting cells are simultaneously growing and dividing, and (2) on the other hand, a rudimentary knowledge of the cell cycle will aid us in developing the necessary models in Chap. 7.

6.2.1. Experimental Methods: Synchronous Cultures

While the interested reader is referred to Mitchison [2] for details on experimental techniques, we should at least outline the principal methods employed for experimental elucidation of the cell cycle. The most obvious approach, that of observation of a single cell as it passes through its cycle, is of limited value. It is time-consuming and the information it offers is restricted because measurements are difficult. Only in special cases is the method of much value.

It would obviously be simpler to perform biochemical analyses on a large number of cells. Unfortunately, if we have many cells in suspension, it is highly probable that at any instant different individual cells are at different stages of the cell cycle; i.e., some of the suspended cells are very young, some are mature and on the brink of dividing, and others are in between. In such a heterogeneous population of individuals, with most conventional methods one measures averages and not character-istics of individuals.

To escape this difficulty, the ingenious technique of *synchronous culture* has been developed. In the modern practice of the *selection-synchrony* version of this method, the first step is collection of cells which are at the same stage of the cell cycle. This can be achieved by attaching cells to a solid support and washing off newborn daughter cells as they are formed. The cells leaving the column at any instant are all of the same age, hence synchronous. Another practice involves density-gradient centrifugation of a cell suspension: cells at different stages of the cell cycle generally separate because of differences in size and density. Then, the *homogeneous population* from a particular density zone is placed in a nutrient solution and allowed to grow.

If we then monitor the number concentration of cells as a function of time, a stepwise increase in cell numbers is expected. This should occur because once every cell cycle, the number of cells should double. The experimental results in Fig. 6.17 illustrate this behavior. For at least the first two cycles, the population of cells appears to remain homogeneous and synchronized: all the cells are at approxi-mately the same point in their cycle at any instant of time. As already observed in Chap. 1, a homogen-eous population is convenient to work with experimentally since it provides a useful amplification (millions of cells instead of one) of individual characteristics. In this manner many features of the cell cycle can be studied in detail.

Another common procedure in synchronous cultures is *induction synchronization*. Here, some environmental conditions are varied periodically in order to force a population into a synchronized stepwise growth pattern. Since the normal metabolic processes of the cells may be disturbed by such treatment, selection synchrony is a preferable practice for probing cellular physiology.

We should note carefully that in Fig. 6.17 synchrony seems to be fading as growth proceeds. The initially stepwise increase in cell numbers gradually shifts to a more continuous pattern of growth. This

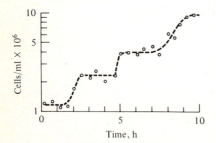

Figure 6.17 Stepwise increases in cell numbers occur during synchronous culture of the yeast *Saccharomyces pombe*. [*Reprinted from J. M. Mitchison and W. S. Vincent, A Method of Making Synchronous Cell Cultures by Density Gradient Centrifugation in I. L. Cameron and G. M. Padilla (eds.), "Cell Synchrony,"* p. 328, Academic Press, New York, 1966.]

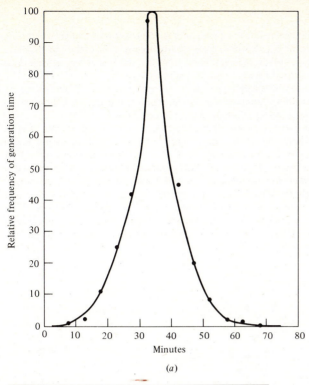

(a)

Range of generation times, min	Number of cells	Range of generation times, min	Number of cells
5–10	0	40–45	45
10–15	1	45–50	20
15–20	11	50–55	8
20–25	25	55–60	2
25–30	42	60–65	2
30–35	97	65–70	0
35–40	65	70–75	5

(b)

Figure 6.18 Different individuals in a population of the bacterium *Aerobacter aerogenes* may have significantly different generation times as the graph (*a*) and table (*b*) show. (*Reprinted from A. C. R. Dean and C. Hinshelwood, Growth, Function and Regulation in Bacterial Cells, pp. 372, 373, Oxford University Press, London, 1966.*)

occurs because the cell population is losing its homogeneity. Behind this tendency toward heterogeneity is the stochastic nature of cell growth and multiplication. While we shall pursue this matter further when analyzing cell populations in Chap. 7, the data presented in Fig. 6.18 illustrate one manifestation of the problem. The length of the cell cycle is not a definite, fixed quantity for any single cell. Thus, two initially identical daughter cells may divide at different times. However, the length of the *average* life cycle of a population of billions of cells is a well-defined number which is very useful in reactor design (Chap. 7).

6.2.2. The Cell Cycle of *E. coli*

Even the simplest of cells may exhibit relatively complex patterns of biosynthesis and growth during the cell cycle. This is indeed the case for *E. coli*, whose life cycle will occupy most of this section. Since most of the available data on the procaryotic cell cycle come from this one bacterium, we cannot be absolutely sure that the features discussed here are typical of all procaryotes.

E. coli is a rod-shaped bacterium whose length provides a reasonable measure of cell volume. As indicated in Fig. 6.19, increase in length is continuous throughout the cycle and appears to be approximately exponential. Total protein increases in a fashion similar to cell volume.

In order to appreciate the complexities of the cell cycle, individual cellular constituents must be examined. First consider DNA synthesis and cell division, events which are coupled in a normal *E. coli* cell. In Fig. 6.9 we observed that the *E. coli* chromosome is a circular molecule and that DNA replication occurs at a fork between parent DNA strands. This situation obtains only when the cell is in an environment which causes slow growth (doubling time = cycle time $\geq$ 60 min). For such a slowly growing bacterium, DNA synthesis begins after division of the mother cell and stops before subsequent division occurs. When better surroundings spur faster growth, there may be more than one DNA molecule per cell and there may be several moving forks on each molecule. This is true for cells with doubling times of less than 40 min. DNA synthesis occurs throughout the cell cycle for these fast-

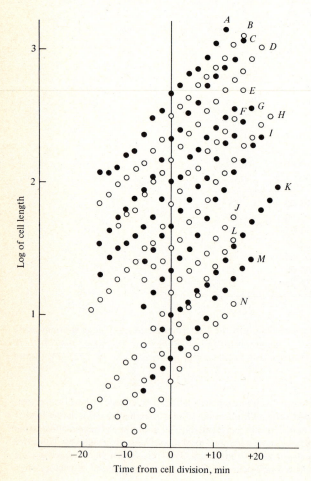

Figure 6.19 Each lettered set of data corresponds to individual *E. coli* cells and their respective daughter cells. Elapsed time between adjacent datum points is 2 min, and the cell length at division ranges from 4.1 to 6.0 μm. (*Reprinted by permission from M. Schaechter et al., Growth, Cell and Nuclear Division in Some Bacteria, J. Gen. Microbiol., 19: 421, 1962.*)

growing cells. After the bacterial chromosome has been duplicated, the daughter chromosomes separate. A transverse wall then forms across the cell, and division follows.

Relatively few data exist at present on RNA synthesis during the cell cycle. The available evidence suggests that it is continuous. However, the composition of an RNA fraction believed to be mRNA has been observed to fluctuate. Consequently, the synthesis of individual proteins may also be expected to oscillate over several cycles.

Such behavior has indeed been observed in a number of experimental studies dealing mostly with inducible and repressible enzymes. Synthesis of a large number of inducible enzymes can be induced in bacteria at any point in the cell cycle. Usually the rate of induced enzyme synthesis shows a stepwise increase in synchronous culture. A typical example of such behavior is shown in Fig. 6.20.

Several possibilities arise in the case of represssed enzymes. If bacteria are grown under conditions where end-product repression is either minimal or maximal, the rate of enzyme synthesis is constant for an interval equal to the cell-cycle time and then doubles. On the other hand, there is a regular periodic pattern of enzyme synthesis for intermediate levels of repression. Several experimental findings suggest that this periodic behavior results from a stable oscillation in the cell's feedback control system (Fig. 6.21). One model consistent with the main biological features of enzyme synthesis

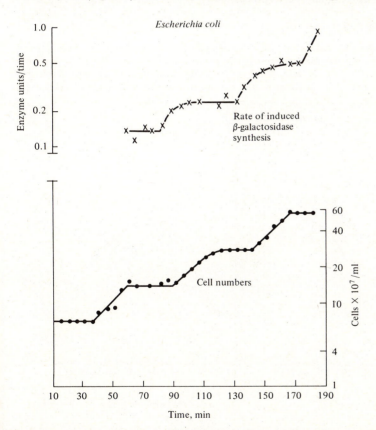

Figure 6.20 The time course of induced β-galactosidase synthesis in a synchronous culture of *E. coli*. The upper graph shows the initial rates of induced β-galactosidase synthesis in aliquots removed from the culture at the indicated times. [*Reprinted from W. D. Donachie and M. Masters, Temporal Control of Gene Expression in Bacteria, in G. M. Padilla, G. L. Whitson, and I. L. Cameron (eds.), "The Cell Cycle," p. 37, Academic Press, New York, 1969.*]

control is

$$\frac{dV}{dt} = kV \tag{6.1}$$

$$\frac{dn_R}{dt} = \beta n_E - \gamma V \tag{6.2}$$

$$\frac{dn_E}{dt} = \frac{\alpha n_G}{K_1 + n_R/V} \tag{6.3}$$

where V = cell volume
n_R = number of repressor molecules
n_E = number of enzyme molecules
n_G = number of genes

By proper choice of the parameters, this model exhibits oscillatory solutions resembling the curves shown in Fig. 6.21. It is recognized, however, that the model is incomplete since it does not explain why the period for variations in enzyme synthesis exactly coincides with the length of the cell cycle. At the present time, the exact nature of this coupling is unknown.

6.2.3. The Eucaryotic-Cell Cycle

At the outset we should recall that there is no such thing as a typical eucaryote. This imaginary cell, however, does serve as a useful device when discussing general features common to many eucaryotes, and we shall employ it in that sense here. In some instances, differences between various eucaryotes will be recognized. For example, consider the growth of mouse cells vs. growth of an amoeba (Fig. 6.22).

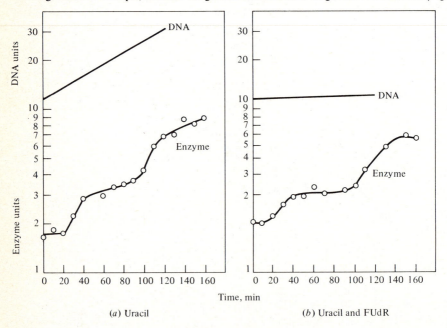

(a) Uracil   (b) Uracil and FUdR

Figure 6.21 A synchronous culture of *Bacillus subtilis* was divided into two separate cultures at time t = 0. (a) Uracil alone was added. (b) Uracil and 5-fluorodeoxyuridine (FUdR), an inhibitor of DNA synthesis, were added. Synchronous synthesis of the enzyme ornithine transcarbamylase is nevertheless very similar in the two cultures. [*Reprinted from W. D. Donachie and M. Masters, Temporal Control of Gene Expression in Bacteria, in G. M. Padilla, G. L. Whitson, and I. L. Cameron (eds.), "The Cell Cycle," p. 37, Academic Press, New York, 1969.*]

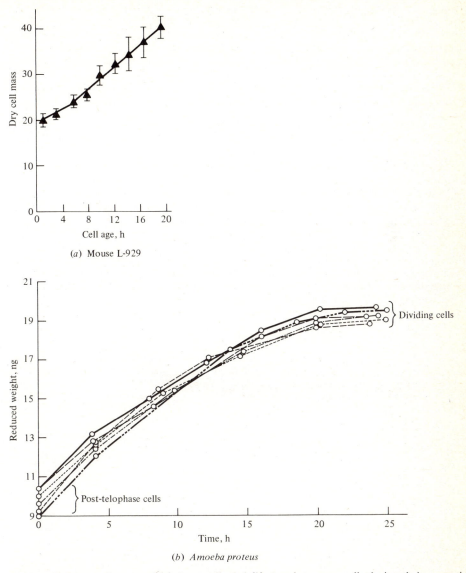

(a) Mouse L-929

(b) Amoeba proteus

Figure 6.22 Increase in mass of (a) tissue cells and (b) Amoeba proteus cells during their respective cell cycles. The different curves in (b) are for six different individual cells. [(a) Reprinted from D. Killander and A. Zetterberg, Quantitative Cytochemical Studies on Interphase Growth. I. Determination of DNA, RNA, and Mass Content of Age Determined Mouse Fibroblasts in vito and of Intercellular Variation in Generation Time, Exp. Cell Res., **38**: 272, 1965; (b) Reprinted from D. M. Prescott, Relations Between Cell Growth and Cell Division. I. Reduced Weight, Cell Volume, Protein Content, and Nuclear Volume of Amoeba proteus from Division to Division, Exp. Cell Res., **9**: 328, 1955.]

While the mouse cells show a linear or exponential increase of dry mass with time, the growth rate of the amoeba declines as the cell grows and is essentially zero during cell division. The second pattern is more typical of eucaryotes in general.

In the eucaryotic cell cycle, there is more differentiation between various parts of the cycle than in procaryotes. The standard pattern of the eucaryotic cycle is given in Fig. 6.23: it is subdivided into four phases, M, G_1, S, and G_2, whose relative durations are indicated in the figure. In the G_1 phase, protein and RNA are actively synthesized, but DNA is not. (Recall that under some conditions DNA synthesis is continuous throughout the procaryote cycle.) In contrast to the procaryotic linear pattern, enzyme synthesis is almost always periodic in eucaryotes. A particular enzyme increases sharply in amount at a definite point of the cell cycle. The available evidence supports sequential transcription of DNA as the mechanism underlying this behavior. The data plotted in Fig. 6.24 show the patterns of α-glucosidase synthesis for different yeast strains containing different numbers of genes coding for the enzyme.

Following the G_1 phase, the chromosomes are duplicated in the S phase. Localization of this process in the latter portion of the overall cycle is evident in Fig. 6.25. Notice that growth of the nuclear volume of the amoeba is greatest at the conclusion of the cycle.

The significance of the G_2 phase is not fully understood at present. After its completion, cell division starts (M phase). Division in eucaryotes proceeds by a well-orchestrated process called *mitosis* (Fig. 6.26). We shall not delve into any details here. Since it is readily observable by a light microscope, the basic features of mitosis have been known for many years and can be found in a good biology or cell-physiology text.

Recently developed instruments called *flow microfluorometers* permit rapid observations of DNA and protein contents of individual cells. After their DNA and proteins have been selectively stained with different fluorescent dyes, cells in suspension flow rapidly (several hundred to several thousand cells per second) through a laser beam and fluorescence excited by the laser is monitored at different wavelengths corresponding to the DNA and protein dyes. Besides numerous other applications, data of this type promise to reveal new details about the cell cycle. For example, the DNA distribution in Fig. 6.27 clearly shows the fractions of a Chinese hamster ovary-cell population in the G_1, S, and G_2 plus M phases.

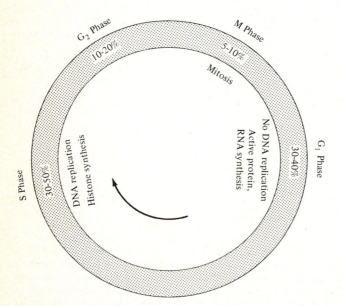

Figure 6.23 The sequence of events during a typical eucaryotic cell cycle.

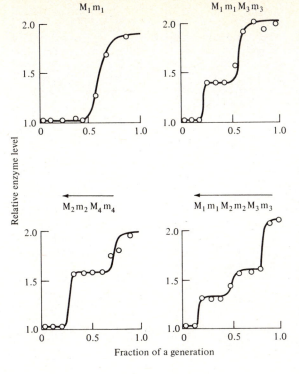

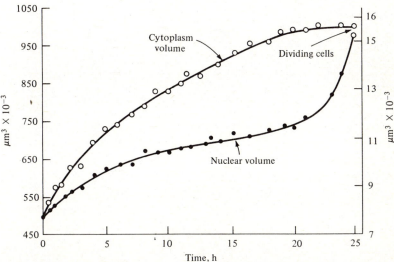

Figure 6.24 Levels of α-glucosidase during one cell cycle of different strains of the yeast *Saccharomyces cerevisiae*. Listed above each plot is a simplified genetic map read from right to left during the cell cycle, which shows the presence of genes (Mm) coding for α-glucosidase in the various strains. [*Reprinted from H. O. Halvorson et al., Periodic Enzyme Synthesis in Synchronous Cultures of Yeast, in "Cell Synchrony,"* I. L. *Cameron and G. M. Padilla (eds.), p. 102, Academic Press, New York, 1966.*]

Figure 6.25 Because different biological macromolecules are synthesized at different times during the eucaryotic cell cycle, cytoplasm (O) and nuclear volume (●) of *Amoeba proteus* follow distinct time courses. (*Reprinted from D. M. Prescott, Relations Between Cell Growth and Cell Division. I. Reduced Weight, Cell Volume, Protein Content, and Nuclear Volume of* Amoeba proteus *from Division to Division, Exp. Cell Res.,* **9**: 328, 1955.)

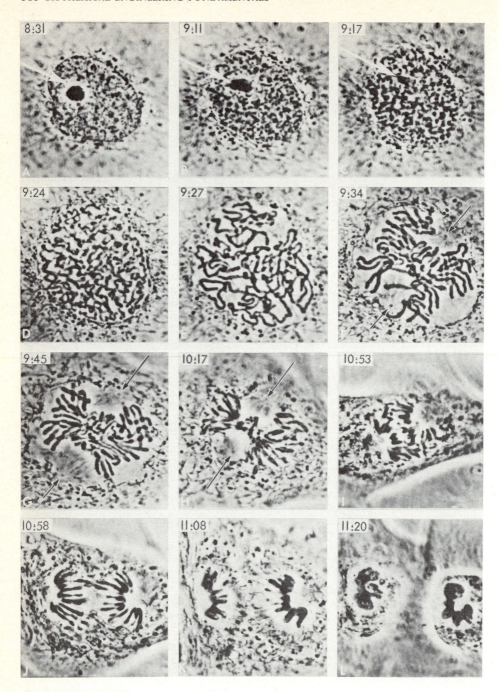

Figure 6.26 Photomicrographs (× 800) showing changes in the nuclear zone of an animal cell during mitosis. (*Reprinted by permission from W. Bloom and D. W. Fawcett, "A Textbook of Histology," 10th ed., p. 72, W. B. Saunders Co., Philadelphia, 1972.*)

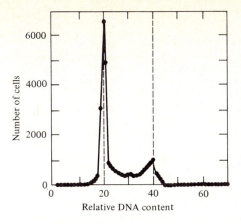

Figure 6.27 The DNA contents of large numbers of individual cells can be rapidly measured using a flow microfluorometer. The first peak in this distribution indicates cells in the G_1 phase, while the second peak contains cells in the G_2 and M phases (Chinese hamster ovary cells). (*Reprinted by permission from J. L. Marx, Lasers in Biomedicine: Analyzing and Sorting Cells, Science,* **188:** *821–823, 874, May, 1975. Copyright © 1975 by the American Association for the Advancement of Science.*)

In this section we have discovered a few of the intricacies of the cell cycle. Actually, disappointingly little is known about it. From our overview of cell metabolism in chap. 5, we realize that an extremely complicated network of chemical reactions must be carefully coordinated in order to utilize nutrients efficiently. While much of the biochemistry of these reactions has been discovered, we have very little idea about how they are executed *in vivo*. Figure 6.28 reveals some intriguing data suggesting that there are indeed variations in the *timing* of the many interacting reaction sequences within the cell. Thus, cellular reactions are often localized not only in space, as in immobilization of an enzyme at a membrane surface, but also in time. In view of these complexities, we must be both careful and humble as we face the task of mathematically representing the kinetic processes of living cells.

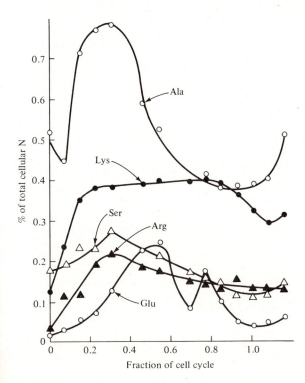

Figure 6.28 Complex variations occur during the cell cycle of *Chlorella pyrenoidsa* in the proportions of free amino acids. (*Reprinted by permission from T. A. Hare and R. R. Schmidt, Nitrogen Metabolism During Synchronous Growth of Chlorella: II. Free-, Peptide-, and Protein-Amino Acid Distribution, J. Cell Physiol.,* **75:** *73, 1970.*)

6.3. ALTERATION OF CELLULAR DNA

In addition to mutation, a variety of processes change the genetic material within a cell. These have both positive and negative implications from the standpoint of industrial microbiology. From the former perspective, any method for altering a cell's DNA content provides an opportunity for development of more productive strains. On the other hand, an unplanned modification of a desired species' DNA can cause expensive failures of industrial microbial processes.

Much of what follows pertains only to bacteria, *E. coli* serving as the subject of most research on these topics. We shall not discuss here the elaborate and complex processes of sexual recombination which operate in many eucaryotic organisms. A brief introduction to this topic within the context of commercial microbes is provided in Elander [11].

6.3.1. Viruses and Phages: Lysogeny and Transduction

We know that viruses are agents of human disease such as common colds, influenza, smallpox, polio, yellow fever, mumps, measles, and many other ailments. Also, several commercial biological processes employing bacteria (including cheese making and antibiotic production) can be severely disrupted by viral infection of the bacteria. The subgroup of viruses which infect bacteria are called *bacteriophages* or just *phages*.

Viruses are constructed of protein and nucleic acid and also may include lipoproteins. The function of the protein is to house and protect the nucleic acid viral component and sometimes to attach the virus to a living cell. The nucleic acid component, which may be either DNA or RNA, is ultimately responsible for the infection and its aftermath. In Table 6.5 the diverse properties of several viruses are summarized; the viral nucleic acid may be single-stranded or double-stranded and is often circular.

Outside a living cell of the proper type, the virus is an inert particle which cannot reproduce by itself. Multiplication of the virus occurs after it has infected a host cell. Thus, viruses are parasites. A brief summary of the basic differences between viruses and living cells is provided in Table 6.6.

Next we shall consider the typical "life cycle" of a virus using as an example the virus called *phage* λ, which infects the *E. coli* bacterium. After attaching to the cell wall, the phage injects its DNA into the cell's interior. At this point two alternative courses are possible (see Fig. 6.29). One option is a state of *lysogeny*, where the phage DNA, now called the *prophage*, is integrated into the bacterial chromosome. If this occurs, the cell lives and reproduces normally, at the same time copying the prophage and creating more lysogenic cells. Phages, like θ, which can enter into a lysogenic relationship with their hosts are called *temperate phages*.

The other possible outcome of temperate-phage infection, called the *lytic cycle*, invariably kills the host cell. The lytic cycle always results from infection by *lytic phages* such as phage T_2 (see Fig. 6.30). During the lytic cycle the phage is said to be in a vegetative state: the phage DNA literally takes over control of the cell. It first directs the ribosomes to synthesize enzymes which destroy the host cell's DNA (this is not a universal feature of viral infections, however) and which will multiply the phage DNA. Then the protein components necessary to create an intact phage particle are synthesized. These proteins spontaneously join with phage DNA to form complete, highly organized bacteriophage particles.

After this self-assembly of many new phages, the enzyme lysozyme is synthesized. As we have already learned, this enzyme attacks the murein cell wall of bacteria. Subsequently, lysis occurs; the cell breaks apart, freeing many phage particles. The electron micrographs of Figs. 6.31 and 6.32 dramatically illustrate this sequence of events.

From this brief survey it should be clear why phage infestation of commercial cultures can be a serious problem. The phage particles are very small, they can multiply rapidly in the right environment, and they can hide in the relatively dormant state of lysogeny.

Occasionally, during reproduction of the phage within its host, some phage particles are formed which contain a small portion ($\sim$ 1 to 2 percent) of the host cell's chromosome. When one of these *transducing phages* injects its DNA into another bacterium, the DNA derived from the first host may cross over with a fragment of the new host's chromosome. In this process, called *transduction*, the

Table 6.5. Characteristics of several kinds of viruses†

Type of virus	Size, nm	Shape, composition, and comment
Animal:		
Cubic symmetry:		
Poliomyelitis	30	Consists of 1 molecule RNA (mol wt 2×10^6) in a spiral, surrounded by protein macromolecules 6 nm diam. arranged as an icosahedron with no retaining membrane; particle mol wt 10×10^6
Helical symmetry:		
Influenza	100	Consists of 1 molecule RNA (mol wt 2×10^6) as a nucleoprotein macromolecule arranged in a helix, the whole coiled and enclosed in a lipoprotein sheath; particle mol wt 100×10^6
Plant:		
Rods:		
Tobacco mosaic	300×15	Whole virus particle rod-shaped, consisting of 1 molecule RNA (mol wt 2×10^6) associated with protein macromolecules arranged in a helix; particle mol wt 39×10^6
Sphere:		
Tomato bushy stunt	30	Icosahedron, consisting of 16% RNA (mol wt 1.6×10^6) and protein; particle mol wt 9×10^6
Insect:		
Silkworm	280×40	Actual virus rod-shaped; DNA constitutes about 8% of dry weight, but in vivo the virus rods are embedded in polyhedral crystalline aggregates of protein 0.5–15 μm diameter
Bacteriophages:		
Double-stranded:		
DNA		
T-even of *E. coli*		Tadpole-shaped phage with DNA (mol wt 130×10^6) confined to head; the protein (some contractile), long tail fibrils involved in attachment to host cell; particle mol wt 250×10^6
Head:	90×60	
Tail:	100×25	
Tail fibrils:	130×2.5	
Single-stranded:		
DNA		
ϕX174 of *E. coli*	22	Dodecahedron with 12 subunits; DNA (mol wt 1.6×10^6) 25% dry weight; particle mol wt 6.2×10^6
RNA	20	Polyhedron containing RNA (3×10^{-12} μg/virus) and
f2 of *E. coli*		protein; nucleic acid content probably similar to that of ϕX174

† S. Aiba, A. E. Humphrey, and N. F. Millis, "Biochemical Engineering," 2d ed., p. 19, Academic Press, Inc., New York, 1973.

Table 6.6 Viral vs. cellular characteristics

Virus	Cell
Only one kind of nucleic acid	Contains DNA and RNA's
Contains only a few enzymes	Contains thousands of enzymes
Reproduced by assembly of nucleic acid and protein components assembled by host cell	Reproduces itself in orderly, controlled fashion

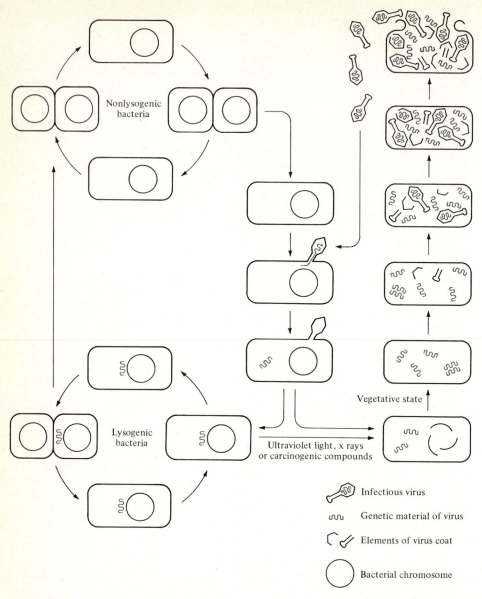

Nonlysogenic
bacteria

Lysogenic
bacteria

Ultraviolet light, x rays
or carcinogenic compounds

Vegetative state

Infectious virus

Genetic material of virus

Elements of virus coat

Bacterial chromosome

Figure 6.29 Possible outcomes of infection of bacteria by temperate phage: lysogeny or cell destruction accompanied by phage multiplication. (*From "Viruses and Genes," by François Jacob and Elie L. Wollman. Copyright © 1961 by Scientific American, Inc. All rights reserved.*)

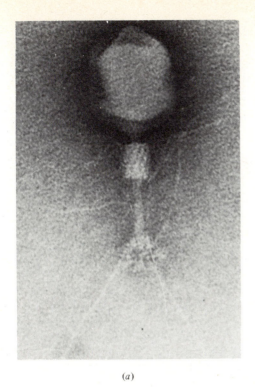

(a)

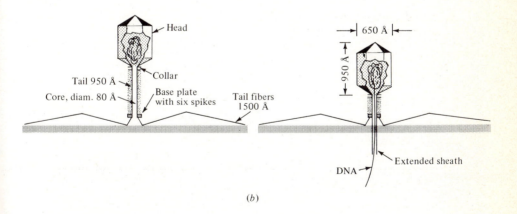

(b)

Figure 6.30 Bacteriophage T2, one of the lytic phages: (a) electron micrograph; (b) schematic diagrams: [(a) *Electron micrograph by R. W. Horne;* (b) *From "Cell Structure and Function," 2d ed., p. 277, by Ariel G. Loewy and Philip Siekevitz. Copyright* © *1963, 1969 by Holt, Rinehart and Winston, Inc.*]

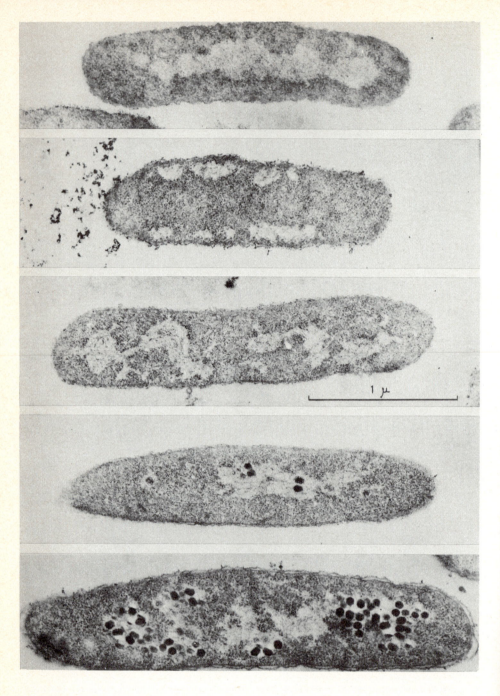

Figure 6.31 Course of infection of *E. coli* by bacteriophage T2 from initiation of infection (top) to 60 minutes later (bottom). (*Electron micrograph by E. Kellenberger.*)

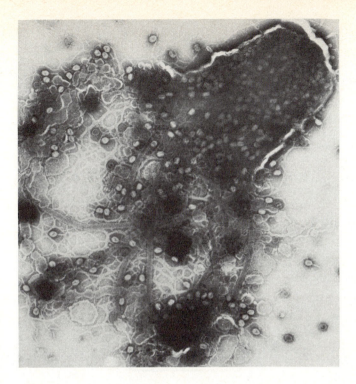

Figure 6.32 Electron micrograph (× 50,000) of the end result of the lytic cycle: the cell has lysed, and numerous new virus particles (white objects) are released. (*Electron micrograph by E. Kellenberger.*)

genetic characteristics of the new host may be altered. Crossing-over generally occurs only between segments which are physically similar and which govern similar characteristics. For example, a transducing phage could carry a small DNA fragment permitting lactose utilization from a lactose-metabolizing *E. coli* strain to an *E. coli* strain unable to use lactose. Figure 6.33 provides a simplified schematic illustration of transduction as well as crossing-over of DNA segments.

Transduction is known to occur with many actinomycetes. In one study involving *Streptomyces griseus*, phage transferred the ability to synthesize the antibiotic streptomycin from a producing to a nonproducing strain. Some of the transduced variants produced more of the drug than the original synthesizing strains.

6.3.2. Bacterial Transformation and Conjugation

Transformation is one of the simplest mechanisms by which genetic material from one cell recombines with another cell's chromosome. Here a double-stranded DNA fragment from the donor cell enters the recipient cell in a process requiring input of metabolic energy. Not all cells will act as recipients: only those in a certain physiological state characterized by synthesis of a particular protein are competent to take up the donor's chromosome fragment. If the donor fragment is similar to a fragment of the recipient's DNA, alteration of the recipient chromosome by crossing-over occurs rapidly.

Besides chromosome fragments, *plasmids* may be introduced into living bacteria by transformation. A plasmid is a DNA molecule that is separate from the bacterial chromosome and duplicates independently. Plasmids are typically relatively small circular molecules with molecular weights of the

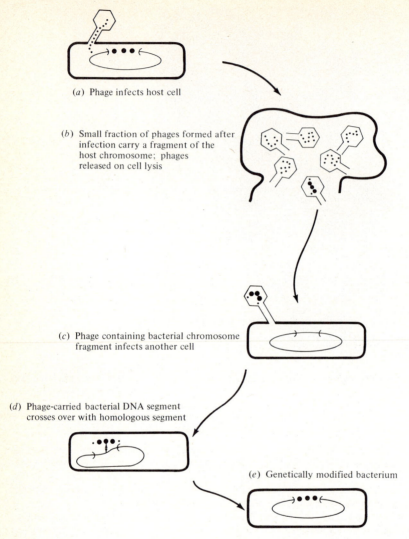

(*a*) Phage infects host cell

(*b*) Small fraction of phages formed after infection carry a fragment of the host chromosome; phages released on cell lysis

(*c*) Phage containing bacterial chromosome fragment infects another cell

(*d*) Phage-carried bacterial DNA segment crosses over with homologous segment

(*e*) Genetically modified bacterium

Figure 6.33 Schematic diagram of bacterial chromosome modification via transduction, a process mediated by bacteriophage. Stages (*a–e*) are shown.

order of 10^6 to 10^8 (see Fig. 6.34). Transformation of plasmids is an integral part of the technique for DNA hybridization discussed in the next section.

While usually nonessential for normal cell function, plasmids can confer useful properties upon the cell they inhabit. For example, plasmids called *R factors* have been identified as the substances responsible for bacterial resistance to antibiotics.

Closely related to plasmids are *episomes*, DNA molecules which may exist either integrated into the cell chromosome or separate from it. One of the better-known episomes is the F, or fertility, factor, which characterizes the partners in conjugating *E. coli* cells.

During *conjugation* of *E. coli* (Fig. 6.35), cells containing the F factor (F$^+$ cells) transmit it to F$^-$ (lacking F) cells, and occasionally some of the chromosome of the F$^+$ partner is transferred to the F$^-$

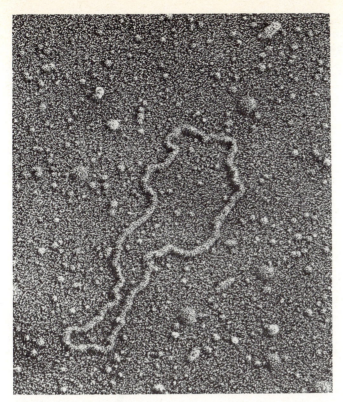

Figure 6.34 A plasmid (pSC101, electron micrograph × 230,000). This circular DNA molecule, which exists and replicates independently from the bacterial chromosome, is an important component in recently discovered methods for recombinant DNA. (*Electron micrograph by Dr. S. N. Cohen.*)

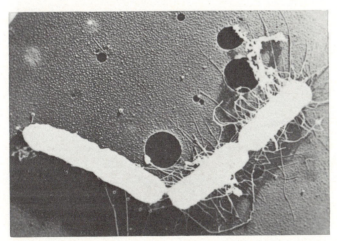

Figure 6.35 Conjugation of *E. coli* cells. The pili (threadlike attachments) of the F⁻ cells clearly distinguish them from the Hfr cells (no pili). F⁻ and F⁻ cells do not conjugate, a process which involves physical attachment of the bacteria. (*Electron micrograph by Dr. F. Jacob.*)

cell. R factors may be transmitted in a similar fashion. Certain F^+ mutants, denoted Hfr, act differently, however; they do not infect the F^- cell with the F factor, but the Hfr partner does donate much of its chromosome. Transmitted segments may then cross over with the F^- chromosome to yield genetically altered strains.

Numerous experiments have revealed that the F^+ chromosome enters the F^- in linear order in the form of a thread. If the transfer is interrupted at different times, chromosomal threads of different lengths are found in the F^- cells. Comparison of the characteristics of recipient populations exposed to different amounts of injected chromosome then reveals the relative locations of genetic information storage on the (F^+) chromosome. In this fashion, detailed genetic maps like that in Fig. 6.36 have been constructed. Notations around this map indicate the relative loci of various mutant characteristics. For example, *thr* refers to a mutant which requires threonine for growth. Many of the other symbols are defined in Watson [1, chap. 7].

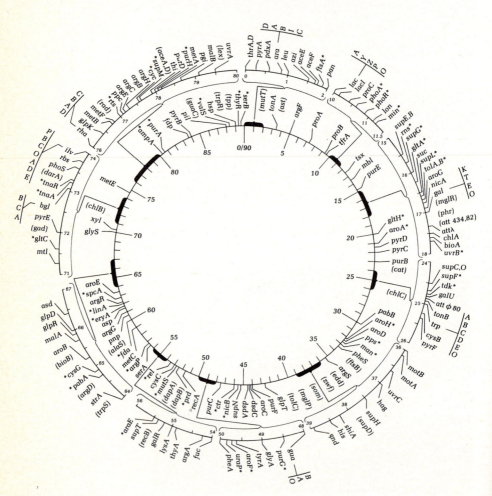

Figure 6.36 A genetic map of *E. coli* K12 showing the relative location of various genes in the circular chromosome. (*Reprinted from A. L. Taylor and C. D. Trotter, Revised Linkage Map of E. coli, Bacteriol. Rev.,* **31:** 332, 1967.)

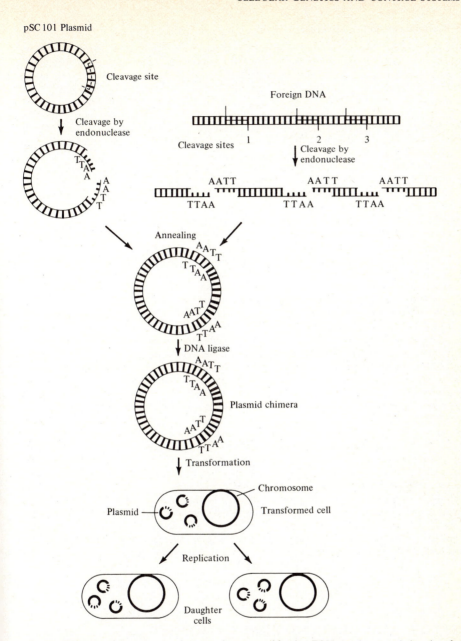

Figure 6.37 Using this sequence of steps, a fragment of foreign DNA can be annealed and sealed into a plasmid which in turn can be introduced into a living cell, where the foreign DNA is replicated and expressed. (*From "The Manipulation of Genes" by Stanley N. Cohen. Copyright* © *1975 by Scientific American, Inc. All rights reserved.*)

6.3.3. Recombinant DNA Molecules

All the processes discussed above occur *in vivo*. There is now another procedure, *in vitro* laboratory manipulation, whereby genetic information from one organism can be transmitted in functioning form to another organism of a different species. Before briefly considering the potential benefits and problems associated with such genetic manipulations, we outline the elements of the technique as it was first applied to introduce DNA from *Staphlococcus aureus* and the toad *Xenopus laevis* into *E. coli*.

The first essential ingredient in constructing a recombinant DNA molecule is a means of breaking and connecting DNA molecules of different origins. Breakage is done with an endonuclease enzyme (Eco R1) which acts at the site of a TTAA base sequence in the DNA molecule. Significantly, both DNA strands are not broken at the same point: one single-stranded fragment containing the bases TTAA is left on one end of the broken DNA, while the other end terminates with the complementary single strand AATT (see upper portion of Figure 6.37). Since these unpaired strand ends contain complementary bases, they are "sticky" and facilitate attachment of other DNA fragments. Similarly sticky pieces of a foreign DNA may also be obtained by cleaving it with Eco R1.

As Fig. 6.37 summarizes, this procedure can be applied using the plasmid of Fig. 6.34 as one DNA component, resulting in the insertion of a segment of foreign DNA into a circular loop. An intact plasmid containing DNA from two sources is then sealed using DNA ligase.

After the host *E. coli* cells are treated with calcium chloride to increase the permeability of the cell membrane, recombinant DNA enter the cells via transformation. Since the replication gene of the original plasmid is not damaged in constructing the recombinant DNA plasmid, the *entire* recombinant plasmid replicates and appears in subsequent *E. coli* generations.

This procedure has been applied to transfer penicillin resistance carried in a *Staphlococcus aureus* plasmid to *E. coli* and to place within *E. coli* a portion of frog gene. In both cases the transmitted genetic information was expressed: the *E. coli* cells in the first example were penicillin-resistant, and the bacteria in the second synthesized RNA from the toad DNA fragment.

Several extensions of this method for genetic recombination have now been developed, expanding the scope of possible manipulations. Presumably these can be used to create bacteria which can synthesize important antibiotics, hormones, and enzymes. Besides these and other positive outcomes, there is a possibility that research with hybrid DNA molecules will produce new infectious species or organisms. Because of this potential hazard, molecular biologists in mid-1974 agreed not to conduct certain types of gene-manipulation experiments. A meeting at Asilomar, California in February 1975 considered the scientific and ethical questions associated with hybrid DNA research, and guidelines for subsequent work in the field were subsequently issued by NIH in July, 1976. At least four major corporations have already initiated research aimed at practical applications of recombinant DNA molecules, one of the most exciting areas or contemporary science.

Clearly a large and ever-expanding array of tools is becoming available for alteration of a protist's genetic characteristics. The increasing ability of the molecular biologist to make such changes in a specific, controlled manner promises to create new generations of biochemical processes. However, at present many of these possibilities remain dreams for the future, with trial-and-error-induced mutations the primary present strategy for strain improvement. The next section summarizes some of the current applications of biological control theory.

6.4. COMMERCIAL APPLICATIONS OF MICROBIAL GENETICS AND MUTANT POPULATIONS

Basic knowledge of molecular biology and cellular control systems have several important practical implications. Some were considered in the previous section; if the location of a desirable or unwanted gene can be determined, a variety of techniques can be applied to attempt modification of that particular gene. Another tactic is induced mutation to a more productive species. The fruits of this approach have already been mentioned in Chap. 1, where we discussed increased penicillin yields based on improved strains obtained by a sequence of mutations. The final portions of this section will outline several other microbial processes where development of special mutant species has played a major role.

Often, improved productivity of a mutant microorganism has a straightforward interpretation in terms of basic cellular control systems. Understanding of the mechanisms controlling biosynthesis of the desired product has another important application: deciding how to formulate and control the growth medium so that productivity is maximized. We consider such environmental manipulations next.

6.4.1. Cellular Control Systems: Implications for Medium Formulation

There are two different approaches to altering cellular productivity via choice of medium composition: (1) add inducers and (2) decrease amount of repressor present. Both approaches are rather obvious on the surface, but in practice some sophistication is required to achieve optimal results.

Beyond the enzyme's substrate itself, nonmetabolizable substrate analogs can be extremely effective inducers for enzyme production. For example, β-galactosidase specific activity in *E. coli* can be increased more than 1000-fold by galactosides. In addition to the isopropyl β-D-thiogalactoside sometimes used for β-galactosidase induction, the following substrate analogs have proved effective (enzyme induced follows in parentheses): *N*-acetylacetamide (amidase), methicillin (penicillin-β-lactomase), and malonic acid (maleate isomerase).

Catabolite repression is known to decrease product biosynthesis in many important processes. Availability of a rapidly consumed substrate such as glucose at high concentrations fosters rapid growth but limits production of the antibiotics penicillin, mitomycin, bacitracin, and streptomycin. Great improvements in product yields can be obtained for penicillin by intentionally fostering diauxic growth (see Fig. 7.7) by using glucose plus a slowly metabolized sugar such as lactose in the growth medium. The desired biomass is grown on the glucose fraction, and a phase of product synthesis follows as the lactose is consumed. An alternative strategy for achieving the same effect is slow addition of glucose to the medium.

Other examples of repressors whose concentrations can be directly controlled by medium formulation include inorganic phosphate (represses phosphatase synthesis in *E. coli* and nuclease production in *Aspergillus quernicus*) and ammonia (represses urease biosynthesis by *Proteus retgeri*). For example, the alkaline phosphatase content of *E. coli* can be increased from about zero up to roughly 5 percent of the cell protein by limiting the phosphate content of the growth medium. In a like manner, avoiding high concentrations of amino acids and sulfate greatly increases protease synthesis by bacteria and *Aspergillus niger*, respectively.

Maintenance of low repressor concentrations is more difficult when the repressor is synthesized by the microorganism. One useful approach in this case is to alter the cell's membrane so that the repressing substance quickly diffuses from the cell's interior to the medium. This strategy is one component of a very productive process for glutamate manufacture using *Corynebacterium glutamicum*. Here, limiting the biotin concentration in the medium permits excretion of glutamate into the medium. Intracellular accumulations of other amino acids are greatest when biotin is present in excess. While a detailed explanation of this effect is elusive, there is some evidence relating biotin effects to phospholipid components of the cell membrane. (Recall Example 2.1: "Modification of Biomembrane Permeability" to give enhanced aspartate excretion.)

In the next section we consider another means of ameliorating the negative effects of repressors.

6.4.2. Utilization of Auxotrophic Mutants

Auxotrophic mutants (see Table 6.3 for selection and detection methods) lack enzyme activity for one or more steps in a biosynthetic pathway. As a result, one or several end products of the pathway are not synthesized. In order for such a mutant to survive, it must be fed the unsynthesized metabolites.

Because the missing metabolites are supplied in the growth medium and are not synthesized by the auxotroph, the engineer rather than the microorganism has control of the metabolites' concentrations. Clearly this can be desirable situation from a practical viewpoint when the unsynthesized end product is a repressor. As Fig. 6.38 shows, such auxotrophic mutants lacking repressor synthesis can be made to overproduce† an intermediate metabolite. By keeping the concentration of repressor E low in the medium, feedback inhibition and repression of pathway enzymes is minimized. The intermediate C (normally the sub-

---

† "Overproduce," as used here, is relative to the optimal allocation of biosynthetic intermediates for cell growth and maintenance. Usually such an optimum from the cell's perspective does not lead to high yields of commercially important end products, hence the motivation for industrial use of mutant species with intentionally imperfect internal controls.

Figure 6.38 Auxotrophic mutants can be used to enhance production of intermediates in a metabolic pathway. In this hypothetical example, the mutant lacks enzyme c and therefore does not synthesize the repressor E. [*Reprinted from A. L. Demain, Overproduction of Microbial Metabolites and Enzymes due to Alteration of Regulation, in T. K. Ghose and A. Fiechter (eds.), "Advances in Biochemical Engineering 1," p. 120, Springer-Verlag, New York, 1971.*]

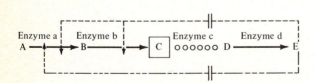

strate for enzyme c, which is absent from the mutant) will then achieve much higher concentrations than in the native organism. As indicated at the close of the previous section, additional manipulations may then be necessary to allow intermediate C to diffuse through the cell membrane into the medium.

The role of auxotrophy in commercial L-lysine manufacture using *C. glutamicum* is illustrated in Fig. 6.39. The productive mutant lacks homoserine dehydrogenase, so that the inhibitory effect of threonine on lysine synthesis (via aspartokinase) is eliminated. Since the auxotrophic mutant does not synthesize threonine or methionine, both these amino acids must be supplied in the growth medium.

Comparison of the aspartate pathway controls for *C. glutamicum* in Fig. 6.39 with the corresponding pathway in *E. coli* (Fig. 5.23) makes an important point: organisms with similar biosynthetic pathways do not necessarily utilize identical control systems. Specifically, notice in Fig. 5.23 that lysine inhibits its own production via the reaction leading to dihydrodipicolinic acid. The corresponding inhibition step is absent from the (unmutated) parent strain of *C. glutamicum*. Also, the aspartokinase system found in the *Corynebacterium* apparently differs from the *E. coli* aspartokinases in that the former is not repressed unless *both* lysine and threonine are present (*concerted* or *multivalent* feedback inhibition), whereas one of the aspartokinase isozymes in the *E. coli* system is repressed by lysine alone. Both these differences confer commercial advantage to the *Corynebacterium*.

Similar strategies have been successfully employed to obtain productive microbial processes for manufacture of the flavor-enhancing purine nucleotides guanosine monophosphate (GMP), inosine monophosphate, (IMP), and xanthine monophosphate (XMP). As Fig. 6.40 illustrates, minimization of AMP and GMP

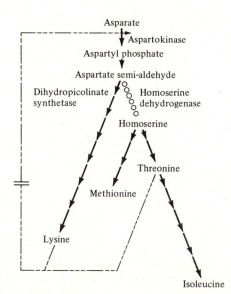

Figure 6.39 A mutant of *C. glutamicum* lacking the enzyme homoserine dehydrogenase permits enhanced production of L-lysine [*Reprinted from A. L. Demain, Overproduction of Microbial Metabolites and Enzymes due to Alteration of Regulation, in T. K. Ghose and A. Fiechter (eds.), "Advances in Biochemical Engineering 1," p. 122, Springer-Verlag, New York, 1971.*]

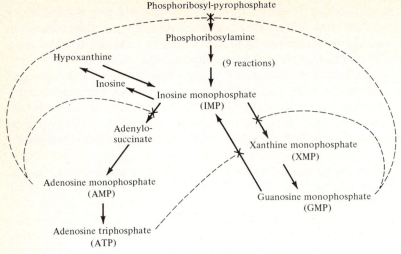

(a) Control of nucleotide biosynthesis

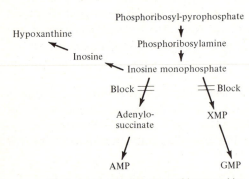

(b) Accumulation of inosine and hypoxanthine

Figure 6.40 (a) In a normal cell, AMP and GMP exert feedback inhibition at several points in the nucleotide biosynthesis reaction sequence. (b) In a mutant which does not synthesize AMP and GMP, the concentration of these components may be kept low by proper medium formulation. This permits enhanced yields of inosine and hypoxanthine. (*Reprinted from S. Aiba, A. E. Humphrey, and N. Millis, "Biochemical Engineering," 2d ed., p. 84, University of Tokyo Press, Tokyo, 1973.*)

concentrations by use of an auxotrophic mutant of *Brevibacterium ammoniagenes* permits accumulation of inosine and hypoxanthine. Appearance of these products in the medium is enhanced by adding a small amount of manganese (Mn^{2+} concentration $\approx 10\mu g/1$) to the medium. Evidence to date indicates that the manganese changes the cell-membrane permeability.

6.4.3. Mutants with Altered Regulatory Systems

Several other types of genetic manipulation and selection have provided commercially superior strains by altering controls at the enzyme and/or at the gene level. Mutants of this type can be used to increase yields of metabolites and enzymes.

To obtain overproduction of a metabolite which acts as an inhibitor and/or repressor of its biosynthesis, we seek mutant organisms whose relevant allosteric enzymes and operons are insensitive to the metabolite's presence. Such mutants are often isolated using antimetabolites, which are toxic analogs of the metabolite in question. Normal cells will not usually grow in a medium containing anti-metabolite since the antimetabolite represses or inhibits biosynthesis of necessary metabolite without serving as a substitute for the unsynthesized metabolite in sub-sequent pathways. On the other hand, strains with deficient feedback controls will not alter their biosynthesis patterns or rates in response to the antimetabolite and will therefore survive in its presence. Table 6.7 lists some of the microbial products whose yields can be enhanced by this technique.

Although resistance to an antimetabolite may involve a variety of control-system alterations, one possibility is elimination or reduction of repression of enzymes in the metabolite's biosynthesis pathway. In this context, antimetabolites can be used to discover mutants with unusually large concentrations of biosyn-thetic enzymes. Such a strategy would be appropriate, for example, in manufac-turing enzymes for subsequent *in vitro* biosynthetic processes. (Sec. 4.6.3).

Other methods can be employed to identify and isolate *constitutive mutants* [7, 9]. In these species, normally induced or repressed enzymes are produced whether or not inducer or repressor is present. Among the enzymes whose yields can be improved with constitutive mutants are β-galactosidase, catalase, phospha-tases, proteases, homoserine dehydrogenase, invertase, histidase, penicillinase, and amidase.

Throughout the previous discussion we have concentrated on various strategies for increasing yield of a desired compound. Additional avenues of appli-cation of such molecular biological principles include biosynthesis of derivatives

Table 6.7. Mutants providing enhanced yields of these end products may be selected using the corresponding antimetabolites†

End product	Antimetabolite used	End product	Antimetabolite used
Arginine	Canavanine	Threonine	α-Amino, β-hydroxy-valerate
Phenylalanine	p-Fluorophenylalanine		
	Thienylalanine	Methionine	Ethionine
Tyrosine	p-Fluorophenylalanine		Norleucine
	Thienylalanine		α-Methylmethionine
	D-Tyrosine		L-Methionine-DL-sulf-oximine
Tryptophan	5-Methyltryptophan		
	6-Methyltryptophan	Histidine	2-Thiazolealanine
Valine	α-Aminobutyrate		1,2,3-Triazole-3-alanine
Isoleucine	Valine	Proline	3,4-Dehydroproline
Leucine	Trifluoroleucine	Adenine	2,6-Diaminopurine
	4-Azaleucine	Uracil	5-Fluorouracil

† A. L. Demain, Overproduction of Microbial Metabolites due to Alteration in Regula-tion, p. 113 in T. K. Ghose and A. Fiechter (eds.), "Adv. in Biochemical Engineering 1," Springer-Verlag, New York, 1971.

of the original products and generation of completely new end products with pharmaceutical and other uses. Brief summaries of these topics are given in Refs. 8 and 11. An important example of such an application is biosynthesis of the tetracycline derivative 6-demethyl tetracycline by a mutant of *Streptomyces aureofaciens*. More stable under acidic conditions than the usual methylated form, this derivative is one of the dominant commercial forms of tetracycline antibiotic.

This completes our foundation in microbiology. With the tools of the past chapters in hand, we are prepared to press on to the engineering of process systems employing living microorganisms. Analysis, modeling, and design of these biological processes will occupy most of our attention for the remainder of the text. In the next chapter, we attack a problem mentioned earlier in Sec. 6.2, the growth kinetics of a single microbial species.

PROBLEMS

6.1. Mutation frequency Mutation processes lead to a new species, the *mutant*. As the DNA material is simply a chemical polymer, it presumably undergoes chemical and physical changes just like any other molecule.

(a) From the data in Fig. 6P1.1a calculate the activation energy of the mutation rate E_m and the preexponential factor k_m°, where $k_m = k_m^\circ \exp(-E_m/RT)$. How do these values compare with those

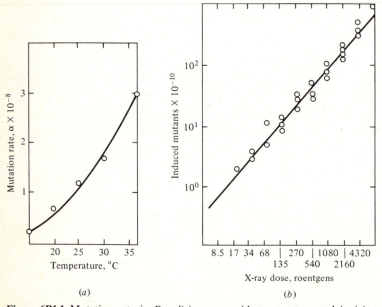

(a)   (b)

Figure 6P1.1 Mutation rate in *E. coli* increases with temperature and ionizing radiation. (a) The effect of temperature on the incidence of mutation *his⁻* (inability to synthesize histidine) to *his⁺* (ability to synthesize it). (b) The number of *met-2⁺* mutants (able to synthesize methionine) induced by exposing *met-2⁺* colonies (nonsynthesizers) to increasingly large doses of x-rays. (R. Sager and F. J. Ryon, "Cell Heredity: An Analysis of the Mechanisms of Heredity at the Cellular Level," John Wiley & Sons, Inc., New York, 1961.)

noted in your freshman chemistry text for simpler chemical species undergoing chemical reactions?

(b) Radiation absorption follows Beer's law: every increment of target absorbs the same fraction of radiation incident upon that increment; therefore $I(x) = I_0 e^{-\alpha x}$. Figure 6P1.1b shows a linear response of mutation appearance vs. initial intensity I_0. Under what condition has all the sample in Fig. 6P1.1b received an equal radiation dose? Would you expect to observe a linearity in the mutant-occurrence–vs.–dose curve when not all cells had received an equal dose?

(c) The mutation rate in Fig. 6P1.1 refers to *E. coli* populations mutated from *his⁻* (cannot synthesize histidine) to *his⁺* (Fig. 6P1.1a), and *met-2⁻* (methionine) to *met-2⁺*. If all DNA alterations occur with the same probability as those deduced here, what is the total mutation rate for DNA of *E. coli*? State your assumptions clearly. Are these mutations of genotype or phenotype?

6.2. Mutation repair and thermodynamics Provided that a large number of individuals are involved, the kinetics of mutation can be usefully described by conventional mass-action kinetics.

(a) In a bacterial population of 1000 l of 3×10^7 cells per milliliter, the mutation of one gene g_1 to a second type g_2 occurs with a frequency of 10^{-8} per cell division, the mutation from g_2 to g_1 occurs at a frequency 10^{-6} per cell division. Calculate the "equilibrium" concentrations of each species.

(b) In cells able to repair damaged DNA enzymatically it is reasonable to suppose that since the enzyme exists, there is an equilibrium between normal and damaged DNA:

$$\text{Normal} \underset{k_{-1}}{\overset{k_1}{\rightleftharpoons}} \text{damaged}$$
$$E_{\text{repair}}$$

Suppose ultraviolet light opens up a second forward path parallel to the first with rate constant k^1. Show that the equilibrium population fraction of damaged DNA rises from $k_1/(k_1 + k_{-1})$ to $(k_1 + k^1)/(k_1 + k^1 + k_{-1})$.

(c) Show that when ultraviolet light is cut off, assuming that the K_m of the repair enzyme is much larger than the total DNA level, the average concentration of damaged DNA returns to its original value at a rate proportional to DNA(damaged)-DNA(damaged)$_{\text{equill, no UV}}$.

6.3. A very simple repressor kinetics model J. Maynard Smith sketches the kinetic control network shown in Fig. 6P3.1.

(a) Assuming steady concentrations of P and gene, develop a kinetic model which could (if solved) predict the time behavior of RNA, M, and Z. Include a loss rate for RNA proportional to its concentration. Proceed as far as possible in obtaining a full transient solution.

(b) From the discussion of this and preceding chapters, indicate whether diffusion should also be considered and what weaknesses this model may have. (This model is considered more completely at the end of Chap. 11.)

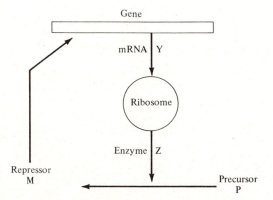

Figure 6P3.1 A simple repressor model. (*From J. Maynard Smith, "Mathematical Ideas in Biology," fig. 30, Cambridge University Press, London, 1971.*)

6.4. Repression of enzyme synthesis (*a*) By inspection of Eqs. (6.1) to (6.3), provide a reasonable justification for the presence and form of each term on the right-hand side of each equation.

(*b*) Again by inspection, comment on what must occur physically for the enzyme level to exhibit periodic behavior of the type shown in Fig. 6.21.

(*c*) The model does not include a time lag to account for diffusion of repressor from its site of synthesis within the cell to its point of action as an enzyme-synthesis repressor. Given the size of a typical cell and a rapid cell-number doubling time of say, 20 min, is this diffusion relatively rapid?

(*d*) Show that if the repressor level is maintained at a constant value by addition to the external medium, the number of enzymes is predicted to increase to a finite limit while V continues to increase. Is this reasonable behavior?

6.5. Cell organization: multienzyme systems Many cellular reactions involve the enzyme-catalyzed formation of an intermediate S_2 between S_1 and S_3, where the equilibrium concentration of S_2 is small relative to S_1 and S_3:

$$S_1 \underset{k_{-1}}{\overset{k_1}{\rightleftarrows}} S_2 \underset{k_{-2}}{\overset{k_2}{\rightleftarrows}} S_3$$

If S_3, etc., are subsequently consumed by the next enzyme-catalyzed reaction, we may neglect the reverse of the second reaction.

(*a*) Assuming $s_i \ll K_i$ for each enzyme, show that the steady-state rate of reaction is given by

$$\text{Rate} = \frac{k_2 s_1}{k_{-1}/k_1 + k_2/k_1} \equiv k s_1$$

if concentrations of S_1 and S_2, E_1 and E_2 are spatially uniform.

(*b*) If E_1 exists on one face of a permeable slab and E_2 on the other, show that the steady-state reaction rate is now

$$\text{Rate} = \frac{k s_1}{1 + k L k_1/k_{-1} \mathscr{D}}$$

and that this is less than $s_{2,\,\text{equil}} \mathscr{D}/L^2$, where L is the distance of separation between planes and $s_{2,\,\text{equil}}$ is the concentration of S_2 equilibrated with the value of s_1.

(*c*) A typical oxygen consumption rate is 10^{-8} mol/(s · cm³). Assuming internal metabolic rates to be of the same magnitude at every step, calculate the *maximum* separation between E_1 and E_2 which would allow this rate for values for $s_{2,\,\text{equil}} = 10^{-4}$ to 10^{-12} mol/cm³ [14].

6.6. Cell organization: ATP production and utilization The intact cell contains localized factories (organelles) which must evidently supply the entire cell with various products. An interesting case which appears usefully examined in one dimension is that of the bull sperm: only the midpiece regenerates ATP, and the attached tail section consumes much of it during motion.

(*a*) Assuming a zero-order reaction for ATP consumption at each point in the tail and a uniform concentration c_0 in the midpiece, show that the ATP profile in the tail is $\bar{c} = \phi^2(z^2/2 - z + \frac{1}{2})$, where ϕ is the Thiele modulus of the tail, z the dimensionless distance from the midpiece, and $\bar{c}$ the dimensionless concentration.

(*b*) Each moving sperm consumes oxygen at 3.7 to 5.0×10^{-18} mol/s for motility. With the aid of the following additional data, calculate the minimum ATP content of sperm needed to maintain diffusive transport to the tail.

ATP diffusivity (corrected for tail water content and
  tortuosity) = 3.6×10^{-6} cm²/s
Midpiece volume = 1.3×10^{-12} cm³   tail length = 5×10^{-3} cm
Tail cross-sectional area (corrected for sheath and fibrous
  matter) = 3×10^{-10} cm²
ATP yield per O_2 consumed = 6

(c) The average observed ATP content is 200×10^{-18} mol per sperm. It has been claimed that one-third to one-half of ATP consumption is used for cell mitochondrial maintenance. Does your calculation in part (b) indicate the need for consideration of other than passive diffusional transport for proper internal distribution of ATP from the mitochondria localized in the midpiece [15]?

6.7. Target theory In an attempt to connect known molecular biology with viral, spore, or cell deactivation, the concept has been developed that "in each cell or in each organism, there are a number of 'targets,' and that changes occur as a consequence of random 'hits' on these targets" [16, p. 87]. Consider the following [16, p. 88]:

Suppose that a cell containing N targets is exposed to a dose of K particles. Let the probability that a particular particle will hit a particular target be p; clearly p is very small.

The probability that a particular target is not hit by a particular particle is $1 - p$.

Hence the probability that a particular target is not hit by any of the K particles is

$$(1 - p)^K \approx e^{-Kp}$$

(a) Noting that if N is large but p is small, $Np \approx (N - 1)p \approx (N - 2)p$, establish the validity of the above approximation for this typical case (large N, small p).

(b) Show that when the proportion of damaged cells is small (damaged means sustaining one or more hits), "the proportion of cells damaged is NKp, i.e., it is proportional to the dose." If there are, in contrast, few undamaged cells, show that for this formulation, the survivor fraction s varies as $\ln s = -NKp$.

(c) When more than one hit is required to damage a cell, the same arguments as used above can be applied. "A bacterium is infected by r similar bacteriophages, each containing N essential genes. Provided that at each of the N loci one gene is undamaged, a functional bacteriophage can be produced by recombination" [16, p. 87]. Show that if "the probability that any particular target escapes is small," $\ln s = N \ln r - NpK$ (here the r targets are the number of virus copies within the cell at the time of irradiation).

(d) Show that your result in part (c) fits the data of Fig. 6P7.1a and evaluate each parameter possible. What phenomena might account for the behavior of Fig. 6P7.1b?

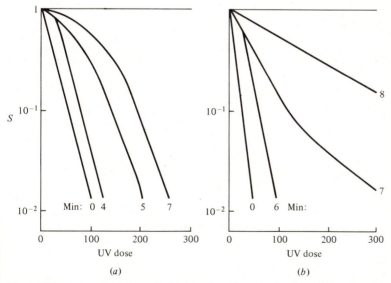

(a)   (b)

Figure 6P7.1 Observed values of infectivity S against dose for bacteria irradiated for different lengths of time after infection: (a) bacteriophage T7; (b) bacteriophage T2. (*After S. Benzer, Resistance to Ultraviolet Light as an Index to the Reproduction of Bacteriophage, J. Bacteriol.,* **63**: 59 1952.)

6.8. Fate of mutant gene in breeding population Mutations occur frequently in populations. The persistence of a mutant gene is surprisingly high, yet the permanence low. For mating diploid populations of N members, at the instant a mutant gene (allele) appears, it has a gene frequency $1/2N$. The total population N is greater than the effective number involved in breeding N_e.

(*a*) For a mutation which is *selectively neutral* (mutant has no advantage or disadvantage vs. original species), the mean number of generations until loss (extinction) of the mutant gene is $t_l = 2(N_e/N) \ln 2N$, and the mean number of generations until fixation (defined as the achievement of gene frequency $= 1.0$) is $t_f = 4N_e$. Here t_l is averaged over all *loss* results only and t_f over only all fixations. Taking $N_e/N = 0.5$ for estimation purposes and a doubling time of 1 h, calculate t_l and t_f for 1 l of mating-cell population with density 10^7 cells per milliliter.

(*b*) The distribution of values for t_f and t_l are relatively small and large with respect to the appropriate mean, as you can show by calculation from

$$\text{var } t_l \approx 4.58 N_e^{\,2} \qquad \text{var } t_f \approx \frac{16 N_e^{\,2}}{N} - \left(2\frac{N_e}{N} \ln 2N \right)^2$$

(*c*) The probability that the mutant becomes fixed in the population is

$$u = \frac{1 - \exp\left[-2(N_e/N)s_1 \right]}{1 - \exp\left(-4N_e s_1 \right)}$$

where s_1 is the selective advantage of the mutant vs. original gene. Show for the typical conditions in microbial populations that $u = 2s_1 N_e/N$. Thus, even when the selective advantage is large ($s_1 = 1$ percent $= 10^{-2}$), the likelihood that it will eventually dominate the culture is small [17].

6.9. Natural selection The relative fitness w of a particular genotype x may be conveniently defined as the ratio of the survival rate for genotype x divided by the survival rate for the mean genotype $\bar{x}$ of the population. (The survival rate is the ratio of occurrence frequency at a point in time divided by the same term evaluated one generation earlier.) Suppose that the fitness of the genotype x varies linearly with the average phenotype expression: $w(x) = 1 - \alpha(\bar{x} - x)$, where α is simply a measure of the *intensity* with which a difference $\bar{x} - x$ affects $w(x)$, $\alpha > 0$. By definition, $g(x)$ is the normalized original genotype distribution $[\int g(x)\, dx = 1.0]$. One generation later, the distribution has become $g'(x) \equiv w(x)g(x)$.

(*a*) Evaluate $\bar{x}'$ in terms of the mean $\bar{x}$ and variance σ^2 of the original population [recall that mean $\equiv \int xg(x)\, dx$, variance $\equiv \int (x - \bar{x})^2 g(x)\, dx$].

(*b*) Show that the change in $\bar{x}$ in one generation is proportional to σ^2, that is, "the rate of evolution is proportional to the genetic variance of the population" (fundamental theorem of natural selection, [18]).

(*c*) When w varies other than linearly with x, Wilson and Bossert [18] claim that the value $\bar{x}' - \bar{x}$ is still "in some way proportional to the genotypic variance." Pick several reasonable forms for $w = f(x)$ and prove or disprove this statement.

6.10. Recombinant DNA Read [13].

(*a*) Outline (from memory) the procedure by which plasmid DNA can be introduced into a bacterium. Repeat for procedure for splicing foreign DNA into pSC101 plasmid.

(*b*) Summarize the potential benefits and problems cited in [13] for future exploration. How does what you have learned in this chapter relate to the material in [13]?

REFERENCES

All the references given in Chaps. 1 and 2 have worthwhile material on various aspects of this chapter. Also recommended:

1. J. D. Watson "Molecular Biology of the Gene," 2d ed., W. A. Benjamin, Inc., New York, 1970. A superb text which, assuming very little biological background initially, carries the reader smoothly into many intricacies of molecular biology.

2. J. M. Mitchison: "The Biology of the Cell Cycle," Cambridge University Press, London, 1971. A good review of the state of knowledge concerning the cell cycle through 1971. The book is well organized and contains frequent summaries emphasizing major results.
3. G. H. Haggis: "Introduction to Molecular Biology," 2d ed., Halsted Press, New York, 1974. Another perspective on many aspects of molecular biology.

Collections of research and review papers dealing with the cell cycle and synchronous cultures which are rich sources of original data:
4. G. M. Padilla, G. L. Whitson, and I. L. Cameron (eds.): "The Cell Cycle: Gene-Enzyme Interations," Academic Press, Inc., New York, 1969.
5. I. L. Cameron and G. M. Padilla (eds.): "Cell Synchrony: Studies in Bio-synthetic Regulation," Academic Press, Inc., New York, 1966.

Review of laser instruments for characterizing and sorting cell populations:
6. J. L. Marx: Lasers in Biomedicine: Analyzing and Sorting Cells, *Science*, **188**: 821, 1975.

Two review papers which examine a variety of practical applications of molecular biology:
7. A. L. Demain: Overproduction of Microbial Metabolites and Enzymes Due to Alteration of Regulation, *Adv. Biochem. Eng.*, **1**: 113, 1971.
8. W. T. Dobrazanski; Microbial Genetics in Pharmacy, *Chem. Br.*, **10**: 386, 1974.

Additional information on a variety of applications, all in D. Perlman (ed.), "Fermentation Advances," Academic Press, Inc., New York, 1969:
9. A. B. Pardee: Enzyme Production by Bacteria, p. 3.
10. K. Veda: Some Fundamental Problems of Continuous L-Glutamic Acid Fermentations, p. 43.
11. R. P. Elander: Applications of Microbial Genetics to Industrial Fermentations, p. 89.
12. A. Furuya, M. Misawa, T. Nara, S. Abe, and S. Kinoshita: Metabolic Controls of Accumulations of Amino Acids and Nucleotides, p. 177.

A fascinating introduction of the topic of recombinant DNA molecules:
13. S. N. Cohen: The Manipulation of Genes, *Sci. Am.*, July 1975, p. 24.

Problems

14. P. B. Weisz: Enzymatic Reaction Sequences and Cytological Dimensions, *Nature*, **195**: 772, 1962.
15. A. C. Nevo, and R. Rikmenspoel: Diffusion of ATP in Sperm Flagella, *J. Theoret. Biol.*, **26**: 11, 1970.
16. J. M. Smith: "Mathematical Ideas in Biology," p. 108, Cambridge University Press, London 1971.
17. M. Kimura, and T. Ohta: "Theoretical Aspects of Population Genetics," Princeton University Press, Princeton, N.J., 1971.
18. E. O. Wilson, and W. H. Bossert: "Primer of Population Biology," pp. 79–83, Sinauer Associates, Stamford, Conn., 1971.

SEVEN

KINETICS OF SUBSTRATE UTILIZATION, PRODUCT YIELD, AND BIOMASS PRODUCTION IN CELL CULTURES

When a small quantity of living cells is added to a liquid solution of essential nutrients at a suitable temperature and pH, the cells will grow. The growth processes of interest to us have two different manifestations according to the morphology of the microorganism involved. For unicellular organisms which divide as they grow, increases in *biomass* (mass of living matter) are accompanied by increases in the number of cells present. This case, which confronts us with a problem in *population* growth, will occupy most of our attention in this chapter. When considering growth of molds, however, the situation is quite different. Here the length and number of mycelia increase as the organism grows. The growing mold thus increases in size and density but not necessarily in numbers.

Associated with cell growth are two other processes, uptake of some material from the cell's environment and release of metabolic end products into the surroundings. As we shall see below, the rates of these processes vary widely as growth occurs. While a general predictive capability will elude us, past experience has shown that among the many microbial processes known, a few general patterns of substrate utilization and product occur frequently. By reviewing them, we shall be better prepared to cope with new problems in microbial application.

Before delving into the available experimental data and associated theories, we should have some appreciation of the difficulties of measuring and monitoring growth processes. Three common techniques, hemacytometer, colony count, and turbidimetric methods, are illustrated schematically in Fig. 7.1. The first two are rather elaborate and require extensive manipulation of the sample. Moreover, the direct-count method does not distinguish between live and dead cells. This disadvantage also detracts from the turbidimetric technique, which is based on the reduction in light transmission as the density of suspended matter increases. Not only do dead cells contribute to the turbidity of the suspension, but in commercial

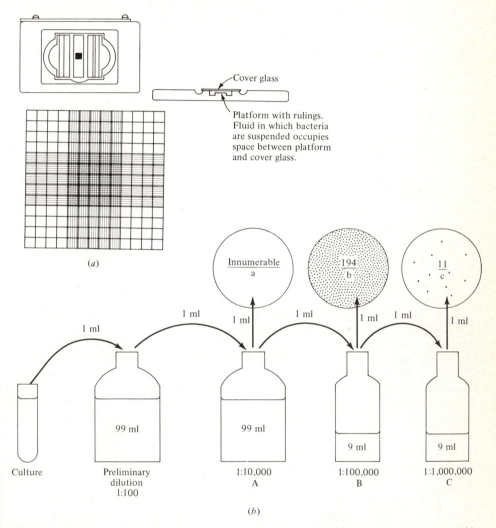

Cover glass

Platform with rulings. Fluid in which bacteria are suspended occupies space between platform and cover glass.

(a)

Innumerable
a

194
b

11
c

1 ml 1 ml 1 ml 1 ml 1 ml 1 ml 1 ml

99 ml 99 ml 9 ml 9 ml

Culture   Preliminary   1:10,000   1:100,000   1:1,000,000
                 dilution   A   B   C
                 1:100

(b)

Figure 7.1 Techniques for measurement biomass concentration. (a) In the hemacytometer a grid facilitates cell counting. (b) In the colony-count method, greatly diluted samples from the original culture are plated on nutrient agar in Petri dishes, where the number of colonies indicates the number of viable cells in the diluted sample. (c) Measurement of turbidity of a cell suspension provides a measure of population density. [*Reprinted by permission from M. Frobisher, "Fundamentals of Microbiology," 8th ed., pp. 46, 47, 49, W. B. Saunders Co., Philadelphia, 1968; (a) also courtesy of Arthur H. Thomas Co., Philadelphia.*]

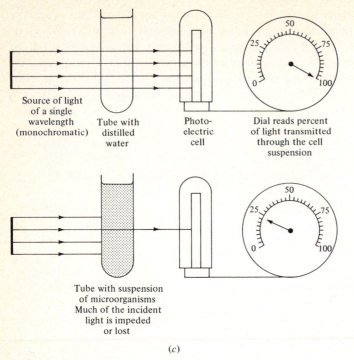

(c)

Figure 7.1 (continued)

practice other debris in solution can cause erroneous measurements. Finally, the optical density of a cell suspension is a linear function of biomass only for low density values (Fig. 7.2). In spite of these difficulties, the turbidimeter enjoys wide application because it is convenient and capable of instantaneous, on-line measurement.

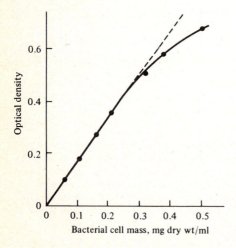

Figure 7.2 Optical density is proportional to cell-mass density only over a limited lower range of values. (*From R. Y. Stanier, M. Doudoroff, and E. A. Adelberg, "The Microbial World," 3d ed., p. 300, 1970. Reprinted by permission of Prentice-Hall, Inc., Englewood Cliffs, N.J.*)

A relatively new approach has shown promise for rapid measurement of *living-cell* concentration. This is an indirect procedure in that it actually measures the light emitted by a luciferase reaction (see Sec. 5.4.2), which in turn depends directly on ATP concentration. ATP concentration is strongly correlated with the number of live cells. In one study a mean ATP content of 4.7×10^{-10} µg per cell was found. Viable cell number measurements by this method agree well with findings of the more time-consuming serial dilution-plate-count approach described in Fig. 7.1.

7.1. GROWTH-CYCLE PHASES FOR BATCH CULTIVATION

Many biochemical processes involve batch growth of microorganisms. After *seeding* a liquid *medium* of appropriate composition with an *inoculum* of living cells, nothing (except possibly some gas) is added to the *culture* or removed from it as growth proceeds. Typically the number of living cells in the culture varies with time, as shown in Fig. 7.3. After a *lag phase*, where no increase in cell numbers is evident, a period of rapid growth ensues, where the cell numbers increase exponentially with time. Although this stage of batch culture is often called the *logarithmic phase*, we prefer the more descriptive term *exponential growth*, which we shall use in the following discussion.

Naturally in a closed vessel the cells cannot multiply indefinitely, and a *stationary phase* follows the period of exponential growth. At this point the population achieves its maximum size. Eventually a decline in cell numbers occurs during the *death phase*. Here an exponential decrease in the number of living individuals is often observed.

Each phase is of potential importance in microbiological processes. For example, the general objective of a good process design may be to minimize the

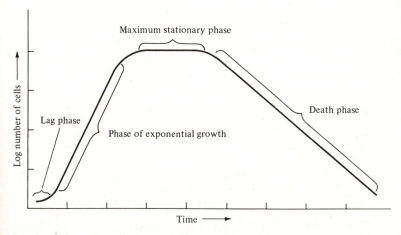

Figure 7.3 Typical batch growth curve of a microbial culture.

length of the lag phase and to maximize the rate and length of the exponential phase, the last objective being achieved by slowing the onset of the transition to stationary growth. To achieve such goals, we should understand the variables which influence the phases of batch growth. Each phase will be discussed individually next, and, where appropriate, mathematical models of the important phenomena will be considered. This discussion will be augmented by Sec. 7.2, where we consider mathematical analyses which bear on the entire batch-growth process.

7.1.1. The Lag Phase

The length of lag observed when a fresh medium is innoculated depends on both the changes in nutrient composition (if any) experienced by the cells and the age and size of the inoculum. The shock of rapid switch to a new environment has several effects on the living cell.

First, recall that the modes of control and regulation of enzyme activity include an adaptive characteristic: when presented with a new nutrient, the cell produces enzymes to allow its metabolization. Thus transfer of a glucose-bred culture on its exponential phase to a lactose medium will necessarily result in a time interval of insignificant cell-division rate while the enzymes and cofactors for the lactose metabolic pathway are synthesized in the cell. (What would happen if a lactose-bred culture were transferred to a glucose-lactose medium?) Similarly, variation in the concentration of nutrients may cause a lag phase. If the new nutrient medium is richer in a limiting nutrient, some time and nutrient will be expended in nonmultiplicative growth while larger concentrations of metabolizing enzymes are created. A decreased nutrient level may result in no lag at all; the exponential rate may resume immediately but at a slower pace.

Many of the intracellular enzymes require activation by small molecules (vitamins, cofactors) or ions (activators) which may have appreciable permeability through the cell membrane. Transfer of a small culture volume or inoculum to a large volume of medium will cause outward diffusion of these requisites for catalysis into the bulk medium if the new medium is lacking in these species or differs appreciably in ionic strength. The rate of growth will fall corresponding to the lower concentrations of such species inside the cell, and again a lag will appear while new machinery to generate such activators is assembled. If essential activators are diluted (vitamins and ions which the cell cannot produce internally), the total level of cell activity must diminish irrevocably. For example, the time of the lag phase of *Aerobacter aerogenes* grown in glucose and phosphate buffer medium increases without limit as the concentration of activator (Mg^{2+}) for phosphatase, a phosphate-transferring enzyme, is decreased to zero, as illustrated in Fig. 7.4.

The age of the inoculum exerts a strong influence on the length of the lag phase. Culture transfer into a new volume of the same medium yields lag-time–vs.–age curves for the bacterium *A. aerogenes* shown in Fig. 7.5a and b. Transfer of young cells into amino acid medium (glucose-phosphate buffer) results in short lag times. By contrast, transfer of an older population (in a slower growth-rate stage due to nutrient depletion and/or toxin accumulation) results in longer lag transients while the population gears its metabolic rates upward in response to induction by higher nutrient concentrations or deinhibition and derepression as toxic and growth-inhibiting substances diffuse out of the older cells into the fresh medium.

The size of the transferred inoculum is a variable, as already seen by the loss upon transfer of such diffusible species as vitamins and activators. Thus while a young cell population shows no lag upon transfer into a medium rich in metabolic intermediates such as amino acids, the same inoculum transferred into ammonium sulfate loses these vital intermediates into solution. A longer lag results for the youngest such species since they will have accumulated only a small amino acid concentration in the original culture medium. With cultures in exponential growth at the time of transfer, their original medium may already contain a reasonable bulk concentration of intermediates, and the dilution on transfer will have a lesser effect, as seen by the minimum in Fig. 7.5b. Again, transfer of an old culture into ammonium sulfate results in a longer lag, just like the transfer into amino acid medium in Fig. 7.5a.

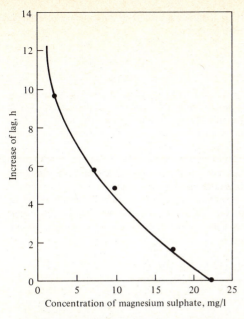

Figure 7.4 Influence of Mg^{2+} concentration on the lag phase in *A. aerogenes* culture. (*Reprinted from A. C. R. Dean and C. Hinshelwood, "Growth, Function and Regulation in Bacterial Cells," p. 55, Oxford University Press, London, 1966.*)

As a hypothesis for modeling, we shall suppose that the lag period is essentially ended when some critical substance *in the cell* reaches a given value c'. For young inocula, Dean and Hinshelwood [3] suggest an equation for the critical-substance concentration *in the cell* as a function of time in the lag phase of the form

$$c = aV + a'n_0 t + a''t \qquad (7.1)$$

where V = volume of old medium transferred

a = concentration of critical substance per unit volume of old medium x (old volume/new volume)

n_0 = number of cells per unit new volume (assumed constant since growth rate is negligible in lag phase)

a' = average increase in cell critical substance (due to production by other cells) per time per cell

a'' = increase in critical substance due to internal cell production

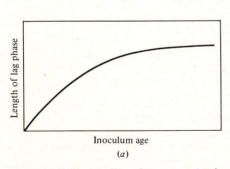

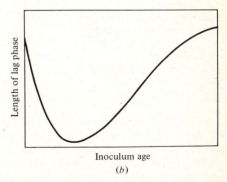

(a)

(b)

Figure 7.5 Influence of inoculum age on *Aerobacter aerogenes* lag phase for (a) an amino acid medium and (b) an ammonium sulphate medium. (*Reprinted from A. C. R. Dean and C. Hinshelwood, "Growth, Function and Regulation in Bacterial Cells," pp. 58, 59, Oxford University Press, London, 1966.*)

Letting t_{lag} denote the lag time and c' the concentration of critical substance at the conclusion of the lag phase, we can rearrange Eq. (7.1) to give the lag-time expression

$$t_{lag} = \frac{c'/a' - aV/a'}{n_0 + a''/a'} \tag{7.2}$$

Transfer of a negligible number of cells, $n_0 \ll a''/a'$, results in a lag time inversely proportional to the productivity a'' of a single young cell (taken to be constant). For large inoculum sizes, t_{lag} varies inversely with n_0. The validity of Eq. (7.2) can be checked by testing the linearity of t_{lag} vs. V at constant n_0 and $1/t_{lag}$ vs. n_0 at constant volume transferred. Such plots are shown in Fig.7.6a and b, again for *A. aerogenes*, where fair agreement with Eq. (7.2) is revealed.

When a new culture is transferred, the order of 99 percent or greater of the cells are alive. In older cultures, account must be taken of the difference between viable or living cell concentrations and the total (biomass) cell concentration in order to determine the lag time quantitatively. The current definition of the lag time is the period before steady (exponential growth) of the living cells is achieved. If the total biomass is the variable actually measured vs. time with such indices as total dry cell weight or total nitrogen level (for protein analysis), allowance for transfer of dead cells must be included in a proper determination of the true lag time.

If we take

$$n_0 = \text{number of live cells in inoculum}$$
$$n_d = \text{number of dead cells in inoculum}$$
$$t_{lag} = \text{true lag time}$$

for $t > t_{lag}$, the total cell number n will be

$$n = n_0 \exp\left[(t - t_{lag})\mu\right] + n_d$$

where μ is the *specific growth rate* (cell rate of formation per number of cells) of the live cells. Therefore,

$$\frac{dn}{dt} = \mu n_0 \exp\left[(t - t_{lag})\mu\right] = \mu(n - n_d) \tag{7.3}$$

Since exponential growth satisfies

$$\frac{dn}{dt} = \mu' n \qquad \text{or} \qquad \frac{1}{n}\frac{dn}{dt} = \mu'$$

the apparent value of μ' (as opposed to the real value μ) is given by

$$\frac{1}{n}\frac{dn}{dt} = \mu' = \frac{\mu(n - n_d)}{n}$$

$$\mu' = \mu\left(1 - \frac{n_d}{n}\right) \tag{7.4}$$

If a very small inoculum is transferred, by the time n has reached a conveniently measurable level, $n \gg n_d$ and $\mu' \to \mu$ (but a large time lag may be involved). Transfer of large aging inocula may result in achievement of exponential growth by the time n is not much larger than n_d. The lag period will have ended, but this fact will not be evident unless correction has been made for the nonviable cells in the inocula since

$$\frac{1}{n}\frac{dn}{dt} = \mu' = \mu(1 - n_d/n)$$

will be changing in time.

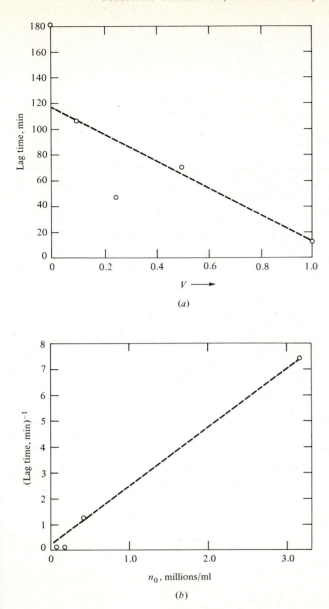

Figure 7.6 To a first approximation, the lag time is proportional to inoculum volume (*a*) and inversely proportional to inoculum cell density (*b*) (*Aerobacter aerogenes* in ammonium sulfate medium). (*Plotted from data in A. C. R. Dean and C. Hinshelwood, "Growth, Function and Regulation in Bacterial Cells," pp. 63–64, Oxford University Press, London, 1966.*)

The extremes of the situation are conveniently tested. A small culture transfer will yield μ independent of the dead-cell population. A large culture transfer will yield an apparent growth-rate constant μ', and the fraction of dead cells at the time of measurement will be

$$\frac{n_d}{n} = 1 - \frac{\mu'}{\mu}$$

The ratio of nonviable to viable cells in the original inoculum is thus

$$\frac{n_d}{n_0} = \left(\frac{\mu}{\mu' - 1}\right) \exp\left[(t - t_{\text{lag}})\mu\right] \tag{7.5}$$

where t and μ' are measured at the same time. Thus by a small-inoculum-transfer ($\mu' = \mu$) experiment and a large-inoculum-transfer ($\mu' \neq \mu$) experiment, measurements yielding μ, μ', and t are made and the ratio of living to dead cells at the time of transfer is determined from Eq. (7.5).

Under favorable circumstances, all living cells transferred exhibit a relatively short lag and then enter the exponential-growth phase. In less favorable cases, the transfer to a new or different medium may utilize so much energy of marginally live cells in the lag phase that a certain death rate occurs in this period. Quantification is difficult, but if the death rate is assumed to be proportional to the number of live cells with a rate constant k_d, the specific rate of cell loss during lag is

$$-\frac{1}{n}\frac{dn}{dt} = k_d$$

and after the lag period the cell number will be

$$n = n_d + n_0 \exp\left(-k_d t_{\text{lag}}\right) \exp\left[\mu(t - t_{\text{lag}})\right]$$

Calculation of the original (nonviable/viable) fraction is now given by

$$\frac{n_d}{n_0} = \left(\frac{\mu}{\mu' - 1}\right) \exp\left[\mu(t - t_{\text{lag}})\right] \exp\left(-k_d t_{\text{lag}}\right) \tag{7.6}$$

The theory just outlined is also applicable to situations where a favored *mutant population* occurs. If the culture medium is unable to support growth of the bulk of the inoculated cells (n_d) but a few mutant cells in the inoculum can thrive, only the mutant cells will proliferate and eventually will dominate the populations.

Multiple lag phases may sometimes be observed when the medium contains multiple carbon sources (Fig. 7.7). This phenomenon, known as *diauxic growth*, is caused by a shift in metabolic

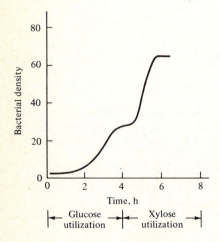

Figure 7.7 In a medium containing initially equal amounts of glucose and xylose, diauxic growth of *E. coli* is observed in batch culture. (*From R. Y. Stanier, M. Doudoroff, and E. A. Adelberg, "The Microbial World," 3d ed., p. 308, 1970. Reprinted by permission of Prentice-Hall, Inc., Englewood Cliffs, New Jersey.*)

patterns in the midst of growth. After one carbon substrate is exhausted, the cell must divert its energies from growth to "retool" for the new carbon supply. A possible explanation for this phenomenon is catabolite repression, discussed in Sec. 6.1.3.

Design to minimize culture and process times normally includes minimization of the lag times associated with each new batch culture. From the previous general discussion and other relevant data (see references), the following generalizations can be drawn:

1. The inoculating culture should be as active as possible and the inoculation carried out in the exponential-growth phase.
2. The culture medium used to grow the inocula should be as close as possible to the final full-scale fermentation composition.
3. Use of reasonably large inocula (order of 5 percent of the new medium volume) is recommended to avoid undue loss by diffusion of required intermediates or activators.

7.1.2. Exponential Growth and the Monod Equation

At the end of the lag phase the population of microorganisms is well adjusted to its new environment. The cells can then multiply rapidly, and cell mass, or the number of living cells, doubles regularly with time. As already mentioned in the previous section, the equations

$$\frac{dn}{dt} = \mu n \qquad \text{or} \qquad \frac{1}{n}\frac{dn}{dt} = \mu \tag{7.7a}$$

with
$$n = n_0 \qquad \text{at} \qquad t = t_{\text{lag}} \tag{7.7b}$$

describe the increase in cell numbers during this period. Thus the rate of increase in n is proportional to n. From the integrated form of Eq. (7.7)

$$\ln \frac{n}{n_0} = \mu(t - t_{\text{lag}}) \qquad \text{or} \qquad n = n_0 e^{\mu(t - t_{\text{lag}})} \qquad t > t_{\text{lag}} \tag{7.8}$$

we can readily deduce that the time interval t_d required to double the population is given by

$$t_d = \frac{\ln 2}{\mu} \tag{7.9}$$

During exponential growth, only a single parameter μ (or t_d) is required to characterize the population. For this reason, the magnitude of the specific growth rate μ is widely used to describe the influence of the cell's environment on its performance. Consider first the influence of temperature: the range of temperatures capable of supporting life as we know it lies between roughly -5 and $95°C$. Procaryotes can be classified according to the temperature interval in which they grow. As seen in Table 7.1, each class has an optimum temperature where growth is maximal and upper and lower temperature bounds beyond which the population cannot survive.

The data in Fig. 7.8 for growth of E. coli dramatically illustrate the strong influence of temperature. Notice in the Arrhenius plot that classical Arrhenius behavior appears at low temperatures whereas there is a rapid decrease in growth rate as the temperature approaches the upper limit for survival of the bacterium.

Table 7.1. Classification of microorganisms in terms of growth-rate dependence on temperature†

Group	Temperature, °C		
	Minimum	Optimum	Maximum
Thermophiles	40 to 45	55 to 75	60 to 80
Mesophiles	10 to 15	30 to 45	35 to 47
Psychrophiles:			
Obligate	−5 to 5	15 to 18	19 to 22
Facultative	−5 to 5	25 to 30	30 to 35

† R. Y. Stanier, M. Doudoroff, and E. A. Adelberg, "The Microbial World," 3d ed., p. 316, Prentice-Hall, Inc., Englewood Cliffs, N.J., 1970.

The similarity of the temperature dependence for growth in Fig. 7.8 with the enzyme-activity–temperature relationship depicted in Fig. 3.38 is unescapable. Apparently at low temperatures the metabolic activity of the cell increases with increasing temperature as the activity of its enzymes rises. When the most thermally sensitive essential protein denatures, however, the cell will die. This hypothesis has been confirmed in several instances by genetic studies in which mutation of a single gene has caused a large change in the maximum tolerable temperature for a microorganism.

Before discussing details of the dependence of growth rates on nutrient supply, we should review the general ideas and practices for construction of cell-culture media. We distinguish two types of media according to their makeup.

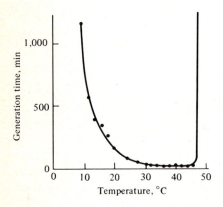
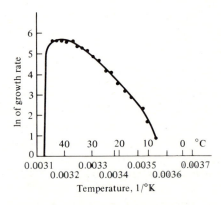

Figure 7.8 (*a*) Up to a point, the growth rate of *E. coli* increases with increasing temperature, but the cells die if the temperature is too high. (*b*) An Arrhenius plot of the data in (*a*). (*From R. Y. Stanier, M. Doudoroff, and E. A. Adelberg, "The Microbial World," 3d ed., pp. 316, 317, 1970. Reprinted by permission of Prentice-Hall, Inc., Englewood Cliffs, New Jersey.*)

A *synthetic medium* is one whose chemical composition is well defined. As Table 7.2 shows, such media can be constructed by supplementing a mineral base with the necessary carbon, nitrogen, and energy sources as well as any necessary vitamins. In addition to providing necessary ions for proper cell function, the mineral base of the medium also contains buffering compounds to reduce large pH fluctuations during the growth cycle. *Complex media* contain materials of undefined composition. For example, in Table 7.2, medium 4 is complex because the exact chemical makeup of the yeast extract is unknown. Other common complex media include beef broth, blood-infusion broth, corn-steep liquor, and sewage.

The general goal in making a medium is to support good growth and/or high rates of product synthesis. Contrary to intuitive expectation, this does *not* necessarily mean that all nutrients should be supplied in great excess. For one thing, excessive concentrations of a nutrient can inhibit or even poison cell growth. Moreover, if the cells grow too extensively, their accumulated metabolic end products will often disrupt the normal biochemical processes of the cells. Consequently, it is common practice to limit total growth by providing only a limited amount of one nutrient in the medium.

If the concentration of one essential medium constituent is varied while the concentrations of all other medium components are kept constant, the growth rate changes in a hyperbolic fashion, as Fig. 7.9 shows. A functional relationship between the specific growth rate μ and an essential compound's concentration was proposed by Monod in 1942. Of the same form as the Langmuir adsorption isotherm (1918) and the standard rate equation for enzyme-catalyzed reactions

Table 7.2. Some examples of synthetic and complex media†

Common ingredients (Mineral base)	Additional ingredients			
	Medium 1‡	Medium 2‡	Medium 3‡	Medium 4§
Water, 1 l	NH$_4$Cl, 1 g	Glucose,¶ 5 g	Glucose, 5 g	Glucose, 5 g
K$_2$HPO$_4$, 1 g		NH$_4$Cl, 1 g	NH$_4$Cl, 1 g	Yeast extract,
MgSO$_4$ · 7H$_2$O, 200 mg			Nicotinic acid,	5 g
FeSO$_4$ · 7H$_2$O, 10 mg			0.1 mg	
CaCl$_2$, 10 mg				
Trace elements (Mn, Mo, Cu, Co, Zn) as inorganic salts, 0.02–0.5 mg of each				

† R. Y. Stanier, M. Doudoroff, and E. A. Adelberg, "The Microbial World," 3d ed., p. 79, Prentice-Hall, Inc., Englewood Cliffs, N.J., 1970.

‡ Synthetic.

§ Complex.

¶ If the media are sterilized by autoclaving, the glucose should be sterilized separately and added aseptically. When sugars are heated in the presence of other ingredients, especially phosphates, they are partially decomposed to substances that are very toxic to some microorganisms.

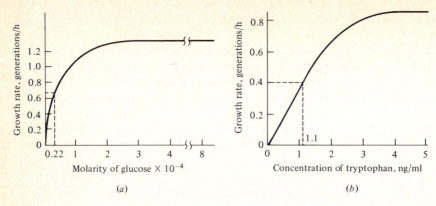

Figure 7.9 Dependence of *E. coli* exponential-growth rate on the concentration of the growth-limiting nutrient: (*a*) glucose medium and (*b*) tryptophan medium (for a trytophan-requiring mutant). (*From R. Y. Stanier, M. Doudoroff, and E. A. Adelberg, "The Microbial World" 3d ed., p. 315, 1970. Reprinted by permission of Prentice-Hall, Inc., Englewood Cliffs, New Jersey.*)

with a single substrate (Henri in 1902 and Michaelis and Menten in 1913), the Monod equation states that

$$\mu = \frac{\mu_{max} c_i}{K_i + c_i} \tag{7.10}$$

Here μ_{max} is the maximum growth rate achievable when $c_i \gg K_i$ and the concentrations of all other essential nutrients is unchanged. K_i is that value of the concentration of nutrient i where the specific growth rate has half its maximum value; roughly speaking, it is the division between the lower concentration range, where μ is strongly (linearly) dependent on c_i, and the higher range, where μ becomes independent of c_i. As shown in Fig. 7.9. K_i values for *E. coli* strains growing in glucose and tryptophan media are $0.22 \times 10^{-4} M$ and 1.1 mg/ml, respectively.

From our earlier examination of the cell's biochemistry, it is apparent that the Monod equation is probably a great oversimplification. As in other areas of engineering, however, this is a case where a relatively simple equation reasonably expresses interrelationships even though the physical meaning of the model parameters is unknown or perhaps does not exist. In some special instances, however, we may attach physical significance to the Monod equation. When discussing active transport in Sec. 5.6.2, we cited evidence that the growth of some bacteria appears to be limited by the rate of active transport of glucose. Active-transport rates, however, usually vary with external concentration of nutrient in exactly the same fashion as enzyme reaction rates—and, of course, just as the Monod equation for μ varies with c_i. In such cases K_i may tentatively be considered as the Michaelis constant for the permease carrier molecule while μ_{max} contains the v_{max} for the permease multiplied by unknown factors reflecting the cell's utilization of the nutrient.

The particular form of Eq. (7.10) is appealing; its simplicity urges several warnings upon the user. First, the value of K_i is quite often rather small. Thus c_i can be $\gg K_i$ rather easily, and the term $c_i/(K_i + c_i)$ may be regarded simply as an adequate description for calculating the deviation of μ from μ_{max} as the concentration c_i begins to diminish. The relation also suggests that when component i is a nutrient (glucose, fructose, etc.), the specific growth rate is finite $(\mu \neq 0)$ for any finite nutrient concentration. Generally, this implied behavior is not well tested for $c_i \ll K_i$.

Since the nutrient level is often initially high in a batch medium $(c_i \gg K_i)$, a true exponential specific-growth rate will be observed in early stages of nutrient consumption. As c_i approaches K_i, the value of μ diminishes and departure from the original exponential law will be predicted. Since K_i is often quite small, however, the stationary phase is often reached before $c_i = K_i$ and the contribution of this correction to the result in Eq. (7.8) is small.

To appreciate the most serious limitation of the Monod equation, we must define the concept of *balanced growth:* growth is *balanced* over a time interval if every extensive property of the growing system increases by the same factor during that interval. As Monod† said of the exponential phase of growth, "it is reasonable to consider ... a ... state ... where the relative concentrations of all metabolites and all the enzymes are constant. *It is, in fact, the only phase of the growth cycle when the properties of the cells may be considered constant* and can be described by a numeric value, the exponential growth rate, corresponding to the over-all velocity of the ... system." This viewpoint is echoed by recent investigators who assert that any attempt to quantify growth of cell populations only in terms of population number or biomass (*unstructured* models) can be complete only for the special case of balanced growth. In any other situation, some aspects of the physiological state of the culture must be included in the analysis if it is to have general validity. The idea here is that populations of the same size may be quite different in, say, the composition of its individuals, depending on the past history of the population. The topic of *structured* models will be considered in Sec. 7.2.

Qualitatively, we may view balanced growth as the situation reached in a constant environment after all necessary cellular adaptations have been accomplished. In a batch system where growth is sufficiently slow and the population is not too large, balanced growth is a reasonable approximation during the exponential-growth phase. The Monod equation describes experimental findings under such conditions quite reasonably.

If, however, nutrient consumption is rapid, the internal cell nutrient levels may be far below the external medium level, which itself may also be changing in time. Under such conditions, the cell's internal control and metabolic regulation may not ever be optimally adjusted since continuous updating of the appropriate enzyme production rates is necessary. Under these circumstances a truly ex-

† J. Monod, The Growth of Bacterial Cultures, *Ann. Rev. Microbiol.* **3**: 371 (1949).

ponential phase may not be found. The Monod equation consequently may break down if growth is rapid. Related forms which have been proposed to model rapid growth rates include

$$\mu = \frac{\mu_{max} c_i}{K_i c_{i0} + c_i} \tag{7.11}$$

and

$$\mu = \frac{\mu_{max} c_i}{K_{i1} + K_{i2} c_{i0} + c_i} \tag{7.12}$$

where c_{i0} is the intial concentration of component i. The objective of such equations is to indicate that as the *initial* nutrient concentration is increased, the observed growth rate shows a deviation from strictly logarithmic growth ($c_{i0} \gg K_{i1}, K_{i2}, c_i$) at progressively higher values of c_i, in agreement with the observed discrepancy between Eq. (7.10) and high-growth-rate experimental data.

Other related forms of specific growth rate dependence have been proposed which in particular instances give better fits to experimental data. For example, Teisser, Moser, and Contois suggest the following models:

Teisser: $$\mu = \mu_{max}(1 - e^{-c_i/K_i}) \tag{7.13}$$

Moser $$\mu = \mu_{max}(1 + K_i c_i^{-\lambda})^{-1} \tag{7.14}$$

Contois: $$\mu = \mu_{max} \frac{c_i}{Bn + c_i} \tag{7.15}$$

The first two examples render algebraic solution of the growth equations much more difficult than the Monod form. The equation of Contois contains an apparent Michaelis constant which is proportional to biomass concentration n. This last term will therefore diminish the maximum growth rate as the population density increases, eventually leading to $\mu \propto n^{-1}$.

The specific-growth rate may be inhibited by medium constituents such as substrate or product. An example due to Andrews [9] proposes that *substrate inhibition* be treated by the form:

$$\mu = \mu_{max} \frac{c_i}{K_i + c_i + c_i^2/K_p} \tag{7.16}$$

Alcohol fermentation provides a nice example of *product inhibition;* the anaerobic glucose fermentation by yeast has been treated by Aiba, Shoda, and Nagatani [7,8] with specific-growth function of the type

$$\mu = \mu_{max} \frac{c_i}{K_i + c_i} \frac{K_p}{K_p + c_p} \tag{7.17}$$

It has often been observed that below some low threshold value of substrate, cells do not proliferate and/or the desired product does not appear in the fermentation. This phenomenon may be due to a *maintenance* process wherein the cell must consume substrate at a small but finite rate just to survive. Possible biological explanations for this minimum level of fuel consumption include work being done to maintain concentrations of various species by active membrane transport (just balancing losses by passive membrane diffusion) and continuous repair or replacement of damaged DNA, RNA, enzymes, and other cellular constituents. The general form of the specific-growth rate *including maintenance* is

$$\mu = \mu_{max} \frac{c_i}{K_i + c_i} - k_e \qquad (7.17a)$$

Most of the experimental data supporting maintenance models derive from continuous culture, which will be considered in detail in Chap. 9.

It is possible that two (or more) substrates may simultaneously be growth-limiting. While few data are available, a Monod dependence on each limiting nutrient may be used, so that

$$\mu = \mu_{max} \frac{c_1}{K_1 + c_1} \frac{c_2}{K_2 + c_2} \cdots \qquad (7.18)$$

In the absence of convincing data for this form, we may regard it simply as a useful indicator that growth depends on several limiting nutrients. Additional models which usefully describe other situations will be discussed in Sec. 7.2 and 7.4 and in Chap. 9.

In closing this section, we recall that for exponential growth

$$\frac{1}{n} \frac{dn}{dt} = \mu$$

describes the behavior of biomass or cell-density variation. Here we have considered *unstructured* models for μ which depend only on external variables such as c_i and T. The past history, age distribution of cells, and other *structural* information which may affect μ are considered in a subsequent section.

7.1.3. The Stationary Phase and the Maximum Population

Deviations from exponential growth eventually arise when some significant variable such as nutrient level or toxin concentration achieves a value which can no longer support the maximum growth rate.

Exhaustion of a particular critical nutrient may appear rather sharply at a given time since the cells are rapidly increasing the *total* rate of nutrient consumption in the exponential-growth phase. To formulate a rough analysis of this event, we suppose that the rate of nutrient A consumption is proportional to the number of living cells until the stationary phase is reached:

$$\frac{da}{dt} = -k_a n \qquad (7.19)$$

Next we assume that exponential growth continues unabated until the stationary phase is reached, and we take the time when exponential growth begins at time zero. Then

$$n = n_0 e^{\mu t} \tag{7.20}$$

where n_0 is the number of living cells when exponential growth starts.

If the concentration of A at time zero is a_0, we can determine from Eq. (7.19) and (7.21) that A is completely consumed when

$$a_0 = \frac{k_a}{\mu}(n_s - n_0) \tag{7.21}$$

where n_s is the size of the population when A is exhausted and the population enters the stationary phase. Consequently, n_s is the maximum population size achieved during the batch culture (review Fig. 7.3). Rearranging Eq. (7.21), we find the maximum population in the case of nutrient depletion to be given by

$$n_s = n_0 + \frac{k_a}{\mu} a_0 \tag{7.22}$$

Linear dependence of n_s on initial nutrient level has been observed experimentally in many cases (often n_0 is so small that it is frequently imperceptible). Figure 7.10 illustrates this behavior. In the plot of n_s for the bacterium *A. aerogenes* vs. lactose concentration, however, distinct variations from Eq. (7.22) are obvious. Apparently other factors can influence the onset of the stationary phase and the size of the maximum population.

If a toxin accumulates which slows the rate of growth from exponential, an equation of the form

$$\frac{dn}{dt} = kn[1 - f(\text{toxin concentration})]$$

is found useful. In the particular case where toxin linearly decreases the growth rate, we have a specific growth rate

$$\frac{1}{n}\frac{dn}{dt} = k(1 - bc_t) \tag{7.23}$$

where c_t is toxin concentration and b is a constant. A plausible assumption is that the rate of toxin production depends only on n and is proportional to it

$$\frac{dc_t}{dt} = qn \tag{7.24}$$

so that $c_t = q \int_0^t n \, dt$ (assuming $c_t = 0$ at $t = 0$) and the growth equation (7.23) becomes

$$\frac{dn}{dt} = kn\left(1 - bq \int_0^t n \, dt\right) \tag{7.25}$$

The instantaneous value of the effective specific-growth rate μ_{eff}

$$\mu_{\text{eff}} = \frac{1}{n}\frac{dn}{dt} = k\left(1 - bq \int_0^t n \, dt\right) \tag{7.26}$$

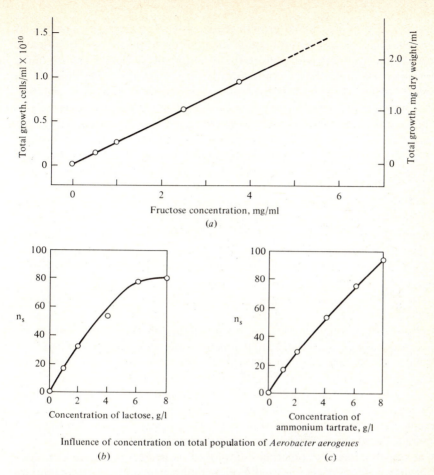

Influence of concentration on total population of *Aerobacter aerogenes*

(b)   (c)

Figure 7.10 Dependence of maximum batch population size on the initial concentration of growth-limiting nutrient: (a) *Pseudomonas* sp. in fructose medium and *A. aerogenes* in (b) lactose and (c) ammonium tartrate media. [(a) *From R. Y. Stanier, M. Doudoroff, and E. A. Adelberg, "The Microbial World" 3d ed., p. 313, 1970. Reprinted by permission of Prentice-Hall, Inc., Englewood Cliffs, New Jersey; (b, c) reprinted by permission from A. C. R. Dean and C. Hinshelwood, "Growth, Function and Regulation in Bacterial Cells," p. 72, Oxford University Press, London, 1966.*]

diminishes more and more rapidly with time, i.e.,

$$\frac{d\mu_{\text{eff}}}{dt} = -kbqn < 0 \qquad \frac{d^2\mu_{\text{eff}}}{dt} = -kbq\frac{dn}{dt} < 0 \qquad (7.27)$$

Growth halts when

$$\frac{1}{bq} = \int_0^t n \, dt \qquad (7.28)$$

Equation (7.23) indicates that growth ceases only when c_t reaches a particular level, $c_t = 1/b$. Dilution of a given toxified medium or addition of a nonnutritive substance which complexes with a given toxin should allow further growth and a consequent increase in n_s, the total number of cells at the time of development of the stationary phase ($dn/dt = 0$). If growth halts due to nutrient exhaustion, dilution with a nonnutritive volume produces no change in n_s. These criteria may be roughly utilized to determine the cause of growth decline and eventual halt. Exact criteria are more difficult to ascertain: growth of nutrient-limited populations does slow somewhat before total exhaustion, as will be seen later, and the growth rate of a poisoned population may become imperceptibly slow long before dn/dt is exactly zero.

The expected dependence of maximum population on the initial level of a given nutrient is sketched in Fig. 7.11. Diminution of nutrient concentration eventually brings the culture to a maximum size which is linear in the initial critical nutrient. Here nutrient depletion apparently causes the cessation of exponential growth. Conversely, a rise in initial nutrient supply may eventually yield an n_s value apparently independent of the nutrient level. This suggests accumulation of toxic products or the existence of some other limiting nutrient as the determining factor.

The regions of dominance of n_s by one medium constituent vs. another are nicely illustrated by glucose and pH influences on *A. aerogenes*, as seen in Fig. 7.12. In amply concentrated glucose-phosphate media, the stationary-phase population is a changing function of pH at all pH values. Considerable reduction of the glucose level yields a region between pH 6.5 and 8.5 where n_s is apparently limited by glucose exhaustion. The dominance of one component over others is rarely totally exclusive: when the acidic portions of the log n_s-vs.-pH curves in Fig. 7.12 are compared, the glucose influence here is still clearly evident.

Just as gowth rates are sometimes employed as measures of nutrient effectiveness, *growth yields* may also be used. For chemoheterotrophs, the total amount of cell material formed per unit amount of substrate provides a measure of the efficiency of the nutrient for supporting biosynthesis. Based on the data for *Pseudomonas* sp. in Fig. 7.10a and the fact that dry cellular material is 50 percent carbon, for example, we can conclude that one-half of the carbon in fructose is

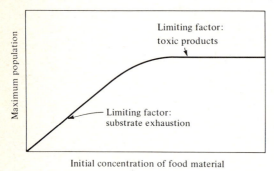

Figure 7.11 Typical relationship between initial nutrient concentration and the maximum cell population in batch culture. When the nutrient concentration is large and has no influence on the maximum population, accumulation of toxic products may be the factor which limits population size.

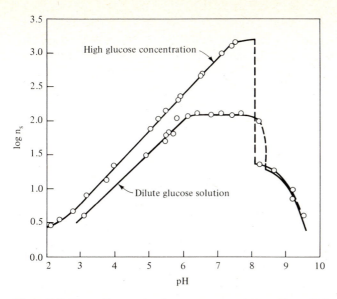

Figure 7.12 Depending on the glucose concentration, medium pH has different effects on the maximum batch population size (*A. aerogenes*). (*Adapted from A. C. R. Dean and C. Hinshelwood, "Growth, Function and Regulation in Bacterial Cells," p. 72, Oxford University Press, London, 1966.*)

eventually incorporated into organic compounds in the cell. In our earlier discussions of growth yields (Sec. 5.3.2 and Table 5.6), we observed that differences in the biosynthetic efficiency of a substrate in anaerobic metabolism ultimately rest on differences in the efficiency of ATP generation from the substrate; current evidence indicates that the ratio of biosynthesized mass to ATP generation is constant at roughly 10 g of dry cellular materal per mole of ATP for anaerobic growth. In an aerobic situation, however, different avenues for ATP utilization become important, and biosynthetic yields are more difficult to anticipate.

7.1.4. The Death Phase

We should not lose sight of the fate of individual cells when examining the population. In general, the population is *not* homogeneous, and the batch growth curve is a gross overview of a very complex system. For example, during the exponential-growth phase, some cells are dividing and giving birth to very young cells at the same time others are growing and maturing. Since cells of different ages generally have distinct sizes and chemical compositions, we could view a cell of a given age as a distinct "species." From this perspective, then, culture of one type of microorganism leads to a population containing a great variety of "species."

The diversity among individual cells becomes increasingly apparent during the stationary and death phases. Some cells are dividing during the stationary phase while others die. Often the dead cells lyse (break open), and the carbohydrates, amino acids and other components freed from the lysed cell are then used as nutrients by the remaining living members of the population. Such cannibalistic events help maintain the population size during the stationary phase. Eventually, due to nutrient depletion and toxic-product buildup, however, the population cannot sustain itself, and the death phase begins.

Relatively few studies have been made on the death phase of cell cultures, perhaps because many industrial batch microbiological processes are terminated before the death phase begins. Usually death of the population is assumed to follow an exponential decay

$$n = n_s e^{-k_d t} \tag{7.29}$$

where now t denotes the time elapsed since the onset of the death phase. This relationship implies that the number of cells which die at any time is a constant fraction of those living.

One physical interpretation of exponential population decay states that there are random lethal events which occur in the culture. When one of these happens to a cell, it dies. An obvious objection to this interpretation is that the past history of the population is neglected. Dean and Hinshelwood [3] suggest that not only do living cells prey on dead ones as the population stabilizes and declines but that competing portions of the cell's interacting metabolic machinery also prey on each other as they compete for scarce intermediates. An extension of their argument leads to a rationalization of the exponential-decay law. Other models of population decline will be discussed in Secs. 7.2 and 7.5.

7.1.5. Variations in Cell Size and Cell Composition during Batch Growth

To increase our understanding of the batch growth cycle and to provide background useful for the forthcoming mathematical representations of cell-population growth, it is worthwhile to review some of the available data on changes in the size and chemical makeup of a population undergoing batch growth. Again we must emphasize that this information does not apply for an individual cell but rather to the diverse collection of cells which constitute the population at any time.

Dean, Hinshelwood, and other investigators made numerous observations of composition variations in cultures of *A. aerogenes* bacteria. Taking the inoculum from a stationary population of a glucose-exhausted culture minimizes the lag phase in a glucose-containing medium. The resulting data are displayed in Fig. 7.13. Notice that early in the growth cycle the average cell size and the RNA/DNA ratio pass through high maxima. Although the amount of DNA per unit cell mass and the ratio of protein to RNA remain relatively constant, when

---

Figure 7.13 Changes in cell characteristics during batch growth of *A. aerogenes*. (*Reprinted by permission from A. C. R. Dean and C. Hinshelwood, "Growth, Function and Regulation in Bacterial Cell," pp. 87–89, Oxford University Press, London, 1966.*)

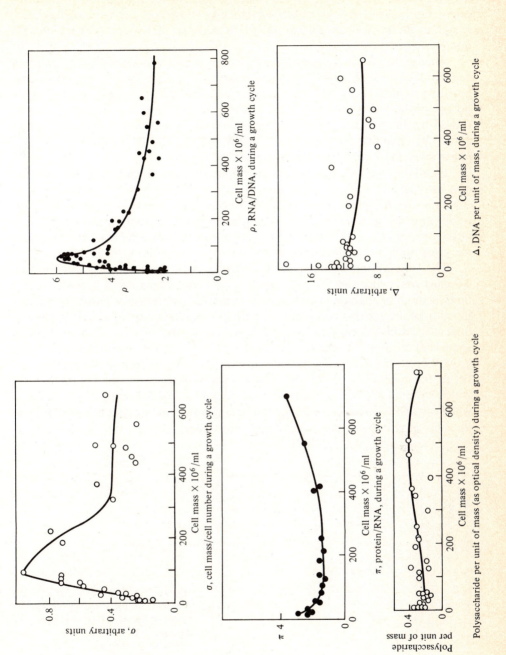

they are taken with the other information, we can see an early picture of the population gearing up its metabolic capacity to exploit the increased nutrient available in its new environment and to generate a larger pool of diffusible intermediates.

Another interesting phenomenon which will be important when we examine end-product rates of formation is evidenced in Fig. 7.14. The average RNA concentration within a cell increases directly with population growth rate. It should be emphasized that this behavior is observed only for growth-rate variations caused by differences in the nutrient medium. If, for example, the growth rate is altered by changing the temperature, RNA concentration during the exponential phase does not seem to be altered significantly.

Finally let us focus closer on the population's characteristics and consider how enzyme activities vary during the growth cycle. While two hydrogenase enzymes in *A. aerogenes* vary little during the exponential growth phase, significant changes in asparagine deaminase specific activity are observed. The initial drop is attributed to dilution of the inoculum while the rise near the end of the growth cycle is believed due to the decrease in medium pH at that stage.

This observation prompts us to emphasize once more that in a batch culture we must never lose sight of the overall *system* comprising a cell population in a fluid medium of changing composition. In a sense we may view batch growth as a function of the starting conditions of both cell and medium: whatever we observe once the batch process has started depends on *both* phases. Thus, it is really somewhat of a misnomer to speak of properties of cell populations in batch growth because these properties are intimately dependent on the interaction between the medium and the population. The technique of continuous culture, which we shall explore in some detail in Sec. 9.1, is consequently better suited in some ways than batch cultivation for learning how growing cells behave, for there the cells' environment does not change with time.

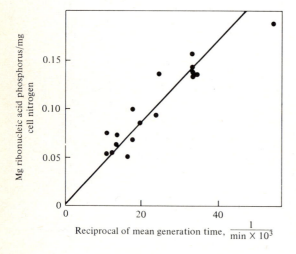

Figure 7.14 These data, which were obtained using a variety of carbon sources, show a direct correspondence between growth rate and the cells' ribonucleic acid content (*A. aerogenes*). (*Adapted from A. C. R. Dean and C. Hinshelwood, "Growth, Function and Regulation in Bacterial Cells," p. 92, Oxford University Press, London, 1966.*)

7.2. MATHEMATICAL MODELING OF BATCH GROWTH

Before further discussion of equations, solutions, and computational results, it is worth reemphasizing the special features of population growth which make mathematical representations difficult. We have already mentioned in Sec. 7.1.2 that some growth models describe the biological portion, or *biophase*, of the system in terms of a single parameter such as cell numbers or biomass concentration. In the terminology introduced by Tsuchiya and coworkers [5], these models are called *unstructured* because they make no attempt to include the physiological state of the population. After first reviewing some unstructured models in Sec. 7.2.1, we shall examine approaches where several variables are used to characterize the biophase.

The growing population consists of a collection of individuals, as we have already emphasized. Thus, the biophase is *segregated* into distinct units, the cells. Seldom is segregation included in growth models, much less the more complex feature of diversity among the separate individual cells. Instead, the biophase is viewed as a continuous medium which is distributed throughout the culture. Thus, we speak of the biophase concentration within a culture vessel or at a "point" inside the vessel just as we ordinarily view, say the concentration of one component in a conventional mixing system even though that component exists as discrete molecules. The validity of the use of the continuum concept in engineering analysis of particulate systems requires that there be a large number of particles at a "point," which is defined as a volume very small relative to the total volume of the system. Such an assumption will become invalid, for example, in the growth of mold pellets, which achieve visible dimensions.

The continuum viewpoint may appear to be most seriously challenged not at the level of cells in the culture but on the finer scale of biological structure. Living cells are extremely small systems, and they do not contain large numbers of molecules of any chemical component. Remembering that we usually deal with numbers of molecules of the order of 10^{23}, the numbers listed in Table 5.9 are quite small.

As an extreme case, since there is only one DNA molecule in a slowly growing bacterium, how can we justify speaking of a "DNA concentration in the cell"? Similar comments could be made about trace ions, the contents of organelles, and many other components within the cell. Although some models described below will employ continuum representations of intracellular events, we must view these as engineering approximations of a typical cell *in* a population of cells as is shown below.

Because of the small numbers of molecules involved, reactions and mass-transfer processes within individual cells should be treated as random events. Direct evidence for the uncertainty typical of cell metabolism has already been given in Fig. 6.19: initially identical individuals exhibit significant variations in generation times. Although stochastic population models can be constructed to attempt to treat such phenomena, these models have not shown significant advantages relative to simpler deterministic ones.

The precision of a deterministic description of a population as small as a typical inoculum (10^5 to 10^6 cells) conveniently illustrates the predictability of cell-population behavior. Suppose that the mean generation time of a population is $\bar{t}$ but that the variation in generation times is wider than that shown in Fig. 6.25. For example, let the distribution of generation times t resulting from sampling *individual cells* be the normal distribution

$$P(t) = \frac{1}{\sqrt{2\pi}\,\sigma} \exp\left[\frac{-(t-\bar{t})^2}{2\sigma^2}\right] \tag{7.30}$$

and take $\bar{t} = 1$ and the standard deviation σ to be 0.5. Then, with 95 percent confidence, we know that the generation time of a single cell is

$$t_{\text{one cell}} = \bar{t} \pm 2\sigma = 1.0 \pm 1.0 \tag{7.31}$$

Now consider a cell suspension containing many cells, say m. Assuming that each of the m cells grows independently of the others, the cell suspension contains m independent samples. Then, the 95 percent confidence limit on generation time for this *population* is $2\bar{\sigma}_{\text{pop}}$, where

$$\bar{\sigma}_{\text{pop}} = \frac{0.98\sigma}{\sqrt{m}} \tag{7.32}$$

Consequently, the uncertainty in the population generation time diminishes rapidly as m becomes large. Taking $\sigma = 0.5$ and $\bar{t} = 1.0$ as before, for example, the 95 percent confidence limits for the population $\bar{t}$ vary with m as given below:

m	95% confidence limits on $\bar{t}$
1	$t = 1 \pm 0.98$
10^4	$t = 1 \pm 0.0098$
10^8	$t = 1 \pm 0.000098$

Such an exercise simply reminds us that, as with other stochastic processes involving relatively large numbers of events (such as fluid-transport phenomena or chemical reactions), the course of a change in population characteristics can be predicted quite precisely even when the standard deviations for individual characteristics are large. Following the progress of an inoculum growing from 10^4 to 10^8 cells per milliliter evidently involves sufficient averaging at all stages to give a well-defined value to the population generation time. In this same sense, we may speak sensibly of the rate of DNA synthesis in a typical cell *in a population* even though in each individual cell only one or two molecules are being assembled at an instantaneous rate which may be quite different from the mean.

7.2.1. The Unstructured Models of Malthus, Verlhurst and Pearl, and Volterra

In the simplest approach to modeling batch culture, we suppose that the rate of increase in cell numbers is a function of the number of cells only. Thus

$$\frac{dn}{dt} = f(n) \tag{7.33}$$

As we shall soon see, such a form does not require us to neglect changes occurring in the medium during growth.

One of the simpler models belonging to the general form given in Eq. (7.33) is *Malthus' law*, which uses

$$f(n) = \mu n \tag{7.34}$$

with μ a constant. We immediately recognize this as the growth rate characteristics of exponential growth, which we have already treated. Malthus' prediction of doom resulting from unbridled population growth has not (yet?) been realized, and transition to a stationary population is generally observed for microbial populations.

Verlhurst in 1844 and Pearl and Reed in 1920 contributed to a theory which included an inhibiting factor to population growth. Assuming that inhibition is proportional to n^2, they used

$$\frac{dn}{dt} = kn(1 - \beta n) \qquad n(0) = n_0 \tag{7.35}$$

a Riccati equation which can be easily integrated to give the *logistic curve*

$$n = \frac{n_0 e^{kt}}{1 - \beta n_0(1 - e^{kt})} \tag{7.36}$$

As illustrated schematically in Fig. 7.15, the logistic curve is sigmoidal and leads to a stationary population of size $n_s = 1/\beta$.

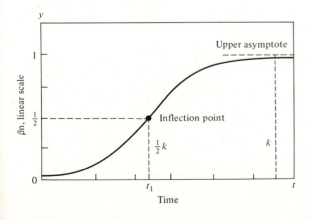

Figure 7.15 The logistic curve $(k > 0, \beta > 0)$.

One possible interpretation of the logistic curve can be formulated by assuming that the production rate of a toxin is proportional to the population growth rate

$$\frac{dc_t}{dt} = \alpha \frac{dn}{dt} \tag{7.37}$$

so that if

$$c_t(0) = 0, \tag{7.38}$$

$$c_t = \alpha(n - n_0) \tag{7.39}$$

In the usual case where n_0 is negligible relative to n, substitution of Eq. (7.39) into (7.23) gives an equation of the same form as (7.35).

A drawback of the logistic equation is its failure to predict a phase of decline after the stationary population has exhausted all available resources. This feature is found in one model developed by Volterra early in this century. In this model, an integral term of the form

$$\int_0^t K(t, r)n(r) \, dr \tag{7.40}$$

is added to Eq. (7.33). Physically we may interpret such a term as a crude recognition that the history $K(t, r)$ of a population influences its growth rate. Through the integral in (7.40), all past values of the population size can influence the growth rate at time t. If K is independent of t, a term such as (7.40) may be viewed as representative of the influence of a component of the culture whose concentration c follows

$$\frac{dc}{dt} = |K(t)n(t)| \qquad c(0) = 0 \tag{7.41}$$

Specifically, let us suppose that K is a constant equal to K_0 and that the history or memory term (7.41) is added to Eq. (7.35). Then the population size is described by

$$\frac{dn}{dt} = kn(1 - \beta n) + K_0 \left| \int_0^t n(r) \, dr \right| \qquad n(0) = n \tag{7.42}$$

As Eq. (7.42) indicates, the sign of K_0 is taken as negative for an inhibitor and positive for a compound which promotes growth. This problem can be solved numerically or on an analog computer, with results as given in Fig. 7.16. So long as K_0 is negative, the population size declines after passing through a maximum.

The unstructured growth models we have just examined have several weaknesses. They show no lag phase and give us no insight into the variables which influence growth. Also, they make no attempt to utilize or recognize knowledge about cellular metabolism and regulation. In the next two sections we shall seek some understanding of how the complex biochemistry of the cell is connected with observed batch-growth phenomena.

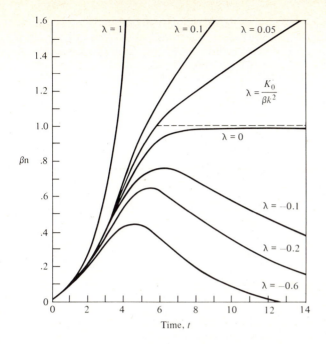

Figure 7.16 is partially labeled within the plot with the following annotations: $\lambda = 1$, $\lambda = 0.1$, $\lambda = 0.05$, $\lambda = \dfrac{K_0}{\beta k^2}$, $\lambda = 0$, $\lambda = -0.1$, $\lambda = -0.2$, $\lambda = -0.6$, vertical axis βn, horizontal axis Time, t.

Figure 7.16 Response of the Volterra model, which includes the population history.

7.2.2. The Kinetic Behavior of Reaction Networks

Conversion of nutrients into metabolic end products often proceeds via long, interconnected sequences, as we have seen in Chap. 5. The study of such networks of reactions can provide considerable insight into some features of cell growth. Moreover, investigation of the kinetics of reaction networks will show that they are to some extent self-regulating.

A common characteristic of many biological networks is the presence of loops with feedback or feedforward of information. As an elementary introductory case, we shall examine the following reaction network discussed by Dean and Hinshelwood [3]:

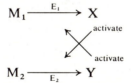

Here $M_1 \rightarrow X$ and $M_2 \rightarrow Y$ denote two enzyme-catalyzed reaction sequences. We shall assume that the concentrations of M_1 and M_2 are constant: either these components are present in great excess, or cellular transport processes maintain them in nearly constant concentrations. In general the level of M_1 and M_2 concentration will depend on the cell's environment.

An information loop arises by assuming that X is an activator for the sequence leading to Y and vice versa. We can write the kinetic equations

$$\frac{dX}{dt} = aY \tag{7.43}$$

$$\frac{dY}{dt} = bX \tag{7.44}$$

where X and Y in the equations denote the amounts (not the concentration) of the species X and Y respectively in the cell. (Why might it be better in principle to employ concentrations? What difficulties would this introduce into the modeling process? This will become clearer in the next section.) Grouped in the rate constants a and b are the M_1 and M_2 concentrations and other factors representing environmental influences. Consequently, if the cell's surroundings change, a and b will usually be altered as well.

The solutions to Eqs. (7.43) and (7.44) are

$$2X = \left(X_0 + \frac{aY_0}{k}\right)e^{kt} + \left(X_0 - \frac{aY_0}{k}\right)e^{-kt} \tag{7.45}$$

$$2Y = \left(Y_0 + \frac{kX_0}{a}\right)e^{kt} + \left(Y_0 - \frac{kX_0}{a}\right)e^{-kt} \tag{7.46}$$

where X_0 and Y_0 are the initial amounts and $k^2 = ab$. After the passage of a time t which is large compared with $1/k$, the amounts of X and Y are described almost exactly by

$$X_e = \frac{1}{2}\left(X_0 + \frac{aY_0}{k}\right)e^{kt} \tag{7.47}$$

$$Y_e = \frac{1}{2}\left(Y_0 + \frac{kX_0}{a}\right)e^{kt} = \frac{kX_e}{a} \tag{7.48}$$

The expressions in the above equations are the exact solution of the following initial-value problem, which consequently can be considered an excellent approximate model for $t \gg 1/k$:

$$\frac{dX_e}{dt} = kX_e \qquad \frac{dY_e}{dt} = kY_e \tag{7.49}$$

$$X_e(0) = X_0' = \frac{1}{2}\left(X_0 + \frac{aY_0}{k}\right)$$

$$Y_e(0) = Y_0' = \frac{1}{2}\left(Y_0 + \frac{aX_0}{k}\right)$$

Thus after a transient time on the order of $1/k$, Y and X maintain a constant ratio $X/Y = a/k$, so that growth is balanced in the sense of Monod (Sec. 7.1.2). Also, as revealed by Eq. (7.47) and (7.48), the amounts of both substances increase exponentially. Depending on the initial conditions (X_0, Y_0), the relative behavior of the transients can be considerably varied. If $X_0 = 0$ and $Y_0 \neq 0$, the initial rate of increase of X is finite while the initial growth rate of Y is zero; the converse is true if $X_0 \neq 0$ and $Y_0 = 0$. If $aY_0 > bX_0$, the initial increase of X with time will be greater than that for Y and conversely. Obviously the conclusions which can be drawn from observation of initial changes following culture transfer will depend very strongly on initial conditions. Eventually, however, a phase of exponential growth is achieved where the growth rate k is *independent* of the initial amounts of X and Y. Still, k may depend on the initial state of the *culture* through the rate parameters a and b.

In a more general situation with a larger serial activation sequence $X_n \to X_{n-1} \to X_{n-2} \to \cdots \to X_1$ the relation

$$\frac{dX_i}{dt} = a_i X_{i+1}$$

holds for all $i \neq n$; the last variable follows the same form

$$\frac{dX_n}{dt} = a_n X_1$$

The transient behavior clearly will depend on the initial values $X_i(0)$, and the exponential-growth phase will again be given by the usual equation

$$\frac{dX_i}{dt} = kX_i$$

where k is the same for all i species in the present example. Whereas $k^2 = ab$ in the earlier example, now the relationship is

$$k^n = a_1 a_2 \cdots a_n \tag{7.50}$$

or

$$k = (a_1 a_2 \cdots a_{n-1} a_n)^{1/n} \tag{7.51}$$

The influence of environmental variations may now be examined. A transfer to a new medium or a new set of conditions might change only a single a_i, say by reducing its value 90 percent, and leave the other a_j $(j \neq i)$ unchanged. The system will undergo a transient response after which, if $n = 10$, exponential growth with a rate constant $k' = (0.10)^{1/10}$ or $k' = 0.8915k$ will occur. Notice that the decrease in growth rate is significantly less (10.85 percent) than the reduction in a_i (90 percent). To understand why this occurs, we employ a generalized form of Eq. (7.48): in exponential growth

$$\frac{X_i}{X_{i+1}} = \frac{a_i}{k} \tag{7.52}$$

Consequently, if a_i is reduced 90 percent, then in the exponential phase, the ratio X_i/X_{i+1} will be decreased about 88 percent. This increase in X_{i+1} relative to X_i tends to compensate for the decrease in the rate constant a_i so that the variation in the overall growth rate is relatively small. Notice, however, that if $a_i = 0$, growth stops.

Branching and rejoining of activation chains are quite common in cell metabolism. As an illustrative example of the kinetics of branched nets, consider the sequence shown schematically below. (The activation again proceeds from larger to smaller subscripts.)

$$X_{j-1} - X_j \begin{array}{c} X_{j+1} - \cdots - X_{j+p} - X_{j+p+1} \\ | \\ Y_1 - \cdots - Y_m \end{array}$$

Then for X_j we have

$$\frac{dX_j}{dt} = a_j X_{j+1} + b_j Y_1 \tag{7.53}$$

at the point where the branching paths converge, and two kinetic equations when they diverge

$$\frac{dY_m}{dt} = b_m X_{j+k} \qquad \frac{dX_{j+p-1}}{dt} = a_{j+p-1} X_{j+k}$$

The $X_j \cdots X_{j+p}$ upper path yields a local growth rate constant

$$k_a = (a_j a_{j+1}, \cdots a_{j+k-1} a_{j+p})^{1/p}$$

The lower $X_j - Y_1 \cdots Y_m - X_{j+p}$ route yields

$$k_b = (b_j b_1 b_2 \cdots b_m)^{1/m}$$

When the k's are viewed as conductances through the parallel network, the local resistance $r_{xy} = 1/k_{xy}$ is given by

$$\frac{1}{r_{xy}} = \frac{1}{r_x} + \frac{1}{r_y}$$

or

$$k_{xy} = k_x + k_y = (a_j a_{j+1} \cdots a_{j+p})^{1/p} + (b_j b_1 \cdots b_m)^{1/m}$$

$$(k_{xy})^p = (a_j \cdots a_{j+p}) \left[1 + \frac{(b_j b_1 b_2 \cdots b_m)^{1/m}}{(a_j a_{j+1} \cdots a_{j+p})^{1/p}} \right]^p$$

and the growth rate constant k for the overall branched network satisfies

$$k^n = \left(\prod_{j=1}^{n} a_j \right) \left[1 + \frac{(b_j b_1 \cdots b_m)^{1/m}}{(a_j \cdots a_{j+p})^{1/p}} \right]^p \tag{7.54}$$

The result again indicates that a change in a particular rate constant, for example in b_3, will cause a new transient behavior until a new constant ratio of components Y_3, Y_4 is reached. Further, the new specific growth rate is again predicted to depart less from the original value for a given change in the rate constants than for an individual activation step.

A useful alternative expression for k can be derived by equating the growth rate for X_j in exponential growth

$$\frac{dX_j}{dt} = kX_j \tag{7.55}$$

with the expression for the same quantity in Eq. (7.53) The result is

$$k = a_j \frac{X_{j+1}}{X_j} + b_j \frac{Y_1}{X_j} \tag{7.56}$$

Initially, suppose that $a_j X_{j+1}/X_j$ is much greater than $b_j Y_1/X_j$. Then

$$k \approx \frac{a_j X_{j+1}}{X_j}$$

If a change of medium is made for which the new value of a_i is essentially zero (where $j \le i \le j + p$) and other kinetic constants remain unchanged, the initial value of k', the new specific growth rate, will not be zero, as seen from Eq. (7.54); evidently

$$k'^{(n-p)} = (a_1 \cdots)^{-p} (b_j b_1 \cdots b_m)^{-m} (a_{j+p+1} \cdots a_n)^{-p} \tag{7.57}$$

or $k' \ne 0$. Since the path from a_j to a_{j+1} has vanished, the new concentrations of Y_1 and X_j along the alternate path will change until

$$k' = b_j \frac{Y_1}{X_j}$$

As before, k' may not differ greatly from k; much of the change is reflected in change of the concentration of Y_1.

While the reaction model just outlined is very simplistic, it may be considered a crude representation of certain real processes. For example, many amino acids (but not all) can be synthesized by some path $\underline{Y}$ in microorganisms. If in the generation of a ribonucleotide, a particular amino acid A is present in the nutrient in sufficient amounts, the synthetic route will not be necessary, and a faster route $X_{j+p-1} \cdots X_{j+1}$ is available which directly incorporates the particular amino acid fragment into the growing ribonucleotide. If transfer is made to a new medium where the amino acid A is completely missing, then some constant $a_j = 0$ and the synthetic machinery in $\underline{Y}$ must be activated to achieve a new exponential growth.

Reintroduction of the culture into the original medium will restore a_j to its initial nonzero value, and the $\underline{X}$ path will be continuous again. Since the flow through the $\underline{X}$ path was originally much greater than the flow through the $\underline{Y}$ path, evidently X_{j+p-1} will be easily increased compared with Y_m by the action of the product of X_{j+p}. The concentration of the X_{j+p} product may then fall toward a new steady value (since less of it may be needed to maintain a given path flux), and since

$$\frac{dY_m}{dt} = b_m X_{j+p}$$

the value of Y_m and all subsequent Y_i $(1 < i < m)$ must also decrease until the original value of k is restored, where Eq. (7.56) is valid and $a_j X_{j+1}/X_j \gg b_j Y_1/X_j$ again.

Thus, transfer of a population to a new medium can *induce* an alternate originally minor metabolic path to become the major material flow route; retransfer to the original solution greatly diminishes the importance of the alternate $\underline{Y}$ route; i.e., the synthesis of amino acid A is *inhibited* in the presence of sufficient amino acid A in the medium. This simple example may serve as a crude model for the *adaptive* behavior of cells; i.e., their ability, within certain bounds, to open or close alternate metabolic routes as medium conditions vary.

The model just described also predicts that if the concentration of A is initially high and falls essentially to zero during the course of batch growth, the relative importance of the metabolic path $\underline{Y}$ would continually increase in time. Under the conditions where the relative rate of disappearance of A is the same order of magnitude as the rate at which the cell is continually trying to upgrade the alternate route $\underline{Y}$, a true exponential growth may not be reached until some time after A has been completely consumed. If A diminishes slowly in time relative to the cell's ability to turn on path $\underline{Y}$, then no lag may be observed and an apparently exponential growth rate may be observed at any instant; in fact, the value of k would be slowly changing.

Several real physical phenomena are omitted from the predictive ability of the previous model. First, the dependence of many interrelationships in such networks are probably more complex than the model reaction system above. For example, a particular activation step may depend on the concentrations of a given substrate, the appropriate enzyme, the presence of a particular vitamin to activate the enzyme, and perhaps a second substrate to which a part of the first will be transferred, e.g., the incomplete ribonucleotide to which part of a given amino acid is enzymatically joined.

Second, the adaptability of a cell, and therefore the appropriate values of the kinetic constants a_i and b_j, may depend on the particular age of the cell, and under some conditions the cell may not survive the change. For example, when the value of X_{j+p-1} decreases greatly in response to the change, the new concentration of Y_m required may be too high or the cell may simply utilize too much of its energy reserve in scaling up the $\underline{Y}$ metabolic machinery.

Also, the analysis just completed does not explicitly recognize the increase in size of the cell and how this process might interact with mass transfer of nutrients and metabolic wastes. Perret's model, discussed next, addresses these points in a qualitative manner.

7.2.3. The Cell as an Expanding Chemical Reactor†

The progression of mathematical treatments accorded any physical phenomenon normally involves a continual increase in the number of relations between the physically important variables which evidently should be considered. The progression continues until the situation is well described or, in less fortunate circumstances, the complexity of the relations becomes unmanageable. Consideration of the life cycle of the cell suggests that a worthwhile extension of the approach of the previous section would fix the feed (or medium nutrient) conditions as before and would include the change in cell volume with time, since on the average, each cell must effectively double its volume in a given mean generation time. Thus, one might choose the following approach: allow cell-volume variation in a rational fashion; include finite reaction rates plus diffusion of feed into changing cell volume (perhaps add some sort of limit or control upon maximum rate of volume change). Also, include diffusion of end products out of the cell.

---

† The material in this section is drawn from C. J. Perret, *J. Gen. Microbiol.*, **22**: 589, 1960; C. J. Perret and H. C. Levy, *J. Theoret. Biol.*, **1**: 542, 1961; and E. M. Williams, *J. Theoret. Biol.*, **15**: 190, 1967. In the last paper the two-component structured model of Sec. 7.2.4 is given and applied to both batch and continuous processes.

Such a description was introduced by Perret in 1960. The model is considered here in two parts: a simple expanding-volume system with linear kinetics, then a means of relating internal rate processes to the cell volume rate of change. The linear system is illustrated in Fig. 7.17. Here a nutrient A diffuses into the cell through the expanding cell membrane. A series of enzyme-catalyzed reactions occur, and a product E leaves through the cell wall.

The scheme in Fig. 7.17 contains two features of interest: (1) through the existence of a number of intermediates, a lag occurs; a step change in A is not seen in E until the effect of the step passes through the intermediates B, C, and D and (2) the product leaving the cell does so by passive diffusion; thus, there must be a gradient of E established across the wall. This gradient means that the final metabolite concentration of E must always be greater in the cell than the medium. If now the walls in Fig. 7.17 are allowed to move slowly outward, solvent will penetrate the membrane and dilute all internal concentrations. The dilution will lower A and increase the flux of A into the reactor by virtue of both an increased surface area and a larger gradient across the membrane. Further, the rate of permeation of E required to maintain a given internal value is also decreased by such a dilution effect.

The fuller characteristics of individual-cell activity were sought by Perret in the more complex reaction scheme in Fig. 7.18. The important features of such a metabolic model include:

1. A major metabolic pathway A-A-B-C-D-E-F-F consuming the primary major nutrient A.
2. An alternate major nutrient J (J̄-J̄-K̄-C̄- ⋯)¯
3. A minor nutrient M and its independent pathway very loosely coupled through O-D to the main pathway.
4. A branching point from -E- with two final products, Y̱ and Ẕ, both of which remain in the cell.

Ignoring both the final -Ẕ-Z step and the production of Y would simply result in a steady accumulation of X and Z in the dead-end reaction sequence E̱-X̱-Ẕ. The first strong pair of assumptions deals with X̱, Ẕ, and Z.

1. The level of Ẕ remains well below that value which corresponds to equilibration with X.
2. The volume of the system (cell) is proportional to the *total amount* of Z.

These assumptions provide that the rate of production of Z will depend on X, which in turn depends on the levels of the *intermediates* in the major metabolic pathway, and that the evolution of cell volume vs. time is dictated by the total quantity of Z within the cell.

If A is fixed, analysis of such a model (excluding Y) yields the following features:

1. The total major nutrient flux per unit volume through the A-F sequence is constant.
2. The level of Ḏ is then constant.
3. The rate of Z production *per unit volume* is constant.
4. A given increase in Z results in a corresponding increase in the volume of the system.
5. The rate of volume increase is proportional to the rate of formation of Z.
6. The rate of increase of the volume (or mass) of the system is at any instant proportional to the volume (or mass). This last statement indicates that the system is growing exponentially (constant growth rate for a given A). If A decreases slowly in time, the specific growth rate must also diminish.

Several features characteristic of living systems result from a few additional assumptions built into the model of Fig. 7.18. First, it is assumed that a later product Y̱ acts as a catalyst for the reaction C̱-D. The system will thus behave autocatalytically; increased Y̱ will increase the C-D reaction, which in turn will increase Y̱. Eventually a point is reached where the rate cannot be further increased by this mechanism: Y̱ has driven the system to its *maximal* exponential rate of growth. Additionally, the

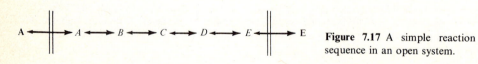

Figure 7.17 A simple reaction sequence in an open system.

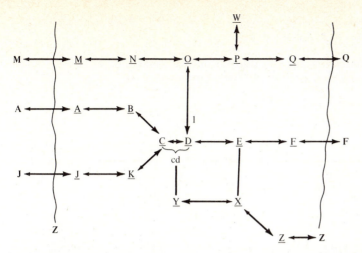

Figure 7.18 Hypothetical reaction network of Perret's model for the bacterial cell.

reaction $A \rightleftharpoons B$ or $B \rightleftharpoons C$ can eventually saturate, being enzyme-catalyzed reactions with Michaelis-Menten kinetics. Thus, at low A concentrations, the growth may be first order in A and will be autocatalytically driven to its maximal rate. A high concentration of A will saturate $A \rightleftharpoons B$ or $B \rightleftharpoons C$ reaction so that a zero-order substrate growth rate results, as is often observed (Fig. 7.9). Thus, for sufficiently large initial A concentrations, a substantial time period may exist within which the specific growth rate is a constant in spite of a continually changing nutrient level A.

Two other features of the model are qualitatively evident. First, if an exponential-growth-phase population is transferred to a new nutrient medium with different levels of A or J, a lag phase may result while the central metabolic pathways readjust their concentrations to a new steady level of expansion. Second, if the new medium contains only M (which can just supply the necessary metabolic energy for the cell to survive), then the main metabolic pathway will empty and the cell will enter a stationary phase of existence in which its volume (or mass) no longer changes with time (very minor $O \rightarrow D$ coupling implies negligible rate of increase of Z).

7.2.4. The Simplest Structured Model: A Two-Component Cell

While Perret's concepts aid us in gaining qualitative insight into cell growth, the model employs many variables which cannot be directly observed and has a large number of parameters. If possible, we would like to construct a much simpler approximate representation of growth kinetics which retains the essential features of real systems. Since in principle the model should be structured in the sense defined earlier, the state of the cell must be characterized by at least two variables. Williams† has constructed a two-component model which does surprisingly well in reproducing many aspects of batch growth dynamics. This section is devoted to a brief outline of this model.

The two components considered by Williams are a synthetic component *R* and a structural and genetic portion *D*. Tentatively, we may regard *R* as RNA and *D* as protein plus DNA. The following interrelationships between these quantities and total biomass *M* are postulated:

1. The *R* portion is produced by uptake of external nutrient *A*. The rate of *A* disappearance is first order in *M* and *A*.
2. The *D* component is fed from the *R* portion.

† F. M. Williams, A Model of Cell Growth Dynamics, *J. Theoret. Biol.*, **15**: 190, 1967.

3. Doubling of the D species is necessary and sufficient for cell division. D synthesis is autocatalytic.
4. Total biomass M is the sum of D plus R.

Translated into mathematical terms assuming kinetics of mass-action form, the equations describing the total *amounts* of the four components (A, R, D, M) in batch culture are

$$\frac{dA}{dt} = -k_1 AM \tag{7.58}$$

$$\frac{dR}{dt} = k_1 AM - k_2 RD \tag{7.59}$$

$$\frac{dD}{dt} = k_2 RD \tag{7.60}$$

$$\frac{dM}{dt} = k_1 AM \tag{7.61}$$

Combining Eq. (7.58) and (7.61) makes it easy to show that total biomass M satisfies a differential equation of the same form as (7.35), and so M will follow the logistic curve given in Eq. (7.36). Consequently an immediate drawback in this case, as in many previous models, is the absence of a death phase. For the many industrial processes requiring microbial growth that are terminated before onset of the death phase, this limitation is not important.

We shall first apply this two-component structured model to the problem of growth initiated by inoculation with cells from a stationary nutrient-exhausted culture. In terms of the model, the stationary culture is characterized by $A = R = 0$ and $M = D$. Thus, the cells are as small as possible and devoid of the synthetic component R. Calculations from Eqs. (7.58) to (7.61) for the resulting batch culture give the curves of Fig. 7.19. Among the features of actual batch growth which are also present in the model are

1. Existence of a lag phase in which cell size increases.
2. An exponential-growth phase where cell size is a maximum.
3. A change in composition of the cells during the growth cycle. Since these changes are evident even during the exponential phase, apparently growth is never exactly balanced.
4. A stationary phase with relatively small cells.

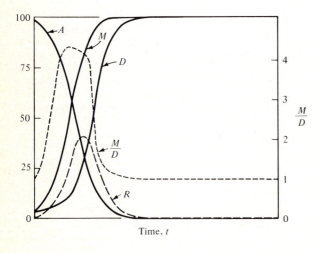

Figure 7.19 Simulation of batch growth using Williams' two-component model. In this case, the inoculum consists of stationary-phase cells. M = total biomass; R = synthetic component; A = external nutrient; D = cell number ($k_1 = 0.0125$, $k_2 = 0.025$, $M_0/D_0 = 1$).

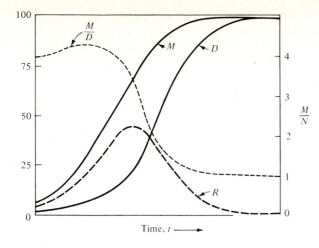

Figure 7.20 Batch growth with a rapidly growing inoculum from Williams' model $(k_1 = 0.0125, k_2 = 0.025, M_0/D_0 = 4)$.

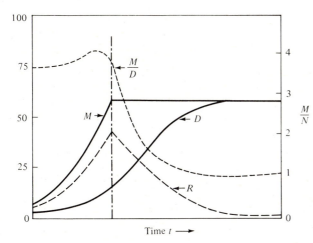

Figure 7.21 At the indicated time, the cells are withdrawn from nutrient solution but continue to divide. Williams' model results, with $k_1 = 0.0125$, $k_2 = 0.025$.

If the inoculum is taken instead from an exponentially growing culture, simulation with the two-component model yields results given in Fig. 7.20. Here the lag phase is absent, but once again composition changes suggest unbalanced growth throughout the cycle. Cells removed from nutrient are known to continue to divide once or twice before becoming stationary; William's model also gives this behavior, as Fig. 7.21 reveals.

7.2.5. Growth of Filamentous Organisms

In the previous portions of this chapter we have been concerned with analysis of microbial populations where increase in biomass is accompanied by an increase in the number of cells. A different situation prevails in molds and other filamentous organisms, where the size and morphology of a mold pellet or pulp varies as growth proceeds. Several experimental studies of batch submerged culture have indicated that biomass increases at a slower than exponential rate with proportionality to

$(time)^3$ providing a reasonable representation of the data. Such a growth pattern can be rationalized from observations of one- and two-dimensional mold cultures. In the first instance, the rate of increase in the colony length is constant, while the radius of the mold colony increases at a constant rate in surface culture (two-dimensional).

Extrapolating to a spherical pellet growing in submerged culture, let us assume that

$$\frac{dR}{dt} = k_g = \text{const} \tag{7.62}$$

where R now denotes the pellet radius. Since the biomass M is

$$M = \rho \tfrac{4}{3}\pi R^3 \tag{7.63}$$

we have from Eqs. (7.62) and (7.63)

$$\frac{dM}{dt} = \rho 4\pi R^2 \frac{dR}{dt} = k_g 4\pi R^2 \tag{7.64}$$

Eliminating R from (7.64) using (7.63) leaves

$$\frac{dM}{dt} = \alpha M^{2/3} \tag{7.65}$$

where

$$\alpha = k_g \left(\frac{6\sqrt{\pi}}{\rho}\right)^{2/3} \tag{7.66}$$

Integrating Eq. (7.65) with an initial biomass of M_0 yields

$$M = (M_0{}^{1/3} + \alpha t)^3 \tag{7.67}$$

Since M_0 is usually quite small relative to M, Eq. (7.67) gives the cubic dependence of M on t mentioned above. The preceding analysis can be placed on a firmer basis by examining transport processes within the pellet. We shall consider this problem further in Sec. 7.4.

The various mathematical models and analyses examined in this section are quite satisfactory in providing a qualitative link between some aspects of basic biochemistry and cell biology and observed patterns of population growth. Although a sound quantitative model is not available at present, improvements in this direction should follow when kinetic data become available for more of the basic steps of cell metabolism. Development of more sophisticated methods for analyzing, simulating, and simplifying very large sets of equations will also contribute to development of more suitable population models. A few more comments on growth modeling will be added in Chap. 9, where we examine continuous culture.

7.3. PRODUCT-SYNTHESIS KINETICS

In this section we shall explore the relationships between product formation rate, cell mass, cell age, and substrate levels. A perusal of some experimental examples will reveal several common patterns of product synthesis kinetics. Unfortunately, however, we do not yet know *a priori* the conditions under which a nutrient–microbial-species combination will yield the maximum amount of desired product. Achievement of this goal requires a reasonably accurate knowledge of cell-metabolism kinetics, an area where current knowledge is woefully incomplete. Because of this limitation, we shall often be forced to resort to empirical or ad hoc design methods for fermentation processes.

In this section and in the remainder of the text, we shall use the term fermentation to mean chemical processes involving microorganisms. As noted in Sec. 5.2, this connotation is standard in industrial microbial utilization although in biochemistry fermentation has a more restricted definition.

Several classification schemes for batch fermentations have been proposed; we shall review below those due to Gaden and Deindoerfer. After examining several product-synthesis patterns in the course of these discussions, we shall be prepared to appreciate an interesting modeling approach which recognizes the diversity and heterogeneity of the microbial population. Other techniques for representing product-formation kinetics will be found in later examples.

7.3.1. Gaden's Fermentation Classifications

In 1955 Gaden [10] proposed the classification scheme of fermentations which is summarized in Table 7.3. Type I fermentations yield products as a direct result of energy metabolism. For the alcohol-fermentation example of Fig. 7.22*a* a reasonable kinetic approximation to the entire course of the fermentation following a brief lag is given by

$$\frac{d(\text{product})}{dt} = -k_1 \frac{d(\text{substrate})}{dt} = +k_2 \frac{d(\text{cell mass})}{dt} \qquad (7.68)$$

Such a product is termed growth-related or growth-associated. Replotting the same data in Fig. 7.22*b* and 7.22*c* indicates that the product appearance rate has a fixed relationship to the rate of nutrient consumption but in fact lags behind the rate of cell growth. The existence of a constant stoichiometric relationship between substrate consumption and product formation, while typical of type I fermentation, is relatively unusual in other classes of fermentations.

The type II scheme is shown in Fig. 7.23, using the citric acid fermentation as an example. Here several maxima for both growth and sugar utilization are evident. The appearance of product exhibits some correlation with the growth rate when both are normalized to the instantaneous cell-mass concentration. Figure 7.23*b* indicates that the first rise in sugar utilization occurs with a renewed growth rate (second maximum of growth curve) but that when the growth rate subsequently diminishes, the sugar consumption remains high for a while and during this latter time the sugar is converted largely into citric acid.

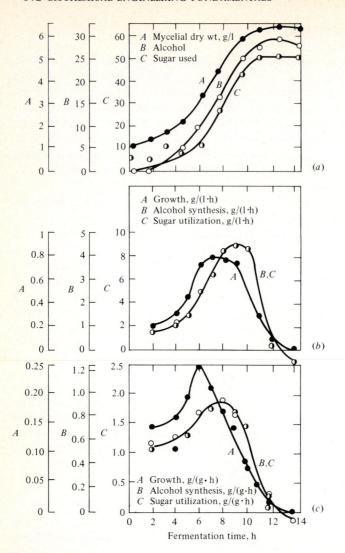

Figure 7.22 The alcohol fermentation, a Gaden type I fermentation: (*a*) time course, (*b*) volumetric rates, and (*c*) specific rates. [*Reprinted from R. Leudeking, Fermentation Process Kinetics, in N. Blakebrough (ed.), "Biochemical and Biological Engineering," vol. 1, p. 203, Academic Press, Inc. (London) Ltd., London, 1967.*]

Table 7.3. Fermentation classifications†

Type I products are simple catabolic products; type III products are complex and result from cell biosynthetic activity; type II products are intermediary

Type	Description
I	Main product appears as a result of primary energy metabolism; desired product may result from direct oxidation of a carbohydrate substrate, e.g., glucose to ethanol, or gluconic acid, or lactic acid; metabolic routes primarily serial; each rate process (sugar utilization, product synthesis, cell mass) exhibits one maximum close to that of others
II	Main product again arises from energy metabolism, but now indirectly; reaction rates complex in behavior; several maxima may appear; as with type I reactions, the overall free-energy change (exclusive of other cell activities) is negative
III	Biosynthesis of complex molecules (class III reactions, Chap. 5) not resulting directly from energy metabolism; cell and metabolic activities reach maximum early; at later stages the formation of desired product becomes most important; oxidative metabolism is low at time of maximum product formation; examples include antibiotic fermentations and biosynthesis of vitamins

† E. L. Gaden, *Chem. Ind. Rev.*, **1955**: 154; *J. Biochem. Microbiol. Tech. Eng.*, **1**: 413, 1959.

Gaden's type III fermentation includes most schemes yielding large biomolecules produced in class III metabolism (Fig. 5.1 and section 5.3.3). The penicillin production example of Fig. 7.24 indicates maximum product yield when the nutrient consumption rate [on a volumetric (grams per liter solution) or cell-mass (grams per grams of cells) basis] is a *minimum* and cell growth has nearly stopped. Figure 7.24c indicates a period of negative specific growth rate where the term

$$\frac{1}{\text{cell mass}} \frac{d(\text{cell mass})}{dt} < 0$$

so that the population has moved beyond the stationary phase for the latter part of the fermentation. Under such conditions, dead-cell lysis may provide a second nutrient source for other cells.

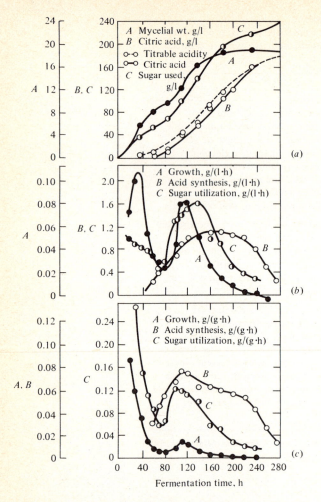

Figure 7.23 The citric acid fermentation, a Gaden type II fermentation: (*a*) time course, (*b*) volumetric rate, and (*c*) specific rate. (*Reprinted from R. Leudeking, Fermentation Process Kinetics, in N. Blakebrough (ed.), "Biochemical and Biological Engineering," vol. 1, p. 204, Academic Press, Inc. (London) Ltd., London, 1967*).

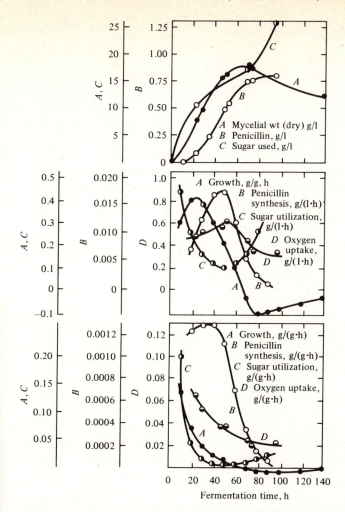

Figure 7.24 Gaden's type III fermentation pattern exemplified by the penicillin fermentation: (*a*) time course, (*b*) volumetric rates, and (*c*) specific rates. [*Reprinted from R. Leudeking, Fermentation Process Kinetics, in N. Blakebrough (ed.), "Biochemical and Biological Engineering," vol. 1, p. 205, Academic Press, Inc. (London) Ltd., London, 1967.*]

7.3.2. Organizing Fermentation Patterns according to Deindoerfer

In Table 7.4, the relation of product formation to nutrient consumption is classified according to a scheme proposed by Deindoerfer [11], which is useful in batch-reactor design. Figures 7.25 to 7.31 illustrate the Deindoerfer patterns.

Figures 7.25 and 7.26 represent examples of simple reactions in batch systems. In the *Aerobacter cloacae* experiment, the data reveal that the ratio of cell carbon formed per glucose carbon consumed does not vary significantly with time. The approximate constancy of this value verifies the assumption used later that the *yield factor* Y ($=$ mass cell produced per mass carbon consumed) is constant for this case. The second example is pertinent to product-formation kinetics. The conversion of glucose to gluconic acid

$$
\begin{array}{ccc}
\text{CHO} & & \text{COOH} \\
| & & | \\
\text{HCOH} & & \text{HCOH} \\
| & & | \\
\text{HOCH} & & \text{HOCH} \\
| & \longrightarrow & | \\
\text{HCOH} & & \text{HCOH} \\
| & & | \\
\text{HCOH} & & \text{HCOH} \\
| & & | \\
\text{H}_2\text{COH} & & \text{H}_2\text{COH} \\
\text{Glucose} & & \text{Gluconic (hexonic) acid}
\end{array}
$$

occurs at a constant rate under the range of conditions reflected in Fig. 7.26. The apparent simplicity of the reaction is deceptive: in *Penicillum notatum*, for example, it is known that the enzyme glucose oxidase catalyses the reaction by transfer

Table 7.4. Deindoerfer's classification of fermentation patterns†

Type	Description
Simple	Nutrients converted to products in a fixed stoichiometry without accumulation of intermediates
Simultaneous	Nutrients converted to products in variable stoichiometric proportion without accumulation of intermediates
Consecutive	Nutrients converted to product with accumulation of an intermediate
Stepwise	Nutrients completely converted to intermediate before conversion to product, or selectively converted to product in preferential order

† F. H. Deindoerfer, *Adv. Appl. Microbiol.*, **2**: 321, 1960.

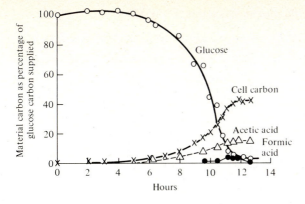

Figure 7.25 Batch growth of *A. cloacae* is a simple reaction. (*Reprinted from S. J. Pirt, The Oxygen Requirement of Growing Cultures of an* Aerobacter *Species Determined by Means of the Continuous Culture Technique, J. Gen. Microbiol.,* **16:** 59, 1957.)

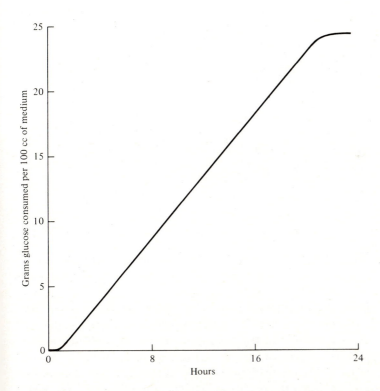

Figure 7.26 Another example of a simple reaction: gluconic acid production from glucose by *Aspergillus niger.* (*Reprinted from A. J. Moyer et al., Fermentation of Concentrated Solutions of Glucose to Gluconic Acid, Ind. Eng. Chem.,* **32:** 1379, 1940).

of a hydrogen molecule to the flavoprotein carrier, En–FAD:

$$\text{Water} + \text{glucose} + \text{En–FAD} \longrightarrow \text{gluconic acid} + \text{En–FADH}_2$$

$$\text{En–FADH}_2 + \text{molecular oxygen} \longrightarrow \text{En–FAD} + \text{hydrogen peroxide}$$

If the second reaction regenerating the H_2 acceptor En–FAD is very slow, the first reaction will occur at a rate equal to that of the second, which may be first order in O_2 and zero order in glucose in a situation of poor but steady oxygen supply. When the regeneration reaction is rapid, the first reaction will depend upon glucose concentration if the glucose level is less than or of the order of the Michaelis constant $K_{glucose}$ for this reaction since

$$\text{Rate}_1 = \frac{k[\text{water}][\text{En–FAD}][\text{glucose}]}{K_{glucose} + [\text{glucose}]}$$

Further, it is conceivable that a reaction network of the above type might involve several intermediates B_i, such that

$$X + A \longrightarrow B_1 + XH_2$$

$$B_1 \longrightarrow B_2$$

$$B_2 \longrightarrow B_3$$

$$B_3 \longrightarrow B_4 \qquad \text{final product}$$

$$XH_2 + O_2 \longrightarrow X + H_2O_2 \qquad \text{regeneration}$$

If the regenerative reaction is slow compared with each of the earlier irreversible reactions, B_4 will appear "immediately" as A disappears and the network will satisfy the kinetic definition of "simple" in Table 7.4 without in fact being a simple reaction lacking intermediates. We have already observed that even a single enzyme-catalyzed reaction may have several intermediates (Sec. 3.3.3).

The simultaneous reaction example in Fig. 7.27 is not merely a simple reaction network of the type

$$B \rightleftharpoons A \rightleftharpoons C$$

where the forward and reverse rate constants for the $A \rightleftharpoons B$ reaction are larger than those for the $A \rightleftharpoons C$ reaction. The cellular population has synthesized various proteins (enzymes) until a satisfactory enzyme level allowing maximum growth is achieved. The remaining sugar is then stored as fat under conditions where the protein-synthesizing system has become largely repressed or inhibited as the cell protein concentration levels off. Earlier discussion of cell metabolism in the absence of an external carbon source suggests that time reversal in Fig. 7.27 would result in qualitatively retracing the cell-fat and cell-protein curves back toward the origin. (How would the sugar curve change?) Sequential substrate utilization was also evident in the citric acid fermentation already mentioned (Fig. 7.23). Finally, we should note from Fig. 7.27 that the length of the phase of greatest growth depends upon the variable measured (total cell mass or cell protein).

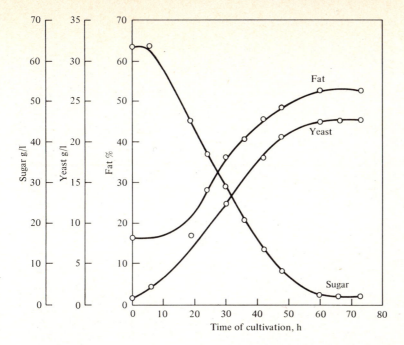

Figure 7.27 Synthesis of cell mass and fat during batch growth of *Rhodotorula glutinis* on glucose, a simultaneous reaction in Deindoerfer's scheme. (*Reprinted from L. Enebo, Microbiological Fat Synthesis by Means of Rhodotorula Yeast, Arch. Biochem.,* **11:** 383, 1946.)

The appropriate interval to measure depends then on the purpose for which the cells are cultured.

The example of sequential, or consecutive, reaction in Fig. 7.28 is reminiscent of the behavior of simple consecutive first-order reactions

$$A \longrightarrow B \longrightarrow C$$

At this point, however, we would do well to remember that only *extracellular* concentrations are reported here, and the concentrations within the cell, where the reactions occur, generally differ from these values. For example, the permeability differences of the cell wall to a ketone vs. a carboxylic acid will also affect the extracellular concentrations measured in the course of the experiment. Indeed, this latter point may relate to a critical aspect of cell metabolism.

In many internal cellular transformations, a *phosphate* form of the particular substrate is transformed to a phosphate form of the product. Since the ionized phosphate moieties have difficulty permeating cell walls compared with the neutral dephosphorylated counterpart, e.g., D-glucose 6-phosphate vs. D-glucose, phosphate intermediates of a reaction sequence will be difficult to detect extracellularly since they are largely contained within the cells. The hydrolyzed product can permeate the cell membrane more easily, but if hydrolysis of an intermediate phosphate is slow relative to its subsequent conversion into a different phosphate

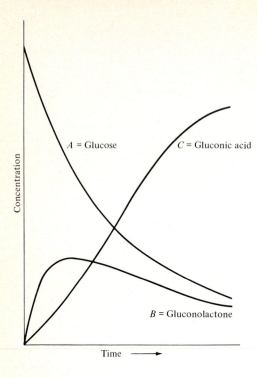

Figure 7.28 A consecutive reaction showing conversion of glucose to lactone to gluconic acid in *Pseudomona ovalis* batch culture. (*Reprinted from S. Aiba, A. E. Humphrey and N. F. Millis, "Biochemical Engineering," 2d ed., p. 113, University of Tokyo Press, Tokyo, 1973.*)

product, the intermediate will be difficult to measure extracellularly. Having a low concentration, this intermediate may not contribute an appreciable lag to the appearance of C in the sequence

$$A \longrightarrow A \text{ phosphate} \longrightarrow B \text{ phosphate} \longrightarrow$$

$$C \text{ phosphate} \longrightarrow C$$

$$D \longleftarrow D \text{ phosphate}$$

The reaction will again appear to be simply consecutive:

$$A \longrightarrow C \longrightarrow D$$

Stepwise reaction examples are illustrated with Figs. 7.29 and 7.30. The first example is the glucose effect mentioned previously; i.e., the presence of glucose represses the inducible production of enzymes which can metabolize other substrates. This stepwise system is then simply two separate reactor systems linked together sequentially in time. Exhaustion of the glucose results in induced synthesis of the appropriate enzymes to metabolize the second substrate (sorbitol in the present example) (recall the network-activation model of Sec. 7.2.2). It may be expected that an appropriate kinetic description would consider glucose only until its concentration reaches (approximately) zero, at which point the cell would

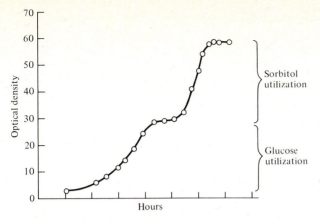

Figure 7.29 Diauxic growth of *E. coli* in glucose-sorbital medium illustrating a stepwise reaction pattern. (*Reprinted by permission from J. Monod, The Growth of Bacterial Cultures, Ann. Review of Microbiol.* **3**: 371, 1949.)

behave as though it were transferred from a glucose medium to a sorbitol medium. A new lag phase would exist while the cell activated the production of sorbitol-metabolizing enzymes and a second exponential phase would eventually be reached (provided the sorbitol was not exhausted in the new lag phase).

The rate data for the stepwise reaction in Fig. 7.30 indicate that the conversion of glucose to gluconic acid is immediately replaced by the conversion of

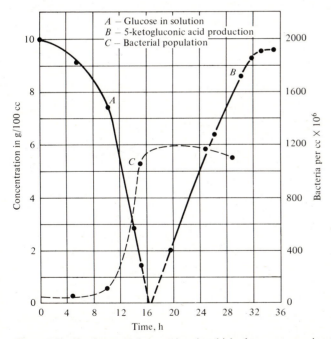

Figure 7.30 Another stepwise reaction, in which glucose conversion to gluconic acid continues until glucose exhaustion. This phase is followed by gluconic acid conversion to 5-ketogluconic acid (batch culture of *Acetobacter suboxydans*). [*Reprinted from J. J. Stubbs et al., Ketogluconic Acids from Glucose (Bacterial Production), Ind. Eng. Chem.,* **32**: 1626, 1940].

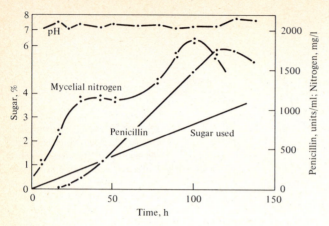

Figure 7.31 The growth rate oscillates in this batch culture with continuous addition of glucose. (*Reprinted by permission from P. Hosler and M. J. Johnson, Penicillin from Chemically Defined Media, Ind. Eng. Chem.,* **45**: 871, 1953. *Copyright from the American Chemical Society.*)

something (gluconic acid?) to 5-ketoglutaric acid. Conversion II begins immediately when conversion I is complete (obviously not all converted glucose becomes glutonic acid; cf. metabolic pathways in Chap. 5).

While the chemical transformations above are similar (oxidative dehydrogenation in each case), the resulting 5-ketogluconic acid is apparently nonmetabolizable and is rapidly excreted from the cell with simultaneous arrest of bacterial growth. If a new time lag exists, it is evidently very brief.

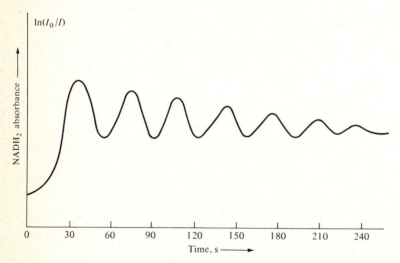

Figure 7.32 Oscillatory kinetic behavior of $NADH_2$ in intact yeast cells under anaerobic conditions (*Saccharomyces carlsbergensis*). (*After B. Chance et al., Cyclic and Oscillatory Responses of Metabolic Pathways Involving Chemical Feedback and Their Computer Representations, Ann. N.Y. Acad. Sci.,* **115**: 1010, 1964.)

The final example in Fig. 7.31 illustrates two features. The presence of intermediates in sizable concentration gives rise to phase lags separating the disappearance of substrate (to form mycelial nitrogen cell mass) and the appearance of product, in this case penicillin. Also, the cell growth and product formation both exhibit oscillatory behavior for this system with constant nutrient addition. Such oscillation in whole single cells was first observed by Britton Chance and his coworkers studying the uptake of glucose by intact yeast cells under anaerobic conditions. Figure 7.32 reproduces the intracellular concentration of $NADH_2$ (referred to as DPNH in earlier literature), which acts as a hydrogen donor in glucose metabolism carrying hydrogen back and forth between reactions catalyzed by glyceraldehyde 3-phosphate dehydrogenase and alcohol dehydrogenase, respectively (see Fig. 5.5). While such oscillatory behavior is not well understood, the existence of oscillatory modes in various nonbiological control systems and electric circuits strongly suggests that oscillation in biological systems is connected with biological control and thus is an inherent characteristic of cell metabolism.

7.3.3. Shu's Segregated Model for Product Accumulation

In 1961, Ping Shu [14] introduced a versatile model for product-formation kinetics which recognizes the segregated, heterogeneous nature of a microbial population. While such an approach proves cumbersome in growth modeling, as we have already noted, it is well suited for representing product accumulation. We include Shu's model because of its potential utility and because it is a somewhat more elegant microbial model than we have considered previously.

To appreciate fully the flexibility of Shu's analysis, we should first mention the now classic study of Luedeking and Piret† on the lactic acid fermentation by *Lactobacillus delbrueckii*. As noted in Eq. (7.68), some fermentations exhibit *growth-associated* product kinetics, where dp/dt is proportional to dn/dt. Figure 7.14 reveals a possible explanation for this phenomenon at the level of basic biochemistry: cells growing more rapidly have a larger number of ribosomes. Presumably this affords more rapid protein synthesis and hence an increase in the overall rates of cellular metabolism, including release of metabolic end products. In other instances, termed *non-growth-associated*, the production rate has been related to the concentration of cells rather than the growth rate.

Luedeking and Piret found that lactic acid accumulation was apparently a combination of the two fermentation types, so that

$$\frac{dp}{dt} = \alpha \frac{dn}{dt} + \beta n \tag{7.69}$$

where α and β were found to be pH-dependent. In the exponential growth phase, where Eq. (7.7) holds, we can write

$$\frac{dp}{dt} = (\alpha k + \beta)n = \left(\alpha + \frac{\beta}{k}\right)\frac{dn}{dt} \tag{7.70}$$

so that during the exponential phase it is not possible to distinguish between growth-associated, non-growth-associated, and hybrid types, as in Eq. (7.69). However, such a model has the advantage of sometimes allowing prediction of product formation rates in *more* that one phase of growth. While

† R. Luedeking and E. L. Piret, A Kinetic Study of the Lactic Acid Fermentation, *J. Biochem. Microbiol. Technol. Eng.*, **1**: 393, 1959.

Luedeking and Piret's two-parameter model is admittedly simple, the concept of combining growth-associated, non-growth-associated, and other influences to model production kinetics has proved quite useful in practice.

To begin examination of Shu's model, we list his basic hypotheses, all of which have some experimental foundations:

1. The population is heterogeneous; i.e., it contains cells of varying ages.
2. There are natural aging phenomena at work on every cell in the population.
3. For a fixed environment, each cells forms product at a rate depending only on the cell's age.

The mathematical embodiment of the last two conditions can be obtained by writing the rate of product formation for a cell of age θ as

$$r_p = \frac{d\mu(\theta)}{d\theta} = \sum_{i=1}^{q} A_i e^{-k_i\theta} \tag{7.71}$$

where $\mu(\theta)$ is the amount of product made by a unit weight of a cell over a period θ of its lifetime and all k_i's are positive. The combination of terms on the right-hand side of Eq. (7.71) represents the several parallel processes which may contribute to product formation: the parameters A_i and k_i are fixed by the organism's characteristics and environment. It is convenient to use the integrated form of (7.71):

$$\mu(\theta) = \int_0^{\theta} \sum_{i=1}^{q} A_i e^{-k_i\theta'} \, d\theta' \tag{7.72}$$

Next, in order to implement hypothesis 1, we introduce an age-density function $\mathcal{M}(\theta, t)$, where

$$\mathcal{M}(\theta, t)\, d\theta = \text{amount of cell mass at time } t \text{ in batch} \tag{7.73}$$
$$\text{with age between } \theta \text{ and } \theta + d\theta$$

Thus, the total amount of product in the vessel at time t is given by

$$P = \int_0^{t} \mathcal{M}(\theta, t) \left(\int_0^{\theta} \sum_{i=1}^{q} A_i e^{-k_i\theta'} \, d\theta' \right) d\theta \tag{7.74}$$

since the range of possible cell ages at that time extends only up to t. (what does this assume about the inoculum?)

Utilization of the model now requires that we determine the age-density function and choose an age-production relationship, as in Eq. (7.71). Suppose that $r(t)$ is the cell-mass growth rate at time t. Now think about the cells which are reaching age θ at time t. All these cells had to be born at time $t - \theta$, and their rate of appearance at *that* time is $r(t - \theta)$. Consequently, the cell mass at time t with ages between θ and $\theta + d\theta$ is $r(t - \theta)\, d\theta$, so that

$$\mathcal{M}(\theta, t) = r(t - \theta) \tag{7.75}$$

During exponential growth for example, $M(t) = M_0 e^{kt}$, so that

$$r(t) = \frac{dM(t)}{dt} = kM_0 e^{kt} \tag{7.76}$$

and

$$\mathcal{M}(\theta, t) = kM_0 e^{k(t-\theta)} \tag{7.77}$$

Using Eq. (7.75), we can easily determine $\mathcal{M}$ for other phases of the growth cycle.

Turning now to the production from a single cell, let us consider the rate function r_p in Eq. (7.71). By employing from one to three terms, several different age-production rate relationships can be represented, as Fig. 7.33 shows. The simple single exponential-decay form is sufficient to describe one set of experimental data rather well. Bakers' yeast was used to convert substrate (α-ketoadepic acid) into lysine. The fermentation was run several times, and the point of substrate addition was shifted. As a result, the total production of lysine varied considerably (Fig. 7.34).

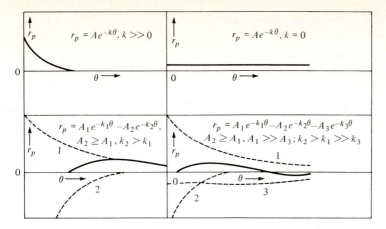

Figure 7.33 Examples of age–production-rate functions for use with Shu's model of product formation.

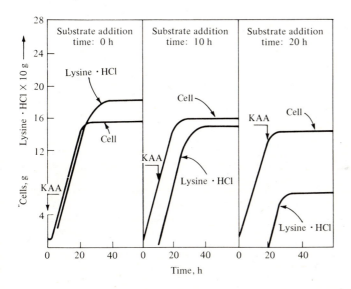

Figure 7.34 Dependence of lysine production on the time of addition of α-ketoadepic acid substrate to a batch yeast fermentation. (*Reprinted by permission from P. Shu, Mathematical Models for the Product Accumulation in Microbiological Processes, J. Biochem. Microbiol. Technol. Eng.,* **3**: 95, 1961.)

Using

$$r_p = 0.0637e^{-0.514\theta} \tag{7.78}$$

as a production model in Eq. (7.59), Shu's model describes not only the time course of a single fermentation (Table 7.5) but the maximum lysine production from several different runs (Table 7.6) quite well.

386 BIOCHEMICAL ENGINEERING FUNDAMENTALS

Table 7.5. Data for lysine production with substrate addition at $t = 0$ and product time course calculated from Shu's model†

Time, h	Experimental conc, g/l		Calc lysine conc, g/l
	Yeast	Lysine	
0	0.9	0	0
5	2.2	0.15	0.22
10	5.9	0.52	0.55
15	10.6	0.83	0.89
20	13.0	1.25	1.23
25	15.3	1.74	1.57
30	16.2	1.64	1.75
35	15.0	1.70	1.77
40	15.7	1.76	1.77
45	15.0	1.82	1.77

† P. Shu, *J. Biochem. Microbiol. Technol. Eng.*, **3**: 104, 1961.

While a quantitative fit was not attempted, a two-term production model provides good qualitative agreement for the citric acid and penicillin fermentations (Fig. 7.35a). Such a two-term form with widely separated k_1 and k_2 values can also reproduce essentially the Luedeking and Piret production pattern expressed above in Eq. (7.69). Figure 7.35b demonstrates how a three-term production model can account for a lag in product formation followed by a decline. Thus it is apparent that many temporal patterns can be represented using Shu's approach.

Indeed, we could criticize the model as being too general—containing enough parameters to fit anything. Also, we could worry about the empirical nature of the production-model function. It would be far more satisfying if it were somehow possible to deduce r_p a priori or at least limit its form using basic biochemical principles. While these doubts and objections are valid, the success of the model in the lysine-fermentation example mentioned earlier is encouraging. It does provide a useful, if simple, means for connecting the time-varying productivity of a single cell with the overall performance of the culture.

Table 7.6. Experimental and calculated maximum lysine accumulations as a function of substrate addition time†

Substrate addition time, h	Maximum lysine conc, g/l	
	Exp	Calc
0	1.82	1.77
10	1.48	1.30
20	0.64	0.45

† P. Shu, *J. Biochem. Microbiol. Technol. Eng.*, **3**: 104, 1961.

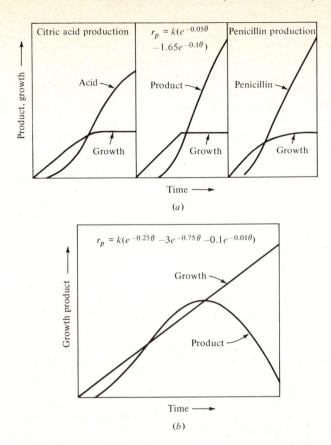

Figure 7.35 (*a*) Shu's model with a simple two term age-production rate function (middle diagram) produces the same qualitative patterns of growth and product yield as observed in the citric acid and pencillin fermentations. (*b*) Another fermentation pattern from Shu's model which includes a lag in product formation and a later decline in product concentration.

7.3.4. Product Yield and Substrate Utilization by Filamentous Organisms

The kinetics of product yield and substrate utilization by molds and other filamentous microbes is usually quite complex. A typical example is the batch fermentation of penicillin. As Fig. 7.24 shows, there are three different regimes of substrate uptake. For the first 20 h, rapid sugar utilization accompanies active growth of mold. As the growth curve passes through the stationary phase, substrate utilization is slow while product-formation rates are at their peak. Following this portion of the fermentation, sugar is again consumed rapidly until it is exhausted.

Several other antibiotic fermentations belong to Gaden's class III and exhibit similar patterns, although there is no general rule applicable to all such fermentations. In streptomycin production, for example, the maximum product synthesis rate is found later in the cycle than for penicillin, and no final rapid sugar-uptake phase occurs. Another antibiotic fermentation involving filamentous organisms, oxytetracycline produced by *Streptomyces rimosus*, is of interest because it introduces the importance of *morphology* for satisfactory performance of the fermentation. Table 7.7 summarizes the major features of this fermentation. Mathematical models to account for cellular differentiation and to relate different states in the hyphae to product-formation kinetics have been proposed but are too complex to be examined here.

Instead, we shall summarize some of the experimental findings on filamentous organisms and product release which illustrate some of the peculiar features of these systems.

1. There is an optimal initial substrate concentration for maximum product yield in a fixed batch time. If there is too much substrate, a "fast fermentation" results, in which little product is formed and substrate is used primarily for biomass production. If there is too little substrate, not enough biomass is formed to manufacture product in the production phase of the type III batch cycle.
2. Product formation is maximized by minimizing the branching in actively growing hyphae in the inoculum. However, the smaller the degree of branching, the

Table 7.7. Stages in batch production of oxytetracycline by *Streptomyces remosus* **in submerged culture†**

No.	Stage	Comment
1	Lag	Lasts about 90 min when the inoculum is small; metabolic activity not measurable
2	Growth of primary mycelium	Depending on inoculum, lasts from 10–25 h; respiration, nucleic acid synthesis, and other metabolic activities at very high level; pyruvic acid concentration at peak; no antibiotic formed
3	Fragmentation of primary mycelium	Lasts about 10 h, during which growth of the mycelium ceases, respiration and nucleic acid synthesis decrease, and pyruvic acid concentration drops to very low value
4	Growth of secondary mycelium	During the 25 h of this phase the secondary mycelium grows to level 2 to 4 times higher than in stage 2; filaments now thin, as contrasted to earlier thick ones; antibiotic production increases rapidly now, nucleic acid synthesis is renewed, but respiration continues to decrease; sugars and ammonia nitrogen rapidly depleted; pyruvic acid concentration may increase slightly
5	Stationary phase	Growth ceases and metabolic acitivity is low as the cycle enters this stage; antibiotic production continues for a time, but on a specific basis its rate is comparatively low

† R. Luedeking, p. 208 in N. Blakebrough (ed.), "Biochemical and Biological Engineering," vol. 1, Academic Press, Inc., New York, 1967; see also J. Doskacil, B. Sikyta, J. Kasparova, D. Poskocilova, and J. Zajicek, *J. Gen. Microbiol.*, **18**: 302, 1958.

longer the lag phase and hence the longer the required batch time. Consequently, there is also an optimal condition for the inoculum.

3. Since fermentations conducted by molds and other mycelial microorganisms are aerobic, it would seem that mixing of a submerged culture should be as intense as possible in order to promote vigorous oxygen transfer to the mold. This is not the case for the penicillin fermentation, where an intermediate degree of agitation has been found to maximize penicillin yields. There are several possible explanations. The morphology of the mold is affected by the applied shear; more vigorous mixing promotes branching of the hyphae, which is believed to decrease product formation. Alternatively, there may well be an optimal rate of oxygen supply, but this is extremely difficult to determine because of the complexity of the system. (Shear effects are reconsidered in Chap. 8.)

7.4. OVERALL KINETICS IN CASES OF CHEMICAL-REACTION–MASS-TRANSPORT INTERACTION

Except for passing mention in connection with Perret's model, we have so far neglected mass-transport effects in our models of microbial kinetics. Mass-transport phenomena appear at several levels, which we might classify as intracellular, cell-membrane–extracellular, and supercellular. The intracellular level should already be familiar from our studies of microbial physiology in Chaps. 1, 5, and 6. Instead of a homogeneous structure, the interior of even the simplest cells contains differentiated regions of varying composition, structure, and function. Transport of material from one region to another—say a protein from a ribosome to the cell membrane—is vital to proper cell function. Although some models dealing with intracellular transport have been formulated, they are still too detailed and too speculative to find practical application for biochemical-engineering problems. Also, Weisz has argued that cellular dimensions are such that internal kinetics is dominated by reaction rather than mass-transfer kinetics, as will be explored shortly. Consequently, we shall not consider in detail any biological-kinetics models incorporating intracellular mass transfer.

This implies that the interior of the cell will be viewed as being well mixed for the purposes of subsequent models. At the next level of our hierarchy, mass transport across the cell membrane and its nearby external environment is of interest. Unless sufficiently rapid, this process can limit cell growth rate, as we shall see in Sec. 7.4.1.

Finally, many microorganisms do not exist as isolated cells. Molds obviously fit into this category since they grow by extending their hyphae rather than by repeated division. The result is a nearly solid, pelletlike structure, into which nutrients must diffuse. Since oxygen is an essential but sparingly soluble nutrient of many molds used industrially, transport of oxygen into mold pellets is an important problem which we shall explore below.

Less familiar but of considerable importance is the propensity of some microorganisms to flocculate or to accumulate at solid surfaces in the form of a film. Many bacteria, for example, are coated on their exterior surfaces by a layer of slimy material, which is predominantly polysaccharides. Upon contact, such organisms tend to stick together, so that particles of agglomerated bacteria develop. The bacteria in the interior of such a floc particle are enmeshed in a matrix of slime material, and they consequently do not experience direct contact with the medium. Instead, nutrients must reach them by diffusing into the floc particle. A similar limitation applies to microbial films. Although the physical situation here differs somewhat from that in molds, the mathematical models usually used for the two cases are essentially identical.

Detailed analysis of this supracellular mass-transfer problem is reserved for Sec. 7.4.2; the general features and assumptions of the standard model are illustrated in Fig. 7.36a. It is assumed that reaction occurs at each point within the pellet and that the kinetics of the reaction is of the same form as observed for isolated cells. (What phenomena might invalidate this assumption?) Mass transport through the pellet occurs via diffusion. The combined effects of these phenomena are gradients of composition within the pellet, each concentration usually varying from a maximum at the exterior surface to a minimum at the pellet center or vice versa. (Which components will have maxima at the outer surface? At the

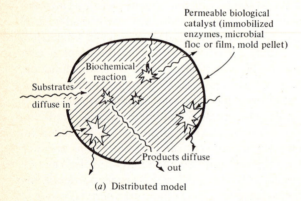

(a) Distributed model

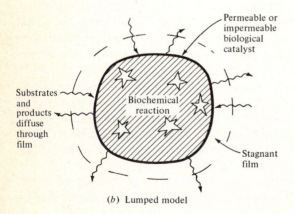

(b) Lumped model

Figure 7.36 Two different models for chemical-reaction–mass-transport interactions in permeable biological catalysts: (a) distributed model; (b) lumped model.

center? Can you imagine a case where a concentration maximum would appear at an intermediate point?) Because both reaction and mass-transfer rate processes are occurring everywhere throughout the pellet, this is termed a *distributed model*. A characteristic feature of distributed models is variation in process conditions as a function of position within the system.

A different situation obtains in the model for cell-membrane-extracellular transport and cell growth, considered in the next section. As Fig. 7.36*b* shows, all concentration gradients are confined to a boundary region in which there is no reaction. Similarly, reaction processes are restricted to the interior, which is uniform in all concentrations. In this kind of model, the mass-transfer and reaction steps are viewed separately; they occur in series rather than in parallel.

Such a view of a physical process is called a *lumped model*. We can arrive at lumped models based on several different rationales: a lumped model may be an excellent approximation to physical reality, as in the case of a well-agitated reactor. On the other hand, we can use lumped models for the sake of mathematical convenience since they are much simpler to manipulate than their distributed counterparts.

7.4.1. A Lumped Model of Substrate Uptake and Utilization by a Microbial Cell

We shall assume that the microorganism in question is spherical, with radius R, and that the relatively stagnant zone surrounding the cell is a spherical shell of thickness ζ. In steady state, the rate of substrate utilization within the cell must be equal to the rate of substrate transport into the cell, so that

$$\frac{4}{3}\pi R^3 \frac{1}{Y_s}\frac{\mu_{max}s_i}{K_s + s_i} = 4\pi R^2 \mathscr{D}_s \frac{ds}{dr}\bigg|_{r=R} \tag{7.79}$$

In writing this balance, we have assumed that the Monod specific-growth rate applies and that the yield factor is constant. The term s_i denotes the intracellular substrate concentration. Substrate diffusion through the stagnant region is presumed to be governed by Fick's Law.

A mass balance on the stagnant spherical cell (see Fig. 7.37) surrounding the cell gives (since no reaction occurs in this film)

$$\frac{d}{dr}\left(r^2 \frac{ds}{dr}\right) = 0 \tag{7.80}$$

which implies

$$r^2 \frac{ds}{dr} = \text{const} = C_1 \tag{7.81}$$

within the film. Subsequent integration of Eq. (7.81) gives

$$s = -\frac{C_1}{r} + C_2 \tag{7.82}$$

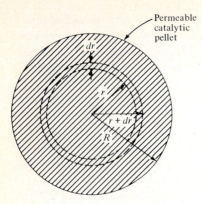

Steady-state mass balance on shell between r and $r + dr$:

$$\left(-\mathscr{D}_s \frac{ds}{dr} \cdot 4\pi r^2\right)\Bigg|_r - \left(-\mathscr{D}_s \frac{ds}{dr} \cdot 4\pi r^2\right)\Bigg|_{r+dr} = r_{u_s} \cdot 4\pi r^2 \, dr$$

Divide by $dr - 4\pi$ and rearrange. Assume effective substrate diffusivity within the permeable pellet $\mathscr{D}_s$ is constant.

$$\frac{\mathscr{D}_s\left(r^2 \dfrac{ds}{dr}\Big|_{r+dr} - r^2 \dfrac{ds}{dr}\Big|_r\right)}{dr} = r^2 - r_{u_s}$$

Take limit as $dr \to 0$,

$$\mathscr{D}_s \frac{d}{dr}\left(r^2 \frac{ds}{dr}\right) = r^2 - r_{u_s}$$

or

$$\mathscr{D}_s\left(\frac{d^2s}{dr^2} + \frac{2}{r}\frac{ds}{dr}\right) = r_{u_s}$$

Figure 7.37 Development of a steady-state balance on a spherical, permeable catalytic pellet using the thin-shell method.

The two constants are evaluated by use of the boundary conditions

$$s = \begin{cases} s_i & \text{at} \quad r = R \\ s_o & \text{at} \quad r = R + \zeta \end{cases} \tag{7.83}$$

where s_0 is the concentration of substrate in the bulk medium surrounding the cell. Since $C_1 = (s_0 - s_i)(R + \zeta)R/\zeta$, Eq. (7.81) gives

$$\frac{ds}{dr}\Bigg|_{r=R} = (s_0 - s_i)\frac{R + \zeta}{R\zeta} \tag{7.84}$$

Using Eq. (7.84) in Eq. (7.79) yields the following relationship between s_i and s_0:

$$\frac{(\mu_{\max}R/3Y_s)s_i}{K_s + s_i} = \frac{\mathscr{D}_s(R + \zeta)}{R\zeta}(s_0 - s_i) \tag{7.85}$$

We seek an expression for substrate-utilization rate in terms of the known medium concentration s_0. Hence, Eq. (7.85) is to be used to express s_i in terms of s_0. If we compare Eq. (7.85) carefully with Eq. (3.103), we note that they are identical except for notation changes. Consequently we can apply the results of Sec. 3.7.2 directly to this problem in its present form. After s_i is written as a function of s_0, the rate r_s of substrate utilization per unit volume of microorganism is

$$r_s = \frac{3}{R} \frac{\mathscr{D}_s(R + \zeta)}{R\zeta} [s_0 - s_i(s_0)] \tag{7.86}$$

which after substitution of $s_i(s_0)$ becomes

$$r_s = \frac{\mu_{max}/Y_s}{2\sigma}\left\{\left(1 + \sigma + \frac{s_0}{K_s}\right) - \left[\left(1 + \sigma - \frac{s_0}{K_s}\right)^2 + 4\frac{s_0}{K_s}\right]^{1/2}\right\} \tag{7.87}$$

where

$$\sigma = \frac{\mu_{max}\, \zeta R^2}{3\, Y_s\, K_s\, \mathscr{D}_s(R + \zeta)} \tag{7.88}$$

E. O. Powell, who originally developed this equation for microbial kinetics applications, compared its goodness-of-fit statistics with those of other growth rate equations, using 13 different sets of experimental data. He concluded that it was superior to the Monod and Tessier forms in nine cases, while the Monod model gave the worst fit in 10 instances. Still, the Monod model neglecting mass transfer is far easier to work with algebraically and contains two rather than three adjustable parameters. Additional details on the derivation and application of this model can be found in Powell's original paper† and in Ref. 16.

7.4.2. Distributed Models for Biological Floc and Mold Kinetics

Deriving the steady-state material balance on substrate now requires formulation of the mass balance on a thin shell within the spherical pellet (see Fig. 7.37). This device permits consideration of all concentrations within the shell as independent of position; in essence we have a very small lumped system. The symbols $\mathscr{D}_s$ and r_{u_s} denote substrate diffusion coefficient and rate of substrate utilization, respectively. Typically $\mathscr{D}_s$ is considered to be constant, while r_{u_s} is a function of s, typically of Monod form:

$$r_{u_s} = +\frac{(\mu_{max}/Y_s)s}{s + K_s} \tag{7.89}$$

If we start from the original shell balance, two subsequent manipulations yield

$$\mathscr{D}_s\left(\frac{d^2s}{dr^2} + \frac{2}{r}\frac{ds}{dr}\right) = +r_{u_s} \tag{7.90}$$

---

† E. O. Powell, The Growth Rate of Microorganisms as a Function of Substrate Concentration, *Proc. 3d Int. Symp. Microb. Physiol. Contin. Cult.*, p. 34, HMSO, London, 1967.

The model is incomplete without boundary conditions on this equation. Since the concentration profile through the pellet will almost always be symmetrical about the center of the sphere,

$$\left.\frac{ds}{dr}\right|_{r=0} = 0 \tag{7.91}$$

We shall assume that the substrate concentration at the external surface of the pellet is equal to the substrate concentration s_0 of the liquid medium bathing the particle. (If an extracellular concentration gradient exists, use the method in Prob. 4.10.) Consequently,

$$\left. s \right|_{r=R} = s_0 \tag{7.92}$$

The *observed overall* rate of substrate utilization r_s by the pellet is equal to the substrate diffusive flux into the pellet. Expressed in moles per pellet volume per unit time, r_s is

$$r_s = \frac{A_p}{V_p}\left(\mathscr{D}_s \left.\frac{ds}{dr}\right|_{r=R}\right) \tag{7.93}$$

where V_p and A_p denote the particle volume and external surface area, respectively. As in our discussion of immobilized-enzyme kinetics (recall Sect. 3.7.2 and Prob. 4.10) we present such rates in a dimensionless form which reveals the extent of diffusional effects. The effectiveness factor η is defined by

$$\eta = \frac{r_s}{r_{u_s}(s_0)} = \frac{\text{observed rate}}{\begin{array}{c}\text{rate which would obtain}\\ \text{with no concentration}\\ \text{gradients in pellet}\end{array}} \tag{7.94}$$

Unfortunately, the effectiveness factor cannot be easily evaluated analytically when r_{u_s} takes the nonlinear form indicated in Eq. (7.89). In this situation, it is necessary to solve numerically the boundary-value problem posed by Eqs. (7.90) to (7.92) and then evaluate r_s using Eq. (7.93). Since such numerical calculations are difficult and time-consuming, we seek to present results in the most compact yet general form. This is achieved here by transforming Eq. (7.90) to (7.91) into an equivalent dimensionless formulation.

Let $\bar{s} = s/s_0$ and $\bar{r} = r/R$; then Eq. (7.90) can be rewritten in dimensionless form

$$\frac{d^2\bar{s}}{d\bar{r}^2} + \frac{2}{\bar{r}}\frac{d\bar{s}}{d\bar{r}} = \frac{r_{u_s}R^2}{\mathscr{D}_s s_0} = \phi^2 \frac{\bar{s}}{1 + \beta\bar{s}} \tag{7.95}$$

where the dimensionless parameters ϕ and β are defined by

$$\phi = R\sqrt{\frac{\mu_{\max}/YK_s}{\mathscr{D}_s}} \qquad \beta = \frac{s_0}{K_s} \tag{7.96}$$

The dimensionless boundary conditions associated with Eq. (7.95) are

$$\bar{s}\bigg|_{\bar{r}=1} = 1 \qquad \frac{d\bar{s}}{d\bar{r}}\bigg|_{\bar{r}=0} = 0 \tag{7.97}$$

The square of the *Thiele modulus* ϕ has the physical interpretation of a first-order reaction rate $R^3(\mu_{max}/YK_s)s_0$ divided by a diffusion rate $R\mathcal{D}_s s_0$. The magnitude of the saturation parameter β provides a measure of local rate derivations from first-order kinetics, with very large values indicating an approach to zero-order kinetics.

In terms of these dimensionless variables, the effectiveness factor is

$$\eta = \frac{3(d\bar{s}/d\bar{r})_{\bar{r}=1}}{\phi^2[1/(1+\beta)]} \tag{7.98}$$

Equations (7.95) and (7.97) reveal that $\bar{s}$ is a function of $\bar{r}$, ϕ, and β only, so that $(d\bar{s}/d\bar{r})_{\bar{r}=1}$ depends only on β and ϕ. This result in conjunction with Eq. (7.98) implies that η depends only upon the Thiele modulus and saturation parameters

$$\eta = f(\phi, \beta) \tag{7.99}$$

A practical problem arises in the use of effectiveness-factor correlations of the form given in Eq. (7.99): the intrinsic-rate parameters μ_m/Y and K_s are frequently unknown. A few simple manipulations readily reveal, however, that uncertainty about the former number need not be a problem. Using Eqs. (7.89) and (7.94), we obtain

$$\phi = \left[\frac{1}{\eta} r_s \frac{R^2}{\mathcal{D}_s s_0} (1+\beta) \right]^{1/2}$$

Substitution of this expression into the right-hand side of Eq. (7.99) gives an implicit relationship between η, β, and a new dimensionless observable modulus Φ, defined by

$$\Phi = \frac{r_s}{\mathcal{D}_s s_0} \left(\frac{V_p}{A_p} \right)^2 \tag{7.100}$$

(Notice that Φ depends on the measurable overall rate r_s; Φ is independent of intrinsic kinetic parameters.) Solution of this implicit equation yields η as a function of β and the observable modulus Φ

$$\eta = g(\Phi, \beta) \tag{7.101}$$

Plots of the η-Φ relationship prescribed by Eq. (7.101) for $\beta \to 0$ and $\beta \to \infty$ are given in Fig. 7.38. Since the effectiveness factors for intermediate K_s/s_0 values lie between the curves for these limiting cases, we see that η is relatively insensitive to the remaining intrinsic parameter K_s/s_0.

Perhaps the most difficult term in Φ to evaluate experimentally is $\mathcal{D}_s$, the substrate diffusivity in the floc or mold particle. Measurements of $\mathcal{D}_s$ have been made by placing a biological floc on a Millipore filter between two well-agitated

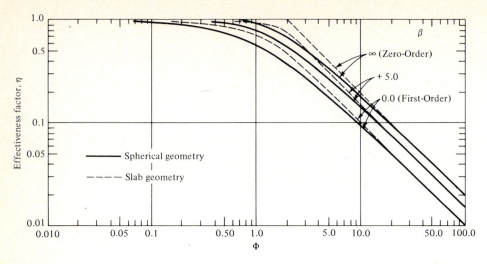

Figure 7.38 Effectiveness factors for substrate utilization with Monod-form kinetics ($\beta = s_0/K_s$). Φ is the observable modulus defined in Eq. (7.100).

vessels containing substrate at different concentrations. Some of the resulting data, which are listed in Table 7.8, clearly show the dependence of $\mathscr{D}_s$ on growth conditions. While the temperature effects revealed here are consistent with the theory of diffusion in liquids, the influence of medium composition is more difficult to anticipate. One possible explanation for the low $\mathscr{D}_s$ values observed with high C/N ratios and glucose is enhanced synthesis of extracellular polysaccharides, which retard oxygen diffusion under these conditions.

The distributed model just described has been applied to respiration rates of *Aspergillus niger* mold pellets. The open circles in Fig. 7.39 are experimental data;

Table 7.8. Dependence of diffusion coefficient $\mathscr{D}_{GB}$ of glucose through microbial floc on growth temperature, substrate, and medium carbon-to-nitrogen ratio (C/N)†

Substrate	Temp, °C	C/N	$\mathscr{D}_{GB}$, cm²/s × 10⁵
Glucose	20	5	0.45
		50	0.05
Methanol		5	0.36
		50	0.36
Glucose	30	5	0.57
		50	0.07
Methanol		5	0.48
		50	0.46

† From W. G. Characklis, J. V. Matson, and D. M. Pipes, Diffusion in Microbial Aggregates, *Proc. AIChE-GVC Joint Meet.*, Munich, September 1974.

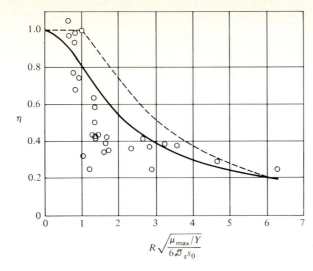

Figure 7.39 Effectiveness factors for oxygen utilization by *A. niger* mold pellets; Experimental data are shown by circles; the solid line is the result of the distributed model with Monod kinetics ($\beta = 1$). The dashed curve results if zero-order kinetics are assumed. (*Reprinted from S. Aiba and K. Kobayashi, Comments on Oxygen Transfer within a Mold Pellet, Biotech. Bioeng.,* **13**: 583, 1971.)

the solid curve was computed using the model assuming $K_s/s_0 = 1$. The agreement is reasonable considering the large degree of scatter in the data. Shortly we shall consider a simplified model susceptible to analytical solutions which leads to the dashed curve in this figure.

Before examining analytical solutions based on approximate kinetics, however, we first seek a formula for η which applies for large Φ (or ϕ) and Monod kinetics. When Φ is sufficiently large ($\Phi \geq 3$, see Table 7.9), diffusion of substrate is slow relative to consumption. In such a situation with diffusion-limited rate, it may be assumed that all substrate is utilized in a thin region within the particle adjacent to its exterior surface, so that the effect of curvature can be neglected in Eq. (7.90). This leaves

$$\mathscr{D}_s \frac{d^2 s}{dr^2} = r_{u_s} \tag{7.102}$$

If the right-hand side of the identity

$$\frac{d^2 s}{dr^2} = \frac{1}{2} \frac{d}{ds} \left(\frac{ds}{dr} \right)^2 \tag{7.103}$$

is substituted for d^2s/dr^2 in Eq. (7.102), integrating the resulting equation with respect to s yields

$$\frac{ds}{dr} \bigg|_{r=R} = \left[2 \int_{s_C}^{s_0} r_{u_s}(s) \, ds \right]^{1/2} \tag{7.104}$$

where s_C and s_0 are the substrate concentrations at the particle center ($r = 0$) and external surface ($r = R$), respectively.

Table 7.9. Criteria for assessing the magnitude of mass-transfer effects on overall kinetcs

Criterion	η value	Limiting rate process	Extent of mass-transfer limitation
$\Phi < 0.3$	~ 1	Chemical reaction	Negligible
$\Phi > 3$	$\propto \Phi^{-1}$	Diffusion	Large

In the diffusion-limited case, $s_C \approx 0$, so that the integral in Eq. (7.104) can be evaluated. Combining Eqs. (7.93), (7.94), and (7.104) gives the desired effectiveness-factor expression valid for sufficiently large Φ or ϕ:

$$\eta \bigg|_{\substack{\text{large } \Phi, \phi \text{ (diffusion} \\ \text{limited overall rate)}}} = \frac{\left[2 \int_0^{s_0} r_{u_s}(s)\, ds\right]^{1/2}}{r_{u_s}(s_0)} \tag{7.105}$$

Evaluation of this formula for the case of Monod kinetics [Eq. (7.89)] yields

$$\eta \bigg|_{\phi \gg 1} = \frac{1}{\phi} \frac{1 + \beta}{\beta} \sqrt{2} [\beta - \ln(1 + \beta)]^{1/2} \tag{7.106}$$

Continuing now our investigation of Monod kinetics, we see that Fig. 7.38 shows that by assuming first-order kinetics

$$r_{u_s} = ks \tag{7.107}$$

($k = \mu_{\max}/YK_s$), a conservative (low) value of η is obtained which is not too inaccurate. This is convenient because the diffusion-reaction model can be solved analytically for the linear rate law of (7.107) (see Prob. 4.10) to give

$$\eta = \frac{1}{\phi} \left(\frac{1}{\tanh 3\phi} - \frac{1}{3\phi} \right) \tag{7.108}$$

where the Thiele modulus ϕ is defined by

$$\phi = \frac{V_p}{A_p} \sqrt{\frac{k}{\mathscr{D}_s}} \tag{7.109}$$

Equation (7.109) can be used to obtain the first-order curve shown in Fig. 7.38 by using the relationship valid for first-order kinetics

$$\Phi = \eta \phi^2 \tag{7.110}$$

At the other end of the spectrum of Monod kinetics is the zero-order approximation, which takes the form

$$r_{u_s} = \begin{cases} k_0 = \text{const} & \text{for} \quad s > 0 \\ 0 & \text{for} \quad s = 0 \end{cases} \tag{7.111}$$

$(k_0 = \mu_{max}/Y)$. Solving the boundary-value problem of Eqs. (7.90) to (7.92) with this substrate-utilization function reveals that substrate concentration s depends on radial position r according to

$$s = s_0 - \frac{k_0}{6\mathscr{D}_s}(R^2 - r^2) \qquad (7.112)$$

This relationship holds as long as s is nonnegative, which will be true from $r = R$ inward to a critical radius $r = R_c$, determined by setting $s = 0$ in Eq. (7.112). This yields

$$\left(\frac{R_c}{R}\right)^2 = 1 - \frac{6\mathscr{D}_s s_0}{k_0 R^2} \qquad (7.113)$$

Equation (7.113) yields a real root for R_c only if

$$R\sqrt{\frac{k_0}{6\mathscr{D}_s s_0}} > 1 \qquad (7.114)$$

Consequently, if this inequality is satisfied, there is an interior portion of the pellet (from $r = 0$ to $r = R_c$) in which $s = 0$ and $r_{u_s} = 0$. Notice that (7.114) is fulfilled for a large number of the oxygen-utilization data points in Fig. 7.39, indicating that in these cases anaerobic metabolism is occurring in the interior portions of the mold pellet. (If s is the only available carbon source, what situation holds in the pellet center?) When inequality (7.114) applies, only an outer shell of the pellet consumes substrate, so that

$$\eta = \frac{\frac{4}{3}\pi(R^3 - R_c^3)k_0}{\frac{4}{3}\pi R^3 k_0} = 1 - \left(\frac{R_c}{R}\right)^3 = 1 - \left(1 - \frac{6\mathscr{D}_s s_0}{k_0 R^2}\right)^{3/2} \qquad (7.115)$$

When condition (7.114) is not fulfilled, a uniform rate of substrate utilization prevails within the entire pellet, giving $\eta = 1$. These results are illustrated by the dashed line in Fig. 7.39.

It is interesting to extend these arguments to multiple-substrate situations. As one substrate is depleted as we move inside a pellet, the metabolic patterns of the interior cells shift to utilize the next best nutrient. Indeed, such interior adaptations could continue further in an elaborate variation of biochemical activities with position within the pellet. In such a case the overall metabolic functions of the pellet become a complex composite of its different internal activities.

Another important feature shown in Fig. 7.38 is the insensitivity of the η-ϕ relationship to particle geometry. For example, the effectiveness factor for a first-order reaction in a slab is

$$\eta = \frac{\tanh \phi}{\phi} \qquad (7.116)$$

This function coincides with the one given in Eq. (7.108) to within about 10 percent over the entire ϕ range. Differences are greatest for ϕ near unity and diminish rapidly for larger and smaller values.

The insensitivity of the η function with respect to both reaction order and particle geometry suggests a useful generalization, which is summarized in Table 7.9 (see Ref. 17 for further details). The criteria summarized are based entirely on the value of the Φ parameter, which, we recall, contains only observable quantities.

Very commonly in nonbiological catalysis η is viewed as a measure of the efficiency of catalyst utilization. Unless η is near unity, greater overall rates can be obtained by subdividing the same amount of catalytic material into smaller pieces, diminishing Φ and thus increasing η. This interpretation is also valuable in the context of biological catalysis, for it indicates how immobilized-enzyme pellet size or mold morphology could be controlled to achieve maximum efficiency.

A fascinating evolutionary concept is that biological structures ranging from cells to organelles to multienzyme complexes have developed in a manner which minimizes diffusional effects on overall rates. Assuming a reasonable oxygen utilization rate of 10^{-8} mol/(s · cm^3), we can use the Φ criterion to compute the

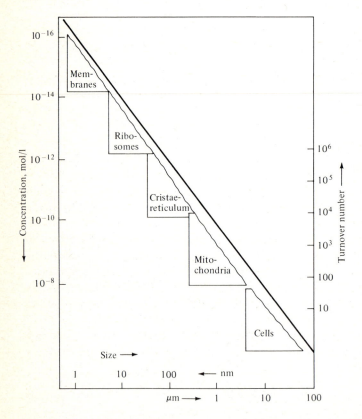

Figure 7.40 The abscissa shows the maximum allowable size for a catalytic apparatus if it is going to process the given small concentrations (ordinate) without diffusional limitations of the overall rate. (Assumes $r_s = 10^{-8}$ mol/s·cm^3, $\mathscr{D}_s = 10^{-6}$ cm^2/s, $s_0 = 10^{-6}$ mol/l). (*Reprinted from P. W. Weiss, Diffusion and Chemical Transformation: An Interdisciplinary Excursion, Science, p. 410, 1973.*)

maximum size of an enzymatic structure which can convert substrate at a given concentration without serious diffusional limitation ($\mathscr{D}_s = 10^{-6}$ cm^2/s is used in this calculation). This information is shown in Fig. 7.40, along with characteristic dimensions of cell components. The turnover-number axis is related to the concentration coordinates using

$$\text{Turnover number} = 60 \times \text{rate/substrate concentration}$$

which assumes that the substrate and enzyme concentrations are approximately equal. (What does this do to the quasi-steady-state assumption?)

This picture is qualitatively consistent with observations that smaller structures are in fact used in nature to process low-concentration intermediates at high rates. The quantitative conclusions of this analysis are also reasonable. For example, assuming that the oxygen concentration in a yeast cell is comparable to the cytochrome c concentration, which is about 10^{-6} mol/l, we conclude from Fig. 7.40 that a yeast cell should be no bigger than 10 μm if mass-transfer effects are to be avoided. Actually yeast cells typically range from 5 to 50 μm in diameter.

7.5. THERMAL-DEATH KINETICS OF CELLS AND SPORES

The activity of microbes in air or liquids can be diminished in any given instance by destruction (heat, radiation, or chemical or mechanical means), removal (filtration or centrifugation), or inhibition (refrigeration, dessication, water depletion, chemicals). The predominant modes for liquid sterilization are heating (industrial) and chlorination (municipal use of water), with particulate filtration the major tool for air sterilization. The latter two processes depend on mass-transfer phenomena discussed in the following chapter; the kinetics of cell and spore inactivation is the topic of this section, and its use in design appears in Chap. 9.

Figure 7.41 illustrates experimental data for vegetative-cell and spore death over a range of temperatures. In general, cells have a greater susceptibility for destruction than spores, consistent with earlier qualitative remarks concerning endospore formation as a survival mechanism of some cells. Similarly, viruses such as the bacteriophages are also inactivated by thermal treatment, so that thermal sterilization in the fermentation industry decreases both the viable microbial population of the liquid-nutrient stream and its virus count, or *titer*.

Before we present some useful equations for estimation of the population reduction rates in sterilization, a few cautionary remarks are in order. The death of a particular cell is probably due to the thermal denaturation of one or more kinds of essential proteins such as enzymes. The kinetics of such cooperative transitions for these molecules may be rather complex in time. Also, the rate of the molecular processes ultimately leading to cell death will depend on environment composition, including the solvent concentration itself. For example, the temperature needed to achieve coagulation (denaturation followed by massive irreversible cross-linking of the denatured protein) of the egg protein albumin increases with decreasing water content as shown in Table 7.10.

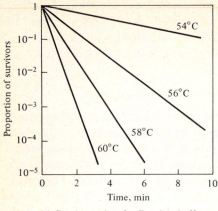

(a) Death rate data for *E. coli* in buffer

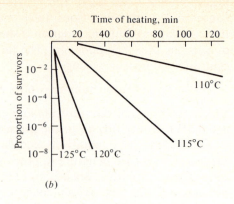

(b) Inactivation of *Bacillus subtilis* spores

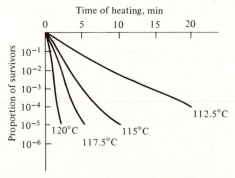

(c) Inactivation of *Bacillus stearothermophilus* spores

Figure 7.41 Experimental data for thermal death of *E. coli* (a) and inactivation of vegatative spores of *Bacillus subtilis* (b) and *B. sterothermophilus* (c). [(a) *Reprinted from S. Aiba, A. E. Humphrey, and N. F. Millis, "Biochemical Engineering," 2d ed., p. 241, University of Tokyo Press, Tokyo, 1973; (b), (c) Reprinted from H. Burton and D. Jayne-Williams, Sterilized Milk, in J. Hawthorn and J. M. Leitch (eds.), "Recent Advances in Food Science, vol. 2: Processing", p. 107, Butterworths & Co. Publishers Ltd., London, 1962.*]

Table 7.10†

Water content, %	Approximate coagulation T, °C
50	56
25	76
15	96
5	149
0	165

† Data from M. Frobisher, "Fundamentals of Microbiology," p. 259, W. B. Saunders Company, Philadelphia, 1968.

These observations and others suggest that it may be more accurate to consider the deactivation of the water-deficient structures such as viruses and spores as requiring some kind of initial hydration, following which the molecular transitions such as denaturation may occur; a similar effect of relative humidity on the survival rate of active bacteriophage is found [22]. A combination of deactivating influences may produce a nonadditive response; e.g., simultaneous desiccation and heat treatment may be less effective than supposed from the individual efficiencies of these processes.

A species population will have a range of cell ages, as discussed previously. Further, the nature of the cell wall and the relative importance of various metabolic pathways to the cell shift with cell and culture age. Thus, the resistance of a species to thermal (or other) destruction will depend on its *history*, a vague term which is difficult to make more quantitative. For example, an exponential-phase population has relatively permeable cell walls, which allow rapid exchange of solutes with the external medium. If these solutes affect protein stability, a consequent difference of exponential-phase mortality from that of a stationary-phase response may be expected. Ultimately, we are left with the old saw that the best data come from measurements on systems approximating those of the application as closely as possible.

To take up analysis of death-rate kinetics, note that the straight lines through the data in the semilog plots are consistent with a first-order decay for the *viable* population level n,

$$\frac{dn}{dt} = -k_d n \tag{7.117}$$

which, for constant k_d, yields

$$n(t) = n_0 e^{-k_d t} \tag{7.118}$$

where n_0 is the concentration of spores or vegetative cells at $t = 0$. The slope of the semilog plots is $-k_d$; a plot of $\ln k_d$ vs. reciprocal temperature provides another straight line, so that the temperature dependence of k_d may be taken as that of the familiar Arrhenius form

$$k_d = k_{d0} e^{-E_d/RT} \tag{7.119}$$

The range of E_d values for many spores and vegetative cells lies between 50 and 100 kcal/mol. The historical term known as the *decimal reduction factor* D_r ($= 2.303/k_d$) is simply the time needed to reduce the viable population by a factor of 10.

These kinetic descriptions work reasonably well provided the concentration of the spores or cells is a large number in the statistical sense of the word (Sec. 7.1). Deviations from such a prediction become more probable as the numbers involved diminish since the standard deviation of, for example, the normal distribution increases as $1/n$. In a deterministic model, the fraction of the population remaining viable (assuming no current growth under the applied lethal conditions) is seen from the previous equation to be

$$\text{Viable fraction} = \frac{n}{n_0} = e^{-k_d t} \tag{7.120}$$

If each organism's fate is independent of the others in that no reproduction occurs, and if the lethal condition is uniformly applied, a stochastic approach to sterilization [21] shows that the probability that at any time t the remaining population contains N viable organisms is given by

$$P_N(t) = \frac{N_0!}{(N_0 - N)!\,N!}\,(e^{-k_d t})^N (1 - e^{-k_d t})^{(N_0 - N)} \tag{7.121}$$

with N_0 the initial *viable number* of organisms in the fluid being sterilized.

The k_d which appears here is the same as the rate constant in Eq. (7.117): in the stochastic model k_d may be interpreted as the reciprocal of the mean life span of the organism. As the number of organisms falls to a low value, the assumptions of a uniform population characterizable by a single k_d value begins to fail. For example, the data in Fig. 7.42 for *Staphlococcus aureus* in neutral phosphate buffer show consistent positive deviations from the random-distribution result of the above equation, suggesting that a minor fraction of the population is more resistant to thermal death than the vast majority.

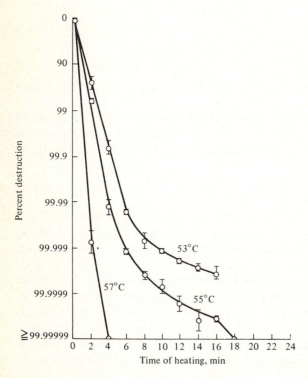

Figure 7.42 Thermal destruction of *S. aureus*. (*Reprinted by permission from G. C. Walker and L. G. Harmon, Thermal Resistance of Staphlococcus aureus in Milk, Whey, and Phosphate Buffer, Appl. Microbiol.,* **14:** 584, 1966.)

While considerable numbers of surviving organisms are acceptable in some applications (viable bacterial concentrations in Grade A pasteurized milk must be less than 30,000 cells per milliliter), more stingent requirements apply, for example, to many pure culture fermentations. In such cases it is important to examine the extinction probability, i.e., the probability that all organisms are inactivated. Setting $N = 0$ in Eq. (7.121) gives

$$\text{Extinction probability } P_0(t) = (1 - e^{-k_d t})^{N_0} \qquad (7.122)$$

from which it follows that the probability of at least one organism's survival is

$$1 - P_0(t) = 1 - (1 - e^{-k_d t})^{N_0} \qquad (7.123)$$

For the usual situation of $N_0 \gg 1$, this becomes

$$1 - P_0(t) = 1 - e^{-N_t} \qquad (7.124)$$

where

$$N_t = N_0 e^{-k_d t} \qquad (7.125)$$

We can interpret $1 - P_0$ physically as the fraction of sterilizations which are expected to fail to produce a microbe-free product.

As we know from our studies of microbial metabolism, living cells often have alternate pathways for tapping an energy source or conducting a biosynthesis. The presence of redundancy alters the chances for organism survival. A statistical analysis shows that the expected value of the population size is

$$E[N(t)] = N_0(1 - \{1 - \exp[\bar{k}_d(t)t]\}^r)^\omega \qquad (7.126)$$

where ω = number of kinds of essential subcellular structure
r = number of units of each structure in each organism
$\bar{k}_d(t)$ = mean specific death rate for each kind of structure

Note that this expression expands into a series of decaying exponentials and is not thus expected to be linear over the major fraction of a semilog plot.

In a similar manner, a cell solution containing m different varieties of microbes (or spores or viruses) may most simply behave as independent deterministic species, the resulting total viable population being the sum of the individual results following Eq. (7.118) with the appropriate individual values of k_{di}:

$$n(t) = \sum_{i=1}^{m} n_i(t) = \sum_{i=1}^{m} n_{i0} e^{-k_{di} t} \qquad (7.127)$$

Again a straight-line semilog plot is not expected. Further, the individual k_{di} values are not routinely available; some methods for estimating the upper and lower bounds on the values of $n(t)/n_0$ from initial data are discussed Hutchinson and Luss.† The validity of these estimates depends on the accuracy of the initial death data and requires evaluation of the second derivative dn^2/dt^2 at $t = 0$.

---

† P. Hutchinson and D. Luss, Lumping of Mixtures with Many First Order Reactions, *Chem. Eng. J.*, **1**: 129, 1970.

In this chapter we have studied the interrelationships between biomass increase, substrate uptake, and product formation in batch culture. From experimental data, the general features of growth for cell populations and filamentous organisms were discerned. Mathematical models of growth helpful for understanding and design of fermentation processes utilizing such organisms have been summarized and reviewed. In Chap. 9, we shall consider design and analysis of a variety of continuous and batch biological reactors. Throughout that discussion we shall rely heavily on the rate expressions presented in this chapter.

Before proceeding to examination of biological reactors, an additional kinetic phenomenon must be studied: mass transfer in macroscopic systems. This subject deals with the transport of oxygen, chlorine, and other gases into liquids for such purposes as aeration and chlorination. A conceptually related problem arises in sterilization of reactor feed streams by filtration of contaminating particles. This topic requires consideration of the mass-transfer phenomena which govern the collection of small particles from moving gas or liquid streams by stationary filter media. These matters and others are examined in the next chapter.

PROBLEMS

7.1. Fermentation batch kinetics Grape juice (natural) is a nearly complete nutrient for yeast. Perform the following batch fermentation.

(*a*) Purchase a simple beer- or wine-making kit or make your own. For the wine you need grape juice (free of settleable solids for this experiment), sugar, a yeast source, a 1-gal vessel, a small beaker, and hose for an airlock.

(*b*) Start-up: suspend active yeast in 100 ml of lukewarm (previously boiled) water and stir well to disperse cells. Add to fermentor (sterilized with boiling water or sodium sulfite solutions) 1 qt grape concentration, 3 lb sugar, and sterilized water (slightly above room temperature) to three-quarters full. Be sure in advance to leave sufficient head space for a small hydrometer (allow for hydrometer introduction through fermenter top). Add yeast suspension, cap tightly, and shake thoroughly for 30 s. Set hydrometer in fermentation liquid. Insert one-hole cork with exit to transfer tube and airlock; set water level in latter ~ 1 cm above hose outlet.

(*c*) *Record* fermentation progress in time (1 wk) by monitoring (1) CO_2 bubble rate (constant hose depth below airlock) (devise a means of measuring volume per CO_2 bubble) and (2) hydrometer reading (do not lose airseal).

(*d*) Plot and discuss data in terms of apparent growth phases and relation between ethanol and CO_2 production (state your assumptions clearly). From knowledge of total sugar added initially (grape + crystallized), and of the dry-cell mass weight at the end of fermentation and your data of part (*c*), demonstrate a carbon mass balance, stating assumptions.

(*e*) It has been said that taste and smell are very sensitive measuring devices. Try your product!

7.2. Temperature variations of growth Johnson, Eyring, and Polissar [23] represent growth of *E. coli* between 18 and 46°C by the following equation for the specific growth rate μ:

$$\mu = \frac{\alpha T \exp\left(-\dfrac{\Delta H_1}{RT}\right)}{1 + \exp\left(-\dfrac{\Delta S}{R} - \dfrac{\Delta H_2}{RT}\right)}$$

where $\quad \alpha = 0.3612e^{24.04}$
$$\Delta H_1 = 15 \text{ kcal/mol}$$
$$\Delta H_2 = 150 \text{ kcal/mol}$$
$$\Delta S = 476.46 \text{ cal/(g mol} \cdot \text{K)}$$

(a) Plot this function as $\ln \mu$ vs. $1/T$.

(b) Show that this equation can be represented as the product of two functions whose form is suggested by the plot in part (a). What explanation(s) rationalize these two individual functions and the value of the above parameters?

(c) In this interpretation of $\mu = f(T)$, what implicit assumptions are made with regard to irreversible deactivations?

7.3. Single- and multiple-substrate kinetics A culture is grown in a simple medium including 0.3% wt/vol of glucose; at time $t = 0$, it is inoculated into a larger sterile volume of the identical medium. The optical density (OD) at 420 nm vs. time following inoculation is given below in column 1 of Table 7P3.1. The same species cultured in a complex medium is inoculated into a mixture of 0.15% wt/vol glucose and 0.15% wt/vol lactose; the subsequent OD-vs.-time data are given in column 2. If OD (420 nm) is linear in cell density with 0.175 equal to 0.1 mg dry weight of cells per milliliter, evaluate the maximum specific growth rate μ_{max}, the lag time t_{lag}, and the overall yield factors Y (grams of cell per grams of substrate), assuming substrate exhaustion in each case. Explain the shape of the growth curves in each case.

Table 7P3.1†

Time, h	OD (1)	OD (2)	Time, h	OD (1)	OD (2)
0	0.06	0.06	4.5	0.44‡	0.43
0.5	0.08	0.06	5.0	0.52‡	0.48
1.0	0.11	0.06	5.5	0.52†	0.50
1.5	0.14	0.07	6.0		0.52
2.0	0.20	0.10	6.5		0.30‡
2.5	0.26	0.13	7.0		0.42‡
3.0	0.37	0.18	7.5		0.50‡
3.5	0.49	0.26	8.0		0.50‡
4.0	0.35‡	0.32			

† Table adapted from D. Kerridge and K. Tipton (eds.), "Biochemical Reasoning," prob. 39, W. B. Benjamin, Inc., Menlo Park, Calif., 1972.

‡ Sample diluted twofold before OD measurement.

7.4. Death or deactivation kinetics: chlorination A number of death or deactivation rate equations have appeared in the literature; an early result was an exponential decay for anthrax spore deactivation by 5% phenol [24]. (The equation for deactivation: $dN/dt = -kN$ is often referred to as *Chick's law*.) Three other forms which have been used are

Logistic: $$-\frac{dN}{dt} = kN + k'N(N_0 - N)$$

t^2: $$-\frac{dN}{dt} = ktN$$

Retardant: $$-\frac{dN}{dt} = \frac{k}{1 + \alpha t} N$$

Thus, for these three, the apparent first-order rate constant changes continually with time.

(a) Integrate each form and sketch the shape of the curve which would be obtained on an appropriate plot in each case, indicating how the parameters in each model would be evaluated from such a graph.

(b) Which law, if any, do the data in Table 7P4.1 for *E. coli* follow?

Table 7P4.1. *E. coli* **survival,** $\%$†
pH 8.5, 2 to 5°C

Cl, mg/l	Time of contact, min				
	0.5	2	5	10	20
0.14	52	11	0.7		
0.07	80	56	30	0.5	0
0.05	95	85	65	21	0.31

† Adapted from G. M. Fair, J. C. Geyer, and D. A. Okan, "Water and Wastewater Engineering," pp. 31–9, John Wiley & Sons, Inc., New York, 1968.

(c) The time to fixed percentage kill or inactivation has been correlated by the form (concentration)^x (time) = const for a given disinfectant. For chlorine (as HOCl), $\alpha = 0.86$ (*E. coli*), 3 (adenovirus), 1 (poliomyelitis virus), 2 (Coxsackie virus A). Why would you (not) expect such variation of α?

7.5. Cell and product kinetics A generalized form of the logistic equation is proposed by Konak [25]

$$\frac{1}{N_\infty^{a+b}}\frac{dN}{dt} = k\left(\frac{N}{N_\infty}\right)^a\left(1 - \frac{N}{N_\infty}\right)^b$$

where $\quad N_\infty$ = stationary-phase population
$\quad a, b$ = const
$\quad N$ = cell mass

(a) Show that the maximum growth rate occurs at $N - N_\infty = a/(a + b)$ and that its value is given by

$$\left.\frac{dN}{dt}\right|_{max} = \frac{kN_\infty^{a+b}a^ab^b}{(a + b)^{a+b}}$$

(b) Indicate by sketches and the previous equation(s) how you would evaluate each parameter of the model.

(c) Show that by replacing N by P on the left-hand side only and choosing a and b appropriately, a product-generating equation results which yields growth-associated, non-growth-associated, or growth- and non-growth-associated behavior analogous to the model of Luedeking and Piret in this chapter.

7.6. Cell networks and product-formation kinetics (a) Summarize why quantification of product formation kinetics is often more difficult than substrate-consumption or biomass-production kinetics.

(b) For an antibiotic, vitamin, or amino acid of your choice, research and write a short paper outlining the microbial synthesis steps, identifying, where possible, slow vs. equilibrated transformations and the presence of enzyme inhibition-activation and induction-repression. As a final portion, propose or discuss (if known) the kinetics of product formation based on the synthesis sequence.

7.7. Thiele modulus of the cell The Thiele moduli of many cells and their inner organelles (membranes, mitochondria, ribosomes, etc.) appear to be at or just below unity [17]. Assuming that the cell may produce enzyme levels well above or below this value, comment quantitatively on why this might (not) reasonably be expected. For simplicity, assume that only one kind of enzyme exists. Include consideration of the rate per volume, the total growth rate, and the cost of enzyme synthesis. Note that the first statement implies that cell kinetics measured in absence of external mass-transfer influences (Chap. 8) will provide true kinetic data, as has been assumed throughout Chap. 7.

7.8. Double-inhibition kinetics The growth of *Candida utilis* on sodium acetate appears to be inhibited by the substrate acetate and also pH [26].

(*a*) With the following assumptions, write down the form for the growth rate in terms of appropriate constants and the hydrogen-ion and total-substrate concentration.

1. The substrate-inhibition function in terms of total substrate is analogous to that for enzyme-substrate noncompetitive inhibition, i.e., that of Andrews, Eq. (7.16).
2. The maximum growth rate μ_{max} is a similar function of hydrogen ion concentration.
3. The inhibitor in the substrate term is the protonated form; i.e., there is an equilibrium between HS (active inhibitor) $\rightleftarrows S^-$ (inactive) $+ H^+$.

(*b*) Show on a Lineweaver-Burk plot for $1/\mu$ vs. $1/s$ how the curve(s) would change with pH. How would you evaluate the parameters of this model?

A number of processes involved in water treatment are similarly affected by pH. An example of anaerobic digester kinetics appears in Sec. 12.4.6.

REFERENCES

1. N. Blakebrough (ed.), "Biochemical and Biological Engineering Science," vol. 1, Academic Press, Inc., New York, 1967. A useful reference for several aspects of biological reactor design. Chapter 6, Fermentation Process Kinetics, by R. Luedeking, is highly recommended.
2. S. Aiba, A. E. Humphrey, and N. F. Millis: "Biochemical Engineering," 2d ed., Academic Press, Inc., New York, 1973. This comprehensive text on fermentation treats batch kinetics in chap. 4 and offers an extended discussion of continuous fermentation in chap. 5. Chapter 9 reviews liquid sterilization.
3. A. C. R. Dean and C. N. Hinshelwood: "Growth, Function and Regulation in Bacterial Cells," Oxford University Press, London, 1966. In addition to discussing the mathematical analysis of reaction networks, this useful monograph presents a large body of experimental information on other aspects of growth and regulation. Highly recommended reading; the only major weakness is undue skepticism about the significance of molecular biology.
4. P. S. S. Dawson (ed.), "Microbial Growth," Dowden, Hutchinson, & Ross, Inc., Stroudsburg, Pa., 1974. A fascinating collection of many of the classic papers dealing with microbial growth. Monod's original paper is reproduced here, as is Powell's work, which is summarized in Sec. 7.4.1.

Review papers on mathematical modeling of cell growth which are especially useful:

5. H. M. Tsuchiya, A. G. Fredrickson, and R. Aris: Dynamics of Microbial Cell Populations, *Adv. Chem. Eng.*, **6**: 125, 1966.
6. A. G. Fredrickson, R. D. Megee, III, and H. M. Tsuchiya: Mathematical Models for Fermentation Processes, *Adv. Appl. Microbiol.*, **23**: 419, 1970.

Papers considering product and substrate inhibition of microbial growth:

7. S. Aiba, M. Shoda, and M. Nagatani: Kinetics of Product Inhibition in Alcohol Fermentation, *Biotechnol. Bioeng.*, **10**: 845, 1968.
8. S. Aiba and M. Shoda: Reassessment of the Product Inhibition in Alcohol Fermentation, *J. Ferment. Technol. Jpn.*, **47**: 790, 1969.
9. J. F. Andrews: A Mathematical Model for the Continuous Culture of Microorganisms Utilizing Inhibitory Substrates, *Biotechnol. Bioeng.*, **10**: 707, 1968.

First presentation of fermentation classification schemes described in the text:

10. E. L. Gaden: *Chem. Ind. Rev.* **1955**: 154; *J. Biochem. Microbiol. Technol. Eng.*, **1,** 413, 1959.
11. F. H. Deindoerfer: Fermentation Kinetics and Model Processes, *Adv. Appl. Microbiol.*, **2:** 321, 1960.

Reviews of oscillations in biological systems:

12. B. Chance, E. K. Pye, A. K. Ghosh, and B. Hess (eds.): "Biological and Biochemical Oscillators," Academic Press, Inc., New York, 1973.
13. J. Higgins: The Theory of Oscillating Reactions, *Ind. Eng. Chem.*, **59:** 18, May 1967.

The original paper proposing the segregated production model discussed in Sec. 7.3.3.:

14. P. Shu: Mathematical Models for the Product Accumulation in Microbial Processes, *J. Biochem. Microbiol. Technol. Eng.*, **3:** 95, 1961.

A first attempt at connecting differentiation in molds with substrate utilization and product-formation kinetics.

15. R. D. Megee, S. Kinoshita, A. G. Fredrickson, and H. M. Tsuchiya: Differentiation and Product Formation in Molds, *Biotech. Bioeng.*, **12:** 771, 1970.

The following sources contain useful discussion of chemical-reaction–mass-transfer coupling. The first concentrates on biological systems, while synthetic catalysts are emphasized in the others. As indicated in the text, however, much of the latter material can be directly applied to biological reactors following minor changes in terminology and notation.

16. B. Atkinson: "Biochemical Reactors," Pion Limited, London, 1974.
17. P. W. Weisz: Diffusion and Chemical Transformation: An Interdisciplinary Excursion, *Science*, **179:** 433, 1973.
18. C. N. Satterfield: "Mass Transfer in Heterogeneous Catalysis," MIT Press, Cambridge, Mass., 1970.
19. E. E. Petersen: "Chemical Reaction Analysis," Prentice-Hall, Inc., Englewood Cliffs, N.J., 1965.
20. R. Aris: "The Mathematical Theory of Diffusion and Reaction in Permeable Catalysts," vol. 1, "The Theory of the Steady State," vol. 2, "Questions of Uniqueness, Stability, and Transient Behavior," Clarendon Press, Oxford, 1975.

An excellent introduction to the literature on sterilization:

21. N. Blakebrough, Preservation of Biological Materials Especially by Heat Treatment, in N. Blakebrough (ed.), "Biochemical and Biological Engineering Science," vol. 2, Academic Press, Inc., New York, 1968.

The factors affecting organisms' susceptibility to sterilization are reviewed in chapters 20 and 21 of Frobisher (Ref. 2 of Chap. 1); consideration of phage destruction:

22. Hango et al.: Phage Contamination and Control, in "Microbial Production of Amino Acids," Kodansha Ltd., Tokyo, and John Wiley & Sons, Inc., New York, 1973.

Problems

23. F. H. Johnson, H. Eyring, and M. J. Polissar: "Kinetic Basis of Molecular Biology," John Wiley & Sons, New York, 1954.
24. H. Chick: Investigation of the Laws of Disinfection, *J. Hyg.*, **8:** 698, 1908.
25. A. R. Konak: An Equation for Batch Bacterial Growth, *Biotech. Bioeng.*, **17:** 271, 1975.
26. J. V. Jackson and V. H. Edwards: Kinetics of Substrate Inhibition of Exponential Yeast Growth, *Biotech. Bioeng.*, **17:** 943, 1975.

EIGHT

TRANSPORT PHENOMENA IN MICROBIAL SYSTEMS

The previous chapters have considered progressively larger scales of distance: from molecular through cellular to fluid volumes containing millions or billions of cells per milliliter. As the sources and sinks of entities such as nutrients, cells, and metabolic products become further separated in space, the probability increases that some physical-transport phenomena, rather than a chemical rate, will influence or even dominate the overall rate of solute processing in the reaction volume under consideration. Indeed, according to the argument of Weisz presented in Chap. 7, many natural cellular systems in their original habitat have Thiele moduli near unity; i.e., they are operating at the maximum possible rate without any serious diffusional limitation. If, in synthetic circumstances, a richer supply of carbon nutrients is created, evidently the aerobic cell will be able to utilize them fully only if oxygen can also be maintained at a higher concentration in the direct vicinity of the cell. This situation may call for extra effort on our part to transfer oxygen, which has sparingly small solubility in aqueous solutions, to the culture at a sufficiently rapid rate.

Evidently, the boundary demarcating aerobic from anaerobic activity depends upon the local bulk-oxygen concentration, the O_2 diffusion coefficient, and the local respiration rates in the aerobic region. This line divides the viable from dying cells in strict aerobes such as mold in mycelial pellets or tissue cells in cancer tumors; it determines the depth of aerobic activity near lake surfaces; and it divides the cohabitating aerobes from anaerobic microbial communities in soil particles. Thus, while the modern roots of biological-process oxygen mass transfer began with World War II penicillin production in the 1940s, its implications are now established to include many natural processes such as food spoilage via undesired oxidation and lake eutrophication due either to inadequate system aeration by natural oxygen supplies or to an excessive concentration of material such as phosphate or nitrate.

A second area of mass transfer is of importance here as well. This concerns the transport and corresponding mechanisms by which colloidal, cellular, and larger than cellular particulate materials in moving air or liquid streams are able to migrate under a variety of forces to the surfaces of very large particles, which act as sinks for the smaller particles. This topic is treated under the general title of filtration theory; it includes mechanical means of removing, by filters, air or waterborne microbial contaminants from streams of interest such as a liquid nutrient before their introduction into a continuous reactor. Other important applications include air filtration by a surgeon's mask or by the cotton plug in a laboratory shake flask, both of which filter the passing air mechanically. The implications of these particle-capture phenomena in biological processes are extremely diverse. For example, the depth to which microbially or virally contaminated water must percolate in soils before becoming potable depends on the filtration efficiency of the intervening soil particles. Even the survival problem of estimating the minimum efficiency with which a larger self-propelled object such as a 200-μm protozoan can catch a 1-μm bacterium is amenable to calculation by this approach, as is the design of hospital and research germ-free rooms.

Strong coupling often occurs between solute diffusion and momentum transport or chemical reactions or (even more complex) both. The case of diffusion and reaction interaction has been considered previously twice: in the analyses of enzyme catalysis in a membrane (Prob. 4.10) and oxygen transfer in a mycelial pellet (Chap. 7). In those circumstances, the Thiele modulus and a saturation parameter K_s/s_0 provide the unifying parameters needed to completely describe the performance, i.e., effectiveness factor, of such systems. Unfortunately, the variety of circumstances under which mass transfer couples with momentum transfer, i.e., the fluid mechanics, is enormous; indeed, it is the substance of a major fraction of the chemical engineering literature. For this text, we content ourselves with fundamental concepts and tabulated formulas for calculation or estimation of the appropriate mass-transfer coefficients for solutes. Particle-filtration theory, excluding electrical effects, is presented quite concisely in relatively simple form, although some of the detailed derivations in the references reveal that the results are deceptive in their simplicity.

A final brief section of this chapter concerns instances where heat transfer may provide an important transport effect which strongly influences the microbial system's behavior through spatial temperature inhomogeneity. Examples here include relatively exothermic fermentation processes, such as trickling-filter operation for wine-vinegar production or wastewater treatment, and that gardner's delight, the compost heap (municipal dump, etc.).

8.1. GAS-LIQUID MASS TRANSFER
IN MICROBIAL SYSTEMS

The general nature of the mass-transfer problem of primary concern in this chapter is shown schematically in Fig. 8.1. A sparingly soluble gas, usually oxygen, is transferred from a source, say a rising air bubble, into a liquid phase containing microorganisms. (Any other sparingly soluble substrate, e.g., the liquid hydrocarbons used in hydrocarbon fermentations, will give the same general picture.) The oxygen must pass through a series of transport resistances, the relative magnitudes of which depend on bubble (droplet) hydrodynamics, temperature, microbial activity and density, solution composition, interfacial phenomena, and other factors. Conceptually, we may imagine four different pathways of solute transport (Fig. 8.1). These arise from different combinations of the following resistances:

1. Diffusion from bulk gas to the gas-liquid interface
2. Movement through the gas-liquid interface

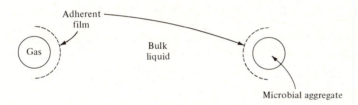

(a) Separate bubble, microbial aggregate

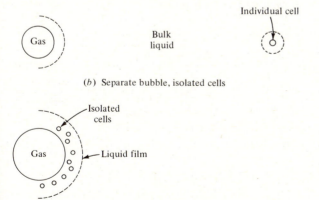

(b) Separate bubble, isolated cells

(c) Microbes concentrated near bubble

Figure 8.1 Gas-liquid-microbe transfer paths: (a) separate bubble, microbial aggregates, (b) separate bubble, isolated cells, (c) microbes concentrated near bubble.

3. Diffusion of the solute through the relatively unmixed liquid region† adjacent to the bubble into the well mixed bulk liquid
4. Transport of the solute through the bulk liquid to a second relatively unmixed liquid region surrounding the microbial species
5. Transport through the second unmixed liquid region associated with the microbes
6. Diffusive transport into the microbial floc, mycelia, or soil particle if appropriate
7. Consumption of the solute by biochemical reaction within the organism

All seven resistances appear in Fig. 8.1*a*. When the microbes take the form of individual cells, the sixth resistance disappears (Fig. 8.1*b*), any internal cell-diffusive resistance being lumped into the apparent microbial kinetics, as described in Chap. 7.

Microbial cells themselves have some tendency to adsorb at interfaces. Thus, as in Fig. 8.1*c*, microbes may preferentially gather at the vicinity of the gas-bubble liquid interface. Then, the diffusing solute oxygen passes through only one unmixed liquid region and no bulk liquid before reaching the cell. In this situation, the bulk O_2 concentration may evidently not always represent the oxygen supply for the respiring microbes, as shown in the following example.

Example 8.1. Effectiveness factor of a microbial monolayer The bubble ($D_b \gtrsim 10^{-1}$ mm) interface is approximately planar with respect to microbial dimensions (order 3×10^{-4} cm). If the dissolved gas saturates the respiratory organelles of the cells, the reaction rate within the microbial film volume is zero order in O_2. Then, if the microbial monolayer is taken to be a uniform planar film (Fig. 8E1.1), the one-dimensional reaction-diffusion equation is

$$\mathscr{D}_{O_2} \frac{d^2 c_{O_2}}{dz^2} - k = 0 \tag{8E1.1}$$

where k = zero-order reaction rate. Rendering c_{O_2} and z dimensionless by $c_{O_2}^*$ (saturation) and D (microbe diameter) leads to

$$\frac{d^2 \bar{c}_{O_2}}{d\bar{z}^2} - \frac{k}{\mathscr{D}_{O_2}} \frac{D^2}{c_{O_2}^*} = 0 = \frac{d^2 \bar{c}_{O_2}}{d\bar{z}^2} - \phi^2 \tag{8E1.2}$$

The Thiele modulus is thus $\phi = D\sqrt{k/\mathscr{D}_{O_2} c_{O_2}^*}$.

The zero-order rate constant k depends on the species. For the actively respiring yeast of the previous discussion,

$$k \approx 3 \times 10^{-2} \frac{g\ O_2}{h \cdot g\ dry\ wt} \frac{1\ g\ dry\ wt}{5\ g\ wet} \frac{1\ g\ wet}{1\ cm^3} \frac{1\ mol}{32\ g\ O_2} \frac{1\ h}{3600\ s}$$

$$= \frac{3 \times 10^{-2}\ mol}{(5)(32)(3600)\ s \cdot cm^3} = 5 \times 10^{-8}\ mol\ O_2/(s \cdot cm^3)$$

---

† For brevity we hereafter refer to such relatively unmixed thin regions as *films* even though the picture suggested is not rigorously correct.

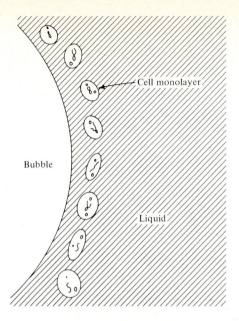

Figure 8E1.1 Schematic diagram of a microbial monolayer near an interface.

Taking $\mathscr{D}_{O_2} = 10^{-6}$ cm²/s (small), $c^*_{O_2} = 1.25 \times 10^{-6}$ mol/cm³ (see Table 8.1 in Sec. 8.1.1) and $D = 3 \times 10^{-4}$ cm gives Thiele modulus

$$\phi = 3 \times 10^{-4} \left[\frac{5 \times 10^{-8}}{(10^{-6})(1.25 \times 10^{-6})} \right]^{1/2} = (3 \times 10^{-4})(4 \times 10^4)^{1/2} \approx 6 \times 10^{-2} \ll 1$$

The assumption that $c \approx c^*_{O_2}$ everywhere within the microbial monolayer is evidently very good since $\phi \ll 1.0$, implying that $c \approx c^*_{O_2}$ everywhere in the film. Since ϕ is proportional to the film thickness L, when L is about 15 times the microbial diameter, $\phi \approx 1.0$ and the concentration of O_2 begins to differ appreciably from $c^*_{O_2}$ at $x = L$. When $\phi \gg 1.0$, the bulk O_2 concentration is quite small, but some of the microbial monolayer will still have c near $c^*_{O_2}$.

Note that in our earlier analysis of coupled diffusion and reaction in Sec. 7.4.2, one of the boundary conditions was $dc/dz = 0$. This applies here at $z = L$ only if no oxygen is utilized in the bulk liquid beyond the cell layer(s). (What is the appropriate boundary condition if O_2 is also consumed in the bulk liquid?)

Similarly, in the microbial utilization of other sparingly soluble substrates such as hydrocarbon droplets, cell adsorption on or near the hydrocarbon-emulsion interface has been frequently observed. A reactor model for this situation is considered in Chap. 9. Finally, the cell may adsorb directly on the interface, without any apparent intervening aqueous phase. Since this situation has been hypothesized but not clearly demonstrated in the literature, it is not considered further here.

The variety of macroscopic physical configurations by which gas-liquid contacting can be effected is indicated in Fig. 8.2. In general, we can distinguish fluid motions induced by freely rising or falling bubbles or particles from fluid motions which occur as the result of applied forces (forced convection) other than the external gravity field. The distinction is not clear-cut; gas-liquid mixing in a slowly

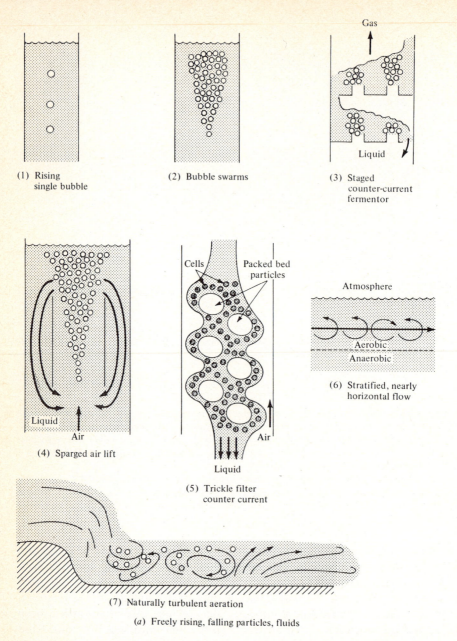

(1) Rising
single bubble

(2) Bubble swarms

(3) Staged
counter-current
fermentor

(4) Sparged air lift

(5) Trickle filter
counter current

(6) Stratified, nearly
horizontal flow

(7) Naturally turbulent aeration

(a) Freely rising, falling particles, fluids

Figure 8.2 Gas-liquid-microbe contacting modes: (a) freely rising, falling particles, fluids, (b) mechanically aerated.

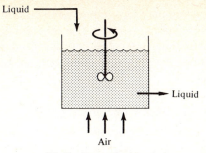

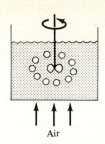

(1) Semi-batch (batch
liquid, continuous
air)

(2) Continuous liquid and air

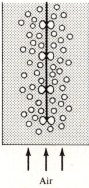

(3) Multiple propeller
(semi-batch or
continuous)

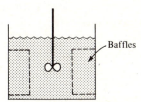

(4) Stirred tank with
baffles (baffles are
often used in designs
(1) and (2) also)

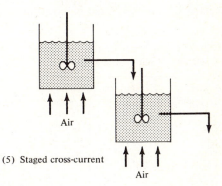

(5) Staged cross-current

(*b*) Mechanically aerated

stirred semibatch system may have equal contributions from naturally convected bubbles and from mechanical stirring. The central importance of hydrodynamics requires us to examine the interplay between fluid motions and mass transfer. Before beginning this survey, some comments and definitions regarding mass transfer are in order.

8.1.1. Basic Mass-Transfer Concepts

The solubility of oxygen in aqueous solutions under 1 atm of *air* and near ambient temperature is of the order of 10 parts per million (ppm) (Table 8.1). An actively respiring yeast population may have an oxygen consumption rate of the order 0.3 g of oxygen per hour per gram of dry cell mass. The peak oxygen consumption for a population density of 10^9 cells per milliliter is estimated by assuming the cells to have volumes of 10^{-10} cm^3, of which 80 percent is water. The absolute oxygen demand becomes

$$\frac{0.3 \text{ g O}_2}{\text{g dry mass·h}} \left(10^9 \frac{\text{cells}}{\text{ml}}\right)(10^{-10} \text{ cm}^3)\left(1 \frac{\text{g cell mass}}{\text{cm}^3}\right)\left(0.2 \frac{\text{g dry cell mass}}{\text{g cell mass}}\right)$$

$$= 6 \times 10^{-3} \text{ g/(ml·h)} = 6 \text{ g O}_2/(\text{l·h})$$

Thus, the actively respiring population consumes oxygen at a rate which is of the order of 750 times the O$_2$ saturation value per hour. Since the inventory of dissolved gas is relatively small, it must be continuously added to the liquid in order

Table 8.1. Solubility of O$_2$ at 1 atm in water at various temperatures and solutions of salt or acid at 25°C†

Temp, °C	Water, O$_2$ mmol/l	Temp, °C	Water, O$_2$ mmol/l
0	2.18	25	1.26
10	1.70	30	1.16
15	1.54	35	1.09
20	1.38	40	1.03

Aqueous solutions at 25°C

Electrolyte conc, M	O$_2$, mmol/l		
	HCl	H$_2$SO$_4$	NaCl
0.0	1.26	1.26	1.26
0.5	1.21	1.21	1.07
1.0	1.16	1.12	0.89
2.0	1.12	1.02	0.71

† Data from "International Critical Tables," Vol. III, p. 271, McGraw-Hill Book Company, New York, 1928, and F. Tödt, "Electrochemische Sauerstoffmessungen," W. de Guy and Co., Berlin, 1958.

to maintain a viable cell population. This is not a trivial task since the low oxygen solubility guarantees that the concentration difference which drives the transfer of oxygen from one zone to another is always very small.

At steady state, the oxygen transfer rate to the gas-liquid interface equals its transfer rate through the liquid-side film (Fig. 8.1). Taking c_g, c_{gi}, c_{li}, and c_l to be the oxygen concentrations in the bulk gas, interfacial gas, interfacial liquid, and bulk liquid respectively, we can write the two equal transfer rates

$$\text{Oxygen flux} = \text{mol } O_2/(\text{cm}^2 \cdot \text{s})$$

$$= k_g(c_g - c_{gi}) \qquad \text{gas side} \tag{8.1}$$

$$= k_l(c_{li} - c_l) \qquad \text{liquid side}$$

where k_g and k_l are the gas-side and liquid-side mass-transfer coefficients, respectively. Note that the mass-transfer coefficients have the dimensions of velocity (centimeters per second).

For sparingly soluble species such as oxygen or hydrocarbons in water, the two interfacial concentrations may typically be related through a linear partition-law relationship such as Henry's law

$$M c_{li} = c_{gi} \tag{8.2}$$

provided that the solute exchange rate across the interface is much larger than the net transfer rate, as is typically the case: at 1 atm of air and 25°C, the O_2 collision rate at the surface is the order of 10^{24} molecules per square centimeter per second, a value greatly in excess of the typical microbial consumption requirements cited above.

Since the interfacial concentrations are usually not accessible in mass-transfer measurements, resort is made to mass-transfer expressions in terms of the overall mass-transfer coefficient K_l and the overall concentration driving force $c_l^* - c_l$, where c_l^* is the liquid-phase concentration which is in equilibrium with the bulk gas phase

$$M c_l^* \equiv c_g \tag{8.3}$$

In terms of these overall quantities, the mass flux is given by

$$\text{Flux} = K_l(c_l^* - c_l) \tag{8.4}$$

Utilization of Eqs. (8.1) through (8.4) results in the following well-known relationship between the overall mass-transfer coefficient K_l and the physical parameters of the two-film transport problem, k_g, k_l, and M:

$$\frac{1}{K_l} = \frac{1}{k_l} + \frac{1}{M k_g} \tag{8.5}$$

For sparingly soluble species, M is much larger than unity. Further, k_g is typically considerably larger than k_l. For example, in completely stagnant circumstances, k_g/k_l is just the ratio of diffusion coefficients in the two phases, which is of the order of 10^4. Under these circumstances we see from Eq. (8.5) that K_l is approximately equal to k_l. That is, essentially all the resistance to mass transfer lies on the liquid-film side. If the bubble rise velocity u and diameter D_b satisfy the inequality $uD_b/\mathscr{D} \gg 1.0$, then k_g/k_l is more like $(\mathscr{D}_g/\mathscr{D}_l)^{1/3} \approx 30$. The conclusion remains unchanged. [The group $uD_b/\mathscr{D}$ is known as the *Peclet* number (see Sec. 8.3.1).]

The oxygen-transfer rate per unit of reactor volume Q_{O_2} is given by

$$Q_{O_2} = \text{oxygen absorption rate} = \frac{(\text{flux})(\text{interfacial area})}{\text{reactor volume}}$$

$$= k_l(c_l^* - c_l)\frac{A}{V}$$

$$= k_l a(c_l^* - c_l) \tag{8.6}$$

where $a = A/V$ is the gas-liquid interfacial area per unit volume and the approximation $K_l \approx k_l$ just discussed has been invoked. Since our major emphasis in this chapter is aeration, we shall concentrate on oxygen transfer and henceforth use k_l in place of K_l as the appropriate mass-transfer coefficient.

It is important to recognize that Q_{O_2} is defined "at a point." It is a local volumetric rate of O_2 consumption; the average volumetric rate of oxygen utilization (moles per time) $\bar{Q}_{O_2}$ in an entire vessel of volume V is given by

$$\bar{Q}_{O_2} = \frac{1}{V}\int_0^V Q_{O_2}\, dV \tag{8.7}$$

In general, $\bar{Q}_{O_2}$ is equal to Q_{O_2} only if hydrodynamic conditions, c_l^*, and c_l are uniform throughout the vessel.

In Table 8.1 we saw that c_l^* is determined by the temperature and composition of the aqueous medium. Often these conditions are beyond our control, or if we can manipulate them, we do so to accommodate the physiological requirements of the cell culture. Consequently, Eq. (8.6) suggests that the main parameters of interest in design for mass transfer are the mass-transfer coefficient k_l and the interfacial area per unit volume a. We shall investigate the variation of each of these parameters with process operating conditions later in this chapter.

8.1.2. Rates of Metabolic Oxygen Utilization

In design of aerobic biological-reactors we frequently use correlations of data more or less approximating the situation of interest to establish whether the slowest process step is the oxygen transfer rate or the rate of microbial utilization of oxygen (or other limiting substrate). The maximum possible mass-transfer rate is simply that found by setting $c_l = 0$: all oxygen entering the bulk solution is assumed to be immediately consumed. The maximum oxygen utilization rate is seen from Chap. 7 to be $x\mu_{max}/Y_{O_2}$, where x is cell density and Y_{O_2} is the ratio of moles of cell carbon formed per mole of oxygen consumed.

Evidently, if $k_l a c_l^*$ is much larger than $x\mu_{max}/Y_{O_2}$, the main resistance to increased oxygen consumption is microbial metabolism and the reaction appears to be *biochemically limited*. Conversely, the reverse inequality apparently leads to c_l near zero, and the reactor seems to be in the *mass-transfer-limited* mode.

The situation is actually slightly more complicated. At steady state, the oxygen absorption and consumption rates must balance:

$$Q_{O_2} = \text{absorption} = \text{consumption}$$

$$k_l a(c_l^* - c_l) = \frac{x\mu}{Y_{O_2}} \tag{8.8}$$

Assuming that the dependence of μ on c_l is known, we can use Eq. (8.8) to evaluate c_l and hence the rate of oxygen utilization.

In general, above some critical bulk oxygen concentration, the cell metabolic machinery is saturated with oxygen. In this case, sufficient oxygen is available to accept immediately all electron pairs which pass through the respiratory chain, so that some other biochemical process within the cell is rate-limiting (Chap. 5). For example, if the oxygen dependence of the specific growth rate μ follows the Monod form, then

$$Y_{O_2} k_l a(c_l^* - c_l) = x\mu_{\max} \frac{c_l}{K_{O_2} + c_l} \tag{8.9}$$

A general solution to an equation of this form was given in Sec. 3.7.2, but here for the sake of illustration we assume that the value of c_l is considerably less than c_l^*. This is not an uncommon situation in biological reactors. Subject to the assumption that $c_l \ll c_l^*$, c_l is easily seen to be

$$c_l = c_l^* \frac{Y_{O_2} K_{O_2} k_l a / x\mu_{\max}}{1 - Y_{O_2} c_l^* k_l a / x\mu_{\max}} \tag{8.10}$$

If the resulting value of c_l is greater than the critical oxygen value c_{cr} (about $3K_{O_2}$), the rate of microbial oxygen utilization is limited by some other factor, e.g., small or low concentration of another substrate, even though the bulk solution has a dissolved oxygen level considerably below the solubility value. The *critical oxygen values* for organisms lie in the range of 0.003 to 0.05 mmol/l (Table 8.2) or of the order of 0.1 to 10 percent of the solubility values in Table 8.1, that is, 0.5 to 50 percent of the *air* saturation values. For the higher values such as obtain for *Penicillium* molds, oxygen mass transfer is evidently extremely important.

Many factors can influence the total microbial oxygen demand $x\mu/Y_{O_2}$, which in turn sets the minimum values of $k_l a$ needed for process design through Eq. (8.8). The more important of these are cell species (Table 8.2), culture growth phase, carbon nutrients, pH, and the nature of the desired microbial process, i.e., substrate utilization, biomass production, or product yield (Chap. 7).

In the batch-system results of Fig. 8.3, the values of oxygen demand and total cell mass vs. time show a maximum in specific O_2 demand in the early exponential phase although x is larger at a later time. A peak in the product $x\mu$ occurs near the end of the exponential phase and the approach to the stationary phase; this is later than the time of achievement of the largest specific growth rate.

The carbon nutrient affects oxygen demand in a major way. For example, glucose is generally metabolized more rapidly than other carbohydrate substances.

Table 8.2. Typical values of $c_{O_2, \text{cr}}$ in the presence of substrate†

Organism	Temp, °C	$c_{O_2, \text{cr}}$, mmol/l
Azotobacter vinelandii	30	0.018–0.049
E. coli	37.8	0.0082
	15	0.0031
Serratia marcescens	31	~0.015
Pseudomonas denitrificans	30	~0.009
Yeast	34.8	0.0046
	20	0.0037
Penicillium chrysogenum	24	~0.022
	30	~0.009
Aspergillus oryzae	30	~0.020

† Summarized by R. K. Finn, p. 81 in N. Blakebrough (ed.), "Biochemical and Biological Engineering Science," vol. 1, Academic Press, Inc., New York, 1967.

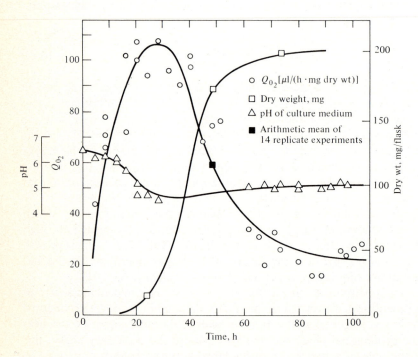

Figure 8.3 Oxygen utilization rate in batch culture of *Myrothecium verrucaria*. (*Reprinted from R. T. Darby and D. R. Goddard, Studies of the Respiration of the Mycelium of the Fungus* Myrothecium verrucaria, *Am. J. Bot.*, **37**: 379, 1950.)

Peak oxygen demands of 4.9, 6.7, and 13.4 mmol/(l·h) have been observed for *Penicillium* mold utilizing lactose, sucrose, and glucose, respectively [2].

The component parts of oxygen utilization by the microbe include cell maintenance, respiratory oxidation for further growth (more biosynthesis), and oxidation of substrates into related metabolic end products. The growth pattern of Fig. 7.30 indicates that glucose disappearance to form gluconic acid proceeds rapidly only when cell growth also proceeds apace, so that the oxygen demand of the *Acetobacter* colony includes the cell-growth price. The subsequent conversion of gluconic acid into 5-ketogluconic acid with resultant stoichiometry

$$C_6H_{12}O_7 + \tfrac{1}{2}O_2 \longrightarrow C_6H_{10}O_7 + H_2O \qquad (8.11)$$

proceeds without substantial cell production, also clearly shown in Fig. 7.30. In this latter phase, the production of Z millimoles of 5-ketogluconic acid per liter-hour would produce an oxygen demand not seriously in excess of the stoichiometric value of $\tfrac{1}{2}Z$ millimoles of O_2 per liter-hour.

A yield coefficient (Chap. 7) may be useful in estimation of the oxygen demand. Let $Y_{O_2/C}$ be the number of moles of O_2 consumed per gram atom of carbon nutrient utilized. In a flow reactor such as Fig. 8.2b, the steady-state carbon consumption is simply the flow rate F (liters per hour) times the substrate concentration difference between inlet and outlet streams $(s_0 - s)$. The overall oxygen demand is then calculable from

$$[V_{\text{reactor}} \text{ (liters)}] \cdot [\text{oxygen demand (mmol } O_2/(l \cdot h))]$$

$$= [F \text{ (l/h)}] \cdot [s_0 - s \text{ (mmol C/l)}] \cdot [Y_{O_2/C} \text{ (mmol } O_2/\text{mmol C)}] \qquad (8.12)$$

For respiring yeast utilizing glucose, $Y_{O_2/C} = 0.4$. The previous example of gluconic acid oxidation would probably give $Y_{O_2/C}$ slightly above 0.5.

More detailed stoichiometries can be formulated by using apparent microbial compositions. With methane as a carbon source, growth experiments using the obligate heterotrophic bacterium IGT-10 [18] can be represented by the overall stoichiometry

$$6.25CH_4 + 7.92O_2 \longrightarrow C_{3.92}H_{6.50}O_{1.92} + 2.33CO_2 + 9.25H_2O \qquad (8.13)$$

where the first term on the right-hand side represents the bacterial mass produced. Here $Y_{O_2/C} = 7.92/6.00 = 1.34$. The value of $Y_{O_2/C}$ was observed to diminish about 20 percent from the methane results for alkanes of carbon number equal to or greater than 4. In general, hydrocarbon substrate fermentations will demand about 2.5 to 3.0 times the oxygen amount consumed per mole of carbon utilized in carbohydrate fermentations. Moo-Young [45] suggests the following general stoichiometries for carbohydrate vs. hydrocarbon fermentations to produce yeasts:

Hydrocarbon:

$$2CH_2 + 0.19NH_3 + 2O_2 \longrightarrow CH_{1.7}O_{0.5}N_{0.19} \text{ (biomass)} + CO_2 + 1.5H_2$$
$$+ 200 \text{ kcal} \qquad (8.14a)$$

Carbohydrate:

$$1.68CH_2O + 0.19NH_3 + 0.68O_2 \longrightarrow CH_{1.7}O_{0.5}N_{0.19} + 0.71CO_2$$
$$+ 1.14H_2O + 80 \text{ kcal} \quad (8.14b)$$

Here $Y_{O_2/C} = 1.0$ (hydrocarbon) vs. $Y_{O_2/C} \approx 0.4$ (carbohydrate); note the linear increase in heat generation with increased $Y_{O_2/C}$.

8.2. DETERMINATION OF OXYGEN TRANSFER RATES

Ideally, oxygen transfer rates should be measured in biological reactors which include the nutrient broth and microbial population(s) of interest. As this requires all the accoutrements for inoculum and medium preparation, prevention of contamination, and environmental control for the microbial culture, it is an inconvenient and troublesome way to conduct mass-transfer experiments. Consequently, a common strategy for study of oxygen transfer rates is to use synthetic systems which approximate fermentation conditions without the complications of a living culture. In such approaches, the major objective is to elucidate the dependence of $k_l a$ on hydrodynamics.

In order for such synthetic media to represent the microbial broth of interest reliably, the following restrictions apply:

1. The *solution viscosity* should remain unchanged, and the rheological behavior of the fermentation liquid should be preserved (see Sec. 8.7).
2. The gas-liquid *interfacial resistance* should remain unchanged.
3. Bubble *coalescense* tendencies should remain the same.
4. Oxygen *diffusivity* and *solubility* should remain unchanged.

In general, the usefulness of a synthetic (nonmicrobial) situation for approximating a fermentation situation depends on the degree to which these conditions are met. Experiments with oxygen absorption into pure water, for example, satisfy few of these criteria. We shall examine the quantitative influence of fluid viscosity, surface-active agents, and nature of mixing shortly. Note also that the microbial oxygen-utilization situations pictured in Fig. 8.1c and d involve the microbial kinetics more intimately with diffusive liquid-side transfer than the examples in Fig. 8.1a and b. These two cases were discussed in Example 8.1 and are considered further in the problems.

8.2.1. Measurement of $k_l a$ Using Gas-Liquid Reactions

Considering now the transport paths in Fig. 8.1a and b, we see that if oxygen is consumed by chemical reaction in the bulk liquid at a sufficiently large rate we will find $c_l \approx 0$. Then the bulk-phase chemical-reaction rate is equal to $k_l a c_l^*$, from which the $k_l a$ value readily follows. A common bulk-phase oxygen sink in many previous mass-transfer studies is the oxidation of sodium sulfite to sulfate in the presence of catalytic metal ions such as Co^{2+}:

$$SO_3^{2-} + \tfrac{1}{2}O_2 \xrightarrow{\text{catalyst}} SO_4^{2-}$$

The kinetics of the rate of oxidation of sulfite solutions to sulfate is complex. The reaction orders for oxygen and sulfite depend on the catalyst used and its concentration, apparently implying a nontrivial series of elementary steps leading to the overall result above. Regardless of the reaction order, the condition sufficient to ensure that the chemical reaction occurs to a negligible extent in the liquid film adhering to each bubble (and thus represents the situations in Fig. 8.1a and b) is a negligible total reaction rate in the film compared with the mass-transfer rate $k_l(c^* - c)$. If ζ denotes the mass-transfer film thickness, this criterion can be restated mathematically as

$$\zeta \times \text{rate}\,|_{\text{film}} \ll k_l(c^* - c) \tag{8.15}$$

The rate in the film will be less than that corresponding to bulk sulfite and saturation oxygen levels, i.e., rate $(c^*, \text{sulfite}_{\text{bulk}})$, and so in terms of these measurable or calculable quantities $(c^*, \text{sulfite}_{\text{bulk}})$ a conservative criterion for negligible film reaction is

$$\zeta \times \text{rate}(c^*, \text{sulfite}_{\text{bulk}}) \ll k_l(c^* - c) \tag{8.16}$$

The "thickness" of the mass-transfer film is given by

$$\zeta = \frac{\mathscr{D}_{O_2}}{k_l} \tag{8.17}$$

Assuming the reaction to be of order α_1 in oxygen and α_2 in sulfite leads to the inequality

$$\frac{\mathscr{D}_{O_2}[k_r(c^*)^{\alpha_1}\text{sulfite}^{\alpha_2}]}{k_l} \ll k_l(c^* - c) \tag{8.18}$$

Thus
$$k_l \gg \sqrt{\frac{\mathscr{D}_{O_2}k_r(c^*)^{\alpha_1}\text{sulfite}^{\alpha_2}}{c^* - c}} \tag{8.19}$$

An illustrative example of Danckwerts' considers an experiment using 10^{-5} M cobalt catalyst (known to give $\alpha_1 = 2$) and sufficient sulfite for α_2 to be 0 (say 0.5 M). For $c \ll c^*$, the above inequality becomes

$$k_l \gg \sqrt{\mathscr{D}_{O_2}k_r c^*}$$

Taking $\mathscr{D}_{O_2} = 1.6 \times 10^{-5}$ cm^2/s, k_r for cobalt catalyst $= 0.85 \times 10^8$ cm^3/(g mol·s), $c^* = 1.35 \times 10^{-7}$ g mol/cm^3 gives

$$k_l \gg 0.01 \text{ cm/s}$$

A less effective catalyst (smaller k_r) would reduce the right-hand side.

Reference to the correlations in this section indicates that this inequality places a *minimum* size on the bubbles which may be used for such an interpretation. Smaller bubbles rising more slowly will have an appreciable reaction rate in the adhering fluid film under these specific conditions. Similarly, larger bubbles in more viscous media than aqueous solutions will exhibit reduced mass-transfer coefficients. However, an enhancement factor E to account for film reaction can be calculated *provided* the reaction-rate constant and reaction order are known [4,5].

A closing series of caveats for sulfite oxidation is illuminating: the rate constant k_r depends on (1) the catalyst and its concentration, (2) the ionic strength of the solution, (3) the presence of catalytic impurities, and (4) the solution pH; for example, in 10^{-5} M cobalt, k_r increases by a factor of 10 between pH 7.50 and 8.50 at 20°C. (The overall reaction generates H^+, so that base must be added to maintain pH constant.)

In spite of these difficulties, the literature contains examples of reactors where $k_l a$ determined from sulfite measurements for a given sparger, stirring rate, etc., correlate closely with the $k_l a$ values observed in an actual fermentation (counter-examples are also evident). Assuming that the configurations in Fig. 8.1a and b represent the fermentation of interest, there is essentially no O_2 consumption in the bubble liquid-side film. Thus, any chemical measure of O_2 absorption attempting to simulate such cell broths must, inter alia, satisfy the fundamental inequality (8.15) above.

8.2.2. Applications of Dissolved-Oxygen Measurements

We can measure oxygen transfer rates in several ways. As noted above, the average volumetric oxygen utilization rate $\bar{Q}_{O_2}$ can be measured chemically in some cases by observing the appearance of an oxidation product. If the experimental system is strictly a batch operation, with no addition or removal of liquid or gas, $\bar{Q}_{O_2}$ is revealed by monitoring the gas volume or pressure changes with time.

Example 8.2. Warburg respirometer An early device for monitoring oxygen consumption is a constant-volume respirometer which Warburg in 1926 devised from earlier similar manometric devices of Barcroft and Haldane (1902) and Brodie (1910). A small volume flask is attached to a U-tube manometer as shown in Fig. 8E2.1. The flask has space for the sample solution, a small open cylindrical reservoir for an alkaline CO_2 absorbent, and a side arm used later to add another component to the initial solution. During a measurement, the manometer fluid in the closed manometer leg rises due to the oxygen consumption of the sample. Periodic changing of the open manometer-tube position restores the closed-leg meniscus to its original level. The resulting height difference between the manometer legs provides a measure proportional to the pressure change of the constant-volume system. The alkali, e.g., KOH, in the separate reservoir communicates only with the gas phase; it rapidly absorbs all the CO_2 liberated from the respiring solution.

The total moles of gas at the start of an experiment are the gas-phase and liquid-phase contributions

$$n_{\text{tot}}(t=0) = \frac{(p - p_{H_2O})V_g}{RT} + \frac{(p - p_{H_2O})MV_f}{RT} \qquad (8E2.1)$$

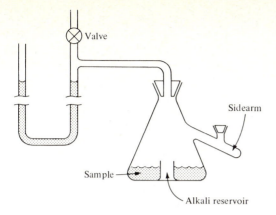

Figure 8E2.1 Schematic diagram of Warburg respirometer.

where p = total pressure over solution

p_{H_2O} = partial pressure of water vapor

V_g = volume of gas

V_f = volume of sample fluid

M = Henry's law constant

At a later time, the gas consumed is reflected in the pressure difference across the manometer leg h in atmospheres, the same units as p and p_{H_2O}. If the liquid solution is still saturated with O_2 at the new oxygen-pressure condition, the total number of moles of gas at this time is

$$n_{tot}(t) = \frac{(p - p_{H_2O} - h)V_g}{RT} + \frac{(p - p_{H_2O} - h)MV_f}{RT} \tag{8E2.2}$$

The net uptake rate $\Delta n_{tot}/t_1$ is thus

$$\frac{\Delta n_{tot}}{t_1}\left(\frac{mol\ O_2}{min}\right) = \frac{h}{t_1}\left(\frac{V_g + V_f M}{RT}\right) = b\frac{h}{t_1} \tag{8E2.3}$$

The term MV_f is typically the order of 1 percent of V_g in a small experimental flask; the term b is the so-called *flask constant*.

The possible oxygen-uptake limitation due to mass transfer of O_2 through the liquid sample is evident; it has been realized for some time that the oxygen uptake rate $\Delta n/t_1$ reflects cell kinetics only if the uptake rate is independent of liquid agitation rate as reflected, for example, in the shaking frequency of the Warburg flasks. As seen clearly in the yeast-uptake example of Fig. 8E2.2a, the minimum shaking rate needed to eliminate gas-liquid mass-transfer resistance increases with the sample mass (at constant V_e). The resulting boundary between appreciable and negligible zones of mass-transfer resistance is plotted in Fig. 8E2.2b. The slope of the line common to all experiments in Fig. 8E2.2a represents milliliters of oxygen per hour per milligram of wet yeast, the specific oxygen demand on a wet cell basis Q_{O_2}, provided that the shaking gives a uniform cellular dispersion and oxygen level in the bulk, i.e., nonsurface liquid sample. Note in all cases that appreciable mass-transfer influence *diminishes* the overall uptake rate for a given mass of cells.

The mechanical shaking stirs the liquid constantly, bringing fresh fluid of lower oxygen content up to the gas-surface interface. Enhanced shaking increases this liquid recirculation, or renewal, at the phase interface. Many theories of mass transfer are developed from an assumed rate of surface renewal due to various scales of liquid motion near the surface. A specific example dealing with mass transfer into stream surfaces appears in Sec. 8.4.

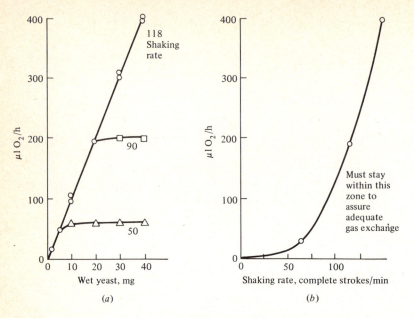

Figure 8E2.2 Total oxygen uptake vs. (*a*) biomass or (*b*) shaking rate in 15 ml capacity flasks. Each diagram shows regions of mass-transfer-influenced uptake and kinetically limited oxygen uptake. Section *a*: Each flask has amount of yeast indicated on abscissa made to 2 ml with 0.02 *M* KH_2PH_4(pH 4.8), and 1 ml 3% glucose; 0.2 ml 20% KOH in center well. Shaken at 50, 90, and 118 complete 2-cm strokes per minute at 28°C. Section *b*: Data from Section *a* plotted to show adequate shaking rate for conditions described. (*Reprinted from W. W. Umbreit, R. H. Burris, and J. F. Stauffer, "Manometric Biochemical Techniques," 5th ed., p. 13, Burgess Publishing Co., Minneapolis, Minnesota, 1972.*)

Agitation, by shaking or stirring, may also change the nature of the fermentation. Bacterial oxidation of sulfur particles appears to require cell-particle contact; sufficient agitation separates the sulfur from the cells, giving rise to an overall drop in oxygen consumption rate.

Similarly, in the fermentation of hydrocarbon emulsions by yeasts, another fermentation involving a particulate nutrient, the cells and emulsion adsorb on each other. It is possible that the violent agitation needed to maintain good aeration, mixing, and emulsion dispersion also dissociates some cells from the substrate droplets. This fermentation is reconsidered later in this chapter, and in Chaps. 9 and 10.

Similar manometer experiments can be used to determine individual aspects of oxygen demand. For example, two flasks containing identical *E. coli* cell mass and medium without carbon substrate each produce the lower-oxygen-uptake line of Fig. 8E2.3, indicative of a *maintenance* oxygen demand (Chap. 7). Subsequent addition of the substrate pyruvate to one flask by use of the flask sidearm produces an increased oxygen uptake rate. Assuming that the maintenance rate is independent of the substrate presence (as was done for some cell growth models in Chap. 7), the *increase* in slope corresponds to a stoichiometric oxygen demand for substrate conversion, provided that the cells were previously adapted to that substrate. (Why is this qualification necessary?) The *difference* in slopes in curves 1 and 2 represents 112 μl O_2 ($\equiv 0.005$ m*M* O_2) per 0.01 m*M* pyruvate added or 0.5 *M* O_2 per mole of pyruvate. From these data, the total oxygen demand (= substrate conversion + maintenance) of a culture can be predicted for other related culture conditions of substrate conversion rate and cell concentrations.

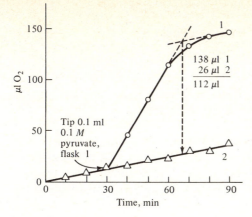

$$\frac{\begin{array}{l}138\ \mu l\quad 1\\26\ \mu l\quad 2\end{array}}{112\ \mu l}$$

Tip 0.1 ml
0.1 M
pyruvate,
flask 1

Figure 8E2.3 Use of Warburg device to determine maintenance O_2 demand and the oxygen consumed per substrate molecule converted. Curve 2: *E. coli*, no substrate present, Curve 1: *E. coli*, pyruvate added at 30 min. is consumed at about 67 minutes. (*Reprinted from W. W. Umbreit, R. H. Burris, and J. F. Stauffer, "Manometric Biochemical Techniques," 5th ed., p. 18, Burgess Publishing Co., Minneapolis, Minnesota, 1972.*)

When gas is continuously added to and removed from the liquid, we use the following O_2 mass balance on the gas phase to determine $\bar{Q}_{O_2}$:

$$\bar{Q}_{O_2} = \frac{RT}{V}[F_{g_{\text{inlet}}}(p_{O_2})_{\text{inlet}} - F_{g_{\text{exit}}}(p_{O_2})_{\text{exit}}] \tag{8.20}$$

Here F_g is the volumetric gas flow rate and p_{O_2} is the partial pressure of O_2. [What assumptions are necessary to justify Eq. (8.20)? Are they generally valid for fermentation processes?]

We would like to use these $\bar{Q}_{O_2}$ values to determine $k_l a$, but, as Eqs. (8.6) and (8.7) and the associated discussion reveal, this requires uniformity of conditions within the vessel, so that the local and average oxygen utilization rates are identical. Consequently, stirred vessels [Fig. 8.2b(ii)] are frequently employed in laboratory mass-transfer studies for biological-reactor design.

When the problem of spatial uniformity has been resolved, $k_l a$ can be extracted from Eq. (8.6) if c_l^* and c_l are known. The first of these is available from solubility data such as Table 8.1, and direct measurement of c_l is now feasible (even in pure microbial cultures) with the polarographic sterilizable electrode.

At the electrode surface, oxygen is reduced according to the overall stoichiometry

$$O_2 + 4H^+ + 4e^- \longrightarrow 2H_2O$$

which may be taken as the sum of the elementary steps

$$O_2 + 2H^+ + 2e^- \longrightarrow H_2O_2$$

and

$$H_2O_2 + 2H^+ + 2e^- \longrightarrow 2H_2O$$

Under neutral or acid conditions and a sufficiently cathodic applied voltage to render the previous two equations irreversible, the current i generated at a bare-metal electrode surface satisfies

$$\frac{i}{2F} = \bar{\mathscr{D}}_{O_2} \frac{\bar{c}_{O_2}(\text{bulk}) f(\text{voltage})}{\zeta} \tag{8.21}$$

where F is Faraday's constant, ζ is the diffusion-layer thickness, and f depends only on the applied voltage. Under steady-state conditions, the current signal is proportional to the local bulk dissolved-oxygen concentration.

The commercial electrode also has a fixed thin oxygen-permeable polymer membrane, which itself has a mass-transfer resistance and thus a characteristic response time to changes of dissolved oxygen level, which may be of the order of 10 s to one min. While the electrodes do not respond instantaneously to oxygen-level variations, a transient method for determination of $k_l a$ is available using an O_2 electrode, as seen in Example 8.3.

Example 8.3. Electrochemical determination of $k_l a$ An oxygen electrode is inserted into a steadily aerated batch or continuous-flow reactor (Fig. 8.2). After a steady electrode response has been achieved, the oxygen-carrying flow is suddenly replaced by an equivalent nitrogen flow, which then strips the oxygen out of solution in a transient manner. Under these circumstances, the voltage response of the probe is given by

$$E(t) = E_0 \left[\frac{\tau^{1/2} \exp(-\beta t)}{\sin \tau^{1/2}} - 2 \sum_{n=1}^{\infty} \frac{(-1)^n \exp(-n^2 \pi^2 \mathscr{D}_{O_2} t/L^2)}{1 - n^2 \pi^2 / \tau} \right] \tag{8E3.1}$$

where $\tau = \beta L^2 / \mathscr{D}_{O_2}$

$\quad E(t) = $ probe voltage of time t

$\quad E_0 = $ probe voltage at time 0

$\quad F_g = $ gas flow rate

$\quad V = $ liquid volume in tank

$\quad M = $ Henry's law constant

$\quad L = $ thickness of probe membrane

$\quad \mathscr{D}_{O_2} = $ oxygen diffusivity in probe membrane $= P_m/L$

$\quad\quad = $ permeability/membrane thickness

and

$$\frac{1}{\beta} = \text{time constant of system (reactor)} = \frac{1}{k_l a} + \frac{VRT}{F_g M} \tag{8E3.2}$$

If the system has a large time constant (small $k_l a$), the value of β is determined from electrode-voltage values for times much greater than the electrode response itself, in which case the series contribution above is negligible. For larger values of $k_l a$ (50 to 500 h^{-1}), Wernan and Wilke [16] suggest using the slope of the electrode response at the inflection point (where the second derivative of E with respect to t vanishes). This method was found to give results as accurate as computer-aided curve fitting using the entire voltage-time response. For various electrode relaxation times, the value of the inverse relaxation time is read from a graph of this parameter vs. the slope of E vs. t at the inflection point.

Evaluation of the desired value, $k_l a$, then proceeds in three steps:

1. From the E-vs.-t electrode-response curve (Fig. 8E3.1) evaluate the slope at the inflection point (steepest slope).
2. From the generalized display of Eq. (8E3.2) in Fig. 8E3.2, obtain β, given the slope and the electrode time constant.
3. Evaluate $k_l a$ from the definition of the system time constant $1/\beta$ above.

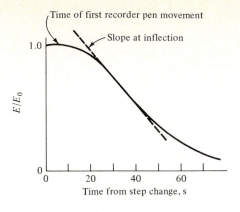

Figure 8E3.1 The inflection-point slope is determined graphically from the electrode response following a switch from sparging of an oxygen-carrying gas to nitrogen sparging. (*Reprinted from W. C. Wernan and C. R. Wilke, New Method for Evaluation of Dissolved Oxygen Response for $k_L a$ Determination, Biotech. Bioeng.*, **15**: 571, 1973.)

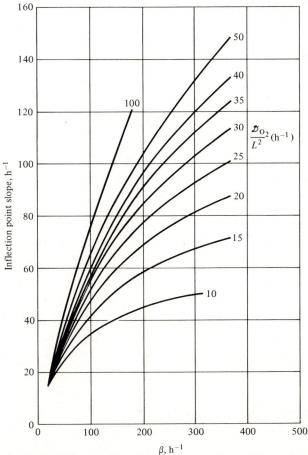

Figure 8E3.2 From this graph, the value of β may be determined from the inflection-point slope and membrane-transport parameters. $k_l a$ then follows from Eq. (8E3.2). (*Reprinted from W. C. Wernan and C. R. Wilke, New Method for Evaluation of Dissolved Oxygen Response for $k_L a$ Determination, Biotech. Bioeng.*, **15**: 571, 1973.)

It is evident that the usefulness of this technique depends on the accuracy with which the electrode response is known, particularly for large $k_l a$ values.

The presence of high solute concentrations may alter oxygen solubility (Table 8.1), the oxygen diffusion coefficient, and the electrode surface itself (thus the electrode response). A complete dissection of these influences clearly requires calibration by independent means.

In many reactor configurations or processes of natural origin, the local oxygen transfer rate varies with position. If such variations occur in the vessel in which mass-transfer rates are measured, the observed vessel-averaged value of $k_l a$ cannot properly be used effectively in *scale-up*, the process of transferring laboratory scale results into larger dimension units. Design methods for such scale-up are considered later in this chapter. Variations of dissolved O_2 within a "homogeneous" phase (bulk fluid, mold pellet, microbial films, etc.) have been examined with miniaturized oxygen probes, the sensing head being of the order of 10 micrometers in diameter. Applications of this instrument include study of local oxygen profiles in mold pellets (recall Sec. 7.4) and the determination of diffusion coefficients in microbial aggregates.

8.3. MASS TRANSFER FOR FREELY RISING OR FALLING BODIES

The rate of material exchange between different regions is governed by the equations of change which describe conservation of mass, conservation of species (such as oxygen), and the momentum balance. When the equations of change are rendered dimensionless in distance, velocity, and concentrations for situations where the density difference between the two contacting phases provides the major driving force for fluid motion, three dimensionless parameters appear in the final expressions. These are the *Grashof, Sherwood*, and *Schmidt numbers*, which, for mass transfer, are defined by

$$\text{Grashof number} = \text{Gr} = \frac{D^3 \rho_g (\rho_l - \rho_g)}{\mu_c^{\,2}} \tag{8.22a}$$

$$\text{Sherwood number} = \text{Sh} = \frac{k_l D}{\mathscr{D}_{O_2}} \tag{8.22b}$$

$$\text{Schmidt number} = \text{Sc} = \frac{\mu_c}{\rho \mathscr{D}_{O_2}} \tag{8.22c}$$

where D is a characteristic dimension. Consequently, we expect mass-transfer-coefficient correlations for such convective motion to involve only these three groups.

8.3.1. Mass-Transfer Coefficients for Bubbles and Bubble Swarms

The mass-transfer coefficient from a bubble, for example, is the proportionality constant between the total bubble flux and the overall driving force, $c_i^* - c_l$. The local flux at the gas-liquid surface is $-\mathscr{D}_{O_2}(\partial c/\partial z)_{z=0}$ (valid for low mass-transfer rates), where z is the coordinate measured into the liquid phase. Thus,

$$k_l = \frac{-1}{c_i^* - c_l}\,\mathscr{D}_{O_2}\,\frac{\partial c}{\partial z}\bigg|_{z=0} \tag{8.23}$$

or, nondimensionally,

$$\text{Sh} \equiv \frac{k_l}{\mathscr{D}_{O_2}}\,D = \frac{-1}{1-\bar{c}_l}\,\frac{\partial \bar{c}}{\partial \bar{z}}\bigg|_{\bar{z}=0} \tag{8.24}$$

Near the gas-liquid interface, the dimensionless concentration $\bar{c}$ has a solution from the transport equations of the form

$$\bar{c} = f(\bar{z},\, \text{Sh},\, \text{Sc},\, \text{Gr}) \tag{8.25}$$

Using this expression to evaluate the derivative $(\partial \bar{c}/\partial \bar{z})_{\bar{z}=0}$ in Eq. (8.24), which is possible in principle, leaves the desired mass-transfer coefficient k_l in the form of the Sherwood number

$$\text{Sh} = \frac{k_l D}{\mathscr{D}_{O_2}} = g(\text{Sc},\, \text{Gr}) \tag{8.26}$$

Thus, the dimensionless mass-transfer coefficient Sh is a function of only the two parameters Sc and Gr.

Correlations for mass-transfer coefficients for falling or rising bubbles, droplets, or solids, have appeared in the literature using other dimensionless groups such as the Reynolds number ($\text{Re} = \rho Du/\mu_c$) or the Peclet number ($\text{Pe} = uD/\mathscr{D}_{O_2}$). In both instances, when an expression for the characteristic velocity u in terms of the density difference $\rho_l - \rho_g$ is substituted, the final result depends only on Gr and Sc according to Eq. (8.26).

For mass transfer from an isolated sphere with a rigid interface, a reasonable approximation to small bubbles in a fermentation broth containing surface-active agents, is predicted theoretically to satisfy Eq. (8.27) provided $\text{Re} = \rho Du/\mu_c \ll 1$ and $\text{Pe} \equiv uD/\mathscr{D}_{O_2} \gg 1$. (Thus $uD/\mathscr{D}_{O_2} \gg 1 \gg \rho Du/\mu_c$, which implies that $\mu c/\rho\mathscr{D}_{O_2} \equiv \text{Sc} \gg 1$. Is the converse true?) In aqueous liquids, since the kinematic viscosity $\nu = \mu/\rho$ is about 10^{-2} cm²/s and $\mathscr{D}_{O_2}$ is of the order of 10^{-5} cm²/s, the Schmidt number is typically of the order of 10^3, justifying use of the equation

$$\text{Sh} = 1.01\,\text{Pe}^{1/3} = 1.01(uD/\mathscr{D}_{O_2})^{1/3} \tag{8.27}$$

For small Reynolds numbers for which this prediction applies, the terminal velocity u_t of a sphere is given by

$$u_t = \frac{D^2 \, \Delta\rho \, g}{18\mu_c} \tag{8.28}$$

Replacing u in Eq. (8.27) with u_t from Eq. (8.28) gives

$$\text{Sh} = 1.01 \left(\frac{D^3 \, \Delta\rho \, g}{18\mu_c \mathscr{D}_{O_2}}\right)^{1/3} = 1.01 \left(\frac{D^3 \rho \, \Delta\rho \, g}{18\mu_c^2}\right)^{1/3} \left(\frac{\mu_c}{\rho\mathscr{D}_{O_2}}\right)^{1/3} = 0.39 \, \text{Gr}^{1/3}\text{Sc}^{1/3} \tag{8.29}$$

[The grouping $(D^3 \, \Delta\rho \, g)/\mu_c \mathscr{D}_{O_2}$ is also known as the *Rayleigh number* Ra.] Notice here that $\text{Sh} = f(\text{Gr}, \text{Sc})$, as expected.

For a larger Reynolds number, the single-bubble result for a noncirculating sphere in laminar flow is

$$\text{Sh} = 2.0 + 0.60 \, \text{Re}^{1/2}\text{Sc}^{1/3} \qquad \text{Re} \gg 1 \tag{8.30}$$

Note that Sh varies as the square root rather than the cube root of the velocity, indicating that a different hydrodynamic regime is present. Again, replacement of u by an appropriate terminal velocity expression will yield $\text{Sh} = f(\text{Gr}, \text{Sc})$.

In many industrial air-sparged reactors (Fig. 8.2 configurations), air bubbles are produced in swarms or clusters of sufficient intimacy that single isolated-bubble hydrodynamics and mass-transfer results fail to describe fluid motion and mass transport accurately in the vicinity of the gas-liquid interface. Calderbank and Moo-Young [1] report that two correlations are sufficient to describe their data for absorption of sparingly soluble gases into liquids which consume the gas chemically. Two distinct regimes of bubble-swarm mass-transfer are evident, the division between them being indicated by a critical bubble diameter D_c. In the absence of surfactants $D_c \approx 2.5$ mm. Bubbles larger than this are typically encountered with pure water in agitated tanks and in sieve-plate columns. Smaller bubbles are frequently found in sintered-plate columns and in agitated vessels containing hydrophilic solutes in aqueous solution.

For $D < D_c = 2.5$ mm,

$$\text{Sh} = \frac{k_l D}{\mathscr{D}_{O_2}} = 0.31 \, \text{Gr}^{1/3}\text{Sc}^{1/3} = 0.31 \, \text{Ra}^{1/3} \tag{8.31}$$

For $D > D_c = 2.5$ mm,

$$\text{Sh} = \frac{k_l D}{\mathscr{D}_{O_2}} = 0.42 \, \text{Gr}^{1/3}\text{Sc}^{1/2} \tag{8.32}$$

Thus, Eqs. (8.29) and (8.31) indicate that in bubble swarms, the mass-transfer coefficient for the same Schmidt and Grashof numbers is reduced about 20 percent compared with the isolated single-bubble case with an immobile surface.

The change of Schmidt number exponent in Eq. (8.32) indicates a changed hydrodynamic regime from Eq. (8.31). For Newtonian fluids, i.e., viscosity = const independent of shearing rate due to stirring speed, bubble velocity, etc., the transition from the $D < D_c$ region to the $D > D_c$ regime is accompanied by a change of bubble shape from nearly spherical (small bubbles) to hemispheric and caplike shapes. For further discussion of bubble hydrodynamics in these swarms, see Ref. 1. The transition value of D varies with surfactant; values as high as 7.0 mm have been reported. In some non-Newtonian fluids, which will be discussed further later, transition with D is much more gradual than the abrupt change observed for Newtonian fluids.

Mass-transfer results for small particles show that as the density difference $\Delta\rho$ diminishes, the Sherwood number approaches 2.0 as a lower limit. For microbial cells, clumps, flocs, etc., as well as for gas oil or other hydrocarbon dispersions, a more accurate form of the Sherwood number is

$$\text{Sh} = \frac{k_l D}{\mathscr{D}_{O_2}} = 2.0 + 0.31\ \text{Ra}^{1/3} \tag{8.33}$$

or

$$\frac{k_l}{\mathscr{D}_{O_2}} = \frac{2.0}{D} + 0.31\left(\frac{\Delta\rho\ g}{\mu_c \mathscr{D}_{O_2}}\right)^{1/3} \tag{8.34}$$

Thus the relative importance of the pure-diffusion result ($\Delta\rho \equiv 0$, $k_l = 2.0\mathscr{D}_{O_2}/D$) vs. the buoyancy term diminishes as particle size increases. For an isolated cell, $2\mathscr{D}_{O_2}/D$ is of the order of 10^{-1} cm/s compared with 10^{-2} cm/s for the Raleigh number term; the mass transfer near its surface therefore resembles that for a sphere in a more or less stagnant medium. Larger diameters due to flocs, films, etc., lead to greater relative contributions from the second term as well as introducing the possibility that diffusion-reaction coupling may influence mass transfer within the larger particles.

8.3.2. Estimation of Interfacial Area

Having evaluated k_l from the appropriate previous formulas, we still must determine the interfacial area a per unit volume. The value of a can be estimated from sparger orifice diameter, overall reactor information, or photographic data, among other means.

A force balance for a bubble leaving an orifice of diameter d predicts departure when the buoyant force $(\pi D^3\ \Delta\rho\ g)/6$ equals the restraining force $\pi d\sigma$:

$$\frac{g\ \Delta\rho\ D^3}{\sigma d} = 6 \tag{8.35}$$

If bubble residence time in the reactor is t_b, volumetric flow rate per orifice is F_0, and total number of (equal) orifices is n, then the interfacial area per unit volume a (neglecting coalescence and change of D with hydrostatic head or absorption) is given by

$$a = \frac{1}{\text{volume}}\ nF_0 t_b\ \frac{\pi D^2}{\pi D^3/6} = \frac{nF_0 t_b}{V}\ \frac{6}{D} \tag{8.36}$$

The quantity $n F_0 t_b$ is the total bubble volume in the reactor. The bubble volume per reactor volume is known as the *holdup* H (volume gas per volume reactor). If it is available from other laboratory, plant, or literature correlations (Example 8.4), it is used directly in

$$a = H \frac{6}{D} \tag{8.37}$$

Example 8.4. Holdup correlations

Agitated tank, diameter $\approx$ height† For $\mathrm{Re}_i^{0.7}(N_i D_i/u)^{0.3} < 2 \times 10^4$,

$$H = \sqrt{\frac{u}{u_t}} \sqrt{H} + 0.015 a_0 \qquad \text{and} \qquad a_0 = 1.44 \frac{P^{0.4} \rho^{0.2}}{\sigma^{0.6}} \left(\frac{u}{u_t}\right)^{1/2}$$

For $\mathrm{Re}_i^{0.7}(N_i D_i/u)^{0.3} > 2 \times 10^4$,

$$H = \frac{a_1}{a_0} \sqrt{\frac{u}{u_t}} \sqrt{H} + 0.015 a_1$$

and

$$\log \frac{2.3 a_1}{a_0} = 1.95 \times 10^{-5} \, \mathrm{Re}_i^{0.7} \left(\frac{N_i D_i}{u}\right)^{0.3}$$

where a_0, a_1 = interfacial area per unit volume of froth
  Re_i = impeller Reynolds number = $\rho N_i D_i^2/\mu_c$
  u = superficial gas velocity (empty-tank basis)
  u_t = bubble rise velocity

and N_i, D_i, P, ρ, σ are as in the text (see Sec. 8.5).

Agitated vessels‡ For air in water (also claimed to reasonably approximate the correlation in Example 8.4), Richards' data can be represented by

$$\left(\frac{P}{V}\right)^{0.4} u^{1/2} = 7.63 H + 2.37$$

where P = horsepower (hp)
  V = ungassed liquid volume, m³
  u = superficial velocity, m/h
  H = volume void fraction (valid for $0.02 < H < 0.2$)

Bubble columns See Fig. 8E4.1 (curve parameter is column diameter).

Laboratory-scale gas-lift column§ Holdup interior to draft tube (sparged):

$$H_1 = \left[(\mu_c - \mu_{\mathrm{H_2O}})^{2.75} + 1.61 \frac{73.3 - \sigma}{74.1 - \sigma} \right] 10^{-4} u^{0.88}$$

holdup in annulus:

$$H_2 = 1.23 \times 10^{-2} \frac{74.2 - \sigma}{79.3 - \sigma} \mu_c^{0.45} \left(\frac{A_{\mathrm{int}}}{A_{\mathrm{ann}}}\right)^{1.08} u^{1.38}$$

† P. H. Calderbank, *Trans. Inst. Chem. Eng.*, **36**: 443, 1958.
‡ J. W. Richards, *Prog. Ind. Microbiol.*, **3**: 143, 1961.
§ M. Chakravarty, S. Begum, H. D. Singh, J. N. Baruah, and M. S. Iyengar, *Biotech. Bioeng. Symp.* 4, p. 363, 1973.

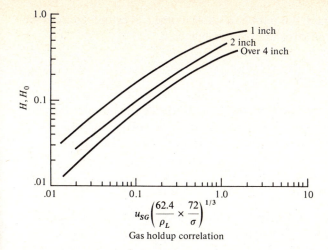

Gas holdup correlation

Figure 8E4.1 Gas holdup correlation. u_{SL} = liquid superficial velocity, ft/s; u_{SG} = gas superficial velocity, ft/s; ρ_L = liquid density, lb/ft^3; σ = surface tension, dyn/cm; $m_s = \dfrac{u_{SG}}{H_0} - \dfrac{u_{SL}}{1 - H_0}$; H_0 = gas fractional holdup at *zero* liquid velocity; and H = gas fractional holdup corrected for liquid flow, if any, $= u_{SG}/m_s$.

Holdup above baffle: $H_3 = 7.5 \times 10^{-3} u^{0.88}$
Total column holdup: $H = 0.03 u^{0.88}$

where μ_c = liquid viscosity at column temperature
$\quad \mu_{H_2O}$ = water viscosity at column temperature
$\quad \sigma$ = gas-liquid surface tension
$\quad u$ = superficial gas velocity
$\quad A_{int}$ = cross-sectional area of draft tube
$\quad A_{ann}$ = cross-sectional area of annulus

   Laboratory-scale gas-lift column† In draft tube:

$$H_d = \frac{u_d}{1.065 u_d + u_\gamma} = \text{void fraction}$$

where u_d is the superficial gas velocity in draft tube and, using volumetric flows,

$$\gamma = \frac{\text{gas flow rate}}{\text{gas + liquid flow rate}}$$

$$u_\gamma = \begin{cases} 32 \text{ cm/s} & \gamma < 0.43‡ \\ 257(\gamma - 0.43) + 32 & \gamma > 0.43‡ \end{cases}$$

 † R. T. Hatch, Ph.D. Thesis in Food Science and Nutrition, p. 150, Massachusetts Institute of Technology, Cambridge, Mass., 1973.
 ‡ $\gamma \approx 0.43$ is the onset of liquid recirculation within the draft tube. γ is defined in terms of gas and net liquid flow rate within the draft tube.

In the absence of such data, the bubble residence time t_b is estimated from the bubble rise velocity integrated over the reactor height h_r

$$t_b = \int_0^{h_r} \frac{dz}{u_b(z)} \approx \frac{h_r}{u_t} \tag{8.38}$$

where in the approximate expression on the right-hand side the bubble rise velocity has been taken to be the bubble terminal velocity. For isolated bubbles, use of the earlier terminal velocity at small Reynolds numbers [Eq. (8.28)] gives

$$t_b = 18 \frac{h_r \mu_c}{D^2 \, \Delta\rho \, g} \tag{8.39}$$

In the case of large, spherical-cap shaped bubbles in Newtonian fluids, the terminal rise velocity to be used in these calculations is

$$u_t = 0.711(gD)^{1/2} = 22.26 D^{1/2} \qquad \text{cm/s} \tag{8.40}$$

For bubble clouds or swarms, calculation of the characteristic bubble rise velocity is more difficult, since neighboring bubbles influence each other's motion and since bubble coalescence and breakup may occur. As a crude approximation, comparison of Eqs. (8.29) and (8.32) suggests that for identical Sc and Sh,

$$\left[\frac{u_t \text{ (bubble cloud)}}{u_t \text{ (single bubble)}} \right]^{1/3} = \frac{0.31}{0.39}$$

so that

$$u_t \text{ (bubble cloud)} \approx 0.50 u_t \text{ (single bubble)} \tag{8.41}$$

For the usual case of dispersions containing a distribution of bubble sizes, the value of D for Eqs. (8.36) to (8.40) is the surface-averaged, or Sauter, mean bubble diameter D_{Sauter}:

$$D_{\text{Sauter}} = \frac{\Sigma m_j D_j^3}{\Sigma m_j D_j^2} \tag{8.42}$$

where m_j is the number of bubbles of diameter D_j.

If photographic studies of the system of interest are available, a transparent triangular grid (Fig. 8.4) can be placed over the photograph and used to determine D_{Sauter}, H (holdup), and a. Let

h = number of hits = termination of a line segment of length L
(Fig. 8.4) in bubble (L = distance in *photograph*)
c = cuts = complete passage of line segment
through bubble
n = number of lines in triangular grid

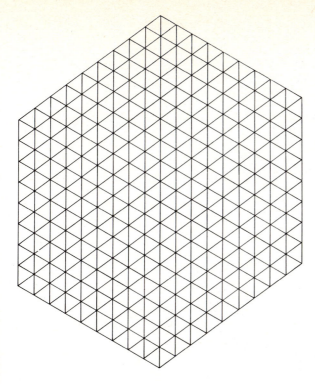

Figure 8.4 Triangular grid for particle size measurements from photographs.

Then

$$D_{\text{Sauter}} = \frac{3Lh}{2c} \quad \text{distance} \tag{8.43a}$$

$$H \text{ (dimensionless)} = \frac{h}{2n} \tag{8.43b}$$

$$a = \frac{2c}{nL} \quad \text{distance}^{-1} \tag{8.43c}$$

8.4. MASS TRANSFER ACROSS FREE SURFACES

Gas transfer through gas-liquid free surfaces (Fig. 8.5) plays a major role in stream reaeration and respiration of aerobic life near the sea surface and in lake communities. Free-surface mass transfer is also important in many industrial microbial processes employing trickle reactors, e.g., wine-vinegar manufacture and waste-water treatment. In the former cases, the depth of oxygen transfer depends on the scale of eddy motions near the liquid surface. Mass transfer into or out of falling-liquid films has been studied frequently, though not often under

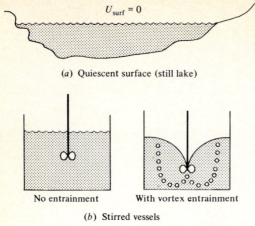

(a) Quiescent surface (still lake)

No entrainment   With vortex entrainment

(b) Stirred vessels

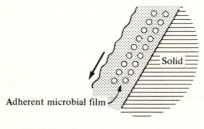

(c) Falling film

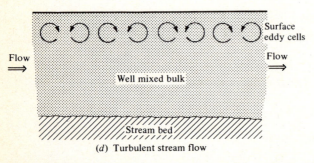

(d) Turbulent stream flow

Figure 8.5 Free-surface aeration configurations: (a) quiescent surface, (b) stirred vessels, (c) falling film, (d) turbulent stream flow.

conditions appropriate to microbial processes. This circumstance is considered first.

The area-integrated absorption rate for a falling laminar liquid film of thickness h, length L, and width W and with zero initial concentration of dissolved gas is given by

$$\text{Integrated absorption rate (moles/unit time)} = WLc_i^* \left(\frac{4 \mathscr{D}_{O_2} u_{max}}{\pi L} \right)^{1/2} \quad (8.44)$$

where $u_{\max}$ is the free-surface velocity. In the derivation, it is assumed that the solute concentration near the solid boundary never departs from zero; i.e., the diffusing solute does not "penetrate" the entire film thickness during the falling-time interval [6].

We define the Reynolds number for the situation as

$$\mathrm{Re} = \frac{u_{\max} R_h \rho}{\mu_c} \tag{8.45}$$

where the hydraulic radius R_h is used as the length scale,

$$R_h = \frac{Wh}{2W + 2h} \approx \frac{h}{2} \qquad \text{if } h \ll W \tag{8.46}$$

From the definition for the mass-transfer coefficient,

$$\text{Integrated absorption rate} = k_l(c_l^* - c_l)WL \tag{8.47}$$

For c_l approximately zero relative to c_l^*, Eqs. (8.44) and (8.47) can be rewritten in the form $\mathrm{Sh} = f(\mathrm{Sc}, \mathrm{Re})$:

$$\mathrm{Sh} = \frac{k_l h}{\mathscr{D}_{O_2}} = 2b(\mathrm{Sc}^{1/2}\mathrm{Re}^{1/2}) \tag{8.48}$$

where h is the length scale for the Sherwood number and $b = (L/h)^{1/2}$. Thus, Sh varies as $\mathrm{Re}^{1/2}$.

Livansky et al. [17] studied CO_2 absorption into aqueous films moving down a slope of known area using water, algal suspensions, and nutrient medium as absorbing fluids. The value of k_l at $\mathrm{Re} = 7$ to 8×10^3 was the same for all three fluids; only the algal suspensions were studied at different Reynolds numbers. (These films may have been turbulent.) Their results can be described by

$$k_l = 4 \times 10^{-5}\,\mathrm{Re}^{2/3} \qquad \text{for } 2000 \le \mathrm{Re} \le 8000 \tag{8.49}$$

In turbulent flowing streams, the scale of circulation is important since this scale determines the depth to which fluid carries fresh, nearly saturated liquid from the surface into the bulk liquid. If we imagine a circulating eddy of length and depth Λ, as shown in Fig. 8.6, the average Sherwood number for mass transfer under turbulent conditions can be defined analogously to Eq. (8.24) as

$$\overline{\mathrm{Sh}} = \frac{\bar{k}_l \Lambda}{\mathscr{D}_{O_2}} = \int_0^\Lambda \left(\frac{\partial \bar{c}}{\partial \bar{z}}\right)_{\bar{z}=0} d\bar{w} \tag{8.50}$$

where $\bar{z}$ and $\bar{w}$ are dimensionless coordinates scaled by Λ, and $\bar{k}_l$ is the *average* value over the eddy length.

The rate of mass transfer is ultimately dependent on $\mathscr{D}_{O_2}$ locally and on the rate at which fluid near the surface is *renewed* by the circulation pattern. An early

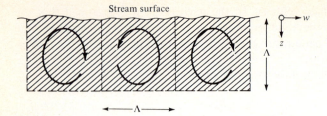

Stream surface

Figure 8.6 Sketch of circulating eddies near a free liquid surface.

derivation due to Higbie† (1935) argued that for fluid elements with identical residence times τ at the gas-liquid surface the mass-transfer coefficient $\bar{k}_l$ should be

$$\bar{k}_l = \sqrt{\frac{\mathscr{D}_{O_2}}{\pi\tau}} \tag{8.51}$$

This form has been extended by Danckwerts for distributions of surface residence times, the result still giving $\bar{k}_l \propto \sqrt{\mathscr{D}_{O_2}}$. A flowing turbulent stream of average flow velocity $\langle u_w \rangle$ has been suggested to have a renewal time τ equal to the ratio of stream depth h to average velocity $\langle u_w \rangle$:

$$\tau_{\text{stream}} = \frac{h}{\langle u_w \rangle} \tag{8.52}$$

thus predicting

$$k_l = \sqrt{\frac{\mathscr{D}_{O_2}}{\pi}\frac{\langle u_w \rangle}{h}} \tag{8.53}$$

If the stream has width W, the interfacial area a per stream volume is

$$\frac{W(1)}{W(1)(h)} = \frac{1}{h}$$

Thus

$$k_l a = \sqrt{\frac{\mathscr{D}_{O_2}}{\pi}\frac{\langle u_w \rangle}{h^3}} \quad \text{(O'Connor-Dubbins)} \tag{8.54}$$

a form which has had reasonable success for describing reaeration of oxygen-deficient lakes and streams. A recent treatment accounting for the variation of k_l with position w in the eddy provides

$$\bar{k}_l = 1.46 \sqrt{\frac{\mathscr{D}_{O_2} u_{\text{rms}}}{\Lambda}} \tag{8.55}$$

where u_{rms} is the rms velocity in the eddy circulation. In general, it appears

---

† R. Higbie, *Trans. AIChE*, **35**: 365 (1935).

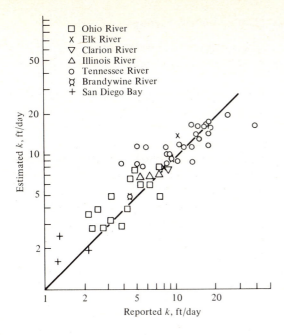

Figure 8.7 Reported reaeration rates vs. rates estimated using Eq. (8.55). (*Reprinted from G. E. Fortesque and J. R. A. Pearson, On gas absorption into a turbulent liquid, Chem. Eng. Sci.,* **22:** 1163, 1967.)

reasonable to take Λ and u_{rms} proportional to mean stream depth and mean stream velocity, respectively:

$$\gamma \times \text{depth} = \Lambda \qquad \gamma \times \text{mean stream velocity} = u_{rms}$$

with the same constant of proportionality γ. A comparison of k_l values estimated from Eq. (8.55) with reported $\bar{k}_l$ values for North American rivers is given in Fig. 8.7.

The motion of waves at air-water interfaces is known to influence gas transfer strongly. Here, in contrast to the preceding treatments, the gas-flow patterns, e.g., average and turbulent velocity components of wind which drives ocean waves, are of major importance. However, the general topic is too complex for this text (see the references).

8.5. FORCED CONVECTIVE MASS TRANSFER

Vigorous mechanical mixing of air-liquid dispersions is often necessary to obtain economic rates of biomass increase, substrate consumption, or product-formation. The concerns of this section are again relationships between appropriate variables allowing estimation of mass-transfer coefficients k_l and/or the interfacial area per volume a. As mechanical agitation has a cost, we shall also consider the sensitivity of $k_l a$ to increased agitation power input.

8.5.1. General Concepts and Key Dimensionless Groups

The functions served by mechanical agitation augment (and in some cases dominate) the influences of convection driven by freely rising or falling dispersed phases:

1. The high shear fields near the impeller tip or other mixer devices produce *small bubbles*, thereby increasing *a* locally. Provided that the rate of bubble coalescence is not correspondingly increased elsewhere in the vessel, the result is an increased value of the volumetric average value of *a*.
2. The fermentation fluid may contain a suspension of solid or other liquid phases which may tend to rise or fall in the vessel. Mechanical mixing provides a more uniform *volumetric dispersion* of these phases in the bulk liquid. For hydrocarbon dispersions, k_l contains a term proportional to the cube root of the phase-density difference $(\rho_{H_2O} - \rho_{HC})^{1/3}$ [recall Eq. (8.33)]; the resulting small mass-transfer coefficient for hydrocarbon-substrate-limited fermentations is increased by agitation.
3. For gas bubbles of given size in vigorously agitated vessels, k_l does not vary significantly with power input since the relative gas or fluid velocity is dominated by density differences. (Why is this true?) The agitator turbulence, however, will decrease D and thus *increase a* for a given holdup; note that this result will change the size of the bubbles and thus k_l through the influence of D.
4. The *maximum* size of loosely aggregated mycelia, microbial slimes, mold pellets, etc., may be diminished by agitation (see Sec. 8.6.4), thus maintaining a smaller microbial Thiele modulus (Sec. 7.4.2) and again rendering the vessel more uniformly mixed with respect to the liquid phase. Examples of *decreased* yields of desired products have been reported at relatively high agitation rates; these may be due to cellular or extracellular enzyme damage, among other possibilities.
5. The liquid-cell suspension may be so viscous that only mechanical agitation provides any degree of bulk-liquid mixing (considered further in Sec. 8.8).

In forced convection, the action of the applied mechanical work produces some characteristic velocity against which other motions can be scaled. For impeller agitation, two scales exist: the rms fluid velocity fluctuation u_{rms}, and the impeller tip velocity u_i, which is proportional to $N_i D_i$, where N_i is the impeller rotation rate in revolutions per unit time, and D_i is the impeller diameter.

For moderate to high mixing intensities, since the local fluid conditions are turbulent, varying rapidly in time, we commonly use time-averaged quantities. While the average velocity locally may be nearly zero, the rms velocity $u_{rms} \equiv \langle u^2(t) \rangle^{1/2}$ reflects the typical average magnitude of the local velocity variations.

In agitated vessels, the value of u_{rms} depends on the power input per unit volume P/V, according to

$$u_{rms} = d \left(\frac{D}{\rho_c} \frac{P}{V} \right)^{1/3} \tag{8.56}$$

where ρ_c is the continuous phase density and d is some constant.

Reduction of the forced-convection balances for total mass, species, and momentum produces the following dimensionless groups:

$$\text{Sherwood number} = \text{Sh} = \frac{k_l D}{\mathscr{D}_{O_2}} \tag{8.22b}$$

$$\text{Schmidt number} = \text{Sc} = \mu_c / \rho_l \mathscr{D}_{O_2} \tag{8.22c}$$

$$\text{Reynolds number} = \text{Re} = \rho_l D u_{\text{rms}} / \mu_c \tag{8.57a}$$

$$\text{Froude number} = \text{Fr} = u_{\text{rms}}^2 / gD \tag{8.57b}$$

Alternately, in stirred systems, the characteristic dimension may be taken as the impeller diameter D_i, and the reference velocity is $N_i D_i$. The subscripts i remind us that the scaling is to the impeller rather than the gas, liquid, or solid particles present in the dispersion. In this case, the appropriate Reynolds and Froude numbers are given by

$$\text{Re}_i = \frac{\rho_l D_i}{\mu_c} N_i D_i = \frac{\rho_l N_i D_i^{\,2}}{\mu_c} \quad \text{and} \quad \text{Fr}_i = \frac{N_i^{\,2} D_i}{g} \tag{8.58}$$

The Froude number has received other definitions. For mass transfer into a suspension of "neutrally buoyant" particles, the following relationship has been suggested:

$$\text{Fr}_L = \frac{N_i^{\,2} D_i^{\,2}}{gL} \tag{8.59}$$

where L is the reactor height. As the Froude number represents the contribution of free-surface dynamics vs. mechanical mixing, the distance of the surface from the tank bottom could logically enter in its description. When the stirred volume refers to the mixing of two phases (continuous and dispersed) of different densities, e.g., hydrocarbon droplets in aqueous phases, the following definition of a modified Froude number may be useful:

$$\text{Fr}_{\text{two phase}} = \frac{\rho N_i^{\,2} D_i^{\,2}}{\Delta \rho \, gL} \tag{8.60}$$

It is clear that close attention should be paid to both the form of correlations and the *definitions* of the groups involved in literature reports; a correlation should never be used or cited without careful group definitions.

8.5.2. Correlations for Mass-Transfer Coefficients

The Reynolds number based on the impeller diameter is convenient since it provides a single parameter over the entire laminar-, transition-, and turbulent-flow range for calculations of power input P. P also depends on N_i and D_i, although in different ways in different regimes. The mass-transfer coefficient of gases depends largely on the hydrodynamics of the liquid film near the bubble; this in turn is dominated by the natural convection buoyancy forces and the turbulent Reynolds number during most of the bubble residence time; thus correlations for freely rising bubbles are most usefully associated with the Reynolds number of Eq. (8.58).

In reactors fitted with baffles to maximize mixing rates within the continuous phase (Chap. 9), the influence of free-surface effects (Froude number) becomes unimportant. In the absence of such surface influences, the dimensionless solutions for the velocity and concentration fields yield

$$\bar{c} = f(\bar{z}, \text{ Re, Sc, Sh}) \tag{8.61}$$

so that Eq. (8.24) gives the dependence of the Sherwood number as

$$\text{Sh} = \frac{k_l D}{\mathscr{D}_{O_2}} = g(\text{Re, Sc}) \tag{8.62}$$

Data of Calderbank's [1] give the correlation

$$\text{Sh (turbulent aeration)} = 0.13\ \text{Sc}^{1/3}\text{Re}^{3/4} \tag{8.63}$$

In terms of power input per unit reactor volume, using relation (8.56) we can show that the variation of k_l(Sh) with P/V is

$$\text{Sh} \propto \left(\frac{P}{V}\right)^{1/4} \tag{8.64}$$

Thus once turbulence has been achieved, so that the previous equation applies, the specific increase in k_l with P diminishes rapidly, as is seen by evaluation of

$$\frac{1}{k_l}\frac{dk_l}{dP} = \frac{1}{4P} \tag{8.65}$$

The power consumption for stirring nonaerated fluids depends upon fluid properties ρ_c and μ_c, the stirrer rotation rate N_i and diameter D_i, and the drag coefficient of the impeller C_{D_i}. The latter is expected to vary with impeller Reynolds number in a different manner for each flow regime: laminar, transition, or turbulent. A well known study by Rushton, Costich, and Everett [24] is summarized in Fig. 8.8 for three impeller geometries. The data are plotted as a dimensionless power input, the power number P_{no} vs. impeller Reynolds number Re_i:

$$P_{no} \equiv \frac{Pg}{\rho_c N_i^3 D_i^5} \tag{8.66}$$

In the turbulent regime, the power input is independent of Re_i,

$$P \propto N_i^3 D_i^5 \qquad P_{no} = \text{const}$$

whereas in laminar flow, the relation is more nearly given by

$$P \propto N_i^2 D_i^3 \qquad \text{or} \qquad P_{no} \propto \frac{1}{\text{Re}_i}$$

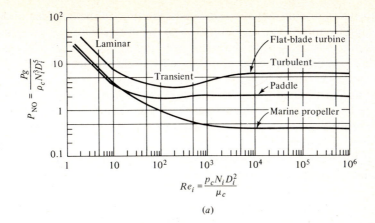

$$P_{NO} = \frac{Pg}{\rho_c N_i^3 D_i^5}$$

$$Re_i = \frac{\rho_c N_i D_i^2}{\mu_c}$$

(a)

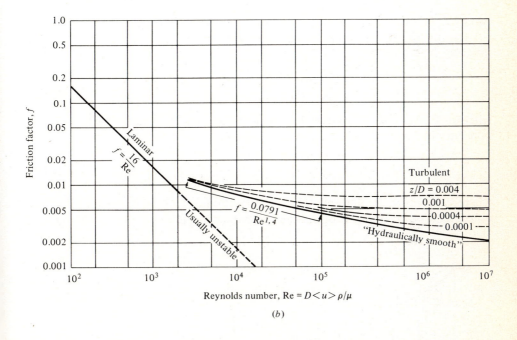

Reynolds number, Re = $D\langle u\rangle \rho/\mu$

(b)

Figure 8.8 (a) Power number vs. Reynolds number (of impeller) for various impeller geometries. (b) Pipe friction factor f vs. Reynolds number, Re. Lower case italic z equals height of surface roughness peaks [(a) *Reprinted from S. Aiba, A. E. Humphrey, and N. F. Millis, "Biochemical Engineering," 2d ed., p. 174, University of Tokyo Press, Tokyo, 1973; modified from J. H. Rushton, E. W. Costich, and H. J. Everett, Power Characteristics of Mixing Impellers, Part 2, Chem. Eng. Prog.,* **46**: 467, 1950. (b) *Reprinted by permission from W. L. McCabe and J. C. Smith, "Unit Operations of Chemical Engineering," McGraw-Hill Book Company, New York, 1954; original curves from L. F. Moody, Trans. ASME,* **66**: 671, 1944.]

the proportionality constant in each case depending on the impeller geometry. It is interesting to note the strong similarity between this figure and the plot of the friction factor in tube flow. In the latter case, the friction factor varies as $1/Re$ in laminar flow (as does the power number vs. Re_i), and, in turbulent flow, f tends to a nearly constant value which has a larger magnitude for pipes with rough walls. Similarly, as the agitator geometry becomes less "smooth," Fig. 8.8b, the power number reaches a higher turbulent plateau value. The latter case is more complicated since the presence of the tank walls and baffles will also exert an effect on the measured power input P.

When the agitated vessel is simultaneously aerated, the power requirements for agitation decrease. The ratio of power requirements in aerated vs. nonaerated vessels, P_a/P vs. a dimensionless aeration rate N_a

$$N_a = \frac{F_g}{N_i D_i^{\,3}} \tag{8.67}$$

(where F_g is volumetric gas rate) has been correlated, as shown in Fig. 8.9.

Except for the most rapidly changing part of the curve, these forms can be fitted to

$$\frac{P_a(N_a) - P_a(N_a = \infty)}{P - P_a(N_a = \infty)} = e^{-m/N_a} \tag{8.68}$$

where $m = $ const. An alternative form which is also useful in turbulent aeration of non-Newtonian fluids is due to Michel and Miller:

$$P_a = m'\left(\frac{P^2 N_i D_i^{\,3}}{F_g^{\,0.56}}\right)^{0.45} = m'\left(\frac{P^2(N_i D_i^{\,3})^{0.44}}{N_a^{\,0.56}}\right)^{0.45} \tag{8.69}$$

where $m' = $ const. In both the above correlations, $P_a(N_a = 0) = P$, that is, the nonaerated power input P of the earlier chapter formulas.

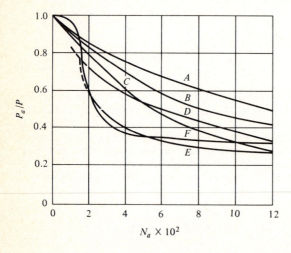

Figure 8.9 Ratio of power requirement for aerated vs nonaerated systems as a function of N_a (see text). [*Reprinted by permission from Ohyama, Y. and Endoh, K. Power Characteristics of Gas-Liquid Contacting Mixers, Chem. Eng. (Japan), 19: 2, 1955.*]

From the relations of the previous paragraph, at constant N_i, D_i, the power input diminishes with increased N_a, that is, increased air flow F_g. This effect appears partially due to the decrease average *density* of the fluid being agitated.

Agitation of air-liquid dispersions affects the interfacial area per volume a more strongly than it does k_l. The reasons for this and some correlation and measurement techniques for the evaluation of a are now discussed.

8.6. SURFACE-AREA CORRELATIONS FOR MECHANICALLY AGITATED VESSELS

Like part of the previous discussion of mass-transfer correlations, this section is based primarily on an extended summary by Calderbank [1]. For a dispersed liquid or gas phase in a continuous liquid phase, the *maximum* size of dispersed-phase diameters for either freely rising (falling) or agitated configurations is due to a balance of opposing forces:

1. The dynamic pressure τ (the sum of shearing and normal stress differences) tends to draw out the droplets into shapes which eventually disintegrate into smaller pieces; this subdivision process is resisted by the following two forces.
2. The surface-tension forces σ/D of the particle tend to restore the droplet to spherical shape (minimum surface-energy configuration).
3. The viscous resistance of the dispersed phase to shearing is proportional to the term $\mu_d D^{-1}\sqrt{\tau/\rho_d}$, where subscript d indicates a dispersed-phase property.

In gas-liquid systems, the term 3 will be negligible compared with term 2. In liquid-liquid contactors the last term should be relatively larger, but a later result of Example 8.9 indicates again that the forces in term 2 appear to predominate even for these all-liquid systems.

The last two restoring forces diminish as $D^{-\beta}$, where β is a positive number. Thus, at some critical diameter D_c, the dynamic pressure, will override the two countering resistances and rearrange the bubble or droplet into smaller portions. At the critical diameter, evidently the following equality holds:

$$m_1 \frac{\sigma}{D} + m_2 \left[\mu_d D^{-1} \left(\frac{\tau}{\rho_d} \right)^{1/2} \right] = \tau \tag{8.70}$$

where m_1 and m_2 are constants.

If the surface-tension forces are much more significant than the viscous forces, as argued, then at the critical drop size

$$m_1 \frac{\sigma}{D_c} = \tau \quad \text{or} \quad m_1 = \tau \frac{D_c}{\sigma} \tag{8.71}$$

The Weber number We is the dimensionless ratio defined by

$$\text{We} = \tau \frac{D}{\sigma} \tag{8.72}$$

Equation (8.71) predicts the existence of a critical Weber number $We_c = m_1$, according to which, for τ dependencies appropriate to the situation of interest (free-rising bubbles, agitation, etc.), we can predict the maximum bubble diameter D_c. The following procedure can be conducted as a thought experiment. If a critical Weber number exists, measurement of D_c (essentially the maximum observed bubble size, though many measuring techniques give D_{Sauter}, not D_c) should provide a correlation for D_c in terms of the surface tension and the variables dominating the dynamic stress. Then assume that resistance term 2 dominates in the determination of τ. From theory, calculate the expected dependence of the dynamic stress τ on the appropriate variables. Compare experiment and theory.

This approach has enjoyed some success, as shown in Example 8.5. For freely rising bubbles or for various agitated or forced-convection configurations, the correlations of D_c or D_{Sauter} with those variables related only to dynamic (term 1) and surface-tension forces (term 2) are surprisingly good. Thus, we conclude that, even for liquids (freely rising bubbles in Example 8.5), this approach is largely valid.

The presence of the additional terms depending on the holdup H (volume of gas per volume of reactor) and viscosity of the dispersed phases μ_d suggest that other forces are also important though not dominant. The bubble diameter observed typically increases with gas holdup, a result to be expected if an additional process such as bubble *coalescense* is acting continuously to diminish the bubble-number density. Coalescence has not been included in the previous discussion; note also that a slight influence of the resistance (term 3) would be expected to produce a weak positive dependence of D_c or D_{Sauter} on the viscosity of the dispersed phase as observed (Example 8.5).

Example 8.5. Maximum D_c or Sauter mean D_{sm} bubble or droplet diameters: correlations and theory

For freely rising bubbles, experimental values[†] are

$$D_c = \left(1.452 \times 10^{-2} \frac{\sigma}{\Delta\rho}\right)^{1/2} \quad \text{cm} \tag{8E5.1}$$

Theory gives $\tau = \frac{1}{6}D \, \Delta\rho \, g$ where τ is the dynamic pressure at the terminal rise velocity. Therefore

$$We = \frac{D^2 \, \Delta\rho \, g}{6\sigma} \quad \text{and} \quad D_c = \left[\frac{(We_c)(6)(\sigma)}{\Delta\rho \, g}\right]^{1/2} = (\text{const})\left(\frac{\sigma}{\Delta\rho}\right)^{1/2} \tag{8E5.2}$$

For agitated vessels, on a power-per-unit-volume basis we list the results of several experiments:

Experiment 1 liquid-liquid:[‡]

$$D_{sm} = 0.224 \frac{\sigma^{0.6}}{\rho_c^{0.2}(P/V)^{0.4}} H^{0.5} \left(\frac{\mu_d}{\mu_c}\right)^{0.25} \tag{8E5.3}$$

† S. Hu and R. C. Kintner, The Fall of Single Liquid Drops through Water, *AIChE J.*, **1**: 42, 1955.
‡ P. H. Calderbank, *Trans. Inst. Chem. Eng.*, **36**: 443, 1958.

Experiment 2 gas–liquid electrolyte:†

$$D_{sm} = 2.25 \frac{\sigma^{0.6}}{\rho_c^{0.2}(P/V)^{0.4}} H^{0.4} \left(\frac{\mu_d}{\mu_c}\right)^{0.25}$$

(8E5.4)

Experiment 3 gas in alcohol solutions:‡

$$D_{sm} = 1.90 \frac{\sigma^{0.6}}{\rho_c^{0.2}(P/V)^{0.4}} H^{0.65} \left(\frac{\mu_d}{\mu_c}\right)^{0.25}$$

(8E5.5)

Theory tells us that

$$\tau \propto \rho_c u_{\text{rms}} = (\text{const})(\rho_c) \left(\frac{PD}{V\rho_c}\right)^{2/3}$$

(8E5.6)

Therefore

$$W_e = (\text{const}) \frac{\rho_c}{\sigma} D^{5/3} \left(\frac{P}{V\rho_c}\right)^{2/3} \quad \text{and} \quad D_c = (\text{const}) \frac{\sigma^{0.6}}{\rho_c^{0.2}(P/V)^{0.4}}$$

(8E5.7)

For agitated vessels, impeller variables [24]

$$P = (\text{const})(\rho_c N_i^3 D_i^5) \quad \text{for } Re_i > 10^4$$

and therefore

$$\frac{P}{V} \approx \text{const} \frac{\rho_c N_i^3 D_i^5}{D_i^3}$$

(8E5.8)

From Eqs. (8E9.6) and (8E9.7), therefore,

$$\frac{D_c}{D_i} = (\text{const}) \left(\frac{\sigma}{N_i^2 D_i^3 \rho_c}\right)^{0.6}$$

(8E5.9)

For turbulent pipe flow [1]

$$\frac{D}{D_{\text{pipe}}} = (\text{const})(We_{\text{pipe}}^{-0.6} Re_{\text{pipe}}^{-0.1}) = (\text{const}) \left(\frac{\sigma}{\rho_c \langle u^2 \rangle D_{\text{pipe}}}\right)^{0.6} \left(\frac{\mu_c}{D_{\text{pipe}} u_{\text{rms}} \rho_c}\right)^{0.1}$$

(8E5.10)

where

$$We_{\text{pipe}} \equiv \left(\frac{\sigma}{\rho_c \langle u^2 \rangle D_{\text{pipe}}}\right)^{-1} \quad \text{and} \quad Re_{\text{pipe}} \equiv \frac{D_{\text{pipe}} u_{\text{rms}} \rho_c}{\mu_c}$$

For flow through an orifice in pipe flow (measured one ft downstream)‡

$$\frac{D_{\text{pipe}}}{D} = 21.6 \left(\frac{D_{\text{orifice}}}{D_{\text{pipe}}}\right)^{3.73} H^{0.121} We_{\text{pipe}}^{-0.722} Re_{\text{pipe}}^{-0.065}$$

(8E5.11)

where We_{pipe} and Re_{pipe} are as in Eq. (8E5.10).

Given D_c or D_{Sauter} and the holdup H, the value of a is calculated from Eq. (8.37). For complex situations like those applying to most of the macroscopic contactor situations in Fig. 8.2, H must be measured directly or obtained from correlations for similar configurations. A representative sampling of such correlations was given in Example 8.4.

† P. H. Calderbank, *Trans. Inst. Chem. Eng.*, **36**: 443, 1958.
‡ J. A. McDonough, W. J. Tomme, and C. D. Holland, Formulation of Interfacial Areas in Immiscible Liquids by Orifice Mixers, *AIChE J.*, **6**: 615, 1960.

8.7. OTHER FACTORS AFFECTING $k_l a$

From the definitions of k_l and a and consideration of the factors responsible for the thickness of the mass-transfer resistance zone near bubble and droplet surfaces, k_l and a will be influenced by alteration of the values of liquid-phase solute diffusivity $\mathscr{D}_{O_2}$, continuous-phase viscosity μ_c, and the gas-liquid interfacial resistance. The liquid "viscosity" may vary with shear rate; this non-Newtonian behavior is of sufficient importance to be discussed separately (Sec. 8.8). The remaining influences are summarized in this section.

8.7.1. Estimation of Diffusivities

The Wilke-Chang correlation is a useful means of estimating (usually to better than 10 to 15 percent) the diffusion coefficient of small molecules in low-molecular-weight solvents:

$$\mathscr{D} = 7.4 \times 10^{-8} \frac{T(x_a M)^{1/2}}{\mu_c V_m^{0.6}} \qquad \text{cm}^2/\text{s} \tag{8.73}$$

where m = solute molecular weight
V_m = molecular volume of solute at boiling point, $\text{cm}^3/\text{g mol}$
μ_c = liquid viscosity

The parameter x_a represents the *association factor* for the solvent of interest; some values for x_a are 2.6 (H_2O), 1.9 (methanol), 1.5 (ethanol), and 1.0 (benzene, ether, and heptane).

The diffusion coefficient will vary with ionic strength (as does solubility, Table 8.1) and with concentration of solutes which change the solution viscosity. Provided that the solute-solvent interactions are not altered in the latter case, the relation

$$\mathscr{D}_1 \mu_{c1} = \mathscr{D}_{\text{ref}} \mu_{c,\,\text{ref}} \tag{8.74}$$

provides a useful scale to correct for changes in solution viscosity from a reference point, say that of pure water.

The diffusion coefficient in microbial aggregates is usually less than that in pure water (2.25×10^{-5} cm^2/s), as summarized in Table 8.3 for oxygen. A recent study examining other variables concluded that the O_2 diffusion coefficient in waste-treatment microbial aggregates decreases from the pure H_2O value with increased aggregate lifetime in the reactor and with increased C/N ratio of the entering waste substrates in the range 20 : 1 to 5 : 1 (see Table 7.8).

Table 8.3. Diffusion coefficients in microbial film

Biomass	Reactor	$\mathscr{D}_{\text{meas}}$, cm^2/s $\times 10^5$	% of H_2O value
Bacterial slimes	Rotating tube	1.5	70
	Submerged slide	0.04	2
Zoogloea ramigera	Fluidized reactor	0.21	8

8.7.2. Ionic Strength

The precise resolution of ionic-strength influences into all pertinent factors appears difficult. An examination of Newtonian fluids gives the physical-absorption result

$$K_l a = \lambda \left(\frac{P_a}{V_l}\right)^n \left(\frac{F_g}{A}\right)^m \frac{\rho_l^{0.533} \mathscr{D}_{O_2}^{2/3}}{\sigma^{0.6} \mu_c^{1/3}} \tag{8.75}$$

where V_l = liquid volume
$\quad P_a$ = power input during aeration
$\quad A$ = reactor cross section perpendicular to flow rate F_g

An empirical fit for λ, n, and m vs. ionic strength $I (= \frac{1}{2}\Sigma Z_i^2 c_i$, i = species, Z_i = species charge) is possible:

$$\lambda = 18.9 - \frac{28.7 I'}{0.276 + I'} \qquad I' = \begin{cases} I & \text{if} \quad 0 \le I \le 0.40 \text{ g ion/l} \\ 0.40 & \text{if} \quad I > 0.40 \text{ g ion/l} \end{cases}$$

$$n = \begin{cases} 0.40 + \dfrac{0.862 I'}{0.274 + I'} & I' = I \le 0.4 \\ 0.9 & I > 0.4 \end{cases}$$

$$m = \begin{cases} \text{monotonic increasing} \\ \text{from } 0.35 & I = 0 \\ \text{to } 0.39 & I \ge 0.4 \end{cases}$$

for $K_l a$ in s^{-1}, P_a in ft·lb$_f$/min, V_l in ft^3, F_g in ft^3/s, A in ft^2, and physical properties in cgs units.

8.7.3. Surface-active Agents

As discussed in Chap. 2, many biochemicals are amphipathic, i.e., contain strongly hydrophobic and hydrophilic moieties which tend to concentrate at gas-liquid and liquid-liquid interfaces. In various phases of fermentations, cells secrete species such as polypeptides which may behave like surfactants, at times leading to foaming tendencies in aerated vessels. Addition of chemical antifoams also affects interfacial resistances to mass transfer, though typically in a manner opposing that of surfactants.

Adsorption of surfactants at the phase interface is a spontaneous process; the interfacial free energy and thus the surface tension σ is reduced vs. the original value. From the correlations in Example 8.5 the values of D_{Sauter} and D_c are expected to decrease, leading to higher values of the interfacial area per volume a.

This tendency for a to increase is countered by the effect of surfactant films on the mass-transfer coefficient k_l. The adsorption of a macromolecular film resembling Fig. 2.1 results in a stagnant, rigid interface. The decreases in k_l discussed below are thought to be due to either or both of two mechanisms: (1) the ease of liquid movement near the interface is reduced due to the decreased mobility of the interface; thus, the variety of mass-transfer theories based on estimating exchange rates of small fluid elements between the surface and the bulk will predict a decreased mass-transfer coefficient (see Refs. 4 and 5 for further discussion); (2) like the cell membrane itself, the molecular film is expected to contribute a resistance of its own, which may cause a departure from the presumed gas-liquid equilibration in the plane of the interface.

Addition of 10 ppm sodium lauryl sulfate (SLS) reduces k_l for oxygen transfer by 56 percent versus pure water. A constant or plateau value of k_l was observed at all higher surfactant concentrations. The surface area per volume a increased slowly throughout the range of SLS concentrations from 0 to 75 ppm, with a minimum of $k_l a$ at about 10 ppm surfactant.

The product $k_l a$ has been observed to increase continuously with surfactant addition in a turbine aerator. Inspection of the data for the reported ratio of a(surfactant)/a(no surfactant) and the corresponding ratio for the product $k_l a$ shows that while a increased 400 percent for addition of 4.0 ppm sodium dodecyl sulfate, $k_l a$ increased only about 15 percent, implying a decrease of about 77 percent in the value of k_l.

This reduction observed in both sets of data is in agreement with results of others reported in summaries by Aiba, Humphrey, and Millis (Ref. 2 of Chap. 7) and Wolynic.† The latter survey for a variety of sparingly soluble gases indicates that the average plateau values of k_l upon surfactant addition corresponded to reductions of k_l by a factor of 60 percent. The sodium dodecyl sulfate data of Benedek and Heideger [30] suggest that turbine-agitated aeration corresponds to the transition region between the two correlations presented earlier in Eqs. (8.31) and (8.32); a similar result appears to be the case for aeration with sieve trays (Danckwerts). This serves to warn the reader that such correlations are useful estimates but should be replaced by experimental values from more pertinent equipment when possible.

8.8. NON-NEWTONIAN FLUIDS

For fluids, the ratio of shear stress τ_s to the velocity gradient du/dy is defined as the viscosity η_v (the subscript v distinguishes the viscosity from the earlier effectiveness factor η). Thus,

$$\tau_s = -\eta_v \frac{du}{dy} \tag{8.76}$$

A plot of τ vs. du/dy for a Newtonian fluid is linear and passes through the origin. A variety of *non-Newtonian* behaviors have been observed in steady flows for liquid solutions of polymers and/or suspensions of dispersed solids or liquids; the features of the more common of these are summarized in Fig. 8.10, where shear rate $\dot{\gamma}$ is used rather than du/dy.

---

† E. T. Wolynic, Ph.D. Thesis, Princeton University, Princeton, N.J., 1974.

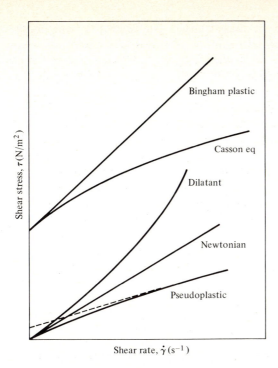

Figure 8.10 Stress vs. shear-rate behavior of Newtonian and common non-Newtonian fluid models. (*Reprinted from J. A. Roels, J. van den Berg, and R. M. Voncken, The Rheology of Mycelial Broths, Biotech. Bioeng.,* **16**: 181, 1974.)

8.8.1. Models and Parameters for Non-Newtonian Fluids

The dilatant, Newtonian, and pseudoplastic behaviors are examples of the general Ostwald–de Waele or *power-law* formulations for fluids

$$\tau_s = -\eta_0 \, |\dot{\gamma}|^{n-1}\dot{\gamma} = -(\text{apparent viscosity})(\dot{\gamma}) \tag{8.77}$$

where $\quad n \begin{cases} > 1 \Leftrightarrow \text{dilatant} \\ = 1 \Leftrightarrow \text{Newtonian} \\ < 1 \Leftrightarrow \text{pseudoplastic} \end{cases}$

Some systems appear not to produce a motion until some finite *yield stress* τ_0 has been applied. For *Bingham plastic* fluids, the form

$$\tau_s = \tau_0 - \eta_0 \, |\dot{\gamma}|^{n-1}\dot{\gamma} \quad \text{where} \quad n = 1, \tau_0 \neq 0, \tau_s > \tau_0 \tag{8.78}$$

is useful.

The last form with finite τ_0 and n less than unity would give a curve similar to the *Casson equation*, which is given by

$$\tau_s^{1/2} = \tau_0^{1/2} - K(\dot{\gamma}^{1/2}) \tag{8.79}$$

In the subsequent discussion of mass-transfer coefficients, power input, mixing, etc., we shall refer to various fermentation or other fluid systems of interest as being (apparently) pseudoplastic, Newtonian, etc., and indicate correlations between the fluid-model parameters and the former quantities of interest. The fluid descriptions in Fig. 8.10 refer to behavior in steady shear flows. Under *unsteady*-state conditions, such as follow a step change in the applied shear rate, time-dependent responses in η_v are often observed, thus demanding a more structured model (in the same sense as the Chap. 7 microbial models) to describe *transient* situations. This latter state may more accurately apply to turbine agitation and turbulent mixing in non-Newtonian systems. These more structured models are a relatively difficult and undeveloped area of mechanics; we simply insert the caveat here that our understanding of the factors responsible for non-Newtonian behavior and their description is weaker than the theories for the previous sections of this chapter.

Non-Newtonian behavior may arise in at least two distinct cases: (1) suspensions of small particles and (2) solutions of macromolecules. It is apparent that the two cases become similar as molecular diameters increase above 50 to 100 Å or as particle diameters fall from the order of micrometer sizes.

8.8.2. Suspensions

Various theories predict that a dilute suspension of spheres should remain Newtonian, the effective viscosity $\eta_{v,\,\text{eff}}$ of the suspensions being given by

$$\frac{\eta_{v,\,\text{eff}}}{\eta_{\text{solvent}}} = 1 + \tfrac{5}{2}\phi + b\phi^2 + \cdots \tag{8.80}$$

where ϕ is the volume fraction solids and b is of the order of 6 to 8. For a bacterial density as high as 10^9 cells per milliliter and a cell diameter of 3×10^{-4} cm, the value of ϕ is about 5×10^{-3}: the effect of the cells is apparently negligible. Higher volume fractions occur in filtration operations such as dewatering (in product recovery from cell broths) and in fermentations producing the mold pellets or mycelia of previous discussions.

At higher volume fractions, solid suspensions may exhibit a yield stress; e.g., aqueous slurries of nuclear fuel particles with diameters in the micrometer range appear usefully modeled by the Bingham plastic model. Mycelial fermentations of *Streptomyces griseus* appear to follow a Bingham form except at very low shear rates; this may be important in aeration. From Eq. (8E5.1) in Example 8.5 note that the maximum stress for a rising bubble is associated with the terminal rise velocity. Thus, in non-Newtonian fluids, a sufficiently small bubble will not exert the yield stress on the surrounding fluid and it will remain fixed in the same fluid element for long times; i.e., it would be expected to circulate in the vessel with the fluid rather than following the usually rising path of larger bubbles.

The Casson equation has been fruitfully applied to descriptions of blood flow. Red blood cells form aggregates, the size of which diminishes with increasing shear forces. The apparent viscosity η_v diminishes with increased shear, as seen from Fig. 8.10 for the Casson equation. In floc-forming fermentations, we may expect stirrer shear forces to control the average size of the microbial flocs; hence the Casson equation may be useful in some such fermentations.

A recent study examined the rheology of penicillin broths vs. time over a large range of shear rates. A turbine impeller was used for viscometry studies rather than a rotating cylinder, the two cited advantages being

(a) the stirring prevents phase separation, which otherwise would affect the reliability of the measurements: without stirring a thin "(cell-free)" layer of liquid is formed at the wall of the cylinder. (b) the shear rate of the impeller is a simple function of the impeller speed and is independent of the rheology of the liquid as has been shown previously [35, pp. 188–189].

This research revealed that the Bingham and power-law models were inadequate to represent the results over the full range of shear rates investigated. A reasonable modification of the Casson equation was derived:

$$(M_\tau)^{1/2} = (M_{\tau_0})^{1/2} \left[1 + \frac{0.69(M_{\tau_0})^{1/2} - 1.1}{(M_{\tau_0})^{1/2}} (N_i)^{1/2} \right] \tag{8.81}$$

where M_τ is proportional to stress: $M_{\tau_0} = 64D_i^3/2\pi K$, N_i is the impeller rotation rate (revolutions per second) and K is the ratio of shear rate to impeller speed (constant). The data agreed well with Eq. (8.81) though a slightly better fit resulted if the second term on the right-hand side was modified to read

$$\frac{0.193(M_{\tau_0})^{0.75} + 0.1}{(M_{\tau_0})^{1/2}} (N_i)^{1/2} \tag{8.82}$$

The importance of the result is threefold: (1) The rheological data were taken over a sufficient range of shear rates to *discriminate* between models which are similar over restricted shear-rate ranges. (2) The results showed that the power law and Bingham laws were most inadequate at low shear rates; this range is likely to be important in determining tank-mixing uniformity; i.e., extrapolation of the Bingham and power laws to this situation would lead to large errors (mixing problems in vessels are discussed in Chap. 9 in connection with fluid residence-time distributions). (3) The modified Casson equation was shown to predict a factor dependent on the mold *morphology* (shape) through the ratio of length to diameter of the filaments. This result (and other theories for nonspherical particles) provide an important potential connection between reactor-design calculations (analysis) and morphology (observation) since the latter subject has been examined for a range of species and conditions in the literature.

The power-law model appears to fit data for mold filaments over the higher range of shear rates. It also describes suspensions of paper pulp; 4 percent paper pulp in water suspensions have power-law parameters $n = 0.575$ and $\eta_0 = 0.418$ $lb_f \cdot s^n/ft^2$. Investigations of oxygen transfer with pulp suspensions have been used to simulate some fermentation conditions: if the overall oxygen uptake is determined by mass-transfer limitations in high shear regions, this may be useful. The previous paragraph casts more doubt on such simulation studies where bulk mixing influences O_2 uptake by the microorganism(s).

In continuous-flow systems with several reactors in series (Chap. 9), the tank number replaces time as an indication of population growth phase, and changes in rheology with tank number may be encountered. For example, two-stage cultivation of *Candida utilis* [36] at large flow rates led to second-tank conditions such that the specific biomass growth rate μ_{growth} became so much larger than specific rate of cell multiplication μ_d that an increase in average *cell size* occurred. A change of cell and aggregate morphology was also noted, indicating again a possible connection between morphology and rheology.

8.8.3. Macromolecular Solutions in Fermentations

Microbes may secrete significant amounts of macromolecular components. The resultant solutions are often non-Newtonian, posing problems not only in aeration and mixing design but also in dewatering schemes encountered in sludge filtration, in waste treatment, and in similar operations in product recovery.

For example, *Aureabasideum pullulans* (*Pullularia pullulans*) 2552 in sucrose medium was found to fit the two-parameter power-law model of Eq. (8.77). The *field apparent viscosity* was evaluated at the condition

$$\frac{du}{dr} = 1.0 \text{ s}^{-1} \tag{8.83}$$

in a Brookfield viscometer. The cell volume fraction in all experiments appears to have reached a maximum of about 4 percent after 5 days. Similarly the maximum yield of polysaccharide synthesized occurred at the same time; it was 60 to 70 percent of initial sucrose. By contrast, the field apparent viscosity often rose from near unity to above 20,000 centipoise (cP) in the course of the fermentation and subsequently decreased by about 80 to 90 percent in the next several days. The value of the power-law exponent n for these runs fell from initial values of 0.67 to 0.78 to about 0.27 at 4 days and subsequently increased to 0.4 to 0.5 at 9 days. It was postulated that secretion of a high-molecular-weight polysaccharide which was degraded in later fermentation stages had occurred. Viscosity experiments with solutions of recovered polysaccharide showed lower n values in the initial secretion periods that the late stages for polysaccharide concentrations of equal weight percents. The n values of such experiments also varied most strongly with concentration in early samples.

The importance of understanding the non-Newtonian nature of fluids concerns several problems which are touched upon in the following sections and Chap. 9. The shear-dependent nature of the fluid renders mixing more difficult for the Bingham, Casson, and pseudoplastic fluids of the previous discussion than for Newtonian fluids. The influence of non-Newtonian behavior also leads to much slower coalescence rates in contrast to results for sulfite oxidation in aqueous solutions. The mixing argument favors a large impeller turning more slowly, the bubble-diameter-preservation argument favors smaller, higher-speed impellers. Finally, in impeller mixing zones, the particles may tend to move away from the high-shear regions, leading to further exaggeration of the inhomogeneity of the reactor.

8.8.4. Power Consumption and Mass Transfer in Non-Newtonian Fermentations

In a batch glucoamylase-producing fermentation using *Endomyces* sp., a similar transient behavior for the power-law exponent was observed: n fell from 1.0 to 0.35 at 70 h then rose to 0.65. An instantaneous impeller Reynolds number Re'_i may be defined according to Calderbank [1]:

$$\mathrm{Re}'_i = \frac{D_i{}^2 N_i{}^{2-n} \rho}{0.1 K_c} \left(\frac{n}{6n+2} \right)^n \tag{8.84}$$

where K_c, the consistancy index, equals the shear stress τ at a shear rate of 1 sec^{-1}.

For the *nonaerated* non-Newtonian broths of *Endomyces*, the power number [Eq. (8.66)] was correlated with the Reynolds number Re'_i, as shown in Fig. 8.11. The curves plotted there can be represented by the relationship

$$P_{\mathrm{no}} = k(\mathrm{Re}'_i)^x \left(\frac{D_i}{D_T} \right)^y \left(\frac{W}{D_T} \right)^z \tag{8.85}$$

where D_i = impeller diameter
$\quad D_T$ = tank diameter
$\quad W$ = impeller width

and k, x, y, and z depend on the range of Re'_i:

	Re$'_i$		
	< 10	10–50	> 50
k	32	11	9
x	−0.9	−0.4	−0.05
y	−1.7	−1.7	−1.2
z	0.4	0.5	0.9

The similarity of Fig. 8.11 to Rushton's data and the pipe friction factor in Fig. 8.8(a, b) is clear.

In aeration studies by the same group, the correlation of Michel and Miller (used earlier for Newtonian fluids) fitted the results in the turbulent regime (Re$'_i$ > 50):

$$P_g \propto \left(\frac{P^2 N_i D_i^3}{F_g^{0.56}} \right)^{0.45} \tag{8.86}$$

The laminar and second regimes (Re$'_i$ < 50) appear reasonably approximated by use of a smaller exponent:

$$P_g \propto \left(\frac{P^2 N_i D_i^3}{F_g^{0.56}} \right)^{0.27} \tag{8.87}$$

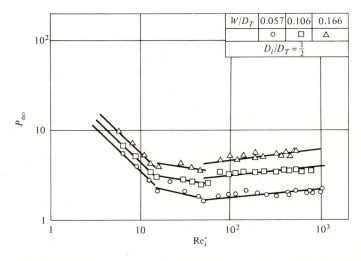

Figure 8.11 Power number vs. impeller Reynolds number, Re$'_i$ (see text) for three different impellers in nonaerated systems. (*Reprinted from H. Taguchi and S. Miyamota, Power Requirement in Non-Newtonian Fermentation Broth, Biotech. Bioeng.,* **8:** 43, 1966.)

The two curves meet at a value of $P^2 N_i D_i^3 / F_g^{0.56} = 2 \times 10^{-2}$, and slightly different proportionality coefficients result for the different impellers used.

The dilution of mycelial cultures by small volume percents of water (10 to 25 percent) produces marked changes in k_l and power-number parameters. The implications for power-input variations with dilution rate in continuous culture may be significant.

The influence of the microbial population depends on the physical situation which applies (Fig. 8.1a to d) and the influence of the microbial particles on the fluid properties. The value of $k_l a$ has been observed to *diminish* 90 percent as the concentration of mycelial *Aspergillus niger* increased from 0.02 to 2.5 percent. *Enhancement* of O_2 mass transfer is observed in the presence of suspensions of *Candida intermedia, Pseudomonas ovalis*, or 0.3-μm alumina particles. Not only do cells and alumina give similar results, but through use of oxidative-phosphorylation (Chap. 5) inhibitors, it is established that the enhancements are typically 40 percent vs. water and are independent of cell viability. It has been argued that the effect of the particles is likely to alter the hydrodynamics near the gas/liquid interface in such a way as to decrease the mass transfer resistance of the adhering fluid film near the bubble.

At the high shear rates needed to mix phases and promote mass transfer, diminutions in microbial activity have frequently been observed. For example, the viability of a relatively large cell, the protozoon *Tetrahymena*, began to be seriously altered by disruption at shear rates > 1200 s^{-1}. In these experiments the maximum shear rate characterized by the impeller tip speed appeared to be a more important variable than the turbulent Reynolds number or power input per unit volume.

The impeller can have other effects: the size of the extant microbial aggregates may be reduced, as mentioned earlier. Taguchi and Yoshida† divided the experimentally observed reduction of mycelial pellet size into two phenomena: (1) chipping small pellicules off the larger pellet and (2) directly rupturing the spherical shape of the pellet. The time evolution of particle diameter D due to the first process appeared to be governed by

$$\frac{dD}{dt} = -k_i (N_i D_i)^{5.5} D^{5.7} \tag{8.88}$$

while the second process could be described as a first-order decay,

$$\frac{dN_p}{dt} = -k_r N_p \tag{8.89}$$

where N_p is the number of nondisrupted pellets remaining. The rate coefficient k_r was correlated with D, N_i, and D_i to give

$$k_r = +(\text{const})(D^{3.2} N_i^{6.65} D_i^{8.72}) \tag{8.90}$$

Assuming that the turbulent Reynolds stress rather than the viscous stress is responsible for pellet rupture and that the pellet resistance depends directly on the measurable tensile strengths of the pellets, Taguchi et al. argue that the last equation may have some theoretical basis.

Regarding the oxygen transfer rate in paper-pulp suspensions, impeller and tank geometry were found to be important, in agreement with the earlier results of Taguchi and Miyamoto [37] for aerated

† H. Taguchi and T. Yoshida, *J. Ferment. Technol.*, **46**: 814, 1968.

and nonaerated *Endomyces* suspensions. In 1.6 percent pulp suspensions, the product $K_l a$ was described by

$$\frac{K_l a}{N_i} = 0.113 \left[\frac{D_T{}^2 h}{WL(D_i - W)} \right]^{1.437} (N_i t)^{-1.087} \left(\frac{D_i}{D_T} \right)^{1.021} \tag{8.91}$$

where t = characteristic mixing time of vessel
  h = liquid height
  W = impeller width

and D_i, D_T and L are as before.

In summary, a number of factors including bubble and cell dimensions, fluid properties and rheology, agitator and tank geometry, and power input determine mass-transfer coefficients and surface area per unit volume. The combination of these estimates with cell kinetics of the previous chapter and notions of mixing and macroscopic reactor configurations of Chap. 9 are important in assembling a complete reactor design. This state of knowledge is obviously wishful; the previous relations in this section are but the beginning of work needed to design confidently such reactors from first principles. As a closing example of some of the complexities yet to be unraveled, we note the time changes of the relations between bubbles, particulate substrate, and cells that accompanies the fermentation of *Candida petrophilum* on *n*-hexadecane, as shown in Fig. 8.12. The description of this gas-liquid-liquid-cell system by the authors† is illuminating:

> During the first period of fermentation, oil droplets are relatively large and cells attach to oil droplets rather than to air bubbles. Air bubbles are unstable and easily renewed. The $k_l a$ value can be kept at the maximum level associated with the fermentor. During the second phase, oil droplets become smaller and cells are adsorbed onto the surface of the oil droplets, forming dense flocs. The flocs tend to attach to the surface of the air bubbles, but they can be easily separated with agitation.
>
> The $k_l a$ value is decreased continuously as fermentation proceeds. The third phase is the last half period of the logarithmic growth phase, in which yeast is growing rapidly, although we cannot observe any oil droplets microscopically in the culture liquid. At this point the $k_l a$ value reaches its minimum throughout the fermentation. . . . [Air] bubbles are covered with yeast cells,

† A. Mimura, I. Takeda, and R. Wakasa, Some Characteristic Phenomena of Oxygen Transfer in Hydrocarbon Fermentation, *Biotechnol. Bioeng. Symp.* 4, pt. 1, p. 467, 1973.

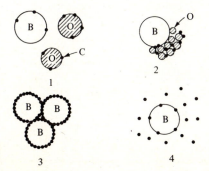

Figure 8.12 Different relationships among air bubbles (*B*), n-hexadecane droplets (0), and yeast cells (*C*) at different stages of batch culture of the yeast *Candida petrophilum*. (1 = lag phase, 2 = first half of exponential phase, 3 = second half of exponential phase, 4 = after n-hexadecane exhaustion.) [*Reprinted from A. Mimuro, I. Takeda, and R. Wakasa, Some Characteristic Phenomena of Oxygen Transfer in Hydrocarbon Fermentation, in B. Sikyta, A. Prokop, and M. Novák (eds.), Adv. Microbial Engineering, Part 1, p. 467, Wiley-Interscience, New York, 1973.*]

and the bubbles come together with cells as an intermedium. They are very stable and float on the surface of the culture liquid. *n*-Paraffin is completely exhausted in the fourth phase. The cells are dispersed uniformly throughout the culture liquid. In this period the nature of the culture liquid may be similar to that in a carbohydrate fermentation, and $k_l a$ recovers its initial levels.

An analytical model including a few of the features of this fermentation type is considered in Chap. 9.

8.9. SCALING OF MASS-TRANSFER EQUIPMENT

As discussed by Oldshue,† the various quantities which may influence the product $k_l a$ in an agitated industrial reaction do not scale in the same way with reactor size or impeller rate.

1. The turbulent Reynolds number Re_t determines u_{rms} and thus bubble mass-transfer coefficients k_l

$$\text{Re}_t = \frac{\rho u_{\text{rms}} D}{\mu} \propto \frac{\rho D}{\mu} \left(\frac{D\ P}{\rho\ V} \right)^{1/3} \tag{8.92}$$

2. The impeller tip velocity $\pi N_i D_i$ determines the maximum shear rate $\dot{\gamma}$, which in turn influences both maximum stable bubble or microbial floc size (Sec. 8.6) and damage to viable cells (Sec. 8.8.4).
3. The power input per unit volume P/V through Re_t determines mass-transfer coefficients and particulate sizes. In laminar and transition regimes of aerators

$$P \propto N_i^2 D_i^3 \qquad \text{from Fig. 8.7}b$$

For turbulent regimes, the power number is constant; thus

$$P \propto N_i^3 D_i^5$$

Then taking V_{reactor} to scale with D_i^3 gives

$$\frac{P}{V} \propto \begin{cases} N_i^2 & \text{laminar, transition aeration} \\ N_i^3 D_i^2 & \text{turbulent aeration} \end{cases}$$

4. The power input *during* aeration is

$$P_a = m' \left[\frac{P^2 (N_i D_i^3)^{0.44}}{N_a^{0.56}} \right]^{0.45} \tag{8.93}$$

Thus,

$$\frac{P_a}{V} \approx m' \left[\frac{P^2}{V^{2.22}} \frac{(N_i D_i^3)^{0.44}}{N_a^{0.56}} \right]^{0.45} \tag{8.94}$$

which will determine the motor size needed during fermentation.

---

† S. Y. Oldshue, *Biotech. Bioeng.*, **8**: 3, 1966.

5. If the vessel liquid is well mixed internally, a characteristic circulation time exists. The liquid recirculation flow rate F_l through the impeller region varies as a cross-sectional area πD_i^2 and tank average impeller velocity varies as $N_i D_i$. Thus

$$\frac{F_l}{V} \propto \frac{N_i D_i^3}{D_i^3} = N_i \qquad \text{time}^{-1}$$

a quantity of importance since it is inversely proportional to the time that fluid may spend away from the homogenizing influence of the impeller.

Oldshue has presented the relationship between these and other variables for the example case with a volume scaling from small (80-l) to large (10,000-l) fermentor (Table 8.4), i.e., a fivefold increase in D_i and a 125-fold increase in V. The verification of the ratios in this table is left as an exercise for the student.

It is apparent that various criteria may be chosen for scaling, depending on the fermentation broth. From a given choice of variable to be preserved on scaling, for example, P/V, the values of k_l and a can be estimated from the equations of this chapter.

The total oxygen consumption rate of the vessel is found by combining $k_l a$ with an appropriate macroscopic description of the vessel. If the bulk liquid composition is uniform and the bubbles are uniformly dispersed throughout the vessel, the mass-transfer rate is simply

$$V k_l a (c^* - c_{\text{liq}})_e = \text{oxygen consumption rate mol } O_2/s \qquad (8.95)$$

where subscript e refers to gas exit compositions (a consequence of the perfect mixing assumption is that exit-stream compositions equal reactor compositions, Chap. 9).

When the bubbles rise in plug flow through the vessel but the impeller still maintains perfect mixing in the liquid phase, c^* varies with position. Over a

Table 8.4. Relationship between properties for scale-up†

Property	Small scale, 80 l	Large scale, 10^4 l			
D	1.0	125	3125	25	0.2
P/V	1.0	1.0	25	0.2	0.0016
N_i	1.0	0.34	1.0	0.2	0.04
D_i	1.0	5.0	5.0	5.0	5.0
V_l	1.0	42.5	125	25	5.0
V_l/V	1.0	0.34	1.0	0.2	0.04
$N_i D_i$	1.0	1.7	5.0	1.0	0.2
Re_i	1.0	8.5	25	5.0	1.0

† S. Y. Oldshue: Fermentation Mixing Scale-up Techniques, *Biotech. Bioeng.*, **8**: 3, 1966.

differential reactor height dz, the instantaneous loss of oxygen from the bubble is

$$HA \, dz \frac{dp_{O_2}}{dt} \frac{1}{RT} = \frac{\text{gas vol}}{\text{reactor vol.}} \left(\frac{\text{differential}}{\text{reactor volume}} \right) \left(\frac{\text{conc. rate of}}{\text{change in bubble}} \right) \quad (8.96)$$

which equals

$$-k_l \cdot a(c_l^* - c_b)A \, dz \quad (8.97)$$

the mass-transfer rate into the liquid. Since $p_{O_2} = Mc_l^*$ (M is Henry's law constant), we have

$$\frac{H}{RT} \frac{dp_{O_2}}{dt} = \frac{HM}{RT} \frac{dc_l^*}{dt} = -k_l a(c_l^* - c_b) \quad (8.98)$$

For constant bubble-rise velocity u_b, $dt = dz/u_b$, and the z variation of c^* is seen to be

$$\ln \frac{(c_l^* - c_b)_{\text{exit}}}{(c_l^* - c_b)_{\text{inlet}}} = \frac{-k_l aRT}{HM} \frac{z}{u_b} \quad (8.99)$$

or

$$(c_l^* - c_b)_z = (c_l^* - c_b)_{\text{inlet}} \exp - \frac{k_l aRTz}{HMu_b} \quad (8.100)$$

The overall mass-transfer rate in the volume Ah is therefore

$$\int_0^h k_l a(c_l^* - c_b)(z)A \, dz = \frac{HMu_b}{RT}(c_l^* - c_b)_{\text{inlet}} Ah\left(1 - \exp \frac{-k_l aRTh}{HMu_b}\right) \quad (8.101)$$

The interaction of mixing, fermentation kinetics, and mass transfer is considered in the remaining text chapters.

8.10. PARTICULATE MASS TRANSFER: FILTRATION

Particulate transport is important in such varied processes as air and water filtration, settling and thickening operations, and coagulation (in addition to the previously discussed influences of particulates on solute mass-transfer coefficients and suspension rheology). Unit operations of settling, clarification, thickening, and coagulation are discussed in other texts commonly found in various engineering curricula and are not further considered here (see references).

This chapter section focuses on the removal of viable organisms from flowing streams by synthetic or natural filtering devices. Treatment of vacuum-filter operation and related techniques for cell harvesting will be discussed very briefly in Chap. 10; further reading for such high cell concentrations is given in the references.

Aerobic fermentation technology requires large continuous flows of essentially sterile air for aeration of appropriate microbial reactors, usually employing one desired species. The removal of airborne contaminants is accomplished primarily by use of mechanical filters, although heating, radiation, and electrically aided removal have been shown to be useful. The dynamics of filtration are assessed by considering the forces acting on isolated particles in flows past filter collector surfaces. Since the contaminant level is usually quite small, the following discussions consider the behavior of isolated particles.

In addition to pretreatment of industrial airflows for aerobic fermentations, air filtration occurs in such common processes as breathing (nasal and lung particle deposition), cigarette-smoke filtration, atmospheric-haze removal by rainfall and, in the health field, the operation of germ-free hospital rooms and the protective action of a surgeon's mask.

The collection mechanisms to be outlined below apply also (in concept) to liquid streams. Thus adsorption of bacteria and viral contaminants onto activated-charcoal packings in tertiary water treatment, soil adsorption from particulate-containing percolating liquids, and even perhaps the mechanics by which protozoa and other motile organisms feed on smaller nonmotile species are susceptible to analysis by considering the mechanisms discussed below.

8.10.1. Single-Fiber Efficiencies and Mass-Transfer Coefficients

For conceptual convenience, consider the flow of a fluid perpendicular to a cylindrical filter fiber, as shown in Fig. 8.13. The primary phenomena leading to particle collection by the filter fiber, assuming for the moment that it is a perfect sink, are the following:

Diffusion As a given fluid element follows its streamline in flow past the collector surface, the small but finite *diffusivity* of the particle leads to a finite collection rate. This particulate movement is termed Brownian motion.

Interception If we assume that the particle center follows the streamline of flow, particles whose streamline passes closer than half the particle diameter, $d_p/2$, to the collector surface will be collected due to their *finite* radius.

Impaction If the particle inertia is sufficiently large, the forces which cause the fluid to change course and move smoothly around the fiber will be insufficient to prevent the particle's *inertia* from carrying it into the collector surface.

Gravitation Due to a density difference, the particle falls (rises) in the fluid, again following a trajectory deviating from the fluid streamlines and, over one-half of the collector surface, resulting in enhanced collection.

Electrical forces Under the influence of an electric field, the particles deviate from streamline flow and are carried to the collector surface. The electric field may be due to particle-collector pairs of opposite charge or to the presence of an external applied field.

Other forces London and van der Waals forces also operate to aid or retard particle collection

The field of fluid mechanics devoted to the motion of particles near solid surfaces is enormous. The detailed discussion of various particle-collection

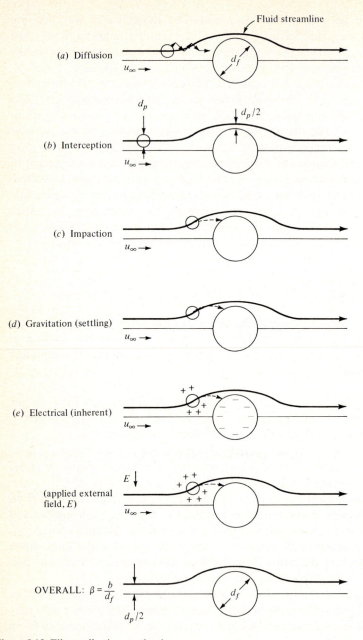

Figure 8.13 Fiber-collection mechanisms.

theories under varying flow regimes is not detailed here but can be found by consulting the references as a start.

Our object in the following discussion is to survey some useful results of single-fiber collection theory and to indicate the relation between such theoretical results and the measurable particle mass-transfer coefficient k_p in a packed bed of filter fibers (or other filter particles). (See Friedlander [12].)

If we consider a very dilute suspension of particles flowing past a filter fiber fixed in space, a fiber removal efficiency β can be defined by

$$\beta = \frac{b}{d_F} \tag{8.102}$$

where b is the width of an approaching flow totally cleared of particles by the filter and d_F is the fiber diameter (Fig. 8.13, bottom illustration). In a unit width of filter mat, the differential number of collector fibers is given by

$$\frac{(1)(\alpha \cdot dz)}{(1)(\pi d_F{}^2/4)} = \frac{\text{differential solid volume}}{\text{volume of fiber of unit length}} \tag{8.103}$$

where z is the distance coordinate in the direction of entering flow and α is the solid volume fraction of filter.

If the fiber surface is a perfect particle sink (no saturation phenomena or unsuccessful strikes), the particles are collected at a rate proportional to their concentration (each particle acts independently). Thus, the differential change in bulk fluid particle concentration N_p is

$$dN_p = -bN_p \frac{\alpha \, dz}{\pi d_F{}^2/4} \tag{8.104}$$

or, rearranging,

$$\frac{dN_p}{dz} = \frac{-4\alpha}{\pi d_F} \beta N_p \tag{8.105}$$

using definition (8.102) for fiber collector efficiency.

In relation to the transport of solutes earlier in this chapter and to the collection of particles by nonfibrous filter mats and other less well-delineated particle collectors, it is also useful for us to define a *particle* mass-transfer coefficient k_p, which satisfies the equation

$$J_p = k_p[N_p(z) - N_p(\text{collector surface})] = k_p N_p(z) \tag{8.106}$$

where J_p is the average particle flux at a given depth z into the filter bed.

Equating the removal rate from the flow to the local (irreversible) collection rate gives

$$\frac{u}{1-\alpha} \, dN_p = -\alpha a_F F \, dz = -k_p \alpha a_F N_p \, dz \tag{8.107}$$

where a_F is the filter surface area per unit filter volume (analogous to a in oxygen-transfer discussion) and u is the superficial, i.e., empty-tube, fluid velocity. Integration of (8.105) and (8.107) leads to

$$\ln \frac{N_p(z)}{N_p(0)} = \frac{-4\alpha\beta z}{\pi d_F} \quad \text{and} \quad \ln \frac{N_p(z)}{N_p(0)} = \frac{-k_p \alpha a_F z(1-\alpha)}{u} \tag{8.108}$$

from which the relation between the *phenomenological* particle mass-transfer coefficient k_p and the single-fiber collector efficiency β is seen to be

$$\beta = \frac{k_p \pi d_F a_F}{u} \frac{}{4}(1-\alpha) \tag{8.109}$$

When all the fibers are identical,

$$a_F = \frac{\pi d_F}{\pi(d_F/2)^2} = \frac{4}{d_F} \quad \text{and} \quad \beta = \frac{k_p}{V}\pi(1-\alpha) \tag{8.110}$$

For either fibrous or nonfibrous filters, the right-hand equation (8.108) describes the initial filter performance. Surface *saturation* effects require the inclusion of a collection history at the fiber surface, resulting in the introduction of a coefficient γ to describe the fraction of particle-collector collisions which result in irreversible particle collection. Filters over time may accumulate substantial loadings of collected particles. Such deposits change the total solid surface of the filter per unit volume a_F and may also seriously alter the apparent geometry of the collection surface. Such particle accumulation often produces an increased pressure drop through the filter. As the general phenomena leading to particle collection are the same in fibrous or nonfibrous filters, we consider only results from theories for single fibers, as these currently provide the more complete insight into filtration.

8.10.2. Functional Forms for Fiber-Removal Efficiency

Some analytical results for Brownian motion, interception, impaction, and gravitational collection from fluids flowing past collectors are presented in Table 8.5. These forms are written in physical variables; since β is dimensionless and proportional to the particle mass-transfer coefficient, we expect that a dimensionless representation of β in terms of dimensionless groups would also be useful.

For diffusion, the particle diffusivity is given by $\mathscr{D}_p = k_B T/3\pi\mu_c d_p$ (this should be corrected for smaller particles, see Prob. 8.15). The Brownian single-fiber collection efficiency can be written

$$\beta_B = 0.9\left(\frac{3\mathscr{D}_p \mu_c}{d_F u_\infty}\right)^{2/3} = m_1 \, \text{Pe}_F^{-2/3} \tag{8.111}$$

where Pe_F is the Peclet number referred to the fiber radius and $m_1 = 0.9(1.5\mu_c)^{2/3}$.

Interception depends, in the first approximation, only upon the ratio of particle to filter diameter, $R_{pf} = d_p/d_f$,

$$\beta_I = 1.5 R_{pf}^2 \tag{8.112}$$

Table 8.5. Functional forms of single-fiber collector efficiency†‡

Brownian motion: diffusion	$\beta_B = 0.9\left(\dfrac{k_B T}{\pi d_F d_p u_\infty}\right)^{2/3}$
Interception	$\beta_I = 1.5\left(\dfrac{d_p}{d_F}\right)^2$
Impaction	$\beta_{Im} = 0.075\left(\dfrac{\rho_p N_p d_p^{\,2}}{9\mu_c d_F}\right)^{1/2}$
Gravitation	$\beta_G = \dfrac{\Delta\rho \, g d_p^{\,2}}{18\mu_c N_p}$

† k_B = Boltzmann's constant, u_∞ = upstream fluid velocity
and ρ_p = particle density.
‡ K. M. Yao, M. T. Habibian, and C. R. O'Melia, *Environ. Sci. Technol.*, **5**: 1105, 1971.

As the phenomena leading to Brownian collection and to interception are not coupled in this analysis, the total collection efficiency for a single fiber can be written as the sum of the two contributions:

$$\beta_{tot} = \beta_B + \beta_I = m_1 \, \mathrm{Pe}_F^{\,-2/3} + 1.5 R_{pf}^{\,2}$$

or
$$\beta_{tot} \, \mathrm{Pe}_F \, R_{pf} = (m_1 \, \mathrm{Pe}_F^{\,1/3})(R_{pf}) + m_2(\mathrm{Pe}_F^{\,1/3} R_{pf})^3 \qquad (8.113)$$

Equation (8.113) has been found to fit data for fibrous air filters when $m_1 = 1.3$ and $m_2 = 0.7$ [12].

The limitations in the theoretical developments leading to β_B and β_I in Table 8.5 are that the Reynolds number referred to the filter must be small,

$$\mathrm{Re}_F = \frac{u_\infty d_F \rho}{\mu_c} < 1$$

the fiber Peclet number must be large,

$$\mathrm{Pe}_F = \frac{u_\infty d_F}{2 D_p} > 50$$

and particle inertia be negligible (impaction not important).

In a real filter, the presence of other fibers alters the flow field neighboring a given fiber and hence also the single-fiber collection efficiency for this fiber. A useful treatment, which includes the volume packing of solids (again assuming identical fibers) α, defines the Kuwabara parameter Ku as

$$\mathrm{Ku} \equiv -\frac{1}{2}\ln\alpha - \frac{3}{4} + \alpha - \frac{\alpha^2}{4} \qquad (8.114)$$

An empirical fit using Ku, an exact solution for β_I in this Kuwabara flow field, and inclusion of an additional combination term for interception-diffusion gives

$$\beta_B + \beta_I + \beta_{BI} = 2.9 \ \text{Ku}^{-1/3}\text{Pe}^{-2/3} + 0.62 \ \text{Pe}^{-1}$$

$$+ \frac{1}{2 \ \text{Ku}}[2(1 + R_{pf}) \ln (1 + R_{pf}) - (1 + R_{pf})$$

$$+ (1 + R_{pf})^{-1} + \alpha(-2R_{pf}^2 - \tfrac{1}{2}R_{pf}^4 + \tfrac{1}{2}R_{pf}^5 + \cdots)]$$

$$+ 1.24 \ \text{Ku}^{-1/2}\text{Pe}^{-1/2}R_{pf}^{2/3} \tag{8.115}$$

The Ku parameter varies from 2.7049 ($\alpha = 0.001$) to 0.034 ($\alpha = 0.5$). For Peclet numbers above 200 and small values of $R_{pf}(< 0.05)$, Eq. (8.115) reduces to

$$\beta_{\text{tot}} \approx 2.9 \ \text{Ku}^{-1/3}\text{Pe}^{-2/3} + \text{Ku}^{-1}R_{pf}^2(1 - \alpha) \tag{8.116}$$

which is the same form as Eq. (8.113) for single fibers. Using the full form of Eq. (8.115) along with the left-hand equation (8.108) allows evaluation of filter performance where only interception and diffusion are important.

At higher flow rates, particle inertia becomes influential in contributing to particle collection as the fluid changes course in passing the filter fiber. The ratio of the initial particle momentum mu_∞ to the work done in pulling the particle with viscous drag over a length characterized by the fiber dimension is the Stokes number (St)

$$\text{St} \equiv \frac{mu_\infty}{3\pi d_p \mu_c d_F} = \frac{\pi(d_p^3/6)\rho_p u_\infty}{3\pi d_p \mu_c d_F}$$

so that

$$\text{St} \equiv \frac{1}{18}\frac{d_p^2 \rho_p u_\infty}{\mu_c d_F} \tag{8.117}$$

Evidently, for St > 1, the viscous forces trying to carry the particle around the fiber will be insufficient, and the *inertia* of the particle carries it into the fiber. The impaction term can be written from Table 8.5 and Eq. (8.117) as

$$\beta_{\text{Im}} = 0.075\sqrt{2 \ \text{St}} \tag{8.118}$$

The gravitation term contains the expression sometimes known as the *gravitation number*, which is simply the ratio of terminal particle velocity to upstream fluid velocity; thus,

$$\beta_G = \frac{\Delta\rho \ g d_p^2}{18\mu u_\infty} = \frac{(u_p)_{\text{term}}}{u_\infty} \tag{8.119}$$

Noting that a typically important term, the interception value $\beta_I = 1.5 R_{pf}^2$, is the order of 10^{-4} when 1-μm particles are collected by 100-μm filters, any appreciable particle-density differences become important at small terminal velocities. This influence will clearly be more important in air filtration than in filtration of liquids.

In the presence of applied electric fields, a neutral particle will be polarized, and an additional particle velocity due to the nonuniform field will arise, leading to further particle collection. The theoretical development here is even more complex than the difficulties underlying the forms of Table 8.5. A useful form for low fiber efficiencies β and small Reynolds number due to Zebel† is

$$\beta_{AF} = 9.66 \times 10^{-4} \frac{d_p^2 (I_N)^2}{d_f u_\infty F(d_p)} \tag{8.120}$$

where I_N is the applied field in volts per centimeter. For $I_N = 10^3$ V/cm, $u_\infty = 10$ cm/s, $d_p/d_F = 0.1$, the values of β are

$$\beta_{AF} = \begin{cases} 5.14 \times 10^{-5} & d_p = 10^{-1} \ \mu m \\ 8.94 \times 10^{-4} & d_p = 1 \ \mu m \\ 9.66 \times 10^{-3} & d_p = 10 \ \mu m \end{cases}$$

The charging of particles and their collection relates to electrostatic-precipitation operation. For detailed discussion of this particle-collection mode see Ref. 11.

The overall collection efficiency β_{tot} depends of course, on the predominant phenomena leading to particle deposition. For example, under conditions where the above electrical effects are unimportant, the domains of influence of the collection mechanisms of Brownian diffusion, interception, impaction (inertia) and gravity are presented in Fig. 8.14a. Equally, Fig. 8.14b provides contours of constant total isolated-fiber collection efficiency (both figures refer to 10-μm-diameter fibers; the formulas in Table 8.5 apply to somewhat larger fiber diameters, but the qualitative results are similar). We notice first that each of the terms of Table 8.5 may be important in air filtration; the same is true in liquids, although the relative influence of the gravitation and impaction terms is much diminished due to decreased $\Delta\rho$ and the larger viscosity of the liquid vs. air.

A most important general result emerges from Fig. 8.14; for any given flow velocity, there exists a valley of relatively inefficient particle removal. This valley spans a wider range of diameters as the air velocity diminishes. In consequence, there is at any velocity a most *penetrating* particle size. We see from Fig. 8.14b that this particle size is characteristically about 0.5 μm for the case illustrated. Reference to Table 8.5 shows that this particle size at minimum single-fiber efficiency changes with fiber diameter, larger fibers being typically less efficient (smaller β). For example, in liquid filtration of neutrally bouyant particles in slow flows (Re $\leq$ 1), the collection efficiency β_T is given to a good approximation by

$$\beta_T = \beta_B + \beta_I = 0.9 \left(\frac{k_B T}{\pi d_F d_p u_\infty} \right)^{2/3} + 1.5 \left(\frac{d_p}{d_F} \right)^2 \tag{8.121}$$

which has a minimum when $\partial \beta_T/\partial d_p = 0$, that is, at $d_p = (0.2 k_B T d_F^2/\pi u_\infty)^{1/4}$.

† G. Zebel, Deposition of Aerosol Flowing past a Cylindrical Fibre in a Uniform Electric Field, *J. Colloid Sci.*, **20**: 522, 1965.

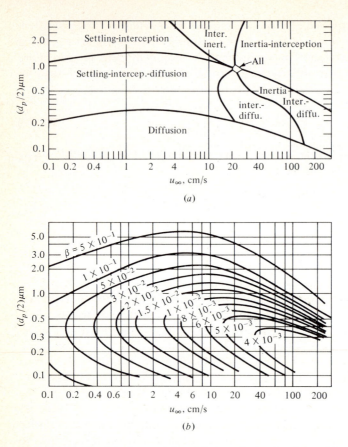

Figure 8.14 (a) Particle size vs. airflow velocity, calculated regimes of dominance for gravity, diffusion, interreption, and impaction (inertia); (b) isoefficiency contours for the same conditions as (a). [*Reprinted from C. N. Davies, "Air Filtration," p. 52, Academic Press, Inc. (London) Ltd., London, 1973.*]

Thus d_p ($\beta = \beta_{min}$) increases with increased fiber diameter.

As the most penetrating particle size is often in the range of diameters characterizing microbial organisms, viruses, and other small particles on which the former may temporarily reside, filter performance should often be evaluated at this worst-case design condition.

The selection of a filter depends upon the individual fiber efficiency, the fiber packing density characterized by α, and the fiber depth L. Given a fiber size and the α for a particular filter, the value of β_T and hence the effective particle mass-transfer coefficient k_p can be evaluated from the equations of this section. The value of z corresponding to a sufficiently deep fiber bed (depth $= L$) to accomplish the desired effluent level of particle numbers $N_p(L)$ is then obtained from the appropriate expression in Eq. (8.108).

Two final aspects of this topic should be mentioned. Nonfibrous filters are geometrically nonuniform assemblies compared with the ideal fibers of this section. Nonetheless, the concepts of the previous discussion apply equally to these newer filter varieties. The pressure drop through filters, especially those which accumulate appreciable particulate loadings, can rise to appreciable and even unmanageable values (see [11]).

8.11. HEAT TRANSFER

In biological reactors, heat may be added or removed from a microbial fluid for the following reasons:

1. It is desired to *sterilize* a liquid reactor feed by heating in a batch or continuous-flow vessel. Thus, the temperature desired must be high enough to kill essentially all organisms in the total holding time (Chap. 9).
2. Since the heat generated in substrate conversion is inadequate to maintain the desired temperature level, heat must be *added*. For example, the reactor is an anaerobic sewage-sludge digestor which operates best between, say, 55 and 60°C.
3. The conversion of substrate generates *excess* heat with respect to optimal reactor conditions for, e.g., maintenance of viable cells, so heat must be removed, as in fermentation of hydrocarbon substrates for single-cell protein production.
4. The water content of a cell sludge is to be reduced by drying.

The first three relate to cell viability and metabolism and will therefore be of concern here. The last example is a unit operation, drying, which is covered in most texts on engineering unit operations.

The heat is transferred between the microbial fluid to or from a second fluid in several ways, i.e., with externally jacketed vessels, coils inserted in a larger vessel, or by evaporation or condensation of water and other volatile components of the cell-containing fluid. Examples of such configurations are shown in Fig. 8.15. Temperature fluctuations between atmosphere and thermally stratified lakes and land also clearly involve heat transfer, the resulting temperatures determining the habitable niches for species. The present section focuses on heat transfer in mechanical reactors (a lake-eutrophication model incorporating daily solar-energy oscillations appears in Chap. 12).

Assuming that transfer rates and changes in other forms of energy are negligible, the fundamental steady-state equation in heat transfer relates the total rate at which heat is generated to its rate of removal through some heat-transfer surface; thus

$$\text{Net generation rate} = \text{removal rate} = h_w A \, \Delta T \qquad (8.122)$$

where ΔT = characteristic temperature difference between fluid and solid surface
A = heat-transfer surface area
h_w = heat-transfer coefficient

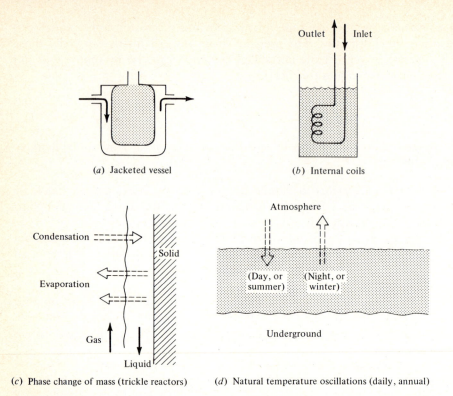

(a) Jacketed vessel   (b) Internal coils

(c) Phase change of mass (trickle reactors)   (d) Natural temperature oscillations (daily, annual)

Figure 8.15 Examples of heat-transfer configurations: (a) jacketed vessel, (b) internal coils, (c) phase change of mass, and (d) natural temperature oscillations.

As with mass transfer, most of the resistance to heat transfer resides in a relatively quiescent thin fluid film near the solid heating-cooling boundary, the bulk fluid being frequently well mixed and thus approximately isothermal. Our main concerns in this chapter are the identification of the magnitude of the microbial heat-generation rate, development of an overall energy balance, and review of useful predictive formulas for h_w for various heater-cooler-sterilizer systems of interest.

The heat-transfer coefficients (Table 8.6) for boiling water and condensing vapors (steam, typically) make such fluids convenient in sterilization "reactors" (Chap. 9). Where lower temperatures are needed, as in heating anaerobic sludge digestors, a nonboiling water stream is useful. Viscous liquids exhibit greater heat-transfer resistances than water; as with mass transfer, this is due to lesser degree of bulk-fluid interchange with wall fluid and also to reduced thermal conductivity (analogous to $\mathscr{D}_{O_2}$).

Inspection of Fig. 8.15 reveals several design problems associated with heat transfer in biochemical reactors. The externally jacketed system has a heat-transfer area A which varies as the tank diameter (say impeller diameter) D_i^2. The

Table 8.6. General magnitude of heat-transfer coefficient h_w † is:

	h_w, kcal/(m^2 · h · °C)‡
Free convection:	
Gases	3–20
Liquids	100–600
Boiling water	1000–20,000
Forced convection:	
Gases	10–100
Viscous liquids	50–500
Water	500–10,000
Condensing vapors:	1000–100,000

† Data from H. Gröber, S. Erk, and U. Grigull, "Wärmeübertragung," 3d ed., p. 158, Springer-Verlag, Berlin, 1955.

‡ Multiplication by 0.204 gives h_w in units of Btu/(ft^2 · h · °F).

volumetric heating or cooling demand of a reactor scales as D_i^3 if the overall microbial reaction rates and power input per unit volume are unchanged. Thus, jacketed vessels, sufficient in rendering laboratory reactors isothermal, must frequently be replaced by reactors with internal coils in heating, e.g., anaerobic sludge digestion, or cooling, e.g. hydrocarbon fermentation for single-cell protein production, of large-scale reactors.

The presence of such internal piping clearly alters mixing patterns, fluid velocities, and perhaps bubble-coalescence rates. The complexity of such a situation reminds us of the need to perform measurements in reactors approximating the desired configurations, as we may expect *a priori* design of such reactors to be more uncertain than for simpler systems. (Problems of vessel mixing patterns are discussed in the text and problems of Chap. 9.) For example, the correlations in Sec. 8.11.2 indicate that the heat-transfer coefficient h_w[Btu/(ft^2 · h) or kcal/(m^2 · h)] changes as we shift configurations from a single cooling coil perpendicular to fluid flow to a set of staggered rows of coils in a tube bundle but still perpendicular to the fluid-flow direction. Thus, the presence of the first row of coils alters the flow pattern past subsequent tube rows.

Heat generation and removal rates are known with sufficient accuracy for some detailed heat-balance considerations to be possible, provided we clearly understand the basis on which such calculations are made and take appropriate precautions in terms of overdesign to allow for some uncertainty. The following section discusses the estimation of the heat-transfer demand (analogous to an oxygen demand); the subsequent sections discuss the conductance h_w.

Before examining those topics, however, we consider an overall energy balance. In a constant pressure system with negligible changes in potential and kinetic energies, the energy balance can be cast in terms of enthalpy changes, i.e., the heats of chemical transformation or phase transformation (e.g., evaporation, condensation), the sensible-heat flow in mass streams, and the heat transfer to or from second fluids acting as heating or cooling devices. Let

Q_{gr} = heat-generation rate from cell growth
Q_{ag} = heat-generation rate due to reactor agitation

$\approx$ power input per unit volume $\dfrac{P}{V}$

Q_{acc} = heat-accumulation rate
Q_{exch} = heat-transfer rate to surroundings or exchanger
Q_{evap} = rate of heat loss by evaporation
Q_{sen} = rate of sensible-enthalpy gain of
streams (exit − entrance)

Then

$$Q_{gr} + Q_{ag} = Q_{acc} + Q_{exch} + Q_{evap} + Q_{sen} \tag{8.123}$$

Cooney, Wang, and Mateles [47] have utilized such a balance to calculate Q_{gr} from measurement of Q_{acc} through monitoring the initial transient temperature rise of a nearly isolated fermentor. Under such an experiment, Q_{evap} and Q_{sen} are quite small, and Q_{exch} represents an important term compared with the *difference* of the larger individual rates $Q_{acc} - Q_{ag}$. As just mentioned, Q_{acc} was monitored calorimetrically, while Q_{ag} was calculated at each gas flow and impeller rate from the correlation of Michel and Miller [Eq. (8.69)].

In a fermentor design, $Q_{acc} = 0$ for a steady-state system [although tempera-ture programming a batch reactor for optimal product yields (Example 10.2) may provide an additional complication]. Q_{ag} is estimated for ungassed or gassed systems using the appropriate power correlations presented earlier.

Neglecting Q_{evap} (which may be an important mechanism in trickle reactors) and Q_{sen} for the moment, the important remaining quantity is Q_{gr}. Methods for estimating or measuring Q_{gr} are discussed in Sec. 8.11.1. In design of larger-scale reactors, the choice of an operating temperature and flow conditions will deter-mine Q_{evap} and Q_{sen}, as the choice of agitator speed and diameter will determine Q_{ag} (corrected for the chosen aeration rate). From expected oxygen or substrate consumption rates, Q_{gr} is calculated by the methods of the following section. The remaining terms are Q_{acc} and Q_{exch}. Whether or not the reactor is operated isothermally, at each instant

$$\underbrace{Q_{exch}}_{\substack{\text{Heater or}\\\text{cooler}\\\text{duty}}} = \underbrace{Q_{gr} + Q_{ag}}_{\text{Generation}} - \underbrace{Q_{acc}}_{\substack{\text{Accumula-}\\\text{tion}}} - \underbrace{(Q_{sen} + Q_{evap})}_{\substack{\text{Removal by other}\\\text{than a solid heat}\\\text{exchange surface}}} \tag{8.124}$$

This last equation sets the heat-transfer magnitude needed to maintain the desired temperature and rate of heat accumulation, if any.

8.11.1. Microbial Heat Generation

From our introductory comments, the problems of liquid-media sterilization, sludge-digester heating, and hydrocarbon fermenter cooling reduce to two. The first process requires knowledge of k_{death} as a function of temperature for the cells (recall Sect. 7.5); the last two demand knowledge of Q_{exch} (kilocalories per hour per volume of reactor) to be transferred in order to maintain the reactor at its desired temperature. The performance of the sterilizer, given k_{death} for the cells, depends upon the wall heat-transfer conductance $h_w A$, where A is the sterilizer heating surface and h_w is the wall heat-transfer coefficient (see Sec. 8.11.2). The microbial reactors involve a heat source arising directly as a *result* of microbial metabolism; methods for estimating this term, Q_{gr}, are our focus here.

The two more fundamental microbial measures which allow a priori calculation of the microbial heat generation rate Q_{gr} from an overall heat balance are estimates based on yield factors Y or heats of combustion data. For the first case, we define a yield factor Y_Δ (grams of cell mass per kilocalorie of heat evolved) analogous to the earlier substrate and oxygen yield coefficients. If Y_s is grams of cell mass produced per gram substrate consumed and ΔH_s and ΔH_c are the heats of combustion of substrate and of cell mass material, respectively (kilocalories per gram), we can write

$$Y_\Delta \text{ (g cell/kcal)} = \frac{Y_s \text{ (g cell/g substrate)}}{(\Delta H_s - Y_s \Delta H_c)(\text{kcal/g substrate})} \qquad (8.125)$$

This arises from approximate energy balances over two pathways, as shown in Fig. 8.16.

Provided that the *predominant* oxidant is oxygen, the heat generation ΔH_s per gram of substrate completely oxidized minus $Y_s \Delta H_c$, the heat obtained by combustion of cells (and extracellular-fluid residue) grown from the same amount of substrate, will reasonably approximate the heat generation per gram of substrate consumed in the substrate fermentation which produces cells, H_2O, and CO_2. More generally, if a cell-maintenance term m is also included [43], we can write

$$\frac{1}{x}\frac{ds}{dt} = \frac{\mu}{Y_g} + k_e \qquad \text{substrate consumption} \qquad (8.126a)$$

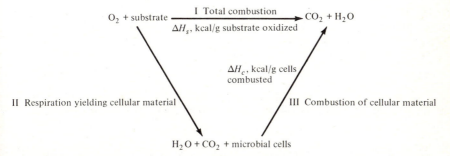

Figure 8.16 Approximate heat balance in substrate consumption.

where x is biomass concentration, so that Y_s is related to the actual growth yield factor Y_g by

$$Y_s = \frac{\mu}{\mu/Y_g + k_e} \qquad \frac{\text{g cell}}{\text{g substrate consumed}} \tag{8.126b}$$

In Fig. 8.16, the heat of combustion of cellular material reasonably represents the energy *stored as biosynthesis* during the generation of cells in step II (*provided that the only metabolic products are* CO_2 *and* O_2). The remainder of the total heat of substrate consumption appears as heat in step II. This latter catabolic quantity represents the heat generation during fermentation.

In the absence of experimental values, the heat of combustion of cells can be estimated from the average cell composition by the following method [43]:

1. A balance for cell combustion is written assuming the products to be CO_2, H_2O, and N_2. For example, for *Pseudomonas fluorescens* growing in a glucose medium

$$\underset{\text{Cells}}{C_{4.41}H_{7.3}N_{0.86}O_{1.19}} + 5.64O_2 \longrightarrow 4.41CO_2 + 0.43N_2 + 3.65H_2O$$

2. The heat of combustion per electron transferred to a methane-type bond is taken to be 26.05 kcal per electron equivalent or, since O_2 has four such electrons, 104.20 kcal per mole of O_2.
3. Therefore the cellular heat of combustion is ΔH_{tot} per weight of "cell." For the *P. fluorescens* example this is

$$\frac{(5.64 \text{ mol } O_2)(104.20 \text{ kcal/mol } O_2)}{(4.41)(12) + (7.3)(1.0) + (0.86)(14) + (1.19)(16)} = 6.44 \text{ kcal/g}$$

4. The value from step 3 is corrected for ash content of the cells, e.g., at 10 wt percent in dried cells, the heat of combustion for *P. fluorescens* from step 3 is

$$(6.44 \text{ kcal/g})(0.90) \approx 5.8 \text{ kcal/g dry cell mass}$$

Therefore

$$\Delta H_c \approx 5.8 \text{ kcal/g dry cell mass}$$

[The heat of combustion of a substrate ΔH_s in Eq. (8.125) can be found in many chemistry handbooks.]

Thus, we can use Eq. (7.7a) to represent the instantaneous mass-generation rate per unit volume in a batch reactor

$$\frac{dn}{dt} = \mu n \tag{7.7a}$$

and the instantaneous microbial heat-generation rate Q_{gr} (heat/time) is evidently given by

$$Q_{\text{gr}} = V_{\text{reactor}} \mu n \frac{1}{Y_\Delta} \tag{8.127}$$

The value of Y_Δ will depend upon both the particular microbial species (through ΔH_c and m) and upon the substrate consumed (through ΔH_s). A sampling of such values appears in Table 8.7. Note in general that hydrocarbons produce more heat than partially oxygenated species $(Y_\Delta(CH_4) < Y_\Delta(CH_3OH),\ Y_\Delta(n\text{-alkanes}) < Y_\Delta(\text{glucose}))$. This concretely reflects the earlier observation that fermentations of hydrocarbons demand greater heat-transfer rates than microbial consumption of carbohydrate at the same mass rate. The corresponding equation for a continuous-flow isothermal reaction at steady state (see Chap. 9) is

$$Q_{gr} = V_{reactor}\,\mu n = (s_0 - s)Y_s F - nk_e \tag{8.128}$$

or

$$\frac{Q_{gr}}{V_{reactor}} = (s_0 - s)Y_s D - \frac{nk_e}{V_{reactor}} \tag{8.129}$$

Recall that, as before, Y_s may depend on the culture age in a batch reactor and upon dilution rate D in a continuous-flow system. The dependence of Y_s (and thus Y_Δ) on cell maintenance appears in Eq. (8.126b).

Some economic estimates from Abbott and Clamen [43] as of 1973 indicate that heat-transfer and mass-transfer (oxygen) operating costs of bacterial cell production from the substrates in Table 8.6 are both appreciable, as seen in Table 8.8.

The accuracy of the assumptions underlying Fig. 8.16 and Eq. (8.125) is exemplified by data for yeast growing on n-paraffins [44]. By experiment, ΔH_s and ΔH_c were found to be 11.4 and 4.7 kcal/g, respectively, giving

$$\frac{1}{Y_\Delta} = \frac{11,400}{Y_s} - 4700$$

Table 8.7. Comparison of yield coefficients for bacteria grown on various carbon sources†

Substrate	Y_s, g cell/ g substrate	Y_0, g cell/ g O_2 consumed	Y_Δ, g cell/ kcal
Maleate	0.34	1.02	0.30
Acetate	0.36	0.70	0.21
Glucose equivalents (molasses, starch, cellulose)	0.51	1.47	0.42
Methanol	0.40	0.44	0.12
Ethanol	0.68	0.61	0.18
Isopropanol	0.43	0.23	0.074
n-Paraffins	1.03	0.50	0.16
Methane	0.62	0.20	0.061

† From B. J. Abbott and A. Clamen, The Relationship of Substrate, Growth Rate, and Maintenance Coefficient to Single Cell Protein Production, *Biotech. Bioeng.*, **15**: 117, 1973.

Table 8.8. Effect of substrate and yield coefficients on operating costs of fermentation†

| Substrate | Cost, cents per pound of cells | | | |
	Substrate	O_2 transfer	Heat removal	Total
Maleate (as waste)	0	0.46	0.75	1.2
Glucose equivalents				
(molasses)	3.9	0.23	0.54	4.7
Paraffins	4.0	0.97	1.4	6.4
Methanol	5.0	1.2	1.9	8.1
Methane	1.6	3.3	3.7	8.6
Ethanol	8.8	0.75	1.3	11.0
Isopropanol	11.6	2.7	3.1	17.4
Acetate	16.7	0.62	1.1	18.4

† B. J. Abbott and A. Clamen, The Relationship of Substrate, Growth Rate, and Maintenance Coefficient to Single Cell Protein Production, *Biotech. Bioeng.*, **15**: 117, 1973.

which is Eq. (8.125) inverted. Direct measurement of Y_s and Y_Δ produced the following results [44]:

| | Y_Δ | |
Y_s, experiment	From Eq. 8.125	Measured
1.25	4400	4640
1.03	6400	6060
1.09	5800	5830

In fermentations or waste-treatment processes the substrates are frequently a mixture of many compounds, each having a different heat of combustion. Writing the standard free energy of combustion as ΔG, Servizi and Bogun noted [46] that the stoichiometric ratio (m_0-moles O_2 needed per complete oxidation of 1 mol of substrate) is proportional to ΔG for a number of compounds. In particular

$$\Delta G = -116 m_0 \qquad \text{kcal/mol substrate} \qquad (8.130)$$

for carbohydrates, TCA cycle intermediates (Chap. 5), and some products of glycolysis, and

$$\Delta G = -104 m_0 \qquad \text{kcal/mol substrate} \qquad (8.131)$$

for aromatics, alcohols, and aliphatic acids. [Hattori (Ref. 4 of Chap. 11) notes that if the chemical oxygen demand (COD) in grams of oxygen per mole is already known, COD $= 32 m_0$ and ΔG for the two groups follows directly by substitution.]

The average free energy released in complete oxidation of multiple substrates is simply

$$\overline{\Delta G} = \frac{\sum_i (M_i\,\Delta G_i)}{\sum_i M_i} \qquad \text{kcal/mole} \tag{8.132}$$

where i = species

M_i = moles of ith substrate

ΔG_i = free energy for complete oxidation of substrate i

Further, it appears that the yield coefficient Y for a reactor containing a number of unicellular species can be defined in the same manner as for the single species (bacteria or yeast) of the preceding discussion. Servizi and Bogan [46] determined the average yield coefficient $\overline{Y}_s$(grams of cell per mole of substrate) for the multiple-species–multiple-substrate system *activated sludge* and found (see Fig. 8.17)

$$\overline{Y}_s = 0.108\,\overline{\Delta G} \qquad \text{g cells/mol substrate} \tag{8.133}$$

Thus, given $\overline{Y}_s$, the other quantities in Eq. (8.125) can easily be determined experimentally, and hence $\overline{Y}_\Delta$ can again be calculated for the microbial heat generation Q_{gr}.

$$Q_{gr} = V_{\text{reactor}}\left\langle\frac{dn}{dt}\right\rangle\frac{1}{\overline{Y}_\Delta} \qquad \text{kcal/h} \tag{8.134}$$

It is reasonable to expect the coefficient in Eq. (8.133) to depend upon the microbial species involved, so that again direct measurement on the appropriate species is desirable. However, the original argument of Servizi and Bogan is that

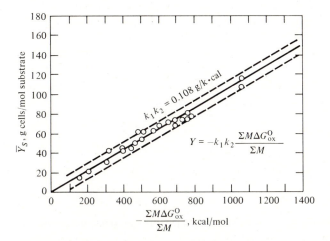

Figure 8.17 Average yield coefficient for activated sludge growth as a function of the average substrate free energy of oxidation. (*Reprinted from J. A. Servizi and R. H. Bogan, Thermodynamic Aspects of Biological Oxidation and Synthesis, J. Water Pollut. Control Fed.*, **36**: 607, 1961.)

$\bar{Y}_s \propto \Delta G$ is based on an assumed linearity between the yield coefficient $\bar{Y}_s$ and the number $\bar{N}_{ATP}$ of molecules of ATP synthesized per mole of substrate consumed. Since $\bar{Y}_s$ = grams of cell per gram of substrate, the coefficient of proportionality between $\bar{Y}_s$ and $\bar{N}_{ATP}$ is simply the grams of cells produced per mole of ATP utilized. This latter value is known to be approximately constant for a wide range of anaerobic species as well as some aerobes (Table 5.6). Thus, Eq. (8.133) may provide a useful estimate in general for mixed-substrate–mixed-population systems in the absence of better data. It is probably accurate only to the order of ± 20 to 30 percent, as can be seen best from the data for the correlation displayed in Fig. 8.16.

A study including *E. coli*, *Candida intermedia*, *B. subtilis*, and *A. niger* growing *separately* on carbon substrates glucose, molasses, or soybean meal also found that the *rate* of heat evolution varies linearly with the *rate* of oxygen consumption:

$$Q_{gr}[\text{kcal}/(\text{l} \cdot \text{h})] = (0.124 \pm 0.003)(\text{rate of } O_2 \text{ consumption}) \qquad (8.135)$$

when the O_2 consumption is in millimoles per liter per hour. In view of the previous equations (8.130) and (8.131) we can expect the proportionality parameter 0.124 to fall about 10 percent for the substrates of Eq. (8.131). The above result was obtained by calorimetry, i.e., by measuring the initial rate of temperature *increase* when the fermentor was thermally isolated from its surroundings.

8.11.2. Heat-Transfer Correlations

From Eq. (8.122) and the overall heat balance [Eq. (8.123)] for heating, cooling, or sterilizing, the general working equation for heat-transfer design is

$$Q_{exch} = h_w A \, \Delta T \qquad (8.122a)$$

In the specific instance of Eq. (8.122), ΔT was the bulk-fluid–wall-temperature difference, A the solid-surface area, h_w the fluid-side resistance to heat transfer. More generally we may know the temperatures of the bulk fluids on either side of the heat-transfer surface and thus only the *overall* temperature difference, $T_{\text{bulk, 1}} - T_{\text{bulk, 2}}$. Therefore we need to develop an expression for the overall heat-transfer coefficient $\bar{h}$, analogous to the earlier overall coefficient K_l for gas-liquid mass transfer. For steady-state heat transfer through a flat wall of thickness l separating the fermentation fluid at $T_{\text{bulk, 1}}$ from a heating or cooling fluid at $T_{\text{bulk, 2}}$, continuity of heat flux demands

$$h_{w1}(T_{\text{bulk, 1}} - T_{\text{wall, 1}}) = k_s \frac{T_{\text{wall, 1}} - T_{\text{wall, 2}}}{l}$$

$$= h_{w2}(T_{\text{wall, 2}} - T_{\text{bulk, 2}}) \qquad \text{kcal}/(\text{cm}^2 \cdot \text{h}) \qquad (8.136)$$

where k_s is the thermal conductivity of the wall in kilocalories per centimeter per second per Celsius degree. In terms of an overall heat-transfer coefficient $\bar{h}$, defined by

$$\text{Heat flux} = \bar{h}(T_{\text{bulk, 1}} - T_{\text{bulk, 2}}) \qquad (8.137)$$

rearrangement of the previous equations yields (8.137);

$$\frac{1}{\bar{h}} = \frac{1}{h_{w1}} + \frac{l}{k_s} + \frac{1}{h_{w2}} \qquad \text{planar wall} \qquad (8.138)$$

In clear analogy with our mass-transfer discussions, the overall resistance $1/\bar{h}$ is the sum of three resistances in series. For heat transfer across a cyclindrical-tube wall in heating or cooling coils, the cross-sectional area for heat transfer changes continuously through the wall. In this instance the appropriate equation for $\bar{h}$ is

$$\frac{1}{\bar{h}_o d_o} = \frac{1}{h_o d_o} + \frac{\ln (d_o/d_i)}{2k_s} + \frac{1}{h_i d_i} \qquad \text{tube wall} \qquad (8.139)$$

where d_i and d_o are the tube inside and outside diameters, respectively. Note the use of subscript o for $\bar{h}_o$ since it reminds us to use the outside tube surface as the basis for a heat-transfer area. The thermal conductivity k_s of the solid depends on the material; e.g., at 100°C, $k_s = 0.908$ cal/(s · cm · K) (copper) and 0.107 cal/(s · cm · K) (steel); k_s increases slowly with diminishing temperature. Appropriate values for different heat exchanger materials are found in standard engineering handbooks.

The analysis of momentum and heat transfer at either fluid-solid interface gives the *individual*-side heat-exchange coefficients (h_{w1}, h_{w2}) or (h_o, h_i) in Eq. (8.138) and (8.139). Where such individual coefficients vary along the heat-transfer surface, an overall *local* heat-transfer coefficient is defined by equations such as (8.138) and (8.139), and a detailed integration over the heat-transfer area is needed to calculate the total heat transferred.

For fluid-wall heat transfer, the important dimensionless groups are the following:

$$\text{Nusselt number} = \text{Nu} = \frac{hd}{k_f} \qquad (8.140a)$$

$$\text{Prandtl number} = \text{Pr} = \frac{C_p \mu}{k_f} \qquad (8.140b)$$

$$\text{Brinkman number} = \text{Br} = \frac{\mu u^2}{k_f(T_{\text{bulk}} - T_{\text{wall}})} \qquad (8.140c)$$

$$\text{Froude number} = \text{Fr} = \frac{u^2}{gd} \qquad (8.140d)$$

and $$\text{Reynolds number} = \text{Re} = \frac{\rho u d}{\mu} \qquad (8.140e)$$

where k_f = thermal conductivity of fluid, cal/(s · cm · °C)
C_p = heat capacity, cal/(g · °C)
d = distance
μ = viscosity
u = velocity
g = gravitational constant

As discussed elsewhere [6], the Brinkman number represents heat production by viscous dissipation divided by heat transport by conduction and may usually be neglected *at the heat-exchanger surface* for our purposes. (In the impeller-tip vicinity, this number becomes important.) Similarly, in a baffled vessel or one with an off-center stirrer, the Froude number is usually negligible.

The heat-transfer coefficient h, rendered dimensionless as Nu, is a function of Pr and Re:

$$Nu = f(Pr, Re) \tag{8.141a}$$

As hydrodynamics may vary with the aspect of the exchange surface, i.e., the length L to diameter d ratio, L/d, correlations are available in the form

$$Nu = f'\left(Pr, Re, \frac{L}{d}\right) \tag{8.141b}$$

Temperature variations induce variations of fluid properties at different points near the heat-transfer surface. As liquid viscosity is the most important of these, the ratio $\mu_b/\mu_0 = $ (viscosity at $T = T_{\text{bulk fluid}}$)/(viscosity at wall temperature) is a useful correlating variable:

$$Nu = f''\left(Pr, Re, \frac{L}{d}, \frac{\mu_b}{\mu_0}\right) \tag{8.141c}$$

As h (and therefore Nu) may be defined as a *local* transfer coefficient or as one which has been *averaged* over the surface in several possible ways, care must be taken to use the appropriate ΔT_{loc} or ΔT_{av} with Nu_{loc} or Nu_{av} from literature correlations.

For fluids of viscosity near that of water, a useful correlation in turbulent flow (heating or cooling) is[†]

$$Nu = \frac{hd}{k} = 0.023 \, Re^{0.8} Pr^{0.4} \tag{8.142}$$

which appears valid when

$$10^4 \le Re \le 1.2 \times 10^5 \qquad \text{turbulent}$$

$$0.7 \le Pr \le 120 \qquad \text{valid for all liquids except molten metals}$$

$$\frac{L}{d} \ge 60 \qquad \text{long tubes}$$

A modification due to Seider and Tate[‡] incorporates an allowance for larger temperature differences; it appears useful for estimating heat transfer with *viscous*

† W. H. McAdams, "Heat Transmission," 3d ed., p. 152, McGraw-Hill Book Company, New York, 1954.

‡ F. E. N. Seider and G. E. Tate, Heat Transfer and Pressure Drop of Liquids in Tubes. *Ind. Eng. Chem.*, **28:** 1429, 1936.

fluids (such as oils):

$$\text{Nu} = \frac{hd}{k} = 0.027 \, \text{Re}^{0.8}\text{Pr}^{0.33}\left(\frac{\mu_b}{\mu_0}\right)^{0.14} \tag{8.143}$$

When natural convection is also important due to the presence of nonuniform fluid density, the Grashof number

$$\text{Gr} = \frac{d^3 g \, \Delta\rho \, \rho_{\text{av}}}{\mu^2} \tag{8.144}$$

appears in the correlation† for liquid flowing in *horizontal* tubes:

$$\text{Nu} = 1.75\left[\frac{d}{L}\text{Pr Re} + 0.04\left(\frac{d}{L}\text{Gr Pr}\right)^{0.75}\right]^{1/3}\left(\frac{\mu_b}{\mu_0}\right)^{0.14} \tag{8.145}$$

(For vertical tubes, the viscosity ratio is replaced by 1.0 and the constant 0.04 by 0.0722.)

When the fluid is known to be non-Newtonian, the forms change. Two equations‡ which have been used for pseudoplastic fluids (Sec. 8.8) are

$$\text{Nu} = \frac{hd}{k} = 2.0\left(\frac{d}{L}\text{Re Pr}\right)^{1/3}\left[\frac{\eta_v(\text{bulk})}{\eta_v(\text{wall})}\left(\frac{3 + 1/n}{3 + 1/n}\right)\frac{1}{2}\right]^{0.14} \tag{8.146a}$$

or

$$\text{Nu} = \frac{hd}{k} = 1.75\left(\frac{d}{L}\text{Re Pr}\right)^{1/3}\left(\frac{3n + 1}{4n}\right)^{1/3} \tag{8.146b}$$

where the power law for pseudoplastic fluids is

$$\tau_s = -\eta_0 \, |\dot{\gamma}|^{n-1}\dot{\gamma} \tag{8.77}$$

and the term η_v in Eq. (8.146) is the coefficient of $\dot{\gamma}$ in Eq. (8.77):

$$\eta_v \equiv \eta_0 \, |\dot{\gamma}|^{n-1}$$

evaluated at the bulk fluid or wall temperature. Note that for a Newtonian fluid ($n = 1$) without large bulk-wall temperature differences, Eq. (8.146) reduces to Eq. (8.145). The viscosity variations with temperature, however, are explicitly represented in Eq. (8.146a) in the same form as Eq. (8.145).

A wide variety of reactor heat-transfer surface and flow configurations are possible. Some correlations for several of these are given in Example 8.6; others can be found in standard heat-transfer texts and in Chap. 3 of Charm (Ref. 22 of Chap. 9). Some example calculations for heat-transfer coefficients and heat-exchanger duties appear in the problem section. Microbial fluids occasionally deposit a residue on the heater surface leading to fouling (time-varying wall heat-transfer coefficient) and a decrease in h for the fluid-side.

---

† R. C. Martinelli and L. M. K. Boelter, *AIChE Meet.*, 1942 (cited in McAdams, op. cit.).
‡ S. E. Charm and E. W. Merrill, Heat Transfer Coefficients in Straight Tubes for Pseudoplastic Food Materials in Streamline Flow, *Food Res.*, **24**: 319, 1959.

Example 8.6. Heat-transfer correlations

Natural convection from vertical plane or cylinder:†

$$\text{Nu} \equiv \frac{hL}{k} = c(\text{Gr Pr})^a \tag{8E6.1}$$

where L is the plate length or cylinder diameter, all parameters are evaluated at $(T_{\text{bulk}} + T_{\text{wall}})/2$, and

$$3.5 \times 10^7 \leq \text{Gr Pr} \leq 10^{12} \qquad c = 0.13, \ a = \tfrac{1}{3} \ \text{(turbulent)}$$

$$10^4 \leq \text{Gr Pr} \leq 3.5 \times 10^7 \qquad c = 0.55, \ a = \tfrac{1}{4} \ \text{(laminar)}$$

Heat transfer in concentric annuli:‡

Streamline flow:

$$\text{Nu} \equiv \frac{hd}{k} = \left(\frac{d_o}{d_i}\right)^{0.8} (\text{Re Pr})^{0.45} \left(\frac{d}{L}\right)^{0.45} \text{Gr}^{0.5} \tag{8E6.2}$$

where d_o, d_i = outside and inside diameters

Turbulent flow:

$$\frac{h}{c_p G} = \begin{cases} \dfrac{0.23 \ \text{Pr}^{-2/3}(\mu_b/\mu_0)^{0.14}}{[(d_o - d_i)/\mu_b]^{0.2}} & \text{outer wall} \\[3mm] \dfrac{(0.023)(0.87)(d_o/d_i)^{0.53}}{[(d_o - d_i)G/\mu_b]^{0.2}} & \text{inner wall} \end{cases} \tag{8E6.3}$$

Gravity flow over horizontal tube surfaces:§

$$h = 65\left(\frac{w}{2\mu_c L d_0}\right)^{1/3} \quad \text{Btu/(h · ft · °F)} \quad \text{if} \quad \frac{w}{2\mu_c L} < 525 \tag{8E6.4}$$

where w = liquid flow rate
L = tube length
d_o = outside diameter

Turbulent flow in tubes:¶

$$\text{Nu} = \frac{hd}{k} = 0.023 \ \text{Re}^{0.8} \text{Pr}^3 \tag{8E6.5}$$

$$b = \begin{cases} 0.4 & \text{for heating} \\ 0.3 & \text{for cooling} \end{cases}$$

Flow perpendicular to isolated cylinder:‖

$$\text{Nu} = \frac{hd}{k} = (\text{Pr})(0.35 + 0.56 \ \text{Re}^{0.52}) \tag{8E6.6}$$

---

† W. J. King, Free Convection, *Mech. Eng.*, **54**: 347, 1932.
‡ C. C. Monrad and J. F. Pelton, in W. H. McAdams, "Heat Transmission," McGraw-Hill Book Company, New York, 1954.
§ W. H. McAdams, et al., Heat Transfer to Falling Water Films, *Trans. ASME*, **62**: 627, 1940.
¶ F. W. Dittus and C. M. K. Boelter, *Univ. Calif. Pub. Eng.*, **2**: 443, 1930 (see McAdams, "Heat Transmission").
‖ S. E. Charm, "Fundamentals of Food Engineering," 2d ed., chap. 4, Avi Publishing, Westport, Conn., 1971.

Flow perpendicular to one row of tubes centered 2d apart:†

$$\text{Nu} = \frac{hd}{k} = 0.21\,\text{Re}_m{}^{0.6}\text{Pr}^{1/3} \qquad (8E6.7)$$

where Re_m is Re evaluated at $v_{\max}$ and $2d$ is defined as

Staggered successive rows of the above type:†

Same as Eq. (8E6.7), but coefficient 0.21 replaced by 0.27 for 3 rows, 0.30 for 5 rows, 0.33 for 10 tube rows or more.

Examples of the overall energy balance, the microbial heat-generation rates, and the sizing of heat exchangers appear in the problems. The subject of transport of heat and mass is, we reiterate, an enormously developed area. The present chapter has provided some conceptual guidelines for estimating the quantities of interest. The literature contains a vast number of references for heat- and mass-transfer correlations under a variety of experimental conditions, as indicated in some of the general references of this chapter. Where possible, use of correlations from experimental configurations most apropos to the situation of interest should be practiced, always taking note of the margin of (un)certainty of the correlation.

PROBLEMS

8.1. Oxygen diffusivities in protein solutions Stroeve [48] noted that an 1881 derivation of James Clerk Maxwell's ("Treatise on Electricity and Magnetism," vol. 1, 3d ed.) for diffusion through a fluid containing spherical obstructions simplified to the form below when the obstructions were impermeable:

$$\frac{\mathscr{D}}{\mathscr{D}_0} = \frac{2(1-f)}{2+f}$$

where $\mathscr{D}$ = apparent diffusivity in suspensions
$\mathscr{D}_0$ = apparent diffusivity in pure fluid
f = volume fraction of obstructions

He found that the form gave reasonable fit to experimental data provided that f was defined as $f = f_p + f_b$, where f_p is the volume fraction of protein and f_b the volume fraction of water physically immobilized on the protein surface. Taking the dimensions of the protein to be those of hydrated hemoglobin (spheroid 65 by 55 by 55 Å), calculate and plot $\mathscr{D}/\mathscr{D}_0$ vs. f_p (not f) assuming no, one, or two monolayers of immobilized water around the protein (range of f_p is 0.1 to 0.5). Compare your results with the following measured values for methemoglobin and comment:

$\mathscr{D}/\mathscr{D}_0$	0.69	0.43	0.17
f_p	0.1	0.2	0.4

† S. E. Charm, "Fundamentals of Food Engineering," 2d ed., chap. 4, Avi Publishing, Westport, Conn., 1971.

8.2. Mass-transfer coefficient Determine k_L for the following conditions:

$$\text{Liquid volume} = 10 \text{ l}$$

$$\text{Turbine impeller diameter} = 10 \text{ cm}$$

$$\text{Vessel diameter} = 50 \text{ cm}$$

$$\text{Speed (rev/min)} = 200$$

$$\text{Air} - \text{medium binary diffusion coefficient} = 0.5 \times 10^{-5} \text{ cm}^2/\text{s}$$

$$\text{Airflow rate} = 2 \text{ l/min}$$

$$\text{Medium viscosity} = 0.01 \text{ g/(cm} \cdot \text{s)}$$

8.3. Oxygen transfer, nonagitated Consider an unstirred aerated vessel with 10 orifices mounted in the bottom. If each is 1 mm in diameter and has an airflow rate of 5 ml/min, what specific cell growth rate will be maintained if oxygen is limiting? Neglect breakup and coalescence and assume the medium is sufficiently dilute for it to behave like pure water.

$$\mu_{\text{max}} = 0.5 \text{ h}^{-1} \qquad K_s = 0.1 \text{ m}M/\text{l} \qquad \sigma = 72 \text{ g/s}^2$$

$$g = 980 \text{ cm/s}^2 \qquad \mu_{\text{gas}} = 2 \times 10^{-4} \text{ g/(cm} \cdot \text{s)} \qquad \mathscr{D} = 0.5 \times 10^{-5} \text{ cm}^2/\text{s}$$

$$\mu_{\text{liq}} = 10^{-2} \text{ g/(cm} \cdot \text{s)} \qquad \rho_{\text{gas}} = 1.4 \times 10^{-3} \text{ g/l} \qquad H_L = 10 \text{ cm}$$

$$Y_{\text{O}/x} = 1 \text{ g O}_2/\text{g cell} \qquad x = 1.0 \text{ g cells/l}$$

8.4. Bubble-column performance (a) Estimate a, H, and k_l for a bubble column under the following conditions:

$$\text{Gas flow} = 20 \text{ std ft}^3/\text{min}$$

$$\text{Liquid flow} = 25 \text{ gal/min (water)}$$

$$\text{Column ID} = 16 \text{ in}$$

$$\text{Average bubble diameter } D = 0.25 \text{ in}$$

(b) An alternate correlation [49] for bubble swarm (liquid or liquid-liquid mass transfer) is

$$\text{Sh} = 2.0 + 0.0187 \left[\text{Re}^{0.484} \text{Sc}^{0.339} \left(\frac{Dg^{1/3}}{\mathscr{D}^{2/3}} \right)^{0.072} \right]^{1.61}$$

Compare the explicit dependence of each physical parameter with that of the text formula. Evaluate k_l again and the percentage difference between the two estimates.

8.5. Stream reaeration (rapids and ponds) A moving stream might be approximated by alternating deep and shallow segments of the same width. If the "deep" and "shallow" segments have depths h_D, h_s and lengths l_D, l_s:

(a) What is the ratio of aeration mass-transfer coefficients for the shallow to deep segments?

(b) Lumping biological activity into a single species, develop an analytic description for substrate utilization by aerobic species of this ponds-rapids configuration, assuming oxygen transfer to be limiting. State your assumptions clearly.

(c) Develop expressions (making simplifying assumptions if needed) for the fraction of total microbial growth occurring in the pond and the fraction of total oxygen transfer occurring in the rapids.

8.6. Simplified stream reaeration: (Streeter-Phelps equation) Stream reaeration can be described as a simple plug-flow phenomenon under conditions where organic sedimentation, sediment reactions, and loss of organic volatiles is unimportant. A balance on organic matter S in a stream of velocity u gives

$$\frac{\partial s}{\partial t} = -u \frac{\partial s}{\partial z} - \mu_{\text{max}} s$$

and the oxygen balance is

$$\frac{\partial c_O}{\partial t} = \underbrace{-u\,\frac{\partial c_O}{\partial z}}_{\text{Gradient}} + \underbrace{\bar{k}_{O_2}(c_{O_2}^* - c_{O_2})}_{\substack{\text{Gain by mass}\\\text{transfer}}} - \underbrace{\mu_{max}\, s\, Y_{O/s}}_{\substack{\text{Loss by microbial}\\\text{oxidation}}}$$

$$+ \text{photosynthesis rate} - \text{algal respiration rate} - \text{sedimentation rate}$$

(a) If oxygen is always in excess for the *microbes* oxidizing the nutrients, show that

$$s(z) = s_{z=0} \exp\left(-\mu_{max}\,\frac{z}{u}\right)$$

(b) At steady state, neglecting photosynthesis, algal respiration, and sedimentation, show that the oxygen-concentration profile satisfies the Streeter-Phelps equation

$$c_{O_2}^* - c_{O_2}(z) = \frac{Y_{O/s}\,\mu_{max}\,s_{z=0}}{\bar{k}_{O_2} - Y_{O/s}\,\mu_{max}} \left[\exp\left(-Y_{O/s}\,\mu_{max}\,\frac{z}{u}\right) - \exp\left(-\frac{\bar{k}_{O_2}\,z}{u}\right)\right]$$

$$+ (c_{O_2}^* - c_{O_2,\,z=0}) \exp\left(-\frac{\bar{k}_{O_2}\,s}{u}\right)$$

where $\bar{k}_{O_2}' = \bar{k}_{O_2}/l$
$\bar{k}_{O_2}$ = eddy-averaged oxygen-transfer coefficient [from (8.38)]
l = stream depth

(c) Establish analytically that for the Streeter-Phelps treatment, $c_{O_2}^* - c_{O_2}(z)$, known as the *oxygen deficit*, has a single minimum at

$$c_{O_2}^* - c_{O_2}(z) = \frac{Y_{O/s}\,\mu_{max}\,s_{z=0}}{\bar{k}_{O_2}'} \exp\left(-Y_{O/s}\,\mu_{max}\,\frac{z}{u}\right)$$

What is the downstream distance z corresponding to this point of maximum oxygen deficit? Repeat this derivation including constant photosynthesis, algal respiration, and sedimentation rates.

8.7. Scaling parameters in aeration (a) Verify the proportions in Table 8.4 by direct calculation.

(b) In agitated aeration, $Sh = \alpha\, Re_i^{m_1} Sc^{m_2}$ according to many correlations. Assuming constant bubble size, show that achievement of identical values of k_l in two different vessels, e.g., small (I) and large (II) requires that the impeller speed in revolutions per minute N_i scale as follows:

$$\frac{N_i(\text{II})}{N_i(\text{I})} = \left[\frac{D_i(\text{I})}{D_i(\text{II})}\right]^{2-1/m_1}$$

(c) Consequently, establish that constant k_l implies

$$\frac{(P/V)_{\text{II}}}{(P/V)_{\text{I}}} = \left[\frac{D_i(\text{I})}{D_i(\text{II})}\right]^{4-3/m_1}$$

For the turbulent correlation in the text, what fortuitous result arises in the previous equation?

(d) For conditions where the bubble size itself is determined by impeller conditions, what relations hold for $N_i(\text{II})/N_i(\text{I})$?, $(P/V)_{\text{II}}/(P/V)_{\text{I}}$?

8.8. Yield coefficients and transfer rates (a) Assuming that cells can convert two-thirds (wt/wt) of the substrate carbon (alkane or glucose) to biomass, calculate the "stoichiometric" coefficients for hexadecane or glucose utilization:

Hexadecane:

$$C_{16}H_{34} + \alpha_1 O_2 + \alpha_2 NH_3 \longrightarrow \beta_1(C_{4.4}H_{7.3}N_{0.86}O_{1.2}) + \beta_2 CO_2 + \beta_3 H_2O$$

Glucose:

$$C_6H_{12}O_6 + \alpha_1' O_2 + \alpha_2' NH_3 \longrightarrow \beta_1'(C_{4.4}H_{7.3}N_{0.86}O_{1.2}) + \beta_2' CO_2 + \beta_3' H_2O$$

(b) Calculate the three yield coefficients Y_s, Y_0, Y_{kcal}. Note that Y_0 and Y_{kcal} for hydrocarbon are below those for glucose even when identical carbon-to-biomass conversion efficiencies are assumed.

(c) Taking results for *Pseudomonas fluorescens* in glucose ($\mu_{max} = 0.63$ h^{-1}, $k_e = 0.25$ g glucose per gram of cell per hour, $Y_0 = 0.667$ g cell per gram of glucose), use Eq. (8.126a) and the above results to calculate oxygen transfer and heat-removal rate requirements vs. time, assuming that $s > K_s$ always, x_0 corresponds to 10^5 cells per milliliter, and $x_{final} = 10^9$ cells per milliliter.

(d) Suppose that a batch aerobic fermenter is run in a cylindrical tank of 6-ft diameter with 130 ft of 1-in-diameter cooling coil arranged on 2-in spacing. Assume that all heat transfer is through the coils and that the bubble sparger always maintains a sufficient $K_L a$ value so growth is not oxygen-limited. If the average fluid velocity perpendicular to the coiled tubes is 10 percent of the impeller-tip velocity, what is the minimum speed in revolutions per minute needed for heat transfer if the average cooling liquid temperature is 18°C and the fermentor should operate no higher than 28°C? Repeat this calculation for cell densities of 10^6, 10^7, 10^8, and 10^9 cells per milliliter (impeller diameter = 4.5 ft, thickness = $\frac{1}{2}$ in, height = 6 in, single paddle).

(e) At 10^9 cells per milliliter, the aeration rate is such that 10 percent of the entering oxygen is consumed by the cells. For a poorly designed sparger, bubble size is too large. What stirrer speed (revolutions per minute) is needed to give adequate bubble size? Can this reasonably be achieved with one large paddle? Would a better design include a second much shorter, high-speed paddle just above the sparger with, for example, $D_2 = 0.2D_1$, $N_2 = 10N_1$?

8.9. Yeast disruption The breakup of yeast cells in a high-pressure homogenizer has been analyzed [50] in a manner parallel to that of bubble breakup. As the cells pass through a very narrow channel, or orifice, with a large pressure drop, fluid forces disrupt the cell wall. The kinetic energy of fluid fluctuations is $K_e = k\rho d_{cell}{}^3 u^2$, and the surface energy of the cell wall is taken to be proportional to $\sigma d_c{}^2$; thus the fluctuating kinetic energy of a liquid volume the size of the cell increases more rapidly than the restraining "surface energy," leading again to a maximum diameter $d_m \propto \sigma/u^2$.

(a) The *distribution* of sizes of yeast cells from a fermentation is given by the cumulative volume fraction $F(d_c)$ (volume fraction of cells with size smaller than d_c) which was observed to have the form

$$F(d_c) = 1 - \exp\left[-\left(\frac{d_c - 1.5}{4.0}\right)^B\right] \qquad d_c \text{ in micrometers}$$

Evaluate the fraction of yeast disrupted (assuming 100 percent efficiency) in the homogenizer as a function of the homogenizer fluid velocity u. The value of B is 1.45.

(b) For flow through narrow homogenizer channels, the turbulent velocity u is proportional to applied upstream pressure (the release pressure in negligible). Sketch the form of the fraction homogenized in a single pass vs. applied pressure on an appropriate graph.

8.10. Dimensionless groups: Buckingham π theorem In heat transfer by forced flow of fluid over a tube surface, the parameters which are physically important in determining the fluid-side heat-transfer coefficient (a conductance h_w) are the characteristic diameter of the pipe D, the fluid velocity v, and viscosity μ (in poise), density ρ, specific heat at constant pressure C_p (in calories per mole per degree), and the thermal conductivity of the fluid k_f (in calories per second per centimeter per degree). The Buckingham π theorem states that "the functional relationship between q quantities whose units can be given in terms of p fundamental units can be written as a function of $q - p$ dimensionless groups. The fundamental units in heat transfer are mass m, length l, time t, and temperature T.

(a) Express h_w, D, V, μ, ρ, C_p, and k in terms of such units, i.e., variable $= m^a l^b t^c T^d$.

(b) Four dimensionless groups are commonly formed from these variables: Nu ($\equiv h_w D/k_f$), Re ($\equiv \rho Dv/\mu$), Pr ($\equiv C_p \mu/k_f$) and Stanton number St ($\equiv h_w/vC_p \rho$). Comment. By inspection, what does the Stanton number represent?

8.11. Heat transfer: anaerobic digester† A well-stirred anaerobic digester (capacity 400,000 lb of sludge per day) is heated externally by pumping sludge through pipes of an external heat exchanger (Fig. 8P11.1). The following information will probably be useful for heat-balance considerations:

---

† Adapted from Metcalf & Eddy, Inc., "Wastewater Engineering," McGraw-Hill Book Company, New York, 1972.

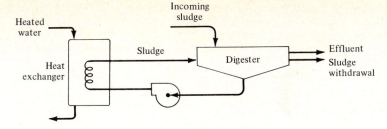

Figure 8P11.1 Anaerobic digester and heat exchanger. (*After Metcalf and Eddy, Inc., "Wastewater Engineering," McGraw-Hill Book Company, New York, 1972.*)

Dimensions (digester): diameter $= 120$ ft   average depth $= 25$ ft

Heat-transfer coefficients:

$$[Btu/(ft^2 \cdot °F)] = \begin{cases} 0.13 & \text{digester side to dry outside earth} \\ 0.15 & \text{digester bottom to wet earth} \\ 0.16 & \text{roof to air} \end{cases}$$

Temperatures: Air $= 16°F$ dry earth $= 32°F$ wet earth $= 40°F$
             Incoming sludge $= 50°F$ digester operating temperature $= 88°F$
Sludge specific heat $= 1.0$

(a) Calculate the steady-state heat-loss rate from the digester.

(b) The maximum heated-water temperature into the exchanger is $130°F$ (higher temperatures may result in sludge-cake development in tubes). Calculate the water flow rate needed if the exit water from the exchanger is at $100°F$.

(c) Taking an average heating-water temperature, calculate the sludge circulation rate needed if sludge return is at $100°F$. If the exchanger tubes are 10-in-diameter tubes each 20 ft long, how many tubes are needed for a superficial sludge velocity of 20 ft/s?

8.12. Power-law fluids: starch hydrolysis Pastes resulting from cooking 1% wt/vol amylopectin in water are pseudoplastic, thus following the power law, $\tau = \eta\gamma^n$, with $n < 1.0$. For batch α-amylase hydrolysis of this paste, the following changes with increasing time were observed [51]:

η, dyn $\cdot s^n/cm^2$	0.32	0.26	0.20	0.14	0.10	0.03
n	0.73	0.75	0.78	0.83	0.85	0.98

(a) Show that these data can be described by

$$\tau_0 = \eta\gamma_0{}^n \quad \text{where} \quad \tau_0, \gamma_0 = \text{const}$$

(b) Establish with an appropriate graph that all τ-vs-γ curves at each degree of hydrolysis pass through a common point.

(c) How would the power input at fixed rotation speed vary with time? Could this parameter be used for on-line batch-process control?

8.13. Power input vs. impeller speed in non-Newtonian fluids For non-Newtonian fluids with power-law indices n less than unity, the shear rate γ may be taken to be proportional to the impeller rotation rate N_i [52]. Show that:

(a) For a fluid between concentric cylindrical surfaces, shear stress on the outer cylinder (as the inner-cylinder rotation speed varies) changes according to $(d\tau/dN_i) \propto N_i{}^{n-1}$.

(b) The Reynolds number, $Re = D_i{}^2 N_i \rho/\mu_a$, where μ_a is the apparent viscosity (shear-rate-dependent), varies as $Re \propto N_i{}^{n-2}$.

(c) For Reynolds numbers defined above, the data for power number ($\equiv Pg_c/D_i{}^5 N_i{}^3 \rho$) vs. Reynolds number fall on or just below that correlation for the Newtonian-fluid values. Thus if P_0 varies as Re^{α}, establish that the power input P varies as $N_i{}^{3+\alpha(n-2)}$.

(d) From the previous information, how would you evaluate the proportionality constant between γ and N_i for a non-Newtonian fluid [52]?

8.14. Air sterilization (a) Determine the length of filter required to reduce the cell concentration to 10^{-8} of its previous value when:

> Filter diameter = 1 m   packing material = cotton
> Fraction of volume occupied by packing = 0.005
> Fiber diameter = 10 μm   cell diameter = 0.5 μm   air velocity = 10 cm/s
> Air viscosity = 2×10^{-4} g/(cm · s)   cell density = 1 g/cm^3
> Particle diffusivity = 0.001 cm^2/s

(b) Repeat if the cells are on dust particles with a size distribution $\rho(r)$ proportional to $\exp[-(r-r_m)^2 a]$, where r_m = mean radius = 2 μm, and there is approximately one cell per particle ($a = 1.0 \ \mu m^{-2}$).

8.15. Particle-slip correction Gas molecules have an appreciable mean free path, i.e., path lengths approximating straight-line trajectories between successive collisions. The Stokes-Einstein diffusivity given by

$$\mathcal{D}_p = \frac{k_B T}{3\pi \mu_c d_p}$$

must be corrected for this noncontinuum behavior when the particle size begins to approach the mean-free-path length λ. A useful formula allowing for this slip factor is a Knudsen-Weber term

$$\text{Correction factor} = 1 + \frac{2\lambda}{d_p}\left(1.257 + 0.4 \exp \frac{-0.55 d_p}{\lambda}\right)$$

Calculate, for the tabulated particle sizes, the percentage error arising from neglect of this factor for:
  (a) The diffusion coefficient
  (b) The single-fiber collection efficiency
  (c) The fibrous filter depth needed to reduce the inlet concentration by 99 percent.

Particles	Size, μm
Small virus	10^{-1}
House dust	5×10^{-1}
Small bacteria, bacterial spores	1
Bacteria	5

The mean free path $\lambda = \mu_c \pi M/2\rho_c RT$, where M = average molecular weight.

8.16. Contributing factors to β_T The results of Fig. 8.14 are due to a solution of equations for isolated fibers using flow fields which neglect the influence of other neighboring fibers; whereas the diffusion-interception result of (8.115) attempts to include neighboring filter effects through the parameter α (solid fraction) and the parameter Ku (see the text).

(a) Using the individual terms in Table 8.5, validate the domains shown in Fig. 8.14a and construct a similar chart for water the carrying fluid. Comment on the similarities and/or differences.

(b) Assuming that $\beta_T = \sum \beta_i$, plot the contours of constant β_T on a d_p-vs.-u_∞ diagram for air; for water.

8.17. Single-fiber ($\alpha \to 0$) vs. finite-volume fraction filters The first two forms of Table 8.5 give $\beta_T = \beta_I + \beta_B$. Assuming (8.115) (which includes the influence of neighboring fibers on collection efficiency) to be correct, construct a calibration chart for air filtration showing contours of constant "error" = $\{\beta(8.115) - (\beta_I + \beta_B[\text{Table } 8.5])\}$ $\beta(8.115)$ on a d_p-vs.-u_∞ plot for $d_F = 100 \ \mu$m.

8.18. Time-varying filtration To estimate the efficiency of a surgeon's mask, suppose the exhalation breathing rate can be approximated by the form $G = A \sin wt$, where $A = 8.0$ l/min and $w = 12$ min^{-1}.

(*a*) Develop an expression for the period averaged ratio of particles escaping a surgeon's mask filter to those entering the mask. Assume the mask has an effective cross-sectional area of 15 cm^2, thickness 5 cm, solid fraction 0.033, and fiber diameter 80 μm.

(*b*) Plot vs. particle size the result of part (*a*) divided by the same result calculated on the basis of average exhalation flow. Comment.

8.19. Filtration theory: predation Some motile organisms consume smaller nonmotile organisms as food. If a 200-μm protozoa can "swim" at 0.1 mm/s, what size should a nonmotile organism be to minimize the possibility of being consumed by the motile predator? State your assumptions.

8.20. Hydrocarbon-fermentation phases (*a*) For the quotation of Mimura et al. (and Fig. 8.12) describing the time course of a particular hydrocarbon fermentation, write a mathematical description of growth and substrate(s) utilization in each phase of the batch fermentation. For each "phase," indicate quantitatively where the controlling resistance(s) to growth may lie, i.e., hydrocarbon solubilization, oxygen transfer, cell metabolism, etc.

(*b*) Postulate various reasons why the cell-hydrocarbon-droplet-air-bubble-solvent system adopts each configuration mentioned by the authors. What obvious experiments are suggested by this direct observation? How would you discriminate between or prove the hypotheses advanced?

8.21. Laboratory procedure Read the "Containment" section of Ref. [53].

(*a*) Summarize the kinds of physical containment, listing especially: use of sterilization, disinfection, and particle filtration.

(*b*) In addition to physical containment, biological containment is also defined. Outline the principles involved, referring to earlier text chapters where appropriate.

REFERENCES

General

Mass-transfer correlations for oxygen transfer:
 1. P. H. Calderbank: Mass Transfer in Fermentation Equipment, p. 102 in N. Blakebrough (ed.), "Biochemical and Biological Engineering Science," vol. 1, Academic Press Inc., New York, 1967.
Oxygen demand of cultures and oxygen solubility:
 2. R. K. Finn: Agitation and Aeration, p. 69 in N. Blakebrough (ed.), "Biochemical and Biological Engineering Science," vol. 1, Academic Press, Inc., New York, 1967.
Oxygen-uptake experiments:
 3. W. W. Umbreit and R. H. Burris: "Manometric Techniques," 4th ed., Burgess Publishing Co., Minneapolis, 1964.
Interactions between diffusion through liquid films and chemical reactions:
 4. P. V. Danckwerts: "Gas-Liquid Reactions," McGraw Hill Book Company, New York, 1970.
 5. G. Astarita: "Mass Transfer with Chemical Reaction." Elsevier Publishing Company, Amsterdam, 1967.
Dimensionless groups in convective mass transfer and many aspects of energy, mass, and momentum transport:
 6. R. B. Bird, W. E. Stewart, and E. N. Lightfoot: "Transport Phenomena," John Wiley & Son, Inc., New York, 1960.
Fermentation fluids:
 7. H. Taguchi: The Nature of Fermentation Fluids, *Adv. Biochem. Eng.* **1**: 1 (1971).
Problems of scaling microbial reactors:
 8. E. L. Gaden, Jr., (ed.): *Biotech. Bioeng.*, **8**: 1966. (Entire volume.)
Mass-transfer examples in microbial reactors:
 9. *Adv. Microb. Eng.*, **1**: 295–580, 1973.

Filtration theory:

10. C. N. Davies (ed.): "Aerosol Science," Academic Press, Inc., New York, 1966.
11. C. N. Davies (ed.): *Air Filtration*, Academic Press, Inc., New York, 1973.
12. S. K. Friedlander: Aerosol Filtration by Fibrous Filters, p. 49 in N. Blakebrough (ed.), " Biological and Biochemical Engineering Science," vol. 1, Academic Press, Inc., New York, 1967.

Water filtration:

13. K. M. Yao, M. T. Habibian, and C. R. O'Melia: *Environ. Sci. Technol.*, **5**, 1105, 1971.

Coagulation, sedimentation, adsorption (not covered in this chapter):

Ref. 10.

14. W. J. Weber, Jr.: " Physiocochemical Processes (For Water Quality Control)," Wiley-Interscience, New York, 1972.

Specific

Batch oxygen demand:

15. R. T. Darby and D. R. Goddard: Studies of the Respiration of the Mycelium of the Fungus *Myrothecoum verracaria, A. J. Bot.*, **37**: 379, 1950.

$k_l a$ by oxygen-electrode transient response:

16. W. C. Wernan and C. R. Wilke: New Method for Evaluation of Dissolved Oxygen Response for $K_L a$ Determination, *Biotech. Bioeng.*, **15**: 571, 1973.

CO_2 absorption into aqueous films:

17. K. Livansky, B. Prokes, F. Dihrt, and V. Benes: Some Problems of CO_2 Absorption by Algae Suspensions, *Biotech. Bioeng. Symp.* 4, p. 513, 1973.

Methane-utilization stoichiometry:

18. D. L. Klass, J. J. Iandolo, and S. J. Knabel: Key Process Factors in the Microbial Conversion of Methane to Protein, *CEP Symp. Ser.*, [93], **65**: 72, 1969.

Bubble-orifice-diameter correlations:

19. D. W. van Krevelen and P. J. Hoftijzer: Studies of Gas-Bubble Formation: Calculation of Interfacial Area in Bubble Contactors, *Chem. Eng. Sci.*, **46**: 29, 1950.

Photographic determination of a, H, D_{Sauter}:

20. P. H. Calderbank and J. Rennie: *Int. Symp. Distill. (Inst. Chem. Eng.)*, 1960.

Shape-velocity dependence of large bubbles in Eq. (8.40):

21. R. M. Davies and G. I. Taylor: *Proc. R. Soc.*, **A200**: 375, 1956.

Free-surface mass transfer to falling film:

Ref. 6, p. 540.

Free-surface mass-transfer to turbulent-stream surfaces:

22. G. E. Fortescue and J. R. A. Pearson: On Gas Absorption into a Turbulent Liquid, *Chem. Eng. Sci.*, **22**, 1163, 1967.
23. D. J. O'Connor and W. Dobbins: The Mechanism of Reaeration in Natural Streams, *J. Sanit. Eng. Div., Proc. ASCE*, **82**: SA6, 1966.

Turbine power number vs. Re (impeller):

24. J. H. Rushton, E. W. Costich, and H. J. Everett: Power Characteristics of Mixing Impellers, pt. 2, *Chem. Eng. Prog.*, **46**: 467, 1950.

Aerated vs. nonaerated power requirements:

25. Y. Ohyama and K. Endoh: Power Characteristics of Gas-Liquid Contacting Mixers, *Chem. Eng. Jp.*, **19**: 2, 1955.
26. B. J. Michel and S. A. Miller: Power Requirements of Gas-Liquid Agitated Systems, *AIChE J.*, **8**: 262, 1962.

Weber number correlations, references:

Ref. 1.

Oxygen diffusivities in microbial films:

27. J. V. Matson and W. G. Characklis: Oxygen Diffusion through Microbial Aggregates, *77th AIChE Meet., Pittsburgh, June 1973.*

Ionic strength influence on $K_l a$:

28. C. W. Robinson and C. R. Wilke: Oxygen Absorption in Stirred Tanks: A Correlation for Ionic Strength, *Biotech. Bioeng.*, **15:** 755, 1973.

Surfactants and mass transfer:

29. W. W. Eckenfelder, Jr., and E. L. Barnhart: The Effect of Organic Substances on the Transfer of Oxygen from Air Bubbles into Water, *AIChE J.*, **7:** 631, 1961.

30. A. Benedek and W. J. Heideger: Effect of Additives on Mass Transfer in Turbine Aeration, *Biotech. Bioeng.*, **13:** 663, 1971.

31. D. N. Bull and L. L. Kempe: Influence of Surface Active Agents on Oxygen Absorption to the Free Interface in a Stirred Fermentor, *Biotech. Bioeng.*, **13:** 529, 1971.

32. S. Aiba and K. Toda: The Effect of Surface Active Agents on Oxygen Absorption in Bubble Aeration I, *J. Gen. Appl. Microbiol.*, **7:** 100, 1963.

33. K. H. Mancy and D. A. Okun: Effect of Surface Active Agents on the Rate of Oxygen Transfer, "Adv. Biol. Waste Treat." p. 111 (1963); see also papers by McKeown and Okun, and Timson and Dunn, and Carver in this same reference.

Rheology of microbial broths:

34. A. Leduy, A. A. Marson, and B. Corpal: A study of the Rheological Properties of a Non-Newtonian Fermentation Broth, *Biotech. Bioeng.*, **16:** 61, 1974 (*A. pullulans* example).

35. J. A. Roels, J. van den Berg, and R. M. Voncken: The Rheology of Mycelial Broths, *Biotech. Bioeng.*, **16:** 181, 1974 (Pencillin broth example).

Morphology change with tanks in series:

36. D. Vrana: Some Morphological and Physiological Properties of *Candida utilis* Growing "Hypertrophically" in Excess of Substrate in a Two Stage Continuous Cultivation, *Biotech. Bioeng. Symp.* 4, p. 161, 1973.

Power-number evolution in non-Newtonian fermentation:

37. H. Taguchi and S. Miyamoto: Power Requirement in non-Newtonian Fermentation Broth, *Biotech. Bioeng.*, **8:** 43, 1966.

Cell density and mass transfer:

38. M. R. Brierley and R. Steel: Agitation-Aeration in Submerged Fermentation, pt. 2: Effect of Solid Dispersed Phase on Oxygen Adsorption in a Fermentor, *Appl. Microbiol.*, **7:** 57, 1959 (*A. Niger*).

39. D. L. Wise, D. I. C. Wang, and R. I. Mateles: Increased Oxygen Mass Transfer Rates from Single Bubbles in Microbial Systems at Low Reynolds Numbers, *Biotech. Bioeng.*, **11:** 647, 1969 (*Candida intermedia* and *Pseudomonas ovalis*).

Agitation and cell damage:

40. M. Midler and R. K. Finn: A Model System for Evaluating Shear in the Design of Stirred Fermentors, *Biotech. Bioeng.*, **8:** 71, 1966.
    Ref. 7.

Paper-pulp suspensions:

41. N. Blakebrough and K. Sambamurthy: Mass Transfer and Mixing Rates in Fermentation Vessels, *Biotech. Bioeng.*, **8:** 25, 1966.

Heat Transfer

Dimensional analysis:
    Ref. 6, pp. 396ff.

42. A. I. Brown and S. M. Marco: "Introduction to Heat Transfer," pp. 85–95, McGraw-Hill Book Company, New York, 1958.

Yield coefficients and heat generation (Y_Δ, Y_s):

43. B. J. Abbott and A. Clamen: The Relationship of Substrate, Growth Rate, and Maintenance Coefficient to Single Cell Protein Production, *Biotech. Bioeng.*, **15:** 117, 1973.

44. M. Kanazawa: The Production of Yeast from *n*-Paraffins, p. 438 in S. Tannenbaum and D. I. C. Wang (eds.), "Single Cell Protein II," MIT Press, Cambridge, Mass., 1975.

45. M. Moo-Young, Microbial Reactor Design for Synthetic Protein Production, *Can. J. Chem. Eng.*, **53:** 113, 1975.

... and multiple substrates, multiple species:

46. J. A. Servizi and R. H. Bogan: *J. Water Poll. Control. Fed.*, **36:** 607, 1961.

Q_Δ and Q_{O_2} correlation:

47. C. L. Cooney, D. I. C. Wang, and R. I. Mateles: Measurement of Heat Evolution and Correlation with Oxygen Consumption during Microbial Growth, *Biotech. Bioeng.*, **11:** 269, 1968.

Problems

48. P. Stroeve: On the Diffusion of Gases in Protein Solutions, *Ind. Eng. Chem. Fundam.*, **14:** 140, 1975.

49. G. A. Hughmark: Holdup and Mass Transfer in Bubble Columns. *Ind. Eng. Chem. Process Des. Dev.*, **6:** 218, 1967.

50. M. S. Doulah, P. H. Hammond, and J. F. G. Brookman: Hydrodynamic Mechanism for the Disintegration of *Saccharomyces cerevisiae* in an Industrial Homogenizer, *Biotech. Bioeng.*, **17:** 845, 1975.

51. Angel Cruz: Kinetics and Shear Viscosity of Enzyme Hydrolyzed Starch Pastes, Ph. D. thesis, Princeton University, Princeton, N.J., 1976.

52. A. B. Metzner, R. H. Feehs, H. L. Ramos, R. E. Otto, and J. D. Tuthill: Agitation of Viscous Newtonian and non-Newtonian Fluids, *AIChE J.*, **7:** 3, 1961.

53. Recombinant DNA Research: Guideline, *Fed. Regis.*, **41**(131): 27902–27943, July 7, 1976.

NINE

DESIGN AND ANALYSIS OF BIOLOGICAL REACTORS

Knowledge of biological reaction kinetics and mass transfer, our primary concerns in Chaps. 7 and 8, is essential for understanding how biological reactors work. In order to assemble a complete portrait of biological-reactor operation, however, it is necessary to integrate these two fundamental phenomena with the gas and liquid mixing and contacting patterns in the unit. Different design and scale-up procedures are required for reactors with different flow and mixing characteristics. Consequently, our major task in this chapter is to blend these various ingredients to obtain a coherent overall strategy for design and analysis of biological reactors.

Most subsequent sections of this chapter deal in part with *continuous* reactors. Here the term continuous means that nutrients and possibly organisms are fed to the reactor continuously and that an output, or effluent, stream is constantly being withdrawn from the system. By contrast, in a *batch* reactor, the contents are not removed until a certain reaction time has elapsed, after which all the reactor contents are harvested. Thus, in the jargon of thermodynamics, a batch reactor is a closed system with respect to exchange of matter with its environment, while a continuous reactor is an open system with mass exchange across its boundaries.

Actually the distinction between batch and continuous operation is not so clear-cut in many biological reactors. In an aerated batch fermentation, for example, gases are sparged continuously into the vessel and exhaust gases are continuously removed. Moreover, due to mass transfer between the gas and liquid phases, this means continuous addition and withdrawal of some components of the liquid phase. Also, small additions of nutrients or precursors may be made during the course of the batch. Conversely, some continuous dialysis biological reactors employ selectively permeable membranes in the exit line so that large molecules and cells are retained in the vessel while the remaining components are removed in an effluent stream. Because of the rather blurred division between batch and continuous biological reactors, it is vitally important that the biochemical engineer go beyond rote design formulas to an understanding of reactor analysis.

Most of what needs to be said about batch biological reactors—stages of batch growth, batch fermentation patterns, product formation in batch systems— has already been presented in Chap. 7. To most engineers unfamiliar with biochemical technology, that discussion of batch growth probably seems overextended and the consideration of continuous processes long overdue. In the modern chemical process industries continuous processing has become almost the rule for large-volume products. It may come as a surprise to learn that almost all fermentations are still conducted batchwise. One reason is tradition: reasonable design procedures for continuous microbial systems did not appear until the 1950s.

There are also unique problems associated with continuous operation of biological processes. While early batch anaerobic fermentations were inherently resistant to contamination by unwanted microbial species, many modern semi-batch (gas-sparged) and continuous fermentations require sterilized feeds of nutrients and gases in order to operate successfully. Thus, continuous fermentor operation also implies continuous sterilization in such cases. An especially troublesome problem in this regard can be infestation of bacteriophage.

In spite of such difficulties, continuous biochemical reactors enjoy many applications, as we shall survey in Chaps. 10 and 12. Also, many natural flows utilizing nonsterile feeds evolve to relatively steady levels of a few dominant populations. For the moment we might mention biological waste-water treatment, food-yeast production, and rivers as examples. Let us turn to consideration of the simplest type of continuous reactor, the perfectly mixed vessel.

9.1. THE IDEAL CONTINUOUS-FLOW STIRRED-TANK REACTOR (CSTR)

Two diagrams of this process are shown in Fig. 9.1a, which illustrates a laboratory implementation, and Fig. 9.1b, which is a schematic diagram of a completely mixed continuous stirred-tank reactor (CSTR). Such configurations for cultivation of microorganisms are frequently called *chemostats*. As this figure suggests, mixing is supplied by means of an impeller, rising gas bubbles, or both. We assume that this mixing is so vigorous that each phase of the vessel contents is of uniform composition; i.e., the concentrations in any phase do not vary with position inside the reactor. The schematic indicates an important implication of this complete-mixing assumption: the liquid effluent has the same composition as the reactor contents (the previously mentioned dialysis reactor being a special counter-example).

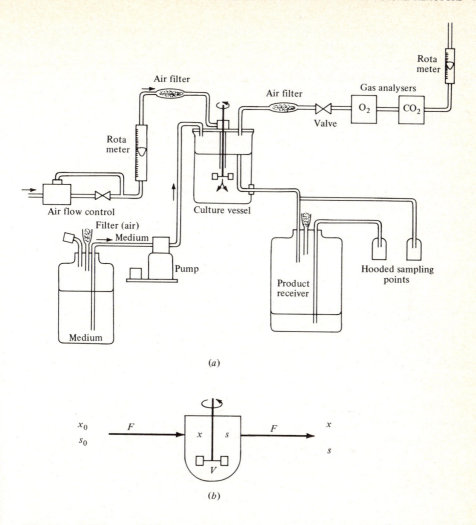

Figure 9.1 CSTRs for continuous cultivation of microorganisms. (*a*) major components of a laboratory CSTR for microbial cultures; (*b*) the notation usually used in modeling and analysis of these reactors.

Because of complete mixing, the dissolved-oxygen concentration is the same throughout the bulk liquid phase. This is of crucial importance in considering aerated CSTRs, for it means that we can often decouple the aerator or agitator design from consideration of the reaction processes. So long as the aeration system maintains dissolved oxygen in the CSTR above the critical concentration, we can view analysis of the microbial-kinetics aspects of the system as essentially a separate problem. Similar logic is also usually applied to the heat-transfer problems which can accompany microbial growth. So long as the vessel is well stirred, has adequate heat-removal capacity, and is equipped with a satisfactory temperature controller, we can assume that it is isothermal at the desired temperature and proceed with investigation of microbial reaction processes. These approaches are implicitly adopted in much of the discussion which follows.

9.1.1. Monod's Chemostat

In our beginning attempts to model and understand continuous cultures of microorganisms, we shall use *unstructured* models: only cell numbers, mass, or concentrations will be employed to characterize the biophase. In the steady state, where all concentrations within the vessel are independent of time, we can apply the following mass balance to any component of the system:

rate of addition to system − rate of removal from system

$$+ \text{ rate of production within system} = 0 \quad (9.1)$$

Letting

x = viable-cell concentration in tank and in effluent stream
x_0 = viable-cell concentration in feed stream
F = volumetric flow rate of feed and effluent streams
V = volume
r_x = rate of cell formation, cells per unit time per unit volume

we have

$$F(x_0 - x) + Vr_x = 0 \quad (9.2)$$

If we neglect cell death and let μ denote the specific growth rate r_x/x, it follows from Eq. (9.2) that

$$Dx_0 = (D - \mu)x \quad (9.3)$$

where

$$D = \frac{F}{V} \quad (9.4)$$

The parameter D, called the *dilution rate*, is equal to the number of tank volumes which pass through the vessel per unit time. It is the reciprocal of the mean holding time or mean residence time, more familiar in chemical engineering reactor analysis. Because D is almost universally employed in the biochemical engineering literature, however, we shall consistently employ the dilution-rate concept and notation.

Often the liquid feed stream to a continuous culture consists only of sterile nutrient, so that $x_0 = 0$. In this case Eq. (9.3) reveals that a nonzero cell population can be maintained only when

$$D = \mu \quad (9.5)$$

i.e., when the culture has adjusted so that its specific growth rate is equal to the dilution rate. When Eq. (9.5) is satisfied, it appears that Eq. (9.3) does not determine x when the feed is sterile. Experiments with continuous culture of *Bacillus linens* have confirmed the indeterminate nature of the population level. After a steady continuous operation was achieved at the 6-h point (Fig. 9.2), two subsequent interruptions of the culture were imposed. In each case, a portion of the reactor contents consisting of cells plus medium was removed and replaced by medium alone. Following each interruption, the system achieved a new steady population of different size.

Such behavior can be expected only when the culture is in a state of exponential growth with the specific growth rate independent of x and s. The indeterminate nature of the system disappears if a particular nutrient is growth-limiting. We can analyze this situation by writing a mass balance on the limiting substrate s and including the dependence of μ on s. In the substrate balance we make use of the yield factor Y, introduced in Sec. 7.1.3:

$$Y = \frac{\text{mass of cells formed}}{\text{mass of substrate consumed}} \tag{9.6}$$

To relate μ to s, Monod's equation (7.10) is frequently used:

$$\mu = \frac{\mu_{max} s}{s + K_s} \tag{9.7}$$

Consequently, Eq. (9.3) becomes

$$\left(\frac{\mu_{max} s}{s + K_s} - D \right) x + D x_0 = 0 \tag{9.8}$$

and the steady-state mass balance on substrate is

$$D(s_0 - s) - \frac{1}{Y} r_x = 0$$

or

$$D(s_0 - s) - \frac{\mu_{max} s x}{Y(s + K_s)} = 0 \tag{9.9}$$

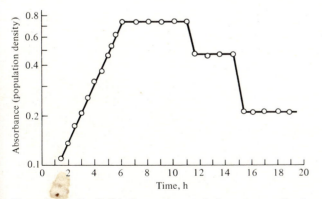

Figure 9.2 As Eq. (9.5) indicates, the population concentration in continuous culture is indeterminate so long as μ is constant (*B. linens* at 26°C with $D = 0.417$ h^{-1}). (*Reprinted from R. K. Finn and R. E. Wilson, Population Dynamics of a Continuous Propagator for Microorganisms, J. Agri. Food Chem.,* **2:** 66, 1954. Copyright by the American Chemical Society.)

Equations (9.8) and (9.9) are often called the *Monod chemostat model*. For the common case of sterile feed ($x_0 = 0$), these equations can readily be solved for x and s to yield

$$x_{\text{sterile feed}} = Y\left(s_0 - \frac{DK_s}{\mu_{\max} - D}\right) \tag{9.10}$$

and

$$s_{\text{sterile feed}} = \frac{DK_s}{\mu_{\max} - D} \tag{9.11}$$

Equations (9.10) and (9.11) contain the explicit steady-state dependence of x and s on flow rate ($D = F/V$). For very slow flows at a given volume, $D \to 0$; thus s tends to zero. Since nearly all feed substrate is consumed by the cells, the resulting effluent cell-mass concentration is $x = s_0/Y$.

As D increases continuously, s increases first linearly with D and then still more rapidly as $D \to \mu_{\max}$. The cell-mass concentration x declines with the same functional behavior: first linear in D, then diminishing more rapidly as $D \to \mu_{\max}$. At some point as D approaches $\mu_{\max}$, x becomes zero. The dilution rate D has just surpassed the maximum possible growth rate, and the only steady-state solution is $x = 0$. This condition of loss of all cells at steady state, termed *washout*, occurs for D greater than $D_{\max}$, where, if $x = 0$ in Eq. (9.10),

$$D_{\max} = \frac{\mu_{\max} s_0}{K_s + s_0} \tag{9.12}$$

The character of these solutions for substrate and cell-mass concentration is shown in Fig. 9.3 for the following parameter values:

$$\mu_{\max} = 1.0 \text{ h}^{-1} \qquad Y = 0.5 \qquad K_s = 0.2 \text{ g/l} \qquad s_0 = 10 \text{ g/l}$$

Notice that near washout the reactor is very sensitive to variations in D; a small change in D gives a relatively large shift in x and/or s.

This sensitivity must be kept in mind if production of cell mass is the objective of the continuous cultivation. The rate of cell production per unit reactor volume is Dx. This quantity is illustrated in Fig. 9.3, and there is a sharp maximum. We can compute the maximal cell output rate by solving

$$\frac{d(Dx)}{dD} = 0 \tag{9.13}$$

where Eq. (9.10) is used to write x as a function of D. Solution of Eq. (9.13) gives

$$D_{\text{max output}} = \mu_{\max}\left(1 - \sqrt{\frac{K_s}{K_s + s_0}}\right) \tag{9.14}$$

If $s_0 \gg K_s$, as is often the case, the value of $D_{\text{max output}}$ approaches $\mu_{\max}$ and consequently is near washout. This situation, evident in Fig. 9.3, may require us to forego maximal biomass production in order to avoid the region of large sensitivity. Inclusion of the practical aspects of sensitivity, controllability, and reliability into optimization problems such as this must not be forgotten in the pursuit of more easily quantified objectives.

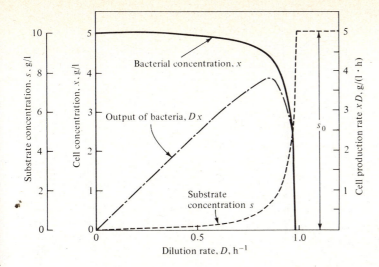

Figure 9.3 Dependence of effluent substrate concentration s, cell concentration x, and cell production rate xD on continuous culture dilution rate D as computed from the Monod chemostat model ($\mu_{max} = 1$ h^{-1}, $K_s = 0.2$ g/l, $Y = 0.5$).

When analyzing the production of end products from a continuous fermentation, we introduce another yield coefficient Y_p, defined by

$$Y_p = \frac{\text{mass of product formed}}{\text{increase in cell mass}} \qquad (9.15)$$

From our earlier classifications of batch fermentations, we are already aware that Y_p is constant for a Gaden type I fermentation but will vary for most other types. Use of the product yield coefficient allows us to write the steady-state product mass balance in the form

$$D(p_0 - p) + Y_p \mu x = 0 \qquad (9.16)$$

so that

$$p = p_0 + \frac{Y_p \mu x}{D} \qquad (9.17)$$

Combined with earlier equations relating μ and x to process parameters, Eq. (9.16) allows calculation of the *concentration* of product in the effluent. The rate of product output is then given by pD, which is maximized for constant Y_p when D has the value specified in Eq. (9.13). Thus, the goal of maximum product output also must be reconciled with sensitivity considerations. (If $Y_p = f(D)$, what value of D maximizes pD?)

An experimental example of continuous-culture behavior is shown in Fig. 9.4. In the culture of the bacterium *Aerobacter aerogenes*, the agreement between experiment and the simple model just outlined is qualitatively correct. However,

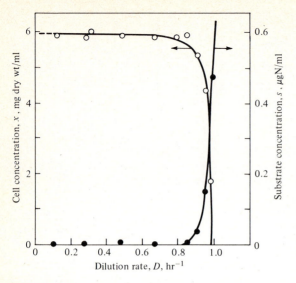

Figure 9.4 Experimental continuous-culture data qualitatively consistent with the Monod model (*A. aerogenes*). (*Replotted from D. Herbert, A Theoretical Analysis of Continuous Culture Systems, in "Society of Chemical Industries Monograph 12," p. 247, London, 1961.*)

both the observed cell-mass production and the observed substrate concentration remain approximately constant over a wider range of conditions than suggested by Fig. 9.3. Other data reveal breakdowns in the above model at very high and very low dilution rates. We shall next examine these two extreme cases in turn and attempt to understand why the Monod model might fail for them.

The data presented in Fig. 9.5 for *A. aerogenes* show a marked decline in cell concentration as the dilution rate is decreased. Similar behavior has also been

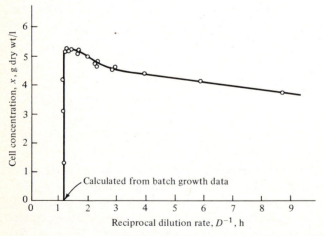

Figure 9.5 Cell concentration decreases as the dilution rate is decreased, a trend contrary to the Monod chemostat model (continuous cultivation of *A. aerogenes* in glycerol medium). (*Replotted from D. Herbert, Continuous Culture of Microorganisms; Some Theoretical Aspects in "Continuous Culture of Microorganisms: A Symposium," p. 48, Publishing House of the Czechoslovakia Academy of Sciences, Prague, 1958.*)

observed for the food yeast *Torula utilis*. This trend, which is contrary to the Monod chemostat model, can be explained by including the possibility of *endogenous metabolism* in the model. By endogenous metabolism we mean that there are reactions in cells which consume cell substance. Thus, we might write

$$r_x = \frac{\mu_{max}sx}{s + K_s} - k_e x \qquad (9.18)$$

to account for this effect. Notice that the additional term in Eq. (9.18) can also be interpreted formally as a cell death rate.

Such a modification of the Monod model is also consistent with other available data. For example, if the rate of respiration of an aerobic culture is proportional to the rate of substrate utilization

$$r_{resp} = \frac{\beta\mu_{max}sx}{s + K_s} \qquad (9.19)$$

then from Eqs. (9.2), (9.18) and (9.19) it follows that the specific respiration rate is

$$\frac{r_{resp}}{x} = \beta(D + k_e) \qquad (9.20)$$

Figure 9.6 displays experimental data which agrees nicely with Eq. (9.20). Observed variations in the yield coefficient Y with D also support a growth rate of the form given in Eq. (9.18). If the rate of substrate disappearance is

$$-r_s = \frac{1}{Y'} \frac{k_s sx}{s + K_s} \qquad (9.21)$$

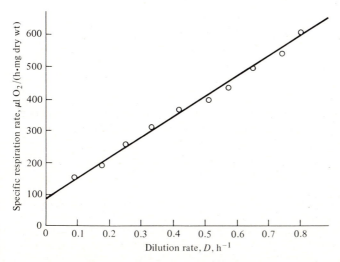

Figure 9.6 The specific respiration rate of the bacterium *A. aerogenes* in continuous culture depends linearly on dilution rate as Eq. (9.20) indicates. (*Replotted from D. Herbert, Continuous Culture of Microorganisms; Some Theoretical Aspects in "Continuous Culture of Microorganisms: A Symposium,"* p. 49, *Publishing House of the Czechoslovakia Academy of Sciences, Prague,* 1958.)

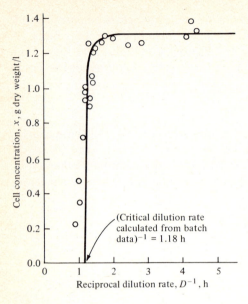

Figure 9.7 This experimental data on continuous growth of *Aerobacter cloacae* shows nonzero cell concentrations at dilution rates exceeding the calculated critical dilution rate. (*Replotted from D. Herbert et al., The Continuous Culture of Bacteria: A Theoretical and Experimental Study, J. Gen. Microbiol.*, **14**: 601, 1956.)

where Y' is the "true" yield coefficient, using Eqs. (9.18) and (9.21) and the definition of Y gives

$$Y = \frac{DY'}{D + k_e} \tag{9.22}$$

Such dependence of Y on D has been verified experimentally for several microorganisms.

At large dilution rates, continuous-culture behavior can deviate significantly from the Monod chemostat model, as Fig. 9.7 shows. Not only is the predicted cell concentration in error near washout, but the population may be maintained at substantially larger dilution rates than theory indicates. Moreover, the yield coefficient decreases as D approaches the critical maximum value. Among the probable explanations for these difficulties is the relatively high substrate concentration which prevails at large dilution rates. Under these circumstances, the substrate may not limit growth, and the members of the cell population may shift their metabolic patterns in recognition of some other limiting factor in their environment. A second possibility is imperfect mixing, a topic explored in Sec 9.1.2.

Example 9.1. Agitated-CSTR design for a liquid-hydrocarbon fermentation† Some microorganisms, e.g., the yeast *Candida lipolytica*, will grow on dodecane and other paraffinic hydrocarbons which are practically insoluble in water. Two alternative mechanisms have been proposed for microbial growth at the hydrocarbon-aqueous-phase interface: (1) cells (characteristic diameter D_c) much smaller than dispersed hydrocarbon droplets (diameter D_h) cluster around the paraffin drops; on the other

† Drawn from M. Moo-Young, Microbial Reactor Design for Synthetic Protein Production, *Can. J. Chem. Eng.*, **53**: 113, 1975.

hand, (2) if $D_h \ll D_c$, we may presume that droplets adsorb onto the outer surfaces of the relatively large microorganisms.

The closing observations in Sec. 8.8 on aerated microbial hydrocarbon fermentations indicate that the adsorption and flocculation relations between air bubbles, cells, and hydrocarbon droplets change over the course of a batch fermentation. For a continuous process, however, we may expect to operate in one particular growth mode, thus a single bubble–cell–hydrocarbon-droplet configuration may be predominant, rendering quantitative description easier.

Moo-Young and coworkers have considered these two situations and have proposed the following modified Monod growth-rate equations which include the effect of surface-area availability:

Case I, $D_c \ll D_h$:
$$\mu = \mu_{max} \frac{s/D_h}{K_s' + s/D_h} \tag{9E1.1}$$

Case II, $D_c \gg D_h$:
$$\mu = \mu_{max} \frac{s/(D_h)^2}{K_s'' + s/(D_h)^2} \tag{9E1.2}$$

where K_s' and K_s'' are modified K_s values. As the Lineweaver-Burk plot in Fig. 9E1.1 indicates, experimental data for the *C. lipolytica*–dodecane system support the second hypothesized mechanism.

Equations useful for CSTR design can be developed by relating input agitator power to dispersed hydrocarbon drop size (Chap. 8). If the diameter D_h of the substrate hydrocarbon droplets is determined by shear at the impeller-tip region, the correlations in Example 8.5 predict that

$$D_h = C\left(\frac{P}{V}\right)^{-0.4} \tag{9E1.3}$$

which was found to fit the experimental data of this study when $C = 0.023$. Substitution of this formula into Eq. (9E1.2) yields

$$\mu = \mu_{max} \frac{s(P/V)^{0.8}}{K_s''' + s(P/V)^{0.8}} \tag{9E1.4}$$

Using this specific growth rate in the ideal CSTR model with constant yield factor gives for sterile feed

$$x = Y\left[s_0 - \frac{DK_s'''}{\mu_{max} - D}\left(\frac{P}{V}\right)^{-0.8}\right] \tag{9E1.5}$$

and
$$D_{max\ output} = \mu \frac{K_s'''}{K_s''' + s_0\left(\dfrac{P}{V}\right)^{0.8}} \tag{9E1.6}$$

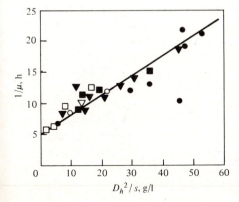

Figure 9E1.1 This double-reciprocal plot of experimental data for growth rates of *C. lipolytica* yeast on dodecane is consistent with the model assuming hydrocarbon droplet attachment to microorganisms. (*Reprinted from M. Moo-Young, Microbial Reactor Design for Synthetic Protein Production, Can. J. Chem. Eng.,* **53**: 113, 1975.)

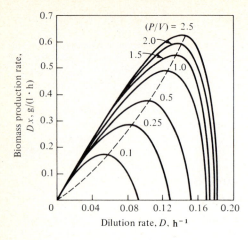

Figure 9E1.2 Biomass productivity for liquid-hydrocarbon fermentation as a function of dilution rate and agitator power input per unit volume P/V (hp/m^3).

Figure 9E1.2 shows the biomass production rate Dx computed using Eq. (9E1.5). Those plots, as well as Eqs. (9E1.5) and (9E1.6), clearly show the importance of considering interactions between biological reactions and fluid mechanics in design and analysis of microbial reactors.

In spite of the changing adsorption patterns in batch hydrocarbon fermentations referred to in Sec. 8.8 and Fig. 8.12, a model assuming that the cells of *C. lipolytica* growing on hydrocarbon droplets adsorb continuously on the hydrocarbon-droplet surface until a cell monolayer is formed gives some agreement with *batch*-fermentation data (Example 10.3).

9.1.2. Incomplete Mixing, Films, and Recycle Effects

Growth of filamentous organisms in continuous culture can be more difficult than continuous growth of yeasts or nonfilamentous bacteria because of the difficulties in handling and transporting mycelia. Some of the limited experimental data are revealed in Fig. 9.8. In this work with *Streptomyces aureofaciens*, many general features of the Monod model are apparent. As in the above discussion, however, deviations from the Monod description are largest at low and high values of the dilution rate.

As this and other mycelial microorganisms may form rather viscous solutions, it may be progressively more difficult to maintain uniform mixing at lower dilution rates. Under such conditions, where slowly circulating sections of the reactor may exhibit high cell-mass concentrations and high solution viscosities, the uniform-mixing (CSTR) assumption clearly becomes invalid. Such nonuniformities may decrease cell yield at low dilution rates (Fig. 9.9) by allowing part of the tank volume to cease circulation. Thus, a lesser working volume with a resultant decreased average residence time results, with lessened cell-mass concentration in the effluent. The presence of cells growing in relatively inaccessible tank regions which interchange only slowly with the bulk of the tank volume will increase the allowable flow rate prior to washout: these cells will survive since they are *not* diluted as rapidly as most of the tank, and these poorly mixed sections will act as low-volume feeds of nonsterile streams entering the reactor.

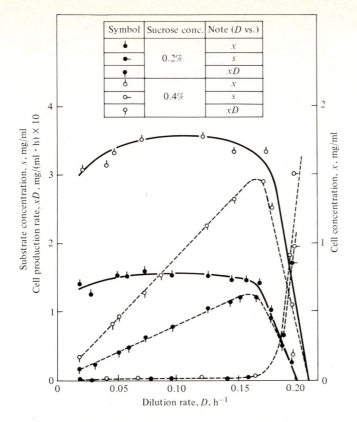

Figure 9.8 Continuous culture of filamentous organisms (here the actinomycete *S. aureofaciens*) often conforms approximately to Monod chemostat behavior, although careful study of these data reveals several deviations from the model. (*Reprinted from B. Sikyta et al., Growth of* Streptomyces aureofaciens *in Continuous Culture, Appl. Microbiol.,* **9**: 233, 1961.)

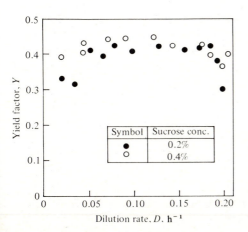

Figure 9.9 Yield factors in continuous culture may not be constant; these data, which were computed from the data in Fig. 9.8, show dependence of Y on dilution rate. (*Reprinted from B. Sikyta et al., Growth of* Streptomyces aureofaciens *in Continuous Culture, Appl. Microbiol.,* **9**: 233, 1961.)

Figure 9.10 shows schematically a simple model of such an imperfectly mixed vessel. Here the reactor contents have been divided into two smaller, completely mixed regions. The feed and effluent streams pass through region 1, whose volume is a fraction α of the total reactor volume. In turn, region 1 exchanges material with stagnant region 2 at volumetric flow rate F'. If we assume Monod growth kinetics with constant yield factor, the following mass balances describe steady-state conditions in this system:

$$x_1 = Y(s_0 - s_1) \qquad \text{region 1 substrate} \tag{9.23a}$$

$$x_2 - x_1 = Y(s_1 - s_2) \qquad \text{region 2 substrate} \tag{9.23b}$$

$$x_2 + \alpha\gamma\mu_{max}\frac{s_1}{K_s + s_1}x_1 = (1 + \gamma D)x_1 \qquad \text{region 1 cells} \tag{9.23c}$$

$$x_1 + (1 - \alpha)\gamma\mu_{max}\frac{s_2}{K_s + s_2}x_2 = x_2 \qquad \text{region 2 cells} \tag{9.23d}$$

where $\qquad \gamma = \dfrac{V}{F'} \qquad D = \dfrac{F}{V} = \text{nominal dilution rate} \tag{9.24}$

The dilution rate at washout for this model is obtained by setting $s_0 = s_1 = s_2$ in Eqs. (9.23) to obtain

$$D_{washout} = \frac{\mu_{max}s_0}{K_s + s_0}\left[1 + \frac{(1 - \alpha)^2\mu_{max}s_0}{K_s + s_0 - (1 - \alpha)\gamma\mu_{max}s_0}\right] \tag{9.25}$$

Since the first expression on the right-hand side is identical to the washout dilution rate for the perfectly mixed system [Eq. (9.12)], the effect of incomplete mixing is to increase the value of $D_{washout}$. If the bracket on the right-hand side of Eq. (9.25) happens to be negative, it indicates that washout is impossible. (What does this mean physically?)

Figure 9.11 shows the effluent cell concentration x_1 and biomass productivity $x_1 D$ as functions of the dilution rate for $\gamma = 0.5$ and a variety of volume fractions α. In this figure, which was computed using the K_s and s_0 values from Fig. 9.7, the change in the x-vs.-D curve as α changes from 1 (perfect mixing) to 0.9 or 0.85 (small stagnant zone) is very similar to the difference between the chemostat theoretical and experimental data given in Fig. 9.7. Moreover, the value $\alpha = 0.9$ corresponds roughly to the vessel volume fraction above the impeller in the experimental reactor used to obtain Fig. 9.7.

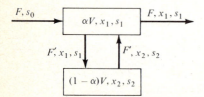

Figure 9.10 Model for an incompletely mixed CSTR with a dead zone.

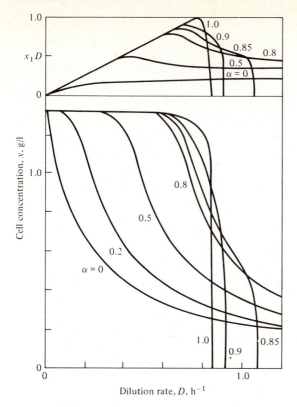

Figure 9.11 Exit cell concentration x_1 as a function of dilution rate D and active zone volume fraction γ for the dead zone model of Fig. 9.10. The upper portion of the figure shows biomass production rate $x_1 \cdot D$ ($\gamma = 0.5$).

This type of model, commonly called a *combined model*, consists of interconnected idealized reactor types. It is but one of a wide class of possible representations of incomplete mixing. Combined models have the advantage of being relatively simple and straightforward to analyze, but experimental evaluation of their parameters can be difficult. Later sections of this chapter will consider other approaches to incomplete mixing.

Another type of nonideality may be involved in preservation of a microbial population at high dilution rates. There may be several solid films of organisms at different points in the vessel. Such colonies can arise, for example, above the liquid level, where splashed droplets have hit the vessel walls, or in crevices and crannies in relatively stagnant zones of the reactor. If we assume that cells on the film at the vessel wall have concentration x_f which is constant with time, reproduction in the film implies addition of cells from the wall into the stirred liquid. In such a situation the steady-state continuous-reactor mass balances take the general form

$$Dx = \mu x + \mu_f x_f \tag{9.26}$$

$$D(s_0 - s) = \frac{1}{Y}\mu x + \frac{1}{Y_f}\mu_f x_f \tag{9.27}$$

where μ_f and Y_f are the specific growth rate and yield factor in the film, respectively. These may differ from the corresponding bulk-liquid parameters μ and Y for a variety of reasons, including diffusion-reaction interactions.

The important thing to notice here is that the $\mu_f x_f$ term in Eq. (9.26) is a source term which is not seriously dependent on D, so that wall growth functions as a second, nonsterile feed which prevents washout. We should note in this connection that laboratory reactors have much larger surface-to-volume ratios than their commercial-sized counterparts, so that in systems involving wall growth, extra care is necessary in scaling up from laboratory data on microbial kinetics.

Another possible source of nonsterile feed is a recycle stream, which can be used to increase biomass and product yield per unit reactor volume. Adopting the notation shown in Fig. 9.12, we take F_0 and F_r as the feed and recycle volumetric flow rates and x_1, x_0, and x as the reactor, recycle-stream, and product-stream biomass concentrations, respectively. These concentrations often differ due to a separator, such as a settling basin, at the point where the reactor effluent stream is split. With $a = F_r/F_0$ and $b = x_0/x_1$, the steady-state biomass-conservation equation for the recycle system is

$$F_r x_0 + \mu x_1 V - (F_0 + F_r)x_1 = 0 \qquad (9.28)$$

so that the overall or external dilution rate D, which is F_0/V, is

$$D = \frac{\mu}{1 - a(b - 1)} \qquad (9.29)$$

Since the microorganisms in the recycle stream are usually more concentrated than in the reactor effluent, $b > 1$. Then Eq. (9.29) reveals that, with recycle, the dilution rate is larger than the organism's specific growth rate. Thus, with organisms growing at the same rate, use of recycle permits processing of more feed material per unit time and reactor volume than in the nonrecycle situation. This feature of recycle is used to great advantage in biological waste-treatment processes, considered in further detail in Chap. 12. (What is the effect of recycle if $b = 1$? What physical interpretation can you provide for your answer?)

Additional important benefits of recycle are revealed by a few manipulations of the system mass balances. Assuming a constant yield factor, the substrate balance is

$$D(s_0 - s) - \frac{\mu x_1}{Y} = 0 \qquad (9.30)$$

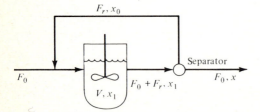

Figure 9.12 Schematic diagram of CSTR with recycle.

Combining this equation with Eq. (9.29), we find that μx_1, the biomass production rate per unit reactor volume, is

$$\mu x_1 = \frac{\mu Y(s_0 - s)}{1 - a(b - 1)} \tag{9.31}$$

This is greater than the nonrecycle production rate by a factor of $[1 - a(b - 1)]^{-1}$. If we assume that μ follows Monod kinetics, we can also show that recycle increases the washout dilution rate by this same factor.

9.1.3. Dynamic Behavior

As in the earlier case of steady-state analysis of continuous culture, we begin our study of the dynamics of these systems with unstructured models. For dynamic studies, the general conservation equation (9.1) must be modified to give the following unsteady-state mass balance:

$$\frac{d}{dt}\text{ (total amount in the system)} = \text{rate of addition to system}$$

$$- \text{ rate of removal from system} + \text{rate of production within system} \tag{9.32}$$

Thus, for a well-stirred culture we have

$$\frac{d}{dt}(Vx) = F(x_0 - x) + Vr_x \tag{9.33}$$

where r_x denotes the rate of biomass production per unit volume. Since V is constant, for sterile feed the mass balance becomes

$$\frac{dx}{dt} = -Dx + r_x \tag{9.34}$$

We shall next explore how the dynamic characteristics of the reactor depend on the growth-rate function r_x. For our purposes *local stability* will be of greatest concern. If a steady state is locally stable, the system will return to that steady state after a small disturbance has acted and moved the system slightly away from the steady state of interest. For an *unstable* steady state, the biomass, substrate, and other concentrations will "run away" from their steady-state values following a small disturbance.

If exponential growth exists in the continuous culture or, in other words, r_x is prescribed by Malthus' law

$$r_x = kx \tag{9.35}$$

then the solution to Eq. (9.34) with the initial condition

$$x(0) = x_i \tag{9.36}$$

is

$$x(t) = x_i e^{(k - D)t} \tag{9.37}$$

Consequently x increases without bound if the organisms grow more rapidly than they are washed out $(k > D)$, and washout occurs in the converse situation $(k < D)$. If $k = D$, the population remains at its initial level; a steady state is achieved only under these conditions, as indicated earlier.

Without examining other specific forms of growth rate function, we can study local stability quite easily when r_x is a function of x only. It has been proved that local stability can be determined by investigation of the linearized form of Eq. (9.34). Proceeding as in Sec. 3.3.2, we can write near a particular steady state x^*

$$\frac{d\chi}{dt} = -\lambda\chi \tag{9.38}$$

where

$$\chi = x - x^* \qquad \lambda = D - \frac{dr_x(x^*)}{dx} \tag{9.39}$$

Since the solution to Eq. (9.38) for $\chi(0) = \chi_0$ is

$$\chi = \chi_0 e^{-\lambda t}$$

we see that the deviation from the steady state at x^* decreases for positive λ and increases if λ negative. Since

$$D = \frac{r_x(x^*)}{x^*} \tag{9.40}$$

from the steady-state form of Eq. (9.34), we are assured of stability if

$$\frac{r_x(x^*)}{x^*} > \frac{dr_x(x^*)}{dx} \tag{9.41}$$

which in turn is true if the derivative of the specific growth rate μ is negative

$$\frac{d[r_x(x^*)/x^*]}{dx} = \frac{d\mu}{dx}\bigg|_{x=x^*} < 0 \tag{9.42}$$

This will be true, it seems, in any biological situation, since violation of condition (9.42) would imply growth at greater than exponential rates in batch culture. This phenomenon has never appeared in any experiments.

It is also known from experiments, however, that continuous cultures are not always stable. Moreover, data on other aspects of continuous-culture dynamics are inconsistent with the one-variable model just explored. Consequently, let us turn to the situation where a single substrate limits growth and examine the dynamic version of the Monod chemostat model. Application of the general unsteady-state mass balance (9.32) to both biophase and substrate and use of Monod's expression (7.10) for the specific growth rate yields

$$\frac{dx}{dt} = D(x_0 - x) + \frac{\mu_{max}sx}{s + K_s} \tag{9.43}$$

and
$$\frac{ds}{dt} = D(s_0 - s) - \frac{1}{Y}\frac{\mu_{max}sx}{x + K_s} \qquad (9.44)$$

For the case of sterile feed ($x_0 = 0$), there are *two* possible steady states, the nontrivial one given earlier in Eqs. (9.10) and (9.11) and the "washout" solution

$$x = 0 \qquad s = s_0 \qquad (9.45)$$

We can determine which of these steady states will exist in a continuous culture by determining their stability. Again, local stability can be studied using the linearized form of Eqs. (9.43) and (9.44). With such coupled, non-linear equations, examination of their linearized approximation is usually the only avenue open for analytical investigations. However, determination of the behavior of even the linearized system is considerably complicated if more than one differential equation is involved. We refer the interested reader to Chap. 11 for details; here we shall only summarize the results of a local-stability study of the Monod chemostat.

	$D > \dfrac{\mu_{max} s_0}{K_s + s_0}$	$D < \dfrac{\mu_{max} s_0}{K_s + s_0}$
Nontrivial steady-state [Eqs. (9.10, 9.11)]	Unstable	Stable
Washout steady-state [Eq. (9.45)]	Stable	Unstable

Another prediction of this analysis is that concentrations cannot approach their steady-state values in a damped oscillatory fashion. As such oscillatory phenomena have been observed experimentally, this substrate-and-cell model is also insufficient to predict all dynamic features of some reactors.

Other weaknesses in the dynamic model of the Monod chemostat are known. It predicts instantaneous reponse of the specific growth rate to a change in substrate concentration: experimentally, a lag is present (see Prob. 10.10). Moreover, growth-rate hysteresis and variations in the yield-factor have been established. Steady cycling has been found in several experimental studies (Fig. 9.13). Consequently, while the Monod chemostat model is quite successful for steady-state purposes in many cases, it has numerous drawbacks as a dynamic representation.

By introducing additional variables into the model, i.e., by giving it more "structure," some of the phenomena unexplained by the Monod model can be accounted for. We could expend another volume on structured dynamic models for continuous culture, but instead we shall only summarize briefly some of the approaches which have been explored. Our emphasis will be on the features of the physical situation which are embodied in the structured model and not on its mathematical details.

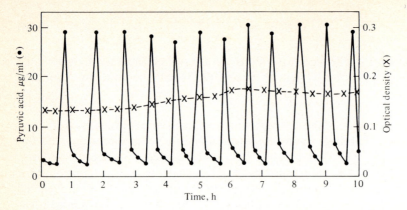

Figure 9.13 These sustained oscillations of pyruvate concentration (•) were observed in continuous culture of *E. coli*. Notice that the cell concentration (crosses) remains approximately constant. (*Reprinted from B. Sikyta, Continuous Cultivation of Microorganisms, Suom. Kemistil.*, **38**: 180, 1965.)

We have already examined the two-component structured model of Williams in Sec. 7.2.4. Applied to continuous culture dynamics, this model reproduces several experimental features not anticipated by the Monod model. Extending the analogy between the Monod growth-rate equation (7.10) and enzyme kinetics, Jefferson and Smith [13] include in their dynamic chemostat model an intermediate species which is an analog of the enzyme-substrate complex. Ramkrishna, Fredrickson, and Tsuchiya [14] include an inhibitor of cell growth in their dynamic model. In a sense, this approach can be viewed as adding more structure to the nutrient phase.

A completely different viewpoint has been taken by Lee, Jackman, and Schroeder,† who consider the influence of flocculation on the overall growth process. We have already observed that the individual cells in many microbial systems form aggregates called flocs. Metabolic processes within such flocs would presumably be different from those in individual dispersed cells; nutrients, for example, would have to diffuse into the floc to reach cells in its interior. Thus, the biophase is viewed as having two components (flocs and individuals) with different kinetics but also with the possibility of interchange of individual cells between the two different morphological forms of biophase. The resulting model exhibits overall yield-factor fluctuations, growth-rate hysteresis, and slower responses than the Monod model—all more compatible with experimental findings than Monod's model.

Yet another conceptual attack is apparent in the model of Young, Bruley, and Bungay [15]. They propose that because of resistances in the mass-transport processes which bring nutrient into the cell, the substrate concentration within the

---

† S. S. Lee, A. P. Jackman, and E. D. Schroeder, A Two-State Microbial Growth Kinetics Model, Water Res. **9**: 491 (1975)

cell is not equal at every instant to the external nutrient concentration, and it is the former quantity which directly influences the cell's growth rate. The model based on this viewpoint exhibits lags in response to environmental changes, as has often been observed experimentally (Prob. 10.10).

In closing our review of chemostat dynamics, we should note another potentially important phenomenon not embodied in the Monod model. In situations where excessive nutrient inhibits growth, the specific growth-rate expression given in Eq. (7.16)

$$\mu = \frac{\mu_{max}s}{K_s + s + s^2/K_p}$$

should be used. A chemostat with this specific growth rate can behave significantly differently from the classical Monod chemostat: now there can be more than one stable steady state for a given set of operating conditions. Dynamic behavior for such a system can be complex, and nonlinear effects not considered in a local stability analysis can be quite important. It has been suggested that this model with its unusual characteristics may help explain the operating difficulties which are common in anaerobic digestion processes. Some aspects of substrate-inhibition effects in chemostats will be explored further in Chap. 12.

9.1.4. Enzyme-Catalyzed Reactions in CSTRs

CSTRs used for enzyme-catalyzed reactions assume a variety of configurations (Fig. 9.14), depending on the method employed to provide the necessary enzyme activity. In the simplest design (*a*), enzymes are continuously added to, and

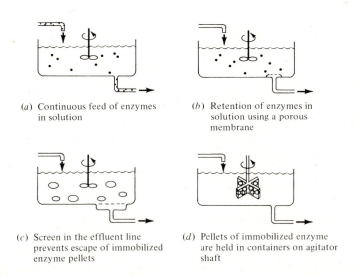

(*a*) Continuous feed of enzymes in solution

(*b*) Retention of enzymes in solution using a porous membrane

(*c*) Screen in the effluent line prevents escape of immobilized enzyme pellets

(*d*) Pellets of immobilized enzyme are held in containers on agitator shaft

Figure 9.14 Schematic diagrams of CSTR designs for enzyme-catalyzed reactions.

removed from, the reactor via the feed and effluent lines. Obviously this approach is practical only when the enzymes are so inexpensive that they are expendable.

Use of more costly enzymes requires that they be retained in the reactor or recycled. Recalling our discussions of enzyme immobilization in Chap. 4 suggests several possibilities. The first of these (Fig. 9.14b) employs a membrane in the effluent stream with pores sufficiently small to prevent escape of the relatively large enzyme molecules in solution. A screen in the effluent line suffices if the enzyme is immobilized on insoluble particles which are suspended in the reaction mixture as a slurry (Fig. 9.14c). Another approach for physical retention of immobilized enzymes within the vessel is shown in Fig. 9.14d, where the enzyme is held in screen baskets attached to the agitator shaft. This configuration, which has also been widely used for study of gas-phase reactions on supported-metal catalysts, is intended to minimize mass-transfer resistance between the liquid phase and the immobilized-enzyme pellets.

Enzyme recycle is feasible only when the enzymes can be readily recovered from the product stream leaving the reactor. Two promising approaches to this problem are containment of enzyme inside liquid-surfactant membranes and immobilization of the enzymes on magnetic supports.

Regardless of which strategy is employed, the common objective is maintenance of the desired enzyme concentration within the CSTR. Assuming that this has been accomplished, we can concentrate our attention on computation of the effluent substrate and product concentrations. The basic principles and general material balances discussed above for microbial growth are applicable, as are additional constraints implied by the relatively simple stoichiometry of these reactions.

For example, for the single reaction

$$S \longrightarrow P$$

1 mol of P is formed for each mole of S which reacts, so that the feed (s_0, p_0) and effluent concentrations (s, p) are related by

$$s_0 - s = p - p_0 \tag{9.46}$$

With Eq. (9.46), reaction-rate expressions $v(s, p)$ which are functions of both s and p can be written in terms of s only, simplifying the necessary algebra. The substrate mass balance in this case takes the form

$$F(s_0 - s) - Vv(s, p_0 + s_0 - s) = 0 \tag{9.47}$$

Table 9.1 gives solutions to this equation for a variety of kinetic forms. These formulas are easiest to use in an indirect fashion: insert the desired substrate conversion into the right-hand side and compute the required residence time and/or enzyme concentration. Design equations for more complicated reactions and kinetics are obtained by similar methods.

Table 9.1. Relationships among substrate conversions $\delta = (s_0 - s)/s_0$, **mean residence time, and catalyst concentration for enzyme-catalyzed reactions in a CSTR†**

Reaction-rate expression for v	CSTR design expression for $V v_{max}/F$
Michaelis-Menten: $\dfrac{v_{max} s}{K_m + s}$	$\delta\left(\dfrac{K_m}{1-\delta} + s_0\right)$
Reversible Michaelis-Menten: $\dfrac{v_{max}(s - p/K)}{K_m + s + K_m p/K_p}$	$\dfrac{\delta\left[K_m + s_0 - \delta s_0 + \dfrac{K_m(p_0 + s_0\delta)}{K_p}\right]}{1 - \delta(1 + 1/K)}$
Competitive product inhibition: $\dfrac{v_{max} s}{a + K_m(1 + p/K_i)}$	$\dfrac{\delta\left[K_m + s_0 - \delta s_0 + \dfrac{K_m(p_0 + s_0\delta)}{K_i}\right]}{1 - \delta}$
Substrate inhibition: $\dfrac{v_{max}}{1 + K_m/s + s/K_i}$	$\delta s_0\left[1 + \dfrac{K_m}{s_0(1-\delta)} + \dfrac{(1-\delta)s_0}{K_i}\right]$

† R. A. Messing, "Immobilized Enzymes for Industrial Reactors," p. 158, table 1, Academic Press, New York, 1975.

9.2. RESIDENCE-TIME DISTRIBUTIONS

Let us now try to imagine what happens to a small clump, or *element*, of fluid after it has entered a CSTR. Because of the vigorous agitation, this fluid will be broken into smaller parts, which separate and disperse throughout the vessel. Thus, some fraction of this fluid element will rapidly find its way to the effluent stream, while other portions of it will wander about the vessel for varying times before entering the exit pipe. Viewed differently, this scenario indicates that the effluent stream is a mixture of fluid elements which have resided in the reactor for different lengths of time. Determination of the distribution of these residence times in the exit stream is a valuable indicator of the mixing and flow patterns within the vessel. As we shall explore in this section and later in this chapter, the residence-time-distribution concept enjoys several worthwhile applications in biological-reactor design and analysis.

9.2.1. Measurement and Interpretation of the $\mathscr{E}$ and $\mathscr{F}$ Functions

We shall consider first an arbitrary vessel with one feed and one effluent line, and it will be assumed for the moment that there is no back diffusion of vessel fluid into the feed line or of effluent fluid into the vessel. In order to probe the mixing characteristics of the vessel, we conduct a *stimulus-response* experiment using a tracer: at some datum time designated $t = 0$, we introduce tracer at concentration c^* into the feed line and maintain this tracer feed for $t > 0$. Then we monitor the system response (in this case the exit tracer concentration) to this specific stimulus. Figure 9.15 shows schematically the general features of this experiment, as well as the shape of a typical exit concentration response $c(t)$.

In practice, a variety of tracers and monitoring techniques can be employed. For example, a dye could serve as a tracer whose concentration could be monitored using a spectrophotometer. Selection of a tracer is limited by the following two requirements: (1) it must not react in the vessel, and (2) it must be detectable at small enough concentration for the local tracer mass-balance equations to be linear in tracer concentration. The second requirement is met so long as the transport and physical properties of the vessel contents are unchanged by the presence of tracer.

Under these conditions, the response to a unit-step tracer input $c_0(t) = H(t)$, where

$$H(t) = \begin{cases} 0 & t < 0 \\ 1 & t \geq 0 \end{cases} \tag{9.48}$$

is obtained by dividing the $c(t)$ function obtained in the above experiment by the tracer feed concentration c^* used in that experiment. The result, the unit-step response of the mixing vessel, is called the $\mathscr{F}$ function:

$$\mathscr{F}(t) = \frac{c(t)}{c^*} = \text{response to unit-step input of tracer} \tag{9.49}$$

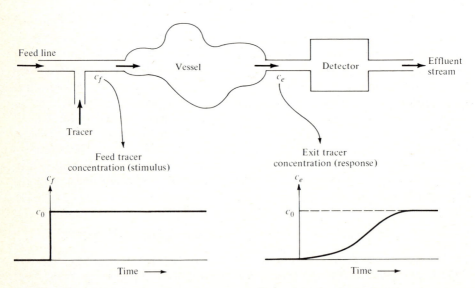

Figure 9.15 Schematic diagram showing experimental measurement of the response ($\mathscr{F}$ curve) to a step tracer input.

What is needed in many cases is not the $\mathscr{F}$ function but the residence-time-distribution (RTD) function† $\mathscr{E}(t)$, which is defined by

$\mathscr{E}(t)\, dt$ = fraction of fluid in exit stream which has been in

$$\text{vessel for time between } t \text{ and } t + dt \quad (9.50)$$

Thus, for example, the fraction of the exit stream which has resided in the vessel for times smaller than t is $\int_0^t \mathscr{E}(\alpha)\, d\alpha$. It follows from definition (9.50) that

$$\int_0^\infty \mathscr{E}(\alpha)\, d\alpha = 1 \quad (9.51)$$

A simple thought experiment will now serve to clarify the relationship between the $\mathscr{E}$ and $\mathscr{F}$ functions. Returning to the stimulus-response experiment of Fig. 9.15, let us imagine the vessel contents to consist of two different types of fluid. Fluid I contains tracer at concentration c^*, and fluid II is devoid of tracer. Consequently, all elements of fluid I must have entered the vessel at some time greater than zero. Then, any fluid I in the effluent at time t has been in the system for a time less than t. On the other hand, fluid II had to be in the vessel at $t = 0$ since only fluid I has entered since then. All fluid II elements in the effluent at time t consequently have residence times greater than t. Assuming that we know the $\mathscr{E}$ function, we can write the exit tracer concentration $c(t)$ as the sum of the fluid I and fluid II contributions:

$$c(t) = c^* \quad \cdot \int_0^t \mathscr{E}(\alpha)\, d\alpha + 0 \quad \cdot \int_t^\infty \mathscr{E}(\alpha)\, d\alpha \quad (9.52)$$

$$\underset{\substack{\text{Fluid I} \\ \text{concentration}}}{} \quad \underset{\substack{\text{Fluid I} \\ \text{fraction}}}{} \quad \underset{\substack{\text{Fluid II} \\ \text{concentration}}}{} \quad \underset{\substack{\text{Fluid II} \\ \text{fraction}}}{}$$

Combining Eqs. (9.52) and (9.49) produces the desired relationship

$$\mathscr{F}(t) = \int_0^t \mathscr{E}(\alpha)\, d\alpha \quad (9.53)$$

which can be differentiated with respect to t to provide the alternative form

$$\frac{d\mathscr{F}(t)}{dt} = \mathscr{E}(t) \quad (9.54)$$

We note first from Eq. (9.54) that $\mathscr{E}(t)$ can be obtained by differentiating an experimentally determined $\mathscr{F}$ curve. Also, the theory of linear systems states that the time derivative of the unit-step response is the unit-impulse response, which reveals that $\mathscr{E}(t)$ can be interpreted as the response of the vessel to the input of a unit tracer impulse at time zero. While an impulse is a mathematical idealization, we can approximate it experimentally by introducing a given amount of tracer into the vessel in a short pulse of high concentration.

---

† Standard terminology from statistics would indicate that $\mathscr{E}$ is a density function with $\mathscr{F}$ the corresponding distribution. The language used above is so firmly embedded in the reaction engineering literature, however, that it would cause confusion to alter it here.

In addition to the experimental methods just described for determining the RTD function, we can sometimes calculate it if a mathematical model of the mixing process is available. Considering the CSTR as an example, the unsteady-state mass balance on (nonreactive) tracer is

$$\frac{dc}{dt} = \frac{F}{V}(c_0 - c) \tag{9.55}$$

To determine the result of an $\mathscr{F}$ experiment for this system, we take

$$c(0) = 0$$

$$c_0(t) = c^* \qquad t \geq 0 \tag{9.56}$$

The solution to Eq. (9.55) under condition (9.56) reveals

$$\mathscr{F}(t) = \frac{c(t)}{c^*} = 1 - e^{-(F/V)t} \tag{9.57}$$

Applying formula (9.54) to the result in Eq. (9.57) reveals that the RTD for a CSTR is

$$\mathscr{E}(t) = \frac{F}{V} e^{-Ft/V} \tag{9.58}$$

Often when dealing with distribution functions such as $\mathscr{E}(t)$, it is helpful to consider the moments of the distribution. The kth moment of $\mathscr{E}(t)$ is defined by

$$m_k = \int_0^\infty t^k \mathscr{E}(t)\, dt \qquad k = 0, 1, 2, \ldots \tag{9.59}$$

Since a unit amount of tracer is introduced into a vessel to observe its $\mathscr{E}$ curve, and since all tracer eventually must leave the vessel, we know that

$$m_0 = 1 \tag{9.60}$$

The first moment m_1 is the mean of the RTD, or the *mean residence time* $\bar{t}$. Under the conditions of zero back diffusion stated at the start of this section, it can be proved [3] for an *arbitrary* vessel that

$$\bar{t} = m_1 = \frac{V}{F} \tag{9.61}$$

which says that the mean residence time is identical to the nominal holding time of the vessel. The second moment m_2 is most often employed in terms of the distribution's variance σ^2,

$$\sigma^2 = m_2 - m_1{}^2 \tag{9.62}$$

which is the average of the squares of deviations from the mean residence time.

9.2.2. Some Applications of the Residence-Time-Distribution Function

Obviously the RTD contains much useful information about flow and mixing within the vessel. One way the $\mathscr{E}$ and $\mathscr{F}$ functions can be used is to assess the extent of deviation from an idealized reactor. For example, we calculated $\mathscr{E}$ and $\mathscr{F}$ above for an ideal CSTR. By comparing these curves with those for a real vessel, we can get an idea of how well the actual system approximates complete mixing.

Different kinds of nonidealities often have distinctive manifestations in the observed response functions. We can gain some appreciation of them by constructing combined models representing various sorts of deviations from the idealized mixing system. Examples are shown in Fig. 9.16: case (*a*) is the ideal CSTR, case (*b*) involves bypassing of the feed stream, and there is a dead volume $(1 - \alpha)V$ in case (*c*). (In contrast to the combined model discussed in Sec. 9.1, this dead volume does not exchange any material with the mixed volume.) When there is bypassing, tracer appears immediately in the $\mathscr{F}$ function, while a dead region results in faster decay of the $\mathscr{E}$ curve than in the ideal case.

We certainly can expect on intuitive grounds that mixing and flow characteristics of a vessel will be critical factors in that vessel's performance as a chemical reactor. In this context, we shall next explore application of a vessel's RTD for predicting the course and extent of reactions within the vessel. This requires further examination of the information provided by the RTD.

Combined model schematic

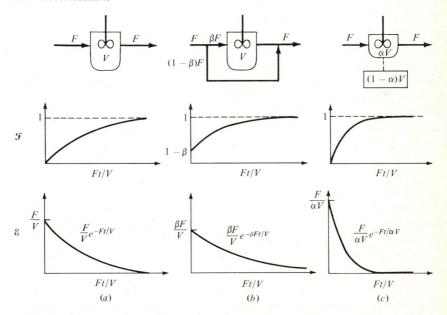

Figure 9.16 $\mathscr{F}$ and $\mathscr{E}$ functions for (*a*) an ideal CSTR, (*b*) a CSTR with bypassing, and (*c*) a CSTR with a dead zone.

Although we cannot discuss all the details here, it is now well established that the RTD does not characterize *all* aspects of mixing (further discussion of this point from a variety of perspectives will be found in the references). The RTD indicates how long various " pieces " of effluent fluid have spent in the reactor, but it does *not* tell us when fluid elements of different ages are intermixed in the vessel.

This point can perhaps be clarified by considering two limiting cases. In the first instance, suppose that fluid elements of all ages are constantly being mixed together. In other words, the incoming feed material immediately comes into intimate contact with other fluid elements of all ages. Such a situation, usually termed a state of *maximum mixedness*, prevails in the ideal CSTR. At the other extreme, fluid elements of different ages do not intermix at all while in the vessel and come together only when they are withdrawn in the effluent stream. In this case, which is called *complete segregation*, reaction proceeds independently in each fluid element: the reaction processes in one segregated clump of fluid are unaffected by the reaction conditions and rates prevailing in nearby fluid elements. Between these two limiting situations falls a continuum of small-scale mixing, or *micromixing*.

Unfortunately for the reaction engineer, the RTD of a reactor is completely independent of its micromixing characteristics. On the other hand, micromixing does influence reactor performance, but not usually to an important degree, as is shown below.

In a reactor with complete segregation, each independent fluid element behaves like a small batch reactor. The effluent fluid is a blend of the products of these batch reactors, which have stayed in the system for different lengths of time. Restating this in mathematical terms, let $c_{ib}(t)$ be the concentration of component i in a batch reactor after an elapsed time t, where the initial reaction mixture in the batch system has the same composition as the flow-reactor feed stream. So long as significant heating or volume change is not caused by the reaction(s), it makes no difference whether one or many different reactions are occurring. A fraction $\mathcal{E}(t)\,dt$ of the reactor effluent contains fluid elements with residence times near t and hence concentrations near $c_{ib}(t)$. Summing over all these fractions gives the exit concentration c_i for the completely segregated reactor:

$$c_i = \int_0^\infty c_{ib}(t)\mathcal{E}(t)\,dt \qquad (9.63)$$

Certainly the individual cells or flocs of cells found in some biochemical reactors are segregated, so that application of Eq. (9.63) seems appealing. However, a rather subtle pitfall exists here. We must remember that living cells contain sophisticated control systems, with which they adapt and respond to their environment. Consequently, the changes which occur during batch culture reflect the combined and interactive influences of the medium and biological phases. Compositions in both of these phases change during the batch in a directly coupled fashion. Consequently, if a cell or floc in a flow system is to behave like the same small batch reactor observed in a batch experiment, it is necessary in general that the cell's environment (the surrounding fluid) also remain segregated in the flow system. If changes in medium composition do not play a critical role in the batch biological reactions, this requirement can be loosened and use of Eq. (9.63) can be better rationalized. Examples of such instances will appear later in this chapter.

Returning now to the influence of micromixing, consider a single half-order irreversible reaction

$$S \longrightarrow P \qquad r = ks^{1/2} \qquad (9.64)$$

occuring in a stirred vessel which is completely segregated but has the same RTD as a CSTR. We might view the half-order reaction as an approximation to the Michaelis-Menten form over a rather narrow range of substrate concentrations. By computing $s(t)$ for reaction (9.64) in a batch reactor and using this with Eq. (9.58) in Eq. (9.63) we find

$$s = s_0 \left\{ 1 - \frac{k\bar{t}}{s_0} \left[1 - \exp\left(-\frac{2s_0}{k\bar{t}} \right) \right] \right\} \qquad (9.65)$$

On the other hand, if the same reaction takes place in an ideal CSTR, which has by definition micromixing at the maximum mixedness limit, the effluent substrate concentration is

$$s = s_0 \left\{ 1 - \frac{(k\bar{t})^2}{2s_0} \left[-1 + \sqrt{1 + \frac{4s_0}{(k\bar{t})^2}} \right] \right\} \qquad (9.66)$$

From a practical design viewpoint, it is fortunate that reactor performance is often not too sensitive to micromixing. For example, for an irreversible second-order reaction in a CSTR, the maximum difference between the conversions at complete segregation and maximum mixedness is less than 10 percent. Thus, even in cases where it is known to be inexact, Eq. (9.63) may provide an adequate approximation of reactor performance. (By series expansion of Eqs. (9.65) and (9.66), develop an expression for the relative difference, $|1 - s\ (9.65)/s\ (9.66)|$.)

In this case and for all single reactions with order less than unity, maximum substrate conversion is achieved at maximum mixedness. Complete segregation provides greatest conversions for reaction orders greater than one. As may be justified using the superposition principle for linear systems, the degree of micromixing does not influence first-order reactions. Thus, for one or more reactions with first-order kinetics, Eq. (9.63) may be used.

Another exploitation of RTD data is the evaluation of parameters in various nonideal-flow reactor models. A variety of such models are considered in the next section along with another ideal reactor, this time one which is always completely segregated.

9.3. TUBULAR AND TOWER REACTORS

There are many continuous biological reactors besides the mixed-vessel type emphasized above. A column can be packed with beads containing immobilized enzymes and used to process amino acids or corn-sugar solutions. In certain waste-treatment processes, waste in solution is removed and oxidized by floc-culated organisms in long, open channels. Another continuous-reactor configuration is the tower fermentor (see Fig. 9.17), in which upward flow of medium and possibly gases suspends flocculated organisms throughout the tower. Such systems are used for continuous beer brewing and growth of yeast or bacteria for use as animal-feed supplements. We begin our investigation of these types of reactors by considering an idealized yet useful prototype.

Figure 9.17 A tower fermentor used for continuous brewing. (*Courtesy of A. P. V. Co., Ltd.*)

9.3.1. The Ideal Plug-Flow Tubular Reactor

When fluid moves through a large pipe or channel, it may approximate *plug flow*, which means that there is no variation of axial velocity over the cross section. If we assume that plug flow prevails in the system, we can formulate the mass balance on the plug-flow tubular reactor (PFTR) easily using the *differential-section* approach. As Fig. 9.18 suggests, the basic steady-state conservation equation (9.1) is applied to a thin slice of the tubular reactor taken perpendicular to the reactor axis. Considering an arbitrary component C, the mass balance on the thin section is

$$Auc\Big|_z - Auc\Big|_{z+\Delta z} + A\,\Delta z\, r_{f_c}\Big|_z = 0 \tag{9.67}$$

where r_{f_c} is the rate of formation of species C in terms of amount per unit volume per unit time. Rearranging and dividing by $A\,\Delta z$ yields

$$\frac{uc\big|_{z+\Delta z} - uc\big|_z}{\Delta z} = r_{f_c} \tag{9.68}$$

Taking the limit of this equation and recalling the definition of the derivative gives the final form

$$\frac{d}{dz}(uc) = r_{f_c} \tag{9.69}$$

So long as the reaction does not cause a change in fluid density (would this assumption be valid for microbial process?), the axial velocity is constant and Eq. (9.69) becomes

$$u\frac{dc}{dz} = r_{f_c} \tag{9.70}$$

The quantity z/u is equal to the time required for a small slice of fluid to move from the reactor entrance to axial position z. If we use this transit time t

$$t = \frac{z}{u} \tag{9.71}$$

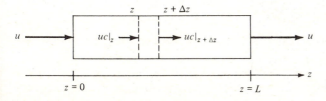

Figure 9.18 Plug-flow reactor.

as a new independent variable, the mass-balance equation (9.70) can be rewritten as

$$\frac{dc}{dt} = r_{f_c} \tag{9.72}$$

which is *exactly the same as the batch-reactor mass balance*. This mathematical demonstration can be supplemented by a physical argument: in plug flow with constant velocity, each thin slice of fluid moves through the vessel with absolutely no interaction with neighboring slices. The system is totally segregated, each thin slice behaving the same as a batch reactor. Consequently, if the initial charge in a batch reactor has the same composition as the feed to the plug-flow reactor, and if the mean residence time L/u for the tube is the same as the batch reaction time, the tube effluent is identical to the batch-reactor product. The boundary condition appropriate for this model is obviously

$$c\Big|_{z=0} = c_0 \tag{9.73}$$

where $z = 0$ denotes the reactor inlet and c_0 is the C concentration in the feed. As an example, we shall assume that the kinetics used in the Monod chemostat are applicable in the PFTR (or the equivalent batch reactor). The mass balances on cells and substrate in the form of Eq. (9.72) are

$$\frac{dx}{dt} = \frac{\mu_{max} xs}{s + K_s} \tag{9.74}$$

$$\frac{ds}{dt} = -\frac{1}{Y}\frac{\mu_{max} xs}{s + K_s} \tag{9.75}$$

with initial conditions

$$x(0) = x_0 \qquad s(0) = s_0 \tag{9.76}$$

On physical grounds or by manipulations with Eq. (9-74) and (9-75), we can see that the s and x concentrations are bound by the stoichiometric relationship

$$x + Ys = x_0 + Ys_0 \tag{9.77}$$

Using Eq. (9.77) to express x in terms of s and substituting this into Eq. (9.74) gives the single ordinary differential equation

$$\frac{ds}{dt} = \frac{\mu_{max}}{Y}\frac{[x_0 + Y(s_0 - s)]s}{s + K_s} \tag{9.78}$$

Integration of this equation subject to condition (9.76) can be achieved analytically, with the result

$$x_0 + Y(s_0 + K_s)\ln\frac{x_0 + Y(s - s_0)}{x_0} - K_s Y \ln\frac{s}{s_0} = \mu_{max}t(x_0 + Ys_0) \tag{9.79}$$

The effluent substrate concentration is the s value corresponding to $t = L/u$, and then x is found with Eq. (9.77). If viewed as the result of a batch reaction, the kinetics of Eq. (9.74) here shows no lag or death phases but does reach a stationary phase.

The physical perspective of the PFTR introduced above readily reveals its RTD. If a tracer pulse is introduced in the feed, it flows through the vessel without mixing with adjacent fluid and emerges after a time L/u. Thus, the tracer pulse in the exit has exactly the same form as the pulse fed into the PFTR, except that it is shifted in time by one vessel holding time. Deviation from such behavior is evidence of breakdown in the plug-flow assumption.

Practical implementation of plug-flow biological reactors can be difficult, especially for aerobic microbial systems. Besides providing adequate aeration, there is the problem caused by microbial growth, which can profoundly influence the flow properties of the culture. Figure 9.19 illustrates one scheme which has been tested experimentally with a *Monascus* sp. mold. Here air, which is fed at 1 vvm (volume of air per volume of vessel per minute), is distributed along the vessel by means of a porous vinyl tube. The RTD shown on the right of Fig. 9.19 was measured using the dye-spectrophotometer approach already mentioned. The tailings on either side of the vessel holding time indicate failure to achieve plug flow: evidently there is some backmixing within the reactor. However, unless operated under conditions very sensitive to details of the RTD, e.g., very high conversion of substrate, this system can be modeled adequately as plug flow.

In contrast to the CSTR, a sterile feed to a PFTR automatically implies a zero biomass concentration in the effluent: plug flow prevents a slice of fluid moving through the vessel from ever being inoculated. One way to circumvent this problem is by recycle, so that the incoming stream is inoculated before entering the vessel.

As indicated above, tubular reactors packed with immobilized-enzyme pellets enjoy several applications. The rate expressions for single particles developed in Chap. 7 must be coupled with other mass balances to produce the necessary rate expressions for use in Eq. (9.70). Considering a single irreversible S → P reaction

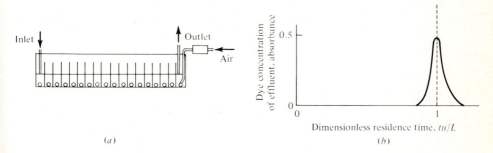

(a)   *(b)*

Figure 9.19 *(a)* Schematic diagram of an aerated tubular fermentor. *(b)* Experimentally observed RTD for the tubular fermentor of part *(a)* (*Reprinted from T. Imanaka et al., Optimization of α-Galactosidase Production in Multi-Stage Continuous Culture of Mold, J. Fermentation Technol. (Japan),* **51**: 431, 1973.)

with intrinsic rate $v = v(s, p)$, for example, the rate of substrate disappearance per unit volume of enzyme pellet at a point in the reactor is

$$v \bigg|_{\substack{\text{overall/unit} \\ \text{volume of pellet}}} = \eta(s_s, p_s)v(s_s, p_s) \tag{9.80}$$

where s_s and p_s are the substrate and product concentrations at the exterior pellet surface at that position inside the reactor. In general the effectiveness factor η, which accounts for intraparticle diffusion, and the rate expression v depend on both s_s and p_s, as indicated.

If mass-transfer resistance between the bulk liquid phase and the pellet surface is next examined, a steady-state material balance on substrate over the pellet gives

Rate of substrate diffusion out of bulk liquid

$\qquad\qquad$ = rate of substrate disappearance by reaction within pellet

Table 9.2. Relationships among substrate conversion $\delta = (s_0 - s)/s_0$ and reactor design parameters for enzyme-catalyzed reactions in a PFTR†

Enzymes in solution‡ or in immobilized pellets with negligible mass-transfer limitations

Reaction rate expression for v	PFTR design expression for $\dfrac{1-\varepsilon}{\varepsilon}\dfrac{Lv_{\max}}{u}$
Michaelis-Menten: $\dfrac{v_{\max}s}{K_m + s}$	$s_0\delta - K_m \ln(1-\delta)$
Reversible Michaelis-Menten: $\dfrac{v_{\max}(s - p/K)}{K_m + s + K_m p/K_p}$	$s_0\left(1 - \dfrac{K_m}{K_p}\right)\left[\dfrac{\delta}{b} + \dfrac{1}{b^2}\ln(1-b\delta)\right]$ $\qquad - \left(K_m + s_0 + \dfrac{K_m}{K_p}p_0\right)\dfrac{1}{b}\ln(1-b\delta)$ where $b = \dfrac{K+1}{K}$
Competitive product inhibition: $\dfrac{v_{\max}s}{s + K_m(1 + p/K_i)}$	$s_0\left(1 - \dfrac{K_m}{K_i}\right)[\delta + \ln(1-\delta)]$ $\qquad - \left(K_m + s_0 + \dfrac{K_m}{K_i}p_0\right)\ln(1-\delta)$
Substrate inhibition: $\dfrac{v_{\max}}{1 + K_m/s + s/K_i}$	$s_0\delta - K_m\ln(1-\delta) + \dfrac{s_0^2}{K_i}\left(\delta - \dfrac{\delta^2}{2}\right)$

† Adapted from R. A. Messing, "Immobilized Enzymes for Industrial Reactors," p. 158, Academic Press, Inc., New York, 1975.

‡ The $(1 - \varepsilon)/\varepsilon$ factor in these equations should be set equal to unity for enzymes in solution.

or
$$4\pi R^2 k_s(s - s_s) = \tfrac{4}{3}\pi R^3 \eta(s_s, p_s)v(s_s, p_s) \tag{9.81}$$

With this equation and reaction stoichiometry, s_s and p_s can be evaluated in terms of the bulk liquid substrate concentration s. Substituting these values into Eq. (9.80) then gives the total rate of substrate disappearance per unit volume of catalyst pellets in terms of the bulk fluid substrate concentration.

This rate factor is converted into a rate for use in the packed-bed-reactor design equations using the void function ε, which is the fraction of reactor volume not occupied by catalyst particles. The expression for the rate of substrate formation per unit void reactor volume is

$$r_{f_s} = -\frac{1 - \varepsilon}{\varepsilon}\eta(s_s, p_s)v(s_s, p_s) \tag{9.82}$$

This formula can now be used in the substrate mass balance, which in the case of plug flow through the packed bed is Eq. (9.70) with c replaced by s. Generally, solution of these equations must be obtained with a computer.

The situation is greatly simplified if intraparticle and external mass-transfer resistances are negligible, since these conditions imply $\eta \to 1$ and $s_s \to s$, respectively. In such circumstances the governing mass balances can be integrated analytically, with results as indicated in Table 9.2 for several types of enzyme kinetics. These equations also obtain for tubular reactors with continuous feed of enzymes in solution. As can be verified by simple operations on the substrate and product mass balances, Eq. (9.46) is applicable here (at steady state) and serves to expedite the calculations.

9.3.2. Tower Reactors

In a fluidized-bed tower reactor like the one shown in Fig. 9.17, liquid flows upward through a long vertical cylinder. Flocculated organisms (or immobilized-enzyme pellets) are suspended by drag forces exerted by the rising liquid; those which are entrained are captured at the top of the tower by expanding the cross section, thereby reducing liquid drag, and fed back into the tower. Thus, by a careful balance between operating conditions and organism characteristics, the flocs are retained in the reactor while the medium flows through it continuously.

Fluidized-bed biological reactors are considerably more complex than the CSTR and PFTR varieties so far examined. For example, in tower fermentors used for continuous beer production, there is a gradient of yeast flocs through the unit. Near the bottom, the organism concentration (centrifuged wet weight per weight of broth × 100 percent) may reach 35 percent, while the yeast concentration drops to 5 to 10 percent at the top of the tower. Moreover, there is a progressive change in medium characteristics along the reactor. Easily fermented sugars (glucose, fructose, sucrose, some maltoses) are consumed first, near the feed point, lowering the medium density. In the middle and upper portions of the tower, the yeast flocs ferment maltotriose and additional maltose. This scenario of rapid initial fermentation followed by slower reactions involving less desirable substrates is consistent with the experimental data shown in Fig. 9.20.

A rudimentary model for such fluidized reactors can be developed by assuming that (1) the biological catalyst particles (microbial flocs or immobilized-enzyme pellets) are uniform in size, (2) the fluid-phase density is a function of substrate concentration, (3) the liquid phase moves upward through the vessel in plug flow, (4) that the substrate-utilization rates are first order in biomass concentration but zero order in substrate concentration, and (5) that the catalyst-particle Reynolds number based on the terminal velocity is small enough to justify Stokes' law (recall Example 1.1). Assumptions (4) and (5) are reasonable for many applications, and (1) to (3) may be adequate approximations.

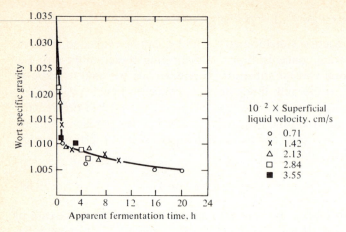

Figure 9.20 Tower fermentor data for yeast growth on a sugar mixture show an initially rapid rate of substrate utilization (as revealed by change in wort specific gravity), followed by a period of much slower utilization rates. (*Reprinted from R. N. Greenshield and E. L. Smith, Tower-Fermentation Systems and their Applications, The Chemical Engineer, **May, 1971**: 182.*)

Under these assumptions, the substrate-conservation equation follows the form of Eq. (9.69):

$$\frac{d(su)}{dz} = -kx$$

or

$$u\frac{ds}{dz} + s\frac{du}{dz} = -kx \tag{9.83}$$

For Stokes flow, the concentration of the suspended biomass can be related to the liquid flow velocity in a fluidized bed by

$$x = \rho_0\left[1 - \left(\frac{u}{u_t}\right)^{1/4.65}\right] \tag{9.84}$$

where ρ_0 is the microbial density on a dry weight basis and u_t is the terminal velocity of a sphere in Stokes' flow (Eq. 8.28). (Should the dry-weight or wet-weight biomass density be used in the terminal-velocity formula?). Note that in the context of the fluidized bed we assume that local biomass concentration is dictated entirely by hydrodynamic factors rather than the biochemical-reaction–metabolism features emphasized for other reactor types. Substituting Eq. (9.84) into (9.83) leaves two unknowns, s and u as functions of position z in the tower.

We complete the model by applying Eq. (9.69) to total mass ($r_f = 0$) to reveal

$$\frac{d}{dz}(\rho u) = 0 \tag{9.85}$$

Expanding (9.85) and using $\rho = \rho(s)$ gives

$$\rho(s)\frac{du}{dz} + \left(u\frac{d\rho}{ds}\right)\frac{ds}{dz} = 0 \tag{9.86}$$

To cast the model in standard form suitable for numerical integration, we may now view Eqs. (9.83) and (9.86) as simultaneous algebraic equations in the unknowns ds/dz and du/dz. Solving this algebraic set, which need not be written out in full here, gives ds/dz and du/dz in terms of s and u: a set of two simultaneous differential equations to be integrated with the initial conditions

$$s(0) = s_f \qquad u(0) = u_f = \frac{F_f}{A_f} \tag{9.87}$$

where A_f is the tower cross section at the bottom. The effluent substrate concentration s_e is $s(z = L)$.

All this is much simplified if we assume that whatever fluid density changes occur do not affect u significantly. With u independent of position, Eq. (9.83) integrates directly, with the result

$$s_e = s_f - k\rho_0 \left[1 - \left(\frac{u}{u_t} \right)^{1/4.65} \right] \frac{L}{u} \tag{9.88}$$

where x from Eq. (9.84) has been inserted and L is the tower height. Such linear dependence of substrate concentration on mean reaction time L/u is apparent in at least portions of Fig. 9.20 (if we also assume a linear relationship between s and ρ).

Unfortunately, however, the data in Fig. 9.20 suggest that our most serious error in this model is lumping of many substrates: the various sugars consumed during the anaerobic alcohol fermentation have been grouped together into a single hypothetical or average substrate in our model. In doing this, we have no way to include the glucose effect, which plays a very important role in the operation of continuous tower fermentors for brewing.

Several commercial processes for single-cell protein production involve aerobic growth of yeast or bacteria in tower fermentors. Since large volumes of gas are utilized in this context, mass transfer and two-phase flow phenomena become dominant features of the system. We learned in the previous chapter that for a sufficient density of rapidly growing aerobic organisms, the overall growth rate is typically limited by the rate of oxygen transfer from the gas bubbles into the liquid phase.

Analysis of this limiting rate process requires knowledge of liquid and gas mixing within the tower. Studies on air-water sparged columns have shown that if (*a*) gas flow rates are large relative to the liquid flows and (*b*) column height L and diameter d_t are of similar magnitude, both liquid and gas phases are well mixed. If $L \gg d_t$, that is, in "long" columns, the gas exhibits plug flow while the liquid is mixed. Consequently, Eq. (9.72) is the appropriate design equation for short towers, and for the more typical long columns, Eq. (8.99) gives the column height L ($= Z$ there) to obtain a desired amount of O_2 transfer.

For the integrated form (8.99) to be valid, it is necessary to maintain the interfacial-area factor a nearly constant along the tower. This in turn requires that the gas remain in bubbling flow. Air-water experiments reveal that the gas bubbles rising through the column will coalesce into slugs if the gas-volume fraction ε exceeds a critical value ε_{max}, which is roughly 0.3. The requirement

$$\varepsilon < \varepsilon_{max} \text{ to maintain bubble flow}$$

can be translated into a design specification for column diameter by noting that at any point in the tower

$$F_g = u_g \, \varepsilon \, \frac{\pi d_t^{\,2}}{4} \tag{9.89}$$

where F_g and u_g are the gas volumetric flow rate and linear velocity, respectively. We may reasonably assume that u_g is the terminal velocity u_t of a single gas bubble in a stagnant liquid and that F_g is roughly the same as the feed-gas flow rate F_{gf}. The latter assumption is rationalized on the grounds that O_2 consumed from the bubbles is at least partially replaced by CO_2. Under these conditions, Eq. (9.89) reveals that ε is smaller than ε_{max} so long as

$$d_t \geq 2 \left(\frac{F_g}{u_t \varepsilon_{max}} \right)^{1/2} \tag{9.90}$$

which can be used to complete sizing of the tower.

Other means of providing small gas bubbles throughout the tower include insertion of perforated plates and/or impellers within the vessel. These internal devices break up any coalesced gas slugs and thereby maintain a large gas-liquid contact area. Two different tower designs with internals are illustrated in Fig. 9.21. In both cases, the objective is to achieve a continuous analog of batch-fermentor behavior.

Some experimental evidence that this has been achieved in a 10-stage cocurrent perforated-plate tower is given in Fig. 9.22b. Here the total mean residence time $\bar{t}_n$ through stage n is defined by

$$\bar{t}_n = \sum_{i=1}^{n} \frac{V_i}{F} \tag{9.91}$$

where V_i is the volume of the ith stage. The changes in *E. coli*, glucose, and RNA concentrations with cumulative residence time are certainly reminiscent of the time course of a batch fermentation. Similar data showing stage-to-stage variations in biomass and substrate concentrations have also been reported for the reactor in Fig. 9.21a.

In both these systems, mixing in each stage is vigorous, so that plug flow is obviously not exactly realized. These are just some of the many biological reactors in which mixing falls somewhere between the completely stirred and plug-flow extremes. Design and modeling approaches for such processes are our primary concern in the next section.

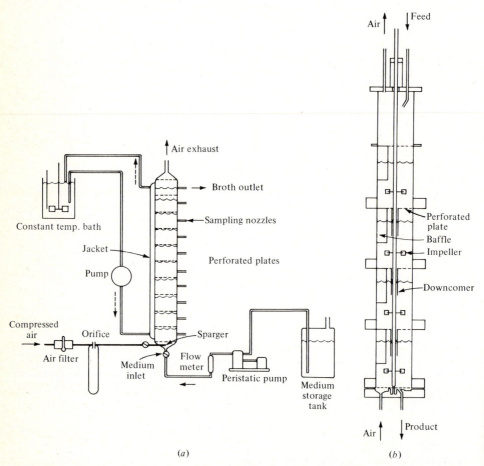

Figure 9.21 Two staged tower-reactor designs. (*a*) cocurrent gas-liquid flow and perforated plates; (*b*) internal impellers, countercurrent contacting, and a liquid downcomer.

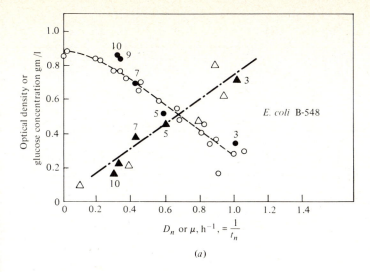

(a)

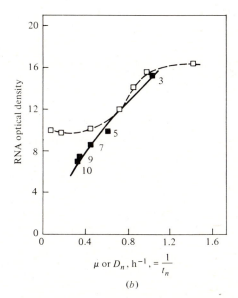

(b)

Figure 9.22 Influence of staging on substrate, biomass, and RNA concentrations. (a) Batch results: (○) *E. coli*: optical density, (△) glucose, (□) RNA, (b) Tower fermentor results: (●) *E. coli*: optical density, (▲) glucose, (■) RNA. Numbers indicate stage number. (*Reprinted from A. Ketai et al., Performance of a Perforated Plate Column as a Multistage Continuous Fermentor, Biotech. Bioeng.,* **11**: 911, 1969.)

9.3.3. Tanks in Series and Dispersion Models for Nonideal Reactors

In cases like the staged-tower reactors just mentioned, the physical configuration of the system suggests a model with a number of ideal CSTRs in series (Fig. 9.23a). These models are also used as empirical representations for other complex mixing and contacting schemes.

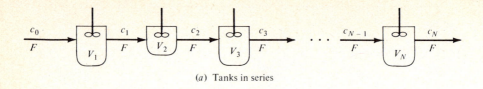

(a) Tanks in series

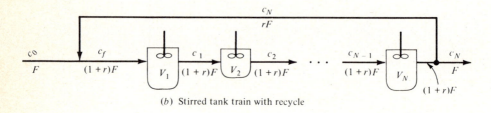

(b) Stirred tank train with recycle

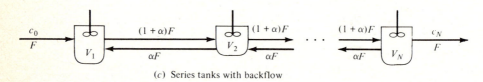

(c) Series tanks with backflow

Figure 9.23 Schematic diagrams of different tanks in series designs and models; (a) tanks in series; (b) stirred tank train with recycle; and (c) series tanks with back flow.

When the notation of Fig. 9.23a is used, the mass balance on an arbitrary component C in the jth tank is

$$Fc_{j-1} - Fc_j + V_j r_{f_c}\bigg|_{\text{tank } j} = 0 \qquad (9.92)$$

For example, if we consider microbial growth in the tanks-in-series system with nonsterile feed ($x_0 \neq 0$), the biomass balances for tanks 1 through N are

$$F(x_0 - x_1) + V_1 \mu_1 x_1 = 0$$

$$F(x_{j-1} - x_j) + V_j \mu_j x_j = 0 \qquad j = 2, 3, \ldots, N \qquad (9.93)$$

These equations can be solved recursively to yield

$$x_1 = \frac{Fx_0}{F - V_1 \mu_1} \qquad X_1 = \frac{\mu_1 V_1 X_0}{\mu_1 V_1 - V_1 \mu_1} \qquad (9.94)$$

and

$$x_j = \frac{F^{j-1} x_1}{(F - \mu_2 V_2)(F - \mu_3 V_3) \cdots (F - \mu_j V_j)} \qquad j = 2, \ldots, N \qquad (9.95)$$

For the case of equal volumes, the simpler form

$$x_j = \frac{\mu_1^{j-1} x_1}{D_1^{j-1} x_1}{(D_1 - \mu_2)(D_1 - \mu_3) \cdots (D_1 - \mu_j)} \qquad \frac{D X_1}{D - \mu_2} = X_2 \qquad (9.96)$$

$$X_j = (\mu_1 - \mu_2)(\mu_1 - \mu_3) \cdots (\mu_1 - \mu_j)$$

results from Eq. (9.95). Here D_1 is the dilution rate of an individual tank ($= F/V_1$). If the feed is sterile ($x_0 = 0$), Eq. (9.93) requires

$$F = \mu_1 V_1 \qquad \text{or} \qquad D_1 = \mu_1 \qquad \text{equal-volume case}$$

These expressions should be substituted into Eqs. (9.94) to (9.96) for a continuous-culture cascade with sterile feed. Obviously $x_{j-1} < x_j$ and additional growth occurs in each stage j unless $\mu_j = 0$.

We can deduce the effect of staging and recycle on the washout dilution rate D_{max} by examining the system of Fig. 9.23b. Assuming Monod growth kinetics with a maintenance term [Eq. (9.18)], sterile feed, equal volumes, and a constant yield factor, it has been shown that D_{max} satisfies

$$\frac{\mu_{max}}{D_{max}} = \frac{N(1 + K_s/s_0)\{1 - [r/(1 + r)]^{1/N}\}}{1 - (k_e/\mu_{max})(1 + K_s/s_0)} \tag{9.97}$$

where as usual D is defined in terms of the overall process

$$D = \frac{F}{V} = \frac{F}{NV_1} \equiv \frac{D^*}{N} \tag{9.98}$$

Setting $r = 0$ in Eq. (9.97) gives the critical dilution rate for the equal-volume cascade without recycle. If in addition to $r = 0$, we take $k_e = 0$ so that the growth kinetics is of classical Monod form, Eq. (9.97) reduces to the familiar expression [see Eq. (9.12)]

$$D^*_{max} = \frac{\mu_{max} s_0}{K_s + s_0} \tag{9.99}$$

This result is expected on intuitive grounds: if the washout occurs in the first tank in the equal-volume train, it will also prevail in the second, third, ..., and Nth tank. (What happens in the non-equal-volume case if $F > V_j D^*_{max}$ for $j = 1, 2, \ldots, k - 1$ and $F < V_k D^*_{max}$?)

Before examining some applications of the tanks in series for the Fig. 9.23a model, we should recognize its relationships with the ideal CSTR and PFTR. In this context it is helpful to consider a fixed total reactor volume V with the number of tanks N as a parameter and

$$V_1 = V_2 = \cdots = V_N = \frac{V}{N} \tag{9.100}$$

Clearly for $N = 1$, we have the single ideal CSTR considered above.

Rearranging Eq. (9.92) and utilizing Eq. (9.100) gives the component C mass balance in the form

$$\frac{c_j - c_{j-1}}{V/NF} = r_{f_c} \tag{9.101}$$

The V/NF term is the holding time of tank j. As $N \to \infty$, this mean residence time per tank approaches zero, as does the difference $c_j - c_{j-1}$: then we can interpret the left-hand side of Eq. (9.101) as the derivative dc/dt, where t denotes the cumulative holding time within the reactor. Consequently, for $N \to \infty$, the C mass balance approaches

$$\frac{dc}{dt} = r_{f_c} \tag{9.102}$$

which is identical to Eq. (9.72). Thus, the tanks-in-series system becomes identical to a plug-flow vessel when N is sufficiently large. This correspondence also makes sense physically. As the number of tanks increases and the holding time per tank falls, there is less intermixing of fluid elements with different residence times.

Two questions are suggested by the previous discussion: How large must N be to approximate plug flow? What type of behavior is expected for N values between unity (CSTR) and this upper asymptote (PFTR)? Information relevant to both concerns is provided by the system's RTD.

We can calculate the RTD by again setting up tracer mass balances on the stirred-vessel cascade and evaluating the response to a unit-impulse input. Carrying out the necessary algebra for the system of Fig. 9.23a subject to the equal-volume condition (9.100) gives

$$\mathscr{E}(t) = \frac{N^N}{(N-1)!}\left(\frac{NF}{V}\right)^{N-1} t^{N-1} \exp\left(-\frac{NF}{V}t\right) \tag{9.103}$$

Plots of this function for a variety of N values are given in Fig. 9.24. The shift in $\mathscr{E}(t)$ with increasing N from an exponential decay to a pulse at $t = V/F$ is apparent.

The variance of the RTD in Eq. (9.103) is

$$\sigma^2 = \frac{1}{N} \tag{9.104}$$

This relationship is useful in developing a series CSTR model for an arbitrary vessel whose RTD has been experimentally determined. Taking the total staged-CSTR system volume V and flow rate F as in the actual process makes the mean residence time for the model match that of the real vessel. Taking N equal to the reciprocal of the experimentally measured variance for the real vessel then ensures that the series-CSTR model RTD has the same second moment as the vessel of interest. If the vessel's RTD has a shape similar to those given in Fig. 9.24, the tanks-in-series model so derived will probably provide an adequate approximation of the real reactor's performance. (Remember that identical RTDs imply identical reactor performances only for total segregation or first-order kinetics).

Fig. 9.25 shows the results of experimental RTD studies of cocurrent perforated plate towers (Fig. 9.21a). In both experiments three plates separated the column into four sections, but in case 1 the plates had 3-mm moles; 2-mm holes were used in case 2, with the total hole volume held at 10 percent in both instances. Comparing the experimental RTDs with the theoretical result for $N = 4$ in Fig. 9.25 reveals that the plates with 2-mm holes provide good staging. Evidently there is some backflow through the plate perforations when they exceed the 2-mm size. Thus, in this case RTD data provide a useful design guideline for preserving the desired segregation in the tower.

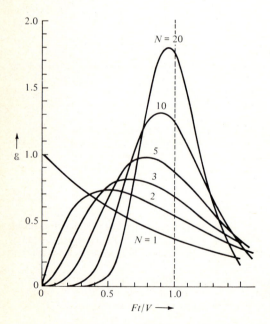

Figure 9.24 RTDs for N ideal CSTRs in series (V = total system volume, individual reactor volumes = V/N).

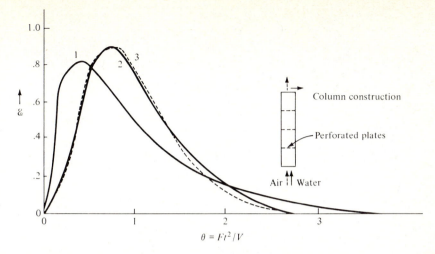

Figure 9.25 RTD data showing flow characteristics of a cocurrent perforated plate column. (1) Hole diameter = 3 mm, plate-area fraction = 0.0981, (2) hole diameter = 2 mm, area fraction = 0.0975, (3) calculated from four CSTRs-in-series model. (*Reprinted from A. Ketai et al., Performance of a Perforated Plate Column as a Multistage Continuous Fermentor, Biotech. Bioeng., **11**: 911, 1969.*)

Sometimes the addition of a growing microbial phase to a reactor dramatically alters its RTD. In the measured RTD of Fig. 9.26a, the eight-plate tower with no growing organisms exhibits a clear staging effect. When bakers' yeast is grown in the same system, however, the RTD measured by a variety of tracers closely corresponds to an ideal CSTR (solid curve in Fig. 9.26b) rather than to an eight-CSTR cascade. This breakdown in staging effect, which is *not* apparent in the *E. coli* data of Fig. 9.22, is apparently due to sedimentation of the yeast suspension. By changing the column design so that four plates had only 3 percent hole area, the RTD with yeast growth became nearly that of a four-CSTR cascade.

RTD data for the countercurrent tower with internal agitation (Fig. 9.21b) suggest that entrained liquid carried upward through the plates represents a significant backflow counter to the liquid flow through the downcomers. After adding a backflow stream connecting adjacent tanks in the model, as in Fig. 9.23c, the experimental RTD can be fitted nicely by appropriate choice of the backflow parameter α (see Fig. 9.27).

A close conceptual analog of this CSTR cascade with backflow is the *dispersion model*. A modification of the ideal PFTR, the dispersion model is derived by considering an axial diffusion process which is superimposed on the convective flow through the tube. Returning to the thin-section model described in Fig. 9.18, we add a dispersion flow $A(-D_z \, dc/dz)_z$ into the section and subtract a similar term $A(-D_z \, dc/dz)_{z+\Delta z}$ for diffusion out of the thin slice on the left-hand side of Eq. (9.67). The same manipulations and limiting processes as followed that equation produce the dispersion-model mass balance

$$\frac{d(uc)}{dz} = \frac{d}{dz}\left(D_z \frac{dc}{dz}\right) + r_{f_c} \qquad (9.105)$$

Usually u and the *effective axial dispersion coefficient* D_z are approximately constant for liquids, so that Eq. (9.105) becomes

$$u\frac{dc}{dz} = D_z \frac{d^2c}{dz^2} + r_{f_c} \qquad (9.106)$$

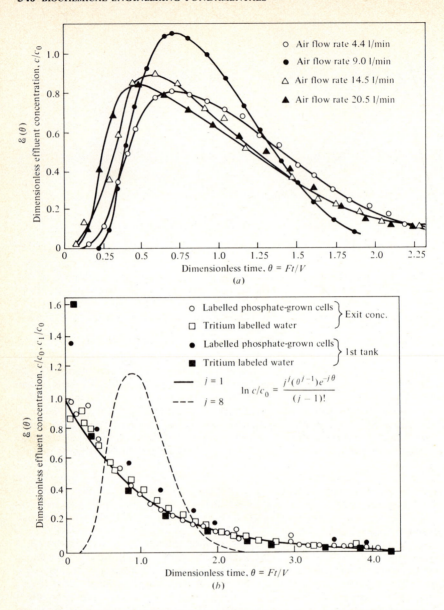

Figure 9.26 Measurement of RTD in an eight-plate tower (hole void fraction = 0.15) by different methods and under different operating conditions. (*a*) Salt tracer data show that the increased aeration flattens and broadens the residence time distribution. (*b*) Yeast cells present in the column; both labeled water and labeled cells used as tracers. (*A. Prokop et al., Design and Physical Characteristics of a Multistage, Continuous Tower Fermentor, Biotech. Bioeng., 11: 945, 1969*).

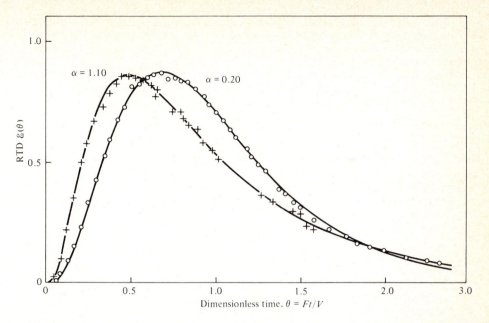

Figure 9.27 Two experimentally measured RTDs ($\bigcirc$ and $+$ points) for tower system of Fig. 9.21*b* with fits by back-flow model of Fig. 9.23*c*. There are four stages. (*Reprinted from E. A. Falch and E. L. Gaden, Jr., A Continuous Multistage Tower Fermentor. II. Analysis of Reactor Performance, Biotech. Bioeng.,* **12:** 465, 1970.)

Since Eq. (9.106) is a second-order differential equation, two boundary conditions are required. The generally accepted ones are

$$uc_0 = \left(uc - D_z \frac{dc}{dz}\right)_{z=0} \tag{9.107}$$

$$\left.\frac{dc}{dz}\right|_{z=L} = 0 \tag{9.108}$$

When D_z is not too large, which is often the case, the complicated conditions (9.107) can be replaced by

$$c\Big|_{z=0} = c_0 \tag{9.109}$$

In this case the variance of the dispersion-model RTD is

$$\sigma^2 = \frac{2}{\text{Pe}}\left[1 - \frac{1}{\text{Pe}}(1 - e^{-\text{Pe}})\right] \tag{9.110}$$

where the axial Peclet number Pe is defined by

$$\text{Pe} = \frac{uL}{D_z} \tag{9.111}$$

Physically Pe may be regarded as a measure of the importance of convective mass transport ($\sim uc$) relative to mass transport by dispersion ($\sim D_z c/L$). Considering the limits of Eq. (9.110) as Pe $\to 0$ and Pe $\to \infty$, we see that the former case gives $\sigma^2 = 1$, which is identical to the ideal CSTR variance. In the limit of very large Peclet numbers, $\sigma^2 \to 0$, which corresponds to plug flow. The transition of the RTD from the CSTR to PFTR extremes as Pe increases is shown by the $\mathscr{F}$ functions of Fig. 9.28. As with the tanks-in-series model, Eq. (9.110) can be used to evaluate Pe for the dispersion model from an experimentally determined σ^2.

Comparison of (9.104) and (9.110) provides a further quantitative coupling between the tanks-in-series model and the dispersion model. For axial Peclet numbers greater than 4, the exponent term in (9.110) is negligible, and Eqs. (9.104) and (9.110) give

$$\frac{1}{N} \text{(tanks in series)} = \frac{2}{\text{Pe}}\left(1 - \frac{1}{\text{Pe}}\right)$$

which shows the similar behavior arising from increased N (decreased size of each backmixed volume) and increased Pe, e.g., by decreasing the dispersion coefficient D_z responsible for backmixing behavior. For large Peclet number, $N = 2/\text{Pe}$.

When Monod kinetics with maintenance is used, the dilution rate D_{max} at washout for the dispersion model is given by

$$\frac{\mu_{max}}{D_{max}} = \frac{\frac{1}{4}\text{Pe}(1 + K_s s_0)}{1 - (k_e/\mu_{max})(1 + K_s/s_0)} \tag{9.112}$$

As the relative influence of dispersion becomes vanishingly small (Pe $\to$ 0), D_{max} decreases to zero. This agrees intuitively with the notion that an ideal PFTR with sterile feed will not support a biological population.

It is important to realize that the axial dispersion coefficient D_z is *not* usually equal to molecular diffusivity. It is a modeling parameter which, when chosen properly, allows the dispersion model to represent some of the effects of several physical phenomena. We shall mention three here: segregated laminar flow in tubes, turbulent flow in pipes, and flow in packed beds.

When fluid moves through a pipe of diameter d_t in laminar flow (Re = $d_t u/v \lesssim 2100$, where v is kinematic viscosity = density/viscosity), the axial velocity u varies with radial position r from the tube centerline according to

$$u = u_{max}\left[1 - \left(\frac{2r}{d_t}\right)^2\right] \tag{9.113}$$

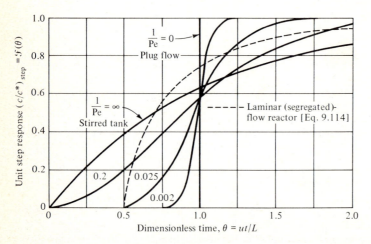

Figure 9.28 $\mathscr{F}$ functions for dispersion (solid curves) and a segregated laminar-flow reactor (dashed curve). Parameters on solid curves are $1/\text{Pe}$.

where u_{max} is twice the average axial velocity $\bar{u} = F/A$. Clearly the residence times of fluid elements located at different radial positions will vary under these circumstances. If we assume that molecular-diffusion effects are negligible (true if $\mathscr{D} \ll d_t{}^2/40L$), examination of the motion of a sheet of tracer injected into the tube inlet at $t = 0$ reveals that

$$\mathscr{E}(t) = \begin{cases} 0 & t \leq \dfrac{u}{2F} \\[2ex] 1 - 4\left(\dfrac{V}{Ft}\right)^2 & t > \dfrac{u}{2F} \end{cases} \tag{9.114}$$

Examining the corresponding $\mathscr{F}$ function in Fig. 9.28 reveals that this situation cannot be represented well by a dispersion model. However, if we include molecular diffusion with diffusivity $\mathscr{D}$ in both the axial and radial directions, a more difficult analysis gives an effective dispersion coefficient of the form

$$D_z = \mathscr{D} + \frac{\bar{u}^2 d_t{}^2}{192\mathscr{D}} \tag{9.115}$$

The dispersion model with D_z given by Eq. (9.115) provides an excellent approximation to the RTD for laminar flow with diffusion so long as the tube is long enough ($L \gg d_t{}^2/40\mathscr{D}$).

In turbulent flow in pipes (Re > 2100), motion of macroscopic eddys of fluid provide an important mechanism for mass, momentum, and energy transport. The effects of eddy transport closely resemble those of molecular-diffusion processes, but turbulent-diffusion fluxes are usually much greater in magnitude than their molecular counterparts. Consequently, the effective dispersion coefficient in this case depends mostly on the state of fluid flow. While the turbulent-flow Peclet number is typically of the order of 3, it falls with decreasing Reynolds number. This trend in turbulent flow as well as a variety of laminar-flow data are displayed in Fig. 9.29. Perhaps the most important biological reactor involving flow in empty tubes is the continuous liquid sterilizer, which is examined in detail in the following section.

In some immobilized-enzyme or mycelial-organism reactors, the tube is packed with particles. These particles are fixed in the bed while the fluid flows around the particles and through the tube. Because fluid is constrained to flow in the interstices between pellets, a fluid element passing axially through the bed undergoes something like a random walk in the radial dimension. The effect of this particle-interrupted sojourn on the RTD of the system can be described by the dispersion model. Theory suggests that Pe_p, defined by

$$Pe_p = \frac{\bar{u}d_p}{D_z} \tag{9.116}$$

is approximately 2 in this instance, where d_p is the pellet diameter. Experimental studies confirm this result under some circumstances, but dependence of Pe_p on Sc and Re_p is apparent in Fig. 9.30.

In closing this discussion of dispersion-producing processes, we should note that there are many additional possibilities, including pipe bends and gradations in depth contours in rivers and streams. Theory is of little help in identifying D_z for these complicated flow situations, and we must consequently rely on measured RTD data to determine an appropriate D_z value.

The tank-in-series and dispersion models are both one-parameter (N and D_z, respectively) nonideal mixing models. In different fashions each spans a continuum of mixing and segregation states ranging from ideal CSTR to ideal PFTR. Thus we are faced with a choice of which model to use. In terms of convenience of computation and analysis, the tanks-in-series representation is usually far superior. Also application of the dispersion model is prone to difficulty when backmixing (σ^2) is large. However, Fig. 9.24 shows that a very large number of tanks is necessary to represent situations near plug flow. Consequently, as a general rule of thumb, the dispersion model is usually preferable for small deviations from plug flow (say σ^2 of the order of 0.05 and smaller), while the tanks-in-series formulation is superior when there is substantial backmixing ($\sigma^2 \gtrsim 0.2$). The second case is typical of fermentors and biological waste-treatment basins. The first is encountered in continuous sterilization of liquids, part of our next topic.

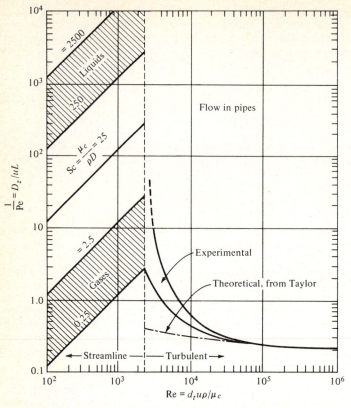

Figure 9.29 Correlations for the axial Peclet number (Pe) in terms of the Reynolds (Re) and Schmidt (Sc) numbers for fluid flow in pipes.

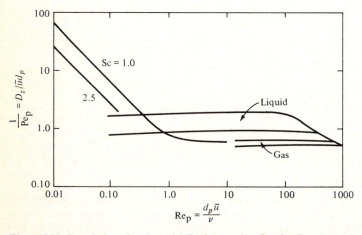

Figure 9.30 Correlations for the axial Peclet number Pe_p for flow in packed pipes. Notice that the dimensionless groups are based upon particle diameter d_p rather than the pipe diameter d_t.

9.4. STERILIZATION REACTORS

Liquids, usually aqueous, can be sterilized by several means, including radiation (ultraviolet, x-rays), sonication, filtration, heating, and chemical addition. Only the last two are widely used in large-scale processes. However, small amounts of liquids containing sensitive vitamins and other complex molecules are sometimes sterilized by passage through porous membranes. In this section we shall concentrate on design of heat-treatment processes.

Requirements for destruction of microbial life in fluids and solids vary widely depending upon the material and its intended use. In some instances, e.g., sauerkraut manufacture and biological waste-water treatment, microorganisms naturally present in the process fluid are responsible for desirable reactions. Inhibitors for the growth of unwanted organisms are rapidly evolved in alcohol, vinegar, and silage production, so that here too sterilization requirements are not extreme. Milk pasteurization involves killing most but not all actively growing microbes. More severe treatment of milk is not practiced because degradation of desired components results. Tradeoffs between destruction of useful compounds and death of unwanted organisms play a major role in choice and design of sterilization and pasteurization equipment.

Pure-culture fermentations, tissue culture, and some food products require more stringent measures. Essentially all contaminating microbial life must be excluded from the system, although the degree of "perfection" also varies somewhat. Economic considerations might indicate, for example, that a contamination probability [$1-P_0$ in Eq. (7.123)] of 10^{-2} is acceptable for a batch-fermentation process. In this case we would expect 1 batch out of every 100 to be lost due to contamination. We could accept this if the loss were comparable to the cost of additional sterilization capacity.

Much more severe requirements hold in the canning industry. A single surviving spore of *Clostridium botulinum* may cause lethal contamination, so that virtually complete elimination is required. Typically, design criterion in this situation specify that the spore survival probability $1-P_0$ be reduced to less than 10^{-12}. This example illustrates the importance of small deviations from essentially complete conversion of substrate (spores and vegetative cells) into products (inactive spores, dead cells) in sterilization reactors. Thus, careful sterilization-reactor design can clearly be critical. Continuous sterilization processes are examined after consideration in the next section of the batch case.

9.4.1. Batch Sterilization

Let us begin by considering a well-mixed closed volume containing a cell or spore suspension. The fluid is to be sterilized by heating, and then cooled to a suitable temperature for subsequent processing. The concentration of surviving organisms resulting from this process can readily be computed starting from Eqs. (7.117) and (7.119):

$$\frac{dn}{dt} = -k_{do}e^{-E_d/RT(t)}n \tag{9.117}$$

where we have explicitly included time variations in the fluid temperature. Separating the variables in Eq. (9.117) and integrating, we find

$$\ln \frac{n_0}{n_f} = \int_0^{t_f} e^{-E_d/RT(t)} \, dt \tag{9.118}$$

where the f subscript denotes final conditions.

Common batch-sterilization designs include one or more of the following heat sources: steam sparging (bubbling of live steam through medium), electrical heating, and heating or cooling with a two-fluid heat exchanger. Deindoerfer and Humphrey† have associated with each heating or cooling mode a particular time-temperature profile; these functions are shown in Table 9.3. The integral on the right-hand side of Eq. (9.118) can be evaluated by segmentation into three intervals of heating, holding, and cooling. A total of four integral forms arise: constant temperature and each of the three transient modes (hyperbolic, linear, and exponential).

Constant temperature:

$$\ln \frac{n_f}{n_0} = \int_0^{t_f} k_{d0} e^{-E_d/RT} \, dt = k_{d0} t_f e^{-E_d/RT} \tag{9.119}$$

Hyperbolic:

$$\ln \frac{n_f}{n_0} = \frac{k_{d0} a (E_d/RT_0)}{(a+b)^2} e^{-(E_d/RT_0)[b/(a+b)]}$$

$$\times \left[E_2 \left\{ \frac{E_d}{RT_0} \left[\frac{1+bt_f}{1+(a+b)t_f} - \frac{b}{a+b} \right] \right\} - E_2 \left(\frac{E_d}{RT_0} \frac{a}{a+b} \right) \right] \tag{9.120}$$

Linear increasing:

$$\ln \frac{n_f}{n_0} = \frac{k_{d0} E_d}{RT_0 a} \left(E_2 \frac{E_d/RT_0}{1+at_f} - E_2 \frac{E_d}{RT_0} \right) \tag{9.121}$$

Exponential:

$$\ln \frac{n_f}{n_0} = \frac{k_{d0}}{a} \left[E_1 \left(\frac{E_d/RT_H}{1+b} \right) - E_1 \left(\frac{E_d/RT_H}{1+be^{-at_f}} \right) \right]$$

$$- \frac{k_{d0}}{a} e^{-E_d/RT_H} \left[-E_1 \left(\frac{E_d}{RT_H} \frac{b}{1+b} \right) + E_1 \left(\frac{E_d}{RT_H} \frac{be^{-at_f}}{1+be^{-at_f}} \right) \right] \tag{9.122}$$

where E_n is the exponential integral

$$E_n(z) = \int_z^{\infty} \frac{e^{-w} \, dw}{w^n} \tag{9.123}$$

---

† F. H. Deindoerfer and A. E. Humphrey, Analytical Method for Calculating Heat Sterilization Times, *Appl. Microbiol.*, **7**: 256 (1959).

Table 9.3. Temperature-time profiles in batch sterilization†

Type of heat transfer	Temperature-time profile	Parameters
Steam sparging	$T = T_0\left(1.0 + \dfrac{at}{1 + bt}\right)$ hyperbolic	$a = \dfrac{hs}{MT_0 \rho C_p} \qquad b = \dfrac{s}{M}$
Electrical heating	$T = T_0(1.0 + at)$ linear	$a = \dfrac{q}{MT_0 \rho C_p}$
Steam (heat exchanger)	$T = T_H(1 + be^{-at})$ exponential	$a = \dfrac{UA}{Mc} \qquad b = \dfrac{T_0 - T_H}{T_H}$
Coolant (heat exchanger	$T = T_{c0}(1 + be^{-at})$ exponential	$a = \dfrac{wc'}{M \rho C_p}(1 - e^{-UA/wc'})$ $b = \dfrac{T_0 - T_{c0}}{T_{c0}}$

where h = enthalpy differences between steam at sparger temperature and raw
medium temperature
s = steam mass flow rate
M = initial medium mass
T_0 = initial medium temperature
q = rate of heat transfer, kcal per unit time
U = overall heat-transfer coefficient, kcal/m²·h·°C)
A = heat-transfer area, m²
T_H = temperature of heat source
w = coolant mass flow rate
c' = coolant specific heat
T_{c0} = coolant inlet temperature
ρ = medium density
C_p = medium heat capacity

† After F. H. Deindoerfer and A. E. Humphrey, *Appl. Microbiol.*, **7**: 256, 1959.

a tabulated function available in many handbooks and computer packages. The result for cooling is obtained by substituting T_{c0} for T_H and using the definitions of a and b appropriate for cooling (Table 9.3).

In each case, the final result yields $\ln(n_f/n_0)$, the logarithm of the ratio of final to initial concentrations. If, for example, electrical heating is followed by holding at an elevated temperature and subsequent liquid-coolant heat exchange, we can write the ratio of final to initial viable-cell concentrations in the form

$$\ln \frac{n_f}{n_0} = \ln \left[\frac{n_f}{n_0(\text{coolant})} \frac{n_f(\text{holding})}{n_0(\text{holding})} \frac{n_f(\text{electrical})}{n_0(\text{electrical})} \right] \qquad (9.124)$$

since $n_0(\text{coolant}) = n_f(\text{holding})$ and $n_0(\text{holding}) = n_f(\text{electrical})$. Rewriting Eq. (9.124) as

$$\ln \frac{n_f}{n_0} = \ln \frac{n_f(c)}{n_0(c)} + \ln \frac{n_f(h)}{n_0(h)} + \ln \frac{n_f(e)}{n_0(e)} \tag{9.125}$$

we see that the overall result $\ln(n_f/n_0)$ is obtained by adding the three appropriate individual solutions above, each evaluated for the particular time interval in that mode of operation.

Another situation which may be usefully examined analytically is thermal sterilization of solids or stagnant fluids. Such processes are important from several perspectives. One significant application is destruction of toxic organisms in sealed food containers. Also, it is necessary to minimize in the closed container all viable microorganisms which could decompose or otherwise spoil the product. Solid particles or microbial aggregates are often found suspended in liquids to be sterilized. We should recognize that these forms tend to protect organisms in their interior from thermal destruction and that more extensive heating is therefore required when such solids are present.

Analysis of both of these processes, at least for simple container and particulate geometries, can be reduced to a two-step recipe analogous to the procedure followed above. First, we solve a transient-heat-conduction problem to determine the temperature in the solid as function of position and time. Assuming constant thermal conductivity k, this problem has the general form

$$\rho C_p \frac{\partial T}{\partial t} = k \nabla^2 T \tag{9.126}$$

where ∇^2 is the Laplacian operator and ρC_p are as defined in Table 9.3. In addition to the partial differential equation (9.126), T is specified as a function of position within the solid at time zero, and the temperature at the external solid surface is known as a function of time for $t > 0$.

Considering a spherical solid, for example, the following equations must be solved to determine the interior temperatures for $t > 0$:

$$\frac{\partial T}{\partial t} = \frac{k}{\rho C_p} \frac{1}{r^2} \frac{\partial}{\partial r}\left(r^2 \frac{\partial T}{\partial r}\right) \tag{9.127}$$

$$T(r, 0) = f(r) \qquad 0 \le r \le R \tag{9.128}$$

$$T(R, t) = g(t) \qquad t > 0 \tag{9.129}$$

where r is distance from the sphere's center and f and g are prescribed functions. If the sphere is initially at a uniform temperature T_0, and if the surface temperature $T(t, R)$ is maintained at a constant value T_1 for $t > 0$, this problem can be solved by separation of variables to obtain

$$T(r, t) = T_1 + \frac{2R(T_1 - T_0)}{\pi r} \sum_{n=1}^{\infty} \frac{(-1)^n}{n} \sin \frac{n\pi r}{R} \exp\left(-\frac{k}{\rho C_p} \frac{n^2 \pi^2 t}{R^2}\right) \tag{9.130}$$

The temperature at the center of the sphere $(r = 0)$ can be deduced from Eq. (9.130) by taking the limit $r \to 0$. The result is

$$T(0, t) = T_1 + 2(T_1 - T_0) \sum_{n=1}^{\infty} (-1)^n \exp\left(-\frac{k}{\rho C_p} \frac{n^2 \pi^2 t}{R^2}\right) \qquad (9.131)$$

Figure 9.31 shows temperature distributions computed from these formulas for a variety of elapsed times. In the context of sterilization, it is critical to note that there is a time lag between the imposition of a high temperature at the surface and achievement of similar temperatures everywhere within the solid. Consequently, the kill of organisms at the sphere's center will generally be less complete than near the surface. This observation remains qualitatively correct for many other geometries and initial and boundary conditions. While we cannot delve further here into the details of transient heat conduction under various conditions, a wealth of additional information, theory, and analytical solutions is available in Carslaw and Jaeger [21].

Once the time-temperature-position relationship has been determined, the resulting destruction of microbes and spores can be calculated. Two different design approaches have been used in the food-processing industry. In the first, the organism concentration at the center of the solid is considered. Since the center heats most slowly, presumably if the center is adequately treated, the remainder of the solid is sufficiently sterilized. To compute the surviving organisms at the center

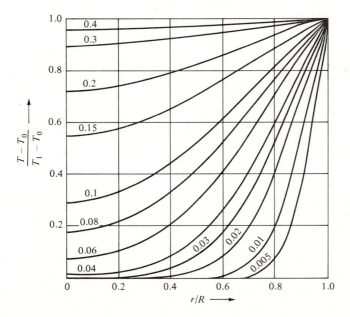

Figure 9.31 Temperature profiles within a sphere of radius R as a function of dimensionless time $kt/\rho C_p R^2$ (given as a parameter on the curves). Initially the sphere temperature is T_0 throughout, and the outer-surface temperature for $t > 0$ is T_1.

we need only insert the temperature-time function evaluated at that position into Eq. (9.118) and compute the integral. For this step, numerical or graphical means are often necessary since $T(t)$ is usually a rather complex expression, such as Eq. (9.131).

Another approach is to conduct the previous calculations many times in order to determine survivor concentrations at each point within the solid. Then by integrating these concentrations over the volume, the number of survivors or probability of survivors can be evaluated. Obviously, this calculation is somewhat tedious although it is straightforward in principle. Consequently, several shortcut design procedures based on this whole-container philosophy have been developed for use in the food industry. Since their explanation requires substantial additional vocabulary and definitions, we refer the interested reader to Charm [22] for a thorough discussion.

While batch sterilization enjoys the advantages of being a relatively simple process, it suffers from several drawbacks. One is the time required for heating and cooling. Related to this disadvantage is another: the extent of thermal damage to desirable components. Many vitamins are destroyed by heating, and proteins can be denatured at elevated temperatures. While the destruction of these components often follows the same kinetics as organism death [Eq. (9.117)], it is important to recognize that the activation energies for these undesirable side reactions are typically much smaller than for the sterilization "reaction." The values listed in Table 9.4, for example, are less than the 50 to 100 kcal/g mol magnitudes which usually characterize spore and cell destruction.

Since the desired reaction here has a higher activation energy than the side reaction, increasing the temperature has the beneficial effect of increasing the ratio of desired rate to undesired rate. This means that if unfavorable Browning reactions or damage to susceptible compounds are to be avoided, the sterilization process should operate at the highest feasible temperature and for the shortest time (HTST = high temperature, short time) which provides the necessary organism death. The slow heating and cooling portions of batch sterilization do not achieve these objectives. Continuous sterilization, considered next, is much better suited for achieving HTST conditions.

Table 9.4. Approximate activation energies for some undesirable side reactions resulting from heat treatment

Reaction	Activation energy E, kcal/g mol
Browning (or Maillard) reaction between proteins and carbohydrates	31.2
Destruction of vitamin B_1	21.0
Destruction of riboflavin (B_2)	23.6
Denaturation of peroxidase	23.6

9.4.2. Continuous Sterilization

Two basic types of continuous-sterilizer designs are shown schematically in Fig. 9.32. In Fig. 9.32a, direct heating is provided by steam injection, with the heated fluid then passing through a holding section before cooling by expansion. The system in Fig. 9.32b features indirect heating using a plate heat exchanger. A number of variations in each type are possible with most featuring high ratios of heat-exchange surface to process volume. Consequently continuous sterilizers provide relatively rapid heating and cooling so that HTST conditions can be realized.

Two different design approaches are available if we assume that the temperature in the continuous sterilizer is approximately uniform. This may be valid at least in the holding sections. In the first method we apply the dispersion model described earlier. When c in Eqs. (9.106) to (9.108) has been identified with n, we take $r_{f_n} = -k_d n$; then the organism concentration throughout from the tubular reactor with dispersion can be determined analytically. The resulting value of n at the sterilizer effluent is given by

$$\frac{n(L)}{n_0} = \frac{4y \exp \text{Pe}/2}{(1 + y)^2 \exp [(\text{Pe})(y)/2] - (1 - y)^2 \exp [-(\text{Pe})(y)/2]} \qquad (9.132)$$

with

$$y = \left(1 + \frac{4 \text{ Da}}{\text{Pe}}\right)^{1/2} \qquad (9.133)$$

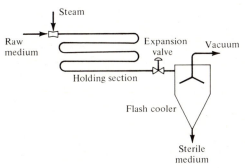

(a) Continuous injection type

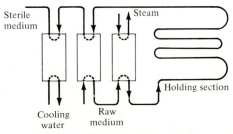

(b) Continuous plate exchanger type

Figure 9.32 Two different continuous sterilizer designs: (a) direct steam injection (b) plate heat exchanger. (*Reprinted from S. Aiba, A. E. Humphrey and N. F. Millis, "Biochemical Engineering," 2d ed., p. 257, University of Tokyo Press, Tokyo, 1973.*)

where here the Damköhler number Da is defined by

$$\text{Da} = \frac{k_d L}{u} \tag{9.134}$$

For small deviations from plug flow (Pe^{-1} small), Eq. (9.132) reduces to the simpler form

$$\frac{n(L)}{n_0} = \exp\left(-\text{Da} + \frac{\text{Da}^2}{\text{Pe}}\right) \tag{9.135}$$

These solutions are conveniently displayed as a plot of remaining viable fraction $n(L)/n_0$ vs. the dimensionless group Da for various values of the Peclet number (Fig. 9.33). From this plot, we can see that as $\text{Pe} \to \infty$ so that ideal plug flow is approximated, the desired degree of medium sterility can be achieved with the shortest possible sterilizer. Consequently, the flow system should be designed to keep dispersion at a minimum.

If flow through the isothermal continuous sterilizer cannot be well represented with the dispersion model, we can make use of general RTD theory. It is difficult to think of a better example of a completely segregated reactor than a suspension of cells or spores subjected to heat treatment. Consequently, provided back diffusion of organisms into the sterilizer feed line is negligible, we can compute the effluent surviving-organism concentration n using Eq. (9.63). Rewriting this equation in terms of organism concentration, we have

$$n = \int_0^\infty n_b(t)\mathscr{E}(t)\,dt \tag{9.136}$$

where $\mathscr{E}(t)$ is the RTD of the continuous sterilizer. Recalling that $n_b(t)$ is the organism concentration at time t in a batch sterilizer with $n_b(0) = n_0$, we can use Eq. (7.118) in Eq. (9.136) to obtain

$$\frac{n}{n_0} = \int_0^\infty \mathscr{E}(t)e^{-k_d t}\,dt \tag{9.137}$$

In some instances evaluation of the right-hand side of Eq. (9.137) is facilitated by noting that it is formally identical to the Laplace transform of $\mathscr{E}$ with the usual Laplace transform parameter s replaced by k_d.

Before leaving the topic of continuous sterilization, we should mention its several advantages in addition to the HTST feature already discussed. First, continuous processing typically requires less labor than batch operations. Also, continuous treatment units provide a more uniform, reproducible effluent than batch sterilization. This is extremely important in the food industry, where small changes in treatment can change the taste of the product.

When practiced in the fermentation industry, batch sterilization is usually done in the fermentor itself (*in situ*). Heating and cooling rates here depend upon the surface-to-volume ratio of the fermentor, and this ratio in turn usually changes during scale-up. The sensitivity of sterilization effects on organisms and medium components to equipment scale can be reduced by using continuous sterilization.

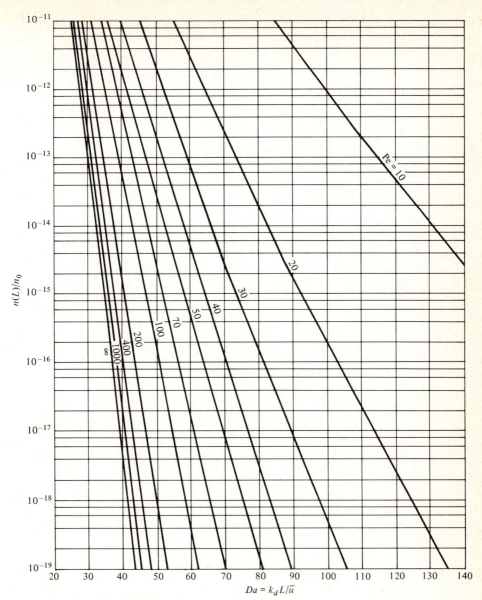

Figure 9.33 Effect of axial dispersion on organism destruction in a continuous sterilizer. (*Reprinted from S. Aiba, A. E. Humphrey and N. F. Millis, "Biochemical Engineering," 2d ed., p. 263, University of Tokyo Press, Tokyo, 1973.*)

If continuous sterilization is used, it is not necessary to design the fermentor also to fulfill the requirements of a good batch sterilizer.

Of course there are some drawbacks to continuous sterilization: direct steam heating can add excess water to the medium, and heat exchangers used for indirect heating or cooling can be fouled by suspended solids. Also, continuous sterilization tends to cause foaming of fermentation media. Additional details on the operation tradeoffs of sterilizers are available in Ref. [21] of Chap. 7. Our attention will now turn to a general comparison of the relative efficiencies of batch and continuous reactors as well as the relationship between kinetic information taken from different reactors.

9.5. RELATIONSHIPS BETWEEN BATCH AND CONTINUOUS BIOLOGICAL REACTORS

It is important to know how various continuous processes compare with batch cultivation with regard to biomass production, substrate utilization, and product yield. Considering biomass production first, we must not overlook the unproductive portions of the batch cycle. As Fig. 9.34 shows schematically, cell growth occurs only during a fraction of the total cycle time t_b. Following the growth phase, cells must be harvested (time t_0), the vessel must be prepared for the next run and inoculated (time t_1), and the lag phase must be passed (time t_2) before the next growth phase is reached. Thus, if t_g is the time of cell growth,

$$t_b = t_g + t_0 + t_1 + t_2 \tag{9.138}$$

Let t_d be the total downtime associated with batch turnaround

$$t_d = t_0 + t_1 + t_2 \tag{9.139}$$

and assume exponential growth during the entire time t_g, so that

$$t_g = \frac{1}{\mu} \ln \frac{x_f}{x_0} \tag{9.140}$$

where x_f is final cell concentration and x_0 is cell concentration in inoculum; then Eq. (9.138) becomes

$$t_b = \frac{1}{\mu} \ln \frac{x_f}{x_0} + t_d \tag{9.141}$$

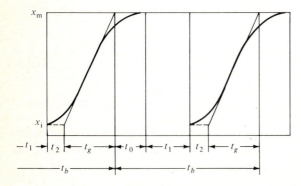

Figure 9.34 Batch process cycle. (*Reprinted from S. Aiba, A. E. Humphrey and N. F. Millis, "Biochemical Engineering," 2d ed., p. 149, University of Tokyo Press, Tokyo, 1973.*)

The holding time t_c required to achieve the same amount of cell growth in a CSTR continuous culture can be determined from Eq. (9.3) to be

$$t_c = \frac{V}{F} = \frac{1}{D} = \frac{x_f - x_0}{\mu x_f} \tag{9.142}$$

so that the ratio of batch to continuous processing times for the same cell concentration x_f is

$$\frac{t_b}{t_c} = \frac{\ln (x_f/x_0) + \mu t_d}{1 - x_0/x_f} \tag{9.143}$$

This quantity is always larger than unity, even if t_d is zero. Consequently, the continuous process always provides a greater yield of cells per unit volume of cultivator vessel than a batch process does.

Such a conclusion is a consequence of the fact that the rate of increase of biomass augments as biomass increases. To put matters in perspective and to establish a foundation for future discussion, it is appropriate at this point to review several basic principles of reaction engineering. Let us consider a reaction scheme where the rate of formation of a single component C, can be expressed as a function of concentration c only:

$$\text{Rate of formation of C} = r_{fc} = f(c) \tag{9.144}$$

For *ordinary kinetics*, which is encountered for most nonbiological reaction systems, $f(c)$ is a decreasing function of c: the more C present, the slower its rate of formation. In an *autocatalytic reaction* such as cell growth, however, $f(c)$ increases with c. Design considerations for these two cases are quite different, as we shall see next.

The mass balances for component C in batch (or PFTR) and CSTR reactors are

Batch (or PFTR): $$\frac{dc}{dt} = f(c) \tag{9.145}$$

Steady-state CSTR: $$F(c_0 - c) + Vf(c) = 0 \tag{9.146}$$

(The formal identity between batch and plug-flow tubular reactors, where $t = z/u$, should be kept in mind throughout the following discussion). From Eqs. (9.145) and (9.146) we can compute the respective holding times t_b and t_c required to convert material at concentration c_0 into concentration c_f:

Batch (or PFTR): $$t_b = \int_{c_0}^{c_f} \frac{dc}{f(c)} \tag{9.147}$$

Steady-state CSTR: $$t_c = \frac{V}{F} = \frac{c_f - c_0}{f(c_f)} \tag{9.148}$$

(we shall neglect the batch turnaround time in this comparison). The batch (PFTR) holding time t_b can best be compared with the CSTR holding time t_c by a geometrical interpretation of t_b and t_c in a plot of $1/f(c)$ vs. c. As Fig. 9.35 depicts

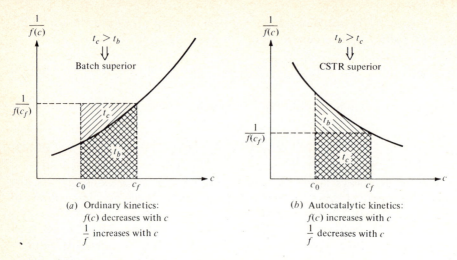

(a) Ordinary kinetics:
$f(c)$ decreases with c
$\frac{1}{f}$ increases with c

(b) Autocatalytic kinetics:
$f(c)$ increases with c
$\frac{1}{f}$ decreases with c

Figure 9.35 Graphical determination for (a) ordinary kinetics and (b) autocatalytic kinetics of required mean residence time for CSTR (////) and PFTR or batch reactors (\\\\).

schematically, t_b is the area under the curve from c_0 to c_f while t_c is the area of the rectangle with sides of length $c_f - c_0$ and $1/f(c_f)$. These constructions are illustrated in the figure for both "ordinary" and autocatalytic cases. Clearly the CSTR reactor is inferior for ordinary kinetics but requires less holding time than the batch system for an autocatalytic reaction. This should reconcile the conclusion in Eq. (9.143) with any contrary experience or intuition which the reader based on "ordinary" reactions.

Many microbial processes are in fact combinations of autocatalytic and ordinary kinetics, as the batch data in Figs. 7.22 to 7.24 indicate. While autocatalytic behavior prevails during the exponential growth phase and its onset, ordinary kinetics is observed as the growth rate decelerates and the stationary and death phases ensue. This suggests that optimal design of some fermentations would involve continuous cultivation in a CSTR followed by a PFTR. As Fig. 9.36 illustrates, a CSTR-PFTR sequence with holding times as indicated will give the most conversion in a given volume (or, conversely, a prescribed conversion in the minimum volume).

These considerations are especially important when substrate utilization or product yield are the dominant design objectives. Returning to Figs. 7.23 and 7.24, we see that, especially in type II and III fermentations, large rates of product synthesis and substrate uptake are exhibited after the growth rate has declined. Thus, while a CSTR might be best from the viewpoint of biomass production, something like the combination CSTR–tubular fermentor will maximize product yield. As we discussed earlier in Sec. 9.3.3, a plug-flow reactor can be approximated by a cascade of well-stirred vessels. Such a cascade is far more practical for most fermentation processes because aeration and pumping of the culture are much easier in a staged operation. Thus, the design method given in Eqs. (9.93) to

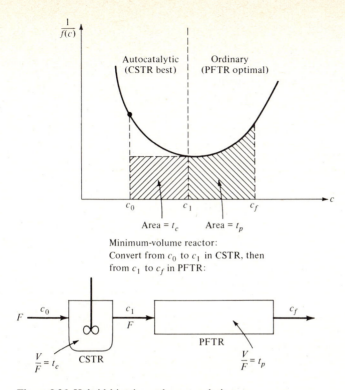

Figure 9.36 Hybrid kinetics and reactor design.

(9.95) can be used to develop continuous processes suitable for a wide variety of process kinetics and design objectives.

In the preceding discussion of the potential advantages of staged continuous fermentation, we referred to Gaden's classification scheme and experimental data for *batch* fermentations. Thus, the implicit assumption was made that there is a connection between batch kinetics and the kinetics of continuous reaction processes. So far we have invoked only a qualitative relationship. Let us see whether a quantitative connection exists, so that data for a batch fermentation can be employed to design a continuous one.

If we assume that the fermentation can be described by only one variable, e.g., cell concentration, we can use Eqs. (9.145) and (9.146) as representative general forms for the batch and CSTR mass balances, respectively. From Eq. (9.145)

Batch: $$\frac{dc}{dt} = f(c)$$

the value of f at a particular concentration c is given by the slope of the batch concentration-time curve at that concentration (see Fig. 9.37a). Consequently from batch data, a graph of $f(c)$ vs. c can be prepared. Rearranging the CSTR mass balance (9.45) into the form

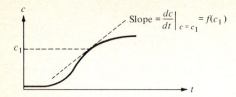

(a) Determination of $f(c)$ from batch data

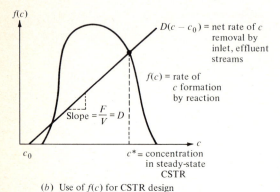

(b) Use of $f(c)$ for CSTR design

Figure 9.37 Use of batch data for (a) determination of $f(c)$ and (b) use of $f(c)$ for determination of CSTR design.

CSTR:
$$f(c^*) = D(c^* - c_0) \qquad (9.149)$$

we see that the steady-state concentration in a CSTR (which has been denoted c^* above) can be determined by drawing a straight line (representing the right-hand side of Eq. (9.149)) with slope D through the c axis at $c = c_0$. As Eq. (9.149) shows, the intercept of this straight line with the $f(c)$ plot gives a CSTR effluent concentration c^* (see Fig. 9.37b). Conversely, this procedure can be used to determine the dilution rate required to achieve a given concentration in the CSTR. Extension of the method to a cascade of two CSTRs is illustrated in Fig. 9.38. While the net-removal-rate line in Fig. 9.36b intersects the $f(c)$ curve three times, the intermediate solution can be rejected because it is unstable. The lowest intersection corresponds to washout, which can be avoided by proper startup. Consequently, the solution of practical interest is the one indicated in the figure.

This design procedure has been tested in an experimental study of the lactic acid fermentation by *Lactobacillis delbrueckii*. Table 9.5 offers a comparison between some of the observed values of bacterial density and dilution rates and the corresponding results from the graphical analysis for a variety of experimental conditions. Apparently the graphical design technique for continuous biological reactors can in some cases be quite accurate, suggesting the utility of the assumption we stated in the previous paragraph.

However, in general we should regard the results of the above technique as preliminary estimates, for such *application of simple batch data for continuous fermentation design is in principle an invalid practice*. We can find objections to this practice on several grounds. First, in a steady-state continuous process with

Schematic of the system:

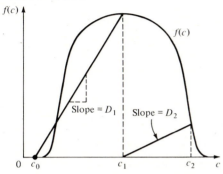

Steady-state mass balances:

$$D_1(c_1 - c_0) = f(c_1)$$
$$D_2(c_2 - c_1) = f(c_2)$$

Graphical solution:

Figure 9.38 Graphical design of a CSTR cascade.

$x_0 = 0$, the properties of the culture—x and s—are not changing with time. Presumably the cell population can better adjust to its environment in this situation. On the other hand, in a batch process both nutrient and population change with time. In this transient case, the cell population is faced with a constantly changing environment and may never be completely adjusted. There is consequently a possibility that batch and continuous cultures may reflect completely different physiological states. (When $x_0 \neq 0$, what transients are continuously experienced by entering cells even though $x = $ constant?)

Table 9.5. Comparison of measured CSTR bacterial concentration with CSTR predictions based on batch kinetic data†

The various bacterial concentrations, given in terms of UOD/ml, correspond to different operating conditions

Experimental measurement	Graphical analysis based on batch data
5.37	5.35
2.4	2.3
8.0	7.15

† R. Luedeking and E. L. Piret, Transient and Steady States in Continuous Fermentation: Theory and Experiment, *J. Biochem. Microbiol. Technol. Eng.*, **1**: 431, 1959.

When developing the procedure for relating batch data to CSTR design above, we emphasized the starting assumption that the process was described by a single variable. Such a description could possibly apply only in the case of balanced growth. In general, cell growth is a process involving a very large number of interactions and independent reactions. Consequently, many variables should be used to model growth, and, in this situation, the connections between batch and continuous operation presumed in the above graphical design procedure easily collapse. As direct evidence of the general necessity for several variables in growth modeling, we need only refer to Figs. 7.3 and 7.22 to 7.24 to see that the rate of formation of cells in batch culture often takes on multiple values for a given cell concentration; e.g., in Fig. 7.3 a horizontal line for constant cell concentration can intersect the growth curve twice. Clearly the volumetric rates shown in Figs. 7.22 to 7.24 are not always in constant proportions, indicating that a model based on a one-reaction representation is likely to fail.

In conclusion, we can say that batch data are useful in a qualitative sense for choosing a processing configuration for a continuous reactor. We can also make preliminary quantitative design estimates for continuous processes based on batch-reaction information. These estimates, however, should be based, where possible, on structured kinetic models and should be checked with laboratory or pilot-scale continuous experiments before completing any large-scale design.

The previous discussion should make us keenly aware that biological-reactor design is a broad and complex subject. In an attempt to bring together many of the elements of this chapter within the context of a single biological process, we shall consider next a rather lengthy example. Also, other illustrations of applications of the principles we have considered in this chapter will be found in the remainder of the book. Still, reactor design is too broad a topic to cover completely in a single chapter; additional details are available in the references.

Example 9.2. Reactor modeling and optimization for production of α-galactosidase by a _Monascus_ sp. mold The enzyme α-galactosidase may be useful in the beet-sugar industry because it can decompose raffinose, an inhibitor of sucrose crystallization. In a fascinating series of papers, Imanaka, Kaieda, Sato, and Taguchi† have investigated production of this intracellular enzyme by a mold they isolated from soil. Their original work deserves serious study: here we summarize some of the major results of their investigations.

As a first step in developing highly productive continuous processes for enzyme synthesis, batch and continuous cultures of the mold were cultivated under a variety of conditions. We list next the major findings from the batch experiments:

1. Among 20 different carbon sources including glucose, fructose, mannitol, and starch, only 4 sugars were effective in inducing high α-galactosidase activity. The strong inducers are galactose, melibiose, raffinose, and stachyose.

† The material in this example is drawn chiefly from T. Imanaka, T. Kaieda, K. Sato, and H. Taguchi, α-Galactosidase Production in Batch and Continuous Culture and a Kinetic Model for Enzyme Production, _J. Ferment. Technol._ (_Japan_), **50**: 633, 1972, and T. Imanaka, T. Kaieda, and H. Taguchi, Optimization of α-Galactosidase Production by Mold, II, III, _J. Ferment. Technol._ (_Japan_), **51**: 423, 431, 1973.

2. Ammonium nitrate gave more enzyme production than the alternative nitrogen sources urea, KNO_3, $(NH_4)_2SO_4$, and peptone. The optimal NH_4NO_3 concentration in the medium is between 0.3 and 0.5 percent bv weight.
3. When grown in a galactose medium, the cell mass is directly proportional to the α-galactosidase activity. When a mixture of glucose and galactose was used as the carbon source, diauxic growth was observed (see Fig. 9E2.1); almost no galactose is consumed until the glucose is nearly exhausted.
4. Figure 9E2.1 also shows that α-galactosidase production does not start until the glucose is almost gone. Separate experiments revealed that glucose concentrations greater than 0.05 percent by weight repress synthesis of the enzyme.

Two different series of steady-state continuous-culture experiments were conducted in a single CSTR. In the first series, the dilution rate was initially at a very low level, and it was increased slowly in a stepwise fashion (shifted up) with observations of steady-state behavior at each D along the way. The data so observed are plotted in Fig. 9E2.2. Especially interesting is the discontinuous jump evident at $D = 0.142$. Below this dilution rate, galactose is being consumed and α-galactosidase is synthesized. When D is increased above 0.142, however, both these activities stop and glucose alone is utilized as the mold's carbon source. Evidently, this jump is a manifestation of the glucose effect, already seen in batch culture of this organism. When cultivated under relatively large specific growth rates (large D's), the mold preferentially feeds on glucose.

Figure 9E2.3 illustrates the results of a similar series of experiments, except that here the CSTR was started up at a high dilution rate. Then, in a sequence of shift-down steps, the dilution rate was gradually decreased. While less sharp than the previous case, another discontinuity occurs, this time around $D = 0.008$. Below that critical dilution rate, enzyme is produced and galactose is assimilated, while no α-galactosidase activity is evident for $D > 0.008$. This is in marked contrast to the shift-up experimental results, where enzyme production was apparent up to $D = 0.142$.

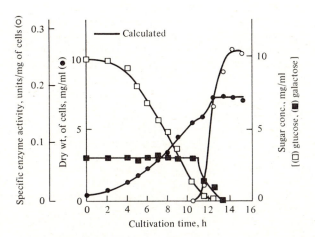

Figure 9E2.1 Results of batch cultivation of a *Monascus* sp. mold in a mixture of glucose and galactose (initial medium composition: glucose 1% (by weight), galactose 0.3%, NH_4NO_3 0.5%, KH_2PO_4 0.5%, $MgSO_4 \cdot 7H_2O$ 0.1%, yeast extract 0.01%). The inoculum was grown in a glucose medium. The initial conditions used in the calculations were $x = 5 \times 10^{-4}$ g/ml, $s_1 = 1 \times 10^{-2}$ g/ml, $s_2 = 3 \times 10^{-3}$ gm/ml, $s_{21} = 0$ μg/mg cell, $r = 0.910$ μg/mg cell, $m = 0$ μg/mg cell, $(rs_{21}) = 0$ μg/mg cell, $e = 0$ units/mg cell. [*Reprinted from T. Imanaka et al., Unsteady-state Analysis of a Kinetic Model for Cell Growth and α-Galactosidase Production in Mold, J. Ferment. Tech. (Japan),* **51**: 423, 1973.]

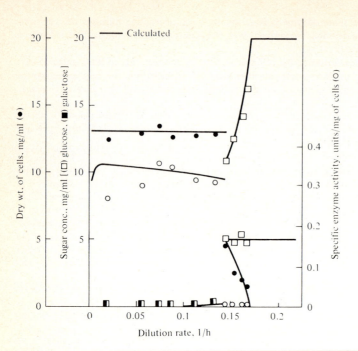

Figure 9E2.2 Steady-state cell and substrate concentrations and specific enzyme activity observed during gradual shift up of dilution rate for continuous culture (30°C). Initially the medium contains 2% glucose and 0.5% galactose. [*Reprinted from T. Imanaka et al., Optimization of α-Galactosidase Production by Mold, J. Ferment. Tech. (Japan),* **50**: 633, 1972.]

Replotting some of the data from the previous two figures in Fig. 9E2.4 clearly shows that this system exhibits multiple, stable steady states between $D = 0.008$ and $D = 0.142$. For dilution rates between these limits, whether or not α-galactosidase is produced depends upon how the reactor is started up. Shifting down into this range results in no enzyme synthesis, while shifting up will provide α-galactosidase production.

Based upon these and other experiments, a mathematical model for substrate utilization, cell growth, and product synthesis was developed. In most respects, the individual model equations in Table 9E2.1 are familiar from our earlier studies: the specific growth rate μ_2 based on galactose is of Monod form, while the specific growth rate of glucose μ_1 includes competitive inhibition by galactose. All the constants in these growth-rate functions were evaluated for two different media from continuous-culture experiments. Parameters labeled G in Table 9E2.2 correspond to a glucose medium (20 g glucose, 5 g NH_4NO_3, 5 g KH_2PO_4, 1 g $MgSO_4 \cdot 7H_2O$, 0.1 g yeast extract in 1000 ml tap water at pH 4.5) while the p subscripts refer to a galactose medium advantageous for enzyme production (5 g galactose, 5 g NH_4NO_3, 5 g KH_2PO_4, 1 g $MgSO_4 \cdot 7H_2O$ in 1000 ml tap water, pH 4.5).

The model for enzyme production is based upon the operon theory of induction, studied in Chap. 6. The specific rate of α-galactosidase synthesis is proportional to the intracellular concentration of mRNA which codes for that enzyme. This mRNA is assumed to decompose by a first-order reaction and is produced provided the intracellular concentration of repressor R is smaller than a threshold value r_c. Below this threshold value, lower r values cause increased specific rates of mRNA synthesis. The repressor is formed at constant specific rate k_2 and decomposes with first-order specific rate $k_4 r$. Repressor concentration is also reduced by complexing with the inducer, intracellular galactose.

The rate of galactose transport into the cell is given by the term in the intracellular galactose mass balance with coefficient U. To take the glucose effect into account, this transport term is set equal to

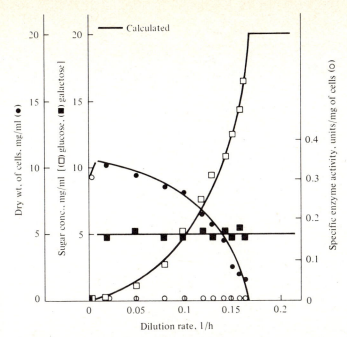

Figure 9E2.3 Steady-state cell and substrate concentrations and specific enzyme activity observed during gradual shift down of dilution rate for continuous culture (30°C). Initially the medium contains 2% glucose and 0.5% galactose. [*Reprinted from T. Imanaka et al., Optimization of α-Galactosidase Production by Mold, J. Ferment. Tech. (Japan),* **50:** 633, 1972.]

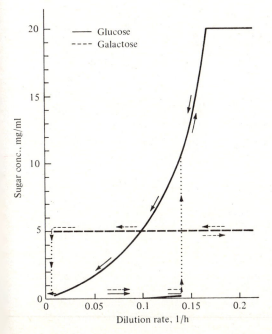

Figure 9E2.4 This replot of some of the curves from Figs. 9E2.2 and 9E2.3 displays steady-state multiplicity and hysteresis in continuous culture. Two stable steady states occur for dilution rates between 0.008 and 0.142. Which is obtained depends on reactor startup. [*Reprinted from T. Imanaka et al., Optimization of α-Galactosidase Production by Mold, J. Ferment. Tech. (Japan),* **50:** 633, 1972.]

Table 9E2.1††‡

Substrate utilization

$$\frac{ds_j}{dt} = -\frac{1}{Y_j}\mu_j x \qquad \text{for } j = 1(\text{glucose}), \ 2(\text{galactose}) \tag{9E2.1}$$

Biomass growth

$$\frac{dx}{dt} = (\mu_1 + \mu_2)x \tag{9E2.2}$$

where

$$\mu_1 = \frac{\mu_{\text{max},\,1}\,s_1}{s_1 + K_1(1 + s_2/K_i)} \tag{9E2.3}$$

and

$$\mu_2 = \frac{\mu_{\text{max},\,2}\,s_2}{s_2 + K_2} \tag{9E2.4}$$

Operon model for enzyme production

Intracellular galactose:

$$\frac{d}{dt}(s_{2I} \cdot x) = Ux\left(\frac{G_2 s_2}{K_{m2} + s_2} - s_{2I}\right) - k_1 s_{2I} \cdot x \qquad \text{for } s_1 < s_{1c} \tag{9E2.5a}$$

$$= -k_1 s_{2I} \cdot x \qquad \text{for } s_1 \geq s_{1c} \tag{9E2.5b}$$

Repressor:

$$\frac{d}{dt}(r \cdot x) = k_2 s - k_3 r \cdot x - k_4 r \cdot s_{2I} \cdot x + k_5(rs_{2I})x \tag{9E2.6}$$

Galactose-repressor complex:

$$\frac{d}{dt}[(rs_{2I})x] = k_4 r \cdot s_{2I} \cdot x - k_5 (rs_{2I})x \tag{9E2.7}$$

mRNA for galactosidase:

$$\frac{d}{dt}(m \cdot x) = \begin{cases} k_6(r_c - r)x - k_7 m \cdot x & \text{for } r_c > r \tag{9E2.8a} \\ -k_7 m \cdot x & \text{for } r \geq r_c \tag{9E2.8b} \end{cases}$$

Enzyme:

$$\frac{d}{dt}(e \cdot x) = k_8 m \cdot x \tag{9E2.9}$$

---

† T. Imanaka, T. Kaieda, K. Sato, and H. Taguchi, *J. Ferment. Technol.* (*Japan*), **50**: 633, 1972.
‡ Concentration variables are $x = $ biomass, $s_1 = $ glucose, $s_2 = $ extracellular galactose, $s_{2I} = $ intracellular galactose, $r = $ intracellular repressor, $(rs_{2I}) = $ intracellular inducer-repressor complex, $m = $ intracellular mRNA, and $e = $ intracellular α-galactosidase. The remaining symbols are kinetic, yield, and transport parameters.

zero whenever the glucose concentration s_1 exceeds a critical value s_{1c}, which is taken to be 2.25×10^{-4} g/ml.

Little information is available for direct evaluation of the rate constants in the operon model. Values for k_3 and k_7 were assigned based on the assumption that the repressor and mRNA half-lives are 40 and 5 min, respectively. The other parameter values listed in Table 9E2.2 were estimated by trial and error to achieve a reasonable fit to the experimental data.

This model is certainly attractive because it includes substantial structure which is heavily based on established biological principles. On the other hand, we could object to the large number of

Table 9E2.2. Parameter values for the mathematical model of α-galactosidase production†

Entries in the right column were evaluated experimentally; the remaining parameters were adjusted to fit the batch and continuous-culture results; G = glucose medium, 30°C, p = galactose medium, 35°C.

$k_1 = 40$ h^{-1}	$\mu_{max,\,1_G} = 0.215$ h^{-1}
$k_2 = 1$ mg/(mg cells $\cdot$ h)	$\mu_{max,\,2_G} = 0.208$ h^{-1}
$k_3 = 1$ h^{-1}	$K_{1_G} = 1.54 \times 10^{-4}$ g/ml
$k_4 = 0.1$ mg cells/(mg $\cdot$ h)	$K_{2_G} = 2.58 \times 10^{-4}$ g/ml
$k_5 = 1 \times 10^{-4}$ h^{-1}	$\mu_{max,\,1_p} = 0.190$ h^{-1}
$k_6 = 1$ h^{-1}	$\mu_{max,\,2_p} = 0.162$ h^{-1}
$k_7 = 8$ h^{-1}	$K_{1_p} = 1.45 \times 10^{-4}$ g/ml
$k_{8_G} = 3.2787$ units/(mg mRNA $\cdot$ h)	$K_{2_p} = 3.07 \times 10^{-4}$ g/ml
$k_{8_p} = 5.0442$ units/(mg mRNA $\cdot$ h)	$K_i = 1.39 \times 10^{-4}$ g/ml
$U = 100$ h^{-1}	$Y_{1_G} = 0.530$
$G_2 = 1$ mg/mg cells	$Y_{2_G} = 0.516$
$K_{m2} = 1 \times 10^{-8}$ mg/mg cells	$Y_{1_p} = 0.377$
$s_{1c} = 2.25 \times 10^{-4}$ g/ml	$Y_{2_p} = 0.361$
$r_c = 0.934$ mg/mg cells	

† T. Imanaka, T. Kaieda, K. Sato, and H. Taguchi, *J. Ferment. Technol.*, **50**: 558, 1972.

adjustable parameters it contains. Several tests can be applied to investigate the suitability of this model. One is based on the following question: Can other models based on different assumptions but containing a similar number of adjustable constants fit the data equally well? If so, we cannot place much confidence in this particular form. Imanaka et al. conducted several such tests, including cases in which (1) intracellular galactose concentration is proportional to galactose concentration in the medium, or (2) repressor formation is proportional to intracellular glucose, or (3) rate of mRNA formation is inversely proportional to concentration of repressor. Any of these modifications in the model resulted in serious discrepancies with the experimental observations.

The other tests of the model which Imanaka et al. considered involved its ability to fit data collected under a wide variety of operating conditions. Indeed, this is the raison d'être for a complex, structured kinetic model, for if the model is sufficiently complete, it can be used to determine optimal operating conditions for the process. Here too the model performed extremely well. The solid curves in Fig. 9E2.1 were computed using this model with initial conditions as indicated in the figure, and the fit is very good.

The model equations for an unsteady-state CSTR can be obtained simply by adding terms of the form D (inlet concentration − concentration in the reactor) to each of the batch equations in Table 9E2.1. Steady-state CSTR mass balances are then obtained by setting all time derivative terms equal to zero. With these equations, the solid lines in Figs. 9E2.2 and 9E2.3 were calculated. The hysteresis and jump phenomena observed experimentally are clearly well represented by the model. As a final test before turning to reactor optimization, the model was used to compute the transient behavior of the CSTR following shift-up. The results predicted by the model as well as experimental data are displayed in Fig. 9E2.5: here the coincidence between measured and calculated responses, including the overshoot of glucose and undershoot of cell concentrations, is quite dramatic.

With the model of Table 9E2.1 thus well established, it was used to compute the enzyme productivity of a variety of continuous-reactor configurations (Table 9E2.3). For each design, the volumes of the various reactors were adjusted to maximize enzyme production. The overall dilution rate $\bar{D}$ indicated in the table is defined as the total medium flow rate into the system divided by the total volume of all reactors in the process.

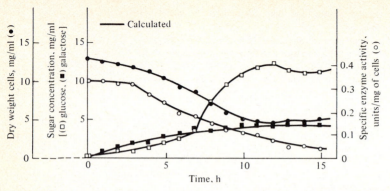

Figure 9E2.5 Transient behavior of CSTR continuous culture following a change in dilution rate from $D = 0.140$ h^{-1} to $D = 0.142$ h^{-1}. The medium contained two carbon sources: glucose (2%) and galactose (0.5%). Initial conditions used in the calculations were $x = 1.30 \times 10^{-2}$ g/ml, $s_1 = 2.23 \times 10^{-4}$ g/ml, $s_2 = 5.01 \times 10^{-5}$ g/ml, $s_{2I} = 2.5$ μg/mg cell, $r = 0.718$ μg/mg cell, $(rs_{2I}) = 1.28$ μg/mg cell, $m = 1.04 \times 10^{-2}$ μg/mg cell, $e = 0.297$ units/mg cell. [*Reprinted from T. Imanaka et al., Unsteady-state Analysis of a Kinetic Model for Cell Growth and α-Galactosidase Production by Mold, J. Ferment. Tech. (Japan),* **51**: 423, 1973.]

We shall examine a few details of systems 3 and 4. In these, as in systems 5 and 6, the basic idea is to grow a large cell concentration in the first part of the system using relatively cheap glucose only. Then a galactose medium is added at a later stage to induce enzyme production. Because enzyme induction takes some time, it is necessary to take into account adaptation of the intracellular reaction systems in the later stage. For this purpose, the induction stage is treated as a completely segregated system, and its effluent enzyme activity is computed using Eq. (9.63), which can be rewritten

$$e_e = \int_0^\infty e_b(t)\mathscr{E}(t)\, dt \qquad (9E1.1)$$

where $e_b(t)$ is the enzyme concentration at time t computed from the batch-reactor model. In the batch calculations, the initial conditions used are the concentrations in the feed to the induction stage.

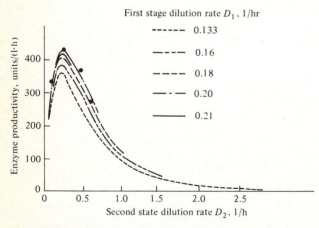

Figure 9E2.6 Calculated and experimental enzyme productivity in a two-CSTR continuous culture (System 3 from Table 9E2.3). [*Reprinted from T. Imanaka et al., Optimization of α-Galactosidase Production in Multi-stage Continuous Culture of Mold, J. Ferment, Tech. (Japan),* **51**: 431, 1973.]

Table 9E2.3. Optimized operating conditions for six different continuous fermentor configurations†‡

System	Type of fermentation process	Optimum operating conditions, D values in h^{-1}	Specific enzyme activity, units/mg of cells	Enzyme productivity, units/(l · h)
1	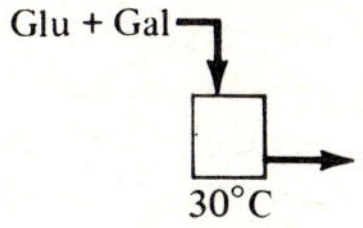 Glu + Gal → [30°C] →	$D = 0.142$ Shift-up system	0.325	554
2	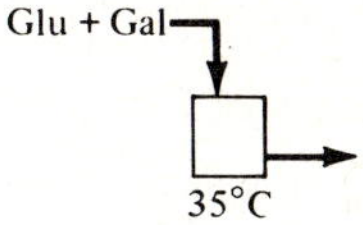 Glu + Gal → [35°C] →	$D = 0.121$ Shift-up system	0.500	559
3	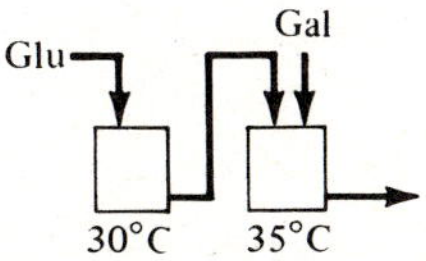 Glu → [30°C] → Gal → [35°C] →	$D_1 = 0.200$ $D_2 = 0.250$ $\bar{D} = 0.118$	0.293	415
4	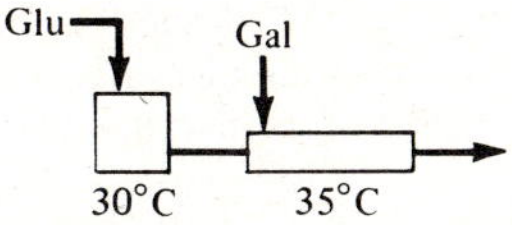 Glu → [30°C] → Gal → [35°C tubular] →	$D_1 = 0.133$ $D_2 = 0.286$ $\bar{D} = 0.097$	0.500	582
5	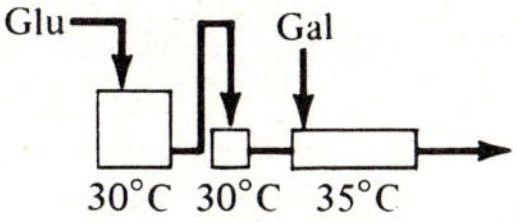 Glu → [30°C] → [30°C] → Gal → [35°C tubular] →	$D_1 = 0.193$ $D_2 = 2.342$ $D_3 = 0.286$ $\bar{D} = 0.117$	0.500	702
6	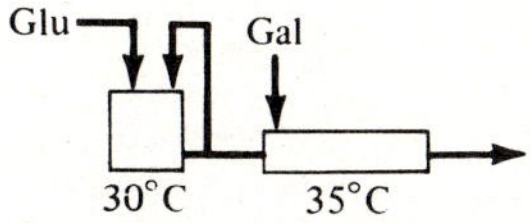 Glu → [30°C] → Gal → [35°C tubular] →	$D_1 = 0.178$ $D_2 = 0.286$ $\bar{D} = 0.117$ $C = 1.34$	0.500	702
		$D_1 = 0.266$ $D_2 = 0.286$ $\bar{D} = 0.145$ $C = 2.00$	0.500	870

† T. Imanaka, T. Kaieda, and H. Taguchi, *J. Ferment. Technol. (Japan)*, **51**: 558, 1973.

‡ Glucose = 2 percent, galactose = 0.5 percent of the total medium.

Using this procedure, the enzyme productivity ($= De_e x$) was computed for the two-stage system (3), with results shown in Fig. 9E2.6. The model revealed that maximum productivity would be obtained with $D_1 = 0.20\ h^{-1}$, and experiments performed under that condition showed nearly exactly the same D_2 dependence as predicted by the model. In system 4, the tubular reactor already shown in Fig. 9.19 was used for the second stage. Again, the model results agreed very well with experimental data for this process (Fig. 9E2.7).

Returning now to Table 9E2.3, we see that well-chosen staged cultures provide substantially greater enzyme production than the best single process. For example, system 6 produces 55 percent more enzyme than the single CSTR. Clearly the availability of a sound kinetic model and the general tools of reactor design have here proved essential ingredients in formulating and optimizing a superior continuous-reaction process.

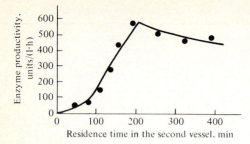

Figure 9E2.7 Productivity of enzyme in a CSTR-tubular fermentor cascade (System 4 from Table 9E2.3). [*Reprinted from T. Imanaka et al., Optimization of α-Galactosidase Production in Multi-stage Continuous Culture of Mold, J. Ferment, Tech. (Japan),* **51**: 431, 1973.]

PROBLEMS

9.1. CSTR kinetics (*a*) Verify each CSTR design equation of Table 9.1.

(*b*) For each CSTR case, indicate graphically how you would evaluate all terms in the reaction-rate expression from an appropriate plot of CSTR performance.

9.2. PFTR kinetics (*a*) Validate the results of Table 9.2 by direct integration.

(*b*) What variables are most convenient to use for each PFTR design equation? Show by sketches how each kinetic parameter in the reaction-rate expression can be evaluated.

(*c*) In general, do CSTRs lend themselves more easily to parameter evaluation than PFTRs? Why (not)?

9.3. Monod kinetics in CSTRs Consider an organism which follows the Monod equation where $\mu_{\max} = 0.5 \text{ h}^{-1}$ and $K_s = 2$ g/l.

(*a*) In a continuous perfectly mixed vessel at steady state with no cell death, if $s_0 = 50$ g/l and $Y = 1$(g cells/g substrate), what dilution rate D will give the maximum total rate of cell production?

(*b*) For the same value of D using tanks of the same size in series, how many vessels will be required to reduce the substrate concentration to 1 g/l?

9.4. CSTR design with inhibition (*a*) Construct a plot of x and s vs. D

$$r_x = \frac{\mu_{\max} s x}{K_s + s + i K_s / K_i}$$

where $s_0 = 10$ g/l
$x_0 = 0$
$K_s = 1$ g/l
$K_i = 0.01$ g/l
$i = 0.05$ g/l
$V_{\max} = 0.5 \text{ h}^{-1}$
$Y_{x/s} = 0.1$ g cells/g substrate

Plot on the same chart the values of x and s when $i = 0$.

(*b*) Suppose a CSTR was designed to operate for inhibitor-free conditions. Develop equations for the ratios (s/s_i) and $xD/(xD)_i$, that is, for the ratio of predicted to observed substrate concentrations and biomass rates. How does the presence of the inhibitor alter reactor behavior with regard to $(xD)_{\max}$ and washout?

(*c*) Repeat parts (*a*) and (*b*) for the case of i increasing with x, taking $i = x/10$.

(*d*) For constant yield factor and $i = x/10$, the growth rate dx/dt is a function of only x and initial conditions. Using the graphical design procedure of Sec. 9.5, what size must two equal-volume tanks in series be to achieve 95 percent substrate conversion in the presence of product inhibition? In the absence of product inhibition?

9.5. Yield coefficient When a negative term is included in cellular kinetics in order to model endogenous metabolism or maintenance energy, the resulting equation (assuming excess substrate) can be written

$$-\frac{1}{Y}\mu x = -\frac{1}{Y_G}\mu x - k_e x$$

where Y_G = growth yield, grams of cell produced per gram of substrate consumed for growth
k_e = grams of substrate consumed for maintenance energy per gram of cell
Y = apparent yield, grams of substrate consumed per gram of cell
Refer to Fig. 9P5.1 for information.

(a) Show that in a CSTR, x and s are related by

$$x = \frac{DY_G}{k_e Y_G + D}(s_0 - s)$$

(b) What relation must hold between k_e, Y_G, and D in this circumstance?
(c) Does the Fig. 9P5.1 for *A. aerogenes* growing on glycerol satisfy this model? Be quantitative.

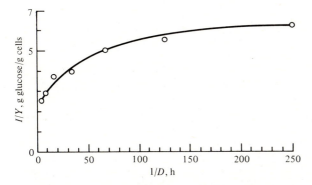

Figure 9P5.1 Apparent yields vs. reciprocal dilution rate. [*From Kirsh and Sykes, Prog. Ind. Microbiol.,* **9**: 155, 1971, *from data of D. W. Tempest, D. Herbert, and P. J. Phipps, in E. O. Powell et al. (eds.),* "*Microbial Physiology and Continuous Culture,*" *HMSO, London,* 1967.]

9.6. Yield coefficients from process data Kanazawa [23] determined the empirical formula $C_{12}H_{23}O_7$ as representative of 330 g of yeast. The *n*-paraffin fermentation was represented as

$$a(CH_2) + bO_2 \longrightarrow C_{12}H_{23}O_7 + dCO_2 + eH_2O$$

(a) Evaluate a, b, d, and e in terms of Y_s (grams of cells per gram of substrate).
(b) In turn, evaluate Y_s in terms of the appropriate parameters for a continuously sparged batch system: aeration rate, oxygen partial pressures of inlet and outlet stream, reactor volume, etc.
(c) What kind of stage-by-stage probes would you use to follow the progress of a continuous multistaged-tower *n*-paraffin fermentation?

9.7. Double-substrate design (both gases) Hamer et al. [24] report that the utilization of methane in the presence of oxygen in a continuous-flow continuously sparged fermenter can be described by the double Michaelis-Menten form of cell growth:

$$r_x = \mu_{max} x \frac{c_1}{K_1 + c_1} \frac{c_2}{K_2 + c_2}$$

where 1 is oxygen, and 2 methane.

(a) Assuming constant yield coefficients, Y_1 and Y_2 (grams of cell per gram of substrate i, $i = 1, 2$), write down the steady-state balances for a sterile- feed system for cells and substrates 1 and 2. Take the overall mass-transfer conductance to be $K_{li}a$ and assume both liquid and gas phases to be completely mixed.

(b) As the dilution rate increases, wash-out will again occur. Show by graphical or analytical evaluation that washout occurs at about $D \approx 0.72$ h^{-1} for the following parameter values: $K_1 = K_2 = 5 \times 10^{-4}$ g/l, $Y_i = 1.25$ g cell/g O$_2$, $Y_2 = 2.0$ g cell/g substrate, $K_{l1}a = K_{l2}a = 100$ h^{-1}, $c_1^* = 0.015$ g/l, $c_2^* = 0.007$ g/l, $\mu_{max} = 0.8$ h^{-1}.

(c) By graphing the appropriate equations, show that the maximum in Dx occurs very close to D(washout) and that if $c_1^* \to$ large, i.e., equivalently if $K_1 \to 0$, the maximum Dx shifts to lower dilution rates.

9.8. Inhibition kinetics If a microbial species is inhibited by a volatile product (such as ethanol), the growth rate can be increased by removal of the inhibiting product via the continuous evolution of the vapor space above the fermentor.

(a) For a batch fermentation following Eq. (7.17), show that the time course of substrate level for the overall reaction: S $\to 0.3$P + cell mass is

$$\frac{ds}{dt} = \frac{-[x_0 + Y_{x/s}(s_0 - s)]}{Y_{x/s}} \mu_{max} \frac{s}{K_s + s} \frac{K_I}{K_I + 0.3(s_0 - s)}$$

where x is biomass (grams per liter) and a constant yield factor $Y_{x/s}$ is assumed.

(b) Integrate this result (by the method of partial fractions, for example) and develop a form for the ratio of biomass densities $x(t$, no inhibitor)/$x(t$, inhibitor).

(c) Evaluate the biomass-ratio parameter of part (b) vs. time when $x_0 = 10^{-6}$ g/ml, $Y_{x/s} = 0.1$ g cell/g substrate, $K_s = 0.22$ g/l, $\mu_{max} = 0.408$ h^{-1}, $K_I = 16$ g/l for two substrate levels: $s_0 = 5.0$ g/l and $s_0 = 70$ g/l, [molecular weight S = 3 (molecular weight P)].

9.9 Graphical design when growth depends on two variables The treatment outlined in Sec. 9.5 allows easy design of continuous systems provided that the cell growth rate is a function of only a single variable x. Suppose the growth rate depends upon two variables, the cell number and the concentration of some soluble species (repressor, inducer, hormone, essential vitamin, etc.)

(a) Show that a three-dimensional analog of the two-dimensional Fig. 9.38 could be used in reactor design for such systems. Sketch the nature of the various functions in this three-dimensional design and indicate the meaning of all intersections between the different functions sketched.

(b) Discuss how you would carry out a series of experiments to obtain the necessary information for the graphical design construction of part (a).

9.10. Homogeneous and film reactor A feed contains a suspension of inert particles as well as substrate for an anaerobic fermentation. The vessel agitation is sufficient to keep the particles suspended and well dispersed. The microbial (single) population partitions itself between the particles and the bulk solution by a linear isotherm:

$$x_s \text{ (cells/cm}^2) = Kx_{bulk} \text{ (cells/ml)}$$

where K has units of ml/cm^2. The absorption process does alter the maximum specific growth rate μ_{max} but not K_s (assuming the Monod form is valid).

(a) Evaluate over all feasible dilution rates when d(inert) = 0.1 mm and volume fraction = 1 percent (1) the ratio of substrate utilization in the presence of inert particles to that occurring in their absence and (2) the influence of the suspended particles on reactor washout.

(b) How would you design a cell-recovery scheme at the reactor exit to maximize biomass recovery?

9.11. Rapid K_s measurement Williamson and McCarty [25] developed a relatively rapid means of determining K_s in microbial kinetics. A small, concentrated feed stream enters the microbial reactor, giving rise to a negligible volume increase over several hours. For values of s allowing less than the maximum possible substrate-utilization rate by the population, $\mu_{max} V/Y$, steady state was achieved in less than 1 h.

(a) Show that a Lineweaver-Burk type of plot (rate vs. $1/s$ at "steady state") allows evaluation of K_s.

(b) Over what range of sampling times is the above analysis valid?

9.12. Glucose effect: serial substrate utilization With multiple substrates, one of which is glucose, substrate consumption patterns in tower or other staged fermentors may be complex, as mentioned in Sec. 9.3.2.

(a) Suppose Fig. 7.7 for diauxic growth typifies cell growth patterns in such cases. From this figure, construct a plot of r_x vs. x (state any assumptions necessary).

(b) From the form of CSTR design for a single tank in Fig. 9.37 it would appear for the curve in part (a) that for low values of D two stable nontrivial solutions exist. Regardless of the state of micromixedness, show that for a *single* tank the second solution cannot exist. Be exact in your reasoning. This example illustrates an invalid application of batch data to continuous-system design.

(c) A tanks-in-series design will give conditions more reflective of batch conditions. Noting that r_x does not vanish between glucose and xylose consumption (Fig. 9.37), use the results of part (a) to design eight tanks in series which (1) consume essentially all the glucose but no xylose; (2) consume as much as possible of glucose and xylose. (The tanks have equal volumes.)

(d) If half the tanks are one-third the size of the other four, what dilution rate and tank sequence would give maximum xylose consumption? Why? Show the design construction for this case.

9.13. Process kinetics *Methylomonos methanolica* grown on methanol at 30°C, pH 6.0 was observed to obey the following kinetic parameter values: $\mu_{max} = 0.53$ h^{-1}, Y_s (methanol yield coefficient) = 0.48 g/g, $Y_{O_2} = 0.53$ g/g, carbon-conversion ($c_{biomass}/c_{methanol}$) efficiency = 0.57 g/g, oxygen quotient = 0.90 mol O_2 per mole CH_3OH, respiratory quotient (RQ) = 0.52 mol CO_2 per mole O_2, k_e = maintenance coefficient = 0.35 g CH_3OH/g, K_s = 2.0 mg/l [26].

(a) Write down the equations for CSTR growth which describe the rates of cell-mass production, oxygen consumption, and CO_2 production vs. dilution rate.

(b) Plot x, xD, and s vs. D (h^{-1}) and locate the predicted maximum values of x and xD when $s_{inlet} = 7.96$ g/l.

(c) Display the variation in the oxygen consumption rate vs. D on the same graph. What stirrer power input per unit volume would be needed to operate the reactor at the maximum productivity $(xD)_{max}$? State your assumptions.

(d) On the same graph, plot the predicted heat-generation rate per volume vs. D.

(e) The specific growth rate μ is observed to be diminished by substrate in an approximately linear fashion from its maximum value of 0.53 h^{-1} at $s = 3$ g/l to 0 at 13 g/l (extrapolated estimate). Repeat part (b) taking this design information into account.

(f) Under what exit conditions would several equal tanks be better than a single tank of the same total volume?

9.14. Liquid sterilization Set up the equation to determine the probability of complete sterility ($n < 1.0$) in a continuous sterilizer if plug flow is maintained and:

$$\text{Specific death rate} = 10 \text{ min}^{-1} \qquad \text{sterilizer volume} = 10 \text{ l}$$
$$\text{Medium flow rate} = 10 \text{ l/min} \qquad \text{temperature of medium} = 131°C$$
$$\text{Collection time} = 5 \text{ min} \qquad \text{original cell concentration } n_0 = 10^3 \text{ l}^{-1}$$

What would you expect to happen to this probability if all the variables given above were held constant but the sterilizer were shortened and widened? Explain your reasoning and show pertinent equations.

9.15. Variable yield coefficient Derive the analog to Eq. (9.14) for the case where Y is given by Eq. (9.22).

REFERENCES

Many of the references given for Chap. 7 contain substantial material on reaction design and analysis. Additional general presentations are available in the following texts:

1. R. Aris: "Elementary Chemical Reactor Analysis," Prentice-Hall, Inc., Englewood Cliffs, N.J., 1970. A highly analytical text with clear insights into the structure of mathematical models. Includes excellent material on reactor dynamics and optimization.

2. K. G. Denbigh and J. C. R. Turner: "Chemical Reactor Theory: An Introduction," 2d ed., Cambridge University Press, London, 1971. A short book fostering intuitive understanding of the workings of reactors and the interplay of physical and chemical features which influence their operation.

3. O. Levenspiel: "Chemical Reaction Engineering," 2d ed., John Wiley & Sons, Inc., New York, 1972. While many other aspects of reactor design are included, this is perhaps the best single source for material on mixing and RTDs.

4. J. M. Smith: "Chemical Engineering Kinetics," 2d ed., McGraw-Hill Book Company, New York, 1970. One of the most popular general texts in the field, made richer by many worked examples illustrating applications to real reactors.

Several international conferences have been entirely dedicated to continuous culture; some of these volumes are not easy to find, but they are all rich in information:

5. "Continuous Cultivation of Microorganisms: A Symposium," Publishing House of the Czechoslovak Academy of Sciences, Prague, 1958.

6. Continuous Culture of Micro-organisms, *S.C.I. Monogr.* 12, Society of Chemical Industry, London, 1961.

7. I. Málek, K. Beran, and J. Hospodka (eds.): "Continuous Cultivation of Microorganisms, Proc. 2d Symp.", Publishing House of the Czechoslovak Academy of Sciences, Prague, 1964.

8. I. Málek and Z. Fencl (eds.): "Theoretical and Methodological Basis of Continuous Culture of Microorganisms," Publishing House of the Czechoslovak Academy of Sciences, Prague, 1966.

9. I. Málek et al. (eds.): "Continuous Cultivation of Microorganisms, Proc. 4th Symp.", Academic Press, Inc., New York, 1969.

A useful paperback:

10. H. E. Kubitschek: "Introduction to Research with Continuous Cultures," Prentice-Hall, Inc., Englewood Cliffs, N.J., 1970.

Incomplete mixing and the effects of films in CSTRs:

11. C. G. Sinclair and D. E. Brown: The Effect of Incomplete Mixing on the Analysis of the Static Behavior of Continuous Cultures, *Biotech. Bioeng.*, **12:** 1001, 1970.

12. J. A. Howell, C. T. Chi, and U. Pawlowsky: Effect of Wall Growth on Scale-up Problems and Dynamic Operating Characteristics of the Biological Aerator, *Biotech. Bioeng.*, **14:** 253, 1972.

Papers on continuous-culture dynamics:

13. C. P. Jeffreson and J. M. Smith: Stationary and Nonstationary Models of Bacterial Kinetics in Well-mixed Flow Reactors, *Chem. Eng. Sci.*, **28:** 629, 1973.

14. D. Ramkrishna, A. G. Fredrickson, and H. M. Tsuchiya: Dynamics of Microbial Propagation: Models Considering Inhibitors and Variable Cell Composition, *Biotech. Bioeng.*, **9:** 129, 1967.

15. T. B. Young, D. F. Bruley, and H. R. Bungay III: A Dynamic Mathematical Model of the Chemostat, *Biotech. Bioeng.*, **12:** 747, 1970.

Tower fermentors: Ref. 16 of Chap. 7.

16. A. Kitai, H. Tone, and A. Ozaki: Performance of a Perforated Plate Column as a Multistage Continuous Fermentor, *Biotech. Bioeng.*, **11:** 911, 1969.

17. A. Prokop et al., Design and Physical Characteristics of a Multistage Continuous Tower Fermentor, *Biotech. Bioeng.*, **11:** 945, 1969.

18. E. A. Falch and E. L. Gaden, Jr.: A Continuous, Multistage Tower Fermentor, I: Design and Performance Tests, *Biotech. Bioeng.*, **11:** 927, 1969; II: Analysis of Reactor Performance," ibid., **12:** 465, 1970.

19. R. G. Ault et al.: Biological and Biochemical Aspects of Tower Fermentation, *J. Inst. Brewing*, **75:** 260, 1969.

20. R. N. Greenshields and E. L. Smith: Tower-Fermentation Systems and Their Applications, *Chem. Eng. Lond.*, May 1971, p. 182.
The standard reference for unsteady-state heat conduction in solids:
21. H. S. Carslaw and J. C. Jaeger: "Conduction of Heat in Solids," 2d ed., Clarendon Press, Oxford, 1959.
Heat-treatment design for food processing:
22. S. E. Charm, "The Fundamentals of Food Engineering," 2d ed., Avi Publishing Co., Inc., Westport, Conn., 1971.

Problems

23. M. Kanazawa: The Production of Yeasts from *n*-Paraffins, p. 438 in S. R. Tannenbaum and D. I. C. Wang (eds.) "Single Cell Protein II," MIT Press, Cambridge, Mass., 1975.
24. G. Hamer, D. E. F. Harrison, J. H. Harwood, and H. H. Topiwala: SCP Production from Methane, p. 362 in S. R. Tannenbaum and D. I. C. Wang (eds.), "Single Cell Protein II," MIT Press, Cambridge, Mass., 1975.
25. K. S. Williamson and P. L. McCarty: A Rapid Measurement of Monod Half Velocity Coefficients for Bacterial Kinetics, *Biotech. Bioeng.*, **17**: 915, 1975.
26. M. Dostolek and N. Molin: Studies of Biomass Production of Methanol Oxidizing Bacteria, p. 385 in S. R. Tannenbaum and D. I. C. Wang (eds.), "Single Cell Protein II," MIT Press, Cambridge, Mass., 1975.

TEN

BIOCHEMICAL REACTORS, SUBSTRATES, AND PRODUCTS I: SINGLE-SPECIES APPLICATIONS

Many fermentation processes practiced in industry employ only a single species of microorganism. In other instances, the action of a single species may predominate even though a mixed microbial population is actually present. In this chapter we shall explore several aspects of single-species fermentations including typical process equipment, operating methodology, and numerous examples of commercial utilization. This discussion will dramatize the importance of the biological and engineering fundamentals emphasized in previous chapters. In addition, consideration of several examples in some detail provides an opportunity for introduction of several new methods and concepts.

10.1. FERMENTATION TECHNOLOGY

While common features are evident among most commercial fermentation processes, significantly different process designs as well as operating practices arise often as a result of varying sensitivity to contamination by undesirable organisms. If it is necessary to avoid any intrusion, the fermentation must be operated on an *aseptic* basis so that a pure culture is maintained. In some situations, e.g., yeast growth at low pH or fermentation of hydrocarbons by carefully selected bacterial strains, aseptic precautions can be relaxed somewhat since operating conditions discourage growth of many potential contaminants.

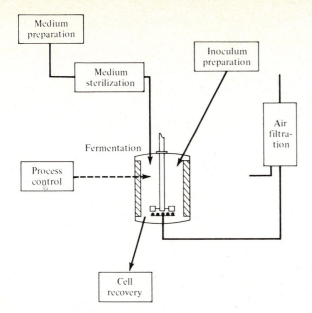

Figure 10.1 The basic operations involved in a typical aerobic fermentation.

Figure 10.1 is a schematic illustration of the important components of a typical fermentation process. Selection of a suitable medium has already been mentioned (Sec. 7.1.2), as have means of sterilizing it (Chapter 9) and any necessary gases (Chap. 8). Additional comments on medium formulation are provided in the following section. Although some influences of the inoculum on process behavior were discussed in Chap. 7, the microbiological problems encountered in this step require further investigation here. Section 10.1.2 will concentrate on the design, instrumentation, and control of the fermentation vessel itself, with a brief survey of selected cell-harvesting and product-recovery techniques following in Sec. 10.1.3.

10.1.1. Medium Formulation

A variety of factors must be considered when formulating a fermentation medium. One relates to cellular stoichiometry and the desired amount of biomass to be produced. The basic concept here is simply a material balance: during the course of cellular growth, small organic and inorganic molecules such as glucose are converted into biomass. Nutrients (reactants) must be provided in sufficient quantities and proper proportions for a specified amount of biomass (products) to be synthesized.

Computation of the necessary amounts of various substrates clearly requires knowledge of the product (biomass) composition. While values vary for specific organisms and for various growth conditions, we may use the data for bacteria, yeasts, and fungi in Table 10.1 as rough guidelines. Thus, for example, a medium to be used to grow 100 g dry weight of a bacterium would require a minimum of 12.5 g of nitrogen, 2 g of phosphorus, 0.2 g of sulfur, and so forth.

Table 10.1. Typical composition of organic and inorganic components of microorganisms

a. Energy-limited fermentations[†]

	% dry weight					
	Bacteria		Yeasts		Molds	
Component	Av*	Range	Av*	Range	Av*	Range
Carbon	48	46–52	48	46–52	48	45–55
Nitrogen	12.5	10–14	7.5	6–8.5	6	4–7
Protein	55	50–60	40	35–45	32	25–40
Carbohydrate	9	6–15	38	30–45	49	40–55
Lipid	7	5–10	8	5–10	8	5–10
Nucleic acid	23‡	15–25	8	5–10	5	2–8
Ash	6	4–10	6	4–10	6	4–10

* Average

b. Inorganic constituents of different microorganisms§

	g/100 g dry weight		
Element	Bacteria	Fungi	Yeast
Phosphorus	2.0–3.0	0.4–4.5	0.8–2.6
Sulphur	0.2–1.0	0.1–0.5	0.01–0.24
Potassium	1.0–4.5	0.2–2.5	1.0–4.0
Magnesium	0.1–0.5	0.1–0.3	0.1–0.5
Sodium	0.5–1.0	0.02–0.5	0.01–0.1
Calcium	0.01–1.1	0.1–1.4	0.1–0.3
Iron	0.02–0.2	0.1–0.2	0.01–0.5
Copper	0.01–0.02	···	0.002–0.01
Manganese	0.001–0.01	···	0.0005–0.007
Molybdenum	···	···	0.0001–0.0002
Total ash	7–12	2–8	5–10

[†] From H. J. Peppler (ed), "Microbial Technology," p. 421, Reinhold Publishing Corporation, New York, 1967.

‡ Values this high are observed only with rapidly growing cells.

§ From S. Aiba, A. E. Humphrey, and N. F. Millis, "Biochemical Engineering," 2d ed., p. 29, Academic Press, Inc., New York, 1973.

Determination of the required carbon from this mass-balance perspective demands some knowledge of the efficiency with which nutrient carbon is incorporated into cellular material. For example, facultative organisms typically incorporate about 10 percent of nutrient carbon under anaerobic conditions and 50 to 55 percent in aerobic environments. Thus, aerobic growth requires a medium with carbon mass about double the desired amount of biomass *carbon*, which (Table 10.1) is 0.48 times the specified total biomass. (How much glucose would be required to grow 100 g of biomass under these conditions?)

Once the elemental requirements have been calculated, choices still remain of the chemical compounds used to supply the necessary elements. Many commercially important microorganisms are chemoheterotrophs whose energy and carbon needs are satisfied by simple sugars. Instead of purified sugars, crude sources such as beet, cane, or corn molasses (50 to 70 percent fermentable sugars) are frequently used as carbon and energy sources in industrial fermentation media. In some instances process wastes like whey and cannery wastes provide cheap yet satisfactory carbon sources for some fermentations. For example, food yeast is grown commercially using a by-product of papermaking, sulfite waste liquor, which contains about 2 percent fermentable hexoses and pentoses.

A variety of possible nitrogen sources are available including ammonia, urea, and nitrate. If the microorganism produces proteolytic enzymes, however, it can obtain necessary nitrogen from a variety of relatively crude proteinaceous sources. Among the possibilities for such crude sources are distiller's solubles, cereal grains, peptones, meat scraps, soybean meal, casein, cereal grains, yeast extracts, cottonseed meal, peanut-oil meal, linseed-oil meal, and corn-steep liquor. Especially important in penicillin fermentation media, corn-steep liquor is a concentrated (50 percent solids) aqueous waste resulting from the steeping of corn to make corn starch, gluten, and other products.

We have mentioned in earlier chapters that some microorganisms require an external source of some amino acids and growth factors. Other microbes which do not have a strict requirement for such medium adjuncts are frequently more productive if nonessential growth factors and nitrogen and carbon sources are provided. In industrial practice growth factors are typically provided by some of the crude-medium components already mentioned, e.g., corn-steep liquor or yeast autolysate. Similarly, these crude preparations often supply many of the minerals necessary for cell function. Other minerals are added to the medium as necessary.

When product formation is the major objective of a fermentation, *precursors* may be added to the medium to improve product yield or quality. Generally the precursor molecule or a closely related derivative is incorporated into the fermentation-product molecule. Specific examples of precursor applications include benzoic acids for production of novobiocins, phenylacetic acid for manufacture of penicillin G, and 5,6-dimethylbenzimidazole for vitamin B_{12} fermentation. In addition to these well-defined precursors, crude media components like corn-steep liquor may also provide useful precursor compounds.

Additional detail on selection and formulation of fermentation media are given in Sec. 6.4.1 and in Refs. 1 and 2. We now turn our attention to other aspects of fermentation technology.

10.1.2. Design and Operation of a Typical Aseptic, Aerobic Fermentation Process

This section will concentrate on aseptic practices since they impose the greatest demands on the ingenuity and thoroughness of the biochemical engineer. Our discussion follows the general sequence of events in the operation of a batch fermentation, beginning with development of an inoculum from a stock culture.

Preparation of an inoculum requires careful proliferation of relatively few cells to a dense suspension of from 1 to 20 percent of the volume of the production fermenter. This involves a stepwise procedure of increasing scale, which is summarized in Fig. 10.2. The starting point is a *stock culture*, a carefully maintained collection of a particular microbial strain. Since the strain may be the result of extensive screening and mutation searches and may constitute a significant competitive advantage in the industry, it is imperative that the integrity of the production species be preserved. The usual strategy for achieving maximal genetic stability in a stock strain is to minimize its metabolic activities during storage. Microorganisms are usually maintained in the desired dormant state by *lyophilization* (freeze-drying) of a liquid cell suspension or by thoroughly drying dispersions of spores or cells on sterile soil or sand. Highly mutated stock organisms are frequently susceptible to *back mutation* and other types of undesirable genetic instability. Thus, constant checks on stock cultures are essential.

If we take lyophilized culture as an example, the next step in inoculum preparation is suspension of the cells in a sterile liquid. A drop of this suspension is then transferred to the surface of an agar *slope*, or *slant*, made by solidifying a sterile nutritive medium in an inclined test tube using agar, a polysaccharide derived from seaweed. After incubation to obtain sufficient growth, the cells are again suspended in liquid and added either to a larger agar surface in a flat-sided Blake or Roux bottle or transferred to a shake flask. These flasks are agitated in machines which shake them in rotary or reciprocal patterns to promote submerged growth with adequate transport of gases to and from the organisms. Several successive steps with increasingly larger flasks usually are required before proceeding to the next step. All the transfers described above must be accomplished under sterile conditions. Rooms especially designed to permit sterilization and maintenance of aseptic conditions and controlled temperature are used in the fermentation industry for these delicate operations.

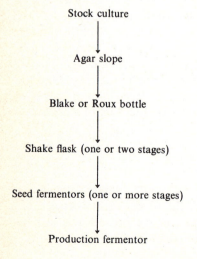

Stock culture

↓

Agar slope

↓

Blake or Roux bottle

↓

Shake flask (one or two stages)

↓

Seed fermentors (one or more stages)

↓

Production fermentor

Figure 10.2 Stages in the propagation of a culture for use in a fermentation.

Further proliferation of the culture is next accomplished in one or more seed vessels, small fermentors with many of the instrumentation and control systems typical of large production units. Conditions in these reactors are chosen to maximize growth of the culture.

Since at this point we have moved from conditions typical of a microbiology laboratory into the plant environment, it is well to pause and consider some of the special design features required to maintain aseptic conditions. First, the system must be arranged to permit independent sterilization of its components. As an example of the extreme degree of care this requires in design and operation, consider the problem of transferring the inoculant from the seed tank to the production fermentor. The schematic in Fig. 10.3 illustrates the required services and sequence of events. All valves in the system must be easy to maintain, clean, and sterilize. For these reasons ball valves are quite popular. Several other general principles of aseptic process design are shown in Fig. 10.3. Specifically, all vessel connections should be steam-sealed: no direct connections between sterile and nonsterile portions of the system should be allowed. Maintenance of a positive pressure on the system ensures that leakage will be outward rather than inward.

The physical characteristics of a typical commercial fermentation vessel are shown in Fig. 10.4. These vessels are usually constructed from stainless steel to minimize corrosion problems and contamination of the fermentation broth by unwanted metallic ions (recall the discussion in Sec. 5.3.3 on the influence of iron ions in the citric acid fermentation). Care must be taken here and elsewhere in the overall process to avoid dead spaces, crevices, and other niches where solids resistant to sterilization can accumulate and where microbial films can grow. All-welded vessel construction with polished welds helps to minimize these problems.

The agitator assembly is designed to meet the mixing and aeration requirements already discussed in Chap. 8. Special attention to the design and maintenance of the aseptic seal is essential to avoid contamination. Although only one

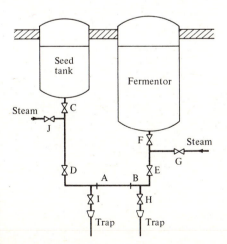

Figure 10.3 Valve and piping configuration for aseptic inoculation of a large scale fermentor. Operation sequence: (1) install pipe section AB; (2) sterilize connection with 15 lb/in^2 gauge steam for 20 min valves D to J open; valve C remains closed; condensate collects in steam traps in branches H and I; (3) cool fermentor under sterile air pressure with valves C, G, H, I, J closed and valves D, E, F left open: sterile medium fills connection; (4) increase pressure in seed tank to 10 lb/in^2 gauge; lower fermentor pressure to 2 lb/in^2 gauge; (5) transfer inoculum by opening valve C; (6) steam-seal fermentor–seed-tank connections by closing C and F and opening G and J; steam and condensate is bled from partially open D and E.

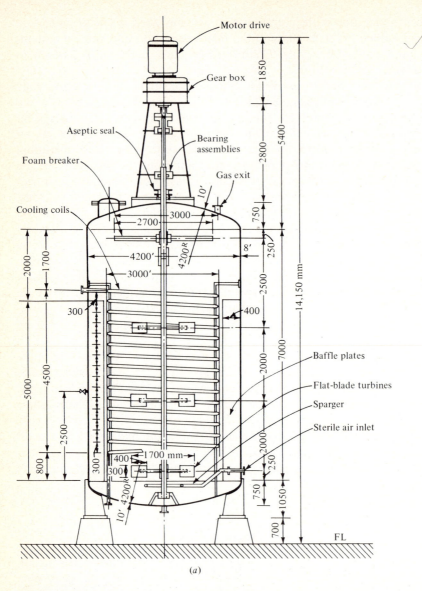

(a)

Figure 10.4 (a) Cutaway diagram of a 100,000 liter fermentor used for penicillin production. (b) photograph looking down into a large production fermentor. [(a) *Reprinted from S. Aiba, A. E. Humphrey and N. F. Millis, "Biochemical Engineering," 2d ed., p. 304, University of Tokyo Press, Tokyo, 1973. (b) Reprinted from R. Müller and K. Kieslich, Technology of the Microbiological Preparation to Organic Substances, Angew. Chem. Int. Edit. Eng.,* **5:** 653, 1966.]

(b)

Figure 10.4 (*Continued*)

impeller is required in laboratory-scale fermentation, several may be necessary in a large commercial vessel.

Typically only 70 to 80 percent of the vessel volume is filled with liquid, with a gas space occupying the top portion of the tank. Often the combined action of aeration and agitation of the liquid promotes the formation of a foam on the liquid surface, especially if the medium contains high concentrations of peptides or proteins. Foams can interfere with the effectiveness of the fermentation by impeding gas mass transfer from the broth to the head space, forcing medium in the shape of foam out of the vessel and contaminating the system when collapsed foam reenters the fermentor. In Fig. 10.4, a supplementary agitator located in the head space serves to destroy the foam. For especially persistent foams, chemical agents called *antifoams* are added to the broth. These compounds destabilize the foam by reducing surface tension. As noted in Chap. 8, the surface-active characteristics of antifoams can decrease the rate of oxygen transfer.

Several functions can be served by the heat-transfer coils within the vessel. If the medium is to be sterilized batchwise within the fermenter, these coils must have adequate capacity for the necessary heating and cooling. The heat-exchange system must also be able to handle the peak process load, which includes the combined effects of microbial activity and viscous dissipation from mixing (Sec. 8.11.1). Typical heat-transfer coefficients for uninoculated medium are about the same as for water, while a dense mycelial broth may exhibit a coefficient more typical of a paste.

Before turning to instrumentation and control of fermentors, we should mention the airlift and tower designs, which are of increasing industrial interest. Both rely on the rising motion of the bubbles of sparged air to provide mixing. In the airlift configuration, air blown in the bottom of the vessel on the outer periphery sets up circulation patterns around the draft tube installed in the center of the vessel's interior. In the tower fermenter, air rises through a tower, sometimes divided into compartments separated by perforated plates. Obviously these de-signs feature smaller power requirements than agitated vessels, and the increased gas-liquid contact time in the tower configuration results in more efficient oxygen utilization. This feature is especially attractive since in the conventional fermentor design outlined above, a large fraction of the oxygen feed, which has been painstakingly sterilized, may escape the system without being used.

Increasingly complex control systems are finding application in commercial fermentations for maximizing process productivity and batch-operating-cycle reproducibility. The limiting factor in the possible sophistication of the control system is the limited availability of steam-sterilizable sensors for parameters of interest. In some processes only temperature is measured, and control of this variable is effected with the heat-exchange facilities already discussed. Other physical factors for which suitable instrumentation exists today include pressure, agitator-shaft speed and power, and foam and liquid level.

Chemical parameters which can be monitored continuously are limited; only the broth pH, redox, NAD level, dissolved O_2 and CO_2 concentrations, and exit-gas composition permit direct measurement. Since other chemical variables such as RNA and ATP concentrations are also of potential importance, an expansion of this list in the near future can be expected. Of all those mentioned now, pH measurement and control is by far the most common. Although the simplest method for limiting pH changes involves the use of an appropriate buffer, this approach is uncertain when a complex medium is used. Also, an inexpensive, nonmetabolizable buffer may not exist in the pH range required for optimum fermentation productivity. Consequently, more modern practice requires controlled metering of an acid or base into the fermentor to maintain the desired pH. Often pH falls during the course of batch operation due to depletion of an ammonia nitrogen source. Use of added ammonia as the controlling base then serves the twofold purpose of maintaining the desired pH and preventing nitrogen limitation of the culture.

Fermentation in the production vessel may take from less than 1 day to more than 2 weeks, 4 to 5 days being typical of many antibiotic manufacturing processes. During this interval, operating conditions must be carefully maintained or varied in a predetermined manner. More details on this operating practice will be provided in one of the examples which follow. Nevertheless, it should be evident from the general review already given that aseptic fermentation is an expensive, time-consuming proposition. Obviously, loss of product from one batch cycle can be extremely costly, so that the emphasis accorded sterilization in earlier chapters is well taken. Another logical conclusion from the above remarks is that only relatively expensive products such as antibiotics, vitamins, and enzymes justify the investment required for strict asepsis. Manufacture of much cheaper compounds will be economical when contamination is relatively unimportant.

10.1.3. Cell Harvesting and Product Recovery

As Fig. 10.5 suggests, recovery of fermentation products can be a complex and multifaceted task. A significant proportion of the overall costs in a fermentation plant often must be spent for product recovery. For example, according to one estimate, the product-recovery stages for amino and organic acid fermentations may constitute up to 60 percent of the plant fixed costs.

However, our examination here of product-recovery techniques is relatively brief for two major reasons: (1) Many of the unit operations involved are common to the nonbiological process industries and thus are treated in many other texts and courses. Readers who have no background in such areas as extraction, adsorption, and crystallization will find brief summaries of process principles and design methods in Ref. 14. (2) Many of the separation methods especially suited for biological products have already been reviewed in conjunction with enzyme purification in Sec. 4.2. As a result, we shall consider in this section only a few selected unit operations commonly employed in the fermentation industry.

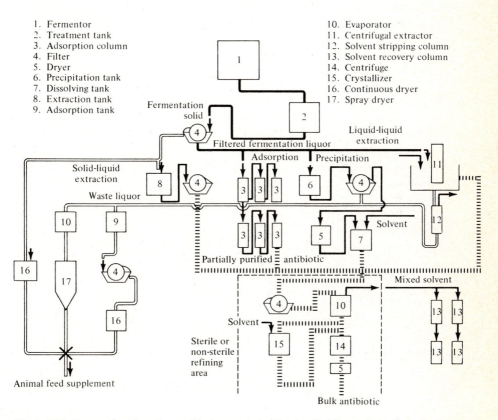

Figure 10.5 Process flowsheet for antibiotic recovery. (*Reprinted with permission from S. C. Beesch and G. M. Stull, "Fermentation," Ind. Eng. Chem. **49**: 1491, 1957. Copyright by the American Chemical Society.*)

The desired product usually resides either in the fermentation medium or the biomass, requiring a separation of these two phases early in the product-recovery process train. Continuous rotary vacuum filters are used for this purpose in antibiotic manufacture; the type schematically illustrated in Fig. 10.6 uses strings to lift off the filter cake (a layer of concentrated solids) which has accumulated on the drum. While string discharge is satisfactory for removing *Penicillium* mycelia, *Streptomyces* mycelia are more difficult to work with and require precoat of the filter cloth with filter aid, e.g., diatomaceous earth (recall Fig. 1.14), and cake removal with a knifeblade to scrape the cake from the rotating drum.

Filtering characteristics of the solid-liquid slurry are often described in terms of the elementary theory of filtration. Assuming laminar flow of filtrate liquid through the cake, we may write [14, p. 19-58]

$$\frac{1}{A}\frac{dV_f}{dt} = \frac{\Delta p}{\mu_c[\alpha(W/A) + r]} \tag{10.1}$$

where A = area of filtering surface

V_f = volume of filtrate collected

t = time

Δp = pressure drop across filter

μ_c = filtrate viscosity

α = average specific cake resistance

W = mass of accumulated dry cake solids = $[\rho w/(1 - mw)]V$, where ρ is filtrate density, w is mass fraction of solids in the slurry, and m is ratio of wet-cake to dry-cake mass

r = resistance coefficient of filter medium

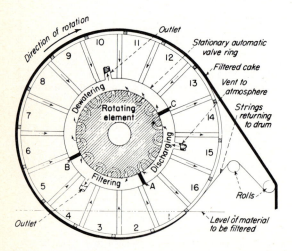

Figure 10.6 Schematic diagram of a string filter in operation. (*Courtesy of Ametek, Inc.*)

Pressure drop across the filter is constant for the rotary vacuum filters commonly used in the fermentation industry. If we assume that the cake is incompressible, α is constant and Eq. (10.1) can be integrated to obtain

$$\frac{t}{V_f/A} = \frac{\mu_c \alpha \rho w}{2\Delta p(1 - mw)} \frac{V_f}{A} + \frac{\mu_c r}{\Delta p} \tag{10.2}$$

indicating that t/V_f is a linear function of V_f under the conditions assumed.

Fig. 10.7a shows a plot of t/V_f against V_f for a *Streptomyces griseus* fermentation broth filtered at various pH values using a cotton cloth, diatomaceous-earth filter aid, and a Δp of 28.4 lb/in^2. Two important features are revealed by this data: the cake is not incompressible since the data for each pH do not fall on a straight line. Generally cells and other organic material from fermentations form compressible cakes.

Also evident in Fig. 10.7 is the strong dependence of filtering properties on broth pretreatment and filtration conditions. Clearly the pH during filtration has a major influence on filtration rates. Figure 10.7b reveals that broth preheating can substantially lower the specific cake resistance, presumably by coagulating mycelial protein.

Centrifugation may be used to remove cells from fermentation broths; yeasts, for example, are sometimes harvested in this fashion. A schematic diagram of one type of continuous centrifuge is given in Fig. 10.8. For dilute suspensions, each cell may be treated as a single particle in an infinite fluid. In this case the analysis of Example 1.1 applies. Such an approach is not valid for concentrated slurries, in which a given particle's motion is influenced by neighboring particles.

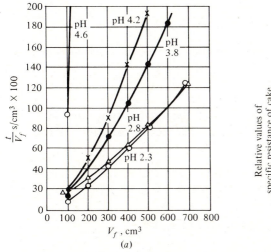
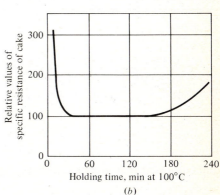

Figure 10.7 (a) pH has a profound influence on the filtration rate of *S. griseus* broth. (b) Heating pretreatment of *S. griseus* broth changes the specific resistance of the resulting cake. [*Reprinted from S. Shirato and S. Esumi, Filtration of a Culture Broth of Streptomyces griseus," J. Ferment. Tech. (Japan)*, **41**: 87, 1963.]

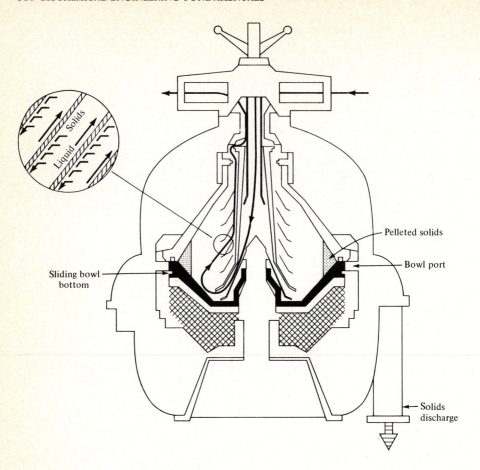

Figure 10.8 In this continuous centrifuge, solid particulates are removed in flow between closely stacked cones. Clarified effluent is withdrawn from the top of the unit.

Correlations of particle velocity u_h in such hindered-settling situations with the single-particle velocity u_0 and the volume fraction of particles ε_p have been developed. Possessing the general form

$$\frac{u_h}{u_0} = \frac{1}{1 + \beta \varepsilon_p^{1/3}} \tag{10.3}$$

the empirical relationships derived between β and ε_p are

$$\beta = \begin{cases} 1 + 305\varepsilon_p^{2.84} & 0.15 < \varepsilon_p < 0.5, \text{ irregular particles} \\ 1 + 229\varepsilon_p^{3.43} & 0.2 < \varepsilon_p < 0.5, \text{ spherical particles} \\ 1\text{--}2 & \text{dilute suspensions } (\varepsilon_p < 0.15) \end{cases} \tag{10.4}$$

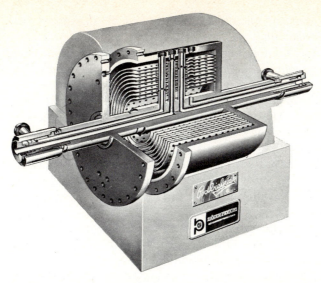

Figure 10.9 Cutaway drawing of the Podbielniak centrifugal-extraction unit which is widely used in the fermentation process industry. (*Courtesy of Baker-Perkins, Inc.*)

Our final topic in this potpourri of unit operations is centrifugal solvent extraction, a process originally developed for penicillin recovery and now widely used in the fermentation industry. To appreciate the advantages of this process, we review some of the difficulties faced by biochemical engineers trying to scale up pencillin production during the 1940s. An early step in the recovery process was conversion of penicillin to acid form by addition of mineral acid to the filtered fermentation broth. Then the penicillin was extracted into an organic solvent. This apparently simple step was confounded by rapid deterioration of penicillin and precipitation of solids under acidified conditions. Consequently, it was necessary to minimize the liquid contact time and to provide some means for solids removal.

These requirements, as well as those of small liquid holdup and countercurrent multistage contacting, were met by the Podbielniak centrifugal extractor (Fig. 10.9). In this unit one liquid phase flows radially inward while the other moves outward radially. The internals, including the perforated annular plates, are rotated at high speed (up to several thousand revolutions per minute) to drive the outward-moving dispersed phase and suspended solids through the continuous phase. Additional details on this unit are given in Ref. 14, p. 21-29.

10.2. PRODUCT MANUFACTURE BY FERMENTATION

As we already know from our studies of microbial metabolism, the processes of substrate utilization, microbial growth, and product formation all occur during batch growth of a culture. In many biochemical processes of commercial importance, however, the predominant activity of interest is product manufacture. This section will review some important microbial products and the organisms and processes used for their manufacture.

When examining these microbiological processes, it is important to remember that they will not be utilized industrially if there is a superior nonbiological process such as straight chemical synthesis. As a general rule, the dividing line between the practicality of synthetic methods and biological conversion depends on the complexity of the product molecule. Although in the late nineteenth and early twentieth century, industrial chemicals such as alcohol, butanol, and acetone were made by fermentation, subsequent developments in nonbiological catalysis led to more economical chemical processes and birth of the massive petrochemical industry. Thus, relatively small molecules are generally cheaper to make today by nonbiological methods. Reversal of this situation can be expected only if future developments cause large shifts in raw-material costs. For example, more abundant supply of carbohydrates made available by new ways to degrade cellulose coupled with scarcity of petroleum-based feedstocks could cause some changes in the positions of the petrochemical and fermentation industries.

Larger, complex molecules such as enzymes and antibiotics can be directly synthesized only with great difficulty if at all. Consequently, biological processes are used exclusively for their production. Since there are very few optically selective nonbiological catalysts, microbial and enzymatic methods usually prevail for making specific optical isomers such as L-amino acids. Similarly, when the cell itself is the final product, as in yeast manufacture (see Sec. 10.3), there is no alternative to a biological process.

10.2.1. Brewing and Wine Making

The most familiar and commercially important products of microbial action are beer, wine, and other alcoholic beverages. A flow diagram of a typical winery process is shown in Fig. 10.10. The feed for wine production may be nearly any ripe fruit or vegetable extract which contains 12 to 30 percent sugar in the juices. The fermentation is accomplished either with wild yeast, present on the grape or fruit skins, or with inoculated selected yeast cultures and pasteurized fruit juices.

Characteristic of many anaerobic fermentation processes, the yeast inoculum is first grown to desired size in the juice under aerobic conditions, generating more cells as well as carbon dioxide; subsequent fermentation to yield 7 to 15% ethanol solutions and further carbon dioxide is carried out anaerobically. Some acids produced during fermentation are degraded by bacteria in postfermentation stages, creating the characteristic bouquet of the wine. The type of wine produced is determined by the fruit used (grape variety, apricots, peaches, etc.), the yeast and bacterial strains involved, possible subsequent fortification, i.e., adding alcohol,

---

Figure 10.10 A process flowsheet illustrating the major operations in modern wine making. (*Reprinted from W. Q. Hull, W. E. Kite, and R. C. Auerbach, Modern Winemaking, Ind. Eng. Chem.,* **43:** 2182, 1951. *Copyright by the American Chemical Society.*)

MODERN WINE MAKING

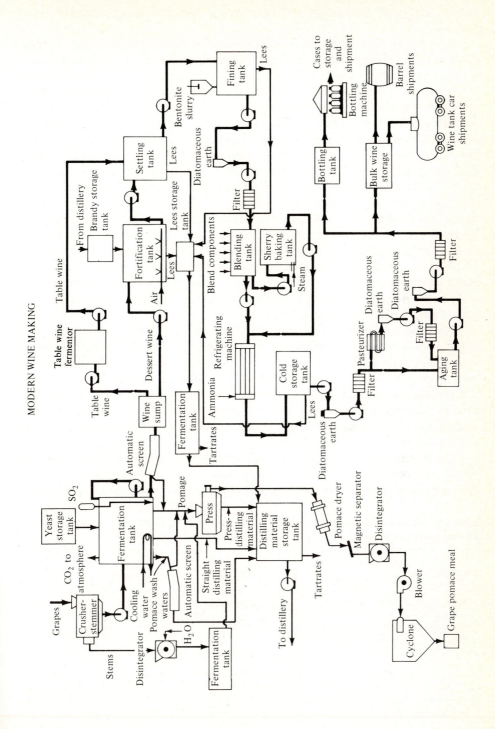

and the series of mechanical and heating, cooling, and aging processes before bottling.

Beer production illustrates several principles of enzyme technology already discussed. To prepare barley for processing, it is incubated for 2 to 6 days so that germination occurs. Recalling Table 4.12, we see that this *malting* step promotes the formation of active α- and β-amylases as well as protease enzymes. The resulting grains are then carefully dried.

Preparation of a nutrient medium called *beer wort* for yeast growth requires *mashing*, which is a carefully controlled warming of an aqueous mixture of malt and starch. Success of the mashing step requires artful exploitation of a complex mixture of substrates and enzymes. Proteins and carbohydrates in the malt-starch mixture must be hydrolyzed, since the yeast involved in the fermentation step can utilize only simple sugars and amino acids as nutrients. On the other hand, total hydrolysis is not desirable because dextrins, peptides, and peptones contribute flavor and body to the beer. Complicating the situation are the characteristics of the hydrolytic enzymes which participate in mashing (Table 10.2). Elaborate cooking recipes with temperatures varying with time over a range from about 40 to 100°C are employed to optimize the complex mashing process. One of these is described in Table 10.3. The end product contains a suitable mixture of yeast nutrients and flavor components.

Next the beer wort is clarified and then boiled for about 2 h with periodic addition of *hops*, ripened, dried cones of the hop vine, which contribute flavor, aroma, and color to the beer and also exert an antibacterial action. The sterilized wort is next cooled and inoculated with a proprietary strain of brewers' yeast (*Saccharomyces cerevisiae*). As in wine making, the yeast is first cultivated under aerobic conditions followed by a switch to an anaerobic environment to cause ethanol and CO_2 production. Finally the product is clarified, pasteurized, aged, and packaged.

If it is not already evident from personal observation, we should now realize that wine and beer are far more than a carbonated, alcohol-water solution. Complex molecules arising from ill-characterized biological and chemical processes

Table 10.2. Three enzyme-catalyzed hydrolyses used in mashing with different components and different rate characteristics†

Enzyme	Substrate (or bond) attacked	Hydrolysis product(s)	Temperature of maximum activity, °C
α-Amylase	Starch (all bonds)	Large-fragment dextrins, amylose	70–75
β-Amylase	Amylose (β-1,4 linkage)	Maltose	57–65
Proteases	Proteins	Amino acids, small peptones	50
		Larger peptides, peptones	60

† Data from Ref. 1.

Table 10.3. Typical mash-production procedure

The process begins with two separate mashes, which are combined

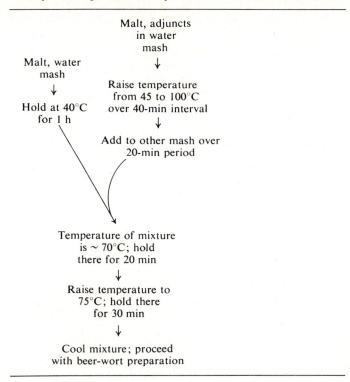

during brewing contribute essential characteristics to the final product. Thus, no synthetic method of wine or beer manufacture seems likely to compete with fermentation methods. Also because of the complicated nature of these products, attempts to make alcoholic beverages by continuous fermentation methods have failed to gain widespread acceptance: apparently the continuous process fails to reproduce all the chemical and physical changes necessary for the characteristic taste, aroma, color, and body of the batch-produced beverage.

10.2.2. Oxidative Transformations

The sour taste of spoiled wine results from oxidation of ethanol to acetic acid by various *Acetobacter* bacteria. The same process is used to advantage for manufacture of *vinegar*, which literally translated means sour wine. By legal definition, vinegar must be made by microbial oxidation of ethanol. Because of this and the special flavor characteristics afforded by bacterial action, direct chemical synthesis cannot compete with the fermentation process.

Since the desired reaction

$$C_2H_5OH + O_2 \longrightarrow CH_3COOH + H_2O \qquad \Delta H = -119 \text{ kcal/mol}$$

$$(10.5)$$

requires considerable oxygen and liberates a substantial amount of heat, special fermentor designs are necessary. In one popular method, which derives from the early nineteenth century, alcoholic liquor obtained by anaerobic fermentation using brewers' yeast is sprayed on the top of a trickling filter (see Fig. 10.11). The bed is filled with packing, typically about 2000 ft^3 of beechwood shavings about 2-in diameter, which serves as a support for an adhering microbial film of one or more *Acetobacter* bacteria. This arrangement permits the efficient heat and mass transfer required to oxidize ethanol while maintaining an active bacterial population. Trickling filters may be viewed as immobilized-cell systems. Extensions of this concept are subjects of current research.

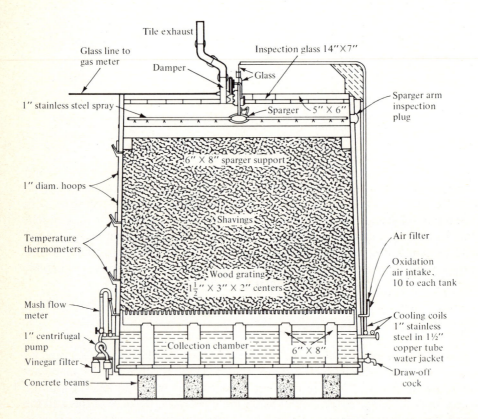

Figure 10.11 Vinegar can be manufactured in a trickling filter where ethanol is oxidized by the microbial film adhering to beechwood packing.

Example 10.1. Reaction rates in microbial films In addition to its use in vinegar manufacture, the trickling filter is also used for biological waste-water treatment (Chap. 12). Microbial films also appear in the rotating-disk contactor for waste treatment, animal-tissue culture, and the leaching of ores by bacteria. Also, new fermentor designs have been proposed which employ microbial films grown on solid particles dispersed in the medium.

In fact, microbial films will arise in almost any system where a solid surface contacts a microbial suspension. Consequently, the presence of films may cause a continuous fermentation to perform otherwise than the simplest theory predicts. Microbial films complicate sterilization: organisms inside the film on other than the heat-transfer surfaces will generally stay cooler than the bulk liquid during the heating cycle (see Chap. 9).

For these reasons and others, microbial films are of considerable significance in biochemical engineering. Excellent reviews touching on film formation, control of film thickness, and film characterization and kinetics will be found in Ref. 9. For the present, we shall examine the rate of substrate utilization in microbial films; product formation rates are sometimes proportional.

The principles underlying this discussion are the same as those already considered in Sec. 7.4.2, except that now we view the film as a plane sheet of thickness L affixed to a planar solid support. Let z denote distance from the solid surface; then a thin-shell balance leads to

$$\mathscr{D}_s \frac{d^2s}{dz^2} - \frac{as}{K_s + s} = 0 \tag{10E1.1}$$

where substrate utilization has been assumed to follow the usual hyperbolic form..

The outer surface of the film is exposed to medium where the substrate concentration is s_0, and the solid surface supporting the film is not permeated by the substrate. Consequently, the boundary conditions for s are

$$s = s_0 \qquad \text{at } z = L$$

$$\frac{ds}{dz} = 0 \qquad \text{at } z = 0 \tag{10E1.2}$$

The boundary-value problem for s posed by Eq. (10E1.1) and (10E1.2) is nonlinear and so in general requires numerical solution. The solution yields s as a function of z: this concentration profile rises from a minimum value at $z = 0$ to a maximum at $z = L$. Consequently, microbes inside the film are exposed to smaller substrate concentrations than cells nearer the outer film surface. Such differences in local cellular environment are believed responsible for a kind of self-regulation of film thickness. As the film grows thicker, substrate concentrations near the solid support fall, while concentrations of metabolic products rise. (What sort of concentration profile would be expected for metabolic end products?) Eventually cells near the support die and undergo autolysis. Such cell disintegration releases a portion of the microbial film, which is swept from the surface into the medium. Then, a new film begins forming on the newly exposed surface.

As in our previous discussions of diffusion-reaction interactions, we shall present overall substrate utilization rates ($\propto (ds/dz)_{z=L}$) for the film in terms of the effectiveness factor. Some results for films (planar geometry) have already been presented in Sec. 7.4.2: the effectiveness factor η in terms of the observable modulus Φ is given by the dashed lines in Fig. 7.38, and Eq. (7.106) gives an analytical expression for η in the limit of small diffusion-rate–reaction-rate ratios. In the following discussion we shall use η_d to denote effectiveness factors computed with the asymptotic equation (7.106).

In computational analysis of reactors containing microbial films, graphical data like that of Fig. 7.38 are inconvenient, and a closed-form expression for the effectiveness factor is desired. From analytical and empirical considerations, Atkinson, Davies, and How† have devised a biological rate

---

† Atkinson, B., Davies, I. G., and How, S. Y.: The Overall Rate of Substrate Uptake (Reaction) by Microbial Film, *Trans. Inst. Chem. Eng.*, London, **52:** 248 (1974).

Table 10E1.1. Rate parameters for reactions in microbial films†
These values for mixed microbial populations (treated in the analysis as a pure culture) were determined experimentally using a biological film reactor

Rate-limiting substrate	a, g/(cm^3·s)	K_s, g/cm^3	$\mathcal{D}_s$, cm^2/s
Glucose	6.66×10^{-7}	6.17×10^{-6}	5.10×10^{-6}
NH$_3$–N	4.94×10^{-8}	2.32×10^{-7}	1.7×10^{-5}

† Computed from data in B. Atkinson, "Biochemical Reactors." pp. 229ff, Pion Ltd., London, 1974.

equation for films which agrees well with the numerical and asymptotic analytical solutions over all ϕ values and for β values from 0.5 to 30. Their result is

$$
\eta = \begin{cases}
1 - \dfrac{\tanh \phi}{\phi}\left(\dfrac{\eta_d}{\tanh \eta_d} - 1\right) & \text{for } \eta_d \leq 1 \\[3ex]
\eta_d - \dfrac{\tanh \phi}{\phi}\left(\dfrac{1}{\tanh \eta_d} - 1\right) & \text{for } \eta_d \geq 1
\end{cases}
\tag{10E1.3}
$$

Some typical values of the parameters required for evaluating overall rates are indicated in Table 10E1.1. While these figures suggest appropriate orders of magnitude, constants for a specific system must be determined experimentally. One method for doing this is the biological film reactor, in which a liquid film in laminar flow passes over an inclined sheet. The thickness of microbial film living on the sheet is controlled by mechanical scraping. Additional details on this technique are described in Ref. [9].

Submerged fermentation for vinegar manufacture had to await special designs for uniform air distribution throughout the vessel. Deprived of oxygen for 10 to 15 s, about one-third of a submerged *Acetobacter* culture will die. Details of the Acetator and Cavitator fermentors developed for this process are given in the references. The Cavitator system is one of the few commercially important continuous fermentation systems.

Several other oxidation processes are conducted commercially with the aid of microorganisms. In our discussion of fermentation kinetics in Chap. 7, we have already seen some of the features of batch production of gluconic acid from glucose (see Fig. 7.28). Gluconic acid and its derivatives enjoy numerous applications in the food, feed, and pharmaceutical industries. A two-step process is involved in its formation: glucose reacts with oxygen to give hydrogen peroxide and δ-gluconolactone, which is then hydrolyzed to yield the acid. Selected strains of the mold *Aspergillus niger* are employed for this fermentation; chemical methods

Figure 10.12 Some steroid transformations mediated by microorganisms. (*Reprinted from M. Frobisher, "Fundamentals of Microbiology, 8th ed., p. 591, W. B. Saunders Co., Philadelphia, 1972.*)

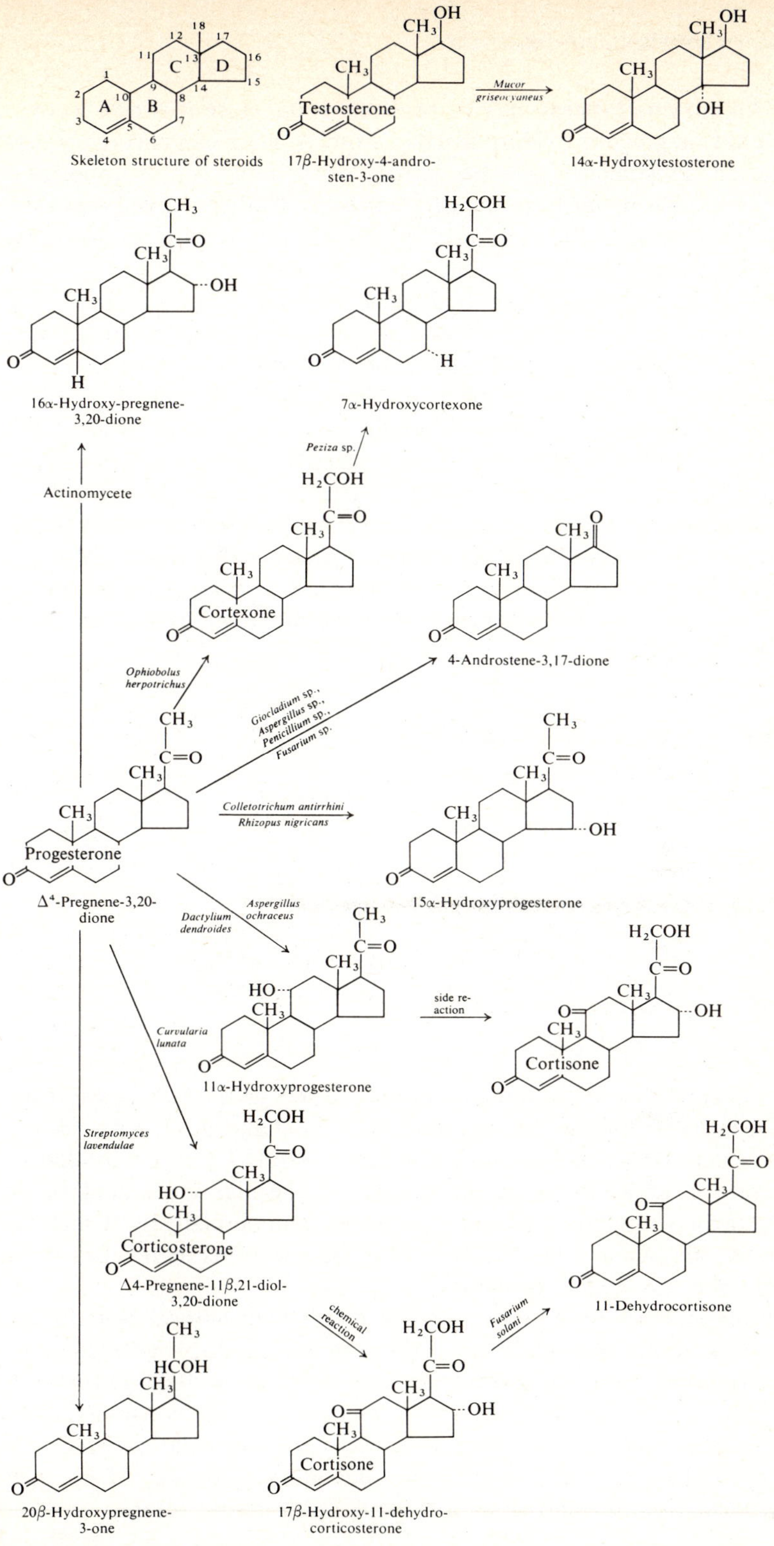

Skeleton structure of steroids

Testosterone
17β-Hydroxy-4-andro-
sten-3-one

Mucor
griseocyaneus

14α-Hydroxytestosterone

16α-Hydroxy-pregnene-
3,20-dione

7α-Hydroxycortexone

Peziza sp.

Actinomycete

Cortexone

4-Androstene-3,17-dione

Ophiobolus
herpotrichus

Gliocladium sp.,
Aspergillus sp.,
Penicillium sp.,
Fusarium sp.

Colletotrichum antirrhini
Rhizopus nigricans

15α-Hydroxyprogesterone

Progesterone

Δ⁴-Pregnene-3,20-
dione

Aspergillus
ochraceus

Dactylium
dendroides

11α-Hydroxyprogesterone

side re-
action

Cortisone

Curvularia
lunata

Streptomyces
lavendulae

Corticosterone
Δ4-Pregnene-11β,21-diol-
3,20-dione

chemical
reaction

Fusarium
solani

11-Dehydrocortisone

20β-Hydroxypregnene-
3-one

Cortisone
17β-Hydroxy-11-dehydro-
corticosterone

595

also are used. An enzyme process may be practical soon, since it is now known that the single enzyme glucose oxidase isolated from *A. niger* can effectively catalyze the oxidation. Presumably catalase activity of the mold is also important, since hydrogen peroxide denatures glucose oxidase and other enzymes. It thus appears that the entire role of the mold in this fermentation route is to produce a few enzymes which permit the desired reaction to occur.

A similar situation appears in microbial transformation of steroids, physiologically important molecules with a common structural skeleton (Fig. 10.12). The unique feature of microorganisms utilized for steroid production is that only a slight transformation of a given substrate is desired, and this change is often accomplished by a single enzyme particular to one microbial species, as indicated in Fig. 10.12. Eventually the particular enzymes effective for these reactions may be isolated, immobilized, and exploited industrially (recall Sec. 4.6.2). Typically, a combination of a chemical step before and/or following a microbiologically based step is required to bring a convenient basis steroid to the desired product. Present uses of such steroid products include treatment of rheumatoid-arthritis symptoms and inflammatory conditions (allergies, skin and eye diseases), as well as anesthetic and antifertility agents.

10.2.3. Organic and Amino Acid Manufacture

Let us return now to several other organic acids produced by the metabolic action of microorganisms on carbohydrates. More specifically, both citric acid and itaconic acid

$$
\begin{array}{ll}
\text{CH}_2\text{—COOH} & \text{CH}_2\text{—COOH} \\
\quad | & \quad | \\
\text{HOC—COOH} & \text{C—COOH} \\
\quad | & \quad \| \\
\text{CH}_2\text{—COOH} & \text{CH}_2 \\
\quad \text{Citric acid} & \quad \text{Itaconic acid}
\end{array}
$$

accumulate in either surface or submerged mold cultures as a result of interruption of the TCA cycle (recall the discussion of Secs. 1.3.3 and 5.3.3; it should be pointed out here that other mechanisms, including possibly C_1 to C_3 condensation, must operate in addition to the TCA cycle for the high citric acid yields observed in practice). Although citric acid is produced commercially by *A. niger*, it appears in almost all living cells. Itaconic acid, on the other hand, has been observed in only two *Aspergillus* species.

The market importance of these two acids parallels their natural abundance; citric acid enjoys widespread application in food preparation, pharmaceuticals, cosmetics, electroplating, tanning, and oil recovery. More than 100,000 tons are produced annually. Several million pounds of itaconic acid are consumed each year, primarily in the resin and detergent industries. The fermentation processes for these two acids share several characteristics including optimum pH < 2 and maximum acid formation during the stationary phase of batch culture.

In contrast to these aerobic processes, lactic acid results from anaerobic metabolism: it is the end product of Embden-Meyerhof glycolysis (recall Sec. 5.2). The chemical is used to provide acidity in food and beverages, to aid in food preservation, to delime hides, to treat fabrics, and to produce resins. It is expensive not because of fermentation costs but because separation from medium impurities is difficult. Bacteria, primarily *Lactobacillus* sp., are the organisms responsible for industrial lactic acid production. In the dairy industry, other lactic acid–producing bacteria are of great importance in both milk spoilage and cheese manufacture. Since mixed populations are typically involved in those processes, we consider them further in Chap. 12.

Numerous amino acids are produced on an industrial scale by fermentation processes. Most important from several perspectives is glutamic acid, the production of which exceeds 100 million pounds annually. A major use of this amino acid is manufacture of the flavor-enhancing agent monosodium glutamate (MSG), which is especially important in Oriental cuisine. Not coincidentally, the breakthrough discovery of the organism *Corynebacterium glutamicum*, which permits single-step glutamate synthesis from glucose in 30 to 50 percent yields, was accomplished in Japan by Kinoshita and coworkers. In addition to isolating this important bacterium, Kinoshita and other Japanese scientists elucidated the importance of altered metabolic controls and membrane permeability (Sec. 6.4.2). The commercial fermentation is conducted in submerged culture with aeration. In Fig. 10.13 the time course of a typical batch run is illustrated (for a different species); notice the use of intermittent ammonia addition to control pH, discussed earlier.

The same *Corynebacterium* species and its mutants have also proved effective in direct synthesis of other amino acids including lysine, valine, homoserine,

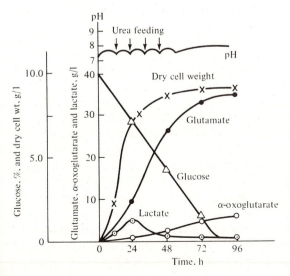

Figure 10.13 Time course of batch production of L-glutamate by *Micrococcus glutamicus* (25°C, initial medium composition: 10% glucose, 0.25% corn-steep liquor, 0.5% NZ amine, 0.1% K_2HPO_4, 0.025% $MgSO_4 \cdot 7H_2O$). (*Reprinted from K. Tanaka et al., J. Agri. Chem. Soc. Japan,* **34**: 593, 1960.)

phenylalanine, isoleucine, and tyrosine. Some of these can also be effectively produced by other microorganisms. Of these amino acids, lysine is next to glutamic acid in commercial production volume. Uses for amino acids include dietary supplements (see Sec. 10.3.1), flavor agents (several applications besides MSG exist), and treatment of disease.

10.2.4. Complex Molecules: Gibberellins, Vitamins, and Antibiotics

We need not discuss enzyme production here since Chap. 4 has already covered that topic. Emphasis in the remainder of this section will be on therapeutic compounds, including vitamins and antibiotics.

First we shall mention the gibberellins, hormones which can stimulate dramatic growth increases in plants (see Fig. 10.14). These compounds all have structures containing the multiple-ringed nucleus exhibited by gibberellic acid

Gibberellic acid

They are produced commercially in aerated submerged fermentations employing the fungus *Gibberella fujikuroi* with yields of the order of 1 g/l after about 400 h. An interesting batch process involving programmed changes in medium composition was patented by Imperial Chemical Industries in the late 1950s. In the first stage of the batch, the fungus grows actively in a nitrogen-limited medium with a C/N ratio from 10:1 to 25:1. The medium used during the final stage contains almost no nitrogen (C/N $\approx$ 40:1), so that growth is arrested. Gibberellic acid accumulates in this phase, and carbon in the medium is replenished by continuous or intermittent additions of a suitable carbon source. Besides providing enhancement of crop yields for cotton, grapes, celery, and other plants, gibberellins are also used to improve α-amylase activity in malt.

Vitamin B_{12}, which is found in almost all animal tissues and products, is apparently produced in nature predominantly by microorganisms. Some animals acquire their vitamin B_{12} supply by absorbing the compound as it is produced by organisms in their natural intestinal flora. Lacking this capability, human beings must acquire their entire B_{12} requirement in the diet. Only recently has chemical synthesis of vitamin B_{12} been accomplished, and the procedure is so lengthy and difficult that it appears to offer no threat to microbial production of the compound, the sole current method of manufacture.

Many protists can synthesize vitamin B_{12}. It appears as an important by-product in the streptomycin and Aureomycin fermentations and has also appeared in significant amounts in the sludge solid residue from biological waste treatment.

Figure 10.14 The growth of peas can be greatly increased by addition of gibberellic acid. (Left to right: greater amounts of gibberellic acid were added.) (*Photo courtesy of Abbott Laboratories.*)

As indicated in Table 10.4, the most productive process reported in the literature through 1967 uses a *Propionibacterium* in submerged culture. A two-step procedure is again employed in batch operation; during the first 2 to 4 days of anaerobic growth, a precursor is produced. This compound is subsequently transformed to vitamin B_{12} in 3 to 4 days in an intensely aerated culture. Although some decrease in yield is apparent, this concept has been extended to a two-stage continuous fermentation, in which the first stage is anaerobic and the second stage operates under aerobic conditions. Recovery of the product is complicated by the need to disrupt the cells, since vitamin B_{12} is an intracellular product. Although many other vitamins can be synthesized by microbes, only riboflavin (vitamin B_2) and the vitamin C precursor L-sorbose are produced by this method in commercial quantities.

Antibiotics are perhaps the most significant products of recent attempts to apply and modify microbial activity to meet human needs. The importance of antibiotics is multifaceted: they combat human and animal diseases and also help increase food supplies; with sales exceeding 1 billion dollars annually, their commercial impact is large; finally, biochemical engineering is in many ways the

Table 10.4. Characteristics of microbial processes for vitamin B$_{12}$ production†

Microorganism	Ingredient of medium	B$_{12}$ yield, mg/l	Comments
Bacillus megaterium	Beet molasses; ammonium phosphate; cobalt salt; inorganic salts	0.45	18-h fermentation (aerated)
Propionibacterium freudenreichii	Corn-steep liquor; glucose cobalt salt; maintained at pH 7 with NH$_4$OH	19	6-day fermentation (3 days aerobic + 3 days anaerobic)
P. freudenreichii	Corn-steep liquor (or auto-lyzed penicillium mycelium); glucose; cobalt salt; maintained at pH 7 with NH$_4$OH	8	Continuous two-stage fermentation; 33-h retention time
Propionibacterium shermanii	Corn-steep liquor; glucose; cobalt salt; maintained at pH 7 with NH$_4$OH	23	7-day batch fermentation (3 days anaerobic + 4 days aerobic)
Streptomyces olivaceus	Glucose; soybean meal; distillers' solubles; cobalt salt; inorganic salts	3.3	6-day fermentation (aerated)
Streptomyces sp.	Soybean meal; glucose; cobalt salt; K$_2$HPO$_4$	5.7	6-day fermentation (aerated)

† From H. J. Peppler (ed)., "Microbial Technology," p. 286, Reinhold Publishing Corporation, New York, 1967.

offspring of the development of antibiotic fermentations. To put this final point into perspective, a short historical digression on the development of the penicillin industry is in order.

Discovered in 1928 by a chance observation of the British bacteriologist Sir Alexander Fleming, penicillin was not pursued for potential therapeutic uses until the fire bombing of England in World War II. This prompted British biochemists to test the drug as an agent for burn treatment. Encouraged by successes in their trials, the English scientists enlisted the assistance of the USDA Northern Regional Research Laboratory (NRRL) in Peoria, Illinois. Several additional fortunate accidents led to dramatic increases in penicillin yields. Corn-steep liquor, a residual process product widely available to the Midwestern laboratory, provided an eightfold yield increase when added to the broth. Also, a worldwide search for more productive strains of the *Penicillium* molds responsible for synthesis of the antibiotic culminated in a moldy cantaloupe found in a Peoria food market. This species, denoted NRRL-1951, has since been further improved by mutations.

To ensure availability of some of the drug during the early 1940s, a simple scale-up practice was used initially. Bottle plants which daily handled about 10^5 1-l bottles, each containing about 200 ml of medium, were used for surface culture of the mold. These plants were built in caves and other environments of suitable temperature. Eventually, air filtration and other aspects of aseptic practice were perfected by engineers, who, in collaboration with microbiologists developing organisms for submerged culture, made possible in 1943 the transition to the first large-scale (originally about 10,000-gal) aseptic, aerated submerged fermentations ever practiced. Many of the process engineering features described in the previous section were developed during this period and have changed little since.

There are a large number of useful penicillin compounds which share the common structure

Basic framework of penicillin

Differences between the various kinds of penicillins derive from differences in the R group. In Table 10.5, several types of penicillins are listed, along with their characteristic R group and other properties. Of those listed, only penicillin G is a natural product. The others belong to the class of semisynthetic penicillins produced mostly after 1959. At that time, the enzyme penicillin amidase was discovered, which could remove the R group from penicillin G, leaving the 6-aminopenicillanic acid core. Extensive tests of synthetically added R groups led to the useful penicillin varieties described in Table 10.5 and many others. Notice in the table that some of these semisynthetic penicillins resist degradation by penicillinase, an enzyme which is produced in an inducible defense mechanism by many bacteria and which destroys natural penicillins. This mechanism possessed by bacteria is the major reason for the extreme degree of asepsis required in the penicillin fermentation. Some of the semisynthetic compounds also exhibit significant activities against gram-negative bacteria, which are unaffected by the natural penicillins. The bacteriocidal action of penicillin is due to interference with bacterial cell-wall synthesis. Since no such synthesis reactions occur in the metabolism of mammals, they are not affected by the drug except for occasional allergic reactions.

Example 10.2. Temperature programming for optimal penicillin production It has been known for some time that high temperatures (30°C) maximize growth rate of the *Penicillium* mold while lower temperatures (20°C) are more favorable for high rates of penicillin synthesis. Although operation at a temperature between these extremes (24 to 25°C) has predominated in past commercial practice, intentional variation of the temperature during the batch, i.e., temperature programming, can conceivably give larger penicillin yields than any constant temperature. Standard mathematical procedures for obtaining the best temperature program have been recently applied to this problem [10] with interesting results which are summarized next.

In order to apply optimal control theory, a mathematical model is necessary. The crosses in Fig. 10E2.1 are experimental data from commercial fermentors operated at 25°C. To avoid any revelation of proprietary-process characteristics, these data were reported only in the nondimensional form shown. These plots do not show the lag phase: the first data point is about 50 h into the fermentation.

Table 10.5. Chemical structures and properties of three different types of penicillin†

R group	Chemical name	Trivial name	Oral absorption	Activity against					Limitations
				Staphylococcus	Penicillinase-producing Staphylococcus	Other cocci	Gram-negative bacilli		
CH_2- (phenyl)	6-(Phenylacetamido)penicillin	Penicillin G	Poor	High	Nil	High	Nil		Variable absorption
(phenyl isoxazole structure, C—C=C, N—O, CH₃)	6-(5-Methyl-3-phenyl-4-isoxazole-carboxamido)penicillin	Oxacillin	High	High	High	Variable	Nil		Variable absorption; narrow spectrum
(phenyl H—C—NH₂)	6-[D(−)-α-Aminophenyl-acetamido]penicillin	Ampicillin	High	High	Nil	High	Medium		Gram-negative bacilli may become resistant during therapy

† From G. T. Stewart, "The Penicillin Group of Drugs," Elsevier Publishing Company, Amsterdam, 1965.

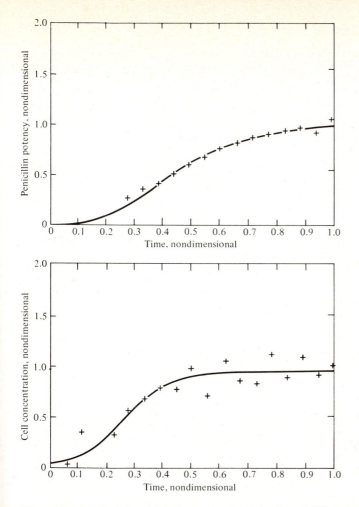

Figure 10E2.1 Experimental data for a commercial batch penicillin fermentation at 25°C. (*Reprinted from A. Constantinides, J. L. Spencer, and E. L. Gaden, Jr. Optimization of Batch Fermentation Processes. I. Development of Mathematical Models for Batch Penicillin Fermentations, Biotech. Bioeng.,* **12:** 803, 1970.)

The trends in these data suggest the following general forms of growth and product-formation kinetics:

$$\frac{dx_1}{dt} = b_1 x_1 \left(1 - \frac{x_1}{b_2}\right) \tag{10E2.1}$$

$$\frac{dx_2}{dt} = b_3 x_1 - b_4 x_2 + b_5 \frac{dx_1}{dt} \tag{10E2.2}$$

where x_1 and x_2 are the nondimensional cell and penicillin concentrations, respectively. Equation (10E2.1) is the logistic equation familiar from Chap. 7, while (10E2.2) is of the Luedeking-Piret type, discussed before, with the addition of a term reflecting penicillin destruction. The need for this term is indicated by the flattening of the penicillin curve in Fig. 10E2.1 for large times. Also other data indicate that penicillin hydrolyses in aqueous solution and that the penicillin synthetic system may also decay with time.

Assuming various sets of values for parameters b_1 through b_5 and comparing the computed x_1 and x_2 time histories with the data show that the sum-of-squares residual for all data points is minimized when $b_5 = 0$ and

$$b_1 = 13.099 \triangleq b_{10} \qquad b_2 = 0.9426 \triangleq b_{20}$$
$$b_3 = 4.6598 \triangleq b_{20} \qquad b_4 = 4.4555 \triangleq b_{40} \tag{10E2.3}$$

The temperature dependence of these parameters can be estimated by combining those values at 25°C with (1) the known optimal temperatures for cell growth rate (30°C) and penicillin synthesis rate (20°C) and (2) experimental evidence showing Arrhenius dependence of the rate constant for penicillin decay with an activation energy of 12 to 15 kcal/mol. These facts suggest the forms

$$b_i(\theta) = b_{i0} g(\theta) \qquad i = 1, 2$$
$$g(\theta) = 1.143[1 - 0.005(30 - \theta)^2]$$
$$b_3(\theta) = 1.143 b_{30}[1 - 0.005(\theta - 20)^2] \tag{10E2.4}$$
$$b_4(\theta) = b_{40} \exp\left[-6145\left(\frac{1}{273.1 + \theta} - \frac{1}{298}\right)\right]$$

where θ is temperature in degrees Celsius.

Utilizing all the above equations, we can now cast the fermentation model in the general form

$$\frac{dx_i(t)}{dt} = f_i(x_1(t), x_2(t), \theta(t)) \qquad i = 1, 2 \tag{10E2.5}$$

with initial values

$$x_1(0) = 0.0294 \qquad x_2(0) = 0 \tag{10E2.6}$$

The objective of maximizing penicillin yield can now be precisely stated mathematically. Find θ as a function of t from $t = 0$ to the final fermentation time $t = T$ such that $x_2(T)$ is maximized. The maximum principle [10] asserts that for the optimal temperature program, which we shall denote θ^*, it is necessary that the Hamiltonian function H

$$H(\theta) \triangleq \sum_{i=1}^{2} \lambda_i^*(t) f_i(x_1^*(t), x_2^*(t), \theta) \tag{10E2.7}$$

be a maximum for $\theta = \theta^*(t)$ for all t between 0 and T. The superscript * in Eq. (10E2.7) denotes values obtained using θ^*, where the adjoint variables λ_i satisfy the differential equations

$$\frac{d\lambda_i}{dt} = \sum_{j=1}^{2} \lambda_j(t) \frac{\partial f_j(x_1(t), x_2(t), \theta(t))}{\partial x_i} \tag{10E2.8}$$

and the conditions

$$\lambda_1(T) = 0 \qquad \lambda_2(T) = 1 \tag{10E2.9}$$

These necessary conditions and associated theoretical results suggest the following iterative algorithm for computing the optimal temperature program:

1. Let $\theta^{(n)}(t)$ denote the nth guess for the program.
2. Solve Eq. (10E2.5) for $x_1^{(n)}(t)$, $x_2^{(n)}(t)$ using $\theta = \theta^{(n)}$.
3. Using $x_1^{(n)}$, $x_2^{(n)}$, $\theta^{(n)}$ to evaluate the time-varying coefficients on the right-hand side of Eq. (10E2.8), integrate those equations backward numerically from $t = T$ to $t = 0$ to determine $\lambda_1^{(n)}(t)$, $\lambda_2^{(n)}(t)$ (this reverse integration is necessary to avoid numerical instabilities often encountered if their integration is begun at $t = 0$).

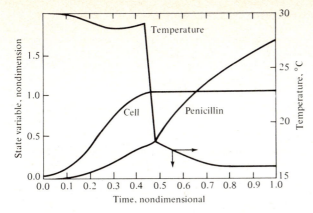

Figure 10E2.2 Computed optimal temperature policy and the corresponding growth and penicillin-production profiles. (*Reprinted from A. Constantides, J. L. Spencer, and E. L. Gaden, Jr., Optimization of Batch Fermentation Processes II. Optimum Temperature Profiles for Batch Penicillin Fermentations, Biotech. Bioeng.,* **12**: 1081, 1970.)

4. Compute

$$\frac{\partial H^{(n)}}{\partial \theta}(t) = \sum_{i=1}^{2} \lambda_i^{(n)}(t) \frac{\partial f_i(x_1^{(n)}(t), x_2^{(n)}(t), \theta^n(t))}{\partial \theta} \qquad (10E2.10)$$

5. Determine the $(n + 1)$st guess for the optimal program using

$$\theta^{(n+1)}(t) = \theta^{(n)}(t) + \varepsilon \frac{\partial H^{(n)}}{\partial \theta}(t) \qquad (10E2.11)$$

where ε is a small positive number.

Constantinides, Spencer, and Gaden employed this procedure with some adaptations to avoid decreasing cell concentrations. The final results are illustrated in Fig. 10E2.2. Notice that the temperature profile is initially large to maximize cell growth and then lower to optimize penicillin formation rate. The increase in penicillin yield provided by optimal temperature programming is quite dramatic: the yield is 76.6 percent larger than that obtained at the best constant temperature of 25°C.

Other models based on different data sets are formulated and optimized in the references mentioned earlier. In those cases penicillin-yield improvements of about 15 percent result from programmed temperature. The temperature variations prescribed by these calculations can be closely approximated in commercial practice with little added cost. Consequently, the combined tools of mathematical modeling and optimization theory should also find fruitful application for other fermentations.

On the order of 50 other antibiotic compounds are produced in commercial amounts by microbial synthesis, mostly by submerged-culture methods. Table 10.6 lists some of these antibiotics. The most important commercially are the streptomycins, tetracyclines, and penicillins. Although most antibiotics are made using molds, a few, including the streptomycins and polypeptide antibiotics (gramicidin, bacitracin), are synthesized by bacteria. In clinical application, the spectrum of antibiotic action must be chosen for optimum effectiveness. For example, while a broad-spectrum antibiotic is effective against very many bacterial species which might cause infection, it may also produce the undesirable effect of killing bacteria in the natural intestinal microbial flora, thereby upsetting digestion and food assimilation. Besides their familiar applications in treatment of

Table 10.6. Several types of antibiotics with source and major characteristic†

Antibiotic‡			
Common name	Trade name	Active against§	Source
Penicillin¶	···	Gram-positive bacteria; *Treponema Neisseria*	*Penicillium notatum*
Fumagillin	···	*Entamoeba histolytica*	*Aspergillus fumigatus*
Paramomycin	Humatin	*E. histolytica*	*Streptomyces rimosus*
Streptomycin	···	Mycobacteria; tuberculosis; gram-negative bacteria	*Streptomyces griseus*
Dihydrostreptomycin	···	Like streptomycin	Streptomycin; also some species of *Streptomyces*
Tetracycline	Achromycin	Broad-spectrum	Chlortetracycline
Oxytetracycline	Terramycin	Broad-spectrum	*Streptomyces rimosus*
Chlortetracycline	Aureomycin	Broad-spectrum	*Streptomyces aureofaciens*
Chloramphenicol	Chloromycetin	Broad-spectrum	*Streptomyces venezuelae*
Erythromycin	Ilotycin, Erythrocin	Broad-spectrum (not Enterobacteriaceae)	*Streptomyces erythreus*
Carbomycin	Magnamycin	Like erythromycin	*Streptomyces halstedii*
Oleandomycin	Matromycin	Broad-spectrum	*Streptomyces antibioticus*
Neomycin B¶	Flavomycin	Mycobacteria	*Streptomyces fradiae*
Viomycin	Viocin	Like penicillin	*Streptomyces floridae, Streptomyces funiceus*
Oligomycin		Fungi of plants	*Streptomyces diastato-chromogenes*
Amphotericin B	Fungizone	*Candida* sp.	*Streptomyces nodosus*
Kanamycin	Kantrex	Broad-spectrum	*Streptomyces kanamyceticus*
Nystatin		Pathogenic fungi	*Streptomyces noursei*
Cycloheximide	Actidione	Saprophytic fungi	*Streptomyces griseus*
Griseofulvin	Grifulvin	Pathogenic fungi	*Streptomyces griseus*
Bacitracin	···	Like penicillin	*Bacillus subtilis*
Polymyxin B¶	···	Gram-positive bacteria	*Bacillus polymyxa*
Pyocyanin††	···	Miscellaneous	*Pseudomonas aeruginosa*

† From M. Frobisher, "Fundamentals of Microbiology," 8th ed., p. 299, W. B. Saunders Company, Philadelphia, 1972.

‡ Several not listed here are valuable commercially, agriculturally, and horticulturally.

§ Not necessarily the only activity.

¶ Several of these antibiotics are in reality mixtures consisting of related compounds such as the penicillins, polymyxin A, B, C, D, carbomycin A and B, etc.

†† Not used medicinally. One of the first known antibiotic substances.

human disease and infection, over 3 million pounds of antibiotics are used in other ways. Added to animal stock feeds, some antibiotics stimulate greatly increased growth. Some antibiotics may also prevent spoilage or control plant diseases.

The cells of specialized animal tissues can often be separated from each other and attached to the surfaces of convenient solids to give biological film reactors known as *tissue cultures*. These cells retain their original functions for a finite number of cell generations, and thus the possibility of in vitro production of extracellular animal or human biochemicals is evident. *Tissue culture* is considered in Sec. 10.3.3 (tissue culture, vaccines); an example of hormone secretion from tissue culture appears in Example 10.4.

10.2.5. Some Undesirable Aspects of Microbial-Product Formation

Several negative effects of microbial products in the food and petroleum industry should be cited. Perhaps the most serious in the process industries is microbial corrosion: a 1957 estimate put the annual loss due to bacterial corrosion at 50 million dollars. Besides damage to pipelines, bacteria cause attack on steel tubing and casing in oil wells. The latter problem is quite serious owing to the difficulty of repair, so that precautions are needed such as bacteriocide injection and use of drilling mud treated to minimize microbial contamination. On the other hand, pipeline corrosion is harder to control but easier to repair.

The general mechanism of corrosion in an aqueous environment involves coupled electrochemical reactions. A corrosion cell is established when anodic and cathodic areas appear on the metal surface. At the anode, metal passes into solution as a positive ion with simultaneous release of electrons, which move through the metal to the cathode, where hydrogen is liberated. These steps are illustrated in Fig. 10.15, which describes a mechanism for anaerobic corrosion caused by the sulfate-reducing bacteria. The FeS produced is cathodic to metallic iron and so promotes growth of the corrosion cell.

Manifestation of anaerobic corrosion depends upon the material attacked. With cast iron, graphitization results: the metal loses much of its iron, leaving a crumbly material which will mark paper. Anaerobic microbial corrosion of steel leaves deep pitting covered by a black deposit rich in FeS.

The sulfate-reducing bacteria are also involved in plugging of oil wells. The permeability of the reservoir will be reduced by the presence of the bacterial cells themselves. Also important, however, is the precipitation of FeS, which may arise from two iron sources, sedimentary rocks or the well casing.

Several different microorganisms can cause corrosion under aerobic conditions. For example, the sulfur-oxidizing bacteria cause formation of sulfuric acid

(1)	$4Fe \rightarrow 4Fe^{2+} + 8e$	anodic reaction
(2)	$8H_2O \rightarrow 8H^+ + 8OH^-$	electrolytic dissociation of water
(3)	$8H^+ + 8e \rightarrow 8H$	cathodic reaction
(4)	$SO_4{}^{2-} + 8H \rightarrow S^{2-} + 4H_2O$	cathodic depolarization (bacteria)
(5)	$Fe^{2+} + S^{2-} \rightarrow FeS$	corrosion product
(6)	$3Fe^{2+} + 6OH^- \rightarrow 3Fe(OH)_2$	corrosion product
Sum:	$4Fe + SO_4{}^{2-} + 4H_2O = FeS + 3Fe(OH)_2 + 2OH^-$	

Figure 10.15 A proposed reaction sequence for bacterial corrosion under anaerobic conditions.

from H_2S

$$H_2S + O_2 \longrightarrow 2S + 2H_2O$$

$$2S + 3O_2 + 2H_2O \longrightarrow 2H_2SO_4 \qquad (10.6)$$

Besides its intrinsic corrosive properties, sulfuric acid produces an anodic cell, which further promotes attack. Sulfuric acid produced by microbial activity is also responsible for corrosion of concrete and stone.

By forming small, knoblike coverings, called *tubercles*, on the metal surface, the iron bacteria create anaerobic regions which are anodic relative to the aerobic environment of the neighboring surface (see Fig. 10.16). Corrosion in this situation can be greatly accelerated if sulfate-reducing bacteria invade the anaerobic microenvironment beneath the tubercle.

Numerous biocides—including oxidants, e.g., chlorine, reducing agents such as acrolein, and heavy-metal compounds—are available for control of these and other undesirable microorganisms. In selecting a biocide, it is important to go beyond its toxicity to the microorganisms and consider secondary effects. For example, if the process water is discharged into an open stream, use of a chlorinated-phenol biocide would be unacceptable.

Microbial spoilage and contamination of foods can take two major forms: either the microbe itself is an undesirable contaminant, or products of its metabolism damage the product. Examples of the former include growth of molds on bread, and the presence of bacteria which can later grow in body tissues and cause disease. Some of the digestive ailments which can be caused by living bacterial contaminants are salmonellosis, bacterial dysentary (shigellosis), perfringens poisoning, and vitriosis.

More common is spoilage of food caused by microbial by-products. Psychro-

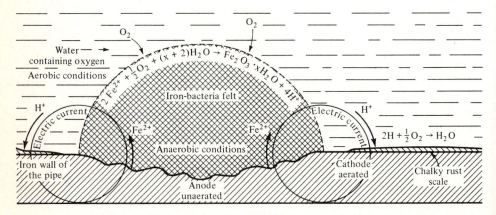

Figure 10.16 Tubercles formed by growth of iron bacteria on an iron surface create an anode-cathode pair and concomitant corrosion. (*After E. Olsen and W. Szybalski, Aerobic Microbiology Corrosion of Water Pumps. I., Acta Chem. Scand.,* **3:** 1094, 1949.)

philic bacteria are the primary causes of meat, fish, poultry, and egg spoilage, while vegetables and fruits are usually spoiled by yeasts and molds. Specific manifestations of microbial spoilage include putrefaction (characterized by unpleasant odors) of meat, development of ropes or slimes in bread and milk, and rots and disintegration of root vegetables and canned fruits.

To prevent spoilage, either freezing or preservative addition can be employed. Freezing apparently injures and kills bacteria by one or both of two mechanisms: (1) an increase in concentration of intracellular solutes as ice separates out, and (2) the intracellular ice crystals themselves cause damage. The action of common salt (NaCl), a familiar preservative, in deterring microbial growth is related to the first of these. By adding NaCl, the availability of water to the microbes is reduced. Already mentioned is the inhibiting effect which reduced pH has on bacterial activity. Consequently, citric and acetic acids are added to many foods to increase their resistance to spoilage. Usually such acid treatment must be supplemented by some other method of preservation, e.g., pasteurization or refrigeration.

Another class of microbial products, the toxins, can be present in food. These are especially serious because they cause acute illness or death. Staphylococcal poisoning, a short-lived but extremely unpleasant ailment, is commonly called ptomaine poisoning. Ham and ham products, in spite of their 2 to 3 percent salt content, are often responsible for transmission of the poison. Apparently, the salt eliminates inhibiting bacteria so that the *Staphlococcus* can grow well. Botulism poisoning, while much rarer, is still of concern because the mortality rate in the United States from such poisoning is about 57 percent. Fortunately, botulism can be well controlled by adequate cooking of all canned foods.

10.3. REACTORS FOR BIOMASS PRODUCTION

Microorganisms themselves are the important end products of several fermentation processes. Obviously fermentation enjoys a monopoly position for these products; no synthetic method yet devised can make living cells. When intentionally cultivated by man, the microbes produced are used either as a supplement to desirable natural microbial activity or as a food source. While we shall concentrate on beneficial applications below, growth of microbial populations can also create problems and costs in oil-production, refining, and waste-treatment industries. Microorganisms which proliferate in the water of cooling towers, for example, must be controlled to prevent clogging of the process. The disposal of excess cell sludge from a biological sewage-treatment process represents an important component of process cost.

An interesting case of Jekyll and Hyde behavior of protists in the context of current technology involves microorganisms using hydrocarbons as a carbon source. Often aqueous liquids are introduced into oil reservoirs in the course of production: certain microbes can proliferate in the aqueous regions, causing increased oil-flow resistance and occasionally even plugging the well. Any attempts to control microbial contamination in oil fields must consider secondary effects carefully. If chlorine is used as a control agent, for example, the petroleum produced later may have undesirable processing properties, since chloride ion reduces the lifetime of platinum catalysts. In a similar manner, microorganisms growing in an aqueous phase within jet-fuel tanks can clog the plane's fuel system. Since several losses of jet aircraft have been traced to microbial contamination, numerous precautions are now applied to ensure that the fuel stays "dry." Also, oil lubricants have been destroyed by microbial attack.

On the positive side, the presence of hydrocarbon-metabolizing bacteria in the soil is a clue to the presence of underlying gas and/or petroleum deposits. Many such deposits leak to the surface trace amounts of gases sufficient to maintain microbial life. Also, yeasts and bacteria for animal food applications are now grown commercially on hydrocarbon substrates. These fermentations, as well as more conventional carbohydrate-based microbial food processes, will be reviewed briefly.

10.3.1. Microbes for Food and Food Processing

One of the oldest examples here, the yeasts, have served the baker since the age of the pharaohs. Originally, desirable strains were maintained in the form of starter doughs, a portion of one dough batch which served to inoculate the next batch. This procedure was obviously not very reliable (why?), and *bakers' yeast* produced by fermentation is now used to prepare each batch of dough. From the latter nineteenth century to the earlier twentieth, bakers' yeasts were obtained from the dual-purpose Vienna process, in which a gently aerated yeast with grain malt yielded both alcohol and yeast. Pasteur's observations of microbial growth in the aerobic and anaerobic modes (recall Chap. 5) coupled with the advent of cheaper, synthetic methods for alcohol manufacture and improved aeration methods for fermentations led eventually to current process practice for manufacture of bakers' yeast.

The fermentation is aerated, typically at rates approaching one volume of air per volume of broth per minute: the maximum oxygen demand is of the order of 2 mmol/(l·min), and the critical oxygen concentration required to avoid alcohol production is about 0.2 ppm. Unfortunately, under aeration rates best for effective yeast propagation, only about one-tenth of the oxygen in air sparged into the vessel is utilized by the growing yeast. The favored medium for the bakers'-yeast fermentation is molasses, which is added to the broth along with mineral supplements during the batch process. The particular organism used is a strain of *Saccharomyces cerevisiae* chosen to optimize both the manufacturing process and yeast marketability. Thus, a suitable yeast should grow rapidly to high yields, should be easy to package in one of several standard formulations (usually compressed yeast cake or active dry yeasts), and should promote vigorous sugar fermentation in dough.

The dough-rising step in baking reveals some interesting interactions between microbial and enzyme activity. Amylases in the flour break down starch and thereby supplement the sugars dissolved in the flour. In utilizing these sugars through the Embden-Meyerhof pathway, the yeasts produce CO_2 and ethanol under the anaerobic conditions which prevail within the rising dough. Biosynthesis of inducible enzymes may be involved here, since sucrose is the primary sugar in the molasses manufacturing medium while maltose predominates in flour. Side products of yeast activity during dough rising include glycerol and higher alcohols, which may contribute to bread flavor.

Dried yeasts are the most common form of microbial food, rich in proteins and B vitamins. The greatest market for these products is for animal-feed supplements although dried yeasts or yeast extracts are sometimes used in moderate amounts as a fortifier for human food. While a typical formula for mixed poultry feed uses 50 lb dried yeast per ton of feed, yeast supplements for human food are usually of the order of 15 g/day for an adult. The amounts in human food are kept small because objectionable tastes and digestive disorders can arise if too much is used, and also because high nucleic acid content of yeasts could raise the level of uric acid, which can lead to gout (uric acid deposition in joints) and urinary-tract uric acid stones. An upper limit of 2 g nucleic acid per day has been suggested.

Because emphasis has been placed on microbial food as a protein source or supplement, it is worthwhile to consider briefly human protein requirements in terms of quantity and quality. (Similar considerations would also apply to animal nutrition.) From nitrogen-balance studies, a joint committee of the Food and Agricultural Organization and World Health Organization (FAO/WHO) has concluded that, for an adult, 0.59 g of protein per kilogram of body weight per day requires replacement; others have recently suggested a value of 0.43 g of protein per kilogram of body weight per day. Several factors must be considered when computing the amount of dietary protein to meet this requirement: individual differences suggest an increase of a 20 to 30 percent safety factor above this average amount in order to meet the protein needs of almost all of the adult population. Larger amounts are necessary for children, pregnant or lactating women, and possibly people doing heavy manual labor in hot climates. (How do these people differ from average? Just consider an unsteady-state nitrogen material balance on the body.)

Most interesting from the aspect of microbial food applications is the matter of protein quality or efficiency. Usually, efficiency is closely correlated with the protein's amino acid proportions. These proportions can vary widely, as shown in Table 10.7 for beef and a vegetable protein. In both cases, the amount of each essential amino acid is shown relative to the amount of that amino acid in whole-egg protein, a very efficient form of protein which serves as a base for this chemical evaluation (see Table 10.8 for egg-protein composition). The essential amino acid in the list with the lowest amount relative to egg protein determines the *chemical*

Table 10.7. Distributions of essential amino acids in beef and wheat†

Essential amino acid	Concentration in whole-egg protein, %	
	Beef protein	Wheat gluten
Histidine	154	94
Isoleucine	84	74
Leucine	87	80
Lysine	141	33
Methionine	84	52
Phenylalanine	69	89
Threonine	90	57
Tryptophan	89	68
Valine	73	61
Chemical score	69	33
Biological quality (BV)	67	42

† From R. I. Mateles and S. R. Tanenbaum (eds.), "Single Cell Protein," vol. 1, p. 35, MIT Press, Cambridge, Mass., 1968.

score for the entire protein; all other amino acids in the protein are presumed underutilized.

Also indicated is the *biological value* (BV) for each protein. This refers to the percentage of ingested protein which is eventually utilized as body protein. While this is the number actually needed to compute dietary protein requirements

Daily protein requirement

$$= \frac{\text{body requirement} \times 1.2 \text{ factor for population diversity}}{\text{BV} \times 0.01}$$

Table 10.8. Comparison of the amino acid distribution in egg, yeasts, and other proteins, mg/g N†

Amino acid	Torula yeast	Brewers' yeast	Casein	Cotton-seed	Soybean	Egg
Tryptophan	86	96	84	74	86	103
Threonine	315	318	269	221	246	311
Isoleucine	449	324	412	236	336	415
Leucine	501	436	632	369	482	550
Lysine	493	446	504	268	395	400
Total sulfur amino acids	153	187	218	188	195	342
Phenylalanine	319	257	339	327	309	361
Valine	392	368	465	308	328	464
Arginine	451	304	256	702	452	410
Histidine	169	169	190	166	149	150

† From R. I. Mateles and S. R. Tanenbaum (eds.), "Single Cell Protein," p. 94, vol. 1, MIT Press, Cambridge, Mass., 1968.

it cannot be determined without extensive, expensive experiments on a human population. Consequently the chemical-score method, which requires only a laboratory analysis of the protein, provides a useful working tool for designing protein diet components.

Notice in Table 10.7 that the low lysine concentration in wheat gluten results in protein waste, unless lysine is provided elsewhere in the diet (e.g., by high lysine corn). Consequently addition of small amounts of lysine would greatly improve utilization of the other components for populations existing almost exclusively on wheat as a protein source. Herein lies a rich prospect for future amino acid fermentations (see Sec. 10.2) as well as impetus for development of single-cell protein (SCP), a term which suggests the use of isolated microbial protein but usually refers to utilization of entire microorganisms as food components. While somewhat of a misnomer, SCP is now widely used in the media and trade literature as a designation for microbial food. Table 10.8 reveals, for example, that torula food yeast is rich in lysine and many other essential amino acids. The importance of the yeast strain employed is indicated by the differences between torula and brewers' yeast. Effects of cultivation conditions will be mentioned later.

In addition to their high protein concentrations (dried yeast, for example, contains about 45 percent protein), microorganisms grow at rates far exceeding macroscopic plants and animals. The classical comparison goes as follows: while a 1000-lb steer can produce 0.9 lb of protein per day and the same amount of soy

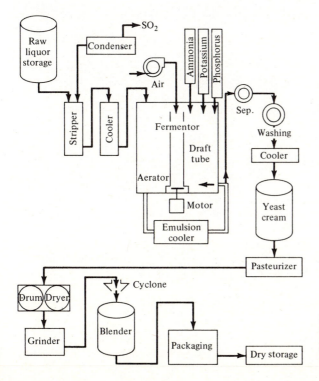

Figure 10.17 Schematic of the continuous process used for torula yeast production on sulfite-waste liquor.

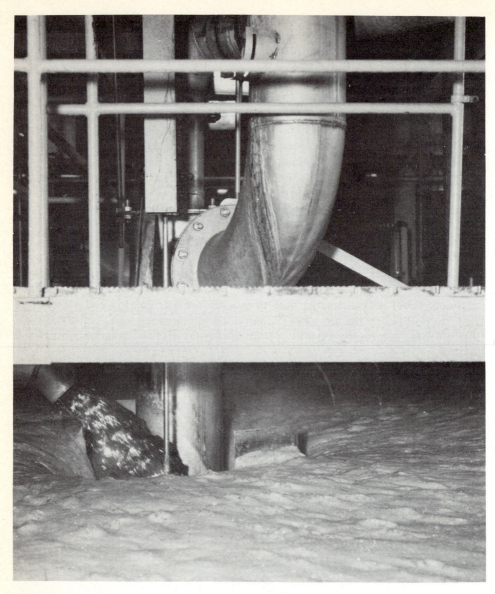

Figure 10.18 Appearance of the broth in a commercial continuous fermentor for torula yeast growth. (*Photo courtesy of the Rhinelander Division of St. Regis Paper Co. Reprinted from L. E. Casida, Jr., "Industrial Microbiology," p. 368, John Wiley & Sons, Inc., New York, 1968.*)

plant gives about 82 lb/day; 1000 lb of yeast can multiply to yield over 50 tons of protein in 1 day under ideal conditions. Even more dramatic differences could be cited based on bacterial populations, some of which can double in mass every 20 min.

As the name suggests, brewers' yeast is a by-product of beer manufacture. Its recovery often involves several washes in order to strip hop resins (causing objectionable flavor) from the yeast. The other major form of food yeast is grown in sulfite-waste liquor. *S. cerevisiae* (brewers' yeast) is not suitable for this process because it cannot effectively utilize the large concentration of pentoses in the sulfite liquor. On the other hand, *Candida utilis*, or torula yeast, grows well on this carbon source and consequently is the organism of choice for the sulfite-waste liquor fermentation. Torula-yeast production is one of the few commercial fermentations operated as a continuous process. Utilizing a Waldhof-type fermentor with an interior draft tube, an external recirculation loop for cooling, and a typical capacity of 60,000 gal, the process requires continuous substrate feed at the rate of 100 gal/min (see Fig. 10.17). The superficial appearance of the broth shown in Fig. 10.18 should accentuate earlier comments about mixing and aeration problems in fermentation systems.

Microbial protein derived from petroleum fractions is of increasing interest, as recent announcements of 50-million-dollar plants with capacities of 10^6 metric tons per year attest. Most processes producing protein from petroleum fractions announced to date employ either yeasts (one is *Candida tropicalis*) or bacteria (often *Pseudomonas aeruginosa*) grown on purified *n*-alkanes containing 10 to 18 carbon atoms; methane-based processes are also known. Providing the paraffin feedstock is highly purified, no toxic or carcinogenic compounds have yet been identified in the product cells. The proposed biochemical mechanisms of hydrocarbon utilization will be found in the references; in any case aerobic conditions are required in order to oxidize the highly reduced carbon in these substrates.

In fact, the oxygen requirements as well as heat-removal duty needed in hydrocarbon fermentations (see Table 10.9) considerably exceed the demands in a

Table 10.9. Oxygen addition and heat-removal needs vs. hydrocarbon fermentations†

Micro-organism	Substrate	Yield, %	O_2 required, g/100 g micro-organism	O_2 factor	Heat evolution, kcal/100 g microorganism	ΔH factor
Yeast	$(CH_2O)_n$	50	67	1	383	1
Yeast	$(CH_2)_n$	85	242	3.6	985	2.6
		100	196	2.9	780	2.0
		115	152	2.3	632	1.7
Bacteria	(CH_4)	100	253	3.8	964	2.5
Bacteria	$(CH_2)_n$	100	172	2.6	780	2.0

† From R. I. Mateles and S. R. Tanenbaum (eds.), "Single Cell Protein," p. 334, MIT Press, Cambridge, Mass., 1968.

carbohydrate fermentation. Approximately 2–3 times as much O_2 is required in the hydrocarbon case. A partially compensating factor is the mixing activity provided by the rising air in such a highly aerated system (about three volumes of air per volume of medium per minute); hence mechanical mixing is not especially advantageous, and the airlift- or tower-fermentor designs are used. Assuming typical fermenter cooling-system design, the heat load indicates a mean ΔT of about 25°F between culture medium and cooling water. Thus, operation of a hydrocarbon fermentation at, say, 86°F in a tropical zone requires refrigeration.

Operating conditions have significant impact on the quality of the microbial product. If the cells are grown at high rates, their nucleic acid content will be larger and cause the toxicity problems mentioned earlier. Oxygen or hydrocarbon

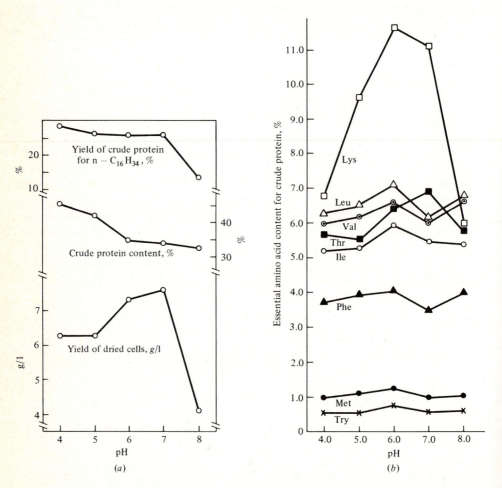

Figure 10.19 (*a*) overall and (*b*) detailed influences of pH on a hydrocarbon fermentation (growth of *Candida tropicalis* on 1% *n*-hexadecane, 0.5% NH_4NO_3 after batch growth for 72 h.) (*Reprinted from J. Takahashi et al., Agric. Biol. Chem.,* **27**: 836, 1963.)

supply can be used as the growth-limiting factor. Usually the latter is preferred since excess hydrocarbon often leads to high fat content and difficulties in separating the organism from the oil and aqueous phases. In Fig. 10.19 several interesting influences of environmental pH on the quantity and quality of petroprotein are apparent. While consistent trends in overall protein content and yield with pH are evident, the data reveal a clear maximum in lysine content and oscillatory behavior for several other essential amino acids. In view of the strong relationship between amino acid mix and protein value, the details of cellular composition are of great practical importance here.

Challenging problems arise in the design and analysis of hydrocarbon fermentations because of the very low solubility of hydrocarbons in water. For example, the solubility of decane is 16 ppb at 25°C. Consequently, the oil phase exists in the medium primarily as dispersed immiscible droplets. Thus, hydrocarbon fermentations involve a four-phase system: aqueous phase, suspended microorganisms, air bubbles, and hydrocarbon droplets. Some elements required for an engineering analysis of such systems are described in the following example.

Example 10.3. A batch-growth model for liquid-hydrocarbon fermentations When a second liquid phase such as a normal-paraffin substrate is introduced into the fermentor, several problems in mathematical modeling arise beyond those already considered. In such a situation we must address the following new questions:

What is the drop-size distribution of the dispersed hydrocarbon phase, and how does it affect microbial growth? Does growth alter the drop sizes?

Where do the growing cells consume the hydrocarbon? In the aqueous phase, where it is scarcely present? Perhaps at the hydrocarbon-water interfaces in the system?

Is the dispersed phase all substrate, or does it contain substantial inert components in which substrate is dissolved?

How does continuous culture change these relationships?

Many of these matters have been explored in an extensive series of papers by Erickson, Humphrey and Prokop [13] and others. Here we shall consider a slightly modified version of their analysis of growth in a batch fermentor when substrate is dissolved in the dispersed phase. Moreover, we shall assume that the substrate concentration s in the dispersed phase is sufficiently small to ensure that its utilization does not alter the interfacial area between the dispersed and continuous phases. Also, the specific growth rate will be presumed to depend on the substrate concentration following the Monod equation (recall Sec. 7.1.2). We shall suppose that substrate is also present in the continuous phase and that equilibrium exists between the continuous- and dispersed-phase substrate, so that

$$s = K_e s_c \tag{10E3.1}$$

where s_c is the substrate concentration in the continuous phase and K_e is an equilibrium constant.

Our final assumptions are that the vessel is well mixed, that the dispersed phase can be adequately represented as a collection of equal-sized droplets with volumes V_d, and that cells exist only at the droplet outer surfaces until these interfaces become completely covered.

We shall let n denote the number of cells in the reactor volume. If the inoculum is small enough for the continuous-dispersed phase interface initially not to be saturated with cells, the first portion of the fermentation is described by

$$\frac{dn}{dt} = \frac{\mu_{max} s n}{K_s + s} \tag{10E3.2}$$

$$\gamma V_d \frac{ds_c}{dt} + V_d \frac{ds}{dt} = \left(\frac{\gamma}{K_e} + 1\right) V_d \frac{ds}{dt} = \frac{1}{Y}\frac{dn}{dt} \tag{10E3.3}$$

where Y is the cell number produced per unit amount of substrate consumed and γ is the volume of continuous phase per unit volume of dispersed phase. Thus, for constant Y,

$$Y = \frac{n - n_0}{V_d(s_0 - s) + \gamma V_d(s_{c0} - s_c)} \tag{10E3.4}$$

where subscript zeros refer to conditions at $t = 0$, the start of the batch process. Using Eqs. (10E3.4) and (10E3.1) to eliminate n and s_c, respectively, from (10E3.3) gives

$$V_d \frac{\gamma}{K_e} \frac{ds}{dt} = \frac{\mu_{max} s[n_0 - Y V_d(1 + \gamma/K_e)(s_0 - s)]}{Y(K_s + s)} \tag{10E3.5}$$

which can be readily integrated to obtain the dimensionless form

$$\frac{\alpha}{\beta + 1}\ln x - \frac{\alpha + \beta + 1}{\beta + 1}\ln\frac{\beta + 1 - x}{\beta} = -\tau \tag{10E3.6}$$

where $x = \dfrac{s}{s_0}$ = dimensionless substrate concentration

$\tau = \mu_{max} t$ = dimensionless time

$\alpha = \dfrac{K_s}{s_0}$ = dimensionless Monod constant

$\beta = \dfrac{n_0}{s_0 \, Y V_d \left(1 + \dfrac{\gamma}{K_e}\right)} = \dfrac{\text{cell number in inoculum}}{\text{maximum cell number produced by growth}}$

The number of cells at the dispersed-phase surface can be computed during this initial stage of growth from Eq. (10E3.4) and (10E3.6). A change in the analysis is necessary, however, when the interface between dispersed and continuous phases is totally covered with microorganisms. If n_f is the number of cells required to cover all interfacial area and t_f the time at which n reaches this number, we have for larger times the mass balances

$$\frac{dn}{dt} = \frac{\mu_{max} s n_f}{K_s + s} + \frac{\mu_{max} s_c(n - n_f)}{K_s + s_c} \qquad t \geq t_f \tag{10E3.7}$$

$$\gamma V_d \frac{ds_c}{dt} + V_d \frac{ds}{dt} = \frac{1}{Y}\frac{dn}{dt} \qquad t \geq t_f \tag{10E3.8}$$

Equaton (10E3.7) takes into account growth in the continuous phase of the cells which reside there.
Letting the subscript f denote evaluation at $t = t_f$, using Eqs. (10E3.1), (10E3.4), and (10E3.7) again in (10E3.8) gives the single ordinary differential equation

$$V_d\left(1 + \frac{\gamma}{K_e}\right)\frac{ds}{dt} = \frac{-\mu_{max} s n_f}{Y(K_s + s)} + \frac{\mu_{max} s[n_0 - n_d + Y V_d(1 + \gamma/K_e)(s_0 - s)]}{y(K_s K_e + s)} \qquad t \geq t_f \tag{10E3.9}$$

Next we integrate (10E3.9) subject to the initial condition $s(t_f) = s_f$ to obtain

$$A_1 \ln\frac{x}{x_f} + A_2 \ln\frac{x - \beta_1}{x_f - \beta_1} + A_3 \ln\frac{x - \beta_2}{x_f - \beta_2} = \tau - \tau_f \qquad \text{for } \tau \geq \tau_f \tag{10E3.10}$$

Here

$$\beta_{1,2} = \frac{\beta + 1 - \alpha \pm \sqrt{(\alpha + 1 - \beta)^2 + 4\alpha(K_e\beta_f + \beta + 1 - \beta_f)}}{2} \tag{10E3.11}$$

where

$$\beta_f = \frac{n_f}{YV_d(1 + \gamma/K_e)s_0} = \frac{\text{cell number to saturate interface}}{\text{maximum cell number produced by growth}}$$

and the + sign in (10E3.11) should be taken for β_1. The other constants appearing in Eq. (10E3.10) are given by

$$A_1 = \frac{\alpha^2 K_e}{\beta_1 \beta_2} \qquad A_2 = \frac{(\alpha + \beta_1)(\alpha K_e + \beta_1)}{\beta_1(\beta_1 - \beta_2)} \qquad A_3 = \frac{(\alpha + \beta_2)(\alpha K_e + \beta_2)}{\beta_2(\beta_2 - \beta_1)} \qquad (10E3.12)$$

Figure 10E3.1 shows batch growth curves computed from Eqs. (10E3.6) and (10E3.10) for the following reasonable parameter values:

$$\beta = 0.001 \qquad \beta_f = 0.05 \qquad \alpha = 0.001$$

The equilibrium constant K_e is the parameter which is varied to obtain the various curves: the largest values of K_e shown are still somewhat small for hydrocarbons, since K_e is of the order of 10^{-8} for C_{12} to C_{18} hydrocarbons. Consequently, we can assume for most hydrocarbon fermentations that growth occurs only at the interfaces between the dispersed droplets and the aqueous phase.

The simplified model based on this assumption describes several different sets of experimental data quite well. For example, the circles in Fig. 10E3.2 are experimental results for batch growth of a yeast, *Candida lipolytica*, on *n*-hexadecane. The solid line is based on the limiting form of the above model for $K_e \rightarrow \infty$ with parameters

$$\mu_{\max} = 0.236 \text{ h}^{-1} \qquad \beta = 0.06 \qquad \beta_f = 0.70 \qquad \alpha = 0.001$$

Notice that the vertical scale is linear rather than logarithmic, so that the straight-line portion of the growth curve between about 9 and 12 h is a stage of *linear* growth.

We encounter linear growth in other systems where growth is limited by surface area. In such cases, unless the surface area changes rapidly, the batch growth rate is approximately constant.

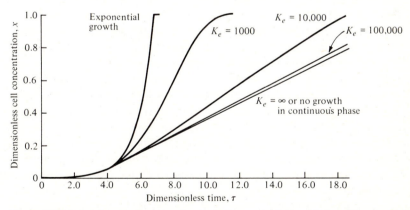

Figure 10E3.1 Batch growth curves for two-phase growth model for different values of limiting-nutrient equilibrium constant. Exponential growth curve is shown for comparison. (*Reprinted from L. E. Erickson, A. E. Humphrey, and A. Prokop, Growth Models of Cultures with Two Liquid Phases. I. Substrate Dissolved in Dispersed Phase, Biotech. Bioeng.,* **11**: 449, 1969.)

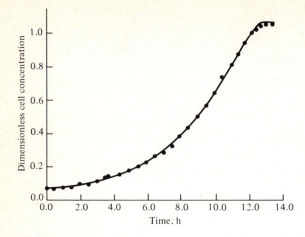

Figure 10E3.2 Comparison of experimental data (●) with calculated behavior (—) based on the assumption that all cell growth is on the oil-drop surface. (*Reprinted from L. E. Erickson, A. E. Humphrey, and A. Prokop, Growth Models of Cultures with Two Liquid Phases. I. Substrate Dissolved in Dispersed Phase, Biotech. Bioeng.,* **11**: 449, 1969.)

10.3.2. Agricultural Applications

The two instances of major concern here involve intentional supplementation of natural microbial activities. By far the most important commercially are the rhizobium bacteria, the organisms which enter the roots of young leguminous plants such as alfalfa, peas, and clover, and cause formation of nodules. Within these nodules the symbiotic action of bacteria and plant serves to fix atmospheric nitrogen into forms usable by the plant. In order to ensure satisfactory nodulation, rhizobium inoculants are now commonly added to the soil, impregnated into the legume seeds, or coated onto them. Different *Rhizobium* species or groupings of various strains are usually required for different types of plants.

Rhizobia are now produced commercially in aseptic, aerated, submerged fermentations. Sucrose or mannitol is provided as a carbon source while yeast extract supplies organic nitrogen and growth factors. Fermentation temperatures are usually from 30 to 32°C, and a 96-h batch run yields a population density of about 5×10^9 cells per milliliter. Although they are obligate aerobes, rhizobia grow well under oxygen tensions down to 0.01 atm. Presumably this low oxygen requirement allows the bacteria to survive within a root nodule.

Since many of the worst insect pests suffer diseases of microbial origin, the intentional use of insect pathogens as insecticides is a logical idea. The fact that only two microbial insecticides have found wide application is perhaps indicative of the stringent requirements on a biological agent for insect control. Among the prerequisites which a biological control candidate must satisfy are virulence against the target insect, rapid action to minimize crop damage, absence of negative effects on beneficial insects, predators, and higher animals, ease of commercial production, and stable storage properties.

Bacillus popilliae, used to control the Japanese beetle, is an obligate parasite and must be produced commercially in Japanese beetle larvae. On the other hand, *Bacillus thuringiensis* can be manufactured using conventional fermentation methods. Over 100 species of *Lepidoptera* and some *Diptera* suffer fatal diseases due to this bacterium. The toxic agent in *B. thuringiensis* is a crystalline protein which forms during sporulation. The high pH value and reducing agents in the gut of susceptible insects cause this toxic substance to dissolve, rendering the gut nonfunctional and causing death within 24 to 48 h.

10.3.3. Immunology, Tissue Culture, and Vaccine Production†

Physiological response to a number of infectious diseases includes the development of specific agents termed *antibodies*. These entities provide a greatly increased efficiency of combat against subsequent invasion by the same infecting organisms, thus conferring an immunity against a specific malady. Because the physiological lag time for production of appreciable amount of antibody is typically much shorter in the second and subsequent contacts with the infectious organisms, e.g., hours vs. many days, there is an enormous general health advantage to be gained by an initial *artificial immunization*. Such artificially initiated immunization generally involves exposure, through an infection or injection, to a mild level of *antigen*, the antibody-stimulating agents of the infecting species. The immunity conferred by physiological response to such mild artificial infection (injection) provides protection against much more rapidly propagating (more virulent) forms of the same or sometimes closely related organisms.

The antigen-containing immunization is usually one of three characteristic preparations:

1. Bacterial *exotoxins* in sterile media
2. *Dead* or *inactivated* microorganisms or virus
3. *Living* organism of *reduced* virulence

The bacterial exotoxins are simply extracellular products of bacterial life; except for extraordinary sterility requirements of the final product, reactor-design principles for such exotoxin production parallel those of other extracellular microbial products. Similarly, the cultivation of organisms of reduced virulence has been accomplished by carrying out growth under relative extremes of temperature or humidity as well as by selection of *attenuated* (milder-virulence) mutant strains. These last methods are difficult areas of study in immunology and genetics, but the principles relating to slow growth in unfavorable conditions and to mutation are those discussed in earlier chapters

For many examples of types 2 and 3, above, the desired product is achieved by propagating the infection in vitro (as opposed to in the live animal, in vivo) through utilization of *tissue cultures*, i.e., animal (or plant) cells which are propagated in artificial reactors. Examples of type 2 include the tissue-cultured poliovirus subsequently inactivated with mild heat and formaldehyde (Salk inactivated vaccine) and the production of the bacterialike but obligate parasites *rickettsias* by propagation of infected live cells from chick embryo (vaccine for Rocky Mountain spotted fever). The Sabin or oral polio "vaccine" is an example of immunization using a live, attenuated tissue culture. (Originally the term *vaccine* referred only to dead or inactivated organisms, but like the term *fermentation*, its general use has broadened.)

Cells from normal tissues possess several general properties which alter the conception of an appropriate reactor design from that for unicellular organisms which have thus far been our main concern:

1. The cells frequently grow only when adhering to a solid surface (successful candidates include glasses, plastics, cross-linked protein, Sephadex beads, titanium, collagen-coated surfaces, filter paper, and polyester films).
2. Through the phenomenon of *contact inhibition*, a feedback mechanism of some sort(s) halts growth upon achievement of a cell monolayer (chemical methods for eliminating such contact inhibition are known).

---

† The various tissue-culture configurations outlined in this section can be found in Ref. 6; specifically, bead supports [6, pp. 372–377]; artificial capillaries [6, pp. 321–327]; spiral coiled sheets [6, pp. 338–344]; and multiplate stacks [6, pp. 333–337].

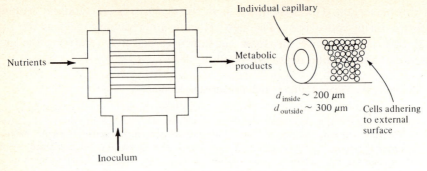

Nutrients →

Metabolic products

Individual capillary

$d_{\text{inside}} \sim 200\ \mu m$
$d_{\text{outside}} \sim 300\ \mu m$

Cells adhering to external surface

Inoculum

(a) Artificial capillary bundles (frequently ultrafiltration devices)

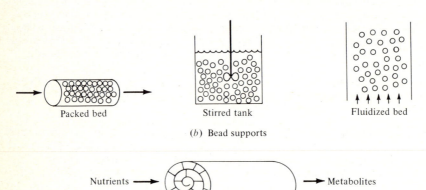

Packed bed

Stirred tank

Fluidized bed

(b) Bead supports

Nutrients → → Metabolites

(c) Coiled sheet (with spacers to maintain sheet separation)

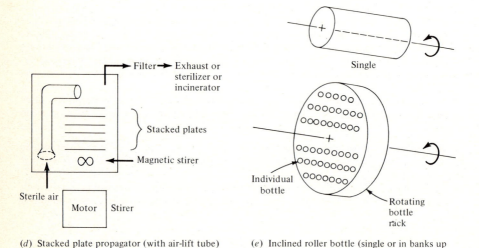

Filter → Exhaust or sterilizer or incinerator

Stacked plates

Magnetic stirer

Sterile air

Motor   Stirer

(d) Stacked plate propagator (with air-lift tube)

Single

Individual bottle

Rotating bottle rack

(e) Inclined roller bottle (single or in banks up to several hundred)

Figure 10.20 High-surface-area configurations for tissue culture.

3. The specialized tissue cells (being derived from a tissue with a specific body function) may often lose much of their original functional ability on repeated multiplication, giving rise to a less specialized, or dedifferentiated, cell.

For our purposes we consider only characteristics (1) and (2). The kinetics of item (3) is largely of concern in the long-term maintenance of in vitro cultures; it clearly poses a major barrier to the use of sustained tissue cultures for biochemical production (antibodies, hormones, enzymes).

Reactor configurations with larger surface-to-volume ratios are shown in Fig. 10.20. The utility and disadvantages of such systems are illustrated by considering the general series of operations for virus production:

1. The inoculum of tissue cells is prepared by chemical [ethylenediaminetetraacetic acid (EDTA)], mechanical, and/or enzymatic (trypsin, collogenase) separation of tissue into individual intact live cells.
2. The inoculation suspension of individual cells is contacted with the high-surface-area apparatus of choice, and the cells become attached to the solid surface.
3. By batch, discontinuous, or continuous addition of nutrient medium to the attached tissue cells, the organisms are cultured to the desired overall cell density.
4. The virus inoculum is mixed into the nutrient reservoir; the virus invades the cell.
5. In the lytic cycle (Chap. 6), the new virus produced is released upon lysis of the invaded cells, the cycle now repeating and eventually consuming all the cells. For cells in the lysogenic state, the final result is the mass of nonlysed cells.

The kinetics of the viral process is thus two serial problems: (1) growth kinetics of the initial attached tissue-cell inoculum to a sufficient density followed by (2) kinetics either of lysogenic cell growth or of the lytic cycle to produced unlysed cell mass or virus-containing filtrate, respectively. In the latter case, the final product is released directly into the liquid medium and is thus easily removed. The former case demands chemical, enzyme (often trypsin), or mechanical (scraping, etc.) removal of attached cells.

We consider the performance of two tissue culture configurations of Fig. 10.20 in the following examples.

Example 10.4. Production of a low to intermediate-molecular-weight product The production of human chorionic gonadotropin (HCG) (placental glycoprotein hormone) by human choriocarcinoma cells (placental tumor) represents the first human hormone for which an in vitro bulk producer has been identified (hamster cheeck pouch cells, to which the original human tumor cells have been transplanted). The rate of HCG production vs. time is displayed for two flow conditions in the artificial capillary (ultrafilter) configuration (Fig. 10.20) in Fig. 10E4.1. The production rates are observed to be logarithmic. The doubling rate of production is 1.2 days for the steeper curve; laboratory monolayer cell-growth kinetics gave similar cell-doubling times of 1.1 to 1.3 days in logarithmic growth. The production HCG may therefore be growth-associated, non-growth-associated, or a hybrid, as we recall from the discussions of Eq. (7.70) for the Luedeking and Piret model. The productivity of the 3-ml actual tissue culture volume equalled that of 217×10^6 cells in monolayer culture technique. As discussed in Chaps. 8 and 9, the interaction between mass transfer (or product) and fluid flow may disguise the apparent kinetics. The outside diameter of the reactor tube is the order of 300 μm $= 3 \times 10^{-2}$ cm. The characteristic relaxation time for species diffusing across this radius into the liquid flowing through the capillaries is thus

$$t_{\mathscr{D}} \approx \frac{(d/2)^2}{\mathscr{D}} = \frac{2.25 \times 10^{-4} \text{ cm}^2}{10^{-6} \text{ cm}^2/\text{s}} \approx 4 \text{ min}$$

assuming $\mathscr{D}_{HCG} \approx 10^{-6}$ cm^2/s. Thus, in the time span of Fig. 10E4.1 (3 to 4 weeks), the data points reflect the "instantaneous" product formation rate. (Note also for Fig. 10E4.1 the complex nutrient media needed. Use with antibiotics penicillin and streptomycin nutrients indicates the medium cannot be heat-sterilized without damage; mechanical filtration is useful for such media sterilization. A survey of extracellular products obtained from tissue cultures is found in Johnson and Boder [7].)

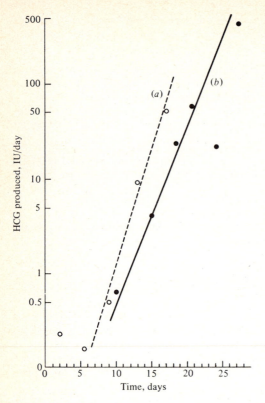

HCG produced, IU/day

Time, days

Figure 10E4.1 Production rate from tissue culture in an artificial capillary. (*a*) No extracorporeal medium replacement. Perfused with 5 ml/min of medium with 83.3 % of Ham's F-10, 13.5 % horse serum, 3.2 % fetal calf serum, 5000 units % aqueous penicillin, 5 mg % streptomycin. (*b*) With periodic extracorporeal medium replacement. Perfused with 0.7 ml/min of medium as in (*a*) plus 0.5 mg % insulin, 0.62 mg % cortisone acetate. (*Reprinted from R. A. Knazel and P. M. Gullino, Artificial Capillaries: An Approach to Tissue Culture* in vitro *in "Tissue Culture: Methods and Applications," p. 321, Academic Press, Inc., New York, 1973.*)

Example 10.5. Cell-growth and virus-propagation kinetics in tissue culture Small beads of porous glass or synthetically derivatized Sephadex (ion-exchange material, Chap. 4) provide high surface area per volume. In the latter case, the use of diethylaminoethyl (DEAE) Sephadex provides a positively charged surface on which negatively charged tissue cells adsorb easily. These nearly neutral buoyant beads require minimal stirring to remain well dispersed in a stirred-tank apparatus (Fig. 10.20*b*). A prior siliconization of the vessel surfaces prevents undesired cell-wall growth. Innoculation of 10^5 cells per milliliter to which are added 8 to 9×10^3 0.2-mm-diameter (wet) beads per milliliter gives a culture surface area of about 9 cm^2/ml [more than five times that of flat monolayer or roller-bottle culture technique (Fig. 10.20*e*)]. The increase in cell number, expressed as the ratio of cell number attached at time t divided by the number of cells inoculated into the bead medium at $t = 0$, is shown in Fig. 10E5.1*a*. The data indicate the absence of an appreciable lag time [the cells adhere over the first 24 h; as these particular cells propagate on solid surfaces *or* in suspension culture (absence of solid surfaces, same situation as typical fermentation), growth apparently proceeds for free and adhering cells] and a nearly logarithmic growth curve for the two oxygen tensions employed. Fig. 10E5.1*b* indicates the achievement of maximal polio virus titer within about 30 h of inoculating two different concentrations of tissue cells adherent to the microcarrier Sephadex beads. The initial cell levels in Fig. 10E5.1*b* appear to be the terminal values of Fig. 10E5.1*b* at ~ 165 h. The termination of virus increase corresponds to the end of the lytic cycle; these data indicate achievement of culture lysis at ~ 30 h but do not illuminate the kinetics of the viral mass appearance between 0 and 20 h (which may be more of a step change than the linear increase shown in this time interval).

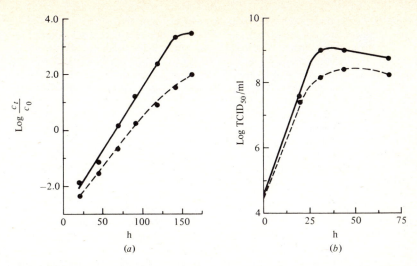

Figure 10E5.1 Production of cell number and viral titer vs. time. (a) c_t (cells attached at time t)/c_0 (cells inoculated at time $t = 0$) vs. time (———75 % O_2 saturation; ------5 % O_2 saturation). (b) $TCID_{50}$ ($\equiv$ virus dose yielding 50 percent tissue culture infection dose per milliliter) vs. time (———13 $\times$ 10^5 cells/milliliter initial cell concentration; ------3.5 $\times$ 10^5 cells/milliliter initial cell concentration). (*Reprinted from A. L. van Wezel, Microcarrier Cultures of Animal Cells, in "Tissue Culture: Methods and Applications,"* p. 372, *Academic Press, Inc., New York, 1973.*)

An example of the kinetics of such complete virus production during a lytic cycle for a poliovirus is shown in Fig. 10.21. During the adsorption and penetration (Fig. 10.21a), the total infectious concentration of virus (reflected in the number of plaque forming units of virus per milliliter of completely ruptured cell medium) drops rapidly since the viral DNA or RNA itself is not infective. In the *eclipse* period, the vegetative system synthesizes quantities of the separate virus parts. During maturation (Fig. 10.21c) assembly begins, the vast bulk of complete virions being assembled in the *latent* interval (Fig. 10.21d). Following completion of virion assembly, the cell is lysed at 21 h by enzymes synthesized under viral direction. Between 6 to 13 h, viral assembly could be expressed logarithmically, reaching its peak at just past 13 h. The kinetics of viral *appearance* in the extracellular fluid (Fig. 10E4.1b) is approximately a step function at 26 h. The propagation of such extracellular virus infection in a culture could thus resemble the kinetics of synchronous cultures (Chap. 7). (Lytic-bacteriophage kinetics often

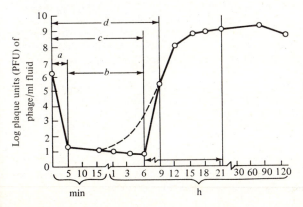

Figure 10.21 Growth of complete infectious virions (poliovirus) vs. time (see text discussion). Time periods: (a) virus adsorption and penetration; (b) eclipse period; (c) maturation; virion assembly beings, (d) latent; completion of assembly. (*Reprinted from M. Frobisher, "Fundamentals of Microbiology,"* 8th ed., p. 212, *W. B. Saunders Co., Philadelphia, 1972.*)

parallels that of Fig. 10.21, the time to cell lysis being the order of 1 h or less). Under other circumstances, virus and cell kinetics are different. Firstly, for the virally infected lysogenic state, the rate of cell replication may be different from that for the normal, uninfected cell. An example in between lytic cycle (virion production, cell death) and lysogeny (viral nucleic acid replication with cell DNA, no virion production) describes some animal-cell kinetics: virions are synthesized and released slowly (days, weeks) by individual passage through the cell membrane. Typically the virus receives its outer envelope from the cell wall in passage.

Of course the live animal (in vivo) itself, offers a convenient (and the earliest) means of vaccine preparation; e.g., smallpox vaccine is the collection of lymph resulting from infection of calfskin by cowpox, and rabies and rickettsial vaccines are producible by direct cultivation in live chick embryos. Such passage of the infectious agent through the cell lines of another (nonhuman) animal is an established method for evolving a modified agent less virulent to human beings (the new agent is frequently more virulent to the new host than the original). Tissue culture (in vitro) has been used to prepare "living" vaccines containing live cells for rabies, yellow fever, measles, and canine hepatitis; the best known example of such a tissue-culture "live" vaccine is the oral polio, or Sabin, vaccine.

The magnitude of the production of biochemicals and vaccines from tissue culture and the possible relation of continuous to batch operating modes are succinctly summarized as follows:†

The pharmaceutical industry is also concerned with a number of other microbiological processes which occupy relatively little fermenter capacity. Among them are the bioconversions of steroids, the production of minor antibiotics, and the production of bacterial enzymes such as penicillin acylase and the vaccines.

Many of these processes have been done in continuous culture at high output rates. The 11-hydroxylation of progesterone has been carried out in the second stage of a two-stage culture of *Rhizopus nigricans* giving a 50–60% yield with a 5-hour contact time. High yields and production rates of tetanus and diphtheria toxin have been obtained in chemostats. Several cell-bound antigens have been produced continuously at rates considerably in excess of the batch culture rates.

The microbiological steps of many of these processes often contribute only a small part of the total process cost. Moreover, the scale of the whole process is often insufficient to warrant the adoption, on the grounds of higher throughput, of a continuous step in what is otherwise a batchwise operation. In these processes conversion yields, product purity, or antigenic quality are more important than output rate. It is possible that the high degree of process control and the short residence times of chemostats will bring about improvements in product quality and thereby justify a continuous cultivation step.

Animal cells in tissue culture can be used for virus multiplication and the synthesis of hormones and other pharmaceutically interesting metabolites. The range of applications expands as medicine increasingly deals with aberrant metabolism on a molecular level. Chemostat techniques have been used in research but, for the present, the use of continuously growing cell lines for the production of material for human injection is prohibited. This prohibition does not apply to veterinary vaccines, though for the reasons given above the small scale of operation makes it unlikely that the continuous processes would be developed for industry. However, the economic potential of cultures of cells of higher organisms, both animal and plant, has hardly been touched and it is possible that new culture methods will need to be developed to exploit their functions.

It is sometimes argued that degeneration (reduction in the productivity of an organism on prolonged cultivation), will preclude the adoption of continuous processes for the synthesis of any product not essential to growth. Almost all of the pharmaceuticals fall into this category and indeed degeneration has proved a problem in antibiotic- and vaccine-producing cultures. Stable cultures are obviously desirable if maximum advantage is to be obtained from continuous

---

† R. G. Righelato and R. G. Elsworth: *Adv. Appl. Microbiol.*, **13**: 339 (1970).

processes and a variety of ways of avoiding the problem are suggested in the literature. Environment manipulation and techniques of reinoculation have proved successful. It is possible that some of these problems can be solved genetically.

Whether or not a process could, to advantage, be run continuously depends in part on the scale of operation. If a batch process fully occupies a plant, then, given a stable market, a continuous process would be worth adopting. The sphere in which continuous operation could prove most valuable is the large-scale antibiotic fermentations. Existing plants could fairly easily be adapted by providing a continuous sterile medium supply and sterile offtake. Most extraction trains already operate more or less continuously and little or no modification would be necessary. In such cases capital already invested in batch plants does not present a major obstacle to the use of continuous fermentation. Indeed the continuous antibiotic processes will not be adopted until substrate conversions and product concentrations compare favorably with the batch fermentations.

Much of the published research on continuous fermentations has not been primarily concerned with developing economic processes but with systems suitable for biochemical or physiological study. If they wish to adopt continuous methods, industrial researchers must be prepared to carry out their own empirical research as they have done with batch cultures, perhaps developing anew both strains and media. Given such effort there appear to be no fundamental reasons why continuous methods should not be adopted for those processes which at present fully occupy tank capacity on a batchwise basis.

We note especially that the successful continuous-culture examples in the second paragraph are microbial; the legal and technical problems in the third and fourth paragraph indicate a more difficult path for continuous tissue culture, the hopeful tone of the last paragraph notwithstanding.

From the topics encountered in the present chapter, it is clear that much of our limited understanding of microbial systems stems from a relative paucity of quantitative data based on fundamental biochemical paths and microbial kinetics. In the same vein but in a different realm, the behavior of reactors containing multiple species (mixed microbial populations) may be better appreciated by the mastery of certain elements of the theories for coupled (chemical or microbial) networks, the subject of the following chapter. As mentioned in the preface, the more informed biochemical engineer will have one ear open to biochemical developments and the other to analysis. The field of this text needs most of all the services of such a "bilingual" practitioner.

We defer discussion of the third predominant facet of microbial activity, namely substrate utilization, for our Chap. 12 review of processes with mixed microbial populations. In the next chapter, we introduce mathematical methods and models especially suited for analyzing mixed populations of microorganisms. Many of the concepts and methods pursued in Chap. 11 go beyond microbial ecology: they enjoy general application for describing interactions of higher organisms and even institutions and organizations.

PROBLEMS

10.1. Fluidized beds of immobilized enzymes (*a*) Chinloy [15] examined the conversion vs. space velocity θ (milliliter per gram of catalyst particles per second) obtained in packed and fluidized beds of protease immobilized on nonporous stainless-steel particles. In both cases, the data fell on the same straight line passing through the origin. Show that the data are not mass-transfer-influenced and that the specific rate constant can be evaluated directly from the data if the enzyme loading per catalyst particle (milligrams E per gram of catalyst) is known.

(b) Gelf and Boudrant [16] studied hydrolysis of benzoylarginine ethyl ester (BAEE) by papain immobilized in porous 170 to 250 μm particles. The following parameter values were reported:

Soluble papain: $K_m = 5 \times 10^{-3}$ M

$$v_{max} = 19 \text{ IU } [\mu\text{mol BAEE}/(\text{min·mg}) \text{ at pH 6.0 and 20°C}]$$

Immobilized papain: K_m (apparent) $= 1.2 \times 10^{-2}$ M

$$v_{max} = 0.05 \text{ IU } [\mu\text{mol}/(\text{min·mg of } support)]$$

The porous support was largely iron oxide particles. From the data in Fig. 10P1.1 determine whether these studies were influenced by external or internal mass transfer. State your assumptions. The catalyst charge in the fluidized bed was 10 g of particles; it was fabricated from a mixture including 100 mg crystalline papain per 30 g oxide.

(c) Discuss how you would design a fluidized-bed reactor of immobilized proteases for hydrolysis of (1) BAEE, (2) casein, (3) 1-μm-diameter gelatin particles.

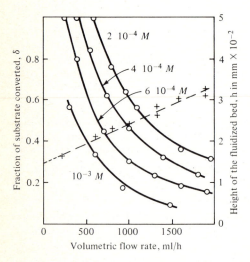

Figure 10P1.1 Fraction of substrate converted (O——O) and height of fluidized bed (+——+) vs. volumetric flow rate. Molarities of inlet substrate concentrations are indicated on the curves. (*From G. Gelf and J. Boudrant, Enzymes Immobilized on a Magnetic Support, Biochim. Biophys. Acta,* **334**: 468, 1974.)

10.2. Optically pure amino acids The process developed by Tosa *et al.* [17] can be represented schematically by Fig. 10P2.1. Assume for the moment that v_{max} is independent of pH. The initial racemic amino acid solution is acetylated by reaction with acetic anhydride; the L-aminoacylase column reverses the acetylation reaction for the L-amino acid, which is then crystallized in alcohol solutions.

(a) Assume that the initial amino acid concentration is $\gg K_m$, develop an expression for the fractional L-amino acid conversion achieved by the enzyme column in plug flow. Repeat including axial dispersion.

(b) The racemization reaction may be taken to be first order reversible, so that the rate is proportional to $c_{D \text{ acid}} - c_{D \text{ acid}}^*$, where $c_{D \text{ acid}}$ is equilibrium D acid level for the solution. If 90 percent of the L acid and 1 percent of the D acid is removed in the wet-crystal stream, along with 10 percent of the entering aqueous phase, what racemization CSTR volume is needed to achieve 95 percent approach to equilibrium?

(c) The enzymatic deacetylation step releases acetic acid into the solution. If the pK's leading to enzyme deactivation were $pK_1 = 5$ and $pK_2 = 8$, what entering pH would give maximum conversion for a 10^{-6} M, 10^{-4} M, or 10^{-1} M feed mixture? (Assume plug flow.) State your assumptions clearly.

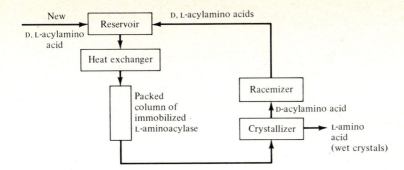

Figure 10P2.1 Catalyzed resolution of amino acids. (*From T. Tosa, T. Mori, N. Fuse, and I. Shibata, Enzymologia,* **31**: 225, 1966.)

10.3. Digestion of insoluble substrates As an example of processes involved with digestion of particulate substrates, the following unit-operation sequence for yeast growth on newsprint has been suggested: mechanical grinding, acid hydrolysis, medium neutralization, addition of additional minor nutrients for yeast growth, yeast fermentor (aerobic), vacuum filtration to separate liquid from cell mass.

(*a*) Sketch the flow scheme above, indicating by arrows points of addition and by circles each unit operation. Include solids conveyers and liquid-pump locations where needed.

(*b*) From any human physiology text, sketch the human food-digestion process in a similar manner.

(*c*) Bionics is the study of natural systems with an eye toward development of synthetic analogs. Discuss similarities and differences between processes (*a*) and (*b*). How might you design a solids handling scheme for part (*a*) using the "conveyer" type in part (*b*)?

10.4. Cell maintenance: instability at small D For some populations, a minimum level of substrate may be needed to achieve a nontrivial steady state. As an example, consider the system with kinetics of the form:

$$r_x = \frac{\mu_{max}\, sx}{K_s + s} - k_e x \qquad r_s = \frac{-1}{Y_s} \frac{\mu_{max}\, sx}{K_s + s}$$

Assuming that the design basis underlying Fig. 9.37 is valid:

(*a*) Show that at substrate level below $k_e K_s/(\mu_{max} - k_e)$ the only steady state in a CSTR system is $x = 0$.

(*b*) If $\mu_{max} = 0.5\,\mathrm{h}^{-1}$, $K_s = 0.2\,\mathrm{g/l}$, $k_e = 0.1\,\mathrm{h}^{-1}$, and $Y_s = 0.6$ g cell/g substrate, plot $(dx/dt)_{batch}$ vs. x and prove by direct solution of the above equations that $dx/dt = 0$ for low s (small D) and for $D > D$ (washout).

(*c*) Considering again the process interpretation of Fig. 9.37, speculate about the effect of small perturbations in x from each steady state. Does the perturbation tend to increase or disappear? While such speculations often are helpful in anticipating stability properties of the various steady states, an exact stability analysis of the type discussed in Chap. 11 reveals possible pitfalls in these speculations. Still, the rigorous analysis indicates that steady states shown to be *unstable* by these simple perturbation arguments are in fact unstable.

10.5. Whey fermentation The fermentation of whey lactose to lactic acid by *Lactobacillus bulgaricus* at 44°C has been observed to fit the model of Luedeking and Piret [Eqs. (7.69) to (7.70)] provided the following modifications are made:

1. The maximal growth rate μ_{max} is $\mu_{max}^{\circ}\left(\dfrac{1-p}{p_{max}}\right)$ for $p_{max} = 7$ percent.

2.
$$\mu^\circ_{max} = \begin{cases} 0.48 \text{ h}^{-1} & p \le 3.8\% \\ 1.1 \text{ h}^{-1} & p \ge 3.8\% \end{cases}$$

3. The value of β is 0.49 in batch experiments, and 0.20 in continuous culture.

(a) Write down the proper equations for s, x, and p in continuous fermentation.

(b) Assuming steady-state behavior, show that at a total retention time of 15 h, two equal stages are better than one but three produce essentially no further improvement in reduction of substrate level. Is the same result true for biomass?

$$s_0 = 4.8\% \qquad x_0 = 0 \qquad p_0 = 1\%$$

$$\alpha = 2.2 \qquad \beta = 0.2 \text{ h}^{-1}$$

$$Y = 0.88 \qquad K_s = 50 \text{ mg/l}$$

(c) Keller and Gerhardt [18] note that when s_0 is less than 5 percent, product inhibition is not particularly strong, thus arguing that "from a practical standpoint, ... cheddar cheese whey (e.g., 4.9% lactose, 0.2% lactic acid) might be fermented adequately in a single stage fermentor, whereas cottage cheese whey (5.8% lactose, 0.7% lactic acid) benefits from an additional stage." Illustrate the magnitude of this benefit by repeating part (b) design using $s_0 = 5.8$ percent, $p_0 = 0.7$ percent.

(d) These authors also point out that addition of sugar would reduce the amount of water which must be removed to get a fixed mass of product. How would sugar addition affect a reactor-design strategy?

10.6. Staged fermentations: hydrocarbons Let us suppose that you have the batch growth curve for the hydrocarbon fermentation described so vividly by Mimura et al. in Sec. 8.6. In Prob. 8.20, you identified the probable controlling resistances of each of the fermentation phases observed. Your company has decided (in your absence) to scale up this fermentation by n tanks in series, where $n \equiv$ number of tanks $\equiv$ number of distinct cell-bubble-substrate configurational phases reported in Ref. 18. For each phase, write down the controlling resistance(s) and discuss quantitatively how you would scale the reactor volume, power inputs, etc., to obtain a scale factor of 5000 from laboratory to process units.

10.7. Propose a problem You are asked to design an interesting and illustrative homework problem which you would be delighted to bequeath to the next class. The problem should include three parts of the following character:

1. A background to the specific process or problem area which includes one or several general references
2. A specific statement of the problem proposed, including any original or reduced data, equations, etc., which would subsequently be useful in the solution of the problem
3. A carefully worked solution to the problem which includes not only the answer but also a comment on the significance of the calculation

Approximate lengths might be 200 to 300 words for Sections 1 and 2.

10.8. Oscillatory feed In waste-treatment plants, the feed rate may vary in time but be periodic with a period length of hours to days. Using the Monod equation for single-substrate limitation, and taking $s_f = $ const and $D = D_0(1 + \alpha \sin wt)$, $\alpha > 0$:

(a) Show that if x and its derivatives are periodic in time,

$$\int_0^{2\pi/w} \left\{ \frac{\mu_{max} s(t')}{K_s + s(t')} - D(t') \right\} x(t') \, dt' = 0$$

The solution $x(t') = 0$ for all t' is the washout result. Indicate the behavior of the quantity in brackets within the integral when a nontrivial solution exists, that is, $x(t') > 0$ for $0 \le t' \le 2\pi/w$.

(b) If $s_f \gg K$ and α is small, near the washout region $s(t')$ will always be close to s_f. Under this condition, show that the nontrivial solution requires $D_0 = \mu_{max}$ and that the variation of x for $0 \leq t \leq 2\pi/w$ is given by

$$x = x(0) \exp \left[\frac{D\alpha}{w} (1 - \cos wt) \right]$$

(c) If $s_f \ll K$, indicate explicitly what equations would have to be solved (by computer) in order to find $x(t)$ for the nontrivial case.

10.9. Lake phytoplankton: oscillating μ A lake contains many microbial species, yet under some circumstances, a single species may be considered in a more or less isolated fashion. An example is that for phytoplankton in a shallow lake (Lake George, Uganda). An initial model is due to Uhlmann, who in 1971 suggested:

$$\frac{dx}{dt} = \mu x + Ix - (Gx + Sx + Ex)$$

Rate of change = growth + inlet − (grazing [i.e., predation] + sedimentation + exit)

(a) Taking $I = 0$ for Lake George, show that this formulation leads exactly to the chemostat equations of Chap. 9.

(b) If over an appropriate interval (day), G, S, and E remain constant but μ oscillates according to the availability of sunlight, say $\mu = \mu_{max} \sin (\pi t/12)$, $t < 12$, and $\mu = 0$, $12 < t < 24$, where $t = 0$ is 7:00 A.M., develop an integral expression for a washout limitation in terms of a period average of appropriate quantities. (This is analogous to Prob. 10.11 except that now it is μ which varies periodically rather than D.)

(c) What assumptions about other species (competitors, predators) are implicit in the above formulation? (For further details, see Ref. 19.) In particular, discuss the interesting interplay between the depth of hydrodynamic mixing and the absorption of sunlight by the phytoplankton [19, pp. 330–331].

10.10. Time lag in transient conditions In the presence of transients, the introduction of a time delay constant ($= 1/\gamma$) for the response of the instantaneous specific growth constant $\mu(t)$ to the changed substrate level has been proposed [20]. The movement of $\mu(t)$ toward the steady-state value μ_0 is given by

$$\mu_0 = \frac{\mu_{max} s}{K_s + s} \qquad \frac{d\mu}{dt} = \frac{\mu_0 - \mu}{\gamma}$$

(a) If s is a function of time $s(t)$, show that the solution is given by

$$\mu(t) = \mu(t = 0)e^{-t/\gamma} + \mu_{max} \int_0^t \frac{e^{-(t-t')/\gamma} s(t') \, dt}{K_s + s(t')}$$

(b) Suppose $s(t) = s_0(1 + \cos wt)$, $\alpha < 1$, and $K_s \ll s(t')$. Obtain the explicit solution for $\mu(t)$ above, plot the ratio $\mu(t)/\mu_{max}$ vs. t for $\alpha = 0.5$, $\gamma = 1$, and $w = 0.1\gamma$, 1.0γ, and 10γ. Explain the differences between these curves. For what sorts of transients ought the time constant γ be included?

(c) If growth $r_x = \mu x s$, write down the equations needed to describe the behavior of a CSTR if $s_f = s_{f0}(1 + \cos wt)$. [Note that this system is quite general, since, for example, any variation $s_f(t)$ can be represented by a Fourier trigonometric series, for example, $s_f = s_{f0} \sum \alpha_i \cos (w_i t + \phi_i)$.]

10.11. Aerobic pipeline reactors "Sewerage systems of many cities comprise both pressure and gravity lines. It has been suggested that these lines may be used as aerobic biological reactors to reduce the biochemical oxygen demand (BOD) on treatment facilities" [21].

(a) Considering the moving liquid phase only, Powell and Lowe [22] derive the result for a plug-flow tubular reactor:

$$\mu_{max} t = \frac{1+\gamma}{\gamma} \ln \frac{\sigma - \gamma}{\sigma_0 - \gamma} - \frac{1}{\gamma} \ln \frac{\sigma}{\sigma_0}$$

where σ = dimensionless substrate level at time 0 (entrance)
$\quad\ \sigma_0$ = dimensionless substrate level at time t (exit)
$\quad\ \gamma = \sigma_0 + x_0$
$\quad\ x_0$ = dimensionless entering cell concentration
$\quad\ \mu_{max}$ = maximum specific growth rate
$\quad\ t$ = detention time

Find the assumed growth law and carefully define each variable in terms of dimensional variables.

(b) Koch and Zandi [21] suggest that an aerobic pipeline reactor may be described as a two-phase flowing system with initial air- and liquid-phase (slurry) volumetric flow rates of Q_a and Q_l. Assuming (1) gas and liquid phases to each be in plug flow with identical velocities and (2) d(total pressure)$/dz = \lambda = $ const, where z = distance in pipeline, write out the two equations describing variation of oxygen concentrations with distance. [Assume that the slurry-phase oxygen consumption rate is linear in cell density, i.e., that the dissolved oxygen level c_{O_2} is always $> c_{cr}$ (Table 8.2).] State any assumptions.

(c) Integrate these equations to find the length of pipeline reactor at which $c_0 = c_{cr}$, that is, the point at which the gas phase should be replenished to maintain aerobic microbial activity. Take $Q_a = 5$ ft^3/s, $Q_l = 500$ ft^3/s, $x_0 = 10$ mg/l, pipe ID $= 120$ in, initial $c_{O_2} = 8$ mg/l, $c_{cr} = 0.5$ mg/l, $s_0 = $ initial BOD $= 150$ mg O$_2$/l, $T = 300$ K, $K_s = 100$ mg O$_2$/l, $\mu_{max} = 0.3$/h, $Y_s = 0.4$ g cell/g BOD, respiration rate $= 0.375$ g O$_2$/(h·g cells), $k_l a = 0.4$/min, $H = 4 \times 10^4$ atm/mol fraction, $\lambda = 0.005$ ft water pressure/ft pipe.

10.12. Bacterial chemotaxis [23] The existence of concentration gradients in natural systems is commonplace. We consider for example the concentration profile of solutes in waters seeping through microbially active soil. Some microbes may *move* toward a richer food supply, and this motion, induced by presence of a solute gradient, is termed *chemotaxis*. The time rate of change of the bacterial density (cells per volume) $n(z, t)$ at a point equals the gradient of the net bacterial flux J of cells per square centimeter per second into the control volume:

$$\frac{\partial n(z, t)}{\partial t} = \frac{\partial J}{\partial z}$$

The flux of cells depends on a passive diffusive motion of cells, $-\mathscr{D}_c\, \partial n/\partial z$, and on an active migration up the concentration gradient. The latter rate may be taken proportional to $n(z, t)$ and to the gradient of ln (substrate), reflecting a decreasing interest in migration with increasing s. Thus,

$$J = -\mathscr{D}_c \frac{\partial n}{\partial z} + kn \frac{\partial \ln s}{\partial s}$$

If the solute profile is assumed to be $s = s_0 e^{-\alpha z}$:

(a) Show that the two above equations reduce to the diffusion equation for a moving medium

$$\frac{\partial n}{\partial t} = m \frac{\partial^2 n}{\partial z^2} + u \frac{\partial n}{\partial z}$$

where the bacterial *motility* $m \equiv \mathscr{D}_c$ and the speed, or *chemotactic velocity*, u depends on the product of the migration rate constant k and the parameter α.

(b) Obtain the full solution of the above equation by separation of variables, assuming no loss of bacteria from the system ($0 \leq z \leq L$).

(c) Show that the maximum value of n is given by $n(t = \infty)_{max} = n_0(1 + Pe_c/2)$ when m is small and by $n(t = \infty)_{max} \approx (n_0)(Pe_c)$ when m is large, where the Peclet number of the problem is

$$Pe_c = \frac{Lu}{\mathscr{D}_c} = \frac{Lk\alpha}{\mathscr{D}_c}$$

(d) Show that the local enrichment, i.e., the number of bacteria which migrate into a peak described by $n(z, t = \infty) > n_0$ (initial uniform distribution) is

$$N = \frac{n_0 L}{Pe_c}\left\{ \frac{Pe_c}{1 - \exp(-Pe_c)} - 1 - \ln Pe_c + \ln\left[1 - \exp(-Pe_c)\right] \right\}$$

REFERENCES

Aiba, Humphrey, and Millis (Ref. 2 of Chap. 7) contains excellent information on design and instrumentation of fermentors. The journal *Process Biochemistry* provides numerous review articles and recent information on microbial technology. More technical information including research data and analytical treatments is presented in *Biotechnology and Bioengineering*, the *Journal of Applied Chemistry and Biotechnology*, and the *Journal of Fermentation Technology*, although many papers in the latter are in Japanese. A more biological viewpoint is evident in the journals *Applied Microbiology* and *Applied Biochemistry and Microbiology*. The annuals *Progress in Industrial Microbiology* and *Society for General Microbiology Symposium* are rich sources of review articles, as are the volumes in the *Advances in Biochemical Engineering* series. The following books have broad coverage of several aspects of microbial technology:

1. L. E. Casida, Jr.: "Industrial Microbiology," John Wiley & Sons, Inc., New York, 1968. A good qualitative review of products, processes, and techniques in the fermentation industry, including patents and economics. Numerous photographic illustrations give a good impression of industrial practice.

2. Henry J. Peppler (ed.): "Microbial Technology," Reinhold Publishing Corporation, New York, 1967. Each of the 18 chapters deals with a specific class of products or microbial processes. The industrial background of each chapter's author provides valuable perspectives on market size, production volume, competitive processes, and historical background.

3. C. Rainbow and A. H. Rose (eds.): "Biochemistry of Industrial Microorganisms," Academic Press, Inc., New York, 1963. Presents some of the biochemical details of commercially important microbial activity. In addition to fermentation processes, the biochemistry of microbial spoilage and sewage treatment is also reviewed.

4. R. Steel (ed.): "Biochemical Engineering (Unit Processes in Fermentation)," Heywood and Company, Ltd., London, 1958. Although somewhat dated, this book offers material on substrates, sterilization, oxygen demand, equipment design, and product recovery for fermentation processes.

5. F. C. Webb, "Biochemical Engineering," D. Von Nostrand Company, Ltd., London, 1964. Besides discussing some microbial products and equipment-design considerations, this reference is unique in including several important subjects, e.g., colloids, emulsions, redox potentials, chemical disinfection, dehydration, radiation, and vaccine manufacture.

6. P. F. Kruse, Jr., and M. K. Patterson, Jr.: "Tissue Culture: Methods and Applications," Academic Press, Inc., 1973. A superb, one-volume summary containing 132 concise contributions covering tissue culture: from enzymes for tissue dissociation to quality control in tissue-culture products.

A useful descriptive introduction to immunology: Ref. 2 of Chap. 1.

Extracellular products from tissue culture:

7. I. S. Johnson and G. B. Boder: Metabolites from Animal and Plant Cell Culture, *Adv. Microbiol.*, **15**: 215, 1973.

8. R. G. Righelato and R. G. Elsworth: Continuous Culture, *Adv. Microbiol.*, **13**: 403, 1970.

References concentrating on specific topics included in this chapter:

9. B. Atkinson and H. W. Fowler: The Significance of Microbial Film in Fermenters, *Adv. Biochem. Eng.*, **3**: 221, 1974.

10. A. Constantinides, J. L. Spencer, and E. L. Gaden, Jr.: Optimization of Batch Fermentation Processes, *Biotech. Bioeng.*, **12**: 803, 1081, 1970.

11. R. Mateles and S. R. Tannenbaum (eds.): "Single-Cell Protein," M.I.T. Press, Cambridge, Mass., 1968.

12. J. B. Davis: "Petroleum Microbiology," American Elsevier Publishing Company, New York, 1967.

13. L. E. Erickson, A. E. Humphrey, and A. Prokop: Growth Models of Cultures with Two Liquid Phases, I: Substrate Dissolved in Dispersed Phase, *Biotech. Bioeng.*, **11**: 449, 1969. (See later volumes of this journal for more papers in this series.)

A general reference for filtration and centrifugation:

14. R. Revvy and C. Chilton, "Chemical Engineer's Handbook," 5th ed., McGraw-Hill Book Company, New York, 1973.

Problems

15. D. Chinloy: Ph.D. thesis, Princeton University, 1976.
16. G. Gelf and J. Boudrant: Enzymes Immobilized on a Magnetic Support: Preliminary Study of a Fluidized Bed Enzyme Reactor, *Biochim. Biophys. Acta.*, **334**: 467, 1974.
17. T. Tosa, T. Mori, N. Fuse, and I. Chibata: Studies on Continuous Enzyme Reactions II. Preparation of DEAE-cellulose Aminoacylase Columns and Continuous Optical Resolution of Acetyl-*de*-methionine, *Enzymologia*, **31**: 225, 1966.
18. A. K. Keller and P. Gerhardt: Continuous Lactic Acid Fermentation of Whey to Produce a Ruminant Feed Supplement High in Crude Protein, *Biotech. Bioeng.*, **17**: 997, 1975.
19. G. G. Ganf and A. B. Viner: Ecological Stability in a Shallow Equatorial Lake (Lake George, Uganda), *Proc. R. Soc.*, **B 184**: 321, 1973.
20. T. B. Young, D. F. Bruley, and H. R. Bungay: A Dynamic Mathematical Model of the Chemostat, *Biotech. Bioeng.*, **12**: 747, 1970.
21. C. M. Koch and I. Zandi: Use of Pipelines as Aerobic Biological Reactors, *J. Water Pollut. Control Fed.*, **45**: 2537, 1973.
22. E. O. Powell and J. R. Lowe: Ref. [7], Chapter 9.
23. I. R. Lapidus and R. Schiller: Bacterial Chemotaxis in a Fixed Attractant Gradient, *J. Theoret. Biol.*, **53**: 215, 1975.

ANALYSIS OF MULTIPLE INTERACTING MICROBIAL POPULATIONS

Until now we have concentrated on systems dominated by a single type of microorganism. Untouched so far are the myriad situations where several different microbial species are important. Among commercial processes, we can cite biological waste-water treatment and cheese manufacture as examples where multiple microbial species are required. Moreover, mixed populations of microorganisms are the rule rather than the exception in natural systems. The natural cycles of carbon, nitrogen, oxygen, and numerous other elements on our planet all require the active participation of many different microorganisms. These applications and others will be pursued more thoroughly in the next chapter.

For the moment, we shall concentrate on the analysis of microbial interactions. We shall seek first to characterize two-species interactions and then extend our analysis to more complex populations. In the following section, four of the six basic types of microbial interactions are considered. Our enjoyment in studying the last two kinds of pairwise interactions is greatly enhanced by analyzing them in detail mathematically. Consequently, Sec. 11.2 presents a general mathematical structure which serves as a springboard for the remainder of this chapter. The final sections consider general and large systems and the development of spatial patterns in mixed populations.

11.1. NEUTRALISM, MUTUALISM, COMMENSALISM, AND AMENSALISM

The first two of these relationships are among the extreme cases possible when two microbial species interact. *Neutralism*, which is relatively rare, means that there is no change in the growth rate of either microorganism due to the presence of the other. Thus, so far as growth rates are concerned, there is no observable interaction. On one extreme from this bland situation is *mutualism*, where both species grow faster together than they do separately. The other extreme, to be considered in Sec. 11.3, is *competition*, where each species exerts a negative influence on the growth rate of the other.

Very few instances of neutralism have been observed. One of these is growth of yogurt starter strains of *Streptococcus* and *Lactobacillus* in a chemostat. The total counts of these two species at a dilution rate of 0.4 h^{-1} were quite similar whether the populations were cultured separately or together. Indeed, it is difficult to imagine many situations in which consumption of nutrients and evolution of products by each species has absolutely no effect on its neighbor. Neutralism can only occur, it would seem, in special environment-microorganism scenarios where each species consumes different limiting substrates and where end products are effectively neutralized or diluted.

Viewed from a different perspective, neutralism implies that the pure-culture behavior of both species is identical to their behavior in mixed culture. The apparent rarity of neutralism casts some shadows on the value of many pure-culture data in describing how mixed populations behave. Prediction of mixed-culture performance from pure-culture studies will be possible only when these studies characterize the relationship between a microorganism and its environment in great detail. Such information on each species can then be hooked together to describe the mixed-population situation.

An analogy with chemical-reactor design may be helpful here: performance of a complex reaction system can be computed only if sufficient thermodynamic and kinetic data are available for *each reaction* which occurs in the system. Once we know how all the intensive variables such as concentrations and temperature affect the reaction rates, heats of reaction, heat capacities, etc., and also how the reactions change concentration, temperature, etc., we can compute (at least for idealized reactors) how the complex system behaves.

Remembering that each cell is an adaptive reactor, we can appreciate the difficulties in extending our reaction-engineering logic to modeling and analysis of mixed populations. Still, we can hope (as is also necessary when dealing with many nonbiological reactors) that models which we know to be gross simplifications of reality will yield proper trends and scaling rules for design.

Mutualism is much more common than neutralism and involves several different mechanisms. One of these is the exchange of growth factors. Such an interaction can be beautifully illustrated by growing a phenylalanine-requiring strain of *Lactobacillus* and a folic acid–requiring strain of *Streptococcus* in a mixed batch culture. Figure 11.1 shows the results with a synthetic medium lacking both

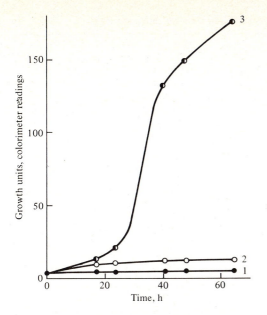

Figure 11.1 Batch growth of pure and mixed cultures of a phenylalanine requiring strain of *Lactobacillus arabinosas* (curve 2 = pure culture) and a folic acid–requiring strain of *Streptococcus faecalis* (curve 1 = pure culture) in a synthetic medium containing neither phenylalanine nor folic acid. Curve 3 shows enhanced growth as a result of mutualism in mixed culture of these two organisms. (*Reprinted from V. Nurmikko, Biochemical Factors Affecting Symbiosis Among Bacteria, Experientia,* **12**: 245, 1956.)

phenylalanine and folic acid. The mixed culture grows well, while separate pure cultures exhibit almost no growth.

Exchange of nutrients may also be involved in mutualistic relationships. Numerous instances are known where mutually beneficial associations exist between aerobic bacteria and photosynthetic algae. While the bacteria use oxygen and carbohydrate, they produce CO_2 and growth factors. The algae, using sunlight as an energy source, convert CO_2 to carbohydrate and also liberate O_2. This system illustrates on a microscopic scale some features of the carbon and oxygen cycles considered further in Chap. 12.

Very close mutualistic ties, such that the partnership is necessary for the survival of one or both species, is often termed *symbiosis*. Microbes are found in many symbiotic relationships with each other as well as with higher organisms. Characteristics of several systems exhibiting symbiosis, some of them quite fascinating, are given in Table 11.1. We shall consider here one example in more detail since it reveals another mode of mutualistic interaction.

Methanobacillus omelianskii, a "bacterium" abundant in anaerobic sludge (see Chap. 12), has recently been discovered to be a mixture of two species. The first converts ethanol to hydrogen and acetate

$$CH_3CH_2OH + H_2O \longrightarrow CH_3COO^- + H^+ + 2H_2$$

but is inhibited by the hydrogen it produces. The second species of the "bacterium" cannot grow on ethanol but consumes hydrogen, yielding methane

$$4H_2 + CO_2 \longrightarrow CH_4 + 2H_2O$$

Table 11.1. Examples of symbiotic relationships involving at least one microorganism

Microorganism	Other organism or site	Comments
Flagellated protozoa	Termites	Protozoa hydrolyze cellulose for termites in exchange for supply of this material, which termites alone cannot digest; flagellates are in turn hosts to bacteria which provide cellulase enzymes
Luminous bacteria	Squid, some fishes	Glands of squid house luminous bacteria, which provide the squid with a recognition device; bacteria accorded protection and nutrients
Rumen micro-organisms	Cattle, sheep, goats	First two stomachs of the cow, the rumen, contain many microbial species, which in exchange for food supply aid the cow in digesting plant material including cellulose, starch, and lipids
Normal microbial flora	Skin, throat, mouth, intestines	Normal flora play an important although ill-defined role in preventing many diseases, as has been shown with studies on germ-free animals
Bacteria	Protozoa	Bacteria live inside the protozoa (an endosymbiosis) and derive nutrients; in one case, at least, bacteria provide their host with needed amino acids and growth factors
Algae	Protozoa	Each protozoan holds 50 to a few hundred algae; algae use light to fix CO_2 and free O_2, which in turn is used by the protozoan to oxidize nutrients, liberating CO_2
Algae	Fungus	Together these form an intimate association called a *lichen;* association is of benefit mostly in very wet or dry environments with scarce nutrients; the alga provides the fungus with organic nutrients; fungal role not well understood
Rhizobium bacteria	Leguminous plants	Bacteria live in nodules formed in plant roots, where they enjoy nutrients provided by the plant; bacteria fix atmospheric nitrogen so that it becomes accessible to plant

Thus, we have a situation in which one species destroys a toxin for its associate, which in turn provides a nutrient for the first. This detoxification type of mutualism may also arise when an aerobe shields an obligate anaerobe from too much free oxygen. Another possibility is maintenance of an advantageous pH by two organisms, one which tends to decrease pH and another which provides the opposite effect.

The final two classes of interactions to be examined in this section involve no significant effect on the first species. In *commensalism*, the second microbe enjoys benefits. The opposite occurs in *amensalism*, where the second species suffers as a result of its interaction with the first.

Several instances of commensalism are similar to the last kind of mutualism we considered. In the commensal version, one species removes a toxin for the second species, but, in contrast to mutualism, the latter organism provides no special benefits for the detoxifier. This type of commensalism is common, as suggested by Table 11.2.

Still more widespread, however, are commensal relationships where one species produces compounds which accelerate growth of another species. The end products serving as bases for commensalism are numerous (see Table 11.3); depending on the particular commensal situation, the produced compound might serve as an energy and/or carbon source for the second species. Such commensal

Table 11.2. Commensal relations: compound removed†

Toxic compound	Details of interrelationship
Concentrated sugar solutions	Osmophilic yeasts metabolize the sugar and thereby reduce the osmolarity, allowing the growth of species sensitive to high osmotic pressures
Oxygen	Aerobic organisms may reduce the oxygen tension, thus allowing anaerobes to grow
Hydrogen sulfide	Toxic H_2S is oxidized by photosynthetic sulfur bacteria, and the growth of other species is then possible
Food preservatives	The growth inhibitors benzoate and sulfur dioxide are destroyed biologically
Lactic acid	The fungus *Geotrichum candidum* metabolizes the lactic acid produced by *Streptococcus lactis*; the acid would otherwise accumulate and inhibit the growth of the bacteria
Mercury-containing germicides	*Desulfovibrio* sp. form H_2S from sulfate, and the sulfide combines with mercury-containing germicides and permits bacterial growth
Antibiotics	Enzymes are produced by some species of bacteria which break down antibiotics; thus the growth of antibiotic-sensitive species is allowed
Phenols	Some bacteria can oxidize phenols, thereby permitting other species to grow
Trichlorophenol	A number of gram-negative bacteria can absorb trichlorophenol in their cell wall lipids and thereby protect *Staphylococcus aureus* from its action

† From J. L. Meers, Growth of Bacteria in Mixed Cultures, p. 158 in A. J. Laskin and H. Lechevalier (eds.), "Microbial Ecology," CRC Press, Cleveland, 1974.

Table 11.3. Commensal relations: compound supplied†

Compound	Species producing compound	Species requiring compound
Purine	*Bacillus subtilis*	*B. subtilis* auxotrophs
Organic acid	*Aerobacter cloacae*	Unnamed bacterium
Isobutyrate	*Corynebacterium diphtheriae*	*Treponema microdentium*
Nicotinic acid	*Saccharomyces cerevisiae*	*Proteus vulgaris*
Vitamin K	*Staphylococcus aureus*	*Bacteroides melaninogenicus*
Nitrite ions	*Nitrosomonas*	*Nitrobacter*
Hydrogen sulfide	*Desulfovibrio*	Sulfur bacteria
Water	*Bacillus mesentericus*	*Clostridium botulinum*
Polysaccharides	Algae	Bacteria
Hydrogen	Rumen bacteria	*Methanobacterium ruminatium*
Methane	Anaerobic methane bacteria	Methane-oxidizing bacteria
Ammonium ions	Many heterotrophs	*Nitrosomonas*
Nitrite	*Nitrosomonas*	*Nitrobacter*
Nitrate	*Nitrobacter*	Denitrifying bacteria
Acetyl phosphate	*Corynebacterium diphtheriae*	*Borrelia vincenti*
Fructose	*Acetobacter suboxydans*	*Saccharomyces carlsbergensis*

† From J. L. Meers, Growth of Bacteria in Mixed Cultures, p. 156 in A. J. Laskin and H. Lechevalier (eds.), "Microbial Ecology," CRC Press, Cleveland, 1974.

relationships are often strung together in a chain so that over time a succession of commensal pairs appears. In a batch system, for example, yeast can convert glucose to alcohol, which serves as a nutrient for *Acetobacter*. This species then produces acetic acid, which in turn is consumed by other microorganisms, and so on.

Also shown in Table 11.3 are several cases where a vitamin or some other growth factor is passed from one species to another. Such situations can readily be constructed by using an auxotrophic mutant as one of the species. Any other species which produces the necessary metabolic intermediate for the first then completes the commensal pair.

Most reported examples of commensalism have been based on batch-culture studies. Indeed, realization of a strictly commensal relationship in continuous culture or other open system is difficult. The problem is avoiding competition: although one organism may aid the other, they both may compete for a nutrient which eventually becomes a limiting nutrient. Therefore, we expect commensalism in open systems only when the species involved differ widely in their nutritional requirements.

Amensalism is the opposite of commensalism: in an amensal relationship, the growth of one species is inhibited by the presence of another. The harmful effects of the offensive species usually are due either to synthesis of toxic products or removal of essential nutrients. Most reported examples of amensalism involve the first mechanism, where the second organism creates an environment within which other species can survive only to a limited degree, if at all.

We have already seen that several antibiotics are produced by microorganisms. One standard laboratory test for microbial antibiotic synthesis is essentially a demonstration of amensalism: an antibiotic-synthesizing and an antibiotic-sensitive species are grown together on an agar surface. The presence of clear zones around the antibiotic-producing colonies is evidence of antibiotic activity (see Fig. 11.2). (Penicillin was discovered by just such an observation.) These clear zones result from diffusion of the antibiotic away from the synthesizing colonies, with resulting inhibition of the susceptible strain.

Antibiotic synthesis by molds and actinomycetes has been emphasized earlier. To appreciate the possible roles of antibiotics in microbial ecology, we must also recognize that algae and other bacteria can also produce antibiotics. Most of these additional bacterial antibiotics are polypeptides. Like penicillin, some of them inhibit cell-wall synthesis while others serve to destroy the permeability barrier provided by the cell membrane.

A related kind of amensalism results because some microbes excrete enzymes which decompose cell-wall polymers. Such organisms derive two different benefits when the lytic enzymes depolymerize cell walls of other species. Possible competitors are destroyed, and the lysed cells release nutrients which can be used by the enzyme-producing microbes. Actually, if the nutrients so released constitute a significant resource for the enzyme-producing species, they benefit from the association. Thus, the ecological relationship ceases to be amensalistic and becomes predation (see Sec. 11.4).

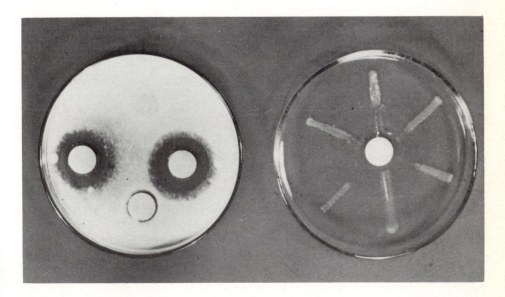

Figure 11.2 Amensalism: diffusion of antibiotic away from the colonies of antibiotic-producing fungus inhibits growth of other organisms on the agar medium. (*Photo courtesy of Charles Pfizer and Co., Inc.*)

Inhibition milder than the above cases may result from microbial synthesis of organic acids. This often lowers pH and inhibits growth of other organisms. More specific inhibitory effects of organic acids are also possible. *Shigella flexneri* is inhibited by formic and acetic acids, and propionate and acetate inhibit growth of the bacteria *Propionibacterium shermanii*. Another example of amensalism involves inhibition of *Nitrobacter* and other chemoautotrophs by small amounts of ammonia produced by other species, especially those which derive energy from amino acids.

Before considering the remaining two kinds of interactions between pairs of different microorganisms, some necessary mathematical diversions are set out in the next section.

11.2. MATHEMATICAL PRELIMINARIES

Since use of vectors and matrices can no longer be avoided, we shall agree that lowercase boldface symbols like **n** denote vectors and that uppercase boldface symbols like **A** denote matrices. Thus, we can compress the notation for a set of differential equations from the cumbersome form

$$\frac{dn_i(t)}{dt} = f_i(n_1(t), n_2(t), \ldots, n_p(t), m_1, m_2, \ldots, m_q) \qquad i = 1, 2, \ldots, p \quad (11.1)$$

to

$$\frac{d\mathbf{n}(t)}{dt} = \mathbf{f}(\mathbf{n}(t), \mathbf{m}) \qquad (11.2)$$

where now **n** denotes a vector of dimension p (with p elements or components; a p-vector) and **m** has dimension q. The function **f**, which depends on **n** and **m** and whose values are p-vectors, will be assumed sufficiently well behaved for Eq. (11.2) to have a unique solution subject to

$$\mathbf{n}(0) = \mathbf{n}_0 \qquad (11.3)$$

where $\mathbf{n}_0$ is a specified p-vector.

From our microbial-ecology perspective, n_i will usually represent the number (or concentration) of species i organisms in the system. The vector **m** contains the parameters which characterize the organisms' environment. For example, m_i might denote medium temperature. Kinetic features of the system are described by the function **f**, which assumes different forms depending on the situation under consideration. Since environmental conditions are often constant or assumed so, explicit dependence of **f** upon **m** is usually not indicated. On the other hand, sometimes differential equations for **m** or at least some of its components are included in the model. This accounts for interactions between the environment and the microbial populations. Representation of population dynamics by an ordinary differential equation as in (11.2) presumes that there are no significant spatial variations in any parameters or variables within the system under consideration. Problems involving spatial dependence of population sizes will not be considered until Sec. 11.6.

We are often interested in the stability characteristics of solutions to Eq. (11.2). Let us consider, for example, a particular steady-state solution $\mathbf{n}_s$. From our earlier definition of steady state, this means that

$$\mathbf{f}(\mathbf{n}_s, \mathbf{m}_s) = 0 \tag{11.4}$$

where $\mathbf{m}_s$ represents a particular set of environmental parameters which does not vary with time. We shall say that the steady state $\mathbf{n}_s$ is *locally asymptotically stable* if $\lim_{t \to \infty} \mathbf{n}(t) \to \mathbf{n}_s$ provided that the initial state $\mathbf{n}_0$ is sufficiently close to $\mathbf{n}_s$. [Our mathematical measure of closeness for vectors is the Euclidean norm, defined by $|\mathbf{n}| = (\sum_{i=1}^{p} n_i^2)^{1/2}$. Then "$\mathbf{n}_0$ sufficiently close to $\mathbf{n}_s$" means that $|\mathbf{n}_0 - \mathbf{n}_s|$ is a sufficiently small real number.] The steady state $\mathbf{n}_s$ is *globally asymptotically stable* if $\lim_{t \to \infty} \mathbf{n}(t) = \mathbf{n}_s$ for any choice of $\mathbf{n}_0$ (except ridiculous ones like negative numbers of a population). If $\mathbf{n}_s$ is an *unstable* steady state, some initial states $\mathbf{n}_0$ arbitrarily close to $\mathbf{n}_s$ lead to trajectories $\mathbf{n}(t)$ which do not approach or stay arbitrarily close to $\mathbf{n}_s$. Thus, in the case of instability, the magnitude of the disturbance, or deviation, defined by

$$\boldsymbol{\chi}(t) = \mathbf{n}(t) - \mathbf{n}_s \tag{11.5}$$

tends to increase from its initial value.

Since the system described by Eq. (11.2) is in general nonlinear, we usually cannot go too far in our analysis without resorting to some approximations. We can attempt to determine behavior near $\mathbf{n}_s$ by expanding the right-hand side of (11.2) in a Taylor's series about $\mathbf{n}_s$ and neglecting all terms of second order and up in the deviations, since they are presumed small. Then we obtain the following linear approximation for our system:

$$\frac{d\boldsymbol{\chi}(t)}{dt} = \mathbf{A}\boldsymbol{\chi}(t) \tag{11.6}$$

where the element a_{ij} in the ith row and jth column of the matrix $\mathbf{A}$ is defined by

$$a_{ij} = \frac{\partial f_i(\mathbf{n}_s, \mathbf{m}_s)}{\partial n_j} \tag{11.7}$$

In the context of mathematical ecology, the matrix $\mathbf{A}$ is often called the *community matrix*. We should emphasize that $\mathbf{A}$ corresponds to some particular steady state. Some systems have more than one steady state for a given $\mathbf{m}$, and this usually implies that a different $\mathbf{A}$ matrix corresponds to each steady state.

The stability properties of the linear approximation are relatively easy to determine since all solutions of Eq. (11.6) usually take the form

$$\boldsymbol{\chi}(t) = \sum_{i=1}^{p} \alpha_i \, \mathbf{c}_i e^{\lambda_i t} \tag{11.8}$$

The quantities $\mathbf{c}_i$ and λ_i are the corresponding pairs of eigenvectors and eigenvalues of $\mathbf{A}$. Thus $\lambda = \lambda_i$ satisfies the characteristic equation

$$\det (\mathbf{A} - \lambda \mathbf{I}) = 0 \tag{11.9}$$

($\mathbf{I}$ is the identity matrix), and the $\mathbf{c}_i$ satisfy

$$(\mathbf{A} - \lambda_i \mathbf{I})\mathbf{c}_i = 0 \qquad i = 1, \ldots, p \tag{11.10}$$

The α_i are constants to be chosen to fulfill the specified initial conditions; consequently they satisfy the linear algebraic equations

$$\sum_{i=1}^{p} \alpha_i \, \mathbf{c}_i = \boldsymbol{\chi}(0) \tag{11.11}$$

Returning now to Eq. (11.8), it is clear that the deviations $\boldsymbol{\chi}(t)$ will all approach zero as time increases so long as all λ_i have negative real parts:

$$\text{Re} \, [\lambda_i] < 0 \qquad i = 1, 2, \ldots, p \tag{11.12}$$

Fortunately, we need not compute all the eigenvalues to verify these inequalities. First, suppose that the determinant in (11.9) has been expanded to provide a pth-order algebraic equation

$$\lambda^p + B_1 \lambda^{p-1} + \cdots + B_{p-1}\lambda + B_p = 0 \tag{11.13}$$

Now we can apply the Hurwitz criterion,† which asserts that all roots of (11.13) have negative real parts if and only if the following conditions are met:

$$B_1 > 0$$

$$\det \begin{bmatrix} B_1 & B_3 \\ 1 & B_2 \end{bmatrix} > 0$$

$$\det \begin{bmatrix} B_1 & B_3 & B_5 \\ 1 & B_2 & B_4 \\ 0 & B_1 & B_3 \end{bmatrix} > 0$$

$$\cdots\cdots\cdots\cdots \tag{11.14}$$

$$\det \begin{bmatrix} B_1 & B_3 & B_5 & \cdots & 0 \\ 1 & B_2 & B_4 & \cdots & 0 \\ 0 & B_1 & B_3 & \cdots & 0 \\ 0 & 1 & B_2 & \cdots & 0 \\ \cdots\cdots\cdots\cdots\cdots \\ \cdots\cdots\cdots\cdots\cdots B_p \end{bmatrix} > 0$$

† C. F. Walter, Kinetic and Biological and Biochemical Control Mechanisms, p. 355 in E. Kun and S. Grisola (eds.), "Biochemical Regulatory Mechanisms in Eucaryotic Cells," John Wiley & Sons, Inc., New York, 1972.

On the other hand, if any of the eigenvalues has a positive real part, there are initial deviations χ_0 which lead to unbounded $\chi(t)$. Thus, instability results if

$$\text{Re}\,[\lambda_j] > 0 \qquad \text{any } j \text{ from 1 to } p \tag{11.15}$$

In the *critical case* where one or more of the eigenvalues has zero real part, Eq. (11.6) admits either oscillatory solutions or nonzero constant solutions.

Since Eq. (11.6) was developed assuming that $\mathbf{n}$ is near $\mathbf{n}_s$, we might expect these results to tell us something about local stability of the original nonlinear system, and our intuition is right. A rigorous proof exists that conditions (11.12) imply local asymptotic stability of $\mathbf{n}_s$ for the original nonlinear system. Also, condition (11.15) guarantees that $\mathbf{n}_s$ is unstable for (11.2) as well as (11.6).† These conclusions remain valid even in the special cases where $\mathbf{A}$ has repeated eigenvalues and a modified form of Eq. (11.8) is needed to represent the linearized response. In the critical case, no general relationship between the nonlinear and linearized system can be asserted. Fortunately, however, this case is relatively rare, so that system linearization provides a powerful strategy for examining local stability of a steady state.

Global stability analysis requires more exotic and intuitive approaches, none of which can be routinely applied. Therefore, let us return to the linearized system, and examine it from the ecologist's perspective.

The interaction between two species near a steady state is characterized by two entries in the community matrix $\mathbf{A}$. Considering for the moment species i and j, we notice that a small increase in the jth population from its steady-state value $[\chi_j(0) > 0]$ contributes a term $a_{ij}\chi_j(0)$ to the initial derivative of χ_i. Thus, if all other populations are initially at their steady-state values $[\chi_k(0) = 0,\ k \neq j]$, the sign of $d\chi_i(0)/dt$ depends on the sign of a_{ij}. We can say that if a_{ij} is positive,

† Ibid.

Table 11.4. **Classification of pairwise interactions based on the signs of the entries** a_{ji} **and** a_{ij} **from the community matrix** $\mathbf{A}(i \neq j)$‡

		Effect of species j on species i (sign of a_{ij})		
		−	0	+
Effect of species i on species j (sign of a_{ji})	−	− − Competition	−0 Amensalism	− + Predation
	0	0− Amensalism	00 Neutralism	0+ Commensalism
	+	+ − Predation	+0 Commensalism	+ + Mutualism

‡ Adopted from R. M. May, "Stability and Complexity in Model Ecosystems," p. 25, Princeton University Press, Princeton, N.J., 1973.

species j has a positive effect on growth of species i and if a_{ij} is negative, that an inhibiting effect is evident. No interaction occurs if a_{ij} is zero. The same argument applies for determining the influence of species i on species j: the effect is stimulatory for positive a_{ji}, inhibitory for negative a_{ji}, and neutral for zero a_{ji}.

All possible combinations of interactions defined in this manner are shown in Table 11.4. We see that the four kinds of pairwise interactions considered in the previous section fit nicely into this scheme. Also, two new types of interactions, competition and predation, appear in the table. Before considering these interactions in detail, we shall explore some implications of the interaction type on population dynamics near a steady state. For the moment, we shall limit this inquiry to two interacting populations.

Example 11.1. Two-species dynamics near a steady state When only two species interact, the community matrix A is 2×2, as seen from the appropriate form of (11.6):

$$\frac{d(n_1 - n_{1s})}{dt} \equiv \frac{d\chi_1}{dt} = a_{11}\chi_1 + a_{12}\chi_2$$

$$\frac{d(n_2 - n_{2s})}{dt} \equiv \frac{d\chi_2}{dt} = a_{21}\chi_1 + a_{22}\chi_2$$

(11E1.1)

or

$$\frac{d}{dt}\begin{bmatrix} \chi_1 \\ \chi_2 \end{bmatrix} = \begin{bmatrix} a_{11} & a_{12} \\ a_{21} & a_{22} \end{bmatrix}\begin{bmatrix} \chi_1 \\ \chi_1 \end{bmatrix}$$

(11E1.2)

The characteristic equation (11.9) can readily be expanded into the quadratic equation

$$\lambda^2 - (a_{11} + a_{22})\lambda + (a_{11}a_{22} - a_{12}a_{21}) = \lambda^2 - (\text{tr } A)\lambda + \det A = 0$$

(11E1.3)

where tr A and det A signify the trace and determinant of A, respectively. Applying the Hurwitz criterion (11.14) to Eq. (11E1.3), we readily conclude that the linearized system is asymptotically stable if and only if

$$\det A > 0$$

(11E1.4)

$$\text{tr } A < 0$$

(11E1.5)

Additional features of the system dynamics are revealed by the solutions to Eq. (11E1.3), which are

$$\lambda_{1,2} = \frac{\text{tr } A}{2} \pm \frac{1}{2}\sqrt{(\text{tr } A)^2 - 4 \det A}$$

(11E1.6)

At least one eigenvalue will have zero real part if either tr $A = 0$ (and det $A > 0$) or det A is zero. Let us exclude this critical case from our considerations and proceed to other possibilities. If

$$\left(\frac{\text{tr } A}{2}\right)^2 > \det A > 0$$

(11E1.7)

both eigenvalues are real and have the same sign: the steady state n_s in question is called a *node*. If det A is negative, there is one positive and one negative eigenvalue and the steady state is a *saddle point*. Complex eigenvalues occur when 4 det A is larger than the square of tr A. In this case the steady state is called a *focus*.

The motivation for these terms, which derive originally from mathematical studies in mechanics, becomes clearer by examining the population trajectories in the phase plane. By definition, the coordinates of the phase plane are χ_1 and χ_2. A specified initial condition $(\chi_1(0), \chi_2(0))$ can be represented by a point in this plane. Likewise, if we plot for all future t the points $(\chi_1(t), \chi_2(t))$ in the phase plane,

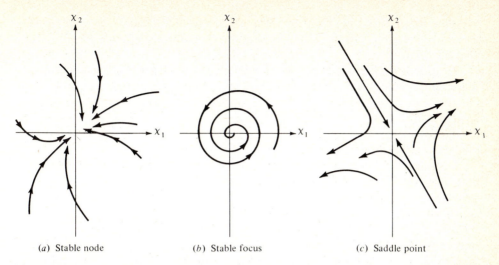

(a) Stable node   (b) Stable focus   (c) Saddle point

Figure 11E.1 Some characteristic phase-plane portraits for the linearized dynamics of two-species ecosystems: (a) stable node, (b) stable focus, and (c) saddle point.

the result will be a continuous curve called a *trajectory*. Thus, plotting trajectories in the phase plane is a convenient device for representing the dynamic behavior of two variables in a single diagram.

Figure 11E1.1 shows typical trajectories for three common cases. Each trajectory shown in the figure corresponds to a different set of initial conditions, and the arrows point in the direction of increasing time. The typical phase behavior of unstable nodes and foci can be obtained simply by reversing the arrows on the trajectories in Fig. 11E1.1a and b. (Why does this work?)

Returning now to the classification scheme summarized in Table 11.4, we see that this scheme refers only to off-diagonal entries of the **A** matrix. Consequently, without further specifications we can say nothing about tr **A** and very little about det **A**.

Let us now consider reasonable assumptions concerning the autonomous specific growth rates a_{11} and a_{22} for some of the interaction types already defined. Mutualism may give rise to larger steady-state populations than possible without the beneficial interaction, so that $a_{11} < 0$, $a_{22} < 0$ obtains. Then condition (11E1.5) certainly holds, but satisfaction of (11E1.4) requires

$$a_{11}a_{22} > a_{12}a_{21} \qquad (11E1.8)$$

If this inequality is not fulfilled, this mutalistic steady state is unstable.

For either commensalism or amensalism in a two-population system, the stability characteristics depend entirely on the autonomous growth rates, for det $\mathbf{A} = a_{11}a_{22}$ and tr $\mathbf{A} = a_{11} + a_{22}$. This reveals immediately that asymptotic stability appears if and only if both a_{11} and a_{22} are negative.

Several of the results just outlined can be generalized to more complex populations, as we shall see in Sec. 11.5.

11.3. COMPETITION: SURVIVAL OF THE FITTEST

Darwin's work on natural selection emphasized the importance of competition between different species. As we shall use it here, the term *competition* refers to dependence of two species on a common factor such as food supply, light, space, or some other resource. Consumption of this common factor by each species limits its availability to the other, so that the growth rates of both organisms are affected negatively.

In a competitive situation, we are interested in discovering whether either species enjoys a natural advantage. A little reflection suggests that the species with the fastest growth rate should do better, since by virtue of its rapid growth it will be able to utilize more of the limiting factor than the slower-growing organism. As we shall see, experimental results as well as mathematical analyses of reasonable models strongly support this idea.

Consider the example shown in Fig. 11.3. Grown separately (open symbols), both *E. coli* and *Staphylococcus aureus* reached similar maximum populations although *E. coli* grew faster. When the two bacteria were cultivated in a batch mixed culture, the *E. coli* growth pattern was identical to that found in pure culture. The *S. aureus* population, on the other hand, was restricted to much smaller maximal levels in the mixed-culture case. Since *E. coli* produces no inhibitors of the *Staphylococcus* organism, this effect can be attributed to the preferential nutrient uptake rate of *E. coli*. (What implicit assumption is involved here?)

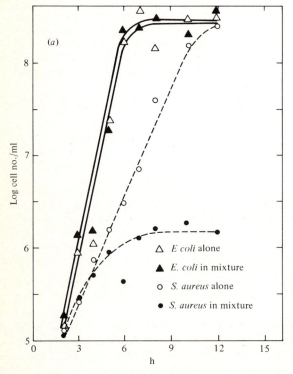

Figure 11.3 Growth of *S. aureus* is reduced when the organism grows in competition with *E. coli* in mixed culture. (*Reprinted from T. R. Oberhofer and W. C. Frazier, Competition of* Staphylococcus aureus *with other organisms, J. Milk Food Technol.*, **24**: 172, 1961.)

11.3.1. Volterra's Analysis of Competition

Growth of two competing species in a closed environment has been analyzed by Vito Volterra, a famous Italian mathematician who established many of the foundations of mathematical ecology. In this nonlinear model, the autonomous specific growth rates μ_1 and μ_2 are both presumed positive. Furthermore, the first species utilizes the limiting nutrient at a rate $h_1 n_1$, and $h_2 n_2$ is the corresponding rate for species 2, where h_1 and h_2 are both positive. Such consumption reduces the specific growth rates of the two species so that

$$\frac{dn_1}{dt} = [\mu_1 - \gamma_1(h_1 n_1 + h_2 n_2)]n_1 \tag{11.16}$$

$$\frac{dn_2}{dt} = [\mu_2 - \gamma_2(h_1 n_1 + h_2 n_2)]n_2 \tag{11.17}$$

The constants $\gamma_1 > 0$ and $\gamma_2 > 0$ in these equations reflect the different effects of nutrient depletion on species 1 and 2, respectively. Turning back to Sec. 7.2.1, we note that this model is simply a two-population generalization of the Verlhurst and Pearl model already discussed.

The ultimate effects of competition can be deduced by multiplying Eq. (11.16) by γ_2/n_1, multiplying Eq. (11.17) by γ_1/n_2, and then adding the resulting equations to obtain

$$\gamma_2 \frac{d \ln n_1}{dt} - \gamma_1 \frac{d \ln n_2}{dt} = \gamma_2 \mu_1 - \gamma_1 \mu_2 \tag{11.18}$$

Integration of this equation yields

$$\frac{n_1^{\gamma_2}}{n_2^{\gamma_1}} = C \exp (\gamma_2 \mu_1 - \gamma_1 \mu_2)t \tag{11.19}$$

where C is a constant of integration which depends on $n_1(0)$ and $n_2(0)$. Let us suppose now that $\gamma_2 \mu_1 - \gamma_1 \mu_2$ is negative. (Does this involve any loss of generality?)

Then Eq. (11.19) reveals that the ratio on the left-hand side approaches zero as time increases. We can readily rule out the possibility that this happens because n_2 grows without bound. Turning to Eq. (11.17), we can see immediately that dn_2/dt is negative for $n_2 > \mu_2/\gamma_2 h_2$, proving that n_2 is bounded. Then, the only conclusion consistent with the limiting behavior demonstrated by Eq. (11.19) is that n_1 approaches zero as time passes.

The net result of this analysis is known as the *competitive exclusion principle*. When two species compete in a *common* limited environment and $\gamma_1/\mu_1 > \gamma_2/\mu_2$, the second species population reaches a limiting size given by $\mu_2/\gamma_2 h_2$ while the first species eventually disappears. This principle can be extended to multiple-species situations described by

$$\frac{dn_i}{dt} = [\mu_i - \gamma_i F(n_1, n_2, \ldots, n_p)]n_i \qquad i = 1, 2, \ldots, p \tag{11.20}$$

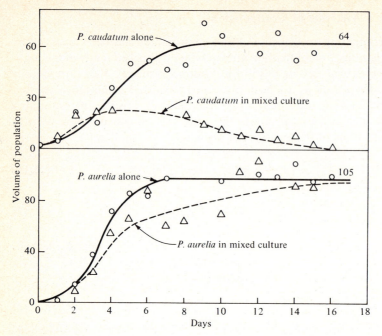

Figure 11.4 Demonstration of the exclusion principle in competitive growth of two protozoa. The slower-growing microbe, *Paramecium candatum*, is eliminated in mixed culture. (*Redrawn from G. F. Gause, Experimental Analysis of Vito Volterra's Mathematical Theory of the Struggle for Existence, Science,* **79**: 16, 1934.)

by arguments identical to the above. Again, the conclusion is extinction of all species but one.

Dramatic experimental evidence illustrating the exclusion principle is shown in Fig. 11.4. Here, in a batch system controlled to maintain constant food supply, a protozoan species which follows the logistic curve [Eq. (7.36)] when grown alone becomes extinct in competition with a faster-growing organism.

The success of the exclusion principle in explaining the outcome of this and numerous other actual instances of competition is somewhat surprising in view of the assumptions behind Eqs. (11.16) and (11.17). It is important to realize, however, that a model of this *form* may be applicable in physical circumstances very different from those assumed in our derivation. In fact, from a strictly mathematical viewpoint, we may view Eqs. (11.16) and (11.17) as approximate representations obtained by expanding a general model with a Taylor's series and retaining only terms through the second order.

11.3.2. Competition and Selection in a Chemostat

Next we shall consider the effects of competition in open systems, using the chemostat as a prototype representative. Extending Eq. (9.37) to the multiple-species situation gives

$$\chi_i(t) = \chi_i(0)e^{(\mu_i - D)t} \qquad i = 1, 2, \ldots, p \qquad (11.21)$$

Thus, whether χ_i increases, decreases, or stays constant with time depends on the relative values of μ_i and D. What happens in the two-species case can be deduced using a rather heuristic argument based on the Monod model.

It is reasonable to assume that the dependence of growth rate on limiting-substrate concentration will rarely be identical for two competing microorganisms. This leaves two possibilities: either the growth rates are related as shown in Fig. 11.5, or the growth-rate curves cross. For the moment, let us assume that the situation depicted in Fig. 11.5 obtains.

Suppose that organism 1 is growing at steady state in a chemostat with dilution rate D. Then, as indicated in Fig. 11.5, the limiting-substrate concentration within the vessel will be s_1. Now, if organism 2 is introduced into the chemostat, it will grow initially at rate μ_2. Since $\mu_2 > D$, the population of species 2 will increase with time, causing the substrate concentration to fall. With $s < s_1$, the specific growth rate of species 1 is smaller than D, so that organism 1 begins to wash out. This trend continues until the substrate concentration reaches s_2, at which point species 2 attains steady state while species 1 continues to wash out.

This pattern of events has been observed in numerous experimental studies. In the simultaneous growth of the bacterium *Aerobacter aerogenes* and the yeast *Torula utilis* in a chemostat, the washout rate of the slower-growing yeast appears to be a function of dilution rate (see Fig. 11.6), as expected from Eq. (11.21). While Fig. 11.6 shows some overshoots and oscillations not included in our simple scenario, other data, such as those plotted in Fig. 11.7, follow our hypothesized chain of events very closely.

If the two functions $\mu_1(s)$ and $\mu_2(s)$ cross, so that which is larger depends on the particular substrate concentration involved, we can conclude by analogy with the argument above that the dilution rate used will determine which species dominates. A dramatic demonstration of such dependence of competitive advantage on dilution rate is revealed in Fig. 11.8. Initially the yeast dominates, but it begins to wash out as soon as the dilution rate is increased. Although the bacter-

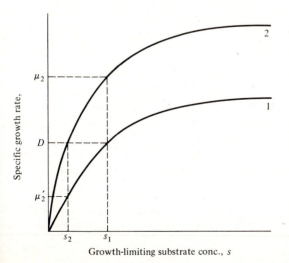

Figure 11.5 Hypothetical specific growth rates of organisms 1 and 2. The text explains why organism 1 will wash out of continuous culture. (*Reprinted from J. L. Meers and D. W. Tempest, The Influence of Extra-Cellular Products on Mixed Microbial Populations in Magnesium-Limited Chemostat Cultures, J. Gen. Microbial,* **52:** 309, 1968.)

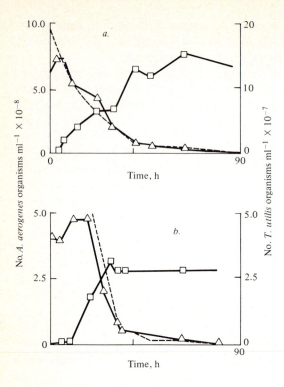

Figure 11.6 Data on competition between *A. aerogenes* (□) and *T. utilis* (△) in a chemostat: (*a*) $D = 0.05$ h^{-1}, (*b*) $D = 0.3$ h^{-1}; $T = 33°C$, pH 6.4, potassium-limited simple-salts medium. Dashed lines are the calculated washout curves for nongrowing organisms. [*Reprinted from J. L Meers, Growth of Bacteria in Mixed Cultures, in A. I. Laskin and H. Lechevalier (eds.), "Microbial Ecology," p. 142. © CRC Press, Inc., 1974. Used by permission of CRC Press, Inc.*]

ium gains ascendency at the higher dilution rate, the roles of the two species are reversed as soon as the dilution rate returns to its starting level.

Before turning to the fascinating class of predation and parasitism interactions, we should point out that the previous discussion also applies to organisms of the same type which have different characteristics. For example, mutant strains

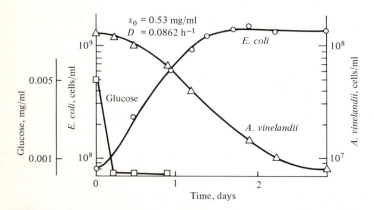

Figure 11.7 Competition between two bacteria (*E. coli* and *Azotobacter vinelandii*) in a chemostat. (*Reprinted from J. L. Jost et al., Interactions of* Tetrahymena pyriformis, Escherichia coli, Azotobacter vinelandii *and Glucose in a Minimal Medium, J. Bacteriol.,* **113:** 834, 1973.)

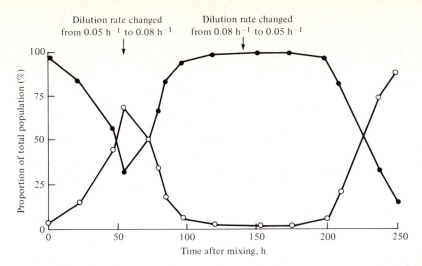

Figure 11.8 Effects of dilution-rate changes on continuous mixed culture of *Bacillus subtilis* (●) and *T. utilis* (○) ($T = 33°C$, pH = 6.4, magnesium-limited simple-salts medium). (*Reprinted from J. L. Meers, J. Gen. Microbiol.*, **67**: 359, 1967.)

especially suited for growth in a particular chemostat environment may develop. These strains may then be viewed as organism 2 in our account of chemostat competition: their population may increase dramatically, eventually completely displacing the original species.

This manifestation of exclusion caused by competition has very important practical implications. Many fermentations have maximal product formation rather than biomass production as their objective. Consequently, strains selected for desirable product yields may well be slower-growing than related mutant strains. Under these circumstances, it becomes increasingly probable that relatively fast-growing mutants will evolve in a chemostat, leading to the washout of the productive strains. As discussed in earlier chapters, part of this problem is avoided in *batch* reactors by use of relatively large inocula of the desired species. As Figs. 11.6 to 11.8 amply demonstrate, such an initial advantage in population density is lost in continuous culture and the contaminant or mutant strain with a larger μ value than the original culture will evidently prevail. As mutants with desirable properties have been produced for batch cultures, we can expect similar efforts in the future to develop productive strains able to compete successfully with useless mutants in continuous reactors.

The effects of wall growth, considered in various lights in Chap. 10, should also be mentioned in the context of microbial competition in open environments. We have already mentioned (Sec. 7.4.1) that a population attached to a surface can survive in a chemostat even when D exceeds the maximal specific growth rate. Similarly, a slower-growing organism, if present in an ecologically advantageous niche such as a film, has been observed experimentally to persist in competition with faster-growing species.

11.4. PREDATION AND PARASITISM

In both predation and parasitism, one species benefits at the expense of the other. We usually distinguish between these two types of interactions by the relative size of the organisms and the mechanism by which one species destroys the other. Predation involves the ingestion of one organism, the prey, by the predator organism. This mode of interaction is relatively common among microbes, with consumption of bacteria by protozoa a prime example. In parasitism, the host, which is usually the larger organism, is damaged by the parasite, which benefits from use of nutrients from the host. Attack of microorganisms by phages, already considered in some detail in Chap. 6, is the best example of parasitism in the microbial world.

Although the physical situations in predation and parasitism differ, their conceptual and mathematical descriptions share many common features. Consequently, for simplicity in our discussion of these forms of interaction, we shall use the generic terms predator and prey in the rest of this section to denote the respective species which benefit and which are damaged by their relationship.

In many instances of predator-prey interactions, the populations of predator and prey do not reach steady-state values but oscillate, as shown in Fig. 11.9 for a bacterium-protozoan system. Other examples of predator-prey oscillations beyond the microbial ones considered in this chapter include the lynx-hare, owl-lemming, and moth–conifer-tree system [1, p. 188]. We can explain such regular fluctuations qualitatively as follows. Starting with a high prey and low predator concentration, the predators ravenously consume the prey, so that the number of predators increases while the prey population declines. These trends continue until there is an overshoot, where the small prey population is inadequate to

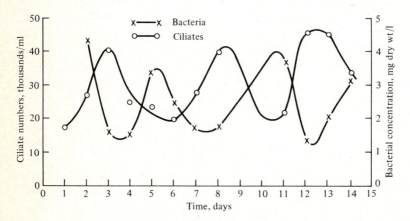

Figure 11.9 Oscillations in chemostat growth of a mixed population of the bacterium *A. aerogenes* and the protozoan *Tetrahymena pyriformis*. [*Reprinted from J. L. Meers, Growth of Bacteria in Mixed Cultures, in A. I. Laskin and H. Lechevalier, (eds.), "Microbial Ecology," p. 151. © CRC Press, Inc., 1974. Used by permission of CRC Press, Inc.*]

support the large predator population. Then the predator population declines due to insufficient food supply while the prey population rebounds. This leads back to the start of the cycle, which may then repeat again and again.

11.4.1. The Lotka-Volterra Model of Predator-Prey Oscillations

A mathematical model for predator-prey interaction that produces such cycles in population sizes was developed by Lotka and Volterra in the late 1920s [8]. In this model, it is assumed that the prey species 1 can multiply autonomously and is consumed by predator at a rate proportional to the product of the numbers of predator and prey:

$$\frac{dn_1}{dt} = an_1 - \gamma n_1 n_2 \tag{11.22}$$

The product $n_1 n_2$ may be viewed as the frequency of encounter between the prey and predator species, and the coefficient γ attached to this term is a measure of how often such encounters lead to death of the prey. It is known, for example, that some bacteria are more resistant to attack by protozoa than others.

The predator species 2 benefits from encounters with prey, and the proportionality constant ε will denote the amount by which the predator population increases per kill of prey. Since the prey is presumed to be the only source of food for predators, the predator population declines in the absence of prey. Thus, the population balance for predators takes the form

$$\frac{dn_2}{dt} = -bn_2 + \varepsilon\gamma n_1 n_2 \tag{11.23}$$

Equations (11.22) and (11.23) constitute the basic Lotka-Volterra model. The nontrivial steady state admitted by the model is

$$n_{1s} = \frac{b}{\varepsilon\gamma} \qquad n_{2s} = \frac{a}{\gamma} \tag{11.24}$$

We can simplify the notation somewhat for our subsequent analysis by introducing new population variables y_1 and y_2, which are scaled by the steady-state values from Eq. (11.24):

$$y_1 = \frac{n_1}{n_{1s}} \qquad y_2 = \frac{n_2}{n_{2s}} \tag{11.25}$$

In terms of these variables, the Lotka-Volterra model becomes

$$\frac{dy_1}{dt} = a(1 - y_2)y_1 \tag{11.26}$$

$$\frac{dy_2}{dt} = -b(1 - y_1)y_2 \tag{11.27}$$

For this model, the phase-plane behavior is readily determined by eliminating time from the pair of Eqs. (11.26) and (11.27) to obtain a single differential equation relating y_1 and y_2. We accomplish this by dividing (11.27) by (11.26) to obtain

$$\frac{dy_2/dt}{dy_1/dt} = \frac{-by_2 + by_1 y_2}{ay_1 - ay_1 y_2} \tag{11.28}$$

then multiplying by $a(1 - y_2)(dy_1/dt)/y_2$, which yields

$$\frac{a}{y_2}\frac{dy_2}{dt} - a\frac{dy_2}{dt} + \frac{b}{y_1}\frac{dy_1}{dt} - b\frac{dy_1}{dt} = 0 \tag{11.29}$$

Integration of this equation gives

$$a \ln y_2 - ay_2 + b \ln y_1 - by_1 = C \tag{11.30}$$

or

$$\left(\frac{y_1}{e^{y_1}}\right)^b \left(\frac{y_2}{e^{y_2}}\right)^a = e^C \tag{11.31}$$

where C is an integration constant dependent on the initial population sizes.

Paths in the phase plane can now be computed from Eq. (11.31) by assuming a value for y_1 and computing compatible y_2 values. Care is necessary when doing this, as zero, one, or two solutions y_2 to this transcendental equation may occur for fixed y_1. In Fig. 11.10 several trajectories are illustrated. Immediately apparent are the existence of closed curves for all initial states except the steady state. (What does this tell us about asymptotic stability of the steady state?) If we transfer this representation to plots of populations vs. time, which requires solution of (11.26) and (11.27), we obtain prey and predator oscillations (see Fig. 11.11) quite reminiscent of data for numerous predator-prey systems, including the information shown earlier in Fig. 11.9.

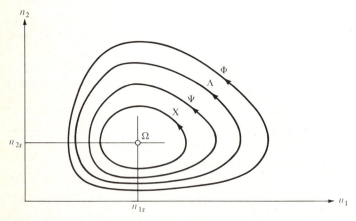

Figure 11.10 Motion in the phase plane as described by the Lotka-Volterra model. Point Ω is the steady state. Oscillatory trajectories labeled X, Ψ, Λ, and Φ correspond to different choices of initial population levels.

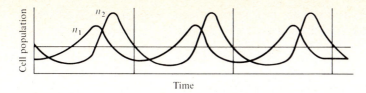

Figure 11.11 Periodic oscillations in the prey n_1 and predator n_2 population sizes from the Lotka-Volterra model.

Before turning to extensions, criticisms, and modifications in the Lotka-Volterra model, let us explore some of its implications a little further. Integrating Eq. (11.26) over one period T of the oscillation, we find that

$$\ln \frac{y_1(T)}{y_1(0)} = aT - a \int_0^T y_2(t)\, dt \tag{11.32}$$

Since y_1 is periodic with period T,

$$y_1(0) = y_1(T) \tag{11.33}$$

so that the left-hand side of Eq. (11.32) is zero. Rearranging what is left gives

$$\frac{1}{T} \int_0^T y_2(t)\, dt = 1 \tag{11.34}$$

or, in terms of the original variables,

$$\frac{1}{T} \int_0^T n_1(t)\, dt = n_{1s} \tag{11.35}$$

Thus, although the periods and the location of the population cycles in Fig. 11.10 depend on the initial conditions, the mean value of prey population, which is given by Eq. (11.35), is independent of the initial conditions and is equal to the steady-state value. Starting with Eq. (11.27) and applying exactly the same approach as that just described, we can easily prove the analogous relation for the mean predator population

$$\frac{1}{T} \int_0^T n_2(t)\, dt = n_{2s} \tag{11.36}$$

The Lotka-Volterra model has some interesting implications for control of microbial or other populations involving predator-prey interactions. Suppose that we introduce an agent which kills both predator and prey in proportion to their numbers. Mathematically, this corresponds to adding terms $-\delta_1 n_1$ and $-\delta_2 n_2$ ($\delta_1 > 0$ and $\delta_2 > 0$) to Eqs. (11.22) and (11.23), respectively. The steady-state concentrations of prey and predator *with the control measure* now become

$$n_{1s} = \frac{b + \delta_2}{\alpha\gamma} \qquad n_{2s} = \frac{a - \delta_1}{\gamma} \tag{11.37}$$

Consequently, the effect of this attempt to control the interacting populations is an *increase* in the mean number of prey and a decrease in the predator average population.

11.4.2. A Multispecies Extension of the Lotka-Volterra Model

Ecologists are interested in the relationship between system complexity and its dynamic behavior. For example, is the system more or less likely to be stable or to exhibit oscillations as the number of different species is increased or as the number and intensity of interactions becomes larger? Although most of our considerations of such issues will be concentrated in Sec. 11.5, we shall consider now a generalized version of the Lotka-Volterra model following the argument of May [5].

The analogous form of Eqs. (11.22) and (11.23) for a system with N predators and N preys is

$$\frac{dn_i}{dt} = n_i\left(a_i - \sum_{j=1}^{N} \alpha_{ij} m_j\right) \qquad i = 1, \ldots, N$$

$$\frac{dm_i}{dt} = m_i\left(-b_i + \sum_{j=1}^{N} \beta_{ij} n_j\right) \qquad i = 1, \ldots, N \tag{11.38}$$

where $n_1, \ldots, n_N$ are the numbers in the N prey populations and $m_1, \ldots, m_N$ are the predator population numbers. Letting $\boldsymbol{\alpha}$ denote the matrix with elements α_{ij}, $\boldsymbol{\beta}$ the matrix with elements β_{ij}, $\mathbf{a}$ the column vector with entries a_i, and $\mathbf{b}$ the column vector with elements b_i, it follows that the steady-state populations $n_{1s}, \ldots, n_{Ns}, m_{1s}, \ldots, m_{Ns}$ for the system described by Eqs. (11.38) satisfy

$$\boldsymbol{\beta}\mathbf{n}_s = \mathbf{b} \qquad \boldsymbol{\alpha}\mathbf{m}_s = \mathbf{a} \tag{11.39}$$

where $\mathbf{n}_s$ and $\mathbf{m}_s$ are vectors with elements n_{is} and m_{is}, respectively.

If we evaluate the $2N \times 2N$ community matrix $\mathbf{A}$ for this system, the result can be partitioned as follows:

$$\mathbf{A} = \left[\begin{array}{c:c} \mathbf{0} & -\boldsymbol{\alpha}^* \\ \hdashline \boldsymbol{\beta}^* & \mathbf{0} \end{array}\right] \tag{11.40}$$

where

$$\alpha_{ij}^* = n_{is}\alpha_{ij} \qquad \beta_{ij}^* = m_{is}\beta_{ij} \tag{11.41}$$

and the $\mathbf{0}$ entries in (11.40) denote $N \times N$ zero matrices. Significant information about the stability of this system can be obtained from the following general formula from matrix theory. For an arbitrary $q \times q$ matrix $\mathbf{C}$,

$$\text{tr } \mathbf{C} = \sum_{i=1}^{q} \sigma_i \tag{11.42}$$

where the σ_i are the eigenvalues of $\mathbf{C}$. Since the trace of the matrix $\mathbf{A}$ in (11.40) is zero, we know from Eq. (11.42) that

$$\sum_{i=1}^{2N} \lambda_i = 0 \tag{11.43}$$

This result leaves two possibilities for the eigenvalues λ_i. First, they are zeros or pairs of conjugate imaginary numbers. Although this is the case for the original one-prey–one-predator Lotka Volterra model, it seems unlikely when N is large. The other alternative is the occurrence of at least some of the eigenvalues of $\mathbf{A}$ in the form $c + id$, $-c - id$ with $c \neq 0$. In this instance, at least one eigenvalue has positive real part and the steady state is unstable. Consequently, this analysis suggests that the effect of additional complexity will be a destabilization of the prey-predator system.

11.4.3. Other One-Predator–One-Prey Models

Inspired by the work of Lotka and Volterra, subsequent investigators have proposed improvements in their model. One deficiency in the Lotka-Volterra approach can be seen by examining Eq. (11.22): in the absence of predator, the prey species n_1 will enjoy unbounded exponential growth. This unrealistic feature can be removed by explicitly accounting for the prey's utilization of the substrate which limits its growth.

Also, several experimental studies have revealed that predator specific growth rates do not vary linearly with prey concentration, as assumed in the original Lotka-Volterra analysis. As indicated by the Lineweaver-Burk plot for a protozoan-bacterium system in Fig. 11.12, an equation of the Monod form

$$\mu_p = \frac{\mu_{p,\,max} n_1}{K_p + n_1} \tag{11.44}$$

will often provide a better representation of predator specific growth rates.

After incorporating both these refinements and assuming constant yield factors Y_s and Y_p for prey growth on substrate and predator growth on prey, respectively, the following populations balances are obtained for a chemostat:

$$\frac{ds}{dt} = D(s_0 - s) - \frac{1}{Y_s} \frac{\mu_{s,\,max}\, s n_1}{K_s + s} \tag{11.45}$$

$$\frac{dn_1}{dt} = -Dn_1 + \frac{\mu_{s,\,max}\, s n_1}{K_s + s} - \frac{1}{Y_p} \frac{\mu_{p,\,max}\, n_1 n_2}{K_p + n_1} \tag{11.46}$$

$$\frac{dn_2}{dt} = -Dn_2 + \frac{\mu_{p,\,max}\, n_1 n_2}{K_p + n_1} \tag{11.47}$$

This model was found by Tsuchiya et al. [12] to describe most of the important features of their experimental studies of predation by the amoeba *Dictyostelium discoideum* on *E. coli* bacteria. After determining kinetic constants for this system (Table 11.5), they computed the solid curves shown in Fig. 11.13. Not only does the model predict oscillations for these operating conditions ($D = 0.0625$/h, $s_0 = 0.5$ mg/ml), but it also does well in reproducing the experimental period and the amplitudes and phase relations of all three variables.

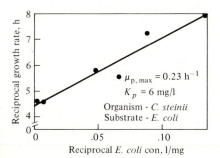

Reciprocal growth rate, h

$\mu_{p,\,max} = 0.23$ h^{-1}

$K_p = 6$ mg/l

Organism - *C. steinii*

Substrate - *E. coli*

Reciprocal *E. coli* con, l/mg

Figure 11.12 A double-reciprocal plot of predator (protozoan *Colpoda steinii*) growth rate and prey (bacterium *E. coli*) concentration. The approximately linear relationship in this plot suggests a growth-rate equation of Monod form (*Reprinted from G. Praper and J. C. Garver, Mass Culture of the Protozoa Colpoda steinii, Biotech. Bioeng.*, **8**: 287, 1966.)

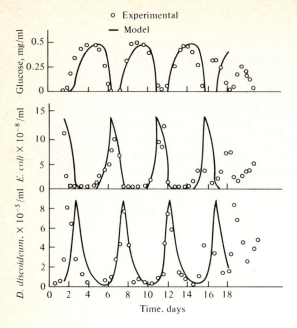

Figure 11.13 Model (—) and experimental (○) results for the prey-predator system *D. discoideum–E. coli* in continuous culture (25°C, $D = 0.0625$ h^{-1}). (*Reprinted from H. M. Tsuchiya et al., Predator-Prey Interactions of* Dictyostelium discoidem *and* Escherichia coli *in Continuous Culture, J. Bacteriol.,* **110:** 1147, 1972.)

Another weakness in the Lotka-Volterra model can now be placed in perspective. We have already seen in Fig. 11.10 that Lotka-Volterra oscillations depend on initial conditions. Viewed differently, as in Fig. 11.14*a*, this means that the Lotka-Volterra oscillations will change period and amplitude following a disturbance to the system. This behavior, sometimes termed a *soft oscillation* by mathematicians, is not very realistic from a physical viewpoint. It means that a small perturbation to the system can alter its dynamic characteristics for all future time.

The observance of regular sustained oscillations in nature, where small upsets occur frequently, suggests that natural predator-prey oscillations are more stable than those obtained from the Lotka-Volterra model. These stability characteristics are found, however, in the model described in Eqs. (11.45) to (11.47). Figure

Table 11.5. Parameter values for the prey-predator model in Eqs. (11.45) to (11.47)†

Organism	Max specific growth rates μ_{max}, h^{-1}	Saturation constant K	Yield coefficient Y
D. discoideum	0.24	4×10^8 bacteria/ml	1.4×10^3 bacteria/amoeba
E. coli	0.25	5×10^{-4} mg glucose/ml	3.3×10^{-10} mg glucose/bacterium

† From H. M. Tsuchiya et al., *J. Bacteriol.,* **110:** 1151 (1972).

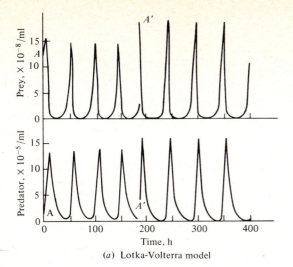

(*a*) Lotka-Volterra model

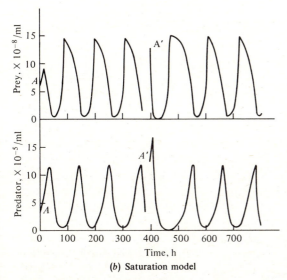

(*b*) Saturation model

Figure 11.14 Comparison of the characteristics of (*a*) the Lotka-Volterra model, which produces soft oscillations dependent on initial conditions, and (*b*) the model of Eq. (11.45) to (11.47), which gives hard oscillations independent of initial conditions. At point *A'* in these calculations, the numbers of predator and prey were shifted to new values. (*Reprinted from H. M. Tsuchiya et al., Predator-Prey Interactions of Dictyostelium discoidem and Escherichia coli in Continuous Culture, J. Bacteriol.,* **110:** 1147, 1972.)

11.14*b* indicates that oscillations predicted by this model are independent of initial conditions. Called *hard oscillations* or *limit cycles*, these oscillations are more reminiscent of actual predator-prey behavior than the soft Lotka-Volterra solutions.

Example 11.2. Model discrimination and development via stability analysis Tsuchiya, Frederickson, and coworkers [12–14] continued the above work and showed how stability analysis could be applied to differentiate between mathematical models of prey-predator systems. They built upon Canale's† results, which show that the chemostat described by Eqs. (11.45) to (11.47) can have

† R. P. Canale: An Analysis of Models Describing Predator-Prey Interaction, *Biotech. Bioeng.,* **12:** 353, 1970.

three different steady states:

Total washout:

$$n_{1s} = n_{2s} = 0 \qquad s_s = s_0$$

Predator washout:

$$n_{1s} > 0 \qquad n_{2s} = 0 \qquad 0 < s_s < s_0$$

Prey-predator survival:

$$n_{1s} > 0 \qquad n_{2s} > 0 \qquad 0 < s_s < s_0$$

Whether more than one of these steady states exists for given operating conditions D and s_0 depends upon the kinetic parameters of the system.

Likewise, stability properties depend upon all these system parameters. Before delving further into a discussion of stability analysis, we take advantage of the special structure of Eqs. (11.45) to (11.47) to reduce the number of dependent variables by one. Dividing Eq. (11.47) by Y_p, multiplying (11.45) by Y_s, and then adding the resulting two equations with (11.46) gives

$$\frac{d}{dt}\left[Y_s(s - s_0) + n_1 + \frac{n_2}{Y_p} \right] = -\frac{1}{D}\left[Y_s(s - s_0) + n_1 + \frac{n_2}{Y_p} \right] \tag{11E2.1}$$

where a term equal to zero $[d(-Y_s s_0)/dt]$ has been added to the left-hand side to make the expressions in brackets identical. Now, integrating Eq. (11.48) shows that

$$Y_s[s(t) - s_0] + n_1(t) + \frac{n_2(t)}{Y_p} = \left\{ Y_s[s(0) - s_0] + n_1(0) + \frac{n_2(0)}{Y_p} \right\} \exp\left(\frac{-t}{D}\right) \tag{11E2.2}$$

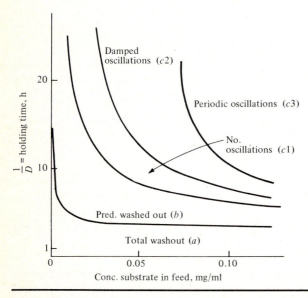

Damped oscillations (c2)

Periodic oscillations (c3)

No. oscillations (c1)

Pred. washed out (b)

Total washout (a)

$\frac{1}{D}$ = holding time, h

Conc. substrate in feed, mg/ml

Figure 11E2.1 Qualitative dynamic characteristics of the prey-predator dynamic model given in Eqs. (11.45) to (11.47) in terms of chemostat operating conditions. (*Reprinted from H. M. Tsuchiya et al., Predator-Prey Interactions of* Dictyostelium discoidem *and* Escherichia coli *in Continuous Culture, J. Bacteriol.,* **110**: 1147, 1972.)

Figure 11E2.2 Dynamic behavior of the glucose–*A. vinelandii–T. pyriformis* food chain in a chemostat. (*a*) Sustained oscillations occur for $D = 0.169$ h^{-1}. (*b*) The oscillations are damped when the dilution rate is reduced to 0.025 h^{-1}. (*Reprinted from J. L. Jost et al., Interactions of* Tetrahymena pyriformis, Escherichia coli, Azotobacter vinelandii, *and* Glucose in a Minimal Medium, *J. Bacteriol.,* **113**: 834, 1973.)

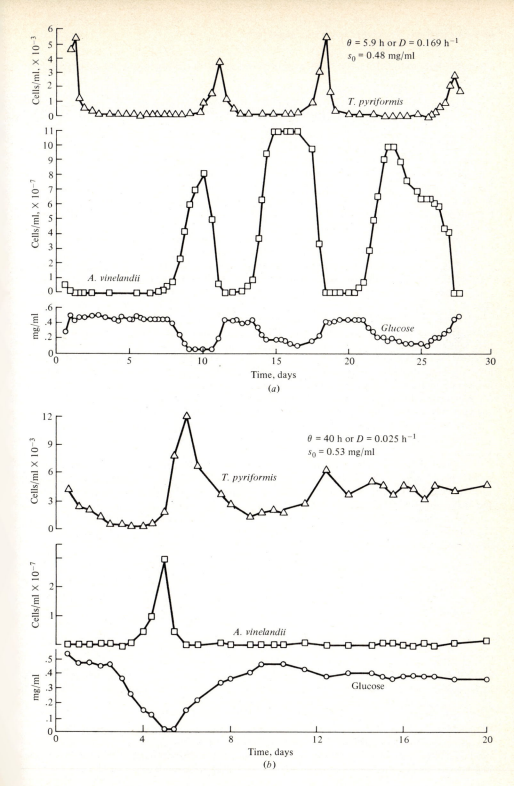

Thus, after three to five chemostat holding times have elapsed, we can say to an excellent approximation that

$$Y_s[s(t) - s_0] + n_1(t) + \frac{n_2(t)}{Y_p} \approx 0 \tag{11E2.3}$$

Using (11E2.3), we can eliminate one of the dependent variables, say $s(t)$, from Eqs. (11.46) and (11.47). Henceforth we shall invoke (11E2.3) and view our problem as a two-dimensional one.

One way to represent the general dynamic features of our chemostat is an *operating diagram* like Fig. 11E2.1. Here for given values of the kinetic constants, regions with different characteristics are labeled in the s_0, $1/D$ plane. Thus, if the point corresponding to given operating conditions s_0 and D lies in region a, for example, only the first of Canale's steady states exists and is asymptotically stable, so that washout of both predator and prey occurs.

Using the methods for local stability already presented, we can show that when steady state c exists, the remaining steady states a and b are always unstable. The three uppermost regions in Fig. 11E2.1 are distinguished by the eigenvalues characterizing steady state c. In region $c1$, both eigenvalues are real and negative: here steady state c is a stable node (recall Example 11.1). Steady state c is a stable focus for operating conditions in region $c2$. Region $c3$ results in instability for steady state c.

This latter possibility apparently poses a dilemma since *none* of the system steady states are stable. This problem is resolved by the theory of Poincaré and Bendixson, which guarantees in this case that limit-cycle oscillations are obtained. Exploitation of this theorem requires that both independent variables have upper and lower bounds, as can be readily shown for this example, and that *the model have only two independent variables*. The latter important restriction has been overlooked in some of the literature on chemical and biological oscillations (see Ref. 10 for further details).

Although the location of the boundary lines in Fig. 11E2.1 will vary somewhat with the choice of the kinetic-parameter values, the general features shown there prevail for reasonable system parameters. One feature of this operating diagram is of great interest in the current context. Suppose that periodic oscillations result for a given feed concentration and holding time. Then sustained oscillations are also expected for all larger holding times and the same inlet concentration of substrate.

This behavior is at odds with the experimental results shown in Fig. 11E2.2 for simultaneous cultivation of *Azotobacter vinelandii* and the protozoan *T. pyriformis* in a chemostat. Although oscillations appear for a 5.9-h holding time, they are damped when the holding time is 40 h. Obviously, the model proposed above is not adequate for this system.

In an attempt to improve the model, we shall add some structure to the description of predator physiology. Introducing two intermediate physiological states N_2' and N_2'' for the predator, we postulate a mechanism

$$N_2 + \alpha N_1 \longrightarrow N_2' \qquad N_2' + \beta N_1 \longrightarrow N_2'' \qquad N_2'' \longrightarrow 2N_2$$

Application of the quasi-steady-state approximation to N_2' and N_2'' gives the following expression for predator specific growth rate μ_p:

$$\mu_p = \frac{\mu_{p,\max}(n_1)^2}{(K_{p1} + n_1)(K_{p2} + n_1)} \tag{11E2.4}$$

According to this model, which reduces to the Monod form (11.44) for n_1 much larger than the smaller of K_{p1} and K_{p2}, the predator growth rate varies as $n_1{}^2$ when the prey concentration is small.

Exactly the same approach as described earlier can be employed to study the stability properties of the revised chemostat model incorporating Eq. (11E2.4). With kinetic parameters obtained from batch studies of *A. vinelandii* and *E. coli* the operating diagram of Fig. 11E2.3 is obtained. This result reveals the disappearance of sustained oscillations when D decreases in accordance with the experimental findings. Thus we have seen here how experimental data plus theoretical stability analysis can find weaknesses in a mathematical model and test an alternative formulation.

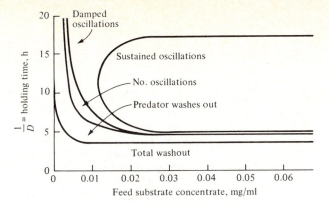

Figure 11E2.3 Operating diagram for the modified predator-prey chemostat model incorporating the multiple saturation predation rate of Eq. (11E2.4). (*Reprinted from J. L. Jost et al.,* Interactions *of* Tetrahymena pyriformis, Escherichia coli, Azotobacter vinelandii, *and* Glucose in a Minimal Medium, *J. Bacteriol,* **113:** *834, 1973.*)

11.5. EFFECTS OF THE NUMBER OF SPECIES AND THEIR WEB OF INTERACTIONS

In this section we shall proceed from specific to rather abstract examinations of how mixed populations with more than two species behave. Before starting our investigation of these sometimes difficult matters, we require some additional terminology from the science of ecology.

11.5.1. Trophic Levels, Food Chains, and Food Webs: Definitions and an Example

Often in nature, there is a hierarchy of food-consumer relationships. For example, the bacteria in the previous section are the consumers of the food glucose, and these bacteria in turn are food for protozoan consumers, which are then consumed by larger organisms. Such hierarchies are called *food chains,* and the successive steps in the chain are designated *trophic levels.* Usually, food chains are not isolated from each other but merge and intertwine to form *food webs.* A simple food web, which has been studied in the laboratory, is illustrated in Fig. 11.15; here, two food chains (substrate → prey 1 → predator and substrate → prey 2 → predator) overlap at the first and third trophic levels.

Before reviewing experimental results for the food web shown in Fig. 11.15 we should recall earlier discussions of its various components. Considering the first and second trophic levels, we have a competitive situation and the exclusion principle applies. In fact, Fig. 11.7 illustrated the validity of the exclusion principle for the first two trophic levels in question. Also, the glucose → prey 2 → predator system of Fig. 11.15 has been studied experimentally and found to yield sustained oscillations under some conditions (see Fig. 11E2.2). Consequently, it will be quite interesting to see what types of dynamic behavior emerge when these and one more prey-predator pair coalesce into the food web of Fig. 11.15.

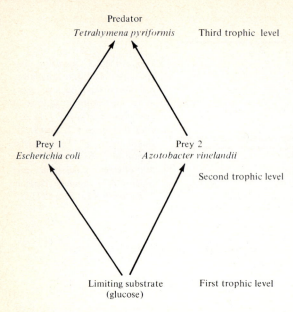

Figure 11.15 Schematic diagram of a simple food web containing three trophic levels. The two bacteria in the second level, which both consume the nutrient of the first level, are both consumed by the protozoan predator in the highest trophic level. (*Reprinted from J. L. Jost et al., Interactions of* Tetrahymena pyriformis, Escherichia coli, Azotobacter vinelandii, *and Glucose in a Minimal Medium, J. Bacteriol.,* **113:** 834, 1973.)

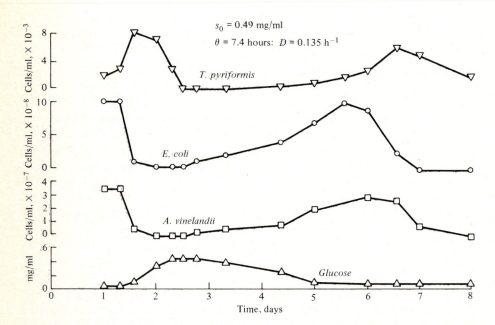

Figure 11.16 Experimentally observed oscillations in the concentrations of members of the food web shown in Fig. 11.15. (*Reprinted from J. L. Jost et al., Interactions of* Tetrahymena pyriformis, Escherichia coli, Azotobacter vinelandii, *and Glucose in a Minimal Medium, J. Bacteriol.,* **113:** 834, 1973.)

Results of an experimental study of this web in a chemostat with $D = 0.1 \ h^{-1}$ are shown in Fig. 11.16. The most notable feature of these results is the *coexistence* of the two bacterial competitors when they share a common predator. Thus, the addition of a predator for both competitors has a stabilizing effect: washout of one of the competitor species is avoided.

Although oscillations are evident in the concentrations of all three populations, the fluctuations are smaller and smoother than in the simpler predator-prey food chain. Unfortunately, the data do not extend beyond 8 days, so that possible damping of the oscillations cannot be ruled out. The experiment was stopped at this point by rapid and extensive growth on the walls of the chemostat.

11.5.2. Population Dynamics in Models of Mass-Action Form

For the remainder of Sec. 11.5, we shall consider abstract methods for analyzing and characterizing complex interacting populations. Note that many of the results cited here will also apply to complex biological reaction systems. Indeed, a frequent starting point in formulating ecological models is the assumption of a set of microbial "reactions" such as those above Eq. (11E2.4). Since the theory outlined in this subsection relies heavily on a chemical-reaction orientation, let us consider an additional example of application of this approach to microbial growth models.

Suppose that the following reactions occur in an isothermal, homogeneous, constant-volume system to which component B is added continuously so that its concentration stays constant:

$$A_1 + B \xrightarrow{\;r_1\;} 2A_1 \qquad A_1 + A_2 \xrightarrow{\;r_2\;} 2A_2 \qquad A_2 \xrightarrow{\;r_3\;} C \qquad (11.48)$$

Moreover, we shall assume that each reaction is elementary as written, so that the kinetics follows directly from the reaction stoichiometry. Consequently, for reaction system (11.48), we have

$$r_1 = k_1 a_1 b \qquad r_2 = k_2 a_1 a_2 \qquad r_3 = k_3 a_2 \qquad (11.49)$$

As a matter of convention, we need not include in our analysis concentrations which are time-invariant. To this end, let us define

$$k^* = k_1 b \qquad (11.50)$$

so that r_1 is

$$r_1 = k^* a_1 \qquad (11.51)$$

The mass balances for A_1 and A_2 can now be written

$$\frac{da_1}{dt} = k^* a_1 - k_2 a_1 a_2 \qquad (11.52)$$

$$\frac{da_2}{dt} = k_2 a_1 a_2 - k_3 a_2 \qquad (11.53)$$

which, upon comparison with Eqs. (11.22) and (11.23), we recognize as the Lotka-Volterra equations.

Now let us progress toward presentation of some very general and powerful theorems which apply to such mass-action kinetic systems. These results often will permit us to conclude, relying only on the *algebraic structure* of the reaction network, that a given system has a single steady state which is globally asymptotically stable. Here the term algebraic structure refers only to the way species are transformed and connected in the reaction network: the results we shall reach apply independent of the specific values of the rate constants or other operating parameters.

Treasures like this theory, developed by Horn, Feinberg, and Jackson [9], rarely come free so we must learn several new concepts and definitions. Our understanding of them will be greatly facilitated by reference to several examples. Some hypothetical mechanisms, as well as two of interest in our biological studies, are listed in Table 11.6. We recognize (3) as an equivalent form of the Lotka-Volterra mechanism, and (6) is one of the proposed models for glycolysis. (Can you identify A_1 through A_7 in terms of some of the compounds in the glycolytic pathway?) The *zero species* 0 appearing in these two mechanisms has the following significance: $0 \rightarrow A_j$ means that component A_j is added to the system at a constant rate. Such a reaction would be used to designate, for example, the addition of substrate to a chemostat via the feed stream. Writing $A_k \rightarrow 0$ means that A_k is removed from the system at a rate proportional to its concentration, as in removal of a species from a chemostat in the effluent stream or the death processes of a species. Thus, the zero species is used to introduce interactions with the system environment: it is the use of this device which permits application of the mass-action theory to open systems.

Next we need to define three integers for a reaction mechanism. These are n_c, the number of complexes in the mechanism; n_l, the number of linkage classes in the mechanism; and n_s, the dimension of the mechanism. Each of these in turn requires additional definitions. First, a *complex* is an entity which appears either at the head or the tail of a reaction arrow. For example, mechanisms (1) and (2) in Table 11.6 have three complexes: $2A_1$, A_2, and $A_3 + A_4$, so that here $n_c = 3$. The Lotka-Volterra mechanism (3) has $n_c = 6$ (A_1, $2A_1$, $A_1 + A_2$, $2A_2$, A_2, and 0), while there are five complexes ($n_c = 5$) for mechanisms (4) and (5): A_1, A_2, $A_1 + A_3$, A_4, and $A_2 + A_5$.

Turning now to linkage classes, let us disregard the direction of the reaction arrows and consider the complexes on either end of an arrow to be *linked* by that reaction. Considering mechanism (1), for example, we see that the complexes $2A_1$ and A_2 are linked directly in this manner. Also, while the complex A_2 is directly linked to the complex $A_3 + A_4$, we can say that the complexes $2A_1$ and $A_3 + A_4$ are linked indirectly: we can get from one complex to the other along some path consisting of direct linkage steps. A *linkage class* is a set of complexes which are all linked to one another either directly or indirectly such that no complex in the set is linked to any outside the set.

From what we have just said about mechanism (1), it has one linkage class $\{2A_1, A_2, A_3 + A_4\}$, so that $n_l = 1$ for this mechanism. Mechanism (2) has the same single linkage class. Mechanisms (4) and (5) have two ($n_l = 2$) linkage classes, which are $\{A_1, A_2\}$ and $\{A_1 + A_3, A_4, A_2 + A_5\}$, and $n_l = 3$ for mechanism (3): $\{A_1, 2A_1\}$, $\{A_1 + A_2, 2A_2\}$, $\{A_2, 0\}$.

Table 11.6. Various reaction mechanisms to illustrate definitions and conclusions of mass-action theory†

† From M. Feinberg and F. J. M. Horn, *Chem. Eng. Sci.*, **29**: 775 (1974).

The concept of *weak reversibility* is related to the linkage notion. Now we consider the direction of the reaction arrows and ask the following question: If there is a directed-arrow pathway leading from one complex to another, is there also a directed-arrow pathway leading from the second back to the first? If the answer is yes for any pair of complexes in the mechanism, we say that the mechanism is *weakly reversible*.

Mechanism (2) in Table 11.6, for example, is weakly reversible. We can move from complex $A_3 + A_4$ to complex $2A_1$ in a single step, and we can return from $2A_1$ to $A_3 + A_4$ by a two-step directed path through A_2. Moreover, similar arguments apply to every pair of complexes in the mechanism. Applying the test underlying the weak-reversibility concept to the other example mechanisms, we conclude that (1), (3), (4), and (6) are not weakly reversible and that (5) is.

One final preliminary matter must be considered before we reach the main theorems. To this end, we define a reaction vector for each reaction, i.e., each arrow, in the mechanism. If there are a total of M different chemical species $A_1, A_2, A_3, \ldots, A_M$ in the mechanism [$M = 4, 4, 2, 5, 5$ for mechanisms (1) to (5), respectively], each reaction vector will have M elements. These elements are determined as follows: the ith element is zero if A_i does not appear in the reaction. If A_i does appear, the ith entry is the stoichiometric coefficient of A_i. This coefficient is assigned a negative sign if A_i is on the tail side of the reaction arrow and positive if the reaction arrow points towards A_i.

For example, mechanism (1) of Table 11.6 has three arrows and four species. Each of these reactions is rewritten below along with the corresponding reaction vectors:

$$2A_1 \longrightarrow A_2 \qquad \mathbf{v}_1 = (-2, 1, 0, 0)$$

$$A_2 \longrightarrow A_3 + A_4 \qquad \mathbf{v}_2 = (0, -1, 1, 1)$$

$$A_3 + A_4 \longrightarrow A_2 \qquad \mathbf{v}_3 = (0, 1, -1, -1)$$

From the set of reaction vectors so obtained, we next determine the maximum number of these vectors which are linearly independent. This number we call n_s, the *dimension* of the mechanism. For instance, $n_s = 2$ for mechanism (1), since $\mathbf{v}_1$ and $\mathbf{v}_2$ above are linearly independent but $\mathbf{v}_3$ is not independent of these two ($\mathbf{v}_3 + \mathbf{v}_2 = \mathbf{0}$).

With the necessary foundation now established, we shall state the *zero-deficiency theorem* [9]. Suppose that for a given mechanism

$$n_c - n_l - n_s = 0 \tag{11.54}$$

Then:

1. For *any* kinetics, mass-action or otherwise, neither a steady state with all M concentrations positive nor periodic oscillations of the concentrations are possible if the mechanism is not weakly reversible.
2. If in addition to condition (11.54) the mechanism is weakly reversible and the kinetics are described by mass-action forms, for all stoichiometrically equivalent positive initial compositions, there is a unique steady state which is globally asymptotically stable. This conclusion applies regardless of the rate-constant values.

Two terms in the last sentence, deserve further explanation. Positive means that each species has a positive concentration. Stoichiometrically equivalent compositions are those which can be transformed into each other by running one or several reactions in the mechanism in the forward or reverse direction. Considering first only $A_1 \rightleftharpoons 2A_2$, the initial concentrations (a_{10}, a_{20}) listed next are equivalent stoichiometrically: (2, 4), (3, 2), (1, 6), and so forth so long as $2a_{10} + a_{20} = 8$. Further, for mechanism (1), the following positive initial compositions $(a_{10}, a_{20}, a_{30}, a_{40})$ are stoichiometrically equivalent: $(2, 1, 1, 1)$, $(3, \frac{1}{2}, 1, 1)$, $(2, \frac{1}{2}, 1\frac{1}{2}, 1\frac{1}{2})$, $(2, 1\frac{1}{2}, \frac{1}{2}, \frac{1}{2})$, etc.

The power of this theorem ultimately depends on how many reaction mechanisms satisfy condition (11.54). Horn has explored this question and found that the vast majority of possible mechanisms are consistent with (11.54). Our examples of Table 11.6 are therefore a somewhat atypical set, since only four [(1), (2), (3), (5)] of the six satisfy (11.54). Let us now consider further the potential applications of this theorem.

Suppose we wish to model an oscillating mixed population by postulating a population "reaction" mechanism involving several elementary steps with mass-action kinetics. The theorem then tells us that for any chance of success, the mechanism must violate either condition (11.54) or weak reversibility. Indeed, this is the case for mechanisms (3) and (6), both of which give rise to sustained oscillations. Since the vast majority of mechanisms we could choose will violate neither, we can use the theorem's conclusions to greatly limit the scope of our search. On the other hand, part 2 of the theorem tells us that for mass-action weakly reversible mechanisms which fulfill (11.54) asymptotic stability in the large is assured regardless of the exact location of the unique steady state or the values of the rate constants. Thus, in such cases, we need not bother with local-stability analyses or search for tricks to verify stability on a global scale.

Part 1 of the theorem also has important ecological consequences. It asserts that if (11.54) is satisfied while weak reversibility is not, some of the concentrations will be zero at steady state. From a population-dynamics viewpoint, then, this theory permits us to discover mechanisms for which at least one species becomes extinct.

Example 11.3. An application of the mass-action theory† In order to illustrate the usefulness of the zero-deficiency theorem, we shall use it to investigate the dynamic behavior of the following mechanism:

$$3A_1 \xrightarrow{k_a} A_1 + 2A_2 \tag{a}$$

$$3A_2 \xrightarrow{k_b} 2A_1 + A_2 \tag{b}$$

$$A_1 + 2A_2 \xrightarrow{k_c} 3A_2 \tag{c}$$

$$2A_1 + A_2 \xrightarrow{k_d} 3A_1 \tag{d}$$

which will be assumed to follow mass-action kinetics. The complexes for this mechanism are

$$3A_1 \qquad A_1 + 2A_2 \qquad 3A_2 \qquad \text{and} \qquad 2A_1 + A_2$$

so that $n_c = 4$. Next we note that the mechanism is weakly reversible. In fact, following the reaction arrows in the sequence (a) to (d) reveals a single directed path which connects all four complexes. Thus, there is only one linkage class and $n_l = 1$.

To find the mechanism dimension, we form four reaction vectors, one for each arrow in our mechanism. Each vector has two components since only two species, A_1 and A_2, are involved in this system. The reaction vectors are

$$v_a = (-2, 2) \qquad v_b = (2, -2) \qquad v_c = (-1, 1) \qquad v_d = (1, -1)$$

Since $-2v_d = 2v_c = -v_b = v_a$, there is only one linearly independent reaction: $n_s = 1$.

Testing condition (11.54) for our mechanism, we find

$$n_c - n_l - n_s = 4 - 1 - 1 = 2$$

so the condition is *not* fulfilled. Consequently, we cannot rule out the existence of oscillations or multiple steady states for this mechanism. Indeed, Feinberg and Horn show that for $k_c = k_d = 1$ and $k_a = k_b < \frac{1}{6}$, we shall observe multiple steady states for this example.

† This example is based on a discussion in Feinberg and Horn [9] and is intended for illustrative purposes only. It does not necessarily correspond to any actual microbial or biochemical reaction system.

11.5.3. Qualitative Stability

In the preceding subsection, we have perused some of the strongest theoretical tools available for a class of nonlinear process models. Our concluding studies of the influences of ecosystem size and complexity will concentrate on the linearized model. First we shall consider what can be said about stability based only on the *signs* of the entries in the community matrix.

If all eigenvalues of $\mathbf{A}$ have negative real parts regardless of the magnitudes of the nonzero elements, we shall say that $\mathbf{A}$ is *qualitatively stable*. Necessary and sufficient conditions for qualitative stability of $\mathbf{A}$ are the following:

1. $a_{jj} \leq 0$ all j

2. $a_{ii} < 0$ for at least one i

3. $a_{ij} a_{ji} \leq 0$ for all $i \neq j$

4. For any sequence of three or more unequal indices $j, k, \ldots, s, t,$

$$a_{ij} a_{jk} \cdots a_{rs} a_{st} = 0$$

5. det $\mathbf{A} \neq 0$

As discussed by May [5], each of these conditions has a direct physical interpretation. Condition 5 ensures that the linear system (11.6) has a unique steady state at $\chi = 0$. While condition 1 requires that the autonomous specific growth of every species must be nonpositive, condition 2 strengthens this restriction by stipulating that at least one of the autonomous specific growth rates be negative. Condition 4 states that there are no closed loops of interactions containing more than two species.

Especially interest in light of the interaction classifications in Table 11.4 is condition 3. Interpreted from an ecological outlook, it states that no mixed population exhibiting a mutualistic (a_{ij} and a_{ji} both positive) or a competitive (negative a_{ij} and a_{ji}) pairwise interaction can be qualitatively stable. On the other hand, all three remaining nontrivial interactions, namely commensalism, amensalism, and predation, satisfy condition 3.

> This mathematically rigorous statement may be plausibly extended into the broader, if rougher, statement that competition or mutualism between two species is less conducive to overall web stability than is a predator-prey relationship. It is tempting to speculate that stability considerations may make for communities in which strong predator-prey bonds are more common than symbiotic ones. This result is not intuitively obvious, yet it is a feature of many real world ecosystems . . . [5, p. 73].

Example 11.4. Qualitative stability of a simple food web† Consider the simple food web sketched schematically in Fig. 11E4.1a. The arrow from species 1 to species 2 indicates that species 2 feeds on species 1, so that $a_{21} > 0$ and $a_{12} < 0$. All the other arrows have analogous interpretations. Consequently, we can determine from the figure that the community matrix $\mathbf{A}_\chi$ corresponding to this web

---

† This example is based on a discussion in May [5].

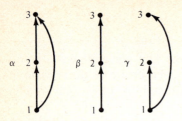

Figure 11E4.1 Three simple food webs, each with three trophic levels.

has entries with signs as indicated:

$$A_\alpha = \begin{bmatrix} - & - & - \\ + & - & - \\ + & + & - \end{bmatrix} \tag{11E4.1}$$

where it has been assumed that all species have negative autonomous specific growth rates. Checking this sign structure against the qualitative stability conditions, we see that, for instance,

$$a_{12}a_{23}a_{31} > 0 \tag{11E4.2}$$

so that condition 4 is violated. Hence there exists some choice of magnitudes for the a_{ij} for which the system is unstable, and we must check the real part of the matrix eigenvalues or an equivalent condition such as (11.14), employing the specific numerical values for the a_{ij}.

In cases β and γ (Fig. 11E4.1), however, the sign structures of the corresponding community matrices take the respective forms

$$A_\beta = \begin{bmatrix} - & - & 0 \\ + & - & - \\ 0 & + & - \end{bmatrix} \quad \text{and} \quad A_\gamma = \begin{bmatrix} - & - & - \\ + & - & 0 \\ + & 0 & - \end{bmatrix} \tag{11E4.3}$$

Obviously conditions 1 to 3 are fulfilled for both these matrices. Checking condition 4, we see that det $A_\beta < 0$ and det $A_\gamma < 0$, so that singularity is no problem. Finally, the physical interpretation of condition 4 and examination of Fig. 11E4.1 immediately reveal that no closed loops of more than two species appear in webs β or γ. This could also be checked by applying condition 4 directly to (11E4.3).

Therefore, food webs β and γ, both obtained by removing one trophic link from web α, are stable regardless of the numerical values of the a_{ij}.

11.5.4. Stability of Large, Randomly Constructed Food Webs

Next we shall review some fascinating work by Gardner and Ashby,[†] May [5], and McMurtrie[‡] on the relationship between stability and ecosystem size and complexity. Suppose that the system contains p different species which if left alone would all exhibit negative autonomous specific growth rates of similar magnitudes. By suitably scaling time, we can then assume that each species contributes an element -1 to the main diagonal of the community matrix.

The effects of interactions will be represented by a matrix **B**, so that the community matrix **A** takes the form

$$A = B - I \tag{11.55}$$

---

† M. R. Gardner and W. R. Ashby, Connectance of Large Dynamical (Cybernetic) Systems: Critical Values for Stability, *Nature*, **228**: 784, 1970.

‡ Paper in preparation cited in Ref. 5.

where $\mathbf{I}$ is the $p \times p$ identity matrix. The extent of interactions and their magnitudes will be specified by two parameters, the connectance C $(0 < C < 1)$ and the strength σ^2, respectively.

$\mathbf{B}$ matrices will now be selected at random (we can use a random-number generator to pick each element b_{ij}) with the following constraints:

Of the p^2 matrix elements b_{ij}, a fraction C will be nonzero. Stated more precisely, the probability that any b_{ij} is zero is $1 - C$.

Nonzero elements have an equal probability of being positive or negative. Thus the mean of these b_{ij} is zero.

The mean square value of the nonzero b_{ij} is σ^2.

For each $\mathbf{B}$ matrix so selected, we can evaluate $\mathbf{A}$ with (11.55) and then pursue standard stability tests. The question of major interest here is this: What is the probability that a particular $\mathbf{B}$ matrix chosen from the randomly selected set gives a stable model? From our formulation it is evident that this probability will depend on the three parameters p, C, and σ^2. May has shown that if $p \gg 1$, the probability that the system is stable approaches unity when

$$\sigma \sqrt{pC} < 1 \tag{11.56}$$

On the contrary, if $p \gg 1$ and

$$\sigma \sqrt{pC} > 1 \tag{11.57}$$

the system is almost certain to be unstable. Figure 11.17 shows results of Monte Carlo simulations which are very consistent with Eqs. (11.56) and (11.57).

Consequently, we have seen that the more species which are present (larger p), and the greater the number which interact (larger C) and the more intense the effect of interaction on growth (larger σ), the more likely it is that the ecosystem

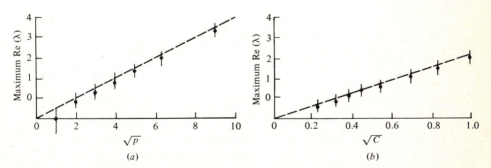

Figure 11.17 (a) The vertical lines are the results of numerical Monte Carlo calculations of the largest real-part eigenvalue of random matrices with $\sigma = 0.5$, $C = 1$. These results and the dashed line, which corresponds to Eqs. (11.56) and (11.57), show that instability becomes more likely as the number of species p increases. (b) Similar results with $\sigma = 0.5$, $p = 40$. Here the destabilizing effect of population interactions is evident. (*Reprinted from R. M. May, Stability and Complexity, in Model Ecosystems, Princeton University Press, Princeton, New Jersey, 1973.*)

will be unstable. While this conclusion is certainly consistent with stability studies of many other types of dynamic systems, we should emphasize here that the current thinking of many ecologists runs in the opposite track. Since the highly stable ecological systems which occur in nature tend to involve many species with a very complex web of interactions, it is often assumed that diversity and complexity in an ecological system make for increased stability. Although empirical observations of natural systems support this view, mathematical analyses like those outlined above do not.

May [5] suggests an idea for resolving this apparent dilemma. Although, on the average, diversity and complexity tend to destabilize a system, the natural world is anything but an "average" system. It has evolved over millennia to its present state, presumably selecting stable interrelationships among populations in preference to unstable ones. This thesis suggests an interesting avenue for future studies in mathematical ecology: find what *special* characteristics a large-scale system should have to be stable. In a sense, the results summarized in the two previous subsections are steps in this direction.

11.6. SPATIAL PATTERNS

Up until now, we have assumed that all populations are uniformly dispersed, or well mixed, within the volume of interest. While convenient for modeling and analysis, this assumption of spatial uniformity is not necessarily correct. Many systems have internal gradients of environmental conditions which tend to force inhomogeneities in population characteristics with position (recall Prob. 10.12).

Even more interesting are the numerous cases where spatial differences in one or several microbial populations arise spontaneously in the absence of any environmental gradients. Although much additional research is required to explore the detailed mechanisms whereby such conditions arise and are maintained, we can safely assume that spatial differentiation is caused by interactions between the individuals involved. These individuals may be members of two different species or cells of a single species at different stages of development.

A vivid example of this phenomenon is presented in Fig. 11.18, which shows regular sporulation of the fungus *Nectria cinnabarina* at 1-mm intervals. Evident here is a spatial periodicity: as we move away from the colony's center, we periodically encounter bands. It is intriguing to speculate that the physicochemical mechanism underlying development of spatial periodicity is the same in concept as that causing periodicities in time. Some mathematical studies point to this conclusion, although it is far from proved in general. Also, there is an actual chemical-reaction system which supports this hypothesis. Left unstirred, the Belousov-Zhabotinskii reaction generates spatial patterns which can also fluctuate with time. Well mixed, this system exhibits periodic oscillations in time.

While much more could be said on this fascinating subject, we have space to consider here only one simple model of spatial differentiation. Access to other work in this field is available in the references.

Following the work of M. E. Gurtin [11], consider two species, 1 and 2, which are confined to a one-dimensional strip from $z = 0$ to $z = L$. The confinement condition means that

$$\frac{\partial n_1}{\partial z} = \frac{\partial n_2}{\partial z} = 0 \qquad \text{at } z = 0 \text{ and } z = L \tag{11.58}$$

We shall also assume that neither species multiplies or dies. As will be proved shortly, this assumption, combined with condition (11.58), implies that the total number of n_1 and n_2 within the region is constant.

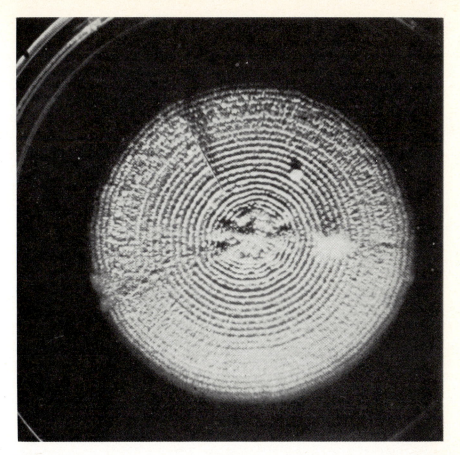

Figure 11.18 This photograph of a 6-cm-diameter mycelial colony of *N. cinnabarina* shows a regular spatial pattern of sporulation which appears here as spore ridges about 1 mm apart. (*Reprinted from A. T. Winfree, Polymorphic Pattern Formation in the Fungus* Nectria, *J. Theor. Biol.,* **38:** 363, 1973.)

Several arguments exists for supposing that the fluxes (numbers per unit time crossing a given position z) $J_1(z)$ and $J_2(z)$ are given by

$$J_i(z) = -D_{i1}\frac{\partial n_1(z)}{\partial z} - D_{i2}\frac{\partial n_2(z)}{\partial z} \qquad i = 1, 2 \tag{11.59}$$

Of course we recognize (11.59) as a general Fick's law relationship for movement of populations. We shall assume for the remainder of this analysis that the flux of species 1 is much more sensitive to the gradient of species 2 than to its own gradient and that species 1 will move in the direction of decreasing density of species 2. This suggests assuming that

$$D_{11} = 0 \qquad D_{12} = D_p > 0 \tag{11.60}$$

A little reflection upon this assumption, perhaps after this analysis is completed, will reveal that it is reasonable in many instances. Making a similar assumption for species 2 gives

$$D_{21} = D_p > 0 \qquad D_{22} = 0 \tag{11.61}$$

Under these conditions, the unsteady-state population balances for organisms 1 and 2 are

$$\frac{\partial n_1}{\partial t} = D_p \frac{\partial^2 n_2}{\partial z^2} \tag{11.62}$$

$$\frac{\partial n_2}{\partial t} = D_p \frac{\partial^2 n_1}{\partial z^2} \tag{11.63}$$

The requisite boundary conditions are already indicated in (11.58), and initial conditions are the starting distributions $n_{10}(z)$, $n_{20}(z)$ of the two populations:

$$n_i(z, 0) = n_{i0}(z) \qquad i = 1, 2 \tag{11.64}$$

Adding Eqs. (11.62) and (11.63) gives

$$\frac{\partial n}{\partial t} = D_p \frac{\partial^2 n}{\partial z^2} \tag{11.65}$$

where n is the total population at a point

$$n(z, t) = n_1(z, t) + n_2(z, t) \tag{11.66}$$

From conditions (11.58) and (11.64), we can readily show that

$$\frac{\partial n}{\partial z} = 0 \qquad \text{at } z = 0 \text{ and } z = L \tag{11.67}$$

and

$$n(z, 0) = n_{10}(z) + n_{20}(z) \tag{11.68}$$

This boundary-value problem for n is of the standard form amenable to solution by separation of variables. Without writing the transient response, we need only recall for the moment that the solution approaches a constant E as time increases:

$$\lim_{t \to \infty} n(z, t) = E \tag{11.69}$$

We can evaluate E by integrating Eq. (11.65) from $z = 0$ to $z = L$ to find

$$\frac{\partial}{\partial t} \int_0^L n(z, t) \, dz = D_p \int_0^L \frac{\partial^2 n(z, t)}{\partial z^2} \, dz = D_p \left[\frac{\partial n(L, t)}{\partial z} - \frac{\partial n(0, t)}{\partial z} \right] \tag{11.70}$$

Condition (11.67) reveals that the right-hand side of (11.70) is zero, from which we deduce that

$$\int_0^L n(z, t) \, dz = \text{const} \tag{11.71}$$

Next, we apply Eq. (11.71) at large times [Eq. (11.69)] and the initial time $t = 0$ to assert that

$$\int_0^L E \, dz = \int_0^L [n_{10}(z) + n_{20}(z)] \, dz \tag{11.72}$$

The left-hand side of Eq. (11.72) is simply EL, and so E follows immediately, permitting us to restate Eq. (11.69) in the form

$$\lim_{t \to \infty} n(z, t) = \frac{1}{L} \int_0^L [n_{10}(z) + n_{20}(z)] \, dz \tag{11.73}$$

Consequently, we have shown that as time increases, the *total* population density approaches a constant equal to the average initial density.

More interesting results are available if we consider the difference w between the two population densities

$$w = n_1 - n_2 \tag{11.74}$$

Combining the equations and conditions on n_1 and n_2 in a manner parallel to the previous development gives

$$\frac{\partial w}{\partial t} = -D_p \frac{\partial^2 w}{\partial z^2} \tag{11.75}$$

$$\frac{\partial w}{\partial z} = 0 \qquad \text{at } z = 0 \text{ and } z = L \tag{11.76}$$

and

$$w(0, z) = w_0(z) = n_{10}(z) - n_{20}(z) \tag{11.77}$$

Here, however, the similarity with the n problem ceases. Those who have studied parabolic partial differential equations will immediately recognize Eqs. (11.75) to (11.77) as an ill-posed problem.

At any point z where $w_0(z)$ is nonzero, the magnitude of w will grow without bound as time advances. We can illustrate this general conclusion with a specific example. When we take

$$w_0(z) = \varepsilon \cos \frac{k\pi z}{L} \tag{11.78}$$

a solution of Eqs. (11.75) to (11.78) is

$$w(z, t) = \varepsilon \cos \frac{k\pi z}{L} \exp \frac{D_p k^2 \pi^2}{L^2} t \tag{11.79}$$

Clearly, $w(z, t)$ approaches either plus or minus infinity as $t \to \infty$ except at the points where $w_0(z)$ is zero. Of course, this result is physically meaningless once either n_1 or n_2 reaches zero, and we must assume that once this has occurred our model no longer applies.

This result plus Eq. (11.73) indicates that unless $n_{10}(z) = n_{20}(z)$ at a point z, one of the species will vanish at that point as time progresses while the other will approach a constant value. The species which survives in a given spatial region will be the one initially present in greatest density. Thus, the end result is total segregation of the two interacting populations.

In this chapter we have explored some of the biological and mathematical principles which underly the behavior of multiple interacting populations. We have seen that experimental studies of these systems are possible under carefully controlled conditions and that mathematical models can be formulated which are consistent with the general features observed in the laboratory. Next, in Chap. 12, we leave the sheltered environment of laboratory, analysis, and computer to consider some of the extremely complicated mixed-population systems which occur in nature and which are exploited industrially. We shall discover that our fundamental studies just concluded are very valuable in these contexts, even though we still lack the necessary insight to accomplish predictive analyses of complex ecosystems.

PROBLEMS

11.1. Competition in batch growth Two competing species A and B are placed in a batch fermenter containing a substrate concentration $s_0 \gg K_A, K_B$, where

$$\mu_i = \frac{\mu_{\max, i} s}{K_i + s} \qquad i = A, B$$

Show that at substrate exhaustion,

$$(n_{Af})^{\mu_{max,A}}(n_{Bf})^{\mu_{max,B}} = \text{const}$$

where n_{Af}, n_{Bf} = cell populations when $s = 0$. (This result was observed by Talling [20] using mixed diatom cultures.)

11.2. Simple mutualism model Meyer, Tsuchiya, and Frederickson [21] have examined the stability of two mutualistic populations with the forms

$$\frac{dn_i}{dt} = -Dn_i + \mu_i n_i \qquad i = 1, 2$$

$$\frac{ds_i}{dt} = -Ds_i + \alpha_i \mu_j n_i - \beta_i \mu_i n_i \qquad i, j = 1, 2; \; i \neq j$$

Here it is assumed that the medium contains all necessary nutrients except s_1 (produced by n_2) and s_2 (produced by n_1).

(a) For each of the growth forms $\mu_i = \mu_{max,i} s_i$ and $\mu_i = \mu_{max,i} s_i / (K_i + s_i)$, show that only two steady-state solutions exist and that the nonwashout solution exists only if $\alpha_1 \alpha_2 > \beta_1 \beta_2$ and additionally, for the Monod substrate dependence, if $D < $ minimum of $(\mu_{max,i}, i = 1, 2)$.

(b) Show by a Taylor's series expansion about the nonwashout solution of part (a) that the solution is unstable and hence that a purely mutualistic interaction of the sort considered above cannot exist.

11.3. Mutualism plus competition Meyer et al. [21] suggest that one resolution to the result of Prob. 11.2 is to add a dependence of the two species on a common substrate, s_3, which appears as a separate factor of the Monod form in each growth equation, for example, Eq. (7.18).

(a) Write out the five equations now needed to describe transient chemostat behavior.

(b) Discover (by graphical or computer means) the solutions when $\mu_{max,1} = 0.3 \text{ h}^{-1}$, $\mu_{max,2} = 0.2 \text{ h}^{-1}$, $K_1 = 10^{-3}$ g/l, $K_2 = 0.002$ g/l, $K_{13} = 0.02$ g/l, $\alpha_1 = 2.0$, $\alpha_2 = 1.0$, $\beta_2 = 4.0$, and the yield factors for s_3 are $Y_{13} = 2.5$ and $Y_{23} = 5$.

(c) Expand the original set of equations in a Taylor's series again, and establish the (in)stability to small perturbations in each variable.

11.4. Mutualism with substrate inhibition A second modification of the conditions of Prob. 11.2 which leads to some stable nonwashout solutions (coexistence) in a chemostat arises when the equation of Andrews (7.16) is assumed to govern species growth:

$$\mu_i = \frac{\mu_{max,i} s_i}{K_i + s_i + s_i^2 / K_i'}$$

(a) Write out the system equations for the time evolution of a chemostat.

(b) Show that four coexistence solutions (not including washout) may exist at steady state provided $\alpha_1 \alpha_2 > \beta_1 \beta_2$ and $D < $ minimum of the maximum absolute values which μ_i ($i = 1, 2$) above can have.

(c) Show by linearized stability analysis that the steady state will be stable if the following conditions are satisfied:

$$\alpha_1 A_1 + \alpha_2 A_2 > 0 \qquad \text{and} \qquad \bar{s}_1 < \sqrt{K_1 K_1'} \text{ and } \bar{s}_2 > \sqrt{K_2 K_2'}$$

or the reverse for *both* inequalities.

$$A_i = \frac{\mu_{max,i} \bar{n}_i (K_i K_i' - \bar{s}_i^2)}{K_i' [(K_i + \bar{s}_i + \bar{s}_i^2)/K_i']^2} \qquad i = 1, 2$$

(The bar over a variable indicates the steady-state value.)

(d) The meaning of the inequalities in the stable solutions is that stability arises only when one species is substrate-rate-limited and the other is substrate-inhibited. Indicate by sketches why such a system is stable to perturbations in s_1, s_2 and why simultaneous substrate limitation or inhibition for two species is unstable.

11.5. *Methanobacillus omelianskii* In the text discussion of mutualism, an anaerobic example was mentioned involving the two species n_1 and n_2 now known to make up "*M. omelianskii*" (Sec. 11.1):

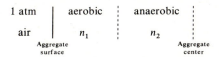

$$CH_3CH_2OH + H_2O \xrightarrow{\quad n_1 \quad} CH_3COO^- + H^+ + 2H_2$$

$$4H_2 + CO_2 \xrightarrow{\quad n_2 \quad} CH_4 + 2H_2O$$

Thus, the first species produces substrate for the second but is now *inhibited* by its own product, H_2.

(*a*) Why is this interaction *mutualism* rather than a form of *commensalism*?

(*b*) Write out appropriate transient system equations assuming reasonable forms for inhibition by H_2.

(*c*) Discuss the system stability using a Taylor's series expansion about a steady state.

11.6. Particles with aerobic and anaerobic zones In soil aggregates, obligate anaerobes may exist in the center of the moist aggregate, and their propagation toward the periphery depends on the reduction of dissolved oxygen to a subcritical level

1 atm	aerobic	anaerobic
air	n_1	n_2
	Aggregate surface	Aggregate center

(*a*) Write down the equations describing the system behavior for the following cases where carbon nutrient s_1 is continuously trickled over the outer soil-particle surface:

1. n_1 requires both nutrient s_1 and oxygen, and n_2 consumes product from n_1 and returns nothing. nothing.
2. n_1 and n_2 compete for the nutrient s_1; no other interaction.

(*b*) How does the nonuniformity of the oxygen distribution, giving rise to aerobic and anaerobic *niches*, change stability concepts from those in Probs. 11.2, 11.3, and 11.4?

11.7. Niches from imperfect mixing Figure 11.8 indicates that the dilution rate will determine the dominant population for a two-species system consuming a common substrate if growth functions μ_1 and μ_2 cross when plotted vs. substrate. Figure 9.10 and the associated analysis for an imperfectly mixed reactor consider a division of the reactor into two well-mixed systems, each with its own dilution streams.

(*a*) Write the system equations for two species in such an imperfectly mixed system.

(*b*) Establish under what conditions one population dominates one vessel and the other the second, i.e., under what dilution rate, etc., a spatial niche stabilizes the unstable coexistence of the two species shown in Fig. 11.8.

(*c*) Why do students and faculty live in separate housing?

11.8. Feinberg-Horn-Jackson analysis For Probs. 11.2 through 11.5:

(*a*) Write down the microbial system description in the terms of the Feinberg-Horn-Jackson analysis (Table 11.6).

(*b*) Identify each complex, each linkage class, and the presence or absence of weak reversibility.

(*c*) Apply the deficiency theorem (11.54) and discuss each result in terms of nontrivial existence and system stability.

(*d*) How would you apply this analysis to a tanks-in-series reactor configuration?

11.9. Intracellular network stability The Feinberg-Horn-Jackson approach applies to any system exhibiting mass-action kinetics, thus to intracellular reaction networks.

(*a*) Write out in appropriate notation the reaction steps involved in nucleotide biosynthesis (Fig. 6.40).

(*b*) Determine from the deficiency theorem whether the normal network (Fig. 6.40*a*) or mutant network (Fig. 6.40*b*) is stable or may exhibit oscillatory behavior.

11.10. Leslie's equations Leslie [22] suggested a predator-prey formulation of the form

$$\frac{dx}{dt} = ax - bx^2 - cxy \qquad \frac{dy}{dt} = ey - \frac{fy^2}{x}$$

The last term in the second equation introduces a ratio of prey to predator.

(a) Plot the equations resulting when dx/dt and $dy/dt = 0$ on a y-vs.-x plot. Choose some simple positive values for a, b, c, e, f, and by occasional evaluation of dy/dx sketch the behavior of the system starting at other than the equilibrium point.

(b) Is the form of the apparent *specific* growth rate for y reasonable as y/x is held constant while x is decreased to small values? Why (not)?

(c) Prove that this system exhibits damped oscillations.

11.11. Statistical mechanics of populations As our essays include mathematical modeling of progressively more complex systems, we often are tempted to draw analogies with apparently similar systems. Read chap. 8 Ref. 23. Write a short summary discussing the similarities and differences between the apparent behavior of molecular vs. cellular populations. Include considerations of open vs. closed systems, equilibrium vs. steady state, equations of motion and their invariants (the conserved quantities).

11.12. Food-web kinetics Consider the sequential food chain $1 \rightarrow 2 \rightarrow 3 \rightarrow 4$, where organisms 2, 3, and 4 are the predators of 1, 2, and 3 respectively.

(a) Assuming a logistic growth rate for each species, and including a predation term, derive the steady-state solutions for this system, showing in particular that the level of 1 depends on *all* the kinetic constants of the system while the level of 4 depends only on the logistic equation for 4 [24].

(b) Since carbon, nitrogen, etc., are obviously recycled, include the concentration of 4 in the growth rate for 1 (representing, for example, mortality by a virulent disease) so that the web is now closed. Show that the steady-state solutions now have identical forms for each species (each is dependent on *all* the web kinetic constants). Discuss quantitatively the stability of these two systems, i.e., the linear vs. cyclic food web.

11.13. Slime-mold aggregation [25–27] When a growing slime-mold amoeba population exhausts its food supply, a relatively quiescent period is followed by an aggregation of the originally uniformly dispersed amoeba into groups spaced ~ 0.1 mm apart. These aggregates eventually produce spores. The aggregation appears to be due to an attractive amoeba response to a gradient in cyclic AMP secreted by the starving amoeba themselves. Let a = amoeba density and ρ = attractant (cyclic AMP) density; then a model can be developed yielding

$$\frac{\partial a}{\partial t} = \frac{\partial}{\partial x}\left(\mu \frac{\partial a}{\partial x} - \chi a \frac{\partial \rho}{\partial x}\right) \qquad \frac{\partial \rho}{\partial t} = fa - k\rho + D \frac{\partial^2 \rho}{\partial x^2}$$

(a) Show from linearization of these equations that assumed solutions of the form $(\sin q)e^{\sigma t}$ for a and ρ lead to stability if $\chi a_0 f < \mu(k + Dq^2)$, $q \neq 0$.

(b) Which wavelength perturbations lead to the earliest instability?

(c) It is argued that the initial system is stable but the starvation changes the system parameter values and produces a situation unstable for a = const. Identify the meaning of each parameter in the model. Which of these might reasonably change in a way to bring the system into instability?

(d) Why is aggregation necessary for spore formation?

(e) Discuss the similarities and differences between this problem and Prob. 10.12.

REFERENCES

Some information on mixed population situations is available in the general microbiology references given in Chap. 1. Reference 3 of that chapter has three chapters on symbiosis. See also:

1. E. P. Odum: "Fundamentals of Ecology," 3d ed., W. B. Saunders Company, Philadelphia, 1971. An extensive and popular textbook dealing with the entire spectrum of ecological science.

2. G. F. Gause: "The Struggle for Existence," Williams & Wilkins Company, Baltimore, 1934. The work that set the stage for most of the future experimental and theoretical studies in microbial

interactions. Many of the concepts, definitions, and outlooks presented in this chapter can be traced back directly to the influence of this classic monograph.

3. A. I. Laskin and H. Lechevalier (eds.): "Microbial Ecology," CRC Press, Cleveland, 1974. The first two chapters deal with microbe-pesticide interactions, and the third chapter reviews the ecology of soil microorganisms. The final chapter, by J. L. Meers, Growth of Bacteria in Mixed Cultures, is outstanding and highly recommended.

4. T. Hattori: "Microbial Life in the Soil: An Introduction," Marcel Dekker, Inc., New York, 1973. A good overview of soil ecosystems, which are very complex and provide many interesting examples of microbial interactions.

5. R. M. May: "Stability and Complexity in Model Ecosystems," Princeton University Press, Princeton, N.J., 1973. A beautifully written monograph on the mathematics of interacting populations which served as the basis for much of Secs. 11.4 and 11.5. The idea that complexity breeds instability is presented in compelling fashion.

6. H. R. Bungay and M. L. Bungay: Microbial Interactions in Continuous Culture, *Adv. Appl. Microbiol.*, **10**: 269, 1968. Although now somewhat dated, still a useful source of ideas and additional reading.

7. A. Rescigno and I. W. Richardson: The Deterministic Theory of Population Dynamics, p. 283 in R. Rosen (ed.), "Foundations of Mathematical Biology," vol. 3, "Supercellular Systems," Academic Press, Inc., New York, 1973. Contains an excellent presentation of Volterra's work on competition and predation and reviews some recent developments in population dynamics, especially the work of Kostitzin.

8. A. J. Lotka: Undamped Oscillations Derived from the Law of Mass Action, *J. Am. Chem. Soc.*, **42**: 1595, 1920. The first work leading to the famous Lotka-Volterra model for predator-prey interactions.

9. M. Feinberg and F. J. M. Horn: Dynamics of Open Chemical Systems and the Algebraic Structure of the Underlying Reaction Network, *Chem. Eng. Sci.*, **29**: 775, 1974. A review paper of the mass-action theory developed by the authors and R. Jackson, summarized in Sec. 11.5.2. (A more recent account is given by M. Feinberg, same volume as Ref. 10, p. 1.)

10. J. E. Bailey: Periodic Phenomena in L. Lapidus and N. R. Amundson (eds.), "The R. H. Wilhelm Memorial Volume on Chemical Reactor Theory," Prentice-Hall, Inc., Englewood Cliffs, N.J., 1977. An introduction to the mathematical theory of spontaneous and forced oscillations set in the context of chemical reaction systems.

11. M. E. Gurtin: Some Mathematical Models for Population Dynamics That Lead to Segregation, *Q. Appl. Math.*, **32**: 1 (1974). For further details on the example presented in Sec. 11.6 as well as related problems.

Papers on predator-prey interactions:

12. H. M. Tsuchiya et al.: "Predator-Prey Interactions of *Dictyostelium discoideum* and *Escherichia coli* in Continuous Culture," *J. Bacteriol.*, **110**: 1147, 1972.

13. J. L. Jost et al.: Interactions of *Tetrahymena pyriformis*, *Escherichia coli*, and *Azotobacter vinelandii* and Glucose in a Minimal Medium, *J. Bacteriol.*, **113**: 834, 1973.

14. J. L. Jost et al.: Microbial Food Chains and Food Webs, *J. Theoret. Biol.*, **41**: 461, 1973.

Further information on spatial patterns:

15. A. M. Turing: The Chemical Basis of Morphogenesis, *Trans. R. Soc. Lond.*, **B237**: 337, 1952.

16. P. Glansdorff and I. Prigogine: "Thermodynamic Theory of Structure, Stability, and Fluctuations," Wiley-Interscience, London, 1971.

17. H. G. Othmer and L. E. Scriven: "Instability and Dynamic Pattern in Cellular Networks," *J. Theoret. Biol.*, **32**: 507, 1971.

18. R. J. Feld: A Reaction Periodic in Time and Space, *J. Chem. Educ.*, **49**: 308, 1972.

19. A. T. Winfree: Rotating Chemical Reactions, *Sci. Am.*, June, 1974, p. 82.

Problems

20. J. F. Talling: The Growth of Two Plankton Diatoms in Mixed Culture, *Physiol. Plant.* **10**: 215, 1957.

21. J. M. Meyer, H. M. Tsuchiya, and A. G. Frederickson: Dynamics of Mixed Populations Having Complementary Metabolism, *Biotech. Bioeng.*, **17**: 1065, 1975.

22. P. H. Leslie: Some Further Notes on the Use of Matrices in Population Mathematics, *Biometrika*, **35:** 213, 1948.
23. J. M. Smith: "Models in Ecology," Cambridge University Press, London, 1974.
24. D. J. Rapport and J. Turner: Predator-Prey Interactions in Natural Communities, *J. Theoret. Biol.*, **51:** 169, 1975.
25. J. T. Bonner: "The Cellular Slime Molds," Princeton University Press, Princeton, N.J., 1967.
26. E. F. Keller and L. A. Segel: Initiation of Slime Mold Aggregation Viewed as an Instability, *J. Theoret. Biol.*, **26:** 399, 1970.
27. C. C. Lin and L. A. Segel: "Mathematics Applied to Deterministic Problems in the Natural Sciences," pp. 22–31, The Macmillan Company, New York, 1974.

BIOLOGICAL REACTORS, SUBSTRATES, AND PRODUCTS II: MIXED MICROBIAL POPULATIONS IN APPLICATIONS AND NATURAL SYSTEMS

The importance of mixed populations of microorganisms can be traced to the beginnings of life on earth. Since then, different species of microbes have played an integral role in the operation of the biosphere and in its evolution. For example, some time between 500 million and 2 billion years ago, the development of primitive algae capable of photosynthesis had reached the point where a significant amount of oxygen (about 1 percent of the current level) was present in the atmosphere. Although microbial life before that time was limited to strictly anaerobic forms, aerobes and facultative anaerobes emerged afterward. Thus the generation of an oxygen atmosphere and subsequent emergence of aerobic life forms on earth can be attributed to primeval photosynthetic microorganisms. In Sec. 12.3, we shall briefly survey contributions of the microbial world to the contemporary global ecosystem.

Mixed microbial populations abound in nature—in the air, soil, and bodies of water. They also grow on and inside higher organisms. Among the most interesting examples are the symbioses with the ruminant animals such as cattle, sheep, and goats (see Table 11.1). In the rumen, which comprises the first two of at least four stomachs, a dense mixed culture of bacteria (about 10^{10} cells per milliliter) and protozoa thrives. In this complicated and extremely diverse population, mutualistic, competitive, amensalistic, and prey-predatory interactions have been observed.

The overall effect of microbial activity in the rumen is decomposition of plant material, including cellulose and other complex carbohydrates, into simpler substances which can be absorbed in the animal's bloodstream. Digestive enzymes are secreted by the ruminant only in stomachs following the rumen. This permits unhindered microbial activity in the rumen and provides for lysis and digestion of ruminant microorganisms in later stomachs. Such a novel design allows ruminants to ingest and utilize much simpler nitrogen nutrients than man and other mammals. Ammonia or urea, for example, are incorporated into combined organic forms such as proteins by rumen microorganisms, and it is the microbially produced nitrogen compounds which are absorbed by the cow in stomachs following the rumen.

Other familiar examples of mixed microbial populations include the natural flora of microorganisms which inhabit the human body. Proper digestion requires the assistance of many bacteria which reside in the intestinal tract. Commensalism and amensalism among these organisms have been observed, and substantial populations of protozoa feed on the intestinal bacteria. Dental plaque consists of several microorganisms, facultative anaerobes being among the most common species. (Why would these be well suited to survive common dental hygienic practices?) The skin is populated by many different protists, with bacteria well entrenched in hair follicles, and yeasts and fungi growing on moist regions. Investigation of how the human body protects itself and often benefits from these numerous cohabitants is a fascinating study but would take us too far afield from our major theme. Consequently, we shall turn now to commercial exploitation of mixed microbial populations.

Here too there is a long history. Ancient artifacts reveal that wine and beer were made by fermenting fruits and grains before 2000 B.C. These processes, at least as conducted then, are excellent examples of applications of *naturally occurring mixed cultures*. In natural mixed-culture processes, inoculation with specific organisms is not practiced. The organisms naturally present when the fermentation begins are responsible for the desired changes. As we shall explore in greater detail below, spontaneous mixed populations are especially efficient in utilizing substrate mixtures.

These examples drawn from naturally arising populations suggest that simultaneous growth of several species offers special advantages and characteristics unattainable in pure cultures. In the next section, we shall briefly review the status of defined mixed-population technology. The remainder of this chapter is devoted to applications of naturally occurring mixed cultures, with heavy emphasis on biological waste-water treatment (Sec. 12.4).

12.1. USES OF WELL-DEFINED MIXED POPULATIONS

The foremost illustration of defined mixed-culture application is cheese manufacture. While the gastronomical benefits of such activities are well known, the economic importance of the cheese and dairy industries is not widely appreciated.

According to one estimate, roughly 4 percent of the United States' Gross National Product is contributed by the dairy industry, with about 15 percent of that figure due to cheese production.

Although indigenous mixed cultures were originally used in making cheese, special starter cultures are now employed to ensure reproducibility of product quality. Cheeses of various types are produced by inoculating pasteurized fresh milk with appropriate lactic acid–producing organisms. The resulting protein-aceous curd precipitated by the acidity of the medium is drained of liquid (*whey*) and allowed to age or ripen by action of bacteria or mold. With hard-curd cheese, an enzyme mixture of rennet (containing the proteolytic enzyme rennin) is added to the inoculated milk after slightly acid conditions are produced. A rubberlike curd eventually results, which is cut into small pieces, heated, drained of whey, and finally milled into shavings and pressed to remove further whey (the cheddaring process). Finally, a second milling, salting, draining, and pressing into molds yields the cheese to be cured. In curing, the slow fermentative (anaerobic) action partially breaks down lipids and proteins to produce additional partially oxidized products such as lactic, butyric, and acetic acids, which, with the continued increase in cheese age, contribute to the sharp taste of the resultant cheese.

Many natural cheeses are similar in initial stages; the final product consistency and taste are predominantly determined in the later curing stages by the organisms used, salt and humidity levels, and the curing temperature. A representative listing of common cheeses and organisms is given in Table 12.1. Many of these organisms synthesize vitamins which increase the nutritional value of the cheese during curing.

Lactic acid bacteria also participate in other defined mixed cultures used in food production. In whiskey manufacture, for instance, a *Lactobacillus* added to the yeast lowers pH to reduce contamination and also contributes to a desirable flavor and aroma. Another example of favorable interaction between a yeast and lactic acid bacterium is ginger-beer fermentation. Also, use of two *Lactobacillus* species increases yield in the lactic acid fermentation.

Consecutive transformation of a nutrient into a desired final product can sometimes be accomplished effectively by a tactic called *dual fermentation*. One such process makes L-lysine from glycerol, with α,ε-diaminopimelic acid (DAP) as an intermediate. In one fermentation, the DAP intermediate is accumulated using an *E. coli* auxotroph. Separately, an *Aerobacter aerogenes* population is grown. This organism synthesizes DAP decarboxylase, an enzyme which acts on DAP to yield L-lysine. Combining the two cultures and adding toluene liberates both DAP and its decarboxylase, so that L-lysine is produced.

There are few cases in which multiple defined cultures are grown simultaneously for commercial use. One is manufacture of β-carotene, where different mating types of the same organism grown together provide about 20 times more product than either type by itself. In the following example, we shall explore how methane utilization for single-cell protein production is improved by using a mixed culture.

Table 12.1. Some cultures used in manufacture of cheese†

a. Cultures used in cheese production

I. Bacteria
  A. Used for lactic acid production primarily:
    1. *Streptococcus lactis* "lactic"
    2. *Streptococcus cremoris* "lactic"
    3. *Streptococcus thermophilus* "coccus"
    4. *Streptococcus durans* (USDA modified cheddar "make" procedure)
    5. *Lactobacillus bulgaricus* "rod"
  B. Used to develop flavor and aroma with or without an effect upon the body
    and texture:
    6. *Brevibacterium linens* "red smear"
    7. *Propionibacterium shermanii* "props"
    8. *Leuconostoc* sp. (flavor associates or citric acid fermenters—"CAFs")
    9. *Streptococcus diacetilactis* (flavor and special uses)

II. Yeasts
  A. Used for maintenance and enhancement of bacterial growth in cultures and
    cheese (*not essential*)
    10. *Candida krusei*
    11. *Mycoderma* sp.

III. Molds (to enhance flavor, aroma, body, texture, composition and appearance):
  A. Externally grown
    12. *Penicillium camemberti*
    13. Miscellaneous species
  B. Internally grown
    14. *Penicillium roqueforti* and variants thereof

b. Use relationship between starter culture types and characteristic cheese varieties

Cheese variety	Type of cultures in use Class I		Classes II and III	How used‡
Brick	A.1–3	B.6, 8, 9	II.A	*x–z*
Camembert	A.1, 2	B.8, 9	III.A.12	*x, z*
Cheddar	A.1, 2, 4	B.8, 9	⋯	*x–z*
Cottage	A.1, 2	B.8, 9	⋯	*x, z*
Cream	A.1, 2	B.8, 9	⋯	*x, z*
Edam	A.1, 2	B.8, 9	⋯	*x, z*
Gouda	A.1, 2	B.8, 9	⋯	*x, z*
Limburger	A.1–3	B.6, 8, 9	II.A	*x–z*
Neufchatel	A.1, 2	B.8, 9	⋯	*x, z*
Parmesan	A.1–3, 5	B.8, 9	⋯	*x–z*
Provolone	A.1–3, 5	B.8, 9	⋯	*x–z*
Romano	A.1–3, 5	B.8, 9	⋯	*x–z*
Roquefort	A.1, 2	B.8, 9	III.B.14	*x, z*
Swiss	A.1–3, 5	B.7–9	⋯	*x–z*
Trappist	A.1–3	B.6, 8, 9	II.A	*x–z*

† Courtesy of G. W. Reinbold, Iowa State University, Ames, Iowa.
‡ Key:
    x = Single-strain cultures; *y* = multiple-strain cultures; *z* = combined mixed-strain cultures (dual purpose, and/or single or multiple strains).

Example 12.1. Enhanced growth of methane-utilizing _Pseudomonas_ sp. due to mutualistic interactions in a chemostat† In Sec. 10.3.1 and Example 10.3, we considered some aspects of single-cell protein production. Substantial recent interest has been focused on gaseous hydrocarbon substrates, particularly methane. Several experimental studies have shown that mixed cultures often have the ability to grow more rapidly on methane than pure cultures of methane-utilizing microorganisms.

A tentative explanation for this observation is a mutualistic interaction like the one shown schematically in Fig. 12E1.1. In this mixed population, studied experimentally by Wilkinson, Topiwala, and Hamer, there are two major classes of bacteria: methane-utilizing _Pseudomonas_ sp. and methanol-utilizing _Hyphomicrobium_ sp. Because methanol is a metabolic end product of the _Pseudomonas_ sp. and also inhibits the growth of those organisms, they benefit greatly from the second

† This example is drawn from T. G. Wilkinson, H. H. Topiwala, and G. Hamer, Interactions in a Mixed Population Growing on Methane in Continuous Culture, _Biotechnol. Bioeng.,_ **16:**41–60 (1974).

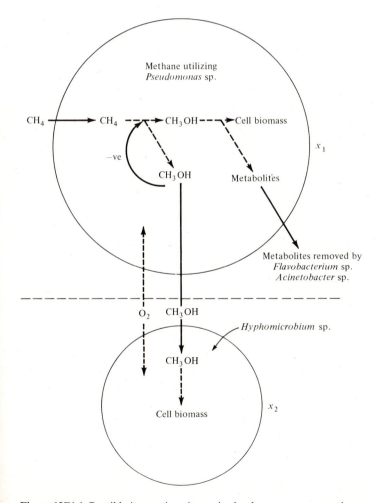

Figure 12E1.1 Possible interactions in a mixed culture grown on methane.

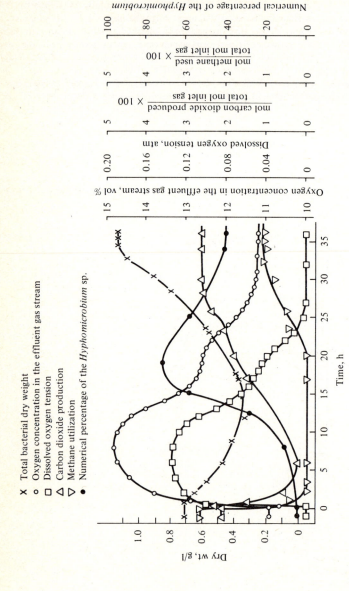

Figure 12E1.2 Transient behavior of a methane-utilizing mixed culture following a shock load to a chemostat. At $t = 0$, the methanol concentration in the fermentor and in the feed was raised to 1.6 g/l. ($D = 0.08$ h^{-1}). (*Reprinted from T. G. Wilkinson et al., Interactions in a Mixed Bacterial Population Growing on Methane in Continuous Culture, Biotech. Bioeng.,* **16**: 41, 1974.)

× Total bacterial dry weight
○ Oxygen concentration in the effluent gas stream
□ Dissolved oxygen tension
△ Carbon dioxide production
▽ Methane utilization
● Numerical percentage of the *Hyphomicrobium* sp.

population, which removes methanol. Clearly, the *Hyphomicrobium* sp. also enjoy the interaction, since they are supplied with a carbon source. The two other types of bacterial species in the mixed population are present in relatively small concentrations and are believed to serve useful functions by utilizing other end metabolites of the *Pseudomonas* sp.

While we shall leave the complete argument for this interpretation to the original paper, one experiment which shows several effects consistent with the scheme is indicated in Fig. 12E1.2. The mixed population was grown in a chemostat, and at $t = 0$ methanol was added to the fermentor and medium reservoir to establish a methanol concentration of 1.6 g/l. As Fig. 12E1.2 shows, this shock causes an immediate and severe drop in methane and oxygen utilization and a gentler decrease in total dry weight of bacteria. Also, the percentage of *Hyphomicrobium* sp. in the fermentor increases. After about 17 h, the methanol concentration in the vessel is near zero, leading to a reversal of the above trends. Presumably at this point *Pseudomonas* activity resumes to a significant extent.

All the component parts for the mathematical model of this system should now be familiar from our previous studies of mass transfer and biological kinetics. Letting x_1 and x_2 denote the concentrations of *Pseudomonas* and *Hyphomicrobium* species, respectively, we see that their unsteady-state material balances for a chemostat are

$$\frac{dx_i}{dt} = -Dx_i + r_f \qquad i = 1, 2 \tag{12E1.1}$$

The forms of the *Pseudomonas* and *Hyphomicrobium* growth rates are chosen to reflect the situation depicted in Fig. 12E1.1. In particular, for oxygen-limited growth of the *Pseudomonas* bacteria, we take

$$r_{f1} = \frac{\mu_{1,\,\text{max}} c_{O_2}}{K_1 + c_{O_2}} \frac{1}{1 + s/K_i} x_1 \tag{12E1.2}$$

Table 12E1.1. Kinetic parameters for a model of methane-utilizing mixed population†

Parameter	Comment	Numerical value
$\mu_{1,\,\text{max}}$	Maximum specific growth rate for methane-utilizing component	$0.185\ \text{h}^{-1}$
$\mu_{2,\,\text{max}}$	Maximum specific growth rate for methanol-scavanging component	$0.185\ \text{h}^{-1}$
K_1	Michaelis constant for oxygen consumption by x_1	$1 \times 10^{-5}\ \text{g/l}$
K_2	Michaelis constant for methanol consumption by x_2	$5 \times 10^{-6}\ \text{g/l}$
K_i	Methanol-inhibition constant (hyperbolic) for x_1	$1 \times 10^{-4}\ \text{g/l}$
Y_1	Stoichiometric yield constant for methanol production by x_1	5.0 g bacteria/g methanol
Y_2	Stoichiometric yield constant for methanol-consumption by x_2	0.3 g bacteria/g methanol
Y_3	Stoichiometric yield constant for oxygen consumption by x_1	0.2 g bacteria/g oxygen
$k_L a$	Oxygen mass-transfer product (specific area · coefficient)	$42.0\ \text{h}^{-1}$
c_{O_2}	Dissolved-oxygen saturation level	0.128 atm (0.008 g/l)

† T. G. Wilkinson, H. H. Topiwala, and G. Hamer, *Biotech. Bioeng.*, **16**: 56, 1974.

where c_{O_2} and s are dissolved oxygen and methanol concentrations, respectively. The inhibition function used here parallels those of Chap. 3 (noncompetitive inhibition of enzyme-catalyzed reactions) and Chap. 7 (ethanol inhibition of yeast fermentation). Since the *Hyphomicrobium* can readily use nitrate as an electron acceptor when dissolved-oxygen concentration is low, we assume that r_{f2} is independent of c_{O_2}, while adopting the Monod dependence on s:

$$r_{f2} = \frac{\mu_{2,\,max}\, s}{K_2 + s}\, x_2 \qquad (12\text{E}1.3)$$

Assuming that yield factors for both species relative to methanol remain constant and that most oxygen uptake is by the *Pseudomonas*, we have the following material balances for s and c_{O_2}:

$$\frac{ds}{dt} = -Ds + \frac{1}{Y_1} r_{f1} - \frac{1}{Y_2} r_{f2} \qquad (12\text{E}1.4)$$

$$\frac{dc_{O_2}}{dt} = k_L a(c_{O_2 s} - c_{O_2}) - \frac{1}{Y_3} r_{f1} - Dc_{O_2} \qquad (12\text{E}1.5)$$

From batch experiments and measurement of oxygen uptake rates, Wilkinson, Topiwala, and Hamer estimated the values of all parameters in this model. Their suggested values are listed in Table 12E1.1.

The steady states computed using this model for various dilution rates are displayed in Fig. 12E1.3, along with some experimental data on this system. The agreement, while imperfect, is adequate. Transient simulations of the effect of methanol addition to the mixed culture produce responses qualitatively similar to those shown in Fig. 12E1.2. However, the real system dynamics are far more sluggish than the model. Apparently, more structure is needed in the model for an adequate reflection of the unsteady-state interactions which occur in the mixed culture.

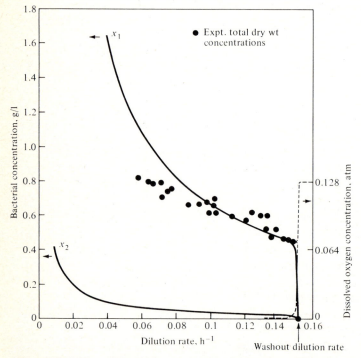

Figure 12E1.3 The solid curves were computed from the model, while the dots are experimental data. (*Reprinted from T. G. Wilkinson et al., Interactions in a Mixed Bacterial Population Growing on Methane in Continuous Culture, Biotech. Bioeng.,* **16:** 41, 1974.)

12.2. SPOILAGE AND PRODUCT MANUFACTURE BY SPONTANEOUS MIXED CULTURES

We turn now to processes where inoculation takes place from natural sources. Under these conditions, the nutrient supply and other environment factors largely determine the resulting dominant mixed cultures. Perhaps the foremost intentional application of this strategy is biological waste-water treatment. To understand this and other natural mixed-culture systems, we must recognize that the microbial world is extremely diverse and dispersed. Thus, we may assume as a working rule of thumb that if there is an environment attractive for growth of a particular microorganism, the chances are high that that particular microbe is growing or will grow there. Consequently, the mixed population which arises in an aerated vessel containing waste water is, by a type of natural selection, especially suited for growing in that environment.

Natural mixed populations are therefore particularly efficient means for utilization of substrate mixtures. While this characteristic is highly attractive in the waste-water-treatment context, it is troublesome when unwanted microbial attack occurs on "substrates" such as wood and food. We shall consider such undesirable activities next, saving discussion of biological waste-water treatment for Sec. 12.4.

Spoilage generally involves decomposition of organic molecules, including polymers such as proteins or carbohydrates. In some instances, e.g., wood rot, it is disappearance of the original substance which causes concern. On the other hand, most undesirable effects of food spoilage derive from the metabolic end products of the attacking microorganism. Both manifestations usually begin with the attack by extracellular enzymes produced by a microorganism.

In the case of wood rot, one or more of the three polymeric constituents of the wood are degraded by a cellulase enzyme system given off by a fungus. Fungi as well as bacteria are implicated in attack on pectin in foodstuffs. This causes disintegration of canned fruits, softening of brined cucumbers, and rotting of vegetables. However, as we saw earlier in Sec. 4.5.2, pectic enzymes also enjoy beneficial applications.

Protein spoilage, another problem in the food industry, can be viewed as a two-step sequence. In the first, called *proteolysis*, whole protein is hydrolyzed to yield peptides and amino acids. Liquefaction of gelatin is a common manifestation of this step. Subsequently, usually under anaerobic conditions, proteins are decomposed, and the amino acids are metabolized to yield foul-smelling products such as putrescine:

$$H_2NCH_2(CH_2)_2CH(NH_2)COOH \xrightarrow{-CO_2} H_2NCH_2(CH_2)_2CH_2NH_2$$

$$\text{Ornithine} \qquad \qquad \underset{\text{decarboxylase}}{\text{ornithine}} \qquad \qquad \text{Putrescine}$$

This process, called *putrefaction*, is evident in the vile odors of badly spoiled meats.

Perhaps the most familiar example of spoilage is sour milk. Although most milk is pasteurized before bottling, this achieves only disinfection, not sterilization. Many sporeforming bacteria survive the process and, with time, cause curdling or putrefaction. Especially interesting is the sequence of microbial populations which typically occur in raw milk at room temperature. Initially lactose (milk sugar) is fermented by streptococci, bacilli, and other bacteria. As a result of this activity, the pH of the milk drops (see Fig. 12.1). This inhibits the original population, and permits acid tolerant species including *Lactobacilli* to gain ascendency. Further reduction in pH to below 4.7 causes curd (a rubbery material consisting primarily of casein, a protein) to form and precipitate.

Next yeasts and molds which can use lactic acid as a nutrient proliferate, increasing the pH. The preeminence of these populations eventually gives way to that of fungi and bacteria, which use fat and casein as nutrients. Eventually oxygen is depleted, and anaerobic bacteria cause putrefaction.

Such successions of microbial species, each enjoying an interlude of dominance during favorable conditions, occur frequently in indigenous mixed cultures. We shall see other examples of this phenomenon in our review below of soil microbiology and in biological trickling filters. Recall the immense practical importance of controlling the initial stages of spoilage: production of curd in milk by bacterial action is the starting point for cheese manufacture, as discussed in the last section.

We find natural fermentation by lactic acid–producing bacteria in other food processes. Pickles are made by a lactic acid fermentation of cucumbers using mixed populations. In the manufacture of sausage and other fermented meats, *Lactobacillus* sp. and other microorganisms produce lactic acid and also accomplish the reduction of nitrate to nitrite, a process that contributes significantly to the development of color and the production of tangy flavors. Batter for sourdough bread is allowed to ferment for 1 or 2 days so that ethanol and organic acids are produced. Sauerkraut is another food prepared using a mixed-culture fermentation.

Although we discussed production of wine, beer, and vinegar in Chap. 10, in some respects those processes belong here. In many cases, the microbial activity responsible for a successful final product arises from a spontaneous mixed culture. Current technology in these areas, however, seems directed at better-defined, more reproducible operations. Consequently, use of carefully screened and preserved inoculum species is increasingly common.

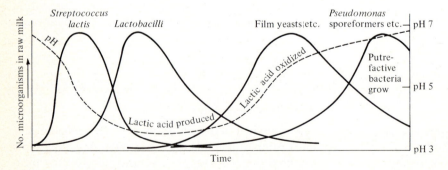

Figure 12.1 Succession of species in raw milk at room temperature.

12.3. MICROBIAL PARTICIPATION IN THE NATURAL CYCLES OF MATTER

Most of our previous discussions of examples of microbial utilization have dealt with processes largely under human operation and control. In order to preserve and improve our environment, it is also important to understand the basic features of natural microbial activities in the biosphere. With such knowledge we may be able to construct useful models for such natural processes as lake eutrophication, biodegradation and water repurification in soils, and stream and estuary ecosystem dynamics. Also, purposeful biological treatment of waste waters, one of the most important tasks microorganisms perform under human direction, in large part mimics components of natural ecological cycles. Thus, a study of microbial activities in nature aids understanding of how life on earth is sustained and also provides a valuable introduction to a critically important application of mixed populations.

Essential to proper function of the earth's ecosystem is cyclic turnover of the elements required for life. Organic matter derived from metabolic wastes or dead organisms must be broken down and converted into inorganic form. Microorganisms are especially suited to play vital roles in this process, which is called *mineralization*. Among the important attributes of the kingdom of protists in this regard are metabolic versatility, high rates of chemical activity, and natural abundance. Taken as a group, microorganisms have the power to decompose every naturally occurring organic compound.

In order to avoid excessive accumulation of any natural organic waste, it is of course not sufficient that degradation reactions occur: these reactions must occur with adequately large absolute rates (specific rate × population density). We have already noted that microorganisms grow at rates far in excess of those for higher organisms. As a consequence, specific rates of substrate utilization are also very large, so that the chemical conversions necessary for mineralization usually occur at rapid rates. The other important factor contributing to the natural importance of microorganisms is their density and widespread distribution. Microbes are extremely numerous in surface waters and topsoil; bacteria have even been discovered in soil samples from Antarctica. It is estimated that 1 acre of fertile soil contains 2 tons of bacteria and fungi in the top 6 in. Combining this density with high metabolic rates, the microorganisms in this amount of soil possess more potential metabolic activity than 10,000 human beings. As much as 90 percent of the CO_2 produced in the biosphere results from respiration in bacteria and fungi.

While microorganisms dominate the mineralization segment of most cycles of matter, they also participate significantly in other portions of these cycles. The particular involvements of protists will be explored in the next two sections. In the first, we consider a gross overview of elemental cycles from a global perspective. Next, some particular better-defined ecosystems are examined.

12.3.1. Overall Cycles of the Elements of Life

Cyclic turnover of the biologically significant elements is often accompanied by cyclic changes in their oxidation state. This is most important, since we are already aware that the biological suitability of an element is often directly linked to its chemical state (see Chap. 5). Although evident in all the element cycles considered below, the linked cycles of carbon and oxygen vividly illustrate the general principles of the cyclic transformation of matter.

The general features of the carbon and oxygen cycles are shown in Fig. 12.2. The major driving force underlying these and the other cycles is photosynthesis, which taps solar energy to reduce CO_2, bicarbonate, and carbonate, the oxidized forms of carbon, while simultaneously liberating molecular oxygen from water (Sec. 5.4). The amount of carbon fixed per year on land and in the oceans is roughly 1.6×10^{10} and 1.2×10^{10} tons, respectively. While green plants are the major contributors to photosynthetic activity on land, photosynthesis occurring in the oceans is almost entirely due to unicellular algae called phytoplankton. Although photosynthesis is the dominant means of CO_2 reduction, it is also conducted by chemoautotrophs. As already noted, mineralization of organic carbon to CO_2 is primarily the consequence of bacterial and fungal metabolic activities.

Carbon is removed or sequestered from the life cycle just described by several mechanisms. Much of the CO_2 released into the atmosphere enters the oceans as bicarbonate ions. There, it can combine with calcium to form calcium carbonate, an insoluble compound which appears in coral shells and limestone. In this form carbon is relatively inaccessible, but much of it is ultimately made available by weathering or by attack of acids. Microorganisms participate in the latter process through synthesis of carbonic, sulfuric, nitric, and other acids.

We are all familiar with sequestered carbon in organic form. *Humus*, an organic residue derived from microbial-resistant plant components, is an important constituent of rich soil. When conditions favor large accumulations of humus, deposits of *peat* are created which, on a geological time scale, can be transformed into coal. Oil and natural gas are other common forms of sequestered organic carbon. Carbon residing in these forms seems destined for eventual return to the biosphere due to man's apparently relentless demands.

Let us next consider how nitrogen cycles in the biosphere and how its chemical state alters in the process. Several basic principles provide useful guidelines for this study; molecular nitrogen in the atmosphere is quite inert; i.e., it is not an acceptable nutrient form for most organisms. Also, its chemical form in living organisms is primarily in a reduced state in proteins.

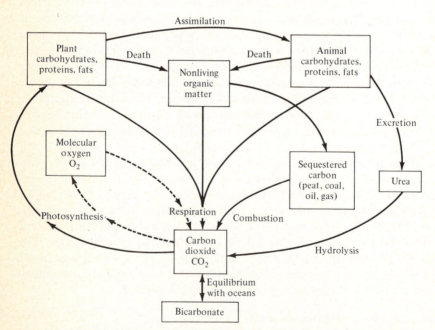

Figure 12.2 Simplified diagram of the carbon cycle. Also shown (dotted lines) is the major component of the oxygen cycle, which is closely linked to the cycle of carbon.

A general overview of the nitrogen cycle is provided in Fig. 12.3. Organic nitrogen is converted into ammonia by microbial action. Although some ammonia escapes into the atmosphere and some is utilized directly by plants and microorganisms, most is oxidized to nitrate (NO_3^-) in a two-step process called *nitrification*. Each half of the nitrification pathway is mediated by a special family of bacteria, primarily the *Nitrosomas* and the *Nitrobacter*, respectively. Both groups are chemoautotrophic and obligate aerobes.

The nitrate thus formed is the best form of nitrogen for plant assimilation; consequently at this point much of the nitrogen flux reenters the pool of reduced organic nitrogen. After incorporation into organic compounds by algae and plants, the nitrogen is in a form suitable for animal utilization. However, microbial activities provide an alternate demand on nitrate. When oxygen is unavailable, as in a microenvironment where respiration has exhausted the supply, many bacteria possess sufficient metabolic dexterity to use nitrate as a hydrogen acceptor. The end result of their activities is denitrification: molecular nitrogen is formed.

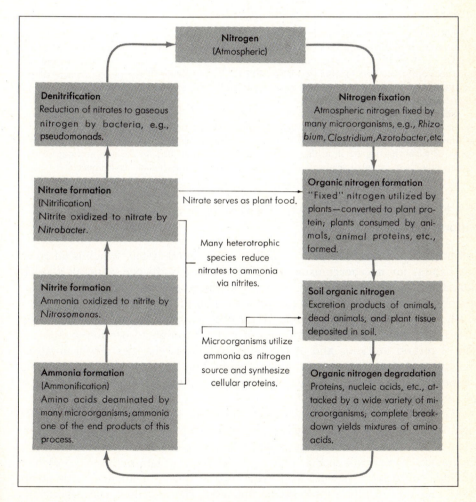

Figure 12.3 Simplified schematic diagram of the nitrogen cycle.

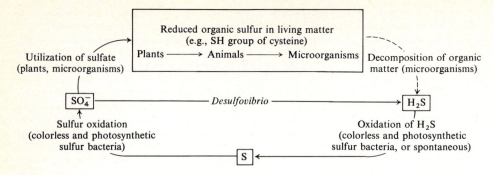

Figure 12.4 Major features of the sulfur cycle.

Fortunately, several specialized microorganisms have the ability to utilize molecular nitrogen and to return it to the biosphere in combined form. Two general types of nitrogen fixation can be identified: in the symbiotic version, mutualistic relationships between *Rhizobium* bacteria and seed plants accomplish nitrogen fixation (recall Sec. 10.3.2). The remaining nitrogen-fixing capacity derives from the nonsymbiotic activity of blue-green algae and a few aerobic (*Azotobacter*) and anaerobic (*Clostridium pasteurianum*) bacteria.

Sulfur also undergoes a cycle of oxidation, reduction, incorporation into, and liberation from, organic matter (see Fig. 12.4). The role of sulfate-reducing and sulfur-oxidizing bacteria in corrosion has already been mentioned (Sec. 10.2.5). For additional information on the sulfur cycle and the organisms which participate in it, the references should be consulted. In the next section, we shall examine how these cycles or segments of them are manifested in particular environments.

12.3.2. Interrelationships of Microorganisms in the Soil and Other Natural Ecosystems

Soil provides a varied and complicated environmental system which is an excellent habitat for microorganisms. It consists of finely divided minerals (largely aluminum silicate compounds), decaying organic residues, and a living mixed microbial population. In addition, water is often present, as is a gaseous phase which may contain N_2, O_2, CO_2, H_2S, NH_3, and other gases. The extensive surface afforded by fine solid granules provides, through adsorption, concentration of certain nutrients and extracellular hydrolytic enzymes. From dissolved minerals and decaying organic material come ions, carbohydrates, nitrogenous compounds, and vitamins. Hence a rich culture medium is available for support of microbial growth.

Syntrophism is a type of relationship in which organisms produce food for each other. Syntrophic relationships are ubiquitous in the soil, and indeed are essential for proper functioning of the cycles of elements. Some examples are illustrated schematically in Fig. 12.5. Notice that one organism (labeled A), which produces cellulolytic enzymes to provide its own nourishment, also feeds others (B, C, and D) from the simple sugars liberated. The metabolic end products of organism A are used by other microbes with different nutritional needs.

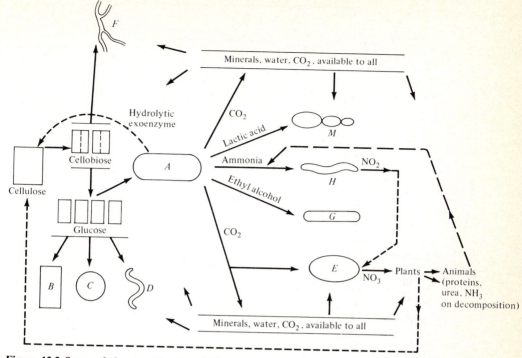

Figure 12.5 Some of the possible syntrophic relationships among microorganisms, plants, and animals in the soil. Metabolic end products of microbe A nourish organisms M, H, G, and E, and cellulolytic enzymes secreted by A hydrolyze cellulose. The resulting simple sugars are nutrients for microorganisms B, C, and D. (*Reprinted from A. L. van Wezel, Microcarrier Cultures of Animal Cells, in "Tissue Culture: Methods and Applications," p. 534, Academic Press, Inc., New York, 1973.*)

Another view of syntrophism, more analogous to the milk-spoilage scenario reviewed in the last section, can be obtained by considering the time sequence of events which follows plowing under of a grass or clover crop. Besides providing many soluble nutrients from the plants, plowing aerates the soil. Conditions consequently favor rapid growth of heterotrophic organisms and facultative autotrophs. Due to the metabolic action of these microorganisms, the temperature within the soil rises, as does its acidity.

After all oxygen has been depleted, strict anaerobes appear, leading to further acid production. Eventually this explosive growth is arrested by nutrient depletion or toxin formation (recall Sec. 7.1.3), and many of the cells die, releasing compounds useful for plant growth. The residual, relatively low-level microbial population consists of species capable of attacking the resistant substances such as those found in humus.

We have only scratched the surface here in examining microbial activities in the soil. For example, almost complete cycles of the essential elements can and do occur under anaerobic conditions. The necessary transformations can all take place in a very small oxygen-depleted ecological niche since all of them are con-

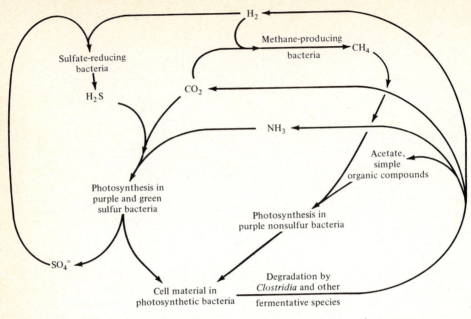

Figure 12.6 Simplified schematic of the cycles of matter in an anaerobic environment.

ducted by microorganisms. Figure 12.6 summarizes the close interrelationships which can exist between photosynthetic, fermentative, sulfate-reducing, and other bacteria in an anaerobic environment. Additional readings in soil microbiology are recommended in the references. We should note in closing this discussion that the vast majority of industrially important microorganisms have been isolated from the soil.

Microorganisms are the primary producers of organic matter in freshwater and marine environments. As illustrated schematically in Fig. 12.7, green plants and large multicellular algae (seaweed) can achieve photosynthesis only in shallow waters near shore. In open water, free-floating unicellular algae called phytoplankton conduct photosynthesis near the surface. Photosynthetic activity becomes difficult and eventually impossible with increasing depth because light is absorbed by water and intercepted by suspended solid materials.

The phytoplankton produced in surface waters feeds an intricate chain of consumers, which begins with animal plankton (zooplankton) and ends with the fishes, whales, and other aquatic animals. The organic wastes resulting from metabolic activities and death in this chain are in turn decomposed by microorganisms. The simple compounds produced in decomposition can be utilized by the phytoplankton, thereby completing a food cycle.

In the example considered next, we shall examine the interactions in a pond in greater detail. By attempting to quantify the qualitative interrelationships in this system, we shall attempt to understand some of the processes leading to *eutrophication*. Here excess nutrient supplies causes explosive growth of algae. Some of the algae produce toxins, and unpleasant odors often accompany the algal bloom.

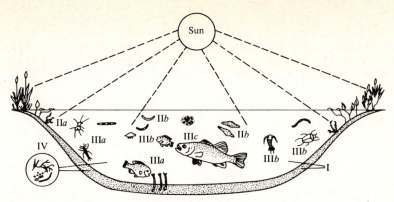

Figure 12.7 Members of the food chain in an aquatic ecosystem which relies on solar radiation as its major energy source. The labels I, II, III, and IV above denote the classes of abiotic components, producers, consumers, and microbial decomposers, respectively.

Eventually much of the algae dies, releasing nutrients for heterotroph consumption. At this point, the net respiration rate exceeds the photosynthesis rate, so that the supply of dissolved oxygen is seriously reduced. This can cause extensive death of fish and many other aerobic organisms, totally upsetting the local balances necessary for survival of the lake or pond ecosystem.

Example 12.2 Mathematical modeling of a natural aquatic ecosystem† Figure 12E2.1 shows a view of chemical and biological interactions in a lake or pond in a form that lends itself directly to mathematical modeling. Before writing mass balances for this complex process, we shall invoke some

† This example is derived from a discussion in Rich [3] of a model by Water Resources, Inc., 1968.

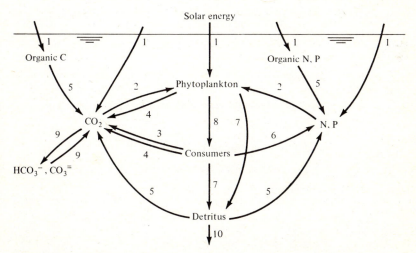

Figure 12E2.1 A view of an aquatic ecosystem which highlights interactions between different segments of this community. The numbers denote external inputs (1), photosynthesis (2), primary respiration (3), endogenous respiration (4), biological decomposition (5), excretion (6), death (7), grazing (8), equilibrium (9) and sedimentation (10). (*From L. G. Rich, "Environmental Systems Engineering," p. 116, McGraw-Hill Book Company, New York, 1973.*)

simplifying assumptions. Phytoplankton will be used to represent the sum of all algal species present, and all consumers of algae are lumped together. The detritus term similarly lumps all dead forms (phytoplankton and consumers); this matter is presumed to settle out of the well-mixed upper lake volume. No horizontal transport is assumed, and no vertical transport mixes underlying water layers with this upper layer under consideration. These mixing assumptions are approximately realized in a stagnant lake in summer.

Balances on each of the nine subsystems shown in Fig. 12E2.1 are now formulated as follows:

Phytoplankton:

$$\frac{dx_p}{dt} = \mu_p x_p(\text{growth}) - k_{rp} x_p(\text{endogenous respiration})$$

$$- k_{mp} x_p(\text{death rate}) - \frac{\mu_c x_c}{Y_{c/p}}(\text{predation by consumers}) \qquad (12E2.1)$$

Consumers:

$$\frac{dx_c}{dt} = \mu_c x_c - k_{rc} x_c - k_{mc} x_c \qquad (12E2.2)$$

Detritus:

$$\frac{dc_D}{dt} = k_{mp} c_p + k_{mc} x_c - k_{dD} c_D(\text{decomposition}) - k_{SD} c_D(\text{sedimentation}) \qquad (12E2.3)$$

Carbon dioxide:

$$\frac{dc_{CO_2}}{dt} = k_{aOC} c_{OC}(\text{from organic carbon})$$

$$+ \frac{1 - Y_{cp}}{Y_{cp}} \mu_c x_c(\text{consumer generation}) + k_{rp} x_p$$

$$+ k_{rc} x_c + k_{dD} c_D + K_L \frac{A}{V}(c_S - c_{CO_2})$$

$$+ f(c_{CO_2})(\text{buffer exchange}) - \mu_p x_p \qquad (12E2.4)$$

Inorganic nitrogen:

$$\frac{dc_N}{dt} = k_{dON} c_{ON}(\text{biological release})$$

$$+ k_{eN} x_e(\text{consumer excretion of nitrogen})$$

$$+ k_{dON} c_D(\text{detritus release})$$

$$+ L_N(\text{inorganic nitrogen addition, external})$$

$$- \gamma_{pN} \mu_p x_p(\text{uptake by phytoplankton}) \qquad (12E2.5)$$

where γ_{pN} = N/C ratio during photosynthesis.

Inorganic phosphorus:

$$\frac{dc_P}{dt} = k_{dOP} c_{OP} + k_{ep} x_c + k_{dpP} c_p + L_p - \gamma_{pP} \mu_p x_p \qquad (12E2.6)$$

Organic carbon:

$$\frac{dc_{OC}}{dt} = L_{OC} - k_{aOC} c_{OC} \qquad (12E2.7)$$

Organic nitrogen:

$$\frac{dc_{ON}}{dt} = L_{ON} - k_{dON}c_{ON} \qquad (12E2.8)$$

Organic phosphorus:

$$\frac{dc_{OP}}{dt} = L_{OP} - k_{dOP}c_{OP} \qquad (12E2.9)$$

Table 12E2.1. Recommended parameter values for the ecosystem model†

Parameter	Value
$\mu_{p,\text{max}}$	$8.33 \times 10^{-2}\,\text{h}^{-1}$
$\mu_{c,\text{max}}$	$8.33 \times 10^{-3}\,\text{h}^{-1}$
K_{Xp}	$0.834\,\text{mmol of C/l}$
K_C	$1.36 \times 10^{-3}\,\text{mmol of C/l}$
K_N	$1.43 \times 10^{-2}\,\text{mmol of N/l}$
K_P	$1.61 \times 10^{-3}\,\text{mmol of P/l}$
K_I	$3.0 \times 10^{-2}\,\text{langley/min}$
z_T	$500\,\text{cm}$
n	10
k_{rp}	$1.25 \times 10^{-3}\,\text{h}^{-1}$
k_{mp}	$1.25 \times 10^{-3}\,\text{h}^{-1}$
k_{rc}	$1.25 \times 10^{-3}\,\text{h}^{-1}$
k_{mc}	$2.08 \times 10^{-3}\,\text{h}^{-1}$
k_{dD}	$4.17 \times 10^{-4}\,\text{h}^{-1}$
k_{sD}	$1.25 \times 10^{-3}\,\text{h}^{-1}$
k_{dOC}	$4.17 \times 10^{-3}\,\text{h}^{-1}$
k_{dON}	$4.17 \times 10^{-3}\,\text{h}^{-1}$
k_{eN}	$1.25 \times 10^{-3}\,\text{h}^{-1}$
k_{dDN}	$4.17 \times 10^{-4}\,\text{h}^{-1}$
k_{dOP}	$4.17 \times 10^{-3}\,\text{h}^{-1}$
k_{eP}	$1.25 \times 10^{-4}\,\text{h}^{-1}$
k_{dDP}	$4.17 \times 10^{-4}\,\text{h}^{-1}$
$Y_{e/p}$	0.4
K_L	$2\,\text{cm/h}$
C_s	$3.0 \times 10^{-3}\,\text{mmol of C/l}$
A	$1.0\,\text{cm}^2$
V	$500\,\text{cm}^3$
K_1	$4.45 \times 10^{-7}\,\text{m/l}$
K_2	$4.69 \times 10^{-11}\,\text{m/l}$
K_w	$1.01 \times 10^{-14}\,(\text{m/l})^2$
γ_{pN}	0.172
γ_{pP}	7.75×10^{-3}
C_z	$2.0 \times 10^{-3}\,\text{m/l}$
α	$2.95 \times 10^{-3}\,\text{cm}^{-1}$
β	$8.20 \times 10^{-6}\,(\text{mmol of C/l})^{-1}/\text{cm}$

† From L. G. Rich, "Environmental Systems Engineering," p. 128, McGraw-Hill Book Company, New York, 1973.

The specific growth rates of phytoplankton μ_p and consumer μ_c are assumed to possess Monod dependencies on each limiting nutrient or energy source. For phytoplankton, the final form depends upon light, carbon dioxide, nitrogen, and phosphorus, whence

$$\mu_p = \mu_{p,\,\text{max}} \frac{I}{K_I + I} \frac{c_{\text{CO}_2}}{K_{\text{CO}_2} + c_{\text{CO}_2}} \frac{c_N}{K_N + c_N} \frac{c_p}{K_p + c_p} \tag{12E2.10}$$

$$\mu_c = \mu_{c,\,\text{max}} \frac{\bar{x}_p}{K_{xp} + x_p} \tag{12E2.11}$$

The daily fluctuations of light intensity I are given by

$$\frac{dI_t}{dz} = -(\alpha + \beta x_p)I_{\text{max}} \sin \pi \frac{t}{t_0} \qquad 0 < t < t_0 \tag{12E2.12}$$

where α = light-absorption coefficient of water
$\quad \beta$ = light-absorption coefficient of algal suspension
$\quad t_0$ = total daylight time
$\quad t_n$ = total night length
$\quad I_t = 0 \qquad t_0 < t < t_0 + t_n$

Simplification in the equations obtains from first averaging μ_p over all values of z. Since the volume is presumed to be homogeneous, only an averaging over $I/(K_I + I)$ is needed. The problem is further simplified if various single substrates are assumed limiting.

The number of parameters involved in this formulation is considerable. It is significant that 22 of the 34 parameters needed were estimated; only 12 values were directly accessible in the literature. Table 12E2.1 lists the values suggested by Rich. As in other modeling areas, the large numbers of parameters here may invite criticism. However, each term in the model equations is clearly identified with a known process. Although more simplified models may be adequate for a single, well-defined environment, we can expect only a more complete and complex model to perform satisfactorily under a wider range of conditions. Presumably, as such models become better validated, we can rely more and more on their abilities to predict the effects of perturbations both natural, e.g., diurnal variation of sunlight intensity, and artificial, e.g., dumping of waste.

12.4. BIOLOGICAL WASTE-WATER TREATMENT

Waste waters contain a complex mixture of solids and dissolved components, with the latter usually present in very small concentrations. In treatment plants, all these contaminants must be reduced to acceptably low concentrations or chemically transformed into inoffensive compounds. The overall system design used to accomplish this varies depending upon the type and amount of waste water to be treated and economic and environmental considerations. Most of the alternatives, however, share enough common features to allow them to be shown schematically on a single diagram like Fig. 12.8. There, the parallel pathways shown for sludge handling and removal of various contaminants represent different options for accomplishing one treatment objective. A typical plant would employ only a few of the many possible pathways from waste water to final effluent.

We consider next the overall purpose of each of the major process trains. In *primary treatment*, the most easily separated contaminants are removed. Taken out here are readily settleable solids (see Fig. 12.9), oil films, and other "light" components. Suspended particles as well as soluble components are removed in *secondary treatment*. In many situations these waste materials are organic, and in these cases use of a biological oxidation process is common. We shall consider these biological processes in further detail later. *Tertiary treatment* is directed at removal of all or some of the remaining contaminants. Among the processes used at this stage are electrodialysis, reverse osmosis, deep-bed filtration, and adsorption.

Wet, concentrated solid wastes, called *sludge*, are removed in primary treatment; cell sludge is generated in secondary biological treatment. We have already mentioned the interplay between substrate utilization and biomass production. Although secondary biological treatment processes, which involve many microbial species, are very efficient in attacking a dilute mixture of organic wastes, they also create biomass. Thus, very small particles and soluble components in liquid wastes are transformed in part into a sludge waste material which is easier to separate out than the original waste. Sludge handling and treatment is consequently an important part of the water-treatment plant. One popular process for sludge-volume reduction in sewage plants is *anaerobic digestion*, where organic material is biologically decomposed in an anaerobic environment.

We should not conclude from this discussion that all three levels of treatment and sludge digestion are always used. Some waste waters are discharged into receiving waters such as streams, rivers, ponds, lakes, and oceans with no treatment whatsoever. In other situations, only primary treatment is applied. However, some form of secondary treatment and sludge processing is the norm for most municipal sewage-treatment plants in the United States, with tertiary treatment only infrequently used at present.

In keeping with the major theme of this text, we shall emphasize the biological components of waste-water-treatment processes. After reviewing the characteristics of typical waste waters in Sec. 12.4.1, we shall consider analysis of activated sludge and related secondary biological treatment methods. Our rapid overview of these important applications of biochemical engineering will conclude with a discussion of anaerobic digestion. Additional information on these processes as well as the physical and chemical operations encountered in treatment plants is provided in the references.

12.4.1. Waste-Water Characteristics

Naturally, the nature and concentrations of contaminants in waste water depend upon its source. There are two main classes of waste waters to consider: industrial effluents and domestic wastes. The latter type is called *sewage*, and it consists of substances such as ground garbage, laundry water, and excrement.

More than 99 percent water, sewage typically contains about 300 ppm (mg/l) of suspended solids and about 500 mg/l volatile material. Much of the suspended solid component is cellulose, and the bulk of organic matter present is in the form of fatty acids, carbohydrates, and proteins in that order. As our earlier discussion of spoilage should suggest, the bad odor of sewage derives from protein decomposition under anaerobic conditions.

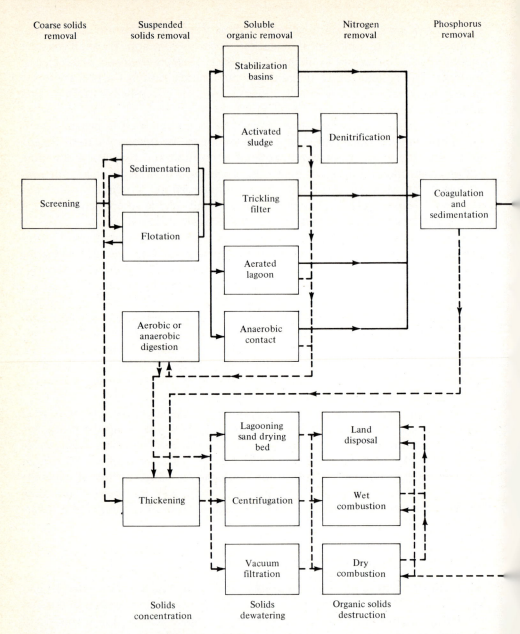

Figure 12.8 Available unit operations for primary, secondary, and tertiary waste-water treatment. (*Reprinted from W. W. Eckenfelder, Jr., "Industrial Water Pollution Control; pp. 6–7, McGraw-Hill Book Company, Inc., New York, 1966.*)

Fine suspended   Bacterial   Trace organics   Inorganic
solids removal   removal   removal   salt removal

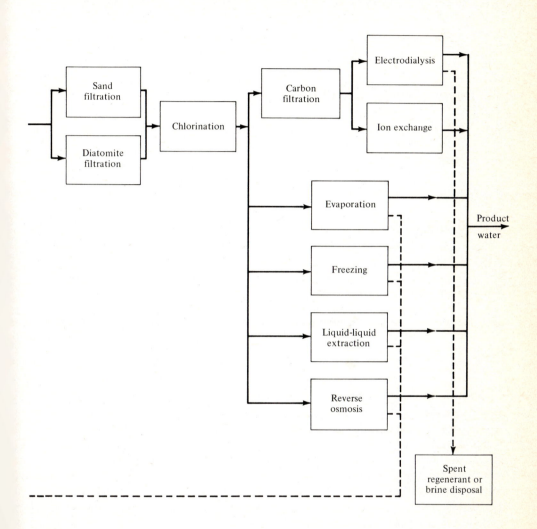

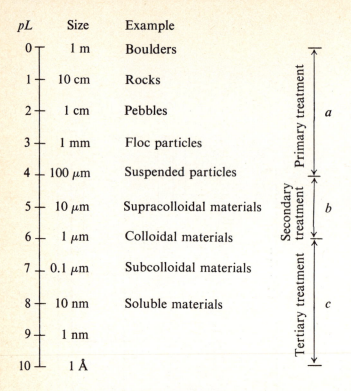

Figure 12.9 Different levels of treatment remove characteristic ranges of particulate sizes from the waste water. (*Adapted from T. Helfgott et al., Analytical and Process Characterization of Effluents, J. Sanit. Eng. Div. ASCE,* **96**: 79, 1970.)

Because of its origins, it should be no surprise that sewage contains a varied population of soil and intestinal microorganisms. Included are aerobes, strict and facultative anaerobes, bacteria, yeasts, molds, and fungi. Since pathogenic organisms and numerous viruses including polioviruses and hepatitis viruses are often present in sewage, it is critically important to isolate drinking supplies and water lines from sewage contamination. The sewage microbial populations provide a continuous mixed-culture inoculum for the biological treatment processes and also supply the metabolic capacity used in the following standard analysis of waste-water composition.

Among the measures of sewage strength or concentration, perhaps the most common index is the *biochemical oxygen demand* (BOD). It is equal to the amount of dissolved oxygen which is consumed by a sewage incubated for a specified length of time at 20°C. The length of the incubation time is often shown as a subscript: thus the BOD determined from a 5-day incubation, which is one of the common intervals, is denoted BOD_5. The amount of dissolved oxygen consumed in an incubation which is continued until carbonaceous biological oxidation ceases is called the *ultimate* BOD (BOD_U). Originally devised in 1898 by the

British Royal Commission on Sewage Disposal, this test was chosen to simulate the conditions of a stream and to provide a relatively direct measure of one of the most damaging effects of sewage discharge: depletion of dissolved oxygen in the receiving waters. A lowered dissolved-oxygen value can quickly lead to death of many aerobic organisms and animals; the result may be a murky, smelly river contaminated with pathogenic microbes.

The *chemical oxygen demand* (COD) is another indication of the overall oxygen load which a waste water will impose on the receiving water. It is equal to the number of milligrams of oxygen which a liter of sample will absorb from a hot, acidic solution of potassium dichromate. Generally, more components of the waste-water sample can be chemically oxidized in this manner than in the standard BOD test. Consequently, the COD value is usually greater than the BOD of the same sample. Although it is less directly related to the polluting effects of sewage than BOD, COD has the advantage of being measurable in about 2 h by conventional methods or in a few minutes using sophisticated instruments.

Both BOD and COD are gross, overall indicators of sewage composition. Nevertheless, they do provide a measure which relates to the environmental damage of the waste water. Moreover, the necessary analyses can be performed with minimal equipment and training in analytical procedures.

Other parameters often used to characterize water quality are phosphorus, nitrogen, and suspended-solids concentrations. In Table 12.2, we see typical values of characteristics for the influent and effluent of a sewage-treatment plant. Among the important contaminants which are not considered in the table are heavy metals and toxic organics such as pesticides, which are often present in small but significant quantities.

The composition of industrial wastes depends strongly on the source. As Table 12.3 reveals, many industrial wastes are far more concentrated than sewage. Also, those derived from processing hydrocarbon materials often contain toxins such as cyanide. We face two related problems with such substances: (1) they are extremely damaging to living organisms of the receiving water, and (2) they may kill microorganisms utilized in aerobic and anaerobic waste treatment. Effective and reasonably economical methods for elimination of such toxic compounds from discharge waters still have not been perfected.

Table 12.2. Some characteristic parameters for water quality
At present effluent standards for nitrogen and total phosphorus concentrations are not always applied

Parameter	Influent raw waste water	Effluent in an acceptable plant
BOD, mg/l	100–250	5–15
COD, mg/l	200–700	15–75
Total phosphorus, mg/l	6–10	0.2–0.6
Nitrogen, mg/l	20–30	2–5
Suspended solids, mg/l	100–400	10–25

Table 12.3. Comparative strength of effluents†

Type of waste	Main pollutants	BOD$_5$	COD
Abattoir	Suspended solids, protein	2600	4150
Beet sugar	Suspended solids, carbohydrate	850	1150
Board mill	Suspended solids, carbohydrate	430	1400
Brewery (bottle washing)	Carbohydrate, protein	550	
Cannery (meat)	Suspended solids, fat, protein	8000	17940
Chemical plant	Suspended solids, extremes of acidity or alkalinity, organic chemicals	500	980
Coal carbonization:			
Coke ovens	Phenols, cyanide	780	1670
Gas works	Thiocyanate, thiosulphate	6500	16400
Smokeless fuel	Ammonia	20000	
Distillery	Suspended solids, carbohydrate, protein·	7000	10000
Dairy	Carbohydate, fat, protein	600	
Domestic sewage	Suspended solids, oil-grease, carbohydrate, protein	350	300
Grain-washing	Suspended solids, carbohydrate	1500	1800
Kier	Suspended solids, carbohydrate, lignin,	1600	3600
Laundry	Suspended solids, carbohydrate, soap	1600	2700
Maltings	Suspended solids, carbohydrates	1240	1480

Table 12.3 (*continued*)

Type of waste	Main pollutants	BOD_5	COD
Pulp mill	Suspended solids, carbohydrate, lignin, sulfate	25000	76000
Fermentation industry:			
Fermentation segment	...	4560	4120
Chemical-synthesis segment	...	960	1580
Formulation, packaging segment	...	145	217
Petroleum refinery	Phenols, hydrocarbons, sulphur compounds	850	1500
Resin manufacture	Phenol, formaldehyde, urea	7400	12900
Starch reduction of flour	Suspended solids, carbohydrate, protein	12000	17150
Tannery	Suspended solids, proteins, sulfide	2300	5100

† Adapted from J. W. Abson and K. H. Todhunter, pp. 318–319 in N. Blakebrough (ed.), "Biochemical and Biological Engineering Science," vol. 1, Academic Press, London, 1967.

In the remainder of this chapter, we shall concentrate on processes and conditions typical of domestic sewage treatment. We should remember, however, that the same biological processes have important applications in treating industrial wastes. Moreover, these two problems often merge, since industrial effluents are in many cases discharged into domestic sewers.

12.4.2. The Activated-Sludge Process

The main component of the activated-sludge process is a continuous-flow aerated biological reactor. As indicated in Fig. 12.10, this aerobic reactor is closely tied to a sedimentation tank, in which the liquid is clarified. A portion of the sludge collected in the sedimentation tank is usually recycled to the biological reactor, providing a continuous sludge inoculation. This recycling extends the mean sludge residence time, giving the microorganisms present an opportunity to adapt to the available nutrients. Also, the sludge must reside in the aerobic reactor long enough for adsorbed organics to be oxidized. Other benefits of reactor recycle were discussed in Chap. 9.

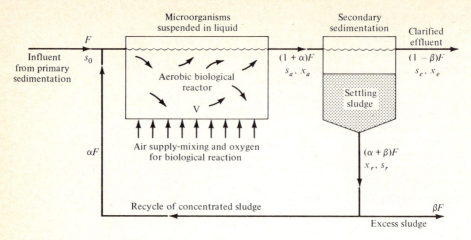

Figure 12.10 Schematic diagram of the activated-sludge process.

To understand the basic mechanisms of substrate removal which operate in this unit, we must examine the nature and morphology of the community of mixed microbes which thrive in the aeration basin. A common bacterium in the activated-sludge population is *Zoogloea ramigera*. Perhaps the most important characteristic of this organism and others in the sludge is their propensity for synthesizing a polysaccharide gel. Because of this gel, the microbes tend to agglomerate into flocs, which are called activated sludge (see Fig. 12.11). A special

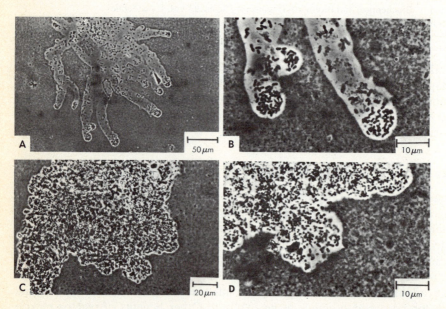

Figure 12.11 Photomicrograph of some of the microorganisms in activated sludge. (*Reprinted from R. F. Unz and N. C. Dondero, Water Research,* **4**: 575, 1970.)

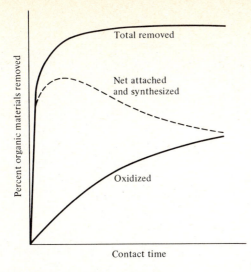

Figure 12.12 Removal of organic material in an aerated, batch activated-sludge system is believed to involve rapid initial physical capture of organics by the sludge, followed by slower biological oxidation.

property of activated sludge is its high affinity for suspended solids, including colloidal materials. Thus, the initial step in removing suspended solids from the waste water is attachment to the floc. Following this, biodegradable components of the adsorbed particulates are oxidized by floc organisms, as indicated in Fig. 12.12.

In order to capitalize on the excellent adsorbent properties of activated sludge, a variation of the conventional process called *contact stabilization* has been devised. As shown in the process flowchart in Fig. 12.13, the recycle settled sludge is subjected to an additional aeration cycle before being mixed with the waste influent in an aerobic tank. In this latter contact basin, organics are removed almost entirely by physical attachment. Biological utilization of these stored organics occurs in the aerated recycle-sludge basin, which also serves the function of restoring the floc's adsorption capacity.

Other versions of the activated-sludge process differ from the conventional design primarily in the contacting pattern of waste water, sludge, and air in the aeration basin. The flowsheets in Fig. 12.13 reveal that in the *step-feed process*, the influent stream is split and introduced at different points of the aeration basin. We appreciate the effects of this distributed feed from our reactor-analysis studies in Chap. 9: the conventional activated-sludge aeration basin is a long, narrow channel which behaves roughly like a tubular reactor with some dispersion. By distributing the feed along the reactor length, the basin is made to behave more like a well-mixed tank reactor.

Actually, a better approximation to a backmixed reactor is achieved by using a circular basin which is vigorously aerated to provide mass transfer and mixing. In such a system, gradients of dissolved oxygen and nutrient concentrations within the reactor are minimized. The activated-sludge population which develops under these conditions is often better suited for dealing with loading fluctuations or with shock loads of toxic material.

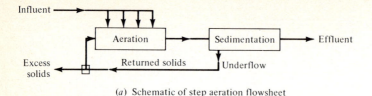

(a) Schematic of step aeration flowsheet

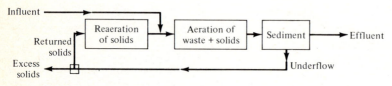

(b) Schematic of contact stabilization (solids reaeration) flowsheet

Figure 12.13 Two alternative flowsheets for biological oxidation processes. The conventional activated-sludge process was shown in Fig. 12.10.

Although the underlying principles of aerator designs for activated-sludge processes are the same as considered in Chap. 8, the aeration systems used can vary widely. Besides the stirred-sparged combination familiar in fermentation applications, the air may be bubbled into the vessel through diffusers on the bottom or sides of the reactor. Alternatively, the surface of the basin may be brushed with rotating blades to create turbulence and promote gas absorption. Other possibilities include the simplex cone, which draws liquid from near the basin bottom and sprays it onto the tank surface. In all cases the aeration and agitation system serves to provide oxygen for microbial respiration, to suspend and mix the sludge and other particulates, and to strip out volatile metabolic products such as CO_2.

Besides possessing the necessary adsorbent and metabolic qualities, a good sludge should settle rapidly. For example, after 30 min in a cylinder, the volume percent of settled sludge should be around 40 times the volume percentage of suspended solids. Much larger values, say 200, indicates a bad sludge, which will tend to overflow from the thickening tank. When the process *bulks*, as this condition is called, it has failed: the effluent will not meet the necessary standards.

While we do not yet understand the causes and mechanism of bulking, examination of the microbial constituents of poor sludge often reveals filamentous bacteria and flagellate protozoa. Healthy sludge, on the other hand, does not contain a significant population of filamentous organisms, and the protozoa present are mainly stalked ciliated species. The protozoa serve a valuable function in the overall process by preying on free, i.e., unflocced, bacteria and thereby clarifying the effluent.

Normally filamentous bacteria and fungi cannot compete with the heterotrophic bacteria found in healthy sludge. Large shocks in influent conditions or improper operation of the unit may, however, create conditions damaging to the desired population, permitting other microorganisms to invade the community. We could conjecture, therefore, that both normal operation and bulking are manifestations of the principles of competition in mixed populations.

12.4.3. Design and Modeling of Activated-Sludge Processes

Although a waste-water-treatment plant for a major city costs in excess of 100 million dollars, the biological reactors they contain are usually designed using extremely simplified and idealized models. Typically, the aeration basin is treated as a perfectly mixed vessel, and sludge is viewed as a single pseudo species whose growth rate follows Monod kinetics with an additional endogenous metabolism decay term. As discussed in Sec. 12.4.2, the substrate or limiting nutrient concentration is usually expressed in terms of BOD.

With the nomenclature indicated in Fig. 12.10, which is consistent with that used in Chaps. 7 and 9, the steady-state mass balance on active solids in the process is

$$(1 - \beta)Fx_e + \beta Fx_r = Vx_a \left(\frac{\mu_{max} s_a}{s_a + K_s} - k_e \right) \tag{12.1}$$

Straightforward algebraic manipulation of this expression shows

$$\frac{1}{\theta_s} = \mu_{max} \frac{s_a}{s_a + K_s} - k_e \tag{12.2}$$

where θ_s is the mean residence time of the activated sludge (ratio of active-solids retention to active-solids effluent rate), sometimes called the sludge age:

$$\theta_s = \frac{Vx_a}{F(1 - \beta)x_e + \beta Fx_r} \tag{12.3}$$

In most sewage-treatment plants, the sludge age is of the order of 6 to 15 days.

Using Eq. (12.2), we can compute the solids-retention time θ_s required to achieve a specified BOD level s_a from a given waste, providing the kinetic constants are known for that system. Typical values for these parameters are given in Table 12.4. We should recall, however, from our earlier discussions of lumping in Secs. 3.4.2 and 9.7.5, that the kinetic "constants" in such simplified models are actually dependent on composition and operating conditions. Consequently, appropriate values for μ_{max}, K_s, k_e, and Y should always be measured for the specific waste of interest.

Assuming a constant yield factor, we can now formulate a steady-state conservation equation for substrate in the aeration basin; recall the recycle biomass balance of Eq. (9.28), which reveals that

$$Vx_a = \frac{YF\theta_s(s_0 - s_a)}{1 + k_e\theta_s} \tag{12.4}$$

Table 12.4. Monod model parameters for utilization of different substrates in aerobic mixed culture†

Organic substrate	Y	μ_{max}, day^{-1}	K_s, mg/l	k_e, day^{-1}	Coefficient basis	Temperature, °C
Domestic waste	0.5	⋯	⋯	0.055	BOD$_5$	
	0.67	⋯	⋯	0.048	BOD$_5$	20
	0.67	3.7	22	0.07	COD	
Skim milk	0.48	2.4	100	0.045	BOD$_5$	
Glucose	0.42	1.2	355	0.087	BOD$_5$	
Peptone	0.43	6.2	65		BOD$_5$	30

† A. W. Lawrence and P. L. McCarthy, Unfied Basis for Biological Treatment Design and Operation, *J. Sanit. Eng. Div. ASCE*, **96:** 768 (June 1970).

This equation provides the active-solids weight Vx_a in the aeration basin in terms of quantities already specified. Next, sludge-settling characteristics must be determined experimentally to find x_r. With this known, the required basin volume V for a given recycle ratio α can be calculated from

$$V = F\theta_s\left(1 + \alpha - \alpha\frac{x_r}{x_a}\right) \tag{12.5}$$

which is obtained by writing an active-solids balance on the aeration basin and using Eq. (12.2).

Since the extreme simplicity of this design procedure differs so dramatically from the complex physical and biological processes which occur in an activated-sludge plant, it is reassuring to see some experimental data supporting this design approach. Figure 12.14 shows the results of cultivation of a mixed population of sewage microorganisms in a chemostat. The solid lines, computed using the same growth model as in Eq. (12.1) and the kinetic parameters shown in the figure, provide an excellent fit to the experimental data. Notice that although the kinetic parameters used here fall within the general ranges of magnitude suggested in Table 12.4, they differ considerably. This underscores the caveat just mentioned: the parameters must be determined for each waste stream, since they will vary with waste characteristics.

Our discussions in Chaps. 7 and 9 suggested that dependence of model parameters on input compositions and operating configuration can be reduced by introducing more structure into the model. Although much remains to be done, an excellent beginning on development of structured models for activated-sludge and other biological treatment processes has been made by Andrews and his colleagues [5–7]. Their work in this area has been motivated by the considerations we have summarized as well as a very important additional one: development of control strategies for waste-water-treatment plants. Every day most plants are subjected to large disturbances which require some adjustment in plant operating conditions. A model with inadequate structure would not accurately reflect the effects of process disturbances and would be useless for simulation studies of plant

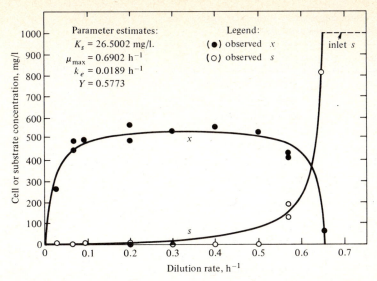

Figure 12.14 parameter estimates shown in plot:

Parameter estimates:
$K_s = 26.5002$ mg/l.
$\mu_{max} = 0.6902$ h^{-1}
$k_e = 0.0189$ h^{-1}
$Y = 0.5773$

Legend:
(●) observed x
(○) observed s
inlet s

Figure 12.14 Comparison of Monod model calculations with experimental data for continuous mixed culture. (*Reprinted from S. Y. Chiu et al., Kinetic Model Identification in Mixed Populations Using Continuous Culture Data, Biotech. Bioeng.,* **14:** 207, 1972.)

dynamics and control systems. We shall return to this point in Example 12.3; now let us examine Andrews' structured sludge model and its characteristics.

In this model, biomass is divided into three components, which are derived from substrate and interconverted according to the scheme

$$
\text{Substrate} \xrightarrow[r_1]{\text{attachment}} \underset{\substack{\text{Stored}\\\text{mass}}}{\text{X}_s} \xrightarrow[r_2]{\text{synthesis}} \underset{\substack{\text{Active}\\\text{mass}}}{\text{X}_A} \xrightarrow[r_3]{\text{residue}} \underset{\substack{\text{Inert}\\\text{mass}}}{\text{X}_i} \tag{12.6}
$$

with respiration occurring above X_s and X_A.

The rate of the attachment step is taken as

$$
r_1 = k_s\left(x_T \frac{f_s s}{s + K_s} - x_s \right) \tag{12.7}
$$

where k_s is a mass-transfer coefficient and x_s is the concentration of storage products in the floc phase. The total mixed-liquor volatile suspended-solids concentration (MLVSS) is denoted by x_T, which in turn is equal to the sum of stored-mass concentration x_s, the active-mass concentration x_A, and the inert-mass concentration x_i:

$$
x_T = x_s + x_A + x_i \tag{12.8}
$$

Finally, f_s is the maximum fraction of the MLVSS which can be storage products, s is substrate concentration, and K_s is a saturation constant. From literature reviews and their simulation work, Andrews' group suggests the parameter values listed in Table 12.5.

Table 12.5 Parameter values suggested by Busby and Andrews for structured kinetic model of the activated-sludge process†

Term	Value	Definition
k_s	3.0	Substrate mass-transfer coefficient $[t^{-1}]$
f_s	0.45	Maximum fraction of MLVSS that can be storage products
K_s	150	Sorption coefficient $[m/l^3]$
μ_A	0.06	Maximum specific rate for conversion of stored mass to active mass $[t^{-1}]$
K_A	80.0	Saturation constant $[m/l^3]$
Y_A	0.66	Mass of active mass formed per unit mass of stored mass converted
Y_i	0.25	Mass of inert mass formed per unit mass of active mass converted
k_i	0.03	Specific decay rate of active mass $[t^{-1}]$

† Adapted from J. B. Busby and J. F. Andrews, *J. Water Pollut. Control Assoc.*, **47**: 1067, 1975.

The specific rate r_2 of the active mass synthesis step is presumed to follow Monod kinetics so that

$$r_2 = \frac{\mu_A x_s}{K_A + x_s} x_A Y_A \tag{12.9}$$

and the constant yield coefficient Y_A, defined as the active mass synthesized per stored mass utilized, characterizes the synthesis step's stoichiometry. Available data suggest that the inert-mass formation rate is first order in active mass:

$$r_3 = k_i x_A Y_i \tag{12.10}$$

Y_i denotes the constant-yield coefficient (mass i formed per mass of A consumed) for this step.

Figure 12.15 shows the results of simulated batch growth using this model. Alternatively these curves could of course be viewed as the concentration in a plug-flow aeration basin as a function of position. An especially interesting feature of these calculations is the great improvement in substrate uptake which occurs with decreasing initial stored-mass concentration. Such behavior is consistent

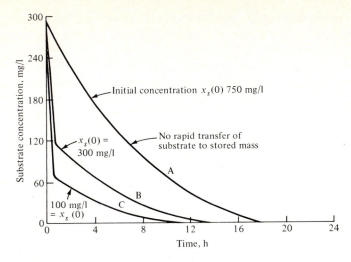

Figure 12.15 The initial stored-mass concentration has a major effect on batch substrate utilization in Busby and Andrews' structured model.

with the increased efficiency provided by the contact-stabilization process, in which the stored mass is largely converted to other mass forms in the aeration tank which follows the clarifier.

Indeed, Andrews and his associates have simulated the steady-state behavior of most versions of the activated-sludge process using this kinetic model. In these calculations mixing in the aeration basin was modeled using four ideal continuous-flow reactors in series. The primary and secondary settlers were represented by models which are presented in detail in the references. Using reasonable loading and sizing parameters, the model gave efficiencies, BOD removal rates, and operating characteristics in good agreement with experimental data on these processes. Some of these results are listed in Table 12.6.

Table 12.6. Performance characteristics of several different biological oxidation-process configurations†
The average influent flow rate was 1000 m^3/h for all cases

Type of process	Removal BOD_U, %	Process loading intensity kg/(kg · day)	Active mass, %	Total tank volume, m^3	BOD_U removed, kg · (unit vol · day)$^{-1}$
Conventional	88	0.5	34	4,500	1.22
Extended air	96	0.15	16	16,000	0.392
Short term	80	0.75	40	2,200	2.21
Contact	88	0.29	28	3,000	1.8
Step feed	91	0.27	28	4,500	1.3

† From J. F. Busby and J. F. Andrews, *J. Water Pollut. Control Assoc.*, **47**: 1055, 1975.

Besides its value in simulating a wide variety of steady-state process configurations, the structured-sludge model just outlined is necessary for analyzing the dynamic behavior of activated-sludge plants. This should be evident from Fig. 12.15, in which the dynamic response of substrate concentration is shown to be highly sensitive to one segment of the total biomass. We discuss dynamics and control later, in connection with the anaerobic digestion process. Those interested in control of activated-sludge systems can gain a good entrée into the available literature from Andrews' review papers [5–7].

Consideration of even more detailed or structured models which explicitly consider population interactions between bacteria and protozoa in activated-sludge units has only recently begun. In a series of papers considering models of activated sludge dynamics, Curds has investigated the food chain shown in Fig. 12.16. As indicated there, only the substrate and bacteria found in the sewage enter the plant in the influent stream. Reexamining Fig. 12.10, however, we recall that the feed to the aeration basin is a mixture of the influent waste and recycle-sludge streams, so that it is important to keep track of organisms which live on the floc phase and so will be included in the recycled sludge. In Curds' model, these species, which are underlined in Fig. 12.16, are the sludge bacteria, the attached and crawling protozoa which prey on the suspended but not flocculated sewage bacteria, and finally the attached carnivorous protozoa, which prey on both free and attached protozoa.

Since by now the mechanics of writing the appropriate mass balances on the system should be familiar, we list only the assumptions needed to formulate Curds' equations:

1. The aeration basin is an ideal CSTR.
2. The specific growth rate of each species shown in Fig. 12.16 depends on the concentration of its nutrient in Monod fashion. The kinetic constants used in the simulation studies are given in Table 12.7; the yield factor for all steps is assumed to be 0.5.
3. The concentrations of floc organisms in the sludge recycle are directly proportional to their respective concentrations in the aeration-basin effluent. Thus, for example,

$$x_{ir} = b_i x_{ia} \tag{12.11}$$

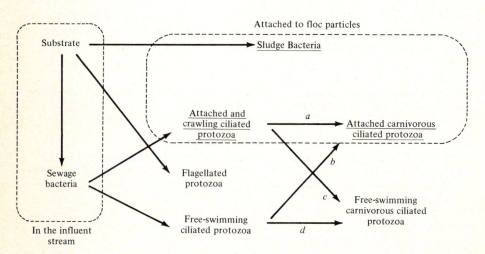

Figure 12.16 The food chain used in Curd's analysis of microbial interactions in an activated-sludge plant.

Table 12.7. Parameters for different microorganisms in Curds' model of activated sludge†

Organism	Concentration in non-variable sewage, mg/l	Range in variable sewage, mg/l	Maximum specific growth rate, h^{-1}	Saturation constant, mg/l^{-1}	Food of organism	Settler concentration factor b
Sewage bacteria	30	15–45	0.5	10.0	Substrate	
Sludge bacteria	...	...	0.3	15.0	Substrate	1.90
Flagellates	...	...	0.4	12.0	Substrate	
Bacteria-consuming ciliates:						
Free-swimming	...	...	0.35	12.0	Sewage bacteria	
Crawling	...	...	0.35	12.0	Sewage bacteria	1.27
Attached	...	...	0.35	12.0	Sewage bacteria	1.90
Carnivorous ciliates:						
Free-swimming	...	...	0.35	12.0	Bacteria consuming ciliates	
Attached	...	...	0.35	12.0	Bacteria consuming ciliates	1.90

† From C. R. Curds, *Water Res.*, 7: 1269, 1973.

where b_i is the *settler concentration factor* for species i. These constants are also listed in Table 12.7. We take $b = 1$ for substrate.

Assumption 3 makes it quite easy to include the sedimentation basin in the model, thereby incorporating the effects of recycle in the simulations.

The presence of free and attached forms of carnivorous protozoa and bacteria-consuming protozoa allows four different prey-predator pairs, labeled a to d in Fig. 12.16. Curds has considered each pair separately, assuming in the analysis that the other three prey-predator interactions at that trophic level are absent. In all cases, the following general trends are observed: in considering the carnivorous ciliates, which have been neglected in many similar analyses, it is found that their presence decreases the concentration of ciliate prey relative to the no-carnivore case. This in turn allows the population of sewage bacteria to increase. Because the sludge bacteria compete with sewage bacteria for a common nutrient, the sludge bacterial population is reduced.

For prey-predator pairs a and b, the ciliate prey is washed out of the system. Interaction c produces oscillations with mild damping which are displayed in Fig. 12.17. Only interaction d, which is relatively rare in activated-sludge populations, leads to stable, nonzero populations of all species at steady state. In all these calculations, the system parameter values assumed are $F = 100$ l/h, $\alpha = 1$, $\beta = 0.05264$, and $s_0 = 200$ mg/l; the influent concentration of sewage bacteria b_0 is 30 mg/l.

One shortcoming of these and many other simulations is the assumption of constant influent flow rate and concentrations. In actual practice the habits of the community lead to sewage variations with time which then appear in the plant influent. To a first approximation, these fluctuations are periodic, with a period of 24 h. The amplitude depends on the upstream system. In plants servicing large cities, for example, the sewage lengths between sewage source and treatment facility vary widely, so that

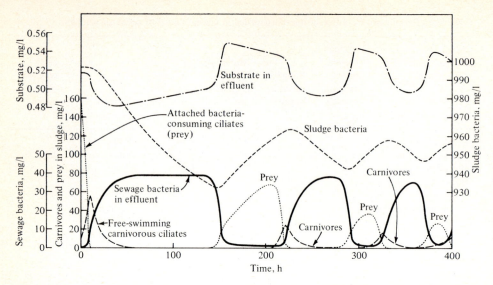

Figure 12.17 Simulation of an interacting activated-sludge population in which free-swimming ciliates prey on attached ciliates. (*Reprinted from C. R. Curds, A Theoretical Study of Factors Influencing the Microbial Population Dynamics of the Activated-Sludge Process—I, Water Res., 7: 1269, 1973.*)

substantial damping of influent load is achieved in the sewer network. Larger variations are expected in smaller plants. The ratio of maximum to minimum value of such oscillations is typically about 3, although this factor can be 10 or greater in some instances.

In one of the first attempts to examine the effects of periodic fluctuations in influent quantity or quality, Curds has considered the behavior of the system outlined above when the feed-stream conditions oscillate according to

$$F = 100 + 50 \sin 2\pi t \qquad \text{l/h}$$

$$s_0 = 200 + 60 \sin 2\pi t \qquad \text{mg/l}$$

$$b_0 = 30 + 15 \sin 2\pi t \qquad \text{mg/l} \qquad\qquad (12.12)$$

where t is time in days and the carnivorous predators have been excluded from the calculations. The resulting behavior is illustrated in Fig. 12.18. This figure shows a succession of dominating populations like those we have seen before in spoilage and soil microbiology. Another interesting feature of these simulations is the survival of attached ciliates, while free-swimming and crawling ciliates are washed out. While free-swimming ciliates are normally not found in healthy sludge, crawling forms are, so that some further refinements in this model are desirable. We should point out in closing that the oscillations seen in Fig. 12.17 are very different from those of Fig. 12.18. In the latter case, the oscillations are *forced* by the cyclic variations in feed conditions. The former oscillations, which occur with constant feed conditions, are *autonomous* and reflect special nonlinear characteristics of the system. Bailey's review (Ref. 10 of Chap. 11) may be consulted for further comments on these matters.

While the structured models developed recently by Andrews, Curds, and others offer great promise for improved design methods and control systems, they have not yet enjoyed widespread application. With further testing and more careful experiments to check their validity, we can expect models of this type to gain wider acceptance in the future.

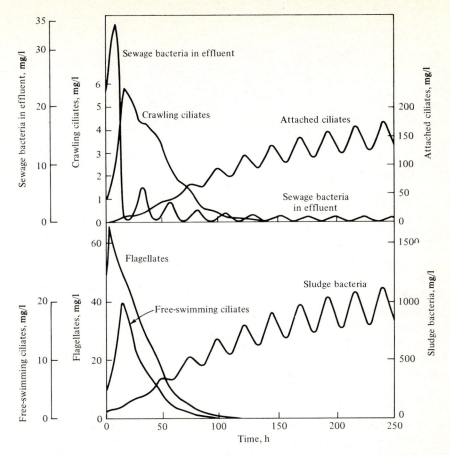

Figure 12.18 Simulation of start up of an activated-sludge unit in which the feed flow rate, feed substrate concentration, and feed sewage concentrations oscillate sinusoidally with a period of 1 day. (*Reprinted from C. R. Curds, A Theoretical Study of Factors Influencing the Microbial Population Dynamics of the Activated-Sludge Process—I, Water Res.*, **7**: 1269, 1973.)

12.4.4. Secondary Treatment Using a Trickling Biological Filter

The so-called *trickling* or *percolating biological filter* is a popular alternative to the activated-sludge process. Although there are significant differences in the design, functions, and microbial flora of the two systems, the basic operating principle of the trickling filter is the same as the vinegar generator discussed in Sec. 10.2.2: a film or slime of microorganisms lives on solid packing which loosely fills a vessel (void fraction about 0.5) designed to permit air to enter the lower portion of the bed. A typical biological-filter design is illustrated in Fig. 12.19.

The use of the term filter to describe this system is in a sense unfortunate, since the mechanism of waste removal is not due to straining but to the same attachment–biological-oxidation sequence which operates in the activated-sludge process. Before examining the microbial populations active in trickling filters, it is

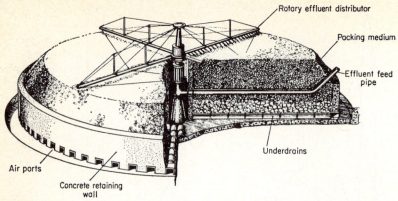

Figure 12.19 A trickling biological filter. (*From J. W. Abson and K. H. Todhunter, p. 326 in N. Blakebrough (ed.), "Biochemical and Biological Engineering Science," vol. 1, Academic Press, London 1967.*)

important to have a clear picture of the engineering and operational characteristics of the unit.

Liquid to be treated is fed to the top of the bed, which typically is 3 to 10 ft deep, either continuously, through fixed nozzles over the bed, or periodically, using a rotating distributor like that in Fig. 12.19. In both cases, the liquid rate must not be high enough to flood the bed. To ensure adequate oxygen supply, the liquid should trickle over the slime-covered packing in films sufficiently thin for the oxygen to continuously supply aerobic organisms in the outer surface of the microbial film. Unlike the activated-sludge process, which often receives forced aeration, air is circulated through the trickling filter by natural convection. The driving force for this convection is the temperature difference generated in the trickling filter by biological oxidation of the sewage; air ports and accompanying ventilation pipes within the filter allow air to enter the bottom and intermediate portions of the bed.

The basic principles of substrate utilization and film growth discussed in Example 10.1 also apply to biological filters. However, the complexities inherent in the mixed-substrate–mixed-population problem have prevented development of analytical design methods for complicated systems such as a biological filter, where there are gradations in concentrations and population densities both locally (within the films) and globally (within the bed from top to bottom). Still, the qualitative understanding of diffusion and reaction in films gained in Example 10.1 is useful. We can anticipate that development of anaerobic regions deep within the film will produce gases which cause portions of the slime to detach spontaneously from the packing. The organisms sloughed off of trickling filters in this fashion are often called *humus*, and this solid debris must be removed in a clarifier following the biological filter. The average film thickness achieved by this spontaneous regulation is a complex function of operating parameters. In a well-operated filter, a film thickness of 0.25 mm is typical.

The usual ranges of loadings and efficiencies of trickling biological filters are listed in Table 12.8. For the conventional process, the hydraulic loadings indicated

Table 12.8. Characteristics of high-rate and low-rate trickling filters†

Feature	Low-rate	High-rate
Hydraulic loading, gal/(day · ft^2)	25–100	200–1000
Organic loading, (lb BOD$_5$)/(1000 ft^3 · day)	5–25	25–300
Depth, ft		
Single-stage	5–8	3–6
Multistage	2.5–4	1.5–4
Dosing interval	Intermittent	Continuous
Recirculation	Generally not included	Always included
Effluent	Highly nitrified, 20 mg/l of BOD$_5$	Not fully nitrified, 30 mg/l or more of BOD$_5$

† From L. G. Rich, "Environmental Systems Engineering," p. 370, McGraw-Hill Book Company, New York, 1973.

result in liquid residence times in the filter of 20 to 60 min. The high-rate option indicated in Table 12.8 is sometimes called a *flushing trickle filter*; because of the higher liquid rate used, slime buildup is limited. On the other hand, this mode of operation flushes out more humus, which must be eliminated in a subsequent settler.

We can draw a useful space-time analogy by taking a Lagrangian trip through the trickling filter; i.e., suppose that we ride through the filter, from top to bottom, in a liquid drop. Then, as we travel through the packed bed, we shall see changes with time in the liquid composition as different components are removed by different microorganisms. In a sense, these changes are similar conceptually to the course of events in milk spoilage (see Fig. 12.1) and in plowed soil. As conditions in the liquid medium change, different species of microoganisms gain ascendency, causing further changes in the liquid and continuing the succession of different microbial populations.

Now let us transfer this observation to a fixed, or Eulerian, frame of reference. What was seen as a sequence in time by a drop moving through the bed is, for a filter in steady state, a pattern in space. Organisms best suited to utilize the feed sewage as a nutrient predominate in the top of the bed, as do tenaciously holding fungi and free-swimming ciliated protozoa. In the lower portions of the filter live stalked ciliate protozoa and nitrifying bacteria. Higher animals are also among the inhabitants of biological filters, with worms and fly larvae the major populations. These animals graze on the slime film which grows on the filter packing, and control of their populations is an important factor in filter operation.

The spatial segregation of organisms in biological filters provides an opportunity for each species to adapt fully to its immediate environment. Because of this, low-rate biological filters usually provide clearer, more highly nitrified effluents than activated-sludge treatment does. Also, experience has shown that filters are less sensitive to shock loads of toxic substances than activated-sludge processes. As indicated in Table 12.9, however, activated-sludge units are in some

Table 12.9. Comparison between trickling-filter and activated-sludge water treatment processes†

Item	Filter	Sludge tank
Capital costs	High	Low
Operating costs	Low	High
Space requirements	High	Low
Aeration control	Partial except in enclosed forced-draft types	Complete
Temperature control	Difficult due to large heat losses	Complete; heat losses small
Sensitivity to variations in applied feed concentrations	Fairly insensitive but slow to recover if upset	More sensitive but recovery quite rapid
Clarity of final effluent	Good	Not as good
Fly and odor nuisance	High	Low

† From J. W. Abson and K. H. Todhunter, p. 337 in N. Blakebrough (ed.), "Biochemical and Biological Engineering Science," vol. 1, Academic Press, London, 1967.

respects superior to trickling biological filters. Choice between the two processes requires careful consideration of waste characteristics, costs, and environmental standards. In some cases, the optimum plant design involves application of both methods. Other options are provided by choice of clarifier and recycle arrangements, as discussed in further detail in the references.

Lagoon systems, while much more primitive than either the activated-sludge or trickling-biological-filter processes, provide another useful method for wastewater treatment. In *oxidation ponds*, which closely resemble natural aquatic-ecosystems, algae free oxygen through photosynthesis, thereby maintaining aerobic conditions for bacteria which consume organic wastes. Oxidation ponds are made quite shallow, typically 2 to 4 ft deep, to avoid establishment of anaerobic zones near the bottom. On the other hand, we find anaerobic conditions or an alternating temporal pattern of aerobic and anaerobic environment in *waste-stabilization lagoons*, which are used for wastes containing settleable solids. Additional data on these processes are provided in Rich [3].

12.4.5. Anaerobic Digestion†

Wastes containing substantial amounts of fermentable organic components can be treated biologically under anaerobic conditions. Although anaerobic treatment has broader applications, a major use arises in treatment of excess sludge solids

† This section is based on the discussion of mathematical modeling of anaerobic digestion in S. P. Graef and J. F. Andrews, *CEP Symp. Ser.*, [136] **70**: 101–127 (1974).

(Figs. 12.10 and 12.13) produced in sewage-treatment processes. As we have discussed earlier, concentrated sludge is produced at several stages of these processes, including waste particulates removed in the screening and primary sedimentation units and also the sludge grown in the secondary biological oxidation process. This material is further concentrated, or thickened, often merely by settling, before disposal, frequently with anaerobic biological digestion as one of the steps in the process.

A simplified schematic of the overall mechanism of anaerobic digestion, which involves a multitude of microbial species, is

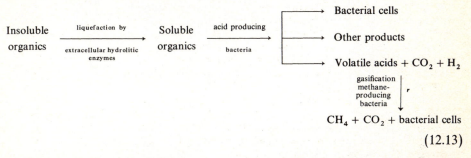

$$(12.13)$$

In the first step, large solid-sludge material is solubilized or dispersed by extracellular enzymes synthesized by a broad spectrum of bacteria. Among the enzymes found in anaerobic digesters are proteolytic, lipolytic, and several celluloytic enzymes. Since solids do not build up in anaerobic digesters, these solubilization reactions apparently proceed fast enough to prevent this step from limiting the rate of the overall reaction sequence (12.13).

Experimental studies of the next portion of the digestion reaction, namely bacterial synthesis of short-chain fatty and volatile acids from soluble organic material, reveal that these steps also occur at a relatively rapid rate. The organisms responsible for these transformations, called *acid formers* for obvious reasons, are facultative anaerobic heterotrophs which function best in a range of pH from 4.0 to 6.5. While the major product of this step is acetic acid, propionic and butyric acid are also produced.

Acetic acid is the most important substrate for the final reaction of the sequence, since about 70 percent of the methane produced has been shown to derive from that component. This gasification step of the process involves methane bacteria, which are strict anaerobes. A narrower range of pH, from 7.0 to 7.8, is optimal for these organisms, which, although difficult to isolate in pure culture, thrive in mixed culture in a properly operated digester. Existing evidence suggests that this conversion of volatile acids to CH_4 and CO_2 is the rate-limiting step in the series of reactions shown in Eq. (12.13).

Figure 12.20 is a schematic diagram of an anaerobic digestion unit. Mixing is provided to prevent high local concentrations of acids from developing. In order to maintain a satisfactory environment for both acid formers and the methane bacteria, digesters are operated at a pH around 7. Also indicated in Fig. 12.20 is an external heat exchanger, which provides an above-ambient temperature in the

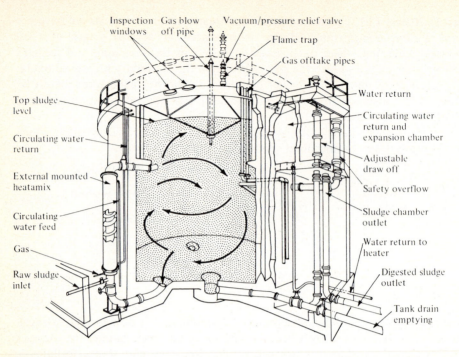

Figure 12.20 Schematic diagram of an anaerobic digestion unit. (*Reprinted from B. Atkinson,* *"Biochemical Reactors," p. 24, Pion Ltd., London, 1974.*)

vessel. At present, the usual practice is operation at the temperature in the meso-philic range which maximizes the rate of sludge digestion, about 90 to 100°F. There is limited evidence that more rapid digestion is available in the thermophilic range, with largest rates at about 130°F. Operation at this temperature level is relatively rare: higher energy cost is one factor which weighs in favor of the mesophilic range of temperatures. The solids-residence time required for anaer-obic sludge digestion at mesophilic temperatures is 10 to 30 days in a well-agitated unit.

Fortunately, the anaerobic digestion process produces a fuel which can be used to reduce energy costs for the waste-water-treatment plant. In some in-stances, the methane produced by anaerobic waste treatment is used outside the plant for heating and power. The gas mixture produced by anaerobic digestion, which is collected from the top of the unit as indicated in Fig. 12.20, is roughly 65 to 70 percent methane, with CO_2 comprising most of the remainder. Hydrogen sulfide, produced by sulfate-reducing bacteria, is present in small amounts, as are H_2 and CO. The digester off-gas has a heating value of 650 to 750 Btu/ft^3 and is produced with a yield of 12 to 18 std ft^3 per pound of organic matter decomposed. Since this gas has a substantially lower Btu value than natural gas (about 1000), it has not been such an attractive product in areas where natural-gas supplies are plentiful. With rising energy costs, however, increasing attention is being devoted

Table 12.10. Effects of anaerobic digestion on sewage sludge†

Fraction	Raw sludge	Digested sludge
Ether-soluble	34.4	8.2
Water-soluble	9.5	5.5
Alcohol-soluble	2.5	1.6
Hemicellulose	3.2	1.6
Cellulose	3.8	0.6
Lignin	5.8	8.4
Protein	27.1	19.7
Ash	24.1	56.0

† From J. W. Abson and K. H. Todhunter, p. 339 in N. Blakebrough (ed.), "Biochemical and Biological Engineering Science," vol. 1, Academic Press, London, 1967.

to anaerobic digestion as a potential fuel source, albeit after the necessary H_2S removal.

As a result of anaerobic digestion, the sludge is in much better condition for further treatment. First, the organic sludge solids are reduced by as much as 50 to 60 percent. Moreover, the composition of the sludge is profoundly changed (see Table 12.10). Because of these alterations, digested sludge is much less putrefactive than raw sludge, and it is also easier to dewater. After dewatering, which is often accomplished with rotary-drum vacuum filtration, the sludge is dried further, then spread on land as a fertilizer, dumped, or incinerated. Figure 12.8 indicates some of the other options for sludge treatment, and others are discussed in the references.

12.4.6. Mathematical Modeling of Anaerobic-Digester Dynamics

Despite production of a gaseous fuel and residual solids with fertilizer value, anaerobic digesters have a bad reputation because they are prone to operational problems. Many digester failures have been documented, with the major known causes classified as hydraulic, organic, and toxic overloading. In the first case, the dilution rate exceeds the growth rate of digester microbes, which are then washed out of the unit. High organic substrate concentrations, on the other hand, cause buildups of volatile acids. Methane bacteria are inhibited, and the digester "sours" as pH falls and failure ensues. When substances toxic to the methane bacteria enter the digester in sufficient amounts, washout of this population causes failure of the overall process.

Since improved operational practices could be of great benefit to enhanced success of anaerobic-treatment processes, there is an obvious incentive for studying the dynamic characteristics of these units and attempting to develop suitable control strategies. We shall review in this section a very interesting mathematical model of anaerobic digestion which was developed by Graef and Andrews. Also, some of the control schemes which they studied will be examined in Example 12.3. This model provides a fitting climax for this text because it involves intricate interplays between physical, chemical, and biological factors. It therefore exemplifies a synthesis of classical engineering skills with basic biological knowledge to achieve a biochemical engineering analysis.

We have already mentioned that conversion of volatile acids by the methane bacteria appears to be the rate-limiting step in the sequence of biological reactions (12.13). Assuming that all volatile acids can be represented as acetic acid and that the composition of methane bacteria can be approximated by the empirical formula $C_5H_7NO_2$, Graef and Andrews [7] develop the following stoichiometry for the gasification reaction:

$$CH_3COOH + 0.032NH_3 \longrightarrow 0.032C_5H_7NO_2 + 0.92CO_2 + 0.94CH_4$$
$$+ 0.096H_2O \tag{12.14}$$

The limiting substrate for this reaction is presumed to be the nonionized volatile acids. The concentration of the nonionized form, HS, differs in general from the total concentration s due to the ionization reaction

$$HS \rightleftharpoons S^- + H^+ \qquad K_a \tag{12.15}$$

where S^- is used as a shorthand notation for ionized substrate. Since $-\log K_a \equiv pK_a$ is 4.5 and digesters operate at pH's above 6, we know that almost all the acid is in the ionized form, so that

$$s^- \approx s \tag{12.16}$$

and

$$(hs) \approx \frac{s(h^+)}{K_a} \tag{12.17}$$

In order to incorporate the known inhibitory effects of high substrate concentration in the model, Graef and Andrews [7] modified the Monod expression for specific growth rate to the form

$$\mu = \mu_{max} \left[\frac{1}{1 + K_s(hs) + (hs)/K_i} \right] \tag{12.18}$$

which is familiar from our kinetic studies in Chaps. 3 and 7. Also included in the bacterial growth rate is a death rate, which is presumed to be first order in toxin concentration [tox]†:

$$r_D = -k_T[tox] \tag{12.19}$$

Based on the available data and Graef and Andrews' estimates, the parameters appearing in these rate expressions have the following values: $\mu_{max} = 0.4 \text{ day}^{-1}$, $K_s = 0.0333 \text{ mmol/l}$, $K_i = 0.667 \text{ mmol/l}$, and $k_T = 2.0 \text{ day}^{-1}$.

---

† To avoid confusion, the concentration of some components are indicated within brackets; others are denoted as usual by lower case italics.

If constant yield coefficients for the ratios (cell mass)/(limiting substrate), $Y_{X/S}$; CO_2/(cell mass), $Y_{CO_2/X}$; and CH_4/(cell mass), $Y_{CH_4/X}$, are assumed, the material balances on the substrate and biological phase in the digester take the form indicated at the bottom of Fig. 12.21. The quantities R_B and Q_{CH_4} denote the rate of CO_2 and methane formation, respectively, due to biological gasification. For conditions typical of anaerobic digesters, the gas density ρ_G is 0.0389 mol/l, and $Y_{CH_4/X}$ and $Y_{CO_2/X}$ are both 28.8 mol/mol. As Fig. 12.21 shows, methane is quite insoluble, so that all the methane formed enters the gas phase.

This is not the case for CO_2, however, which exists in the liquid phase in two forms as well as in the gas phase. The rate of mass transfer of CO_2 from the gas to liquid phases is given by the familiar form from Chap. 8:

$$T_G = k_L a([CO_2]_D^* - [CO_2]_D) \tag{12.20}$$

where $[CO_2]_D$ is the concentration of dissolved CO_2 and $[CO_2]_D^*$ is this concentration at equilibrium. From Henry's law

$$[CO_2]_D^* = K_H p_{CO_2} \tag{12.21}$$

where p_{CO_2} is the partial pressure of CO_2 in the gas phase. Graef and Andrews suggest values of 100 day^{-1} and 3.25×10^{-5} mol/(l/mmHg) for $k_L a$ and K_H, respectively.

Another pathway by which CO_2 can appear in the liquid phase is bicarbonate association, according to

$$HCO_3^- + H^+ \rightleftharpoons H_2O + CO_2 \qquad \text{equilibrium constant} = 1/K_1 \tag{12.22}$$

If we let R_C denote the rate of this reaction, the mass balance on bicarbonate in the liquid phase is

$$V\frac{d[HCO_3^-]}{dt} = F([HCO_3^-]_0 - [HCO_3^-]) - VR_C \tag{12.23}$$

We can obtain an independent expression for $d[HCO_3^-]/dt$ using the requirement of electroneutrality, which can be written

$$[H^+] + c = [HCO_3^-] + [S^-] + [OH^-] + a + 2[CO_3^{2-}] \tag{12.24}$$

where c is the total cation concentration, including contributions of calcium, sodium, magnesium, and ammonium, and a is the total anion (chlorides, phosphates, sulfide, etc.) concentration. For a digester operating in the normal range of pH 6 to 8 $[H^+]$, $[OH^-]$, and $[CO_3^{2-}]$ are negligible, and Eq. (12.24) becomes

$$z = [HCO_3^-] + s \tag{12.25}$$

where z, the net cation concentration, is defined by

$$z = c - a \tag{12.26}$$

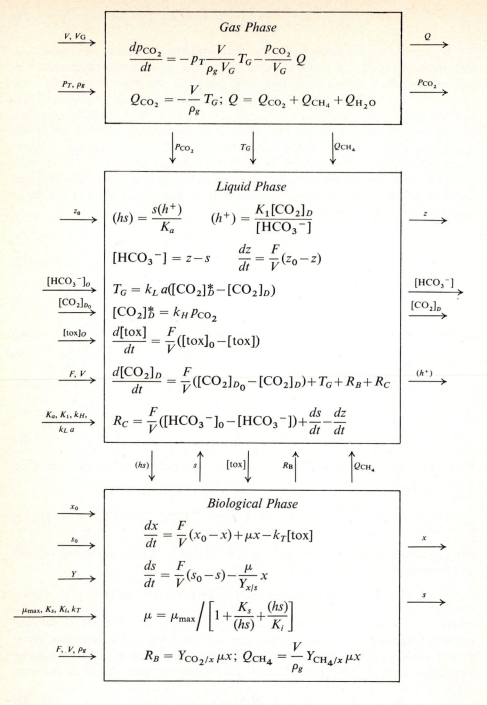

Figure 12.21 Summary diagram of the mathematic model of anaerobic digestion. The arrows indicate information flow between subsystems and interactions with the external environment. (*Reprinted from S. P. Graef and J. F. Andrews, Mathematical Modeling and Control of Anaerobic Digestion in G. F. Bennett (ed.), "Water–1973" CEP Symp. Ser. No. 136, vol. 76, p. 101, 1974.*)

and s^- has been replaced by s, as discussed earlier. If sulfide concentration is not too large, z corresponds approximately to ammonium-ion concentration. We shall suppose that the Z mass balance may be taken as

$$\frac{d(z)}{dt} = \frac{F}{V}(z_0 - z) \tag{12.27}$$

Now, differentiating Eq. (12.25) with respect to time gives

$$\frac{d[HCO_3^{-}]}{dt} = \frac{d(z)}{dt} - \frac{d(s)}{dt} \tag{12.28}$$

Eliminating $d[HCO_3^{-}]/dt$ from Eqs. (12.28) and (12.23) gives

$$R_c = \frac{F}{V}([HCO_3^{-}]_0 - [HCO_3]) + \frac{d(s)}{dt} - \frac{d(z)}{dt} \tag{12.29}$$

This rate is included in the material balance for liquid-phase CO_2, as indicated in Fig. 12.21.

The gas-phase mass balances are relatively straightforward, as Fig. 12.21 shows. Notice how the various gas-production rates computed elsewhere in the model are used to obtain the total effluent flow rate.

Next let us review some of the simulation results obtained using the model. Table 12.11 lists the parameter values and standard steady-state conditions employed in these calculations. Simulation of a batch digester, which is achieved simply by setting $F = 0$ in the continuous model, shows that increasing initial organism concentration, increasing initial pH, and decreasing initial substrate concentration all lead to smaller batch digestion times. The same trends have been observed in the field.

Another situation considered by Graef and Andrews is digester start-up. They showed that the model predicts: (1) a decreased time for start-up if initial pH is increased or the feed-sludge concentration is increased, (2) failure if initial pH or feed-sludge concentration is too low, and (3) alleviation of possibility of digester failure during start-up by slowly raising the digester loading to its final value. Again, field units show similar characteristics.

The model summarized above also exhibits the three modes of failure discussed at the start of this section. Simulation results for two of these cases, organic and hydraulic overloading, are displayed in Figs. 12.22 and 12.23, respectively. In all these calculations, a step change in a process input occurs at $t = 1$ day. If the magnitude of this change is sufficiently small, say less than 35.7 g/l in feed-substrate concentration and less than 2.5 l/day in the case of a hydraulic-flow-rate disturbance, the digester attains a new, stable steady state in the vicinity of the original operating state. Substantially larger step changes in these inputs cause the process to run away: pH and methane product drop precipitously while the effluent-substrate (volatile acids) concentration rapidly climbs. Computed results for a pulse of a toxic agent also reveal digester failure if the toxic overloading is too large.

Table 12.11. Standard steady-state conditions and parameter values used in simulation of anaerobic-digester dynamics†

Influent variables		Steady state conditions		Parameters and constants	
Term	Value	Term	Value	Term	Value
s_0	167 mmol as acetic	s	2.0 mmol as acetic	V	10 l
	10 g/l as acetic		120 mg/l as acetic	V_G	2 l
z_0	50 meq/l	(hs)	0.0112 mmol as acetic	μ_{max}	0.4 day^{-1}
F	1.0 l/day		0.672 mg/l as acetic	p_T	760 mmHg
$[CO_2]_{D_0}$	0 mmol	$[HCO_3^-]$	48.0 mmol	D	25.7 l/mol
x_0	0 mmol		24,000 mg/l as $CaCO_3$	I	0.1 mol/l
$[HCO_3^-]_0$	0 mmol	x	5.28 mmol as $C_5H_7NO_2$	$k_L a$	100 day^{-1}
$[tox]_0$	0 mmol		597 mg/l	K_s	0.0333 mmol/l
		z	50 meq/l	K_i	0.667 mmol/l
		$[CO_2]_D$	9.0 mmol	$K_1\ddagger$	6.5×10^{-7}
		p_{CO_2}	273 mmHg	$Y_{X/S}$	0.032 mol organisms/mol substrate
		μ	0.1 day^{-1}	$Y_{CO_2/X}$	28.8 mol CO_2/mol organisms produced
		pH	6.91	$Y_{CH_4/X}$	28.8 mol CH_4/mol organisms produced
		Q_{dry}	6.35 l/day		
		Q_{CH_4}	3.91 l/day		

† From S. P. Graef and J. F. Andrews, *CEP Symp. Ser.*, [136] **70**: 130, 1974.

‡ At 38°C; $I = 0.1$.

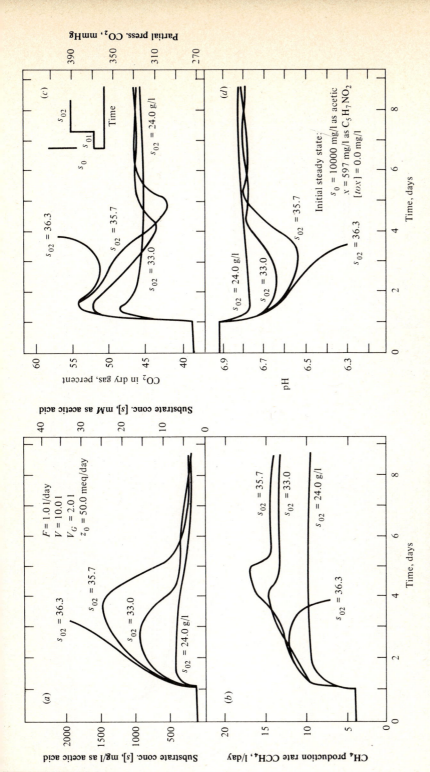

Figure 12.22 Different sized step changes in feed substrate concentrations s_0 to an anaerobic digester produce different patterns of dynamic response. For a step change to $s_0 = 36.3$ g/l, the process runs away. (*Reprinted from S. P. Graef and J. F. Andrews, Mathematical Modeling and Control of Anaerobic Digestion, in G. F. Bennett (ed.), "Water–1973" CEP Symp. Ser. No. 136, vol. 76, p. 101, 1974.*)

733

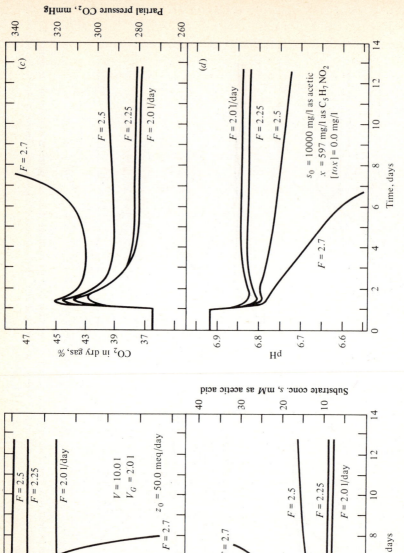

Figure 12.23 Calculated anaerobic digester response to step changes in hydraulic flow rate $F(F_1 = 1 \text{ l/day})$. (Reprinted from S. P. Graef and J. F. Andrews, Mathematical Modeling and Control of Anaerobic Digestion, in G. F. Bennett (ed.), "Water–1973" CEP Symp. Ser. No. 136, vol. 76, p. 101, 1974.)

734

Consequently, we see that the model outlined above characterizes the qualitatively dynamic features of anaerobic digestion quite well. Further study of this model and subsequent refinements will greatly aid our understanding of these complex processes and aid in development of design and operational strategies for improved performance. In fact, several suggestions for control design and possible flags for impending failure are discussed further in Andrews et al. [5–7]. One of the most interesting control schemes studied by these authors is reviewed in the following example. We should emphasize that a dynamic model which retains the known essential dynamic features of the real process is an indispensable ingredient for such investigations of controller design.

Example 12.3. Simulation studies of control strategies for anaerobic digesters The following four methods of feedback control of anaerobic digesters were considered by Graef and Andrews: (1) gas scrubbing and recycle, (2) base addition, (3) organism recycle, and (4) flow reduction. Since the first approach is the most unusual, and since it uses the ionic-equilibria portions of the model, we shall concentrate on control via gas scrubbing and recycle in this discussion. [See Prob. 12.12 for control via (3) or (4).]

A schematic diagram illustrating this control configuration is provided in Fig. 12E3.1. As indicated there, some of the effluent gas from the digester is scrubbed to remove CO_2, and then this gas is recycled to the digester. How much CO_2 is removed by passage through this loop is determined by the pH within the digester. When digester pH falls below a threshold value, gas flow through the CO_2 scrubbing loop is increased. Removing CO_2 from the digester gas phase will cause the carbonic acid concentration in the liquid to drop, thereby creating an increase in pH.

This rather subtle approach to pH control has several potential advantages relative to conventional techniques, which require addition of a base. If strong alkali is added, it may create, at least temporarily, localized regions of very high pH without effectively raising pH in the total vessel. Moreover, metal cations in the alkali can be toxic to the microbial population of the digester. Lime addition, another possibility, has the problem of creating an insoluble carbonate precipitate.

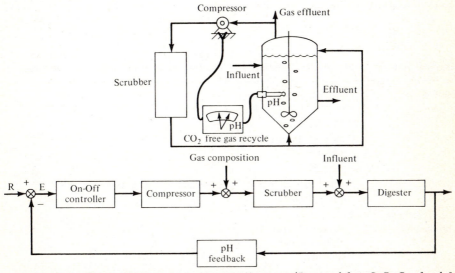

Figure 12E3.1 Off-gas scrubbing and recycle control system. (*Reprinted from S. P. Graef and J. F. Andrews, Mathematical Modeling and Control of Anaerobic Digestion, in G. F. Bennett (ed.), "Water– . 1973" CEP Symp. Ser. No. 136, vol. 76, p. 101, 1974.*)

After gaining experience with this control system, Graef and Andrews concluded that a multilevel control provided best results. In this scheme, there are two overlapping on-off bands for the gas-recycle flow rate Q_R:

$$Q_{R1} = 0 \qquad \text{pH} < 6.75$$

$$Q_{R1} = Q_1 \qquad 6.75 \leq \text{pH} \leq 7.00$$

$$Q_{R1} = 0 \qquad \text{pH} > 7.00$$

Second stage Q_{R2}:

$$Q_{R2} = 0 \qquad \text{pH} < 6.65$$

$$Q_{R2} = Q_2 \qquad 6.65 \leq \text{pH} \leq 7.00$$

$$Q_{R2} = 0 \qquad \text{pH} > 7.00$$

with
$$Q_R = Q_{R1} + Q_{R2}$$

Figure 12E3.2 shows the response of the digester with gas-scrubber feedback control to organic overloads. Notice that the controlled system survives a step increase in feed substrate of 40 g/l, while this overload is more than sufficient to cause the uncontrolled digester to fail. Unfortunately, this mode of control could not prevent failure when toxic or hydraulic overloading occurred. The successes and shortcomings of other control strategies are summarized in Table 12E3.1. From the simulation results indicated there, it appears that a multivariable control scheme is necessary for stable digester operation in the face of all three kinds of overload. With such a control system, digester variables, e.g., pH and methane production rate, would be measured, and in response to all these measurements more than one variable (such as gas and sludge recycle and residence time) would be manipulated. Development of a satisfactory multivariable control strategy remains a worthwhile problem for further research.

Table 12E3.1. Summary of control strategies for anaerobic digester control†

In each case, the manipulated variable was switched in an on-off fashion to maintain the measured variable near its set point value

Measured variable	Manipulated variable	Control effective in preventing digester failure in the event of:		
		Organic overloading	Toxic overloading	Hydraulic overloading
Digester pH	Flow rate through gas-recycle line with CO_2 scrubbing	Yes	No	No
	Flow rate of caustic added to digester	Yes	No	No
	Liquid flow rate in and out of digester	Uncertain	Uncertain	Yes
Rate of methane production	Rate of recycle of sludge collected from digester liquid effluent	No	Yes	No

† Adapted from S. P. Graef and J. F. Andrews, *CEP Symp. Ser.*, [136] **70**: 101, 1974.

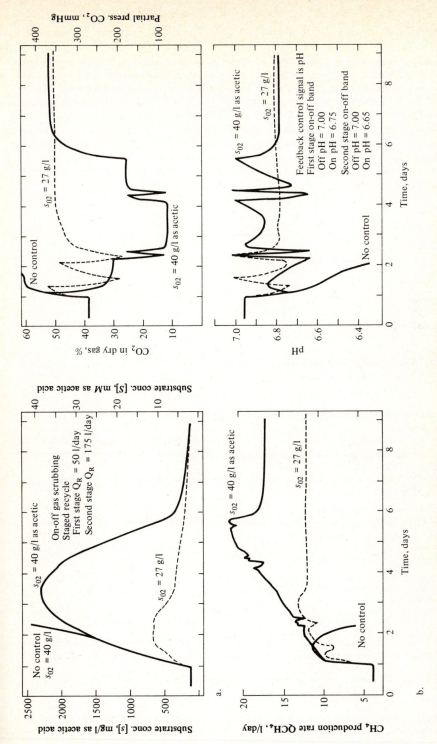

Figure 12E3.2 Effectiveness of the off-gas scrubbing control policy in counteracting an organic overload. (*Reprinted from S. P. Graef and J. F. Andrews, Mathematical Modeling and Control of Anaerobic Digestion, in G. F. Bennett (ed.), "Water–1973" CEP Symp. Ser. No. 136, vol. 76, p. 101, 1974.*)

PROBLEMS

12.1. Time scales: two-species systems [10] Experiments in continuous culture with lactate as the growth-limiting nutrient have revealed that the K_m for *Pseudomonas* is approximately 6 mg/l while that for *Spirillum* is approximately 12 mg/l. If the maximum growth rate for *Pseudomonas* is 0.6 and that for *Spirillum* is 0.8, discuss the effect of lactate concentration in a mixed continuous culture containing bacteria. How do the two time scales for growth ($= 1/\mu$) change with lactate concentration?

12.2. Activated-sludge kinetics: time lag and recycle For the process in Fig. 12P2.1 obtain an explicit result for the time variation of biomass x when growth depends upon s in a Monod function dependence but also includes a time lag, as discussed in Prob. 10.10. Add a death rate $= k_d x$ to the previous model, take $F_0 = $ const but $s = s(t)$, according to Prob. 10.10. (For computer calculations of more complex models, see Ref. 11.)

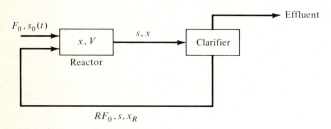

Figure 12P2.1 Simple activated-sludge scheme.

12.3. River self-purification Wuhrman divides stream recovery into the four categories shown in Table 12P3.1 (next page).

(a) Set up the appropriate kinetic, flow, and conservation equations and carefully indicate the data you would need to describe analytically (1) particle sedimentation, (2) reactions in water mass, and (3) chemical transformations in the sediment.

(b) What backgrounds would you look for in people being interviewed to set up a laboratory to study river purification?

12.4. Two-tank anaerobic digestor Pohland and Ghosh [12] discuss the interaction between acid formers and methane formers in anaerobic digestors which is conceptually similar to that of Prob. 11.5. The detailed stoichiometry proposed is

Acid formation:

$$4C_3H_7O_2NS + 8H_2O \longrightarrow 4CH_3COOH + 4CO_2 + 4NH_3 + 4H_2S + 8H^+ + 8e^-$$

Methane formation:

$$8H^+ + 8e^- + 3CH_3COOH + CO_2 \longrightarrow 4CH_4 + 3CO_2 + 2H_2O$$

These workers suggest the possibility of an easier environmental control for the methane formers if two tanks in series are used. The growth parameters measured on a glucose feed to the first reactor were:

Acid formers: μ_{max}(glucose) $= 1.25$ h^{-1} $K_g = 22.5$ mg glucose/l

Methane formers: μ_{max}(acid) $= 0.14$ h^{-1} $K_a = 600$ mg acetic acid/l

(a) Taking the kinetic forms for the two species to be those suggested in Prob. 11.5b, write out the steady-state equations governing two tanks.

Table 12P3.1†

Overall reaction	Main mechanism
1. Transport and incorporation of compounds from the water mass into deposits	*Sedimentation* (eventually after flocculation reactions) of inorganic or organic particulate matter
2. Reactions within the water mass and on surfaces of suspended matter (eventually transport of reaction products into the sediment)	*Chemical* Acid-base reactions, oxidation-reduction processes, adsorption, precipitation, etc. *Biochemical* Dissimilation of inorganic or organic metabolizable compounds by organisms; death or other inactivation of parasitic organisms
3. Exchange reactions of volatile compounds in the water mass with the atmosphere	Loss of volatile matter to atmosphere Equilibration reactions of dissolved gases with atmosphere (loss or uptake of O_2, CO_2, N_2, etc.)
4. Transformation of sediments from reduced to more oxidized states, including destruction of organic materials	Chemical and biochemical oxidations within sediment

† K. Wuhrman, Stream Purification, p. 122 in R. Mitchell (ed.), "Water Pollution Microbiology," John Wiley & Sons, Inc., New York, 1972.

(*b*) Prove that for dilution rates in the first tank in excess of 0.14 h^{-1}, the methane formers are washed out. What does this imply about using separate tanks to (largely) separate species in a food chain?

(*c*) Suppose the flow rate is greater than 0.14 h^{-1}. Solve for the outlet conditions in the first and second tanks when $V_2 = V_1$ and when $V_2 = 5V_1$ and sketch the outlet waste concentration (unconverted substrate) as a function of D. What trade-off is apparent in using increased D to favor species separation?

(*d*) In actual waste-treatment processes, settled cells and unconverted solid wastes would be recycled from each reactor exit to a settler-separator back to the same reactor inlet. Write out the substrates and species balances for such a two-tank system.

12.5. Industrial vs. waste-treatment reactors A comparison between industrial and waste treatment microbial reactors is summarized in Table 12P5.1 (next page).

(*a*) Comment on the characteristic similarities and differences, including when the critical design criteria are likely to be similar or different.

(*b*) $k_l a$ has been observed to vary with $(P/V)^n$, where $n = 1.33$ (sewage), 0.72 (yeast broth), and 0.5 (endomyces or mycelia fermentation), and 0.4 (hydrocarbon fermentation). Attempt to rationalize these exponents. How do they affect the results of part (*a*)?

12.6. Bifunctionalism in bacterial synergism When two distinct species carry out distinctly different chemical transformations, the result may be termed bifunctionalism, by analogy with bifunctional

Table 12P5.1

Property	Waste treatment	Industrial fermentation
Temp, °C	10–30	20–50
Variation	± 5	± 0.5
Rheology	Newtonian	Newtonian, non–Newtonian
Viscosity variation, cp	1–10	1–1000
Substrate concentration, g/l	0.1–5	5–40
Loading (BOD), ppm	100–5000	5000–40,000
Reactor size, gal	50,000–400,000	250–40,000
Power per unit		
volume, hp/10^3 gal	0.05–0.5	1–20
Growth rate, h^{-1}	0.05–0.1	0.1–1.0
Loading, lb BOD/1000 ft^3	10–1000	

† C. L. Cooney and D. I. C. Wang, *Biotech. Bioeng. Symp.* 2, p. 63, 1971.

catalysis involving two distinct kinds of catalytic sites, e.g., dehydrogenation and isomerization. An example in the microbial context is

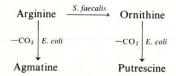

(a) Assuming Monod forms for each of the three conversions, write down the system equations for a chemostat.

(b) At low arginine levels, sketch the system outlet behavior as the dilution rate is increased continuously.

(c) If μ_{max} for *E. coli* is the same for both steps but larger than μ_{max} for *S. faecalis*, what should be the relative size (V_1/V_2) to maximize putrescine production? (fixed total volume). Repeat for μ_{max} (*E. coli*) $< \mu_{max}$ (*S. faecalis*).

(d) What design strategy, assuming a nonsterile feed, minimizes putrescine production? Which strategy, (c) or (d), would please the neighbors?

12.7. Nitrification in soil McLaren [13] has suggested that a first approximation to nitrification by soil microbes is given by considering a serial reaction:

$$NH_4^+ \xrightarrow{k_1} NO_2^- \xrightarrow{k_2} NO_3^-$$

In laboratory soil-enrichment cultures, the first and second conversions can each be associated with a single species, *Nitrosomonas* and *Nitrobacter*, respectively.

(a) At steady state, suppose that maintenance and cell replacement consume these nutrients by reactions which are zero order in NH_4^+ and NO_2^- with the rate constants k_1, k_2 above proportional to local microbial concentrations. If the respective biomass concentrations x_1, x_2 are independent of depth z, evaluate the vertical-concentration profiles of NH_4^+, NO_2^-, and NO_3^- [normalized to NH_4^+ $(z = 0)$] if the downward fluid velocity is u and ion-exchange effects of nutrients with the soil are neglected. Plot the results for the cases $k_1/k_2 = 0.1$, 1.0, 10.0 using dimensionless distance (zk_1/u). Include the two situations: (1) nitrogen uptake for new cellular material can be neglected; (2) a separate zero-order reaction with rate constant βx_1 or βx_2, respectively, allows for NH_4^+ utilization by *each* species for maintenance and replacement.

(b) If NH_4^+ is indeed the limiting nutrient for *Nitrosomonas* maintenance and replacement, evaluate from part (a) the dimensionless depth at which the assumption of $x_1 = $ const everywhere fails. Experimentally, the NH_4^+ oxidizer profile has been measured in one instance to have the following values: $2 \times 10^5/cm^3$ (surface water), 2×10^5 per gram of soil (0 to 1 cm depth), 2×10^3 per gram of soil (1 to 3 cm depth), 2×10^2 per gram of soil (3 to 5 cm depth) [1, and references therein].

(c) The assumption of constant biomass level is an obvious convenience. Show that the model above does not allow determination of depth variations of x_1 and x_2.

(d) Discuss how you would devise a model for unambiguous prediction of *Nitrosomonas* and *Nitrobacter* profiles, keeping in mind the last two chapters and a quotation from 1923:

[It] is shown that the soil is normally inhabited by a very mixed population of organisms, varying in size from the smallest bacteria up to nematodes and others just visible to the unaided eye, on to larger animals, and finally earthworms, which can be readily seen and handled. These organisms all live in the soil, and therefore must find in it the conditions necessary for their growth [14].

12.8. Stability of activated-sludge interactions Apply the Feinberg-Horn-Jackson analysis to the calculated results for the activated-sludge pairwise species interactions a, b, c, and d in Fig. 12.15. Do the Feinberg-Horn-Jackson predictions agree with the explicit calculations from Curds' equations?

12.9. Activated-sludge reactor The inflow to an activated sludge reactor has a 5-day BOD of 220 mg/l, the outlet must be below 15 mg/l (Table 12.2). Given the following conditions, complete the design of an activated-sludge reactor as indicated below:

$$\text{Active solids concentration} = 3000 \text{ mg/l}$$

$$\text{Recycle ratio} = 0.46 \qquad F = 4.5 \times 10^6 \text{ gal/day}$$

Y, μ_{max}, K_s, k_d from Table 12.4. All nutrients except BOD are in excess.

(a) Calculate

(1) Sludge age, (2) reactor volume needed, (3) concentration of active solids in recirculation line, and (4) daily aeration rate, assuming 7.5 percent oxygen utilization. Equation (12.5) indicates that minimal V can be accomplished by increasing x_r/x_a. Clearly achievement of increased x_r demands a larger secondary sedimentation unit V_s. While settling of sludge is a complex process which proceeds in several stages (see, for example, Ref. 15), we can approximate the return active solids concentration by

$$x_r = x_{r,\,max}(1 - e^{-\beta t}) + x_a e^{-\beta t} \qquad x_{r,\,max} = x_r \text{ at } t = \infty$$

where $t(\equiv V_s/F)$ is the mean settling time elapsed and β the characteristic constant > 0.

(b) Evaluate x_r(optimum) if all operating costs are divided between the digester and the settler, and the ratio of such costs (1 ft^3 digester/1 ft^3 settler) is γ ($1.0 \le \gamma \le 10.0$). (Neglect construction costs.)

12.10. Simplification of lake models: dissolved O_2 Where there is strong interest in only one parameter, e.g., dissolved-oxygen level, more complex models can often be simplified to provide a convenient, steady-state representation. Treating the scheme of Fig. 12E2.1 in combination with Fig. 8.6 and Eq. (8.54), (1) lump all oxygen-generating processes (gas-liquid transfer + phytoplankton activity) into an upper layer, (2) select an incoming BOD level which mixes uniformly into the upper and lower levels at one end of the lake without settling, etc., and (3) lump all organic carbon-utilizing species (as appears valid for Table 12.3) into a single species with a given minimum dissolved oxygen requirement, etc.

(a) Develop a model, stating assumptions, which predicts when this latter single species will exist only in the upper lake level or in both levels. (The simplest case would regard each level as a CSTR with appropriate intertank transfer rates of O_2.)

(b) From the critical O_2 levels cited in Chap. 8 and from μ_{max}, K_s in Table 12.4, for what flow rates does the habitation transition (upper level only $\rightarrow$ upper + lower level) of aerobic organic carbon consumers occur?

(c) In many rivers and some lakes, increased feed flow rates also bring increased feed BOD concentrations. If the latter varies linearly with inlet flow, how would your conclusions (not) be changed?

12.11. Predator-prey kinetics A generalized predator-prey model formulated by Rozenweig and MacArthur [16] can be cast in the form

$$\frac{dx}{dt} = f(x) - y\phi(x) \qquad \frac{dy}{dt} = -ey + ky\phi(x)$$

where x and y are prey and predator densities, respectively.

(a) Compare this form with that of Lotka-Volterra [Eqs. (11.22), and (11.23)], and list as many physical or biological circumstances as you can conceive where the forms of $\phi(x)$ and $f(x)$ would differ from those implied by the Lotka-Volterra model. Be specific. What is the meaning of $\phi(x)$? Of $f(x)$?

(b) In experiments with the predator-prey system, *Paramecium* and *Didinium*, some 1973 results of Luckinbill are described by Smith (Ref. 23 of Chap. 11, p. 33): (in each case, the recorded densities for prey never approached their self-limiting density):

1. "... there is first a rapid increase of prey, followed by an increase in the predators, which capture all the prey and then starve."
2. "Prolonged coexistence was achieved by adding methyl cellulose to the medium; this renders the medium viscous and slows down the swimming of both species. However, there was still an oscillation of increased amplitude, ending in the extinction of the predator."
3. "Persistent coexistence was achieved by adding methyl cellulose and at the same time halving the concentration of food for the prey species."

Rationalize these observations. What further experiments would you perform to confirm or disprove any assumptions which you have made?

12.12. Anaerobic digester control Write out appropriate equations to describe anaerobic digester dynamics when (a) organism recycle or (b) flow reduction (increase) is used as the control method, Include a dynamic equation for the appropriate control variable, stating your rationale for its form.

REFERENCES

Substantial material on mixed microbial populations, including additional detail on the natural cycles of matter, can be found in Refs. 1 to 3 of Chap. 1. Also useful are Refs. 1 and 2 of Chap. 10 and most references in Chap. 11. Other readings on applications of mixed populations.

1. T. Hattori: "Microbial Life in the Soil: An Introduction," Marcel Dekker, Inc., New York, 1973. A broad-based introduction to soil microbiology which emphasizes quantitative treatment wherever possible. Besides chapters on soil microbes, including their physiology, interactions, and roles in plant growth and geochemical transformations, this text offers substantial readings on the soil environment.
2. R. Mitchell: "Introduction to Environmental Microbiology," Prentice-Hall, Inc., Englewoord Cliffs, N.J., 1974. A broad survey of microbial action in the biosphere. Nutrient cycles, eutrophication, community ecology, and waste treatment are among the many examples discussed.
3. L. G. Rich: "Environmental Systems Engineering," McGraw-Hill Book Company, New York, 1973. This environmental-science text emphasizes the application of modern modeling and systems methods to processes, including waste treatment.
4. J. W. Abson and K. H. Todhunter: Effluent Disposal, chap. 9 in N. Blakebrough (ed.), "Biochemical and Biological Engineering Science," vol. 1, Academic Press, London, 1967. A good brief review of industrial waste-water treatment. Chapter 10 by R. F. Wills, dealing with sedimentation and flocculation, provides an excellent complement to waste-water-treatment aspects emphasized here.
5. John F. Andrews: Review Paper: Dynamic Models and Control Strategies for Wastewater Treatment Processes, *Water Res.*, **8**: 261–289, 1974. An excellent summary of the state of the art.

Detailed information on topics of special interest:

6. J. B. Busby and J. F. Andrews: A Dynamic Model and Control Strategies for the Activated Sludge Process, *J. Water Pollut. Control Fed.*, **47**: 1055, 1975.
7. S. P. Graef and J. F. Andrews: Mathematical Modeling and Control of Anaerobic Digestion, *CEP Symp. Ser.* [136] **70**: 101–127, 1974.
8. C. R. Curds: A Theoretical Study of Factors Influencing the Microbial Population Dynamics of the Activated-Sludge Process, I: The Effects of Diurnal Variations of Sewage and Carnivorous Ciliated Protozoa, *Water Res.*, **7**: 1269–1284, 1973.
9. S. Y. Chiu, L. E. Erickson, L. T. Fan, and I. C. Kao: Kinetic Model Identification in Mixed Populations Using Continuous Culture Data, *Biotech. Bioeng.*, **14**: 207–231, 1972.

Problems

10. C. T. Calam and E. W. Russell, Microbial Aspects of Fermentation Process Development, *J. Appl. Chem. Biotech.*, **23**: 225, 1973.
11. H. W. Blanch and I. G. Dunn: Modelling and Simulation in Biochemical Engineering, *Adv. Biochem. Eng.*, **3**: 159–162, 1974.
12. F. G. Pohland and S. Ghosh: Developments in Anaerobic Treatment Processes, in Biological Waste Treatment, *Biotech. Bioeng. Symp.* 2, p. 85, 1971.
13. A. D. McLaren: *Soil Sci. Soc. Am. Proc.*, **33**: 55, 1969.
14. E. J. Russel: "Microorganisms of the Soil," Longmans, Green and Co., London, 1923.
15. R. P. Canale and J. A. Borchardt: pp. 120–121 in W. J. Weber, Jr. (ed.), Physiocochemical Processes for Water Quality Control, Wiley-Interscience, New York, 1972.
16. M. L. Rozenweig and R. H. MacArthur: Graphical Representation and Stability Conditions of Predator-Prey Interactions, *Am. Nat.*, **97**: 209, 1963; **103**: 81, 1969.

$671135473